PERVASIVE LOOPS

PERVASIVE LOOPS

A New Framework for Studying Origin of Life
and Consciousness

Muralidhar Ravuri

Contents

FOUNDATIONS

BRAIN AND CONSCIOUSNESS

SENSING THE EXTERNAL WORLD

Contents

ORIGIN OF LIFE

DIVERSITY OF LIFE

Dedicated to

My parents Subbarao and Indiradevi,
my wife Sridevi,
my kids Chaitanya and Agastya,
my sister Udayalakshmi
and my brother Vidyasagar

Preface

I started writing this book as a journal of ideas that I have been collecting over the last 20 years on the problems of origin of life and the structure of consciousness. My goal initially has been to understand how large-scale structured dynamical systems work. However, as I started taking detailed notes after reading several research papers over the years, it became increasingly difficult to create a coherent view and, more importantly, to validate any of the ideas I was proposing. The range of problems are so wide that it was very easy to find counterexamples for any reasonable hypothesis I had.

I needed to create a framework to ensure that I was not going in endless circles without making significant progress. That's when I realized that any idea I propose should be constructive. It is not enough to just analyze and explain how these systems could have worked if no one can construct and verify them. The fact that these are some of the most complex systems we have ever encountered is not an excuse to not construct them. In fact, the single most important guiding principle that I tried to follow can be stated as follows: If you are in a lab or in front of a computer along with this book, can you build or simulate the ideas proposed for these large-scale structured dynamical systems from scratch? Furthermore, if you are not present, can nature create these systems using the same ideas? If you or nature cannot do this, then the ideas must be necessarily incomplete.

Clearly, this book is incomplete in the above sense because I have not provided such detailed constructions. However, this principle has forced me to strongly critique myself and question everything I am claiming here. For example, when I describe how free will works, I ask myself if I can construct a system that will appear to operate in a way that is indistinguishable to a human exhibiting free will. Similarly, we can ask if the ideas proposed can construct synthetic life, synthetic consciousness, systems that can sense space and the passage of time and so on. Living beings exhibit numerous such features that are nontrivial to construct. One cannot be content until the ideas yield valid constructions. This simple principle had given me an iterative process to refine the ideas over the years – constantly guided by a need construct such large-scale structured dynamical systems. Whether other readers see the same way as I do or not, this approach has given me an honest way to see the gaps. So my next 20 year effort, now that the book has taken a decent shape, is to actually work towards constructing these systems. It will not be an easy task, but it is worth a try.

Let me begin by briefly describing the problems considered in this book to get a better sense of the above challenges. For this, let me divide the last 4 billion years into three periods and discuss each of them below. Firstly, the objective of my theory would be to explain the creation of life when there was no life on earth, say, 4 billion years ago. By this, I do not mean looking for specific conditions for life to form on earth. In fact, I do not know these conditions at all. I want to claim that life is not so unique and special that we should think these conditions are critical (cf. existing theories that claim that under the correct set of conditions life was created just by random chance). One of the first things I want to show is how to create life in a 'controlled' way and, if possible, using synthetic materials besides carbon-based compounds. Next, I want to show how nature can create life *by itself* without any help from an intelligent being using organic materials. This includes explaining how hundreds of distinct and necessary chemicals come together in a small space at the correct time in the correct order without relying purely on random chance. In addition, we should explain the *creation* of these features as well. If special conditions, catalysts or locations are important, we should explain what laws of physics help 'create' these conditions repeatedly. I call this transition period, from nonlife to life, as *time period I*. Until now, there is no known theory that explains this satisfactorily i.e., if we exclude the only two solutions offered for now: (a) random chance mechanisms that rely only on abundance of chemicals and large amount of time and (b) on a Creator. One reason to exclude these two solutions is that they are not repeatable. We cannot use these 'solutions' to go to a laboratory and create life from either inanimate objects or from dead cells.

Secondly, after the first life forms appeared on earth, they have evolved to take different shapes and forms during the course of 4 billion years, until humans came into existence. The goal of the theory would be to explain how to create such a variety of life forms and how to explain their interactions with each other. There are two special cases worth highlighting here. The first one relates to how the internal complexity within a single cell arises or with how multiple cells coordinate to perform complex tasks in a multi-cellular organism. The second one is about how different species work together as part of a healthy ecosystem. I call this period, with a diversity of life, as the *time period II*. Darwin's theory of evolution is one of the successful theories that seem to address this period quite well. However, in chapter 71, I will discuss several shortcomings of Darwin's theory (see also Behe [8]). Unfortunately, the alternate theories proposed as a solution (which is primarily a Creator) are not as complete or satisfactory either. Both Darwin's theory and Intelligent Design fall at two extremes, namely, pure randomness followed by survival of the fittest and a Creator respectively. They hide the elegant structure that makes large-scale structured dynamical systems work. In this book, I will outline this unique structure in considerable detail.

Thirdly, the existence of human beings introduces enormous complexity because of their highly evolved brain. Human brain displays powerful features like intelligence, memory, language, emotions, free will and consciousness. Surprisingly, the architecture of the cell and the brain are significantly different. The DNA in nerve cells of the brain does not play a direct role in our everyday tasks (like with thinking and planning) unlike a somatic cells' everyday tasks (like digestion and transport of chemicals). Instead, the internal architecture of the entire network of our brain is directly responsible for these abilities. The objective of the theory is to explain what consciousness, feelings, free will and other human abilities are, in a concrete way.

Furthermore, I want the theory to suggest designs for constructing synthetic conscious systems. I call this period, with complexity within our brain, as the *time period III*. There is no existing theory that explains how consciousness and other higher human abilities work, let alone explain it with constructive designs. Just as I claim that the creation of life is not special, I claim that consciousness is not special either.

Recently, however, we have seen a resurgence in AI using techniques like deep learning. One of the most recent and amazing results in this field is from DeepMind's reinforcement algorithm called AlphaGo Zero that does not depend on supervised data to learn how to play the game of Go. In fact, it started just from random initialization and the basic rules of the game to learn and eventually beat the previous winner AlphaGo Lee 100 games to zero (and, hence, indirectly the previous world champion Lee Sedol in Go). The implications and extensions of such deep learning algorithms to other fields are only just beginning. With such rapid progress in this field and with the applications so wide reaching, it still remains to be seen if this truly captures synthetic consciousness.

I have personally been working in deep learning for a few years now and have recently applied several deep learning techniques to self-driving, specifically in object detection and tracking. I am amazed to see how well it works. However, I have not had the opportunity to relate deep learning techniques to the theory proposed here. This is an interesting area of future work for me. There, definitely, seem to be similarities between the two and I have constantly been inspired by the amazing progress made by the researchers in deep learning. I would be thrilled to see if deep learning evolves to be a more general form of AI as well, using and merging some of the ideas proposed here (or otherwise too). Nevertheless, an interesting aspect of the ideas presented in this book is that it tries to cover all three time periods in a cohesive way.

Now, three common questions come to mind when someone claims that they are working on such diverse set of problems. They are: (i) why do you need a new theory – can you not use an existing theory like, say, Darwin's theory and extend it somehow, (ii) why do you need to work on all these problems – can you not just pick one or a few specific problems instead and (iii) why publish the entire work in one piece – why not publish it incrementally, get feedback and iteratively improve on the ideas?

For the first question, unfortunately, it is not possible to extend the current form of Darwin's theory or other existing theories. If it was that easy, somebody would have done it already. Let me pick on Darwin's theory for now. Darwin's theory relies heavily on the existence of DNA. However, in period *I*, there is no DNA. Instead, we need to explain how DNA and life was created naturally in the first place. In period *III*, DNA does not play any direct role for consciousness. Is what you are going to do next (like cook, eat or watch a movie) encoded in the DNA? Even in period *II* when Darwin's theory is well applicable, it does not explain how to handle complexity with the highly coordinated and precisely timed chemical reactions that occurs within your body right at this instant. If these perfectly coordinated reactions do not happen, you simply die (as an individual and eventually as a species). Therefore, we need a replacement to Darwin's theory that can handle all of these cases effectively. The theory presented in this book is one such theory.

For the second question, since these are large-scale complex systems, it is easy to create an impression that you are looking at some toy problems, taking a simplistic view or ignoring the real difficult problems. Experts in these fields will, especially, feel that you are just scratching the surface and there are far more intricate issues to consider. Besides, it is not clear if your solution generalizes beyond the problems you have considered. As a result, the approach will feel incomplete and less convincing. A unified theory, though not a requirement, would be nontrivial to discover by looking at just a subset of problems. If we manage to discover a unified theory, it would not only be valuable, but will reveal new insights into the underlying structure of our universe itself, as we will see in this book.

For the third question, it is important to realize that not all problems can be improved on iteratively. For problems like finding a cure for cancer, there is a natural notion of error (from the true solution). This error helps us solve the original problem incrementally with partial solutions along the way. We keep improving this error to eventually find a cure for cancer. On the other hand, for problems like free will and consciousness, a notion of partial solutions is ill-defined. I cannot say that I have built a system that is 50% closer to the solution of free will or is 40% self-aware and I will improve it incrementally bringing it to 60% self-aware, 70% self-aware and so on over a period of time. Therefore, to claim that we have a theory of large-scale structured complex systems, we should include a minimum set of problems within our theoretical framework. This minimum set must include problems I have mentioned earlier (within periods *I, II* and *III*).

Now, you should not get the impression, from the way I have been describing here, that I have already solved some of the hard problems. In fact, this is so far from the truth. In chapter 72, I have listed several problems that still need answers. It is best to view the theory presented here as just the beginning of a formal study of large-scale structured dynamical systems. Until now, complex systems have been studied with a heavy emphasis on the fundamental laws. This has produced a slow progress in our understanding of systems that have billions of highly interacting systems. The new theory offers a fundamentally different way to study complex systems by looking at the structure of stable parallel loops (to be discussed later in the book). It surely opens up more new questions than I have answered. Nevertheless, I claim that some of the most difficult questions, which I have referred to as the minimal set from the three time periods *I, II* and *III*, can been satisfactorily answered. This should give us hope that the approach presented here can indeed lead to a formal study of large-scale structured complex systems.

Now, it is useful to have an open mind when reading the new theory presented here. There is typically a strong negative bias against any proposals for the origin of life, consciousness and other features of our brain. This is rightfully so since most proposals tend to go into philosophical details with less constructive solutions. While the philosophical implications are exciting, I have chosen to postpone them where possible and, instead, try to address the constructive aspects of these ideas. Large-scale structured dynamical systems have deep and interesting structures even without getting into philosophical discussions. Since the basic human curiosity to understand our existence is what drives us, my hope is that this book helps you get a step closer in your personal journey.

Sunnyvale, California
October 2017

Muralidhar Ravuri

Acknowledgements

I would never have been able to do this work if not for the absolute and complete support from my family for more than twenty years. The constant and everlasting support from my parents Subbarao and Indiradevi have been extremely critical throughout my life. They have always encouraged me to try something new and challenging.

My wife Sridevi has given me the most incredible freedom, love and affection in every possible way I can ever expect. During the past 15 years, she has also contributed in very significant ways to guide the theory to be correct and complete. Whenever I present my theory to her, like every few months for the last 15 years, she would ask what seems to be the most basic and simple questions, but my investigation would lead to very important results. She ensured that I did not lose sight of the big picture. She was extremely instrumental in identifying several holes in my theory over these years, forcing me to patch them in unique ways.

Watching our sons Chaitanya and Agastya grow, while presenting a glimpse of my childhood that I do not remember, has both inspired me and has given me a hope towards a possible solution to this problem. In fact, watching every little detail of their lives for the last several years made me believe that my theory has value, at least, to me. I have presented several examples in this book from my personal experiences based on watching them grow and generalizing them as scenarios common to most children.

My sister Udayalakshmi and my brother Vidyasagar have constantly encouraged me to try the most challenging problems. Since they were always available to support me emotionally, I felt that I had no obstacles to overcome. My family environment has been the most powerful force for my continued interest in this problem for more than two decades and I cannot thank enough for this. Their true belief in me even though this problem presented a zero chance of any useful solution in my lifetime, created the perfect and happiest atmosphere at home. Finally, watching life itself and its complex possibilities has been the ultimate motivation.

1
Introduction

1.1 Synopsis

There is a quest to explain how the brain works, how humans developed consciousness, how the very life came into existence in a world devoid of all of these. This quest has inspired a number of scientists, philosophers and almost every human being to understand our very existence in this universe. How will the ultimate answers to these questions look like? Will we be able to comprehend them? Or will they be so complex that it will always remain a mystery?

In this book, I will introduce three universal concepts that provide a foundation to answer these questions clearly and completely. The first of these three concepts is that we should analyze all large-scale structured interacting dynamical systems as loops – not just sequentially, with a cause producing an effect, but the effect in turn producing a cause and so on – as loops. The dynamical events being analyzed as loops can occur at any scale, from a quantum level to astronomical scales. Note, however, that in most cases, looped dynamics is not natural because we cannot have a perpetual motion machine. They should be accompanied by an active source of external or internal energy.

The second one is that all dynamical interactions inherently are parallel. A given dynamical system is capable of exhibiting life-like patterns only when we include the parallel interactions and the accompanying serialization that occurs within its subsystems. Our thought processes follow serial step-by-step analysis like, say, when analyzing a problem or when planning an event. This makes us believe that all conscious thoughts are either effectively serial or can be treated as serial if the interval of our perceived timeline is made extremely small. However, we will see that such serial viewpoints can never clarify life, consciousness and other features of large-scale structured dynamical systems. I will show, for example, that parallelism, though nontrivial to keep track of, is necessary to build features like free will (see section 1.7.3 for a short introduction).

The third one is the familiar notion of stability. When we work with systems involving millions of interacting dynamical subsystems like cells and the brain, just ensuring the stability of the entire system for a sufficiently long period is a hard problem. The complex time-varying interactions, the imperfect timing of interactions, wear and tear, unexpected damages, uncertainties and others tend to destabilize a large-scale dynamical system (at least, synthetic ones) quite easily. In this book, I will show how to *design* a special class of large-scale structured dynamical systems that remain stable for a sufficiently long time in spite of these uncertainties.

These three notions will now force a radical change in our thinking and our approach to problem solving. It will require us to break out of the conventional linear and serial analysis of the world and, instead, study it using loops and parallel approach. Everyone has seen or experienced these three concepts individually in everyday life. It is my thesis that they form the underlying connection between all living and large-scale non-living systems in our universe.

At first, it will not appear convincing that three simple concepts, namely, stable parallel loops, can provide answers to such difficult and diverse set of questions. Indeed, I will introduce additional ideas unique to each problem, which will later be combined with these three concepts to formulate complete answers. Nevertheless, what will come as a surprise is that these three concepts form the basis of a new theoretical as well as a constructive framework for problems as diverse as the origin of life from inanimate objects to consciousness in human beings.

If we look at all existing theoretical and experimental approaches to these problems, it is interesting to note that they do not work well with each other. A proposal for the origin of life from inanimate objects does not fit well with a proposal for free will or consciousness. A proposal for how you design a car or a computer does not fit well with how life itself is created (either for the first time or from an embryo subsequently). A proposal for free will does not fit well with a proposal for sensing space or the passage of time and so on. Darwin's theory is one example of a framework that tries to address these questions, though not completely (like, for example, it does not address the periods before the existence of life and after the existence of consciousness). In other words, we can say that existing approaches do not offer a unified theoretical framework that works for all cases of large-scale structured dynamical systems.

The theory proposed in this book, on the other hand, offers a unified view to all of these questions and several more, among both living and nonliving large-scale structured dynamical systems. This theory is radically different from all existing theories. It is, therefore, entirely nontrivial to be convinced of and to convince that the concepts introduced here do apply to a wide range of scenarios. I will, therefore, spend considerable amount of time analyzing most cases from a theoretical or a logical perspective. I will also provide constructive designs for synthetic life and synthetic consciousness with guarantees, at least at a conceptual level. The existence of a unified theoretical framework to study all large-scale structured dynamical systems ranging from the creation of life from inanimate objects to the creation of consciousness is the deepest result of this book. The constructive designs, though discussed in detail, will need more work, especially as we try to build them. The emphasis on constructive designs, which is also unique, has forced the theory developed here to be non-philosophical and complete. In comparison, most researchers working on the origin of life and features of consciousness rarely offer constructive designs that do not require other conscious beings or a significant amount of random chance. This has created a huge gap in the reasoning that was nontrivial to address until now.

Therefore, in this book, we have two broad goals: (i) understand the origins (of life and consciousness) and (ii) show how to recreate life and consciousness naturally and synthetically. To achieve this, I will make the theory *invariant* to cells, neurons, organic materials and current living beings by decoupling the abstractions introduced in the book from their actual constructions within living beings. We will do this while still gaining insight into the properties and features of living beings and conscious humans. This is akin to 'systems biology' approach. Such an approach will suggest that features like life and consciousness, though rare and slow to create, are not unique or special. They become properties of a special class of dynamical systems (as we will see).

The approach I will take in this book is as follows. Using the above three notions, I will first present a special class of physical and chemical dynamical systems called stable parallel looped (SPL) systems. Within this class, I will show how to create an *infinite family* of stable dynamical systems that have millions of interacting subsystems with *any specified complexity* (chapter 2). The existence of this infinite family is the reason why I have singled out stable parallel loops as the foundational concepts for the problems of origin of life and consciousness. Equipped with such an infinite family, it will then become easier to construct or choose dynamical systems with special life-like properties.

Specifically, I will show how specially constructed *physical SPL systems* can exhibit features like consciousness (chapters 10-23), emotions (chapter 49), feelings (chapter 22), sensing space (chapter 35), passage of time (chapter 40), free will (chapter 28) and other higher human abilities (chapter 50). I will also show how special *chemical SPL systems* can be used to create primitive life forms from inanimate objects entirely naturally (chapters 52-68), multi-cellular organisms from unicellular organisms for the first time (chapter 70), eukaryotes from prokaryotes (chapter 70), guarantee a direction in evolution in spite of randomness (chapters 53-55) and others. For each of these problems, it may not be obvious how to use the infinite family of physical and chemical SPL systems. However, as we develop the theory and work out several examples, we will begin to see the wide range of features that can be built using SPL systems.

Let me start by describing what we mean by large-scale structured dynamical systems. I will loosely refer to them as structured complex systems. Even though conventional complex systems include both structured and unstructured ones (like chaotic systems), in this book, I only consider structured systems. I will discuss several difficulties we face when working with large-scale structured dynamical systems. Next, I will discuss a few problems we face when we start applying existing approaches to these complex systems. This suggests that we need to develop a new theory. I will discuss why we need an SPL framework to address these issues and not some other framework instead. After motivating the new theory, I will present an intuition on why and how the SPL framework specifically addresses the problems of origin of life from inanimate objects, the diversity of life and consciousness in human beings.

1.2 Large-scale structured dynamical systems

The classes of systems considered in this book are 'large-scale structured dynamical systems'. I prefer to use this term instead of 'complex systems' because the latter term is less descriptive and vague. Our goal is to propose a new theory using which we want to understand features exhibited by these systems. Examples of these systems are living beings and our brain. Examples of features are creation of order from disorder, direction in evolution and consciousness. Our hope is to extend the new theory beyond the examples and features considered here as well.

What are large-scale structured dynamical systems? How do they differ from typical dynamical systems we encounter in our everyday life? It is difficult to propose a precise definition for large-scale structured dynamical systems because of the wide range of features they exhibit. Therefore, I will follow a similar approach taken when defining other difficult concepts like life and consciousness. The idea is to list a set of properties these systems exhibit. For example, life is defined as a system that exhibits properties like reproduction, growth, adaptation, structured organization of subcomponents, homeostasis and others (see McKay [101] and Davison [31]). However, notice that properties like reproduction, growth and others are not precisely defined either and yet most people understand them. For example, it is not clear how many components should be reproduced (like < 30%, > 60%, or DNA should be 100% copied but proteins and other molecules can be present in much lower numbers). Similar issues exist with other properties like growth and adaptation used in the definition of life. Therefore, in a sense, we use these properties primarily to identify systems that are *not* living beings.

The situation is similar when defining consciousness in terms of properties like self-awareness, free will and others. These difficulties will always exist when defining concepts or features exhibited by systems with millions of components arranged in a structured manner. Yet, I will define a large-scale structured dynamical system in terms of properties it should exhibit, in the same spirit as above.

Even after specifying a set of properties, one additional difficulty is when we build an *inorganic* system that exhibits each of these properties. For example, would we be willing to call an inorganic or a synthetic system exhibiting all the above features of life and consciousness as valid? There may still be philosophical objections in this case. I hope that such objections are only temporary.

Given this, let me start by defining what a system is. A **system** is any collection of interacting components and relationships that we want to study as a whole. Physical objects like tables, chairs, cars, air molecules in a room, chemicals in a test tube and others are some examples of systems. The conscious person observing an event and the set of features he is interested in, typically, plays a role in the choice of the system, its components and their sizes. For example, a geneticist studying cellular locomotion may pick the collection of genes, enzymes and other chemicals as the system to study. An engineer, on the other hand, may pick larger mechanical components within the cell as the system.

A **dynamical system** is a system that evolves along a single timeline according to a fixed set of laws. In general, different components of the system evolve at different rates ranging from slowly varying to fast varying. A 'subsystem' is considered static (the opposite of dynamic) if its timescale of evolution is too slow compared to the timescales of other components within or even outside the system. For example, continent evolution is considered static within the shorter yearly timescales of a living being whereas they are treated as dynamic at much larger timescales (like thousands of years).

Using these two definitions, we can now define a **large-scale structured dynamical system**, loosely referred to as a **structured complex system**, as one with the following properties. Several of these properties were borrowed from the definition of life (see McKay [101] and Davison [31]). The goal here is to choose those properties that encompass living beings and our brain, at minimum, because they are the subject of study in this book.

a) *Large-scale and interactions:* The system should be composed of a large number of interacting dynamical subcomponents i.e., physically moving and interacting parts. The number of components is sufficiently large like, say, in the range of thousands or more to call them as large-scale.

b) *Organized structure:* The components are expected to be well-organized (cf. with the same requirement in the definition of life).

c) *Adaptive dynamics:* The system should change over time in response to the environment. The timescales considered are short enough to treat the variations as dynamical or time-varying (versus components that are mostly static). In addition, the dynamics should not be purely internal, but should correlate with the environment i.e., the system should respond to the environment as well.

d) *Growth:* The structure and interactions between the components should grow over time while still maintaining stability. In other words, the growth should not be detrimental to the system. The growth should happen more broadly across the entire system and not just locally through the accumulation of components.

e) *Behaviors:* The system as a whole should exhibit one or more features or behaviors considered nontrivial. The presence of these features (like free will, ability to sense the passage of time and others) within some systems, which are difficult to explain using the current laws of nature, prompts us to treat these systems as special. Some examples of these features are emergent features (i.e., those that are not the 'sum of its parts').

It is hard to be precise where the boundaries are when ascertaining structured complexity within dynamical systems. Yet, the hope is that it is not hard to tell apart if a specific system presented to you satisfies the above properties to classify it as a large-scale structured dynamical system or not.

Clearly, all living beings from unicellular to multi-cellular organisms, our brain, large societies and populations of species (viewed as part of an ecosystem) do satisfy all the above properties so we can call them as large-scale structured dynamical systems. They are well-organized, have large number of interacting subcomponents, adapt, grow over time and exhibit nontrivial emergent behaviors. Among man-made systems, consider a computer as an example. It certainly has a large number of transistors, resistors and other electronic components arranged in a structured way. The dynamics can be viewed as electric current flowing through these components.

However, a computer does not grow and is not adaptive to changes in the environment (cf. our brain which also has a large number of neurons arranged in somewhat static network, although the difference is that it grows autonomously by forming new connections between neurons over time). Also, the behaviors exhibited by a computer are not considered emergent (at least for now). Therefore, I would not treat a computer as a large-scale structured dynamical system as defined above. This is not to say that a computer is less interesting. Rather that the theory presented in this book is not directly applicable to such systems.

Most man-made systems like cars, tables and chairs do fall in the same category i.e., they are not large-scale structured dynamical systems in the above sense and, therefore, the theory presented here would not apply to them. They are certainly important systems to construct and study, though not in this book. Similarly, chaotic systems are not considered as large-scale structured dynamical systems in the above sense and will not be discussed in this book. In the next section, I will discuss several unique difficulties we face when studying large-scale structured dynamical systems. Understanding these unique difficulties will also help us distinguish large-scale structured dynamical systems from traditional ones.

Given a system, we can ask if it exhibits the above features (a) – (e) and if it faces the unique difficulties discussed in the next section. If the answer is yes for most of these questions, it is likely that the theory presented in this book is applicable to them. Otherwise, it is likely that there is an alternate way to study these systems. I use the term 'large-scale structured dynamical systems' to represent this difference between the two types of systems. The hope is that the term *large-scale structured dynamical system* used in this book becomes clearer, not just through the above definition but through the examples (like living beings and our brain) and through the other facets like the difficulties we face when studying them.

1.3   Difficulties with structured complex systems

Large-scale structured dynamical systems are inherently difficult to study and understand. In this section, I will discuss several of them that I have personally struggled with over the last 20 years (relating to the origin of life and consciousness). Let me start by discussing the obvious difficulties that follow directly from the definition of large-scale structured dynamical systems. Firstly, these are large-scale systems. This makes it nontrivial to keep track of all components, theoretically and computationally, at all times. Secondly, these systems are structured. As a result, their components cannot be studied entirely probabilistically. When you perform a certain action, it is not useful to say that this action is one of many possible random chance outcomes. This is unlike studying unstructured chemical reactions in a statistical or thermodynamical way by considering random collisions of, say, an Avogadro number of molecules. The features exhibited by these systems do not stem from random chance encounter of components, rather through structured, reliable and causal sequence of events. Thirdly, these systems are dynamical. The components are interacting at fast enough timescales that we cannot pause to analyze it. Sometimes, a long cascading sequence of events occurs while we are still analyzing just part of the system. Fourthly, combining all of these aspects, namely, structure, large-scale and dynamical interactions, compounds the overall difficulty compared to when these aspects were present individually.

In this section, I will discuss issues unique to large-scale structured dynamical systems beyond the ones discussed above. We tend to typically ignore several of these when studying everyday dynamical systems. However, we cannot avoid these questions if we ever want to discover a complete theory of large-scale structured dynamical systems.

1.3.1 Building large-scale structured dynamical systems

The first difference between large-scale structured complex and non-complex dynamical systems is with the mechanisms and processes used to build them. Recall that the main reason for building a dynamical system is to verify experimentally if the theoretical assertions are valid. However, since a structured complex system has thousands of organized and yet moving parts, how can we be sure that the system being studied would indeed exhibit the theoretical behaviors proposed, if we never construct it? The theoretical analysis typically assumes relatively simple models with several approximations. It appears that these approximations can easily cause divergence between the model and the reality.

Given this, let us look at the following fundamental question. Why does nature need to build a large-scale structured dynamical system? The first requirement is stability. The system should not collapse at least until a conscious being can verify its existence and its features. Let me compare the problem of stability for: (a) man-made simpler systems like cars or computers and (b) large-scale synthetic or natural systems like single cellular, multi-cellular or conscious beings. Let us assume that we start with the most basic raw materials in both cases. These are, for example, chemical elements like carbon, hydrogen, oxygen, iron, silicon and others – *not* transistors, resistors, nuts, bolts, enzymes, DNA or other complex organic compounds.

Preventing collapse of structured complex systems during their creation: To build any dynamical system, it is natural to start with small-order systems first and then add new features over time. As we do this, we would need to ensure basic stability (or survival) even under disturbances, measurement noises, structural defects, abnormal growth and so on. For simple systems, we try to achieve this by carefully picking suitable materials and designs. Once we have a feasible design, we save the creational or the assembly process within, say, a few schematics.

However, with living beings, thousands of chemical reactions are coordinated to attain stability when performing even basic functions like breathing, eating, walking and others. A similar schematic-based approach of specifying its construction is not sufficient. The construction of a living being is a highly dynamical process. We need, additionally, the correct sequence of steps, the correct set of external conditions, when, in what order and how to change them and so on. For example, imagine we have all chemicals needed for a living cell, but placed in separate test tubes (like 20 amino acid test tubes, 4 nucleic acid test tubes, glucose test tube and other test tubes, one for each of the thousands of organic compounds present in a cell). A unique challenge to address here, beyond the static schematic design, is how we can combine these chemicals while maintaining stability *during* the process of creation.

For large-scale structured dynamical systems, the creation process is disturbed by the partially created system since several subcomponents start to operate before the entire system is fully created. For example, a subcomponent like a heart operates, pumping blood and nutrients within the body, even as the rest of the organs are being formed before birth. This self-disturbing feedback loop can lead to potentially unstable situations quite easily. With simpler man-made systems like an airplane, a partially incomplete airplane does not operate (i.e., fly or run its engines) and disturb its own construction. With the creation of a living being, the existing chemicals poured from the test tubes begin to react right away before the complete cell is created. How can we prevent these hundreds of reactions from reacting abnormally, thereby avoiding a collapse (cf. with cancerous growth or miscarriages)?

The laws or mechanisms needed for creating an airplane are Newton's laws, Maxwell's laws, choosing strong materials, using scaffolds and other support structures to withstand forces, pressures and so on. These mechanisms can be carefully controlled in a factory since they are static. For large-scale structured dynamical systems, the presence of moving or chemically reacting parts during creation adds a new layer of complexity. No laws or mechanisms have been presented that addresses this difficulty. In this book, I will show that by using a special class called stable parallel looped (SPL) dynamical systems, we can avoid the above difficulties. I will show that SPL systems can be incrementally constructed to arbitrary complexity in a dynamical fashion (see chapter 2 and beyond). The operation and construction are not disjoint from each other, at least in the sense that both are dynamical steps (unlike the usual case where former is dynamic and the latter is static).

1.3.2   Coordinating a complex system during creation

As mentioned earlier, for a simple system, the construction and operation are disjoint and distinct phases. This is not the case with a structured complex dynamical system. Given this, how can nature create a large-scale structured dynamical system like a living being by itself? Even with help from conscious humans, this self-operational loop discussed above introduces unpredictability, uncontrollability and severe time constraints. Take, for example, the existing order or coordination within living beings. If you are in deep sleep or you are knocked unconscious, most parts of your body behave like dead weights – you cannot lift your hands or legs against gravity even though there are no damages to your body or brain (if I raise your hand and let it go while you are asleep, it falls down as if it were a nonliving component). The only difference to a conscious state is that there is a disruption in the correct distribution of chemicals, their concentrations, the sequences of chemical reactions and their timing. Similarly, a simple cut on your hand triggers a long sequence of chemical reactions to seal and heal the wound. A living being easily dies if such correct sequences of steps do not occur.

We seem to take these sequences of steps for granted as if they were memorized within the DNA. However, naturally coordinating such large number of cascading sequences of chemical reactions is thermodynamically unfavorable (since we seem to create order from disorder). If the localization and timing of accumulation of chemicals is not correct, the cascading chemical sequence stops or proceeds in a different pathway even if the DNA is undamaged. Therefore, DNA is not sufficient for the problem of coordination. It does memorize the set of molecules to create, but not the sequence of steps necessary for coordination.

Except living beings, we do not see such degree of coordination with other systems that contain thousands of ordered interacting components. If nature is capable of achieving coordination seemingly easily, what laws and mechanisms allow this to happen? In this book, I will show that for stable parallel looped (SPL) systems, the above problems become easier. Specifically, I will show that a linear cascading sequence of steps in an SPL system can be equivalently represented as disturbance propagation in a suitable SPL sub-network (see chapter 58).

1.3.3   Creating order from disorder naturally

Large-scale structured dynamical systems, as defined in section 1.2, are considered to have a sufficient degree of order. We have already seen several issues when constructing such large-scale dynamical systems. Let us now look at issues related to who will build these systems and how. There are two ways to build any system: (a) naturally, relying on just the standard physical laws and (b) synthetically, with help from conscious beings.

The former approach requires creation of order (i.e., the structured complex system itself) from disorder. At first glance, such a process seems to contradict the second law of thermodynamics. Second law states that in a closed system, the entropy or disorder keeps increasing over time, unlike what happens in a living being. Yet, to reconcile with the second law, people argue that the creation of a living being cannot be considered occurring within a 'closed' system as there is external energy supplied by the Sun. Indeed, people refer to sugar crystals, snowflakes, cave formations (stalagmites and stalactites) and other simple structures as natural examples in which order is created from disorder. Unfortunately, the laws and mechanisms underlying these simple examples do not generalize to the creation of single cellular, multi-cellular organisms or conscious beings from inanimate objects. Besides, the non-closed-system argument does not help us understand why and how order is created from disorder, albeit in an open system. Answering this for truly large-scale complex systems is one of the primary objectives in this book.

Consider the latter approach of creating complex systems synthetically (item (b) above). Here, the design and analysis involves considerable planning prior to building them. For now, these steps are hidden within a human brain (i.e., the designer). We use the term creativity with no reference to the underlying physical and chemical laws to express this design phase.

Direction in evolution: One other related question is with the creation or evolution of order across thousands of generations within a given species, across multiple species or within the entire ecosystem. Examples of these are evolutionary changes to create short-beaked and long-beaked finches in the Galapagos Islands (across generations), evolving from limbs to wings (across generations and across species), evolutionary changes between predator-prey, parasitic and symbiotic relationships (between species), evolution of multi-cellular organisms from single cellular organisms (changes within the entire ecosystem) and others. Given that we do not want to invoke a designer or Creator to guide the evolutionary process, how does nature guarantee a direction in evolution? Why should evolution proceed in a direction to keep living beings alive, discovering chemicals to maintain the coordination at all stages of life as well as across generations? When living beings are subject to any stress, they could have simply died. Yet, living beings not only struggle to survive but also evolve new features to adjust better with the environment. Any other collection of molecules even if taken from living beings (like from dead cells or fetus from a miscarriage) never evolves to the same extent under such stresses. What structure and internal dynamics distinguish these two cases (evolution in live versus dead cells), especially since the total collection of molecules (and mass) are almost identical between living and dead cells? If we cannot explain creation of life from dead cells, how can we accept creation of life from inanimate objects through random chance or random self-assembly four billion years ago even if we pick favorable conditions?

Darwin's theory implicitly assumes that the basic survival is guaranteed (i.e., by assuming features like transport of correct molecules, timely breakdown of molecules thereby avoiding cancerous growth and others). If we assume basic survival at every step, Darwin's theory seems to suggest that survival of the fittest is capable of creating a direction in evolution. However, we have seen that the coordination of hundreds of chemical reactions using natural mechanisms is a highly nontrivial problem (see section 1.3.2). The set of non-living configurations using a given collection of atoms and molecules are astronomically larger than the configurations that guarantee life. Maintaining feasibility of life at all stages is 'the' biggest challenge for surviving, adapting and discovering new evolutionary features. The transition from living to dead (or order to disorder) has much higher probability than from dead to living. This is also favored by the second law of thermodynamics. What then are the laws and mechanisms that favor the opposite i.e., from disorder to more order?

When we exclude a designer, Darwin's theory only identifies random mutations as the main source for generating new features. How can they *guarantee* a direction in evolution? I will show that, in addition to random mutations, we need specific properties of SPL systems in order to 'extract' a definite direction in evolution (chapter 54). If these specific properties are not present, we cannot guarantee a direction in evolution within a species (like short-beaked and long-beaked finches in Galapagos Islands), across species (like predator-prey, symbiotic and parasitic evolutionary features), across multiple generations and in the entire ecosystem. Once again, we will see that SPL systems allow the possibility of guaranteeing a direction in evolution without a designer.

1.3.4   Laws of nature obeyed when discovering designs

Consider the problem of building a man-made system (like say, a car or a camera). Two distinct phases in this process are worth highlighting. The first phase is to come up with a feasible design (i.e., a design that functions as a camera, say). This design phase relies primarily on human creativity. During this phase, the inventor would first need to be innovative and would need to craft a clever design, choose a set of components to use, decide where to place them, how to assemble and so on. How does one get these ideas? What laws of nature should be applied in order to create the design of a car (not the car itself – just its design)? Can we apply Newton's laws, Maxwell's laws, quantum mechanical laws and Einstein's general theory in some specific way to give rise to a shape of a car and its design? The second phase is to actually build the system (like the car) and see that it works according to the original design specifications. For most systems, these two phases are iterative in nature.

One of the problems with existing theories, discoveries and inventions is that they do not capture the above design phase. They do, however, capture how these systems operate after they are built. In fact, the design phase does not fit within the realm of science (which we anecdotally refer to as "design is an art rather than science"). We do, however, recognize and accept that the discovery process is hidden within a human's brain as creativity. We acknowledge the inventor's intelligence and creativity, but hardly address or attempt to identify the physical laws that led to this discovery. Note, however, that the system being designed, after it is built, must obey the standard laws of nature.

Creativity requires consciousness and awareness, at least of the external world to some degree. This applies even to animals like when a bird builds a nest. The origin of creativity must be hidden somewhere within the true dynamics, namely, just physical and chemical reactions of our brain (of at least the original designer). To see why this is the case, note that all human beings started from a single fertilized egg composed of just physical and chemical reactions. Nothing meta-physical was introduced during pregnancy or after birth, at least in an obvious or measurable way until we have a fully formed brain. Continuing further, the rewiring within our brain through our everyday experiences relies once again on just physical and chemical dynamics. All through this process, nature seems to have relied only on existing known physical laws like Newton's laws, Maxwell's laws, quantum mechanical laws and others. Given this, will we be able to express even creativity (needed for designing systems) in terms of just the same set of laws?

This gap should be addressed by any theory that attempts at explaining the origin of life and our consciousness. The first single-cellular organism created 4 billion years ago has considerable structure, much more complicated than a car, that we can say it has a certain special 'design'. If we want to exclude (a) randomness and (b) a Creator as the 'laws of nature' responsible for this design (of the first life form), it becomes imperative that we bring the problem of design into the realm of science from the realm of art. This is a necessary requirement that all existing theories have essentially ignored until now.

For the design of a car, we were able to get away by attributing the design to human creativity. However, for the problem of origin of life, we have no choice because there are no other conscious life-forms present during that period. We have to solve the design problem using the same laws of nature as the ones that will be used later when the design yields a construction and the constructed system begins to operate in our world. For otherwise, we would have to resort to pure randomness or a Creator to perform the design phase. In this book, I will show that *physical SPL systems* can be used to express features related to human consciousness and our brain including creativity. On the other hand, *chemical SPL systems* can be used to express features related to the origin of life and creation of order from disorder naturally. Therefore, SPL systems act as a bridge to create and understand all features from the origin of life to human consciousness. SPL systems, as we will see, are capable of encapsulating the design phase itself into the standard laws of nature.

1.3.5 Controlling causes and effects

Newton's second law states that when a force F is applied to an object O of mass m in a specific direction, the object moves with an acceleration a in the same direction according to $F = ma$. In this case, the cause is the force F and the effect is the motion of mass m. The important question to ask here is: who applies the force F? Excluding the four fundamental forces of nature, the only answers are: (a) random disturbances like wind, volcanic eruptions, tornadoes, meteor strikes, earthquakes, rain, landslides and others or (b) living beings like a human pushing the object O or a dog kicking it. The former sources are not directed and are less predictable while the latter ones are well controlled and repeatable.

Imagine what the world would look like if there were no life forms on the planet. How would objects move from location A to location B? Who would apply forces to make an object move according to Newton's second law in a precise path like from A to B as shown in Figure 1.1? If the path from A to B is not a straight line, then specific forces do need to be applied at specific times along specific directions to make the

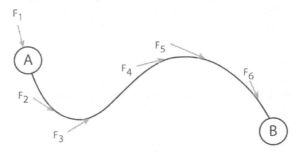

Figure 1.1: Forces applied to change the direction of motion of an object. If an object moves from A to B along a curved path, forces F_1-F_6 need to be applied at specific locations at specific times. Who applies these forces? Such complex pathways are difficult to occur naturally and repeatedly. Yet, the problem of origin of life requires such directed pathways.

object change directions and follow the path from A to B. In this case (with no life forms), the only causes are random disturbances of type (a) mentioned above. Forming interesting motion patterns on such a planet would be minimal or coincidental. Forming large-scale structured complex patterns would be almost impossible.

On earth, every rearrangement of living or nonliving objects that occurs every day like transportation of objects across cities, migration patterns of living beings, shuffling objects in your house and others, are primarily caused by living beings. If not for living beings, most objects on this planet would rarely move around enough in, say, a person's lifetime (except from random weather patterns) that it is worth observing. It is like gazing at the moon looking for structured complex patterns (or at the clouds to see a human face pattern appear occasionally). We humans with free will (as well as other animals) are capable of setting up suitable conditions and applying necessary forces in definite directions to make an object move with high precision.

Controlling causes and effects in a structured manner are possible, primarily, by living beings. This is, in fact, needed for all living beings to exist. This is an important point rarely emphasized. All nonliving systems simply obey the laws of nature thereby generating only simpler patterns compared to what we observe in a living world. Why is this important? Clearly, the problem of origin of life requires us to start with conditions without any life forms and eventually transition to a world with living beings. The set of chemical reactions that occur within a living being are highly structured, occur in a specific sequence and result in directed actions. For this to happen, nature needs to ensure that the forces F applied on different objects at different time instants be of a definite magnitude and direction. This, as we have just seen, is nontrivial when there is no life on the planet.

Let me elaborate this further with a simple example. Consider four objects A, B, C and D located at different places as shown in Figure 1.2. These objects could be molecules like H_2O, glucose, a protein, ATP or other chemicals that a living being requires. They may or may not be present in abundance. Consider the situation in which we want A and C to combine to form AC, followed by transportation of AC to a different and sufficiently distant region R_1 where B should also arrive at about the same time to react with AC to form ACB (see Figure 1.2). After ACB is formed, we want it to move to region R_2 where it combines with D (after D arrives there as well) to form $ACBD$. Such a sequence of steps is not unusual within living beings. For example, when a hormone arrives at a cell, a series of cascading reactions occur at the surface of the cell, which eventually continues inside the nucleus as well by actively transporting chemicals into the nucleus. When neurotransmitters arrive at the dendrites of a neuron, a series of steps occur eventually generating an action potential that propagates along the axon to distant regions causing the release of neurotransmitters through the synaptic cleft to the neighboring neurons. I will discuss several examples of these later in the book. In fact, cellular biology is filled with examples of these sorts and I refer the reader to textbooks like Kandel et. al [82] and Garrett and Grisham [48].

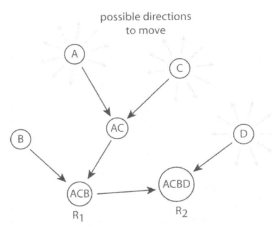

Figure 1.2: Controlling the direction of motion to follow a chemical cascade of reactions. In this example, each molecule has multiple directions to move along. However, they need to pick specific directions at specific time intervals in order to run a specific set of chemical reactions. Each such change in direction requires a new force to be applied. Who applies these forces when there is no life yet on the planet?

The question once again is: who directs the motion of objects A, B, C and D in the sequence specified above to eventually form $ACBD$? We cannot argue that such a precise sequence of steps is not important. This is because life ceases to exist if not for such precision. For example, during embryogenesis, if the accumulation of chemicals does not occur in a specific sequence, it is possible to grow multiple eyes, limbs at abnormal locations or not form entire organs (see Nüsslein-Volhard [115]). If you cut your finger, your body needs to follow a precise sequence of steps for both blood clotting and for fighting infections. Otherwise, we can easily die.

Unless specific forces are applied, the objects A, B, C and D do not move in the desired directions. If A and C are moving in different directions, unless specific forces are applied, they do not change direction to meet and form AC. If random disturbances are the only sources for applying forces (i.e., of type (a) above), then the number of possible directions for object A and B are infinite as shown in Figure 1.2. Among these possible directions, the number of possible paths that cause both A and C to approach each other are too few. Note that we are not talking about a single molecule A and C moving to a given region, but a large collection of them. Therefore, the probability of A and C approaching each other if the only mechanisms are purely random is close to zero. Of course, the standard counterargument suggested here is that we wait long enough and that we have a large number of objects A and C. With large number of objects and a sufficiently long time, A and C can indeed approach each other. However, within a short period, we would want B to approach AC, followed by D to approach ACD at a different region and after another short period. And this is just for a single simple chemical cascade of reactions. A living cell has hundreds of such chemical cascades happening every day. On the other hand, if humans with free will exist, the probability of chemical cascades can be made equal to one. A human could apply the necessary forces to change direction at will and make A and C move together, followed by moving B and others in the sequence specified above (like, say, a human gets the correct test tube of chemicals and mixes them in the correct order at the correct time).

Given that humans did not exist 4 billion years ago, but we still want A, B, C and D to move as described above, is the creation of life a purely random chance event in this universe? Is our existence that lucky? Or is there a clever way to convert this seemingly zero-probability event into 100% guaranteed event? In this book, I will show how to achieve such guarantees using what are called minimal structures (see chapters 57 and beyond). The minimal structures are those based on SPL systems that can occur naturally with finite probability i.e., without help from living or conscious beings. These minimal structures do not identify precise set of chemicals, chemical reactions and environmental conditions. This is because if we did that, the question of who finds these chemicals and sets up these conditions will still be unanswered. What is the intuition behind this new solution?

Let me now try to motivate the solution at least partially. Let us change the problem a bit (just temporarily) by considering A, B, C and D to be magnetic materials instead. Objects A and C move randomly as before. The number of possible pathways is still infinite. However, the instant A and C are reasonably close enough, the magnetic attractive (or repulsive) force suddenly eliminates all of the infinite pathways and picks the direction of motion to just a single pathway – either A and C move towards each other or away from each other.

In other words, what we observed with this example is the following. We started with a system whose components move around randomly and we have imparted additional structure to it. This structure was the inclusion of standard laws of nature like magnetic materials and electromagnetic force. This structure was enough to improve the probability of A and C getting together from zero to near certainty (almost 100% probability) when they are close enough. The same situation applies when A and C are electric charges subject to electric force or even heavy masses subject to gravitational force.

Of course, these specific laws of nature do not generalize when A, B, C and D are molecules like NH_3, CO_2, CH_4 and other organic molecules (because they do not have magnetic properties or other special attractive properties). We need additional structures, laws or properties to cover all cases. Nevertheless, there is hope with this approach. We can look for new but simple structures or laws that we want to impose on a random system in order to improve the overall probabilities. Each of these newly introduced structures should be simple enough that nature has a finite probability of constructing them or using them (similar to how we used magnetic force only when the objects are close enough). These structures should themselves be able to apply forces and change direction of motion of objects in a somewhat predictable and repeatable way even in the absence of living beings. These are the *minimal structures* mentioned earlier and they will be based on SPL systems that will be discussed in detail in this book (see chapter 57).

The difficult part is to take the above simple examples and minimal structures and show how to apply them to systems as complex as living beings. We want to identify and specify 'all' details at least in a theoretical sense instead of just stating them in vague terms (cf. sugar crystals, snowflakes and other simple patterns as common examples stated for creating order from disorder that simply do not generalize to the creation of systems as complex as life itself). To do this, we once again need to show how to build an infinite family of large-scale structured dynamical systems, specifically, both physical and chemical SPL systems of arbitrary complexity, as we will see.

1.3.6 Analysis versus synthesis of features

For large-scale structured dynamical systems, the explanations people offer when analyzing a given feature is, for now, disjoint from the corresponding explanation given to synthesize the same feature. This is not the case with simpler man-made systems. For example, analyzing how a car works versus building it is not too disjoint from each other. The components, the static and dynamical arrangement of them, the mechanisms and principles both for creating and for understanding are quite similar.

However, the explanations given for understanding how a single cellular organism works versus how to create it from inanimate objects have no connection. For example, the DNA sequence, the transcription and translation mechanisms needed to *understand* the role of different enzymes for a given feature are not related to how nature *discovers and creates* the necessary special and long-chained DNA sequences or the enzymes. For now, the initial synthesis of DNA sequences are only through random mutations (according to Darwin's theory), but the mechanisms to understand the operational features are not random. In fact, they are the standard laws of nature like quantum mechanical laws, Maxwell's laws and others. However, these laws do not explain how multiple molecules would come together in a small region in space and in time so they get a chance to react (like within a cell). In fact, this even appears thermodynamically improbable because it is analogous to expecting air molecules to reach one corner of a room. Yet, this is required at minimum for creating new molecules including the correct DNA, RNA or amino acid sequences. What laws and mechanisms bring the correct sequence of monomers in the correct order at the correct location and at the correct time to create the correct sequenced polymer, reliably and repeatedly? With man-made systems like a car, this coordination is done by conscious humans.

The situation is similar with features of our brain like consciousness, self-awareness, free will and others. A simple test to determine whether the explanations for analysis and synthesis are disjoint is by asking the following generic question: can we use the explanation used for analysis to construct the same feature? For example, can we use the arguments given to help us understand free will to build a system with free will as well? How about understanding versus creating a system that senses the passage of time, understanding self-awareness versus creating a system that becomes self-aware and others? Similarly, can we use the mechanisms discovered to help us understand how an enzyme is created from DNA to explain how nature created the entire process of creating the enzyme? Such questions expose the gap in the original reasoning, if present. In this book, the SPL framework proposed will be used for both analysis and synthesis of specific features.

1.3.7 Error accumulation from idealized assumptions

When studying large-scale structured dynamical systems, it is common to make idealized assumptions, at least, initially. During this phase, we ignore a few undesirable features like friction, heat of dissipation, electrical resistance, saturation, delays and others. The hope is that these factors can be included later without significantly altering the analysis. However, with thousands of interacting dynamical subsystems, keeping track of the errors is as hard as keeping track of the desired signals themselves.

While this difference exists during our analysis, conscious humans have a choice to try out different idealized models. However, nature trying to create large-scale structured dynamical systems like cells and brain does not have such a choice between idealization and reality. When an electrical signal propagates along the axon of a neuron, there are undesirable properties like electrical resistance, saturation and delay that decrease the speed of propagation of the neuron spike. Yet, nature would need to take advantage of them instead of fighting to eliminate them. As an interesting example, I will show how electrical resistance, which is usually considered undesirable because it generates unnecessary heat, is directly *responsible* for giving rise to neurons' information carrying ability (see chapter 5).

In this book, we will not make any idealized assumptions as we develop both the theory and propose designs for large-scale structured dynamical systems. Instead, I will formulate principles, concepts and mechanisms that incorporate these seemingly undesirable features right within the definition. For example, loops are not idealized with perfect periodicity, fixed time period T of repetition, indefinite repetitions or reaching identical states during each repetition Rather, loops already factor these pragmatic cases within its definition (see chapter 2).

In essence, I push the above seemingly undesirable features into the axioms and definitions in order to ensure that the subsequently derived results and relationships are rigorous and well defined. This is in contrast to the traditional approach in which we try to be extremely precise with the axioms and definitions. Unfortunately, such a precision forces the subsequently derived analytical models to be detached from reality because of the idealized assumptions. Such an approach rarely generalizes to large-scale structured dynamical systems because of the magnification of ignored effects. These result in a huge gap in the frameworks used by the theorists compared to the experimentalists, say, in biology and in complex systems. Experimentalists argue that the models chosen by theorists are not realistic and that they do not capture the scale of complexity observed within living cells. The new approach based on physical and chemical SPL systems avoids these issues, as we will see.

1.3.8 Causality within parallel dynamics

When we study large-scale structured dynamical systems, it is inevitable that we need to keep track of dynamics of several components simultaneously. For example, in a single cell, several chemical reactions occur in parallel. Different types of proteins are constantly created several of which are broken down, cellular respiration (Kreb's cycle) occurs continuously, different molecules are transported in and out of the nucleus and so on (see textbooks like Alberts et. al [2] for several detailed examples). Even though the pathways are not visible, several of these reactions are directed. There is both simultaneity of occurrence of dynamics across a collection of seemingly unrelated chemical reactions (i.e., parallel dynamics) and at the same time occurrence of several cascading linear sequence of chemical reactions (i.e., a serial flow).

In the case of our brain, we have a well-defined network of pathways. This makes the parallel dynamics and the serial flow mentioned above a bit more obvious to observe. Several neurons in our brain are simultaneously excited like, for example, several cone cells fire on your retina simultaneously when you look at, say, a tree. This is the parallel dynamics. At the same time, the firing of a given neuron releases neurotransmitters to the neighboring neurons causing them to fire action potentials as well like, when say, the firing pattern from the retina flows through the optic nerve to the primary visual cortex. This is the serial flow of signals within the network of our brain.

In addition to the serial flow and the parallel dynamics that occurs within large-scale structured dynamical systems, humans who analyze these systems introduce another layer of complexity by bringing in what are called abstractions. These abstractions are typically sentences in a language. Humans create logical or causal statements distinct from the raw dynamics mentioned above. For example, we say 'as you smelled freshly baked pizza, you started feeling hungry'. This statement is simple and it feels like a causal statement with the cause, 'smelling freshly baked pizza', producing an effect, 'hunger'. However, the true causality in this statement originates from the raw dynamics that occurs within our brain and our body. The raw dynamics in this example is far too complex to keep track of – we need to track dynamics of thousands of neural firing patterns originating from our olfactory sensory neurons until eventually it produces a feeling of hunger. Other examples of causal-looking sentences are: 'since you were a bit early at swinging your bat, you missed hitting the ball', 'as you were walking by, you briefly looked down and noticed a dime on the sidewalk' and so on.

Almost every sentence has certain logic to it. This makes us believe that most of our sentences are causal. They also encode and store the true causal dynamics that occurs in the external world into a suitable form in our brain. Sentences, especially of the form 'if A, then B' appear causal. Here A is the cause and B is the effect. However, we need to realize that the true source of causality exists only within the raw dynamics of the event. We have discovered language as a convenient form of communication that compresses an enormous amount of parallel dynamics occurring within, say, our brain or our body (like the neural firing pathways within our brain network). If the external causality (of dynamics occurring in the real world like a ball rolling down the plane) is not mapped correctly in our brain, then it is difficult to trust anything we see around us. The presence of parallel dynamics raises several difficult questions, most of which we tend to overlook because of our belief in our language abstraction. Let me pick a simple example to illustrate one of these issues.

Consider looking at a red table followed by looking at a blue chair. The causality here implies that we have looked at the blue chair 'after' we have looked at the red table. This seems so obvious that we simply take it for granted. However, let us briefly look at the true dynamics occurring within our brain that is responsible for creating this impression of causality. Our brain is a complex interconnected network of neurons. A given neuron fires an action potential to the neighboring connected neurons when the membrane potential at the axon hillock exceeds a threshold voltage. We will discuss this in detail later as well (see also Purves [122]). When the image of a red table falls on the retina, a large collection of cone cells arranged in the shape of an inverted table fires. These signals proceed into the brain through the primary visual cortex through a complex network of pathways. If we see a blue chair, a different set of cone cells in the shape of an inverted chair fire (different from inverted table) and follow through a different set of pathways. The pathways are different because: (i) several disjoint set of cone cells fire for a table versus a chair and (ii) the timing of firing even for the common set of cone cells can still be different.

Note that there are thousands of pathways to keep track of for both table and chair images originating from the retina. They proceed through a complex network – some are along short pathways and others are along long pathways. There is no way to know ahead of time which ones are the shorter ones, where they are destined to go and when they should reach there. It is quite likely that an earlier pathway is shorter than the later one because of this complex parallel dynamical network. This can be a problem because it is easy to violate causality, when, say, the later pathway for a blue chair is shorter and has reached a given destination (i.e., the one responsible for recognition) earlier than the earlier pathway for a red table, especially if the time lag between viewing red table followed by viewing blue chair is short. We would then incorrectly conclude that we saw the blue chair before the red table.

The parallel dynamics within a complex network has essentially eliminated all guarantees for causality for a collection of neurons. Note that causality is guaranteed for each individual pathway, by definition (because of the cascading serial flow). However, it is when we track thousands of neural pathways as a group (like the shape of an inverted table or a chair), we can no longer guarantee causality. Yet, we rarely make mistake in causality with such large groups of neural pathways.

There are other problems too. The blue color could have reached a 'recognition' state at the same time as the table instead of the chair (because of longer pathways). This can make it seem that the table is blue instead of red and the chair is red. The legs of the table, its surface texture, its edges, angles and other patterns could have taken different paths from the top of the table making the whole problem of combining them into a single entity (i.e., a red table) a difficult one. In a parallel network, it is natural to expect each of the neural pathways for these subcomponents to take different paths. However, it becomes an almost impossible problem to coordinate and recombine them later after each subcomponent is individually recognized.

Who will coordinate this effort? Which pathways will wait and how, given that these are chemical reactions? What if you blink and did not receive a signal for a brief period? What if during this period, the order of signals has reversed? How would our brain know what is the correct order and what is the incorrect one? What if there are delays in processing like when you are distracted for a brief period? Should our brain drop some of the signals if they become inconsistent? How would our brain even know that some of the signals are inconsistent? Who is validating all of this information flow through the system? Are there any markers with, say, timestamp information to suggest how to reconstruct the data later (similar to how we do with internet packets that are transmitted across different routers)? Whenever we split information and send across different channels, it is a nontrivial problem to reconstruct the original message. How is our brain solving this problem? Does our brain even care about solving these problems? Since our brain simply obeys the laws of nature, there is no obvious notion of preferred pathways unless some other natural mechanisms ensure such preference.

Our brain solves these problems with almost no errors. We never see a red table and blue chair in a non-causal order (i.e., blue chair followed by red table instead). We never see a red chair or a blue table in a mismatched order instead of the correct form, namely, a red table and a blue chair. Every causal sentence like those mentioned earlier (i.e., of the type: if A, then B), though involves 'groups' of neural pathways, which can potentially run into the same issues mentioned here, are never in a non-causal order (i.e., if B, then A). Even though we hardly think of these issues for every logical sentence we formulate, we can no longer dismiss them as easy problems.

In this book, I will show how to use the SPL architecture of our brain to address these issues entirely naturally. One of the key ideas is to transmit only those types of signals through the network that is invariant and robust to losses. For example, we cannot transmit the true information, namely, blue or red color directly through the network (which is, unfortunately, what we do with computer networks). If we did, we would need to keep track of this 'useful' information and then later coordinate/combine them in the correct order, which we know is a hard problem. We need a suitable robust encoding based on standard neural spikes and a common set of neurotransmitters, while instead using a unique way to reconstruct the original signal that guarantees causality and accuracy. I will discuss how we can do this, thereby avoiding the above problems, in the later chapters using SPL systems.

1.3.9 Theory versus experiments

With most scientific approaches, progress in theory and experiments go hand-in-hand. Sometimes new theoretical models are proposed to explain a given experimental result. This model then serves to predict outcomes in new situations, prompting new experiments to be developed. If the predictions are accurate, we gain confidence in the theoretical model. If not, we look for an alternate theoretical model. In this sense, experiments are used to ultimately validate any proposed theory.

There is also a two-way interaction between theory and experiments. Sometimes experiments guide the development of new theories (like the black-body radiation, Young's double slit experiment and others to guide the development of quantum mechanics). At other times, the theory guides new experiments as well. Some of these experiments are designed to validate the theory (like observing bending of a ray of light by gravitational mass to verify Einstein's theory of relativity), but others are seen as new inventions prompted primarily because of the theory (like, say, radio transmission built using Maxwell's theory).

What is the interaction between theory and experiments when studying large-scale structured dynamical systems? Since there are practical difficulties with keeping track of thousands of interacting dynamical components in real-time, the theoretical models tend to make severe approximations. As mentioned in section 1.3.7, these idealized assumptions make the theory deviate from experiments considerably to the point that they are no longer useful. Similarly, if we perform experiments involving thousands of dynamical components (like, say, simulating a brain network or cellular processes), the amount of data collected will be so large that it is not obvious what patterns to look for. Without a valid theory that identifies the set of useful patterns to study, there is still too much information to process.

Additionally, there are fundamental theoretical difficulties with the topics discussed in sections 1.3.1-1.3.8 like the chicken-and-egg problem with the origin of life, how a system knows about itself to become self-aware and others. It is not obvious how to take simple theoretical patterns for basic components and then generalize them to create large-scale structured dynamical systems. For example, having a detailed model for a single neuron does not suggest how billions of neurons work to create features such as consciousness, free will and others. In this sense, proposing a theoretical model is more important, at least initially, to resolve these fundamental issues before designing experiments to validate the theory.

Another reason for proposing a theory first is the following. For some problems involving large-scale structured dynamical systems (like for finding a cure for cancer), we can find solutions in an incremental manner. Here, there is a natural notion of error that we can keep improving over time. For these problems, even though theoretical models do help, experiments play a more critical role at each step. However, there are other problems for which incremental solutions do not make much sense. For example, we cannot build a system that is 50% self-aware or is 40% capable of sensing the passage of time and then we look at improving this system iteratively until it acquires 100% self-awareness. For these types of problems, we need a complete theoretical solution in one-step because there is no notion of error that we can keep reducing if we tried the experimental approach. It is more critical to have a self-consistent 'theory' before we device an experiment. What experiments should we try when we do not know what 100% self-aware or 100% sensing passage of time means in a theoretical sense? How do we even know when to stop i.e., that we have built the desired feature completely?

In this book, I will focus mostly on the theoretical aspects to address the questions raised in the previous sections. At the same time, I will propose an infinite family of physical and chemical SPL systems as a way to construct large-scale structured dynamical systems. This infinite family will serve as a proof of existence of realistic large-scale structured dynamical systems (i.e., with almost no idealized assumptions). I will show how to use them to construct each feature of life and consciousness.

1.4 Why a new looped systems framework?

From the above discussion, it is clear that we need to overcome several new challenges unique to large-scale structured dynamical systems. The previous section provided just a glimpse of questions that I have personally struggled with over the last 20 years. Throughout the book, I will present several more such questions as we study specific topics (and offer solutions). However, do we need a new framework to address these issues? Can we not use existing theories and laws instead? In this section, I will discuss why we need a new SPL framework (at an intuitive level) and how it differs from existing approaches. I realize that we have not yet defined stable parallel looped (SPL) systems. I will define them later in chapter 2. For now, let us assume that the terms stable, parallel and loops are just what they mean in the colloquial sense. You will see later in chapter 2 that these colloquial meanings are fairly close to the definitions we will use in this book, as one would expect.

The current research on the problems of origin of life and consciousness is too vast to describe here in a few pages. Yet, it is safe to say that most of the above issues are not addressed by existing approaches to a satisfactory degree. It is well accepted that these problems are unsolved and that the underlying structures identified are far from complete. It is also not easy to compare the new theory with existing theories because there is no existing theory that spans both the origin of life from inanimate objects and the problem of consciousness. For this reason, I will only compare specific topics with the corresponding existing proposals. For example, in chapter 10, I will briefly compare the new theory with other existing theories on consciousness. Later, in chapter 71, I will compare the new theory with Darwin's theory of evolution. As a result, the comparison in this chapter will not comprehensive. I refer the reader to books like Wills and Bada [166] and Jablonka and Lamb [77] to pursue more detailed study of the existing research.

In this section, I will argue that the causal space-time dynamics and active loops are both necessary and sufficient when analyzing and building large-scale structured dynamical systems. I will only provide an intuition for now and will limit the discussion to living beings as the exemplary class of large-scale structured dynamical systems. The goal is to show the broad applicability of the new framework and the unified principles that bind diverse set of topics ranging from the origin of life from inanimate objects to consciousness in human beings.

1.4.1 Using randomness to create definite phenomena

Consider the problem of origin of life. One of the questions here is how we guide molecules and/or objects to move to the correct location, at the correct time, in the correct order, for a correct period of time and do this repeatedly for a sufficiently long time to make multiple chemical reactions work together to produce useful products. Since there is no agent to help us, we need to explain these steps using just random mechanisms. As we can see, there is a nontrivial amount of coordination needed across hundreds of different types of chemicals.

To address this problem, we need to first identify laws that allow us to start from random initial conditions and let us converge to a definite nonrandom state (like the correct place or the correct time). If we can do this, there is hope that we can apply the same laws repeatedly to help coordinate a large number of steps. Let me give a simple example of a dynamical system that indeed exhibits this property. Consider a pendulum or a swing that contains a single asymptotically stable point (its rest point). These systems eventually converge to their unique and nonrandom state, namely, the asymptotically stable point, even for a large set of random initial conditions. For example, a pendulum converges to the resting state for most initial conditions. In this simple example, even if we started from random initial conditions, we know where the system will settle down to after a long time, namely, its asymptotically stable point. Our objective is to find more such systems and conditions.

In real physical systems (as opposed to idealized mathematical dynamical equations), the presence of interactions like friction, delays, saturation and other nonlinear phenomena makes it less likely to have asymptotically stable points. However, one pattern that is common in most nonlinear and practical dynamical systems is what is called as a *limit cycle* (see chapter 14). Limit cycle is not a single point like an asymptotically stable point. Rather, it is a set of points in the state-space of the dynamical system with the property that if the system enters a state corresponding to a point on the limit cycle, then it continues to stay on this limit cycle. The system that has a 'stable' limit cycle has the additional property that when you have random initial conditions close enough to the limit cycle, the system converges to the limit cycle (see Figure 14.3).

Limit cycles manifest in the real world as periodic phenomena, oscillations or other types of repetitive phenomena. Such repetitive phenomena are more common in the real world than the asymptotically stable points. If, in addition, we relax the requirement that the repetitive phenomena should repeat for a very long time (like idealized limit cycles) or that they should have a fixed period of repetition (like perfect periodicity), then the repetitive phenomena are even more common in our everyday life.

Therefore, given that purely random phenomena can converge to such repetitive limit cycles at least for a short period before it gets disturbed, it implies that we have several dynamical systems and conditions that can exhibit non-randomness from randomness. This is important because we do not require external agents like God or other conscious humans to create or to converge to a limit cycle for such a class of dynamical systems i.e., we can create order from disorder (in a limited sense as discussed).

The question then is whether nature can create more order from disorder utilizing these limit cycles in unique ways. If that were the case, the system evolves and converges over time to a large set of interesting non-random phenomena even if they start with randomness. Living beings are examples of such convergence to non-randomness from randomness by increasing the set of limit cycles they can exhibit over time (as we will see in chapters 53 and beyond).

These limit cycles are what I have been calling in more general terms as loops in this book. I also require stability of these loops (or limit cycles) and the fact that there are a large number of parallel sets of loops for systems to exhibit interesting phenomena. These are what I call as stable parallel looped (SPL) systems, at least from an intuitive level. I will present more details on these systems in chapter 2.

If there is no such convergence to stable parallel loops for large-scale structured dynamical systems, then these systems tend to have considerable randomness (like air molecules in a room) to exhibit interesting patterns. The more they evolve towards SPL systems (to be discussed later), the more chance they will have to exhibit interesting patterns as well as to offer guarantees of non-randomness from randomness. Therefore, the main objective of this book and for the problems of origin of life until the creation of consciousness in humans is to identify how to build a large, possibly, infinite family of SPL systems (with large collections of limit cycles). We want these systems to be easy to build both in the physical and in the chemical world without the help of any external agents like God or conscious humans (see chapter 2 that discusses how to create such an infinite family). We will see that such creation of non-randomness happens iteratively over millions of years. This is a simple intuitive view of the importance of physical and chemical SPL systems.

1.4.2 Necessity and sufficiency of active SPL dynamical systems

In this section, I want to argue that active stable parallel looped dynamical systems (terms used in a colloquial sense for now) are necessary and sufficient for constructing life and consciousness without the help of other conscious beings. Looped dynamical systems are those that have some sort of reliable and repeatable behavior. We will see that loops are not just convenient but they are necessary for creating large-scale structured dynamical chemical systems. For example, if you have a large set of chemical reactions that should coordinate naturally, this is possible only if they are part of some looped reactions. For otherwise, the probability that the correct chemicals be present at a given location and at a given time is almost zero. Repeatability of looped reactions is necessary in that sense.

Let me elaborate on this point by considering the following simple analogy. Consider a set of molecules moving freely before the existence of life as analogous to traffic flow with people driving around arbitrarily along different pathways. To extend the analogy, two molecules (CO_2 and H_2O) reacting to form a new molecule (H_2CO_3) as part of a chemical reaction is analogous to two people meeting at a given location and sharing a ride. These two people travel together now. A molecule (H_2CO_3) breaking down into two molecules (H_2O and CO_2) is analogous to splitting the ride into two people and letting them travel in different directions. A generic chemical reaction ($2H_2 + O_2 \rightarrow 2H_2O$) would require mixing two or more people to create new people (here is where the analogy becomes weak) and letting them move freely.

One of the questions we want to answer for the problem of origin of life is how simple molecules like NH_3, CO_2, H_2O and others can eventually combine to form a unique structure which we regard as life. As mentioned previously, one question to answer is how these freely moving simple molecules come together at the correct place at the correct time in the correct order, converting to the correct chemicals and doing so repeatedly while also running a well-defined ordered sequence of chemical reactions (let's call this temporarily as the *coordination problem*). This is analogous to seeing how certain specific people, who were traveling arbitrarily, can congregate at some region at the correct time in the correct order repeatedly, while combining, sharing rides and so on. Observing such unique traffic patterns among randomly moving people seem almost impossible. The required organization appears to be too perfect and there is nothing obvious that seems to guide towards such perfection (excluding Creator or analogously people communicating and planning together to congregate). Answers based on chance (or without clear explanation based on physical laws) combined with long waiting period or choosing an abundant number of molecules (or people) seem less convincing for the type of structure we expect to create in a living cell. This is one reason why the proponents of Intelligent Design claim that life was not created naturally and instead it was designed by some intelligent system, possibly, a Creator.

While the above objections are valid, we do not need to go to one extreme and offer Creator as the solution. There is a better and simpler solution that we can borrow from the traffic flow analogy. To see how to extend the analogy, let us first understand why the coordination problem appears so difficult. It is because we have neither imposed nor identified any 'reasonable' structure within the motion of molecules. We have assumed them to move completely arbitrarily and independently. Therefore, seeing the emergence of a highly ordered system naturally from such a state of disorder seems unconvincing. What then is a reasonable structure we can identify or impose on this system? One of my answers is loops (see chapter 52 and beyond for other structures), but let me briefly explain what I mean using the above analogy.

If we want to coordinate 10 different groups of randomly moving molecules to reach a given region, the chances of this happening naturally is close to zero (cf. air molecules reaching one corner of a room). However, if we breakdown the problem into several smaller problems (one for each group of molecules) and impose the restriction that each group moves in looped pathways as shown in Figure 1.3 (the pathways can be more complex than depicted as long as they are loops), then the chances of all 10 groups reaching region R simultaneously has a finite probability. In other words, by imposing a looped structure, we have effectively increased the probability from near-zero to a finite value. This is the key idea – just make each group of molecules as part of some physical and chemical loops. How do we do that? Who does that? Is this requirement even reasonable i.e., is it naturally possible? These are some of the questions I will answer in later chapters.

But let's look at whether this idea is practical or if it is just a theoretical suggestion. I claim that looped networks are quite common around us. Traffic flow, weather patterns, economic systems, periodic systems, oscillatory systems and several others are examples of looped systems. We probably did not realize the pervasiveness and/or the importance of loops. Let me return to the traffic flow analogy to elaborate on the idea discussed in Figure 1.3. Consider a situation where people want to go from San Jose to San Francisco. This can happen in one of two ways: (a) everyone takes

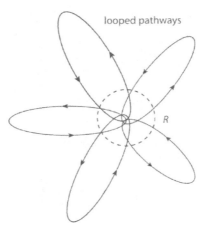

Figure 1.3: Imposing a requirement of looped pathways for molecules. Here, we have molecules moving independently but in looped pathways. The actual looped pathways could be much more complex than those depicted here. Imposing this looped requirement, however, makes it easy for all groups of molecules to still meet at a given region R with finite probability compared to the case when the same groups of molecules move independently and randomly (like air molecules in a room).

their car and drive themselves independently or (b) people mostly use public transportation like a combination of trains, buses and, possibly, even cabs. Approach (a) is analogous to the less structured random motion of molecules while approach (b) is much more structured and fits with the idea of Figure 1.3 because several *loops* were prebuilt into the transportation system. For example, the cab service has some approximate loops (like going from P to Q and back – this is not a precise loop, but the point is that the cab is not wandering off i.e., there is some repeatability). The trains and buses clearly operate in a looped manner – they go from P to Q and back along a specific path on a regular basis. They also operate independently of peoples trips. We have several choices for looped pathways like Muni, Caltrain and BART for trains/buses and with multiple lines (like red line, blue line and so on). With buses, we have several numbered buses, one for each specific looped route. If the train or bus frequency is high enough, then if you miss your connection train at a location X, you only need to wait for a few minutes until the next bus arrives.

One way to interpret, as a loose analogy, is that most existing research tries to address the problems of origin of life and consciousness using approach (a) while this book shows an alternate solution using approach (b). I show how to build large 'public transportation' networks (which are analogous to physical and chemical SPL networks).

Of course, the issue with approach (b) is that someone has to build all of these looped routes. They should operate regularly and continuously (i.e., they should not shut down for extended periods of time) and with sufficient frequency (i.e., there should be a train every few minutes). I want to argue that this looped network idea extended to the chemical world is not a stringent requirement even before the existence of life. In the prebiotic world, it is conceivable that there are special pockets like hydrothermal vents and hot springs where richer set of chemicals existed as loops. This surely needs experimental verification. Yet, to give us confidence, it is worth looking at some examples like simple reversible chemical reactions, complex reactions like Belousov-Zhabotinsky reaction and others, reactions with the presence of catalysts or enzymes, multi-stage looped pathways like Kreb's cycle (see Figure 2.8) and more general metabolic networks (see Jeong et al. [80]; Nicholson [114]).

Once we show that looped physical and chemical networks are pervasive (which will take the rest of the book!), Figure 1.3 suggests that the problem of coordination has a finite probability. By identifying and sometimes even imposing a looped network, the problem of coordination became easier. There are other advantages too. You let the existing looped network (maintained by somebody else like the transportation authority) 'work' for you. You hop on and hop off at specific locations and ride along part of the loop. You are not 'consuming energy' except during these hops. If you want several people to move towards San Francisco at peak time, say, you introduce a disturbance in the form of increasing the frequency of number of trains and buses in one direction (or one half of the loops – going towards San Francisco). This 'disturbance' will transport more people in that direction. In more general looped networks too, disturbance propagation is a useful way to produce a cascading sequence of steps naturally (like, say, when you have a cut on your finger, a cascading sequence of steps should occur to seal and heal the wound – we can achieve this using disturbance propagation within a suitable looped network instead of worrying about questions like how to memorize these steps, who created, planned and executed them in the correct order and others). I will discuss disturbance propagation in detail in chapter 58. In other words, looped networks are indeed pervasive and useful patterns to study.

Even in the case of our brain, looped dynamics is fundamental for forming memories, sensing passage of time and self-awareness. If you look at a table, turn away and look back at the same table, how do you know that you are looking at the same table? Repeatable events such as these have to be interlinked within looped dynamics of our brain to form a membrane of truths (more details later). The external events can then be correlated to the internal dynamics of our brain to ensure consistency and one-to-one correspondence between the two types of dynamics. This is necessary for consciousness, as we will see. If not for the sustained looped dynamics within our brain, every memory would be just a fleeting one, making it impossible to relate causal events in a continuous manner.

In the chemical world, another way to see why looped reactions are necessary for creating life is by considering Figure 1.4. It shows the energy change for a few sample reactions in abstract terms. Some reactions proceed naturally from an uphill energy state to a downhill state. They can be either entropy driven or enthalpy driven reactions – see Borgnakke and Sonntag [15] that describes how Gibb's free energy can be used to determine whether a process or a chemical reaction proceeds spontaneously in either the forward or the reverse direction. After reaching the

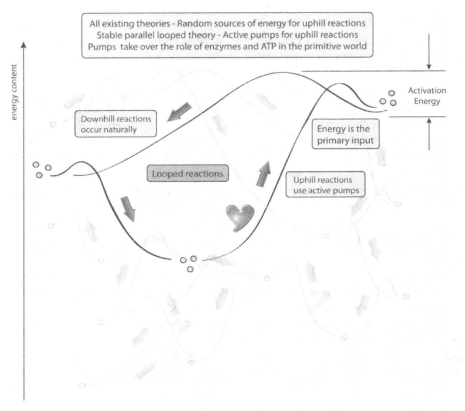

All existing theories - Random sources of energy for uphill reactions
Stable parallel looped theory - Active pumps for uphill reactions
Pumps take over the role of enzymes and ATP in the primitive world

energy content

Activation Energy

Downhill reactions occur naturally

Energy is the primary input

Looped reactions

Uphill reactions use active pumps

Figure 1.4: A looped network with uphill and downhill reactions. If we represent different molecules at different heights based on their energy content, we can view the downhill reactions as ones that occur naturally. However, reactions proceeding uphill require either passive random energy or an active source of energy. All existing theories on origin of life assume a random energy source for uphill reactions while the new theory suggests using special SPL pumps instead (see chapter 59).

downhill state, if the molecules with lower energy state do not react and turn back into higher energy state by consuming energy (through other chemical reactions), the system, as a whole, would eventually settle down at the lowest energy states i.e., near absolute zero temperature. The only way to avoid this situation is by supplying external energy (to help a different set of chemical reactions to proceed in the upward direction), thereby completing loops (Figure 1.4). There are multiple ways to bring the molecules uphill: (a) using passive random-chance mechanisms, (b) using enzymes, replicases or other catalysts, which are 'non-minimal' structures (see section 1.5.3 that briefly describes what minimal structures are) used by all existing theories and (c) using active looped mechanisms with pumps i.e., minimal structures used by the new theory in this book (see chapter 59).

There are several other advantages to looped dynamics like self-sustaining property, minimal energy consumption, ability to coordinate well, ability to explain coincidences, self-replication across a collection of molecules, cooperation across systems and others, which I will discuss in chapter 3 (and later as well).

However, note that looped dynamics does not occur naturally in most situations, especially if the patterns we want to observe are more structured and more interesting. The primary reason for this is that we cannot have a perpetual motion machine. Therefore, energy must be supplied at approximately regular intervals if the looped dynamics were to continue for a long time. There are multiple ways to supply energy. In the primitive world when there is no life, the primary mechanisms are passive ones like energy from the Sun, volcanic eruptions, hydrothermal vents, hot springs, wind gusts and others. These sources of energy though useful are not structured enough to ensure repeatability of any specific pattern. Another way to say this is that the system of interest is relying on energy from the environment. If the system of interest were to exhibit repeatable dynamics, then relying on the environment for energy is less than ideal. Instead, what we need is an active source of energy within the system itself. Relying only on the environment is equivalent to relying on random chance. The structured complexity of living beings is almost impossible to form purely on random chance.

Therefore, an active source of energy within the system is necessary to create life. The active source of energy can be viewed as generated by a special subsystem within the system of interest. This subsystem captures energy from the environment, stores it internally in a form that the rest of the system can use, as needed. A natural good design for such a subsystem is another SPL system itself. In chapter 59, I will propose several variations of active material-energy looped systems (i.e., a kind of a primitive pump) as examples of such subsystems. I will show how to create such subsystems naturally as well. These subsystems would give a better control for the system, thereby providing an opportunity to exhibit large-scale structured dynamics that is also repeatable.

Next, it is important to note that parallel dynamics is inherent in any living being. The set of chemical reactions that occur within a cell do so in parallel. There is no notion of waiting – no component or chemical reaction waits for other chemical reactions to occur. We do, however, see a cascading sequence of reactions that occur within cells. Sometimes they are triggered by a hormone and other times they are triggered by an influx of chemicals like, say, when an ion channel opens. In general, there are multiple ways to trigger a linear cascade of reactions. This seems to suggest that we do have serial dynamics. Nevertheless, parallel dynamics is prevalent and natural in physical systems (simply because no system waits for another dynamics). Serial dynamics is just a consequence of cause and effect whenever they are properly aligned.

Another reason why serial dynamics is not a good assumption in physical systems (except for systems that are already conscious and self-aware to a degree) is that such an assumption makes it difficult to explain how to construct the same long cascading sequence of steps repeatedly. Such long sequences of steps suggest the presence of a conscious designer who has planned them ahead of time. Standard physical laws do not naturally allow a specific sequence of steps to occur *repeatedly* in most cases. In chapter 58, I will show how to translate a long sequence of steps equivalently into disturbance propagation in a suitable SPL network (see section 1.6.2 for a short introduction). In other words, serial dynamics emerges naturally out of SPL networks when you apply a suitable disturbance. There is no reason to introduce a conscious designer in such SPL systems. This avoids one of the major hurdles, namely, how nature can memorize long sequences of steps without the help of a conscious being, when trying to explain the origin of life.

Parallel dynamics is also necessary for features of our brain. For example, parallel dynamics and a suitable grouping across, not along, the parallel dynamics are fundamentally necessary to create free will, as we will see (see section 1.7.3 for a brief introduction). When a self-sustaining dynamical membrane can be formed (more details later), such a grouping can obey seemingly arbitrary laws, giving rise to a notion of free will. Serialized analysis or viewpoints can never explain the root cause of free will.

Now, each of the above concepts i.e., active source of energy, parallel and looped dynamics are useful only if the resulting system is stable for a sufficiently long time. Necessity of stability of the combined system is, therefore, obvious. The trick is to show how to maintain stability even if the structured complexity grows rapidly, as is the case within any living being. The basic survival of a living being, viewed as a generalized form of stability, is not obvious. In this book, I will show how the infinite families of SPL systems allow us address this issue quite naturally. This establishes the necessity of active stable parallel looped dynamics, at least, in an intuitive sense. There are several topics I have just glossed over. The rest of the book will cover them in detail (which are not as simple as one would have thought).

To show the sufficiency of active SPL systems for the problems of origin of life and consciousness, we do need to construct specific SPL systems exhibiting life-like properties. The objective with the rest of this book is to demonstrate this aspect too. For now, I will narrow down the focus to the problems of origin of life, diversity of life and consciousness. I will pick specific issues within these classes of problems to provide additional intuition on how SPL framework is suitable. The rest of the book will expand on these ideas.

1.5   Origin of life from inanimate objects

The goal of this section is to provide an intuition for why the concepts of stability, parallelness and loops (used in a colloquial sense) are critical for the problem of origin of life from inanimate objects. Consider a simple thought experiment in which we have sealed room filled with several types of simple molecules like CH_4, NH_3, H_2O, H_2O_2, H_2S, CO_2 and others. Let us assume that it is supplied with sufficient amount of randomly distributed energy for as long as we are performing this experiment. Initially, the molecules in the room are in a state of complete disorder. The question we ask now is: what would we expect to see in this room after, say, 100 million years? If the same set of molecules stays in the same state of disorder, we clearly do not require any further explanation. On the other hand, if we see considerable order like

L-form amino acids, D-form sugars, DNA, enzymes and even primitive life forms, we have a difficult question to answer. How could such high degree of order be created from the initial disorder? What natural laws and mechanisms make this happen? The only answer for now is random chance-based mechanisms, which is clearly unsatisfactory for the high degree of order. The purpose of the new theory is to present a rigorous alternative by identifying new structures, dynamics and less chance-based mechanisms to create order from disorder.

Take, for example, the familiar notion of stability of molecules. Those molecules that are stable under a given set of environmental conditions (like temperatures and pressures) remain after a long time while the unstable molecules disintegrate and disappear (or become trace elements like H_2O_2 compared to H_2O). Therefore, in the above sealed room, we see that stability alone is capable of producing order from disorder. After 100 million years, only those molecules from the initial set that are stable (under those conditions) will remain. We want to know if there are additional patterns and properties beyond stability that can bring further order within this sealed room.

If we place highly ordered systems like enzymes and bacteria into the sealed room as the initial set, it is not surprising that more order will be created subsequently (see Darwin's theory). Therefore, the above problem of evolution of order is interesting only if the initial structures in the sealed room are the most elementary ones. I refer to them as **minimal structures**, using which we want to explain how nature creates highly ordered systems.

We can now formally state the objective of the problem of origin of life as follows: *describe the structures, patterns and natural mechanisms within existing physical laws using which the probability of creating the first life forms from inanimate objects keeps improving over time.* The term 'first life forms' (or primitive life) refers to systems that satisfy the commonly accepted definition of life, namely, specifically organized systems capable of self-replication, growth and adaptation, at minimum. The belief is that we can expand these primitive life forms to create all other life-like features using Darwinian evolution.

1.5.1 Common issues

If we arrange the timeline as shown in Figure 1.5, we see that we have inanimate objects composed primarily of CH_4, NH_3, H_2O, H_2S, CO_2 and others at the lowest level (about four billion years ago). There is no apparent structure at this level, except random distribution and random collisions of the above molecules. However, at the highest level, we have life with intricately detailed structures and a diverse set of simple and complex organic molecules. The objective is to explain how the transition occurred from such a disorderly state to a highly ordered one using entirely natural means.

Towards this end, Miller-Urey experiment (Miller [106] and Miller and Urey [108]) is one of the first one to show that amino acids and other organic compounds, necessary for creating life, can be spontaneously created from water (H_2O), methane (CH_4), ammonia (NH_3), hydrogen (H_2) and a simulated lightening through electric

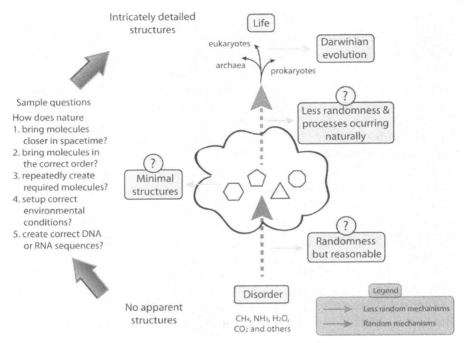

Figure 1.5: The problem of creating life from inanimate objects. Our goal is to explain how we start from an initially disordered state of simple molecules like CH₄, NH₃, H₂O and CO₂ to create intricately detailed structures like living beings. The transition from initial random mechanisms to less random mechanisms happens through the creation of *minimal* structures. These minimal structures should be simple enough that they can occur easily and naturally (i.e., simpler than specific RNA sequences that can both self-replicate and self-catalyze). Our task is to propose mechanisms for each of the steps indicated by '?' by answering several questions listed here.

sparks. Since then, several other experiments have been performed to show how to create other organic compounds critical for life like purines and pyrimidines under prebiotic conditions (see Miller and Cleaves [107], which summarizes a large number of experimental results for creating almost all critical monomers necessary for life under prebiotic conditions).

Next, to create a highly ordered system from these monomers naturally, we need to answer a set of questions unique to the generation of order. Restricting ourselves to the problem of origin of life, some of these questions are (see Figure 1.5):

(a) What laws or structures bring the prerequisite molecules in a small spatial region (cf. air molecules reaching one corner of the room) to allow a chemical reaction to occur? This seems to oppose the second law of thermodynamics.

(b) How can they accumulate within a short period? Otherwise, they may disintegrate or move away before they can be part of a long cascading chain of chemical reactions? For a few reactions, abundance of molecules or chance mechanisms as an answer may seem satisfactory (like the spontaneous self-assembly of phospholipids to form a closed membrane). However, this is unsatisfactory for most processes of life.

(c) What laws help setup the correct conditions like temperatures, pressures, pH and specific catalysts for a given reaction to occur? This seems to be external to the system of interest.

(d) What laws help coordinate a long sequence of reactions to create a given structure or to perform a given task?

(e) What laws or structures direct the creation of a specific DNA or RNA sequence for the first time?

(f) If structures like DNA or enzymes are created by chance, how can they be created reliably and repeatedly so the order can be recreated in the event they are starved (like during ice ages or meteor strikes) or when they disintegrate? The mere presence of all these prerequisite molecules is not enough to create high degree of order (cf. dead cells or all required chemicals dumped into a beaker). We need a natural sense of regulation or coordination at a large-scale to avoid collapse.

In spite of these issues, the only scientific mechanism proposed until now is that life was created by chance. Some, as yet undiscovered, chemicals acted as precursors to DNA and enzymes. These chemicals were hypothesized to have favorable properties like self-assembly, self-catalysis and self-replication and were more stable under primitive conditions. There have been several proposals for these chemicals like ribonucleic acid (RNA), peptide nucleic acid (PNA), threose nucleic acid (TNA), glycol nucleic acid (GNA), polycyclic aromatic hydrocarbons (PAH) and others.

One popular theory for the origin of life is the RNA world hypothesis (see Woese [173]). It assumes that the precursors to DNA are RNA instead. Their goal then is to create RNA replicases i.e., RNA sequences catalyzing the replication of RNA from an RNA template. RNA world hypothesis seems to address the chicken-and-egg problem between DNA and enzymes (i.e., DNA is needed to create enzymes and enzymes are needed to replicate DNA) if we can discover RNA replicases. An estimate of the length of RNA replicase sought for is comparable to that of the Spiegelman monster. For example, it was first shown that a 4500 nucleotide bases sequence of Qβ virus exhibited both self-catalytic and self-replication properties i.e., it is a RNA replicase (called Q-Beta Replicase). This sequence was later reduced to a new replicase sequence of only 218 bases after 74 generations using Darwinian selection (Kacian et. al [81] and Fry et. al [47]) and later to only 48 or 54 nucleotides long (Oehlenschläger and Eigen [116]).

The mechanisms provided for creating RNA replicases are random chance and spontaneous self-assembly. They are combined with favorable locations (like hot springs, hydrothermal vents and others) and favorable conditions (like reasonable temperatures, pressures and catalysts like clay montmorillonite that are found under primitive conditions). This seems unreasonable even if the number of nucleotides needed is just 48 or 54.

Even though I was arguing against random chance mechanisms, it should be noted that all theories, including the one introduced in this book, have a component of randomness, especially at the lower levels (Figure 1.5), if we were to avoid invoking a Creator. However, since random chance mechanisms do not 'improve' the probability of creating life (for example, probability of bringing molecules close in spacetime is no better in the second million years after randomly trying for the first million years), our objective is to minimize creating complicated structures and complex sequences of steps using such mechanisms. We will rely on a different type of randomness to achieve this called repeatable randomness, which I will discuss next.

1.5.2 Repeatable randomness

The type of randomness assumed in this book is different from arbitrary randomness assumed in the current evolutionary theories. I call it as *repeatable randomness* (see sections 52.3 and 71.3). Let me highlight the difference briefly. Some events like the formation of an enzyme for the first time, creation of self-catalytic RNA and the creation of cellular processes are random with almost zero probability of occurrence (since the number of possibilities of creating a sequence of given length are exponential). Other simpler events like rainfalls, wind blowing in a given direction and formation of bubbles is random as well, but is repeatable, though not reliably. Hence, they have nonzero and sufficiently high probabilities. Random events of the first type i.e., the non-repeatable and near-zero probability events are ruled out within the new framework. Instead, the approach is to only work with repeatably-random and sufficiently high-probability events.

The approach is as follows (see chapters 52-68). We start with simple repeatably-random events with a nonzero probability of occurrence. Using these events, we create an initial set of structures called the *minimal* structures (see Figure 1.5). We will then show how to combine them to create more complex repeatably-random systems, in such a way that the probability of occurrence of newer systems is actually increased beyond original systems. The processes used to combine the systems themselves will be reliable and repeatable as well, i.e., they are not low probability processes. We will continue this process in an iterative way where, at each time, we keep building systems that have higher and higher probability of occurrence. Eventually, we will show how to reach a state where the probability of creating a system becomes 100%. Such a threshold system is precisely the first primitive single cellular organisms capable of Darwinian evolution using which we can create all modern life forms like archaea, prokaryotes and eukaryotes (see Figure 1.5). In this way, it becomes possible to create the first life forms from inanimate objects (see chapters 52-68).

The differences between each theory stem from (i) the choice of minimal structures, (ii) the justification for randomness to create them and (iii) the natural nonrandom mechanisms for the creation of the threshold system capable of Darwinian evolution. As an example, the minimal structures for RNA world hypothesis are RNA replicases inside a closed compartment (see Woese [173]). This system is capable of Darwinian evolution. Other theories suggest structures based on polycyclic aromatic hydrocarbons (see Ehrenfreund et. al [35]), peptide nucleic acids and others as possible precursors to RNA. The minimal structures in RNA world hypothesis and others are still too complex as discussed earlier. They do not reveal the true structure and laws necessary to create life. What, then, are the minimal structures of the new theory proposed in this book? Let me now identify a few of them here.

1.5.3   Minimal structures

Minimal structures are the ones that are easy to form naturally under prebiotic conditions using which we can create more complex structures reliably. Some examples of minimal structures are already well known and others are new. Firstly, *stability* of molecules is already identified as a cause for the existence of nonrandom collections of molecules. Stable molecules with long half-life under the current conditions remain while the reactive ones disappear. Since we would not be able to specify the exact collection of chemicals without knowing the precise conditions, we pick special geological locations like hydrothermal vents, hot springs and others that maintain similar conditions over long periods to ensure steady accumulation of these stable molecules. Some of these chemicals are CH_4, NH_3, H_2O, H_2, HCN and others used in prebiotic experiments like Miller [106], Miller and Urey [108], Oró [119] and others (McCollom et. al [100], Hargreaves et. al [66] and Miller and Cleaves [107]).

The second simplest structure identified is a *closed compartment*. Reactions occurring within a closed compartment have better conditions like temperatures and pressures than the ones occurring in the open environment. Using the abundance of molecules and energy, their random distribution, random collisions between molecules and the laws of chemistry, it was shown that the synthesis of lipids and fatty acids could occur under prebiotic conditions (Oró [119], McCollom et. al [100], Hargreaves et. al [66] and Miller and Cleaves [107]). It was also shown that vesicles (closed compartments) do form under prebiotic conditions using fatty acids, organized as micelles (Hanczyc et. al [63] and Hanczyc and Szostak [64]). Clay montmorillonite catalyzes the vesicle formation and the vesicles are stable for days to months. In addition, it was shown that when fatty acid micelles are added, the vesicles even grow and split using a process called flipping (Hanczyc et. al [63] and Hanczyc and Szostak [64]). These two minimal structures are indeed used by most theories on origin of life.

In spite of an improvement when using a closed compartment over an open environment, the products can only be created if all reactants enter this small spatial region within a short period. What mechanisms bring the reactants close together to form the above fatty acids? Unfortunately, passive and random diffusion mechanisms do not offer better than random chance (analogous to air molecules reaching one corner of a room). If multiple products must be created in a cascading linear sequence, as is the case for most tasks performed by living beings, coordinating all of the corresponding reactants in space and time through passive random chance to enter the closed compartment is unlikely. We need additional structure.

The third minimal structure is an *active* power source, not just the equivalent of

ATP, which powers internal reactions, but more importantly an ability to suck inputs into the closed compartment from the environment. All problems relying on a collection of specific inputs require this feature, at minimum. Indeed, we see that all living beings have retained this feature of actively fetching the necessary inputs or food to self-sustain for a long time. A simple example is with breathing. We actively suck air into our lungs. We do not let air diffuse into our body. Figure 59.1 described later is a simple 'pump' design called an active material-energy loop that captures the essence of breathing or of actively seeking food.

The first important change a simple pump design brings is an ability to run a chemical reaction actively, not just passively. Until now, a chemical reaction occurs passively through random collisions if the correct environmental conditions and chemicals are present at the desired locations and at the correct time. However, with several active suction-based pumps, chemical reactions no longer need to wait for all reactants to passively accumulate in a small spacetime region by chance or wait for better environmental conditions to appear randomly. These active systems like pumps are the first set of structures that let the new theory deviate from all existing theories.

Next, even if favorable chemicals (like O_2, ATP and others) are naturally created using the three structures identified so far, how can we address the steady accumulation and repeatable creation of these molecules? The current models account for this by trying to identify unique conditions when the half-life is high. For example, at 100 °C, half-life of decarboxylation of alanine is 19,000 years (White [162]). However, molecules like nucleosides, peptide bonds and phosphate esters only have a half-life ranging from a few seconds to a few hundred minutes (White [162]). Therefore, expecting the same conditions to work for most molecules is unlikely. Yet, life requires not only the stable molecules, but the less stable and low half-life ones as well.

Therefore, the fourth structure needed is a way to regenerate molecules continuously, namely, a *looped* set of reactions. Each reaction within a chemical loop can be assisted by a collection of active pumps mentioned above. When the loop is completed, all chemicals involved self-sustain and have steady concentrations. When we have a network of chemical loops, it becomes easy to regenerate the entire network even if some of the chemicals are lost. In other words, recreation of these looped chemicals is a repeatably-random event. This makes looped reactions the most favorable for forming large-scale structured dynamical systems. The system as a whole requires abundant energy and chemicals as inputs. The physical loops like circulation, water cycle, seasonal cycles and vortices cause steady mixing to generate diversity. These looped reactions are the next structures that let the new theory deviate from all existing theories.

We see that living beings have retained the looped dynamical property through the creation and growth of complex metabolic networks (Nicholson [114]). In fact, physical and chemical looped structures exist at all layers. In the primitive world, I will show that active pumps will facilitate the formation of chemical loops, while in the current living world specialized enzymes perform the same function. Since all current life forms share a threshold looped network (Nicholson [114] and Jeong et. al [80]), if we can gradually and continuously keep increasing the complexity of looped networks, the central objective of improving the probability of creation of life over time can be satisfied.

Note that concepts similar to 'chemical' looped systems like hypercycles were studied previously and applied to the problem of origin of life, though they were specialized to autocatalytic systems with self-replicative catalysts (Eigen and Schuster [36]; Hofbauer and Sigmund [70]). However, the creation of an infinite family, unification of both physical and chemical SPL systems and design of SPL systems using minimal structures like pumps instead of self-replicative units like enzymes and replicases, all of which will be discussed in this book, is new. In section 1.4.2, we have already seen that loops are necessary for creating order (see Figure 1.4).

Using the four structures identified here, namely, stability, closed compartments, active pumps and loops, the primitive world starts to exhibit detailed chemical networked structure over time. However, as the looped network begins to create polymers, a new set of problems appear. For example, creating a random protein or an RNA sequence from a pool of amino acids and nucleic acids has little purpose unless the polymer sequence is precise. What laws or structures direct the creation of a specific sequence for the first time and repeatedly, which for RNA world hypothesis is an RNA replicase? If the mechanisms are based on 'parallel' random collisions, creating a randomly permuted RNA sequence and a correct sequence become indistinguishable even though the functional differences are significant. The usefulness of a polymer sequence does not make it any easier to be created reliably and repeatedly. Therefore, we need a 'serial' mechanism using a template sequence to create the same polymer sequence repeatedly without caring about the usefulness of the created sequences. How can nature do this without the equivalent of an RNA replicase or enzymes? In this book, I will identify additional structures (helical geometry and complementary nucleotide bases – chapter 62) to solve these polymer-specific questions.

The rest of the book aims at showing how to create a primitive system capable of Darwinian evolution using these minimal structures. Since the new approach is constructive, we need to provide designs and mechanisms for the entire process starting from inanimate objects to the creation of the first primitive life forms. As expected, this is easier said than done. Let me itemize a few tasks below to see why this is nontrivial.

- Create a variety of monomer molecules using prebiotic processes similar to those discussed in Miller and Cleaves [107],
- Identify mechanisms using chemical SPL processes to create new collections of organic chemicals,
- Create a catalyst like a primitive enzyme without using amino acids and DNA,
- Create a random polymer chain like RNA, DNA and others using the primitive enzymes,
- Replicate polymers like RNA and DNA reliably,

- Create a primitive genetic code,
- Create primitive amino acid based enzymes,
- Transition from primitive genetic code to modern genetic code,
- Replicate, transcribe and translate DNA to mRNA,
- Create life-like systems that actively search for food, sense the external world and explore, and
- Create the first complete primitive life forms.

Each of these tasks will require identifying additional structure within chemical SPL systems. In this book, I will discuss them in considerable detail.

1.6 Diversity of life

The goal of this section is to provide an intuition for why the concepts of stability, parallelness and loops (used in a colloquial sense) are critical for the problem of diversity of life. Consider the period after the first life forms were created. There was considerable diversity of life in next 3.5 billion years. Nature had developed several new features and a diverse set of species. Darwin's theory of evolution is regarded as a well-understood and well-tested theory to explain this diversity. Therefore, in this book, I will discuss just a few topics like how nature guarantees a direction in evolution; how physical laws of nature create competition and cooperation within and across systems and how organisms memorize a large sequence of steps.

In addition, there are other causes for creating diversity of life on earth. Two prominent events are: (a) when prokaryotes transitioned to eukaryotes and (b) when the first multi-cellular organisms were created. A few difficulties that I will discuss only briefly when trying to explain the origin of multi-cellular organisms or how to create multi-cellular features like a heart, an eye and an ear are: how does a single cell encode information about the construction of every organ and subsystem for the first time, how does a cell encode the relative physical locations, their functional relationships, their physical and chemical coordination, the sequence of steps for using every feature, their physical and chemical adaptations, their growth, the creation of new organs and how, together, they all behave as a single entity (see chapter 70).

1.6.1 Direction in evolution

Evolution of life has a natural direction in which more order is created with time even without the help of intelligent and conscious beings. This progression in order happens within a single organism like when it grows to become an adult, within a population in a given species, within multiple species and within the entire ecosystem. The order may not continue indefinitely, but its existence for long periods deserves an explanation. There are three related problems here: (a) creating the order for the first and each subsequent time, (b) sustaining the order as the system executes its dynamics and (c) adapting it over time.

One of the challenging problems unique to living beings is to explain how such detailed internal structure can be created in a natural way. Such level of order is simply nonexistent among naturally occurring nonliving systems. Examples of spontaneous order seen in nonliving systems like crystals, occurring purely randomly, are far too trivial to compare with living beings. In chapter 53, I will describe an iterative mechanism using the second law of thermodynamics to address creation of order. This mechanism combines a few repeatably-random systems to create new systems which are increasingly reliable and which have a higher probability of occurrence. The idea is to take two systems, one that is stable and the other that is 'worn out' due to the second law effects, and merge them to create a system that is, at least, as stable as the first one. Working with two copies with one that is stable, instead of only one copy, ensures that the newer system is better than both systems. If, instead, you only had one system, the second law is bound to damage it by introducing disorder rather than help rebuild it to form a better system. This mechanism is termed as the stable-variant merge loop pattern (see chapters 53-55).

One common variation of this is with living beings that reproduce sexually. Consider a new stress introduced within such a population. Imagine you take one parent whose genes are stable (or more fit) and another parent whose genes are less fit and, hence, subject to more random mutations, and you let them reproduce sexually. The resulting offspring with two copies of DNA will be, at least, as fit as the stable parent and, yet, has a much higher probability to create new features to overcome the stress using the mutations of the less fit parent (see section 54.3). In other words, the stable parent ensures a direction in evolution while the less fit parent ensures that new variations or features are introduced. In comparison, Darwin's theory proposes random mutations as a way to create new features. However, Darwin's theory does not *guarantee* a direction in evolution. There is a much higher chance of failure with pure random mutations of Darwin's theory whereas with the stable-variant merge loop, the fitness is not severely affected because of, at least, one stable copy of the DNA from one parent.

Another variation is one in which horizontal gene transfer takes the role of sexual reproduction in the above stable-variant merge loop pattern (see section 55.1). In chapter 55, I will also discuss other important variations of stable-variant merge loop across entirely different species. Some of these are predator-prey relationships, parasitic, symbiotic and more complex relationships within our ecosystem. Each of these stable-variant merge loop patterns guarantees a direction in evolution. Such guarantees are not possible with Darwin's theory.

The advantage of this mechanism is that it is applicable even among nonliving systems i.e., even before life exists. However, this is more pronounced among systems that use SPL architecture. With these SPL systems, the second law introduces more degrees of freedom as it attempts to damage the original system from, say, wear-and-tear. Another stable system with looped dynamics can now combine with the damaged system to ensure the creation of a higher-order and a feature-rich system (see section 53.4).

1.6.2   Memorizing a long sequence of steps

Most problems involving creation requires discovering and executing a large number of rather special sequences of steps. For example, to create a single enzyme, the sequences of steps followed are the transcription of DNA to mRNA followed by translation of mRNA to a protein. If you have a cut on your hand, long sequences of steps are followed like for clotting the blood, regenerating the skin to seal the wound and for fighting any infections. These steps need to be 'memorized' as well. It is not sufficient just to copy the DNA to the offspring.

While we did make considerable progress towards discovering the sequence of steps followed for tasks like cell division (mitosis and meiosis), embryogenesis and others, we rarely seem to ask how the cell memorizes these steps. We have little information on how nature discovered and memorized them for the first time so the cells could continue to follow them almost flawlessly each time. The number of sequences of steps memorized and followed during your lifetime is enormous. If such sequences of steps are not memorized, it is very easy to die even from a simple cut. What is the source of memory for all of these well-coordinated linear sequences of steps?

Note that the commonly accepted source of memory i.e., the DNA is for creating a list of enzymes. It is not a source of memory for following a sequence of steps for a given task. Consider a simple example like deep sleep. The sequences of chemical reactions during deep sleep versus when you are awake are subtly different. This difference is sufficient to lose complete control of your own limbs. The DNA is the same in both cases. However, the sequences of steps (i.e., the chemical reactions) that are executed are different. Therefore, DNA is not the source of memory of a sequence of steps. What is it then? As you may have noticed, Darwin's theory does not explain this memory.

One simple explanation is that the execution of a sequence of steps is simply inevitable. They are just embedded within the laws of nature, analogous to how a ball inevitably rolls down an inclined plane if you accept gravity and other physical laws. There is no special memory needed to explain how waves propagate in an ocean or to explain other physical phenomena after you accept the standard laws of nature. Even in your body, it could be hypothesized that we do not need any additional source of memory to execute a sequence of steps other than the standard laws of nature.

However, such an explanation does not work. This is because we want reliability and repeatability of the same sequence of steps. We do not need any special memory to explain how a ball rolls down an inclined plane. However, if we want an approximately same path to be followed by the ball as it rolls and collides with several neighboring balls, which together produce an unique pattern and we want this to occur repeatedly like, say, 10 times, the standard laws of nature, as a valid explanation is unconvincing. Similarly, the sequences of steps that occur within our body and our brain are quite repeatable and reliable. Otherwise, basic survival of the species itself is at stake. This is what I mean when I say living beings need to 'memorize a sequence of steps'.

This seems like an impossible task without a designer. In the nonliving world, we rarely see more than 2-3 well-coordinated sequence of steps occurring without the help of an intelligent being. A conscious being sets up the correct conditions, arranges them in a special way, and corrects them if the steps start to deviate. If there were no conscious beings, but we still do see a long sequence of steps occurring naturally, we will quickly dismiss it as a coincidence (cf. an occasional appearance of a human face in the clouds – who has arranged the clouds in this form? – the only satisfactory answer is just random chance). Existing physical laws do not provide simple answers for the *creation* of such structured and well-coordinated patterns.

In this book, I will show that the problem of creating and executing a well-coordinated sequence of steps, necessary for the existence of life, is equivalent to disturbance propagation in an appropriate network of loops (see chapter 58). If you want to design a chemical process with a long sequence of steps, a natural and sustainable way to achieve this is first by creating an appropriate network of looped reactions and then introducing a disturbance. A disturbance would then propagate naturally to execute the required sequence of steps as we will see. Therefore, nature's task is simply to design a suitable SPL network and to keep increasing its size and complexity over time. The actual memory of a sequence of steps will automatically follow if we introduce a suitable disturbance into this SPL network. This is a new source of memory identified in this book beyond the memory from the DNA. When the SPL network reaches a threshold complexity, it would be possible to follow most chemical sequences of steps entirely naturally. I will discuss more details in chapter 58.

1.6.3 Competition versus cooperation

From a philosophical point of view, there seems to be two main 'forces' that drive innovation and diversity we see in the multi-cellular world. They are the competitive and cooperative elements. I want to identify briefly which physical laws are responsible for creating competition and cooperation between systems. We will see that loops once again play a critical role to create cooperation naturally.

The notions of competition and cooperation are so prevalent that we use them even when discussing large-scale man-made systems as well. For example, we do say that competitive and cooperative forces shape our economy. The presence of a healthy competition drives innovation within an industry compared to a monopoly. We also see cooperation between industries, say, as some sort of balance or conservation like the balance between supply and demand. The cooperation also manifests in different forms like, for example, one company provides a design or a specification while the other company builds a part that meets this specification. One company requires a service with certain quality and another company provides that service. These services could be quite broad like building services, janitorial services, electricity, legal, accounting, email, telephone and so on. If we look at the entire economy, the cooperative elements of the above type are quite prevalent.

With multi-cellular organisms too, cooperation is extremely common. A multi-cellular organism is, by definition, a system that has strong cooperation between its cells and organs. The various organs cooperate so the entire organism can survive even at the expense of sacrificing a few components in dire situations. The various organelles cooperate in a eukaryotic cell to perform basic cellular functions. In fact, there is one popular theory to explain the origin of a few organelles in a eukaryotic cell, especially the mitochondria and plastids (like chloroplasts). This is the endosymbiotic theory (see Margulis [133]). Based on detailed microbiological evidence, the theory suggests that some specific primitive prokaryotes were engulfed by other larger prokaryotes to attain a symbiotic relationship. These later became as mitochondria and plastids. The challenge is to show how to extend this notion of cooperation beyond these examples. For example, when the world initially had single cellular organisms operating independently, what drove these systems to cooperate and, subsequently, create a combined cooperative system as a new multi-cellular organism? Each system working independently had to compromise on certain functionality in order to cooperate. Why would they do that, especially when they do not have a level of awareness (or the concept of greater good)?

To a conscious being, the concepts of competition and cooperation seem obvious. Given two systems, we can both identify and create either cooperation or competition between them using our awareness of the external world. Our objective is to see if nature can do the same when there is no consciousness to guide it i.e., we want to see if cooperation and competition can automatically emerge from the standard physical laws of nature (before consciousness existed).

Consider the origin of competition first. As we already know, Darwin's theory focuses primarily on competition to explain the diversity of life. This is captured in the notion of 'survival of the fittest'. To see the root cause of competition generalized to nonliving and non-conscious living systems as well, consider several similar dynamical systems that have an active source of energy. If all of these systems share a common resource (cf. multiple companies selling a similar product to the same set of customers), then the fact that they are actively consuming inputs suggest that they are competing to get a larger share. This happens quite naturally i.e., without conscious awareness. Some systems consume more inputs than the others do. Since there is no notion of consciousness in these systems, we can say that they are not intentionally trying to compete. Yet, the fittest will survive in such a situation. Even if a monopoly exists (i.e., a single system consumes almost all of the inputs), it is possible to disturb this state to create competition by introducing stresses like unforeseen environmental variations (weather patterns, earthquakes, wind gusts, rainfall, volcanic eruptions and others). Other examples that can disturb a monopoly are systems that are agile and better fit at a smaller scale. The more dynamical the environment is, the higher the chance that the fitness shuffles among these systems. Therefore, competition naturally emerges when multiple systems share a common resource. Such a competitive advantage is the underlying hypothesis of Darwin's theory to create new and diverse set of features in living beings. However, it is interesting to note that Darwin's theory focuses less on cooperation.

Let us now look at how we have cooperation in physical systems, again especially among nonliving and non-conscious living systems. If we see multiple streams of water travelling downhill, we never see cooperation between them to create a specific pattern. When multiple clouds coalesce, we never see cooperation between them to form the shape of, say, a human face. Even in cases when there are clear patterns like with tornadoes, hurricanes and others, we rarely see cooperation between components to form this pattern. It seems like a bit of a stretch to say that nonliving beings are cooperating towards a common goal even if we include attractive and repulsive forces within electrical, magnetic, gravitational and other forces. On the other hand, there is a great degree of cooperation among non-conscious living beings. For example, the mere existence of a multi-cellular organism shows an enormous degree of cooperation between various organs within its body. How can we then explain this difference?

Specifically, what is the root cause of cooperation within nonliving systems? The answer is the existence of looped dynamical systems. Previously, I had mentioned that if we have two systems that cooperate, they typically do so in the following way: the output of one system is the input to the other system and vice versa. This is an example of a simple looped system if we combine both systems into a single larger system. Let me elaborate this further by picking a simple looped system, like the one in Figure 2.1 (i.e., a system with a ball rolling down an inclined plane and a human placing it back on the inclined plane to complete the loop). Such a looped system can be split into two parts. In one part, the ball rolls down the inclined plane and stops. In the other, a human brings the ball and places it on the inclined plane to complete the loop. When this happens, the first part can occur once again. If this looped dynamics continues for a long time, we can say that both parts are working cooperatively. This is especially true because the natural state of any system is *not* to continue the looped dynamics (since we cannot have a perpetual motion machine). In fact, we do need a constant supply of external energy to continue any looped dynamics, which does not happen easily among nonliving systems.

Therefore, given that looped dynamics does not occur naturally often, we would need to have cooperation between two or more systems to create a loop. One system will cause a downhill reaction while another system will cause a corresponding uphill reaction by consuming energy. Together they complete the loop (see Figure 1.4). Downhill reactions can occur naturally, but if the corresponding uphill reactions do not occur to form a loop, then the world will settle down to a low energy state with almost no possibility of life. We have already seen that this was the basis for the origin of life i.e., the physical and chemical SPL systems.

Even within non-conscious living beings, we see cooperation because of the presence of looped processes. Plants consume energy from the sun, CO_2 from animals (and H_2O) to produce O_2 and starch, which are consumed by animals to produce CO_2 and H_2O needed by plants. Together, they complete a loop so life could continue. This loop, split between two systems (plants and animals), is clearly an example of cooperation. Even within a cell, we have several chemical reactions that consume ATP (an energy source generated in the mitochondria) to produce ADP, while the mitochondria consume ADP to generate ATP, thereby completing the loop. Each such chemical loop in the metabolic network (see Jeong et al. [80]; Nicholson [114]), though occurs in a single cell, is well regulated. We can identify several chemical reactions managed by different subsystems within our cells for each such chemical loops. We can view all of these subsystems as cooperating to achieve regulation of the chemical concentrations. Even if we disturb the system by introducing additional chemicals, the loops return to the original states by propagating the disturbance within the SPL network, as we have briefly seen before (see details in chapter 58).

Therefore, if a looped chemical reaction, a looped physical process or a combination of both occurs within two or more subsystems causing a loop to occur for a long time, we can say that these subsystems are cooperating to allow a possibility that would otherwise have not happened naturally (namely, the continued occurrence of the looped dynamics). The fact that looped processes *cannot* occur naturally in most cases (since we cannot have a perpetual motion machine) implies that two or more systems must cooperate to make this happen. In cases like earth revolving around the Sun, we accept the looped process as caused by the fundamental force of gravity. In almost all other examples of looped processes, the mere existence of looped dynamics becomes the root cause of cooperation between a few subcomponents.

In this chapter, we have been arguing that active looped processes are necessary for explaining both the origin of life and consciousness. What we are seeing just now is that the same looped processes are also responsible for creating cooperation within nonliving (and living) systems, at least in a philosophical sense. Therefore, our goal once again is to build large-scale stable parallel looped networks over time to promote cooperation. In other words, the current discussion provides yet another intuitive reason to study SPL systems for understanding the evolution of diversity on earth.

1.7 Discovering structure of consciousness

The goal of this section is to provide an intuition for why the concepts of stability, parallelness and loops (used in a colloquial sense) are critical for the problem of consciousness. During the period life existed on earth, we see that nature has created the most interesting set of patterns in our universe. Among these patterns, only in the last several thousands of years did a species, namely, human beings, had an opportunity to see, feel, understand and express this beauty in an elaborate language, even though life existed for about 4 billion years on earth. No other species demonstrated this ability as clearly as human beings do. This ability is what we call, in a very broad sense, as consciousness. How did human beings develop this ability?

Only a conscious person can ponder on questions like what is the purpose of life, how I know that I exist, what I am doing with my life, how long am I going to live, what is happening around me and so on. Consciousness gave humans the ability to realize life itself. Each of these questions are completely irrelevant whenever we are not conscious (like during deep sleep, under general anesthesia and others), when humans did not develop consciousness or with other animals that do not have a similar degree of consciousness.

Among animals, we notice that some exhibit memory of a few local events and some others seem to understand even death. From these observations, most people tend to accept that consciousness, emotions and other unique human abilities are present in other species as well, but perhaps to a much lesser degree. Such a gradational representation of feelings and consciousness is considered the least contradictory from a logical point of view, even though there is only a limited, if at all, scientific evidence. Given such gradations, what is a suitable definition of **minimal consciousness**, i.e., the lowest degree of consciousness below which the concept has no relation to consciousness?

The word 'consciousness' is loosely defined and, yet, very clearly understood by people who are already conscious. If we ever attempt to explain what it means to a 5 year old, we quickly realize the limitation with words or a language. Yet, when he becomes a 15 year old, all that is needed is his collection of life experiences and a little introspection to clarify what consciousness means, in a way no language could have ever done satisfactorily. These fundamental limitations with language seem to pose a major hurdle towards a scientific theory of consciousness.

When we think of the word consciousness, we see that it is related to self-awareness, free will, our ability to know things or events, our ability to sense space and the passage of time, perception, cognition, human memories and even intelligence. Consciousness is a combination of all of these well-defined features, though we may pick one or more of them to be more closely related than the others. One reason why we do not want to exclude these features from consciousness is that there are no examples of human beings where the exclusion is unambiguous. Any human being who we would consider as conscious exhibits every one of these features to some well-defined degree. For example, a person who just senses space or just senses the passage of time and is still not conscious does not exist. Yet, we do not want to say that all of these features put together are precisely what consciousness is. For this reason, in this book, we will look at a notion of minimal consciousness.

The typical objectives when defining minimal consciousness would be (a) to develop a rigorous theory, (b) to understand the physical and dynamical structure of our brain to allow us develop consciousness, (c) to extend it to cover all other notions commonly attributed to human consciousness like self-awareness, free will and feelings and (d) to create consciousness synthetically. As we dwell deeper into each of these questions, several other topics will appear to be highly intertwined with one another. It becomes necessary to study each of them as well. It is this entangled nature of the problem of consciousness that makes it impossible to try to understand it purely from a philosophical point of view (see Figure 1.6).

Therefore, it becomes necessary to rely on human consciousness and compare the differences between conscious and unconscious states. The examples I will use throughout the book for the unconscious or less conscious states are the following seven commonly accepted ones – *deep sleep, under general anesthesia, knocked unconscious, coma, severely drunk, sleepwalking and even a newborn baby*. These seven examples will let us evaluate statements about consciousness in a logical and rigorous way. They serve as counterexamples when analyzing a statement. In the next few subsections, I will provide intuition for at least a few such statements.

1.7.1 Necessary and sufficient conditions

There have been several attempts at providing new frameworks for consciousness (Baars [6]; Dennett [32]; Chalmers [19]; Crick and Koch [28]; Tononi [154]). Each of

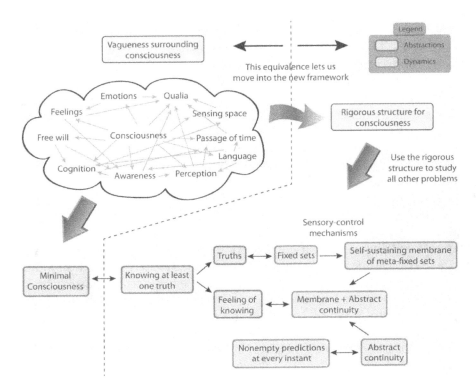

Figure 1.6: The central result is that minimal consciousness is equivalent to knowing at least one truth. This result lets us move away from the vagueness surrounding consciousness instead to the study of truths, the structure and relationships between them and the feeling of knowing using a dynamical framework of stable parallel loops. Specifically, this approach reveals that minimal consciousness is equivalent to the existence of a self-sustaining dynamical membrane of meta-fixed sets made of abstract continuous paths.

these recognizes individual aspects within the complexity of our brain network as the source of consciousness. A few examples are: identifying the neural correlates for consciousness (Koch [86]), taking an information theoretic approach to consciousness by studying the connectivity in graphs (Tononi [154]), looking at the origin of consciousness within humanity (Jaynes [79]), sensorimotor account for visual consciousness (O'Regan and Noë [117]), quantum mechanical view for consciousness (Hameroff [62]) and evolutionary views on consciousness (Edelman [34]; Dennett [32]). However, none of these attempts stated clear *necessary and sufficient* conditions for consciousness.

For example, Tononi [154] suggested that "consciousness is integrated information". Necessity of integrated information i.e., that which is beyond the sum of its parts is well justified. However, why is integrated information sufficient? What is the 'nature' of this integrated information that would make it sufficient? When we build a machine in which the information is well integrated, what are we integrating and why would we regard it as conscious? Besides, the integrated information in terms of purely static structural properties of the brain network (for example, static structural regions like fusiform gyrus in area V8 identified as neural correlate of color – Crick and Koch [28]) cannot be sufficient. This is because the static network and regions such as fusiform gyrus are structurally the same, even if manually stimulated, both when we are conscious and when we are not (like during deep sleep and under general anesthesia). Therefore, static structures cannot be the complete set of equivalent conditions. Dynamical properties must necessarily be included for consciousness. Yet, there is rarely a mention of what physical and chemical dynamics are necessary and sufficient.

As another example, O'Regan and Noë [117] have mentioned that visual consciousness is the outcome of sensorimotor contingencies. Here again, necessity of actions and sensing is well justified. However, why and what sensorimotor interactions would make them sufficient for consciousness? If we build a machine, how many sensorimotor contingencies should be included and in what order? When such a machine becomes complex, what is the nature of this complexity and what resulting structural and dynamical properties make it sufficient? We can similarly evaluate each of the existing frameworks on consciousness and they all fall short with either necessity or sufficiency.

In this book, I deviate from existing approaches by providing both necessary and sufficient conditions for each concept introduced, even if it appears trivial at first glance. For example, one of the central results is the following (Theorem 12.1):

A system is minimally conscious if and only if it knows at least one truth.

To understand this statement, note that truths are necessary because whenever we are conscious, we know many truths like, say, we know our hands, legs, our emotions, tables and others. Even the equivalent contrapositive statement is true – if we do not know anything like during deep sleep, under general anesthesia, when severely drunk, when in a coma, when sleepwalking or when knocked unconscious, then surely we cannot be conscious. Truths are also sufficient because the moment we know a single truth, we are conscious or, equivalently, the contrapositive implies that when we are not conscious like during deep sleep, under general anesthesia and others, we do not know even a single truth (more details in chapter 12).

This central result has given us one set of necessary and sufficient conditions. It allows us to study the problem of consciousness in a much more systematic way (see Figure 1.6). At first glance, this equivalence sounds either trivial or that we have pushed the problem of consciousness into the problem of knowing. The objective then is to clarify and show how to build a dynamical systems framework to address both truths and knowing. How do we define the abstract concept of truth in a dynamical systems framework? How would we use such a dynamical notion of truth to study the problem of knowing? These questions are not easy to answer, as we would expect. Nevertheless, we can show that the special family of stable parallel looped (SPL) dynamical systems to be discussed later provides a good framework. For example, using this family, I will show that (Theorem 23.2):

A system is minimally conscious if and only if it has a threshold self-sustaining dynamical membrane of meta-fixed sets.

There are several terms in this statement that are unclear at this stage and it would take nontrivial amount of time to define them. Rather than attempt defining these terms, I will assume that you already understand a few terms (like self-sustaining, membrane and threshold) in a colloquial sense. Assuming such a level of understanding, I will state a few less-rigorous statements with the hope that it gives an intuition nevertheless. Loosely speaking, if the membrane exists, you are minimally conscious and if it 'tears down', you are not conscious. For each of the seven unconscious states mentioned earlier like deep sleep, under general anesthesia, coma and others, we notice a lack of self-sustaining dynamical membrane (more details in chapter 19). Furthermore, the above Theorem 23.2 allows us to (a) build synthetic conscious systems and (b) study human consciousness empirically as well (as we will see later).

One interesting consequence of the above result is that knowing a single truth is equivalent to knowing an entire membrane of truths. In other words, we cannot know a single truth like 'does the table in front of me exist?' by itself. To know just one such truth, we have to know an enormous number of truths (about space, passage of time, geometric properties, our own body and others) simultaneously. These are all part of the threshold membrane of meta-fixed sets. This is both a nontrivial and less intuitive result (more details later).

The above result also addresses the standard philosophical hurdle on consciousness – the infinite recursive nature of questions when we look at the seat of consciousness. This is sometimes referred to as the mind-body problem. The infinite recursion comes from asking which system controls your body, which system controls that system that controls your body and so on *ad infinitum*. Another way to view this problem is by asking how a dynamical system, that simply obeys the laws of nature, is capable of knowing *about* the very laws of nature. It is like asking how a car, that simply moves on the road obeying the natural laws, steps back to see itself and understand that it is indeed on a road. What representation in the physical world would give a system such an ability? We will see later that the self-sustaining membrane of meta-fixed sets is one such representation. The membrane allows us to both sense and control the external world. It may not be obvious to see the connection until we understand clearly what meta-fixed sets are, which collection of meta-fixed sets are included in the threshold membrane, how this collection changes over time i.e., how it expands or shrinks (even tearing down during deep sleep and reconstructing it during REM sleep) and so on.

The next step to use the equivalent statement of Theorem 12.1 is to formulate necessary and sufficient conditions for when a system can know facts. To see the importance of this, consider why we understand sentences even if we have never heard them before (like sentences in this book). Is it because of grammatical correctness? Why? Universal grammar (Chomsky [21]) is not sufficient for *understanding* language, though some representation of it appears necessary. This is because you need to be conscious first before you can understand anything (Theorem 12.1). Besides, it is not clear why a machine with a universal grammar would 'understand' language. Even among humans, for example, we understand every word in 'eat to like I pizza' and, yet, not the sentence. A simple rearrangement of this sentence to 'I like to eat pizza' completely changes our ability to understand. Analogously, a jumbled arrangement of individually understandable pieces of a jigsaw puzzle does not make us understand the whole picture. As we start to rearrange it correctly, we do understand it. When we forgot who or where we saw a given person, we can eventually know this by recollecting and rearranging the sequence of events in a 'correct' order (analogous to grammatical correctness). In each of these cases – correct arrangement of words, correct arrangement of jigsaw-like puzzles, correct arrangement of sound variations, correct arrangement of events and so on – what are the necessary and sufficient conditions to understand them? I will show that we need two structures – the above dynamical membrane and a notion of continuity that generalizes to abstractions i.e., a concept called abstract continuity (more details in chapters 21-22).

Notice that with the above necessary and sufficient conditions, we do not answer the question of 'why' you have these properties like consciousness or feeling of knowing. It only states that if a set of conditions are satisfied, you will "always" have a feeling of knowing and vice versa. The term "always" here is modulo the known set of examples of unconscious states that I mentioned earlier like deep sleep, under general anesthesia, coma, severely drunk and others. From an experimental verification point of view, these are the primary examples we have to perform experiments and validate the necessity and sufficiency of the statements discussed. As we discover more examples, it may be that we have to revise these statements. However, for now, I have tried to use as many known counterexamples as possible to justify these statements. This is similar to other scientific questions like with gravity, mass or charge. We typically address how things work, but not why they exist. They just do. Sometimes such questions are not within the realm of this theory. Our goal is to define concepts and build a comprehensive set of relationships between them. It is purely a relativistic theory in that sense.

The important point to keep in mind here is that we need to study the problem of consciousness by introducing necessary and sufficient conditions for each new concept introduced. Indeed, I will present several new ideas and concepts later in the book. Figure 1.6 summarizes these steps and suggests the new alternate way. It suggests that the difficulties with studying consciousness and its interactions with other topics like free will, sensing space and time can be simplified using the necessary and sufficient conditions proposed in this book. What will give us confidence in the new theory is from: (a) the constructive approach we take where I provide designs and mechanisms for each new concept and (b) ensuring we have necessary and sufficient conditions for each concept. However, this is not an easy task. Several problems need mechanisms that do not require other conscious beings. Let me list a few of them here that I will discuss later in the book (see Figure 10.1).

- Propose a new model for neurons and memory
- Discuss abstract continuity – an ability to understand sentences you have never seen before

- Specify necessary and sufficient conditions for attaining a feeling of knowing
- Describe mechanisms to create several common fixed sets
- Show how to build and use sensors reliably without any help from the external world
- Describe how we learn to control our own body (actuators) without the help of an external world
- Study the dynamics of serial versus parallel systems
- Identify the origin of free will and show how to create a free-will-like system
- Show how a human can be sure of the existence of the external world
- Show the existence of an empty space and other geometric properties like size, distance, shape and others
- Show how we sense three dimensional space i.e., how we have a cohesive perception
- Describe how we can sense the passage of time
- Discuss how a system becomes self-aware
- Discuss our ability to compare conscious systems – dual equality
- Provide necessary and sufficient conditions for realizations and emotions
- Discuss several special abilities of the human brain like intelligence, capturing gist, language, mathematics and others
- Build a synthetic conscious system

Each of these problems are not only theoretically hard to analyze, but are also difficult from a practical standpoint. For this reason, we devote a large part of this book to address each of these topics.

1.7.2   Minimality – knowing

In the above discussion, I have picked 'knowing truths' as the minimal set for consciousness. Why should we pick them? One reason is that knowing is so fundamental that most of us treat it as synonymous to consciousness. For example, knowing yourself is self-awareness. Knowing the existence of the external world involves knowing the existence of space, knowing the passage of time, knowing the external objects and having a cohesive perception. Knowing your interactions with the external world includes your free will. Knowing your feelings is related to knowing your subjective quality of your experiences (*qualia*). Therefore, it appears that if we can formalize the concept of knowing (Figure 1.6), we can extend it to the above specific forms of knowing.

Conversely, if we do not include knowing into the minimal set, the resulting definition does not appear to have any relationship to the notion of consciousness. For example, if you indeed do not know absolutely anything, why would you consider yourself conscious? This situation indeed occurs in each of the seven less conscious states mentioned above (like deep sleep, coma and under general anesthesia). *If someone says that 'nothing exists in this universe', then even though the statement appears obviously incorrect, we can guarantee that it is false if and only if we are conscious.* To convince ourselves, consider one of the seven less conscious states, like, say, when we are in deep sleep. Even though I and the rest of the universe continue to exist physically, I, as a system, am incapable of 'knowing' my existence or the existence of the universe. This is true with any fact for which an adult conscious human can evaluate truth. Even if you are repeatedly performing the same task (like an newborn drinking milk every day), you do not know or realize what you were doing, why you were doing and that you were even doing it, to the same degree as adult conscious humans. This statement is at the heart of every problem related to knowing anything in the universe. This is the reason why I use it to define *minimal consciousness*. I will discuss the details of minimal consciousness and extensions to other features like free will and passage of time later in the book.

The work ahead of us, in spite of accepting the importance of knowing is, considerably nontrivial. One reason is that knowing is an abstract concept while we need to convert it into a dynamical systems framework. We also need to provide constructive mechanisms along the way. I will discuss all of these details later in the book.

1.7.3 Consciousness – a property of SPL systems

Given that we are taking a constructive approach and that we are looking for necessary and sufficient conditions, it turns out that consciousness will become a property of a special subset of SPL systems. This is a point worth emphasizing. As discussed briefly, the special class of SPL systems allows the possibility of creating an overall stable and, yet, complex system with millions of interacting dynamical subsystems (see chapter 2). With such large-scale structured dynamical systems, we can create several unique dynamical patterns that are reliably repeatable as well.

In this book, I will show that these patterns will look very similar to the features of our brain like, say, consciousness, emotions and other feelings. From this perspective, it is clear that these patterns, however complex they appear to be, are purely properties of SPL dynamical systems that can be synthetically generated as well. I will then argue that such a synthetic system is, indeed, indistinguishable from a human being with respect to properties like free will, space, passage of time and others. This makes consciousness itself a property of SPL dynamical systems. In other words, we do not require neurons, organic compounds, DNA and others to create consciousness (and life as well). This is a surprising result and may be hard to accept. Hopefully, if and when you finish reading the book, this statement will become clearer.

What exactly is a property of dynamical system? If we consider a simple pendulum as a dynamical system, it has a property called asymptotically stable point corresponding to the vertical rest position. Several other oscillatory systems have different types of stable points as well. Similarly, there are other properties called unstable points, saddle points, limit cycles for nonlinear dynamical systems, attractors, periodic orbits and others. From these examples, we can say that a property of a dynamical system is a unique and distinct state that can be reliably and repeatedly created, not just in a single dynamical system, but also in a sufficiently large class of dynamical systems. In this book, the classes of dynamical systems we consider are stable parallel looped systems.

The existence of these dynamical properties can be guaranteed simply by creating the stable parallel looped system in a certain specific way (described in detail in chapter 47). The fact that we are working with large-scale structured dynamical systems, fortunately, helps in creating these properties in multiple different ways. For example, no two human brains are identical and, yet, they all exhibit the same set of properties like consciousness and emotions. There is considerable flexibility in the sequence of steps for creating such complex systems just by comparing them with the billions of human beings who live on earth. Once the correct architecture is chosen and an approximately correct set of steps are followed, consciousness, for example, becomes inevitable within these dynamical systems. The system does not have a choice, at this point, whether to become conscious or not.

Intuition for the root cause of free will: Let me consider free will as a nontrivial example of such a property. I will provide a brief intuition here deferring the details to chapters 26-28. One central problem of free will is to design a system that is composed of a large number of components, each of which obey the standard laws of nature, but the system as a whole can obey seemingly arbitrary laws. For example, a human being can perform seemingly arbitrary tasks whenever he wants (within reasonable limits) even though every molecule or cell within a human obeys just the standard laws of nature. Nonliving systems or less complex living beings do not seem to have such ability. In fact, their actions are quite predictable when we use the standard laws of nature (like Newton's laws, Maxwell's laws and others).

On the other hand, I cannot predict what you will do next even though I know that each component within your body obeys the same laws of nature. The answer is not mere complexity in the form of a large number of components that I need to have access to (which I don't). For example, I do not know what you will do next while you know it well, even though you do not have access to the same large number of the components of your own brain and body. Also, a computer has millions of transistors and other components and yet it is predictable. Even a single cellular organism with much more complexity than a computer is quite predictable (at least to a degree). Yet, something unique within the structure of a human body (and brain) has given rise to nonpredictability. What is this unique structure i.e., what is the root cause of our free will? How do we construct such systems?

Two other problems come to mind when talking about free will, namely, the existence of choice and our ability to know that we have a choice. The former problem can be best understood only after we know how to design a system that obeys seemingly arbitrary laws. The latter problem is closely related to the problem of consciousness itself. Indeed, you know about your choices only if you are already conscious. I will, therefore, postpone the discussion on these two problems.

The key insight to the problem of creating a system obeying seemingly arbitrary laws comes from the observation that the dynamics *along* a flow is uniquely different from the dynamics *across* the flow. Let me explain what this means using a few examples. If we have 10 balls rolling down an inclined plane, we notice two distinct patterns: (a) the dynamics of each ball along the flow (i.e., rolling down the inclined plane individually) and (b) the change in shape of all 10 balls viewed together at every second, say, i.e., the dynamics across the flow. The laws obeyed by each ball in the former case (along the flow) are the standard Newton's laws. However, the laws obeyed in the latter case (across the flow – the pattern formed by these balls) can be made sufficiently arbitrary.

Another example is with a CRT (cathode ray tube) television screen. The beam of electrons falling on the screen (along the flow) obeys the standard laws. However, the movie pattern on the screen (across the flow) can be as arbitrary as we want i.e., they can be any sequence of time varying images. A similar example is the firing pattern triggered from an image falling on the retina. The neural firing pathways along the optic nerve into the primary visual cortex are the dynamics along the flow, whereas the 'fixed sets' (to be discussed later) triggered for a pattern like a table, a chair, a tree and others corresponds to the dynamics across the flow. The former dynamics obey the standard laws (action potentials travelling along axons and so on) whereas the latter dynamics is sufficiently arbitrary. As another example, take a look at a smoke ring. Every molecular motion within the smoke ring obeys the standard laws of nature (along the flow). However, the shape of the smoke ring (across the flow) wiggles and jiggles quite arbitrarily. It can obey arbitrary laws. All of the above examples are real physical examples that exist in nature. They are not abstract mathematical concepts.

These examples highlight one simple point: the laws obeyed *across* the flow are sufficiently arbitrary even though the laws obeyed *along* the flow are just the standard laws of nature. Such a requirement is precisely what we were looking for in a free will system. Our task now is to figure out how to exploit this idea to build a system that, as a whole, obeys sufficiently arbitrary laws. This suggests that the system we want to build should contain collections of components, with each collection having both dynamics along the flow and across the flow. Operations like sensing and control within this system should always be across the flow and never along the flow. This is not entirely trivial but can be done with relative ease using active SPL systems, as we will see in chapters 26-28.

These examples also highlight the root cause of free will (at an intuitive level). Free will stems directly from parallel dynamics and not from serial viewpoints. Of course, we still need a clever design to satisfy the above requirement. The self-sustaining membrane of meta-fixed sets mentioned previously (see Theorem 23.2) is precisely one such example of an SPL system that satisfies this requirement (as we will see). Such an SPL system can both sense and control itself. In chapters 26-28, I will discuss more details. In other words, we can see that free will emerges as a property of SPL systems. In fact, in this book, we will see that most topics related to consciousness can similarly be shown as properties of SPL systems.

1.8   Organization of the book

In summary, we can say that SPL systems are an important class of systems to study for the problems of origin of life, diversity of life and consciousness, at least intuitively speaking. This (long) introduction highlighted several new questions to focus on when studying large-scale structured dynamical systems and presented an intuition for why and how active stable parallel looped systems can address them clearly, completely and constructively. Let me now start exploring each of these topics in detail.

In the following chapters, I will discuss the topics in the reverse timeline i.e., the problem of consciousness first, then the diversity of life followed by the problem of origin of life. One reason for doing this is that the problem of self-construction of the brain and neurons is not as critical for the problem of consciousness as understanding how and what specific structures give rise to the features of consciousness. In addition, physical systems, the primary components critical for consciousness, are easier to understand than chemical systems (because assembling physical systems to create a structure is common in our everyday life compared to assembling structures using chemical reactions).

The problem of origin of life deals primarily with the issue of how nature self-constructs large-scale structured dynamical systems like a living cell. Not dealing with such self-construction issues is easier to understand in the beginning. We can first focus on the properties of stable parallel looped systems instead. Once we understand these properties, we will be in a position to ask and understand how nature can self-construct such large-scale structured dynamical systems as well.

In spite of the reverse timeline approach taken here, the reader can pick the three main topics discussed in this book i.e., consciousness, diversity of life and origin of life in any order because they are sufficiently decoupled from each other. The advantage is that the core concepts for all three topics are still based on SPL systems.

2
Foundations

In this chapter, I will introduce the three concepts central to this book, namely, loops, parallelism and stability. I will use simple examples to provide an intuition behind these concepts. While we are already familiar with these concepts, I will present them in a form that highlights the viewpoint needed for this book.

The goal of this chapter is to address the following two basic questions: (a) how do we construct large-scale structured dynamical systems that do not collapse for a sufficiently long time and (b) how does nature construct such large-scale systems without help from other conscious beings? The approach I will take is as follows. I will start with a simple physical (as opposed to chemical) system and evaluate how to construct it naturally. It will turn out to be not easy enough. I will identify aspects that are difficult and try to eliminate some of them. This will give rise to a second example. I will then discuss additional difficulties with this second example to propose a third example and so on. Using this iterative procedure, I will show how to create a seemingly simple and yet powerful infinite family of physical dynamical systems in which the latter ones will have more intricate structure but nicer properties than the former ones.

However, it will turn out that all these designs, while answering (a), would not answer (b) cleanly i.e., we would still need a conscious being to create systems from this infinite family. To truly eliminate a conscious being from these designs, we need to switch to chemical systems. Towards this end, I will present a similar infinite family of large-scale chemical systems using an analogy with the above infinite family of physical systems. The resulting chemical designs will also be iteratively constructed with more intricate chemical structure. Together, the two infinite families of physical and chemical systems offer the necessary foundations for studying origin of life and consciousness, as we will see later in this book.

Another consequence of these infinite families is that we will now have a way to create stable dynamical systems of arbitrary complexity. In section 1.3, we saw that creating stable dynamical systems with millions of moving parts is a nontrivial problem. As it happens, the infinite family proposed here, though looks almost trivial, is the only known way to create arbitrarily complex large-scale structured dynamical systems, at least for now. Their existence unifies origin of life and consciousness into a single framework and will make the study of other complex systems simpler and systematic, as we will see. In contrast, existing theories on structured complex systems are incomplete because no reliable and repeatable way of creating 'real' large-scale structured dynamical systems has been presented. As a result, any behaviors they would predict are not easy to verify experimentally on a synthetic system.

2.1 Loops

Consider two simple examples to illustrate the familiar notion of loops: (1) a swing in a playground and (2) a ball rolling down an inclined plane as shown in Figure 2.1.

In the case of a swing, you are at one end of the swing. You push the swing when it reaches you. The swing moves towards the other end and comes back to you. I call this a physical loop. In the ideal case, when we write the dynamical equations and plot the phase-space of position versus momentum, we will see a loop for each initial condition. You can continue this back-and-forth motion several times with a push in each 'loop'.

In the case of a ball rolling down a plane inclined at an angle θ, you place the ball of mass m and radius r on the top (at a height h) and let it go. The ball rolls down the inclined plane and comes to a halt after a while. This, by itself, does not form a loop. However, when a human brings the ball back to the top of the inclined plane, we have completed a physical loop. You can start the motion once again by letting the ball roll thereby repeating these loops.

From these two examples, we identify a *physical loop* by visualizing a particle in the system that moves along a physical path from a given point and returning to the same point after a period T. With chemical systems, we have a collection of molecules and a corresponding set of chemical reactions causally connected such that a subset of the products of one chemical reaction become the reactants of another chemical reaction. We say that such a series of chemical reactions form a causal *chemical loop* if and only if starting from an initial set of chemicals, we can have a series of chemical reactions producing several intermediate chemicals, which eventually produce the original reactants after a period T. I will discuss examples of chemical loops later in section 2.6. Generalizing physical and chemical loops, we can define looped systems as follows:

Definition 2.1 (loop): A dynamical system exhibits a loop if and only if there exists a small neighborhood N of states with the property that the system enters neighborhood N both at time t and at time $t + T$.

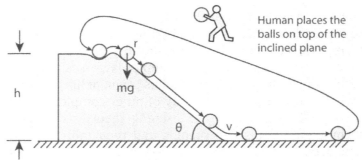

Human places the balls on top of the inclined plane

Figure 2.1: A simple physical SPL system. As balls roll down the inclined plane and come to a halt, a human picks it up and places it on top to continue the looped motion.

The looped process defined above repeats itself for a while. There is a clear sense of direction of propagation of information or dynamics in a looped system. This definition generalizes the notion of loops to support a large collection of examples by relaxing the requirements as follows:

1. *Paths nonidentical in each loop*: The path of a physical loop, as described above for physical and chemical systems, is not necessarily the same during each repetition. In other words, we do not expect the system to repeat the exact same dynamics in every loop. We make this assumption because we want the system to have an ability to move away from its current behavior using just minor variations in each repetition. Had we assumed precision in repetition in each loop, the behaviors exhibited would be quite limiting and would not capture reality well enough.

2. *State only approximately same in each loop*: The components of the physical loop need not return to the exact location in the next iteration. It can stay in a small neighborhood of the initial point. More generally, we reach an *almost same* state as the initial state and not the *exact* same state before continuing the next loop. We impose this requirement because *macroscopic repeatability* is more important for the behaviors of interest than the extreme precision in repetition at a microscopic level.

3. *Time period T(t) varies across repetitions*: The period T need not be constant across repetitions. The subsystem executing a loop can accelerate the first time around, go slowly in the second loop and so on. This is again important to capture reality. The exact time period is not as important as the existence of a time period $T(t)$ that can possibly vary over time. For example, for the problem of origin of life, the repeatable creation of a collection of molecules is more important than whether the repetitions occur exactly after time intervals $T, 2T, 3T, 4T$ and so on.

4. *Repetitions need not occur forever*: The looped paths do not have to repeat forever, i.e., for an infinite amount of time. I will call the system a looped system until it keeps repeating the looped dynamics. Later, the system can collapse from this state because of several reasons like dissipation of energy, wear and tear, destruction due to excess energy, severe damages and others. This assumption is needed to avoid mathematical or other types of idealizations.

5. *System not collapsing for a sufficiently long time*: The system should be stable sufficiently longer than the average time period $\bar{T}$ of repetitions, say, at least 10 times longer. This assumption is needed to restrict ourselves to large-scale structured dynamical systems that exhibit interesting behaviors long enough for someone else to verify. This ensures a sense of reliability and belief in the observable dynamics of our systems. It also eliminates the possibility that the system exhibited a behavior merely by chance or as a lucky event.

6. *Existence of sufficient energy and material inputs*: There exist energy and material inputs that can be supplied in small bursts with a time-varying period such that the system can remain stable while repeating the loops for a

sufficiently long time. This assumption can be thought of as a restatement that we cannot have a perpetual motion machine. However, it is stated in a less restrictive form in which we only need energy and material inputs in small bursts – not continuously and not for extended periods of time.

In chapter 3, I will discuss additional properties of loops in further detail. Let me first briefly apply these assumptions to the above two examples. The swing, for example, need not return to the exact same point. The ball does not have to be placed at the exact same location each time. The swing can wobble differently or the ball could take different trajectories in each loop. The time it takes to complete the loop for the swing or the ball need not be a constant. The swing oscillations or the rolling ball repetitions need not occur indefinitely. Furthermore, external energy would need to be supplied to continue the looped dynamics (thereby avoiding a perpetual motion machine). A human supplies this energy in small bursts in these examples.

In this book, I will continue to make the above assumptions whenever I discuss looped systems. With these assumptions, we are now able to capture a large set of examples like dynamical systems that are oscillations around an equilibrium point in physical systems, dynamic equilibrium in chemical reactions, limit cycles in physical and chemical reactions (which are usually far from equilibrium), periodic and cyclic systems, strange attractors like Lorenz attractor, which are considered chaotic.

It is common to separate the dynamics in examples like these into two groups – system and environment. In the case of the swing, one way to choose a system is to pick the swing together with the rope tied to it. Then, the structure to which the swing is attached, the atmosphere and you pushing the swing, would be considered as the environment. In the rolling ball case, one choice is to include just the ball itself as the system. The inclined plane, the ground beyond the inclined plane on which the motion of the ball continues, yourself bringing and placing the ball back on top of the inclined plane would then be the environment. If the system is composed of several subsystems (see parallelism in section 2.2), it is possible that only a subsystem rather than the entire system exhibits a loop as defined above.

What are some of the key similarities and differences in these two examples? In the case of the swing, we, the environment, need to supply only a small amount of energy in each loop unlike the ball-rolling case. Secondly, the swing can continue its dynamics for a sufficiently long time with small bursts of energy. The ball rolling down the inclined plane, on the other hand, requires a lot of energy from the environment to continue the loops. The human needs to walk to the location where the ball stopped, pick it up and walk back to the initial state, in order to continue its loop. A restatement is that the energy lost by the swing in each loop is sufficiently low compared to the rolling-ball case.

If we exclude ourselves from the environment in both cases, the chances of the rolling-ball system occurring as a loop naturally is almost zero because it is unlikely for the ball to get to the top of the inclined plane. It is a dynamical system, nonetheless, whose dynamics 'ends' when the ball comes to a halt. The swing can, however, continue as a loop for a longer time naturally even when you are no longer pushing it. A small input as random energy can be imparted occasionally, say, whenever a strong

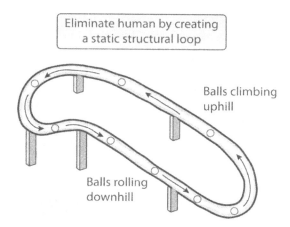

Figure 2.2: A physical SPL system with one static structural loop. We can eliminate the human, who places the ball to complete the looped dynamics in Figure 2.1 by structurally closing the loop. Balls roll downhill naturally while they climb uphill occasionally if external energy is supplied.

wind blows. The swing may not oscillate to the same height as before but can have small oscillations each time a strong wind blows. The wind energy trapped by the swing is effectively used because it is lost slowly compared to the same wind energy trapped by the ball that pushes it on the ground from a rest position (because of higher friction losses). Also, if you look at the system a year later, you would notice that the ball would have rolled off to a distant place because of wind and other disturbances whereas the swing is possibly oscillating at the same place. The information that the ball was once placed on the inclined plane is completely lost and is not relevant anymore.

Since the system in Figure 2.1 requires a complex dynamical system like a conscious human to complete the loop, let me propose a simpler alternative. Consider a new structure for the environment as shown in Figure 2.2. It is modified in such a way that it already forms a 'structural' loop by curling the path back to the top of the inclined plane. Let us assume that the loss in energy through friction, as the ball rolls on this structure, is sufficiently low. In this case, when the ball reaches almost the top of the inclined plane, it only needs a small extra amount of energy to continue the loop again. This can be supplied, say, by an occasional gust of wind (provided the timing is correct as well).

We can get a rough estimate of the new height h_1 that the ball reaches after it rolls down to the bottom from a height h and continues its journey upwards as shown in Figure 2.2. We use the conservation of energy as follows. Consider the ball of mass m and radius r at height h (see Figure 2.1) rolling down the inclined plane. Its potential energy is mgh where g is the acceleration due to gravity. When the ball reaches the bottom, its potential energy is mostly converted into kinetic energy. From the conservation of energy, we have:

$$\frac{1}{2}mv^2 + \frac{1}{2}I\omega^2 + E_L = mgh \qquad (2.1)$$

Here, I is the moment of inertia of a spherical ball $= \frac{2}{5}mr^2$, ω is the angular velocity of the ball, v is the linear velocity of the ball when it reaches the bottom and E_L is the total loss in energy in the form of heat and friction when the ball reaches the bottom of the inclined plane. If we assume that the ball rolls without slipping, then $v = \omega r$. Substituting for I and ω as above and simplifying equation (2.1), we get:

$$\frac{7}{10}mv^2 + E_L = mgh \qquad (2.2)$$

Therefore, on the journey downhill, the ball's potential energy was mostly converted into kinetic energy. If the ball now continues to move upwards as in Figure 2.2, it will reach a height h_1, lower than h. To calculate h_1, we use the conservation of energy once again. This time, the kinetic energy will be converted back to potential energy. Assuming the same amount of energy loss E_L, we now have:

$$\frac{7}{10}mv^2 = mgh_1 + E_L \qquad (2.3)$$

Simplifying the two expressions, we get new height h_1 as:

$$h_1 = h - \frac{2E_L}{mg} \qquad (2.4)$$

The energy losses E_L prevent the balls from reaching the top of the inclined plane to form a loop in one direction (cf. the ball oscillating back-and-forth until it stops at the bottom). We need to supply an extra amount of energy $2E_L$ to continue the looped dynamics in a single direction. This is much lower and simpler than creating a human and then requiring him to place the ball back on top of the inclined plane for each loop. This new structure of the ball and the environment makes the likelihood of the continued dynamics longer than our initial example (Figure 2.1).

Even with these advantages, the chance of creating such a structural loop (Figure 2.2) by nature itself is still quite low. As a result, it appears as though we have not achieved much by redesigning the system. However, there is an important difference when we evaluate which of these designs would 'survive' i.e., run their dynamics longer. In Figure 2.1, the structural requirement of creating an inclined surface is simple, but the dynamical requirement of the ball getting on top of the inclined plane, which must happen in *every* loop, is unlikely through natural means. On the other hand, in Figure 2.2, the structural requirement, of creating a looped structure, is unlikely, but it needs to happen *only once*. Subsequently, the dynamical requirement becomes relatively easier because the little extra energy needed to continue the looped dynamics could be supplied, say, from wind and other random disturbances in the nature.

We have thus reduced the large degree of randomness that must occur in every loop (in the form of creating a human and using him to continue the looped dynamics) to large randomness occurring once combined with smaller randomness in each loop. In section 2.4, I will show how to make this initial random event less unlikely as well. In fact, I will show how to create an infinite family of dynamical systems of any specified complexity incrementally and iteratively with each new design eliminating some of the inefficiencies of the former design, but in turn adding new and intricate structural requirements.

In summary, what we have seen here are two examples: a swing that is a loop and a rolling ball on an inclined plane that is not a loop (if the human is excluded), but can be altered to form a looped dynamical system by modifying the structure. We can generalize these examples to say that if the system and the environment can combine naturally, by itself, to create a loop, then this combined system tends to remain and sustain its dynamics for a relatively long time. If we look back in time, all dynamical systems that survived are precisely these. For example, electrons, protons, atoms, molecules and others sustain themselves because their energy requirements from the environment and the corresponding losses are very low. This low amount of energy is easily trapped from a mere random disturbance that occurs in nature and is lost slowly as well. The formation of the structural loop itself is the primary less probable event and not the ability to run the dynamics.

2.2 Parallelism

The second important concept in this book is parallelism. To understand this better, consider the opposite concept, namely, that of serial dynamics within physical or chemical systems. This is typically the standard viewpoint taken when analyzing simple dynamical systems.

When we pick a dynamical system to study, we first identify the critical entities of interest. We make approximations with respect to sizes and timescales as necessary. For example, when studying the dynamics of billiard balls, we consider the balls themselves as entities of interest, not the individual atoms and molecules within each ball. When studying traffic flow, we consider cars and other vehicles as entities of interest, not their components like wheels and engines.

The next step is to define the state of this system using the entities identified above. The state is typically specified using a vector $X(t)$ of variables (typically, position and momentum) that evolves along a single timeline t. The dimension n of the state vector $X(t)$, sometimes referred to as the number of degrees of freedom, is as large as is necessary to capture the features of interest. At any given instant $t = t_0$, the state vector $X(t_0)$ is considered as a full specification of the system relative to the features of interest. The objective is to determine $X(t)$ for all time t. One common model chosen is to specify the time evolution of the state vector $X(t)$ as a set of ordinary or partial differential equations. Then, we can compute the time evolution of $X(t)$, at least in theory, by solving these equations.

This approach of studying a dynamical system is what I call as the ***serial view***. The reason I call it a serial view is because there is a single timeline variable t for the entire state vector $X(t)$. My claim is that such a serial analysis will not fully clarify features like the origin of life and consciousness in large-scale structured dynamical systems.

The n components $x_1(t), x_2(t), \dots, x_n(t)$ of the state vector $X(t)$ come from different physical entities within the system, in general. For most systems, these entities are interconnected, thereby allowing them to exchange forces and energy. In the differential equation model, this interrelationship or coupling manifests itself through cross-terms (like $x_1(t)x_2(t)$ or $x_3(t)x_5(t)$). For example, the differential equation for a component $\dot{x}_1(t)$ can have cross-terms that contains $x_2(t), x_3(t)$ and others. The serial view of the physical system is now modeled as a collection of coupled differential equations.

The spatial locations (and momentum) of each of the entities within the system change as the single timeline variable t changes. When we solve these coupled differential equations, we compute these spatial locations and momentum at each instant. In essence, we are taking spatial snapshots of all entities within the system at each instant t, analogous to creating movie frames at each instant t. Specifically, if we divide the time interval $[t_a, t_b]$ into m discrete instants $t_1 = t_a, t_2, t_3, \dots, t_m = t_b$, the state vector $X(t_i)$ for $i = 1, 2, \dots, m$ represent the analogous movie frame snapshots. Such a movie-like representation is the serial view of a dynamical system. The accuracy and the mathematical correctness of this model can be improved by increasing m, thereby decreasing the sampling gaps $\Delta T_i = t_i - t_{i-1}$ for $i = 1, 2, \dots, m$.

When the dimension of the state vector is small, which is typically the case with most systems we study, the computational complexity of the above serial view is manageable. We can either solve the differential equations in closed form or numerically in a computer.

When studying large-scale structured dynamical systems like a single cell or the human brain, we generalize the above serial formalism in a conceptual sense. While this is logically correct, the serial view offers little help in understanding the behaviors exhibited by such structured complex systems. The corresponding differential equations are both difficult to write and difficult to solve both theoretically and computationally. The equations are highly coupled. The dimension of the state vector m is in the thousands or millions for such systems.

We, therefore, resort to approximations to the model in order to reduce the dimensionality of the state vector. Unfortunately, there are several issues with a lower-dimensional model as discussed in section 1.3.7. For example, the lower-dimensional model no longer captures critical aspects of the true large-scale structured dynamical system. The approximate model only addresses a very small subset of the behaviors exhibited by the complex system. The errors in approximation are magnified easily resulting in a divergence from reality. Furthermore, the lower-dimensional model is less causal and more probabilistic since the small dimensionality only yields average behaviors.

In other words, while the serial view for large-scale structured dynamical systems is a conceptual generalization of the corresponding view of smaller dynamical systems, it offers little insight into how a structured complex system exhibits features like life, free will, consciousness and self-awareness. With serial view, we are taking snapshots of the entire system along a single timeline. However, for features like consciousness, the entire system is not critical at all times. Rather, different subsystems play important role at different times. It is the combination of these subsystems and their interactions that result in features like consciousness, as we will see. A different approach is to keep track of dynamics within subsystems along its own timeline and then synchronize the dynamics only when the subsystems interact. This is what I refer to as the ***parallel view*** of large-scale structured dynamical systems.

Whereas the serial view attempts to use a single timeline for the entire system, a parallel view splits the large-scale structured dynamical system into several subsystems that keep changing over time and tracks the causal dynamics individually within its own timeline. Recall that a large-scale dynamical system has several entities and a large number of degrees of freedom. It is also typical that a given entity interacts with a very small fraction of entities within the system. For example, a given neuron in our brain connects to about 1000 other neurons on average, which is very small compared to a total of 100 billion neurons. The same applies to a cell. A given molecule can participate in very few chemical reactions compared to the total number of possible reactions that occurs within a cell.

Given this, it is better to track the space-time dynamics of each individual entity in its own timeline as a 'flow' event as long as there is minimal interaction between the entities. Each flow event is a *causal* dynamical event and, therefore, has its own natural timeline. The dynamics within each event can be viewed individually as a serial system as described above. Since we plan to choose different timelines with the parallel view, it is necessary to synchronize these timelines, at least whenever the causal events do converge or intersect, which does happen quite frequently. In fact, exciting things happen when the events exchange information, mass or energy during this period of convergence. They may later diverge into their individual timelines. The subsystems that are part of a given flow event need not be the same for all time. They do change with respect to the observer's timeline. They are also different for each feature of interest. This flow based causal view is what I term as the parallel view of a large-scale structured dynamical system.

A simple example to describe this is by looking at the television screen where electron beams in the cathode ray tube hits a flat screen to produce an image (see Figure 2.3). Here you can imagine each dot on the screen to be independently excited with time-varying intensities and properties to produce different colors at different times. The independent excitation is, however, not true in a real cathode ray tube, but there is no loss of generality in assuming it conceptually. In this example, there are two types of time-variations that occur: (a) the flow-based time variations of the properties of each individual dot on the screen and (b) the time-variations of the entire screen, i.e., the frames of the movie itself.

There is one temporal variable for each dot, whose intensity and other properties vary over time in a causal manner. The causality is guaranteed from the corresponding ray in the electron beam (i.e., the flow event). There is also a two-dimensional spatial pattern spanned by the entire television screen composed of thousands of such dots. This pattern corresponds to a movie or a television show. The spatial pattern also changes in the same time dimension. The spatial pattern represents objects like trees or cars. The variations of these patterns are entirely arbitrary i.e., they do not obey any laws of nature. These two types of variations, namely, individual parallel variations on each dot and image variations on the entire screen (as a serial movie with a single timeline as perceived by a conscious human) are essentially orthogonal to each other.

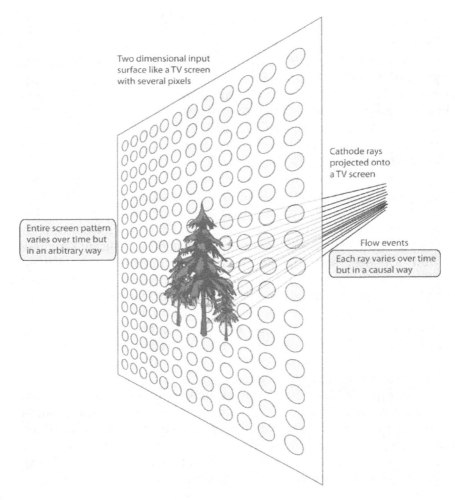

Two dimensional input surface like a TV screen with several pixels

Cathode rays projected onto a TV screen

Entire screen pattern varies over time but in an arbitrary way

Flow events

Each ray varies over time but in a causal way

Figure 2.3: A screen projecting an image. Parallel flow dynamics of each ray is causal whereas time-varying dynamics of a screen pattern is entirely arbitrary.

How is it possible to explain and analyze serial time-variations in the two-dimensional (2-D) spatial pattern in terms of parallel variations in the patterns of the individual dots/pixel? The laws obeyed by the parallel variations of individual dots are the standard causal laws of physics. What laws are obeyed by the time evolution of the 2-D spatial patterns on the entire television screen? The answer is that they are entirely arbitrary. Namely, the time evolution of 2-D spatial patterns on the television screen can take the form of any arbitrary movie we like. I will expand on this idea later in chapter 28 to create a system that obeys seemingly arbitrary laws even though the individual components obeys the standard laws of physics. The resulting system will appear to exhibit a feature similar to free will.

Now imagine your retina as a television screen with images of external objects exciting the rods and cones in your eye. These excitations are transmitted along the neurons into the inner regions of the brain. As with the television screen, we have the above two types of variations to keep track of. Firstly, we need to keep track of each individual pathway of neuron excitations, starting from the retina to deep inside the brain. This is the flow event for a single neural excitation pathway. Here we keep track of dynamics 'along' the flow (analogous to the cathode ray flow). Secondly, we need to keep track of an entire spatial pattern of excitations on the retina (of an image like a tree) across the complex network of the brain. Here, we want to keep track of dynamics 'across' the flow (i.e., spatial dynamics along an observer's timeline cross-sections, say, at time t_1, t_2, t_3 and so on). Initially, the parallel excitations from retinal neurons corresponding to, say, a tree propagate along the optic nerve is a well-defined linear path. However, as they spread across the visual cortex and other regions of the brain, the linear order that existed at the optic nerve is lost because of the cross-pathways and loops.

Given the importance of both these variations (along and across the flow), it is no longer useful to analyze our brain as a serial system by discretizing or sampling the time intervals to sufficiently small scales using the observer's timeline. The serialization only keeps track of *across the flow* dimension of the entire system, not just the relevant subsystems. The physical dynamics, the laws of nature and the notion of causality is well defined *along the flow* and not across the flow as mentioned above (Figure 2.3). The patterns across the flow have a sufficient degree of variability, making it seem that the time-evolution of these patterns do not obey physical laws. The timeline of the movie images on the screen can be called 'abstract' (i.e., not obeying physical laws) and can be made as arbitrary as we want. In chapter 28, we will see how this unique difference can be exploited to create a system exhibiting a feature similar to free will. In fact, we will see that this unique difference between serial and parallel view is the root cause of free will.

The parallel view discussed here introduces new challenges as well. One such challenge is with causality as discussed in section 1.3.8. Consider, for example, looking at a tree followed by looking at a house. How does our brain infer the causal connection (at least the time sequence aspect) i.e., that the house is viewed 'after' the tree? At the input source (the retina), the parallel set of neural firing paths does obey causality. However, subsequently, the parallel neural firing paths corresponding to a tree potentially disperses to different regions within our brain compared to those corresponding to a house. Since there are cross-pathways and loops, it is possible that the neural firing pathways corresponding to a house moves faster/further or reaches a 'recognition' state earlier than the tree. If this happens, the *inference* of causality – that the tree is viewed before the house – will be broken (not the causality of physical dynamics itself).

To make matters worse, continuous adaptive changes to our structural neural network through our everyday learning causes further disruption to our ability to infer the causal connection. Furthermore, the dynamical firing patterns are also not identical each time you see the same image. This is because the time taken for a neuron to fire action potentials is (a) different among different neurons, (b) can be altered chemically both naturally (say, through the release of neurotransmitters like endorphins after a prolonged workout) and synthetically (say, by consuming alcohol) and (c) varies significantly with time because of the past state. The ion channels are blocked accordingly, resulting in an altered timing of neural firing patterns. This can result in a temporary distortion in the inference of causality.

The above problem is further amplified when several images keep falling on the retina in a specific timeline order at almost every instant. The streams of retinal firing patterns are transmitted continuously along the optic nerve at every instant. The firing patterns coming later (in the observer's timeline) may take pathways that are already excited than the firing patterns before them. This may not happen for long periods, but they do happen sometimes. With such mismatched linear time-order at the neural layers, how can we guarantee coherence and synchronize the information to achieve a clear *serial* thought in our mind instead of producing random gibberish recognition outcomes?

Furthermore, we control and direct our thoughts in such a way that this parallel neuron firing patterns that were spread to distant parts of the brain are conceptually converged in a specific way to solve or understand a given topic. It produces a continuity of thought that can be tracked effectively. You are able to start, stop and switch between various tasks without ever going into infinite loops. What is the relationship between a topic that we understand (at the spatial pattern level i.e., an abstract human language level) versus what happens at the massively parallel and detailed neuron firing patterns? In chapters 28 and 39, I will discuss several mechanisms for consciousness that addresses these issues resulting from parallelism in our brain.

The patterns that emerge from multiple timelines or a parallel view has good coordination within large-scale structured dynamical systems like cells and our brain making them worthwhile to study closely. In the nonliving world, we see smoke rings or the appearance of a human face in the clouds as some examples of such patterns. Of course, these patterns are considered as either designed or random. In this book, I will show how to create and self-sustain such complex patterns in stable parallel looped systems entirely naturally and in a reliably repeatable way. In fact, if we can construct a system that is capable of producing and controlling the formation of patterns like a smoke ring in both space and time (i.e., wherever and whenever – not just by random chance), such a system has similarities to free will, as we will see in further detail in chapter 28.

In summary, parallel systems have several highly interacting and dynamical temporal and spatial patterns that give rise to interesting behaviors and coordination's. They also lack a linear order. While these properties make them difficult to analyze, a rigorous and systematic study is nevertheless possible provided we combine parallelism with loops introduced earlier, as will be shown in this book.

2.3 Stability

The third concept important in this book is that of stability. The state of a system is considered stable whenever the system returns to this state when subject to small perturbations. The opposite notion is that of an unstable system. An example of an unstable state is an acrobat balancing on a rope. What will happen when a physical system does indeed become unstable? There will be a point of apparent discontinuity in its dynamics and typically, at this point, the system breaks down. If it does not completely break down, we sometimes view it as a bifurcation. Other similar notions applied to a single state are asymptotic stability and saddle points. For systems that have limit cycles and strange attractors, we can evaluate their stability as well.

In the case of a swing, the resting position is considered stable. With the system in Figure 2.2 (ball on an inclined plane modified to form a structural loop), the position of the ball when it comes to rest is a stable position. When these systems reach a stable state, they need extra, sometimes a large amount of energy to start the motion once again, unlike the case when they are already in motion as loops. This is similar to the activation energy needed to make a chemical reaction proceed. This energy is higher to start a chemical reaction than to continue producing products.

Another common term associated with stability is equilibrium. Some examples of equilibrium systems are with thermodynamic equilibrium, reversible chemical reactions reaching a state of dynamic equilibrium and others.

Yet another way to describe stability is as follows: *a system is in a stable state S if it resists small changes that try to alter its state S*. In this case, the system will resist the perturbation and tries to return to the unperturbed state. For example, if the system appears to be in a state of rest, it prefers to be in this state. If the system is in a state of motion, it resists when being stopped. In the case of reversible chemical reactions in a state of equilibrium, if you change the temperature, pressure and concentrations by a small amount, the reaction responds by proceeding in a direction that opposes this change. This is the familiar Le Chatelier's principle.

This form of the definition is a generalization of the previous one. It avoids a few ideal assumptions. These are similar to the assumptions made when defining loops in section 2.1.

1. *No need to define state in precise terms*: The new definition does not assume that a state can be precisely defined. This allows us to model the behavior of stability for a large number of 'real' dynamical systems without a loss of qualitative precision. We do lose mathematical precision. However, for large-scale systems of interest here, with millions of subcomponents, the mathematical precision only introduces computational complexities, which at the present time, is preventing us from capturing true relationships within such systems.

2. *Stable state not fixed in time*: The new definition does not require the stable state to be fixed in time. We do need this because most real dynamical systems interact and integrate with a number of other systems over time. These adaptive changes produce interesting evolutionary features in which the stability continues to hold albeit in an altered state. The ability to study stability of such time-varying systems is one of our objectives.

3. *Stable state returned after a perturbation is only approximately same*: When resisting a small perturbation, the new definition does not assume that the

returned state has to be the same state as before. We only need the state to be similar in a macroscopic sense. Furthermore, most systems alter structurally under perturbations (like, say, they bend, twist or deform with some plasticity). The dynamics exhibited by the deformed system is no longer the same as the original system. This results in a change in the stable state (cf. changing stable states when a child grows into an adult). We would like to allow the new definition to encapsulate such real possibilities as well.

There is a detailed mathematical theory of dynamical systems that describes the notion of stability formally (see Khalil [84]). For example, in the case of continuous dynamical systems in two dimensions, Poincaré-Bendixson theorem characterizes the long-term behavior (see Khalil [84]). It essentially states the following: *dynamics that stays in a bounded region in two dimensions approaches either a fixed point or a limit cycle*. In higher dimensions and in discrete dynamical systems, there can be strange and chaotic attractors. It is tempting to think that loops are inherently stable. However, there are notions of unstable limit cycles, saddle points where only oscillations along certain directions are stable and not along others. Other examples include resonant states that are usually unstable (to small perturbations) and so on.

In summary, stability of a dynamical system introduces a state S in which the system resists changes to small perturbations in any direction. When combined with loops and parallel dynamical components, the dynamics of the physical system exhibits several discrete states in which the system can be locked into (analogous to being trapped in a local minimum instead of a global minimum). The system also transitions between these discrete stable states. When this happens, the system exhibits new and interesting emergent behaviors, as we will see next.

2.4   Stable parallel loops

Having introduced all three concepts (loops, parallelism and stability), it is possible to propose examples where only two of the three concepts play a role, like stable loops, parallel loops or stable parallelism. I will skip these cases and, instead, discuss systems that have all three properties. I call these systems as stable parallel looped (SPL) dynamical systems. They are a special class of dynamical systems that allow the creation of a seemingly simple and yet powerful infinite family of stable dynamical systems of any specified complexity, as we will see. In this section, I will show how to construct them within both physical and chemical world (unlike existing theories). Let me begin by defining stable parallel looped dynamical systems.

Definition 2.2 (stable parallel looped system): A dynamical system is said to be a stable parallel looped (SPL) system if and only if the following conditions hold: (a) *existence of dynamical loops* – there exists parts of the physical or chemical dynamics of the system that recur approximately the same with a time-varying period $T(t)$ for a sufficiently long time and, if necessary, when supplied with energy and/or material inputs repeatedly, (b) *parallel interacting loops* – the system has one or more loops that exchange inputs and energy with one other and (c) *stability* – there exists sufficiently small disturbances for which the system does not collapse (like, say, bounded inputs producing bounded outputs).

In order to exclude systems that die down quickly, I will typically impose the following condition as well – an SPL system should be stable sufficiently longer than the time period of the slowest looped dynamics (say, at least 10 times longer). Furthermore, I only require approximate repeatability and periodicity with no necessity to continue the dynamics indefinitely (see also the other assumptions specified in sections 2.1-2.3 when defining loops, parallelism and stability).

The simplest example of an SPL system is a swing in a playground considered earlier. A person pushes at regular intervals to supply small energy to continue the oscillatory (or looped) motion. The other two conditions in the definition are trivially satisfied. Other examples are limit cycles, periodic, cyclic and oscillatory dynamics (like pendulums and waves). With chemical systems, examples of SPL systems are chemical oscillatory reactions like Belousov-Zhabotinksy reaction, Briggs-Rauscher reaction and interconnected cycles like Kreb's cycle, Calvin-Benson cycle and more general metabolic networks (see Nicholson [114] and Jeong et. al [80]) within living beings.

Our objective now is to show how to create large-scale physical and chemical SPL systems starting from the simplest SPL systems. In the next few sections, I will present an infinite family of 'synthetic' SPL designs of any specified complexity. You can define a partial order (a binary relation '≤' that is reflexive, antisymmetric and transitive) among this family of dynamical systems using a metric like the amount of self-sustaining time over a long period. In chapter 53, I will show how to redesign this family so they can form 'naturally'. Together, they will be seen as a way to create more order from disorder (seemingly violating the second law of thermodynamics, but in a local sense for open systems – see chapter 53).

2.5   Physical SPL systems

Let us revisit the examples discussed in section 2.1. We started with an SPL system shown in Figure 2.1. We modified it to create a new SPL system (Figure 2.2) by eliminating the costly requirement of a human being. There are still three problems with the new design in spite of the improvement from Figure 2.1 to Figure 2.2.

Firstly, the direction of motion along the loop is not fixed. The balls can move in both forward and backward direction. If you imagine a number of balls moving along this path with different levels of energy, some balls will start moving in the reverse direction as soon as they slow down. They will collide with balls moving in the forward direction hindering the overall motion (see Figure 2.2). This is analogous to resistance observed with the propagation of water waves in non-looped versus looped water pipes leading to a quick dampening of the disturbance. The final state is where all balls end up at the bottom i.e., the lowest-energy state. If, however, there is a single direction of motion, the perturbations along the direction of the loop could have helped in maintaining the loop longer and would have been more stable.

Secondly, there is only one stable point in this system, which is at the bottom of this path. This is partly responsible for reverse motion as the balls are attracted to

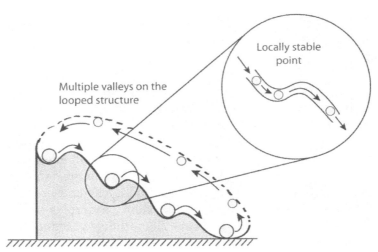

Figure 2.4: A physical SPL system with multiple locally stable points within the static structural loop. One stable point in the SPL system of Figure 2.2 is now replaced with multiple stable points. Balls can now settle down at one of the valleys instead of all of them reaching at the bottommost point. Multiple valleys can be present in the uphill direction as well, though not shown in this figure.

this stable point from both directions. As shown in equation (2.4), the height h_1 is smaller than h causing the balls to reverse direction instead of continuing forward to complete the loop. Only when an extra energy $2E_L$ is supplied to the ball at the correct time would it continue its looped journey in the same direction.

Thirdly, with only one stable point we need to supply a sufficiently large amount of energy (mgh for each ball) to make the system resume its looped dynamics after all the balls reach this stable point. Resuming the looped motion has become costly when the dynamics comes to a halt.

Let me attempt to fix these problems, at least partially, by redesigning the system as shown in Figure 2.4. In Figure 2.4, you see that there are several hills and valleys along the way. Each of the valleys is a stable point, giving rise to multiple stable points instead of just a single one. The valleys are at different heights with respect to each other. Different balls moving along this system can now end up in different valleys, instead of all of them at the bottom.

Let us look at the advantages of this design. Firstly, reverse motion is now possible only locally at each valley instead of once for the entire system. This is a slight improvement because the average number of collisions has decreased. Secondly, adding multiple stable points gave rise to more dynamical patterns within this system. Finally, the balls that stay in each of these stable points only need a small amount of extra energy to take them to the next higher stable state (i.e., much smaller than mgh – for example, $mgh/100$ if we have 100 stable points placed equidistantly from top to bottom). The average amount of energy needed by the balls for resuming the loop motion is lower compared to the system shown in Figure 2.2. For example, if we supply a little more than $mgh/100$ to the ball at the topmost stable point, then it will tip over and potentially roll all the way down before settling in a different local stable point. On the other hand, supply the same amount of energy ($mgh/100 + \Delta$) to a ball in Figure 2.2 will only cause local oscillations. Surely, the friction losses will be higher in Figure 2.4 than in Figure 2.2. Nevertheless, the 'interesting' dynamics and the reshuffling of state (rearrangement of balls, in this case) is quite different and arguably better in Figure 2.4 than in Figure 2.2 even with less external energy (like $mgh/100$).

Also, we can say that the energy received from the external environment is now effectively stored at each of these valleys as potential energy. If this system receives energy from the environment in random bursts, the balls move and collide with neighboring balls to push them out of a stable state A to the next higher stable state B, effectively capturing the external energy for later use. In Figure 2.2, however, when they reach the bottom-most state, all of the energy is effectively lost.

We could once again use the conservation of energy as we did with equations (2.1)-(2.3) to compute the amount of energy stored at each of these valleys, the likelihood of different balls at different heights, the net extra energy needed to restart the dynamics of N balls and compare this with the total energy $Nmgh$ needed for Figure 2.2 when all the balls reach the stable points. I will skip these details.

The system in Figure 2.4 certainly exhibits interesting dynamical states than the system in Figure 2.2, especially if there is a sufficient amount of random excess energy available in the environment. Can we improve the system further to make it trap the external energy better, create additional interesting states and, potentially, run its dynamics for a longer time? If we analyze Figure 2.4, we will notice that this system runs quite efficiently as long as it is already in motion. If the external energy level drops for some reason and the balls come to a halt, the total input energy needed to restart the motion is still quite high (even though it is not as much compared to Figure 2.2). The energy needed for a dynamical system to start its motion can be termed as activation energy analogous to that of a chemical reaction. In Figure 2.4, the activation energy is still a bit high.

We should, therefore, avoid the need to activate the system by continuing to keep it in motion. One way to do this is by lubricating the system so the balls do not lose energy quickly through friction. As a result, the system does not come to a halt and continues its motion for a long time. Another improvement is to replace the stable *points* with stable *loops* with a small height h_l of each loop. Figure 2.5 is the next design that does this. If the structural loops are well lubricated, the balls that enter a given loop will keep going around for a long time (instead of oscillating back and forth as in Figure 2.4). Furthermore, as the balls in the loops are in constant motion locally, the energy needed to make them move from one stable state A to the next higher stable state B is much lower than when they are at rest. Going in a loop is also unidirectional especially if the height h_l is small. It avoids the reverse motion like the back-and-forth oscillations we see for a stable point. As long as the local looped dynamics occurs, the corresponding collisions that hinder the motion are eliminated. Of course, the above discussion assumed certain ideal conditions. Keep in mind that more than the

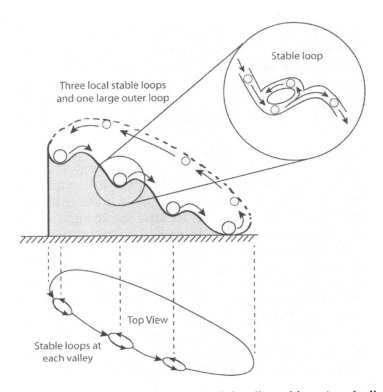

Figure 2.5: Physical SPL system in which locally stable points (valleys) are converted to stable loops. Each locally stable point of Figure 2.4 is converted into a stable loop. Some of the balls continue dynamics in each of these local stable loops and others continue along the outer loop. Though not shown in the figure, not all stable points need to be converted into stable loops and multiple loops can be present on the uphill direction as well.

advantages of Figure 2.5 compared to Figure 2.4, the creation of a new system is the primary takeaway point. Also, the fact that we can convert stable points into stable loops and vice versa (as can be seen with Figure 2.4 and Figure 2.5) is a good way to create more and more intricate SPL systems.

Furthermore, the dependency on the external energy is relaxed further with this design. With the previous designs, the external energy is expected to be more or less constant if we want the dynamics to keep running. With this design, however, small random bursts of external energy from the environment, which always exist from wind and other disturbances, can fluctuate. Yet the new system utilizes this energy effectively to have continued dynamics. The extra energy needed to kick the balls to the next higher level is not just from the local bottommost location as in Figure 2.4, but from other intermediate locations as well, due to their continued motion within the looped path. This is an important point to remember – as the ball is moving in the local loop, when a random external energy burst occurs, the ball may already have been more than halfway higher on its upward journey. If that were the case, the external energy requirement to complete the local looped dynamics got reduced by half. If more energy was imparted, the balls can either orbit around each of these tiny loops faster, or jump to the next higher energy state to trap the external energy. When the external energy levels drop, the balls lose the stored energy slowly (because of good lubrication).

Therefore, the expectation that the external world is a reliable source of energy is not as critical. Instead, the energy can be supplied quite intermittently. Interestingly, this is a unique property with all living beings – accumulating or storing food for later use and not relying on the external world to supply food constantly.

It is not correct to assume stability of a complex parallel looped system even if each of the smaller sub-loops is individually stable. When we changed the system to Figure 2.5, it was important to coordinate all these parallel loops to have a single direction. This is to ensure that the balls do not collide in opposite direction thereby hindering the motion, especially when they start moving from one state to the next higher state. If each of these loops is uncoordinated or misaligned, the dynamics within them simply collides with each other. The entire dynamics could collapse as a result. Fluid circulation (or water pipes) where dynamics can proceed in both directions is not very efficient (cf. open versus closed circulatory systems in living beings). A clear separation as arteries and veins with flows along only one direction is quite desirable. Human brain is another example of a physical structure where the looped dynamics occur only in directed one-way paths. This avoids the problem of achieving overall stability resulting from insufficient coordination with multiple parallel loops. In chapter 5, I will describe the internal details of a neuron and present a mathematical model that highlight several properties needed for good coordination in our brain.

If there is a lot of external energy available, there is less need to capture and use it efficiently. However, if the amount of energy available for a system is less, the ability to use the energy efficiently becomes critical for continued dynamics of the system. As we see with the above examples, the easiest way to achieve this is by having stable parallel looped designs. SPL designs are quite effective at trapping the available external energy and using them at a possibly different time when the external energy is scarce. This can be seen as an effective redistribution of available external energy. There are several situations when there is a shortage of energy. For example, when several similar or competing systems share a given amount of energy, you tend to have an overall shortage of energy or food resources. External factors also introduce a genuine shortage of available energy, like during ice ages. Systems with designs similar to Figure 2.5 like atomic orbital models with similar transitions between levels and molecular orbital dynamics tend to self-sustain longer even under such dire situations.

So far, we have taken one isolated example and modified it to make it more intricate and effective (in some sense). With our approach, the new system effectively utilized the resources by aligning the system and the environment to form an integrated stable parallel looped system instead of keeping them disjoint from one another. The expectation was also that the resulting system is simpler, rather than

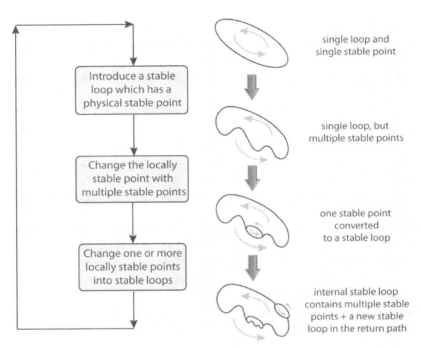

Figure 2.6: An iterative process to create an infinite family of physical SPL systems of arbitrary complexity. Starting from a single looped SPL system, it is possible to create a multitude of physical SPL systems depending on how we modify subcomponents of it at various stages of evolution.

harder, to occur naturally. This may not always be the case. However, this system is actually part of a larger system in the universe. It needs to interact with a number of other systems in order to be effective. We can use the above designs once again to create much larger SPL systems and iteratively improve them to eliminate operational inefficiencies. The key idea used (when you compare Figure 2.5 and Figure 2.4) is the ability to replace some or all of the stable points in an SPL system into stable loops and vice versa. The generic process shown in Figure 2.6 is as follows:

1. Introduce a stable loop, which typically has one physical stable point (like Figure 2.2).
2. Change the single physical stable point into multiple stable points (like Figure 2.4).
3. Change one or more stable points into stable loops (like Figure 2.5).
4. Continue step (1) for each stable loop iteratively.

Figure 2.6 shows the evolution of one SPL design after just two iterations. At any stage, a stable point can be converted into a stable loop, which in turn can be converted into multiple stable points, *ad infinitum*. By continuing this process over several iterations, we can create much larger physical SPL systems. It is also possible to merge (Figure 2.7(a)), link or chain (Figure 2.7 (b)) multiple physical SPL systems in an infinite number of ways to create new SPL systems with an increased set of loops

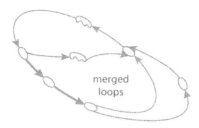

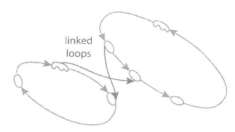

Figure 2.7: Mechanisms to create an infinite family of physical SPL systems. (a) We can merge two or more physical SPL systems to create a new SPL system with different features and properties. (b) Linking two or more physical SPL systems is yet another way to create a new SPL system. Using different ways of merging and linking, we can now create an infinite family of physical SPL systems.

and an increased ability to exhibit interesting behaviors. During this process, we need to ensure that the parallel loops are coordinated to have a single direction of flow for achieving stability for a long time. Blood circulation systems and our brain networks are examples of similar large-scale physical SPL systems with directed flows. In this way, we can iteratively create *physical* SPL systems of increasing order and any specified complexity. In chapters 52-68, I will describe a 'bottom-up' approach for creating large-scale *chemical* SPL systems. The bottom-up approach and chemical SPL systems are necessary for the problem of origin of life i.e., for creating life-like systems from inanimate objects using only natural means.

One point to note with Figure 2.6 is that there cannot be waste products for structured complex systems if they were to be effective. Every part of the system would need to be recycled, if necessary, by including it in a much larger system, i.e., as complete loops.

If we apply the iterative process of Figure 2.6 for several iterations and combine it with the merging and linking process of Figure 2.7, it is possible to generate a structured and yet arbitrarily complex SPL system. The resulting interconnected stable parallel looped system now starts to look more like how cells or our brain is organized. To compare, our brain has a complex interconnected network of neurons similar to the inclined plane and rolling ball structures discussed. Structurally, both the rolling ball structures and the brain network are not significantly different. They both have a large number of looped pathways within their respective networks.

However, the operation of looped dynamics in both of these cases are considerably different. In the case of our brain network, what propagates within the loops is an electrical disturbance. Inputs in the form of light, sound and pressure, for example, cause chemical variations within sensory neurons which open ion channels. This causes an influx of sodium ions into the nerve cell. This disturbance changes the membrane potential causing the neuron to fire action potentials along its axon releasing neurotransmitters at the synapses. The neurotransmitters now open ion channels on the post-synaptic neurons. The process now repeats causing the disturbance to spread from one neuron to the next to very distant regions within our brain.

I will discuss this process in considerable detail in chapter 5. For now, it should be clear that the balls rolling along looped structures discussed here is analogous to the propagation of electrical disturbance across a similar looped brain network. The one difference is that there is physical motion of balls within the looped pathways in Figure 2.2, Figure 2.4 and Figure 2.5 whereas in the brain network, it is the *disturbance that propagates* within the looped pathways (see section 5.3), not the ions themselves. Nevertheless, both examples sustain looped dynamics while consuming energy along with an ability to increase their SPL network complexity.

Ecosystem cycle is another example of such well-coordinated linked systems. Complex societies and economies are other such examples that evolve by linking several efficient looped systems to form a more stable and much larger complex looped systems. For now, the connection between SPL systems and living systems like cells or brain does not appear entirely clarified. I will discuss more on this connection in the later chapters. Specifically, I will show how SPL systems are capable of exhibiting a number of new and interesting patterns similar to living beings.

Most classes of interconnected dynamical systems have the property that the higher the complexity, the easier it is to destabilize them. SPL systems are the only known class for which the opposite can be satisfied – higher complexity can make them easier to self-sustain longer (cf. examples in the previous sections 2.1 - 2.4). However, as you may have noticed, the above physical SPL systems cannot be created naturally i.e., without help from other conscious beings, with some guarantees. Even though there are several possible variations to these designs, is it possible to create specific structures or let them evolve naturally in specific directions? To address this question, we need to switch to chemical SPL systems.

2.6 Chemical SPL systems

Having seen how to construct an infinite family of physical SPL systems, we can show the 'existence' of a large family of chemical SPL systems via a mapping specified in Figure 2.9. However, before I describe this, let me briefly revisit the notions of stability, parallelism and loops, but applied to chemical systems.

Firstly, the notion of stability is already well understood for reversible chemical reactions in terms of Le Chatelier's principle (see section 2.3) and chemical equilibrium. Secondly, parallelism, as discussed for physical systems, continues to be applicable for several interacting chemical reactions within living beings. They produce a cascading chain of reactions. Even a single chemical reaction occurs with parallel interactions between several molecules of reactants and products. Lastly, looped chemical reactions are quite prevalent among organic chemical reactions within cells, though they are not emphasized as much. Nicholson [114] included all the metabolic pathways involved in the ATP metabolism in mitochondria and chloroplast as a map. This chemical network clearly contains several parallel

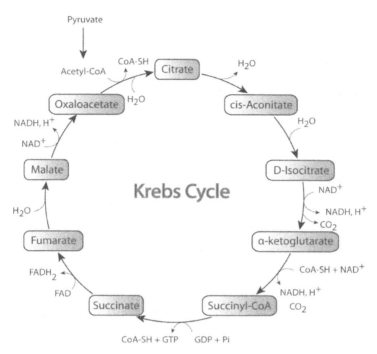

Figure 2.8: Krebs cycle, which is central for energy generation in most animal cells under aerobic conditions.

chemical loops. Nicholson has created a number of other minimaps, maps of inborn errors of metabolism, animaps and others, which also illustrate chemical loops at a detailed level. As another example, Jeong et. al [80] represented all metabolic reactions involving small molecules of a yeast cell. The presence of complex network of parallel loops is again quite apparent.

In Figure 2.8, we have one of the most important chemical loops, namely, Kreb's cycle that is central for energy generation in most animal cells under aerobic conditions. Glucose is first converted to pyruvate using glycolysis. Pyruvate is oxidized to acetyl-CoA in the presence of oxygen (aerobic conditions), which then begins a series of chemical reactions that is part of the Kreb's cycle. In the absence of oxygen (anaerobic conditions), pyruvate is reduced and proceeds instead with fermentation. Under aerobic respiration, the theoretical yield is 36 molecules of ATP, the primary storage of energy in cells, versus 30 ATP molecules of actual yield (as there are leaks in protons in the inner mitochondrial membranes) for each molecule of glucose.

Enzymes assist each of the reactions within the loop. For example, the conversion from citrate to *cis*-aconitate uses aconitase enzyme and conversion from *cis*-aconitate to isocitrate relies on isocitrate dehydrogenase enzyme. There are at least eight enzymes required for the complete loop of reactions to occur (for Kreb's cycle). In the case of plant cells, the corresponding energy generation is through Calvin-Benson cycle (see Raven and Johnson [126] for more details).

In the case of inorganic reactions, chemical oscillations (or loops) were discovered only in the early 1900's. One of the most well studied and visually observable oscillations is Belousov-Zhabotinsky reaction (see Belousov [10] and Zhabotinsky [177]). Belousov had discovered this reaction in an attempt to find an inorganic analog of Kreb's cycle. It was later explained using autocatalysis. The overall reaction is represented by:

$$3 CH_2 (CO_2H)_2 + 4 Br O_3^- \rightarrow 4 Br^- + 9 CO_2 + 6 H_2O \qquad (2.5)$$

It is broken down into the following three reactions to explain how the oscillations occur.

$$Br O_3^- + 5 Br^- + 6 H^+ \rightarrow 3 Br_2 + 3 H_2O \qquad (2.6)$$
$$2 Br O_3^- + 12 H^+ + 10 Ce (3+) \rightarrow Br_2 + 6 H_2O + 10 Ce (4+) \qquad (2.7)$$
$$Br_2 + CH_2 (CO_2H)_2 \rightarrow Br CH (CO_2H)_2 + Br^- + H^+ \qquad (2.8)$$

The mechanism involves two processes in which reaction (2.6) is dominant when bromide concentration is high. This causes reaction (2.8) to proceed, resulting in a decrease in bromide concentration. At low bromide concentration, reaction (2.7) becomes dominant. When reaction (2.7) proceeds, it starts to increase the bromide concentration once again, thereby producing oscillations. For a more detailed analysis, reactions (2.6) and (2.7) are subdivided into several other reactions (see Field and Noyes [43] and Field et. al [42]).

Note that a chemical reaction under constant temperature and pressure proceeds spontaneously when Gibb's free energy (ΔG) is negative. This makes the system proceed to a thermodynamic equilibrium at which point ΔG becomes zero. The oscillations described earlier are, therefore, quite far from thermodynamic equilibrium and the second law of thermodynamics is not violated.

A number of other mechanisms have been proposed for chemical oscillations like, say, Lotka-Volterra (predator-prey model), the Brusselator and the Oregonator – all based on autocatalysis. Mathematical models based on such oscillations like hypercycles and replicator equations were studied extensively (see Hofbauer and Sigmund [70]). Other sources of chemical oscillations have been discovered like bistability and strange attractors or chemical chaos (see Atkins [4]). In other words, all three concepts introduced in this book are equally applicable for systems with chemical reactions.

Let me now describe how to create a large family of *chemical* SPL systems using the infinite family of *physical* SPL systems described previously. Consider the following mapping shown in Figure 2.9. Here, we map all the balls as 'different' types of molecules. With physical SPL systems, we treated all balls as identical. For chemical SPL systems, each ball is potentially a different molecule. Represent different heights of the inclined plane as different energy levels. A specific molecule with a corresponding energy state represents a ball at a given height. The act of balls rolling downhill represents a chemical reaction in which reactant molecules of higher-energy state transition to product molecules of lower-energy state (Figure 2.9). Although the balls in physical SPL systems do not change, the corresponding representation as molecules in a chemical SPL system does change with each

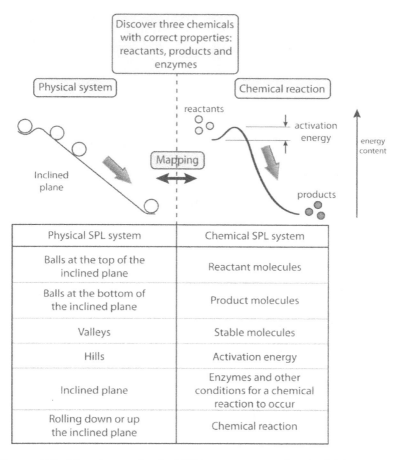

Figure 2.9: Mapping a physical SPL system to a chemical SPL system. Balls rolling down an inclined plane from one energy state to another are mapped to an uphill or a downhill chemical reaction. As shown in the table, the activation energy, the enzymes and other conditions that assist a chemical reaction are mapped, provided reactants, products and enzymes with suitable properties exist or can be discovered.

transition. The valleys correspond to special locally stable molecules of a given energy level. The hills represent the activation energy that a chemical reaction needs to overcome.

There are multiple ways to map the static physical structure of inclined plane in the chemical world. In one representation, they correspond to dynamical enzymes. Enzymes assist a given uphill and downhill chemical transition. While the balls absorb the external energy directly to produce the motion along the inclined plane, with chemical SPL systems, the enzymes (analogous to the inclined plane) also absorb the energy. Their unique shape and size cause the uphill or downhill transition to create new products (cf. induced-fit model of enzymes). In another representation, they correspond to suitable conditions like temperatures, pressures and energies that

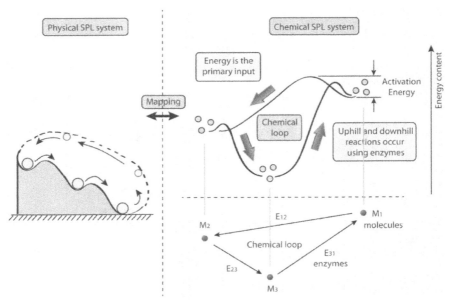

Figure 2.10: A sample chemical SPL system mapped from a physical SPL system. Using the mapping of Figure 2.9, a physical SPL system is mapped into a corresponding chemical SPL system. There are three locally stable points in both the physical and the chemical SPL systems. Most metabolic pathways within living beings (like the urea cycle and the Kreb's cycle) are much more complex than this sample chemical SPL system. Each reaction requires the discovery of a suitable enzyme both in the uphill and downhill directions to complete a loop.

allow a given chemical transition. Other representations include closed compartmental structures, catalysts, chemical pumps and a combination of these. Each of these representations operates in a similar way, though the degree to which these structures assist the uphill and downhill reactions differ compared to enzymes.

Figure 2.10 shows one example of a chemical SPL system utilizing the above mapping. It contains two downhill reactions and one uphill reaction to complete a chemical loop. The chemical loop itself is virtual unlike a physical loop. The downhill reaction tends to occur easily and naturally, while the uphill reaction requires considerable external energy and several favorable conditions, both in the downhill and uphill directions, to form a simple chemical loop. An example of a complex organic chemical SPL system involving multiple enzymes all of which are nontrivial to setup naturally, reliably and repeatedly is Kreb's cycle. In Figure 2.10, three enzymes assist in converting molecules in the loop.

From Figure 2.9 and Figure 2.10, mapping a physical SPL system into a chemical SPL system is quite simple in an 'abstract' sense. However, creating a 'real' chemical SPL system is nontrivial because we need to discover different types of chemicals and enzymes with several relative constraints between them. As an example, consider Figure 2.10. We need to represent three reactions for the mapping of the physical SPL system. From Figure 2.10, each reaction requires three sets of chemicals – reactants,

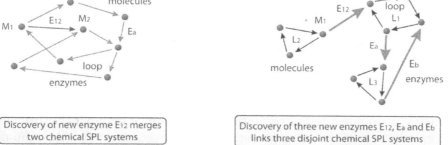

a. Converting stable points to stable loops

b. Merging chemical SPL systems

c. Linking chemical SPL systems

Discovery of new enzyme E_{12} merges
two chemical SPL systems

Discovery of three new enzymes E_{12}, E_a and E_b
links three disjoint chemical SPL systems

Figure 2.11: Mechanisms to create an infinite family of chemical SPL systems. (a) A chemical reaction with a stable point can be converted into a new SPL system with a chemical loop with the discovery of several new chemicals and enzymes like with Kreb's cycle. (b) Two or more chemical SPL systems can be merged to create a new SPL system. (c) Two or more chemical SPL systems can be linked to create a new SPL system. Using each of these processes, we create an infinite family of chemical SPL systems. Each of these processes requires the discovery of new enzymatic pathways.

products and enzymes. For three reactions, we need a total of nine types of chemicals. However, since the physical and chemical reactions are interconnected, they impose constraints – the products of one reaction = reactants of another reaction. This implies we only need three chemicals (M_1, M_2 and M_3) and three enzymes (E_{12}, E_{23} and E_{31}). Furthermore, the relative energy levels should also be such that $M_1 > M_2 > M_3$ (to align with the three valleys of the physical SPL system). Besides, the activation energies of each reaction (which are altered by enzymes) should be comparable to the corresponding depth of valleys in the physical SPL system that is being mapped.

For more complex physical SPL systems (like even Figure 2.7), the practical difficulty of discovering a real chemical SPL system that satisfies all constraints of the correct mapping is compounded. Having a repository of metabolic networks (Nicholson [114] and Jeong et. al [80]) from each species greatly helps us in finding chemicals with required relative relationships. Each design in Figure 2.4 - Figure 2.7 can now be represented as chemical SPL systems using the above mapping with a suitable choice of chemicals and enzymes, if they exist.

Merging, linking, chaining and the generic process mentioned earlier (Figure 2.7) for physical SPL systems continue to be applicable here (Figure 2.11). They help generate a large collection of chemical SPL systems when we choose appropriate enzymes, organic reactants and products. This collection can be ordered with increasing complexity using, say, the degree of the chemical network as the measure. As an example, replacing stable points (Figure 2.4) with stable loops (Figure 2.5) is equivalent to replacing the formation of H_2O and CO_2 from glucose and O_2 with a stable looped reaction, namely Kreb's cycle (Figure 2.11(a)). More generally, if we pick all the chemicals from the metabolic network (Nicholson [114]) of a given organism, we get a chemical SPL system. Therefore, *each living being itself can be viewed as a chemical SPL system*. This gives rise to an infinite family of chemical SPL systems of arbitrary complexity (by using either, say, one SPL system for each living being or by using the above generic merging, linking and chaining mechanisms). The above mapping of physical to chemical SPL systems is direct with organic reactions. Yet, we can represent even inorganic chemical loops like Belousov-Zhabotinsky and Briggs-Rauscher reactions as interacting SPL systems using the above mapping.

2.7   Designing large-scale SPL systems

A complex system like a cell or our brain has millions of dynamical subsystems. It is quite common to *analyze* such structured complex systems to understand different behaviors they exhibit (like an ability to search for food, struggle to survive, replication, free will, consciousness and others). However, how would we *design* such structured complex systems? If we look at any example of a structured complex system, it always starts as a simple system. Over a period, as the system scales, several features with multiple interactions between them will be introduced. This scaling brings unforeseen complexity and errors. We try to fix the errors as much as possible, sometimes by completely redesigning the system, to make it robust. In other words, it is common that a structured complex system results as an outcome rather than through a direct or intentional design. In fact, the notion of designing a system with an aim of having millions of subcomponents seems not only unnecessary but also less clever. Why would a designer need millions of subcomponents to create a given feature? Can he not achieve the same by building it in a simpler way? On the other hand, it feels reasonable when a system evolves and grows gradually into a million subcomponent structured complex system with wider usage, like with cells or the human brain.

Therefore, if we were to design a system that will eventually evolve into a million-subcomponent system, how should we start the first version of the design? What should our architecture and approach be for subsequent iterations? The new design paradigm I proposed in the previous sections was to create structured dynamical systems by ensuring that they form stable parallel loops with the rest of the system. We will try to satisfy this requirement during each iteration and when we introduce new features. Such an approach directs the evolution of a structured complex system in a clear path of progressively well-interconnected systems (which we will see in further detail later in the book). This is analogous to building a transportation looped network of trains and buses and using them to get from location A to location B instead of the alternative approach of driving from A to B (which we briefly discussed in section 1.4.2).

Let me now briefly explore the possibility of designing a structured complex system like a single cellular organism starting from simple designs. To begin with, let me identify a few unique behaviors of living beings that should be modeled in the initial designs. For example, several common behaviors exhibited by living beings are: a cell wants to live, it struggles when it nears death, it fights back, it fetches its own food, it responds to the external environment as if it senses the environment, it grows and reproduces and so on. When designing a cell, we would like to understand how these unique behaviors emerge. Nonliving systems typically do not exhibit these behaviors. However, with our initial designs, we do not require all of these features. Which one of these should we include, at minimum?

There are two types of designs we are interested in here: one that uses a conscious and intelligent being like a human and the other without any help from an intelligent being i.e., by nature itself. In this section, I will focus only on the former designs. Starting from chapter 52, for the problem of origin of life, I will discuss the latter natural designs in detail. The design paradigm I will take is to create structured complex systems using physical and chemical SPL systems discussed in the previous sections.

Among the list of features mentioned above, two of the most important ones are (a) that a system wants to live and (b) that it fetches its own food to survive. The terms 'want', 'live', 'survive' and 'its own' have an associated subjectivity and an emotional feeling in our colloquial usage. However, with the designs discussed here, I am only using them in a mechanical sense. For example, survival, in our first version of design, means that we want the dynamics to sustain for as long as possible. Similarly, 'wanting to live' can be interpreted for now as the systems' ability to resist its own destruction in some reasonable sense. Also, 'fetching its own food' means that the system has active mechanisms to seek materials and energy needed for its continued dynamics. In other words, materials and energy are not passively supplied to the system through random collisions, wind disturbances and others.

The reason we pick these two features is the following. If we observe a system that (i) exhibits these two features, (ii) 'looks' like a cell or a known living being and (iii) for which we do not know its internal structure, then there is a good chance that we would believe this system to be alive. Such a system would appear autonomous i.e., move on its own and fetch its own food. We would also observe that as we threaten the systems' survival, it would fight back. No nonliving system behaves in this manner. Even though we would not know if such a system can grow and/or reproduce, it would appear alive at first glance. It would be like watching a robot that stops what it is doing and starts searching for a power outlet when it is low on battery power. Also, this robot would fight back if you prevent it from reaching a power source as if it understands that it would deplete its energy and would be turned off i.e., it appears as if the robot is struggling to survive. We do not have such autonomous robots yet.

In chapter 59, I will show how to create simple naturally occurring designs that exhibits these two features (a) and (b), but in a mechanical sense mentioned above. I will also show how to extend such designs to create all other life-like features including growth and replication (see chapter 63). In other words, *I will show that all properties of life can be created from inanimate objects provided we have initial designs that captures just features (a) and (b) above, even if only in a primitive and mechanical sense*. Therefore, we can say that these two features generate all other life-like features including replication.

These simple designs are what are called the minimal structures (see chapter 57). The central result for the problem of origin of life is that we do not require creation of highly improbable complex structures like replicases that can both self-replicate and self-catalyze. Instead, we can use a collection of minimal structures based on physical and chemical SPL designs and then iteratively create more structured complex systems (see chapters 53 - 68).

Let us look at feature (b) closely. The SPL systems discussed until now (in sections 2.1-2.6) are passive. For example, a pendulum as an oscillating system continues its dynamics as long as external energy (using, say, batteries) is supplied to the system. Otherwise, it slows down and the oscillations come to a halt. The pendulum system does not 'seek' energy by itself to maintain the oscillations for a long time. This is true with most nonliving systems. Energy either is consumed randomly from external sources or is supplied by us, conscious human beings, to keep them working (through batteries and other power sources). The source of energy is external to the system.

On the other hand, all living systems seek energy and drive themselves actively and autonomously. What is the source of this drive? Clearly, we need an active source of energy – a mechanism that generates energy – within the system. But, how does a source of energy translate into a drive to survive? For example, having solar cells combined with rechargeable batteries can be considered as a good active source of energy. However, machines with such a setup do not have a drive to survive (or to continue their dynamics). In addition to having batteries that do not run out for a long time, the system needs additional mechanisms to recharge the batteries as needed, by itself i.e., we conscious beings should not be deciding when to recharge them.

For example, we breathe air ourselves utilizing our internal sources of energy. Similarly, our heart pumps continuously for like 60-70 years. We feed ourselves to replenish the energy consumed. With these examples, we not only have an active source of energy, but additional mechanisms to detect low energy and, therefore, seek food. Together, these are best represented as looped mechanisms at various levels like with materials or energy production, their consumption, detection of low and high levels of materials or energy and others, as we will see. Note that human level of consciousness and emotions are not needed for these drives because bacteria, plants and even primitive animals have been seeking energy for billions of years now.

Let us design the first version using our breathing mechanism as a guide. Our

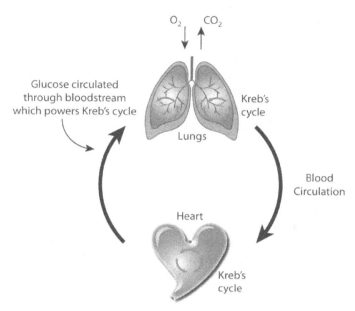

Figure 2.12: A simplified loop-based representation of our breathing mechanism. O_2 enters the system and CO_2 leaves the system. Within the system, we have blood circulation that carries, glucose, O_2 and CO_2 to different regions. Kreb's cycle is the primary source of energy generation to power all components like heart and lungs.

breathing is an active process. Air does not diffuse through our nose or mouth. Rather, it is actively sucked into our lungs. How do we even do this without being consciously aware of it? Note that we breathe even when we are unconscious like during deep sleep. In fact, if we try to hold our breath, we will be forced to let go and breathe normally, not because we want to live. Rather, it is an involuntary reaction because there seems to be a strong backpressure generated internally to eventually force us to breathe. This is an important mechanism to capture in our first design.

Figure 2.12 includes just the minimal set of components (like lungs and heart) abstracted in a suitable form to capture our *active* breathing mechanism. In Figure 2.12, we have an intake/exhaust system at which inputs (like oxygen and glucose) enter and outputs (like carbon dioxide) exit. The intake and exhaust mechanism is assisted by a subsystem like the expansion and contraction of the lungs. The external inputs then enter a circulatory system (like our blood circulatory system). The circulatory system is powered by a pump-like system (like a heart). Energy is genera-ted by taking the external inputs and by performing a series of chemical reactions (like O_2 and glucose converted into CO_2 using glycolysis followed by Kreb's cycle). A simpler mechanical example of this first version is an internal combustion engine that takes fuel and air to produce energy and work. For simplicity, let us assume that the energy generation process occurs in both the lungs and heart subsystems. The circulatory system helps transmit the external inputs and waste products to the appropriate locations.

Constructing real physical systems with just the components shown in Figure 2.12 is quite easy. We can use either man-made components or chemicals from living beings for each of the subsystems. The main components are a pump P (represented by heart) and an elastic membranous system L (represented by lungs) that can contract and expand. For the latter, we can use a spring loaded system, for example. Internal combustion engines and synthetic nanomotors (see Laocharoensuk et. al [91] and Fennimore et. al [39]) are examples of such designs.

The system operates as follows. As L expands, it sucks external inputs into the system. This generates more energy in L to keep the expansion and contraction process to continue for a while. When L contracts it pushes the outputs through the exhaust. At the same time, the external inputs move through the circulatory system to P. At P, a similar energy generation process occurs that powers the pump to continue the circulation. The cycle now repeats for as long as external inputs exist.

There are several coupled looped processes in this system: (a) expansion and contraction at L, (b) circulation of inputs and outputs through the system and (c) cyclic operation of the pump P. For the system and the looped processes to continue for a while, we need to first 'jumpstart' the system. Afterwards, the system operates the above looped processes as just described.

This system has a few important features. Let us first assume that the energy generated at L and P is much larger than what is needed to power all the subsystems. For otherwise, the dynamics will come to a halt. Whenever this condition is satisfied, we have a coupled loop – the intake of materials generate energy and the generated energy causes further intake of materials. The cycle now repeats until the external inputs are available.

I call this as the *active material-energy looped system*, with Figure 2.12 as our first example. The existence of the above coupled loop between materials and energy interaction is the reason for choosing the name 'active material-energy loop' (see chapter 59). Another example with a mechanical system is an internal combustion engine in a car. Fuel and air are actively sucked in during the compression and expansion strokes of the engine. These two examples are analogous to our active breathing mechanism, though simplified.

In chapter 59, I will show how to build other active material-energy looped systems *naturally*. I will use them to explain the creation of several complex organic molecules, show how to replicate the DNA without using DNA polymerases and other enzymes though less efficiently (i.e., a primitive DNA replication process before the discovery of amino-acid based enzymes) and the ability to carry out several chemical reactions even without the existence of amino-acid based enzymes. Using such active SPL designs iteratively, I will show how to create most other life-like features.

The advantage of the system in Figure 2.12 is that it is autonomous. It does not rely on the external systems (like humans and other living beings) to supply energy needed for continuing its dynamics, at least for a while. Rather, it actively sucks the necessary inputs as long as they are freely available in the environment. This is an important change from the designs in Figure 2.4 – Figure 2.7. These previous designs needed to rely on random energy sources for the balls to get to the top of the inclined plane (or any local peak). The active suction mechanism discussed here avoids this problem and gives a better control over the environment.

The second important feature of this system is that if you choke the system at the intake I, the system resists this by generating sufficient backpressure. For example, if you choke the intake of inputs (like oxygen when breathing), then when the available inputs are consumed, it creates a partial vacuum and hence a pressure difference. Additionally, a compression powered by the energy at L (like lungs) also generates pressure to suck inputs. Similarly, if you choke the exhaust, the system generates a pressure to push outputs (like carbon dioxide) out and clear the pathway. In the case of breathing, we can try this by holding your breath during either the inhale or the exhale process.

Rephrasing these features using words reserved for living beings, we can say that the system in Figure 2.12 is (1) autonomous and is trying to seek the raw materials it needs actively in order to 'survive' or sustain its dynamics, and (2) it resists 'death' through choking, at least, at a single location within the system (i.e., at intake I). We have essentially added life-like features to a mechanical system. The notion of resisting choking in this example is really a generalized stability argument. If you disturb any of the two normal states, inhale or exhale, by closing them, the system resists and brings it back to a normal state i.e., inhale and exhale states are stable. The feature 'resisting death' is only true at the intake I in this example. In the later modifications to this design, it is possible to incorporate additional sensors and actuators to resist death from damages to locations other than intake I as well.

The principles behind the active material-energy looped system described here are generic enough to be applicable for a subset of features in living beings including bacteria and plants, even without visible circulation. For example, the material loop consists of transfer of molecules in and out of the plasma membrane. The energy loop is, say, the conversion of ATP to ADP and vice versa. In addition, the mechanisms described for these simple designs do not have to be as efficient as they are for living beings. Rather, the only requirement is that the energy produced is more than the energy consumed and/or wasted.

It is now possible to include more life-like features into this simple system. I will discuss them briefly, especially since we have built several machines using a similar approach. Here are some modifications to this design.

- We add more energy generation loops using chemical, mechanical, electrical and other systems.
- Since the energy generated is more than the energy consumed, we can add more subsystems like hands and legs (cf. with robots). When combined with active energy seeking ability, this system can be designed to 'search' for its own source of energy as the net energy level drops to low values (Figure 63.1 discusses one such design).
- Currently, the system resists choking only at one location, namely, at the intake I. It is possible to add 'sensors', one at a time, that detect damage at more locations. This sensory information can be coupled with actuators that either

resist damage or make the system move away – fight-or-flight response (cf. with sensory feedback loops with electronic control systems). The system now resists 'death' for a large variety of damages analogous to living beings.

- We can add additional sensors to detect, say, light, sound, pressure and special chemicals (both harmful and useful ones). This lets the system process more information from the environment.

In this manner, we can continue adding other features, one at a time. This is analogous to how cells improve in each generation. The difference here is that we rely on human intelligence and consciousness (cf. how robots and other machines with human-like features are built) whereas cells evolve entirely autonomously. Another difference is that the new approach uses stable parallel looped dynamics with active mechanisms instead of static assembly and passive mechanisms of the traditional approach.

The above discussion on how to build extensions to Figure 2.12 synthetically is intentionally brief because (i) it is not new and (ii) my objective is to explain how nature creates such systems, by itself, not with the help of intelligent and conscious human beings. How does a system evolve to introduce its own construction within itself (like within the DNA)? How does a system develop consciousness and understand its very own existence? My objective is to address such questions in detail using the new active stable parallel looped dynamical framework introduced in this chapter (see chapters 10 - 50).

3
Pervasive Loops

In the previous chapter, I introduced an infinite family of large-scale physical and chemical dynamical systems based on a new stable parallel looped architecture. These systems can be built with any specified complexity. The focus in the later chapters is to identify, design and build the correct subset of systems from within this infinite family that exhibit a specific feature. Examples of features we are interested in building are those exhibited by living beings and humans, like memory (within human beings), free will, consciousness, emotions, perception of space, sensing passage of time, origin of life from inanimate objects, replication of DNA without the presence of enzymes, guaranteeing a definite direction in evolution, creation of the first multi-cellular organisms from just unicellular organisms and others.

Take, for example, the feature free will. Not every SPL system within this infinite family exhibits free will. How would we tweak the SPL designs so that the resulting system does indeed exhibit free will? Similarly, if we want to build a system that senses the passage of time, how would we choose an appropriate SPL (or any other) design? It is possible that a system sensing the passage of time will not exhibit free will and/or vice versa. One-size-fit-all designs, though desirable, are less informative. Rather, it is useful to identify those subsets within this infinite family that give rise to a given feature as it would help us understand its root cause.

The rest of the book takes this approach with each of the above features individually. In this chapter, I will identify additional properties of looped dynamical systems that make them suitable for studying large-scale structured dynamical systems. Recall the definition of loops in section 2.1 along with the assumptions stated. These assumptions allow us to generalize definition 2.2 for an SPL system to cover a large set of examples like dynamical systems that are oscillations around an equilibrium point in physical systems, equilibrium in chemical reactions, limit cycles in physical and chemical reactions (which are usually far from equilibrium), strange attractors like Lorenz attractor, periodic and cyclic systems.

Loops are quite pervasive in nature. Here are a few examples that we have been encountering in our everyday life: pendulums and swings, atomic oscillations, orbital motions, planetary motions, seasonal cycles, water cycle, oscillations in quartz crystals, waves, reversible chemical reactions, ecosystem cycle, Kreb's cycle, Na-K pumps in neurons, oscillations for transmitting neuron spikes along the axon, ice age rhythms, water fountains, motors, pumps, flywheels and wheels, roller coasters, predator-prey models, demand and supply dynamics and so on.

Even though looped dynamical systems are pervasive in nature, it is not obvious how to show that they can be used to create features like consciousness, life from inanimate objects and others listed above. It may be reasonable to feel that looped dynamical systems are necessary for life and all properties exhibited by living beings.

However, it is entirely nontrivial to show that they are sufficient as well. The goal of this book is to show the sufficiency, in precise terms (at least theoretically) by giving constructions based on the infinite family of physical and chemical SPL systems of the previous chapter. In this chapter, I will begin by listing some of the advantages and unique properties of looped dynamical systems that make them pervasive.

3.1 Long term dynamical behavior

If a nonlinear physical system is subject to random energy inputs at various times that are too small in magnitude individually to produce macroscopic change, what will the system's state look like after a sufficiently long period? There are a few possibilities. One possibility is that the system can come to a state of rest in a macroscopic sense, but at a microscopic level (i.e., at an atomic level, say), there will be significant dynamics because of the random energy inputs. An example of this is the state of dynamic equilibrium with reversible chemical reactions. At a microscopic level, several molecules of H_2 and O_2 continue to produce H_2O and vice versa even when the reaction is under equilibrium. Only in a macroscopic sense, the concentrations of H_2, O_2 and H_2O do not change.

The second possibility is that the system has a periodic motion like limit cycles or loops that will sometimes produce macroscopic motions as well. Examples of this are water waves in an ocean, ecosystem cycles, water cycle and others. These macroscopic motions can sustain for sufficiently long time if the environment cooperates by minimizing the loss of energy that is consumed through its normal dynamics (cf. Figure 2.4 – Figure 2.6 in chapter 2).

Yet another common state is one in which a system exhibits random motion. Some examples of this are random molecular motion in a room, turbulence diffusion and others. The other possibilities are less probable in nature. It is the second possibility above (of loops) that is important for creation of life on earth and its sustainability for billions of years, as we will see in the later chapters. The other possibilities do occur, though they are more common among nonliving systems and lifeless planets.

3.2 Self-sustaining property – loops versus non-loops

Consider designing two systems, one with many loops and another with almost no loops. If we now supply energy to both these systems, the system with loops will sustain longer while exhibiting interesting features than the one without loops, as previously explained (see examples discussed in chapter 2). The primary reason is that the available external energy is trapped effectively in a looped system and can be re-used later when the external available energy is lower. The looped system self-sustains either from occasional random external energy or from an internal energy source. For example, consider a disturbance in the form of a pressure wave in a water pipe. If the water pipe is closed as a loop, the wave disturbance circles around the pipe several times without diminishing its amplitude. There are no cross-interactions and collisions between the forward and the backward disturbance dynamics.

If the pipe were not a closed loop however, the wave would bounce back and forth, as it reaches either end. The backward flow of disturbance opposes the forward flow. This resistance causes the energy to dissipate quickly unless there is a resonance, which for a complex non-looped dynamical system, is rare. In other words, most non-looped systems start and end their dynamics in a short period unlike a system with loops. Non-looped systems can only run for long periods if they are very slow processes (like formation of continents). Such slow processes, though interesting in geological timescales, do not produce structured patterns similar to those exhibited by living beings.

This ability to self-sustain the dynamics for a long time also implies that if events are occurring by random chance, then if they are looped events, the probability of their occurrence becomes quite high because of repeatability. On the other hand, if the events do not have loops, their probability of occurrence is low. Note that the probability of occurrence of an event is proportional to the number of times an event occurs over a long period of time. For a non-looped event, the total number of times an event occurs is low compared to a looped event if counted over a long period.

3.3 Consuming minimal external inputs

Most systems with loops are self-contained and they re-use most of their subcomponents. They typically require external physical inputs in small quantities. This is because the dynamics repeats itself. One input they require in larger quantities for the continued loop dynamics is energy (since we cannot have a perpetual motion machine). Take, for example, refrigeration cycle where the refrigerant CFC is re-used throughout the cycle. The energy needed to move the refrigerant through the various components like evaporator, condenser and compressor is the only constant source of input. When designing a system (man-made or natural), it would be ideal if the system takes energy as the only input at several possible locations. Sometimes, this is possible only when we extend the system by including additional dynamical systems.

For example, plants and animals together form a looped system in which the plants consume CO_2 and H_2O from animals to produce glucose and O_2. Animals, in turn, consume these outputs. Together, they form a loop. The only external input is sunlight. This iterative process of including more systems as part of a larger system can be used with other examples as well. We can use this approach until we feel all loops are well coordinated and closed. When this happens, the only input for the larger system will be energy.

As another example, consider the ecosystem cycle shown in Figure 3.1. Plants use solar energy along with other inputs like water, which acts as food for herbivores. Carnivores eat herbivores, which are decomposed by microbes and bacteria. They get broken down to basic elements like carbon dioxide, oxygen and nitrogen, which are used by plants. Together, they form the ecosystem cycle. There are several smaller loops within the ecosystem as well. However, when nature includes all of them as part of bigger loops, we have a very stable ecosystem that is capable of surviving for a long time on earth. The overall ecosystem cycle exist in good harmony, with sunlight as the main source of external energy input.

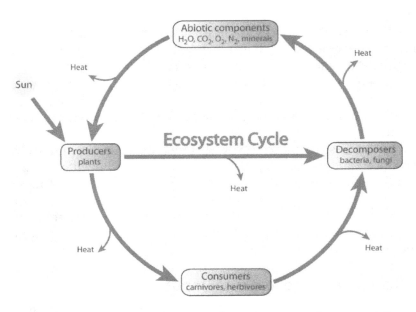

Figure 3.1: Ecosystem cycle. Sun is the main source of energy while the rest of the materials and living beings are recycled.

3.4 Traversing across a chain of loops

If we have a complex network of physical loops that intersect one another (cf. loops in, say, Figure 2.7(a)), a particle can move in a path starting on one loop, traversing a short while and then hop onto another loop. These hops happen quite easily. Examples of these structures are the ones we see in science museums that have complex roller-coaster like designs on which balls roll and move in interesting patterns after they were raised to a certain height and let go. Other examples of such complex system of loops are the brain's neural network and blood circulation. Therefore, a given particle reaches distant location along such paths (see Figure 2.7(a) as an example of physical SPL system).

Besides, the number of such possible paths is large. The net outcome is that these systems produce interesting patterns and behaviors. Even though the particles appear to be moving randomly from one location to another, there is, in fact, visible (like, say, in Figure 2.7(a)) or invisible loops (as physical fields) in the environment.

Living beings, in particular, have the ability to control the very networks on which the particles move just as easily as they control the motion of the particles themselves. These are unlike nonliving systems controlled by force fields. In other words, living beings are proactive whereas nonliving systems tend to be reactive and move under the influence of externally applied force fields (sometimes with help from conscious beings).

3.5   Energy consumption in a complex multi-loop system

The amount of energy spent by particles in a large-scale structured network of loops as they traverse from one location to another is *minimal* relative to a non-looped system of similar complexity (in terms of, say, the total length of the paths), in the following sense. When particles move within a physical loop, there is an uphill and a downhill motion (in terms of energy – see Figure 2.4). The downhill motion occurs naturally while the uphill motion requires energy. In addition, either the supporting structure on which the particles move help supply energy or there is an internal source of energy. The natural dissipation of energy of the particle turns out to be the most significant loss of energy compared to the dynamics itself. Furthermore, the particle spends its energy whenever it needs to hop from one loop to another (as this involves a change in state – cf. Newton's first law of motion). The total energy consumed by this system is the sum of energy consumed within each loop plus the sum of energies consumed for each hop along the entire path.

Let us compare this to the energy consumed by a non-looped complex system of similar complexity (say, in terms of the total length of the pathways). In a non-looped system, every particle needs to start its motion from rest requiring the equivalent of 'activation energy' (cf. activation energy of a chemical reaction). The dynamics stops after a while since this system has less or no repetitions. However, starting and stopping consumes a lot of energy (cf. higher activation energy necessary to 'start' a chemical reaction versus one that is already occurring). This activation energy is an additional energy requirement beyond what is needed by a looped dynamical system i.e., energy consumed along the path (like loops and non-loops) + energy consumed during hops.

Due to this extra high-energy requirement, we would need to design a number of heat sinks to cool the amount of heat generated in the non-looped system. Otherwise, the thermal fluctuations from the heat can hinder the system's ability to continue its dynamics for a sufficiently long time. Therefore, we can say that the design that best utilizes the total available energy while minimizing the loss of energy through heat or friction is by having a large number of well-coordinated loops. This is what I referred to as the *minimal* energy consumption by looped dynamical systems. The description given here is somewhat vague, especially the term 'non-looped system with similar complexity'. Yet, I hope that the central point, namely, that the external energy is effectively trapped and re-used by looped system compared to a non-looped system, is clear. In addition, the requirement of supplying additional energy for activation is lesser for a looped system compared to a non-looped system. For structured complex systems like the brain and the cells, a looped approach is the only one that can ever work because of the above energy requirements. Otherwise, the energy losses and the heat generated would simply destroy the system.

3.6   Coordination and coincidences with loops

We have seen that the energy trapped and re-used by a looped system is quite efficient. A given amount of energy generates dynamics that sustains for a longer time compared to a non-looped system of similar complexity. In addition, the probability of integrating a looped system as part of a larger system is also higher for such systems. This becomes important when we explain the origin of life from inanimate objects starting from chapter 52. The reason is as follows.

If we have a system that is not a loop, then it has a short time frame when the dynamics starts and ends. If we want to create a better system by merging features of this system with several other ones, we need to ensure that the dynamics of all these systems occur during the same period, at the same place and at the same rate. If one system is faster or slower than the other it is less likely to exchange information between the two systems. This makes it very unlikely for structured complex behaviors to occur by nature itself. The favorable opportunity exists only for a brief time and only in an ideal scenario (with respect to location, time period and rates).

However, for a system with loops, the available energy is consumed more slowly and the behavior of the system repeats itself a large number of times. This repetition makes it likely to exchange information or interact with one another in spite of the fact that different systems have different speeds or frequency of dynamics. For example, if three molecules A, B and C missed to meet each other in a given region and a given time interval and if each is part of a, possibly independent, chemical loop, then surely they will get several opportunities to meet in the future as their dynamics repeat with time-varying periods. Coincidences do have a finite probability in a stable parallel looped system. If we need hundreds of such coordination's to occur (as with living systems), repeatable looped systems offers a great advantage. In fact, there is no known way to achieve such coordination naturally if they are non-looped non-repeating complex systems.

We can convince ourselves about the possibility of coincidences within independent repeated processes in an idealized setting as follows. Assume that we have 10 waveforms, each reaching a given region R (similar to Figure 1.3) in one period of repetition, but at different times. Let us represent them as 10 square waves, each with values ranging between from 0 to 1, with distinct time periods equal to multiples of $\sqrt{7}$, say, and distinct phase shifts equal to multiples of $\sqrt{3}$. For example, in Matlab, we can create these waveforms using the following equation: $y_i(t) = 0.5 + 0.5 * (square(\frac{2\pi t}{i*\sqrt{7}} + i * \sqrt{3})$ for i = 1, ..., 10. We assume that the crests of these waveforms correspond to when the object reaches region R and the trough corresponds to when object is away from region R. The problem of coordination corresponds to the question of whether 'all' 10 waveforms reach the same region R simultaneously. To determine this graphically, we first add all of the waveforms: $y(t) = \sum_1^{10} y_i(t)$. If we plot $y(t)$, we will see that the values range from 0 to 10. A value of 10 at a given time corresponds to when all 10 waveforms meet at region R whereas a value less than 10 like, say, 6 corresponds to when 6 of the waveforms meet at region R. From the plot, it is easy to see that all integer values from 0 to 10 are attainable. In more realistic cases, the periods of repetition and the phases will be time-varying as well. Even in these cases, we would observe all 10 waveforms reaching the region R with a non-zero chance. The same coordination problem is not easy for non-looped pathways.

3.7   Self-replication of a collection of molecules

The term self-replication is typically reserved for polymer molecules like DNA and RNA. These are the only known molecules for which an elaborate replication process has been identified relying on complex enzymes like DNA polymerases. The existence of DNA molecules that can replicate reliably and the corresponding transcription property to create enzymes is the central reason for the sustainability and evolution of life for billions of years. Understanding how to discover these mechanisms and the molecules naturally is one of the biggest unsolved mysteries with the problem of origin of life. In chapters 60 - 62, I will show how this can be achieved without the help of amino-acid based enzymes both reliably and repeatedly using suitable SPL systems (like Figure 59.1). These simpler SPL designs are an alternative to RNA world hypothesis and other proposals.

Before we address replication of DNA molecules, it is important to realize that there is another more important replication problem that is largely ignored. This is with the replication of a *collection* of molecules as opposed to a single polymer molecule. The reason this is equally critical for the origin of life is because the existence of life would not be possible unless a large number of organic molecules beyond DNA, RNA and enzymes are present. These are molecules from several functional groups like carbonyl, carboxyl, aldehydes, ethers, esters, alcohols, ketones, amides, amines, sulfhydryl and others. Unless these molecules are also present at the same time and at the same location where the DNA and enzymes are, for a sufficiently large period of time, the cells will not be able to perform any useful functions and life would cease to exist.

How can we guarantee self-replication or the continued existence of all of these molecules before and after DNA is discovered by nature? If these molecules are not present or if they disintegrate before a self-replicating DNA, an RNA replicase or other equivalent molecule is discovered, even DNA would not be useful. While the properties of DNA and its replication process is well understood, how can we guarantee replication of the entire collection of these other organic molecules, each of which cannot be individually replicated easily?

The answer to this nontrivial problem lies with the formation of chemical SPL systems. To see this, consider a collection of molecules and let us ask which subset remain after a long time. The typical answers are (a) inert, less reactive chemicals and ones with high half-life, (b) abundantly available chemicals and (c) chemicals that switch repeatedly and randomly between states. However, with the discovery of chemical SPL systems (see section 2.6), we now have a 'dynamical' case as well.

Claim 3.1: Any collection of molecules that form a chemical SPL network while consuming (i) freely available external inputs/energy or (ii) molecules that are themselves part of other external loops, will continue to exist with high probability after a long time.

To see this, consider molecules within a looped subset. They reuse raw materials while consuming only abundant energy or chemicals unlike a non-looped subset, which produce non-reusable products. These molecules also self-sustain longer compared to non-looped dynamics, which die down quickly. Additionally, non-looped chemicals arriving within a small neighborhood at the correct time as coincidences so they could react, has near-zero probability (cf. air molecules reaching one corner of a room). Looped chemicals, on the other hand, are repetitively produced. As their concentrations keep increasing in a given region, the chances of coincidences improve. Therefore, the probability of looped subsets remaining after a long time is higher relative to non-looped subsets.

If we view the static part of the chemical SPL network as a directed graph (see metabolic networks – Nicholson [114] and Jeong et. al [80]), we can partition it into a collection of strongly connected components (Harary [65]). Each strongly connected component has at least one loop. A looped network is said to be generated by a subset S of chemicals if and only if the entire network can be recreated using S and the corresponding collection of enzymes. With the above partition, the subset S needs to have only one chemical present to regenerate each strongly connected component. Therefore, replicating the entire chemical SPL network is easy as there are several choices for subset S of chemicals (for example, three strongly connected components with four chemicals in each, has 64 choices). A simple random split of the chemical SPL system into two parts will regenerate the entire SPL network in each part.

Even though no individual type of chemicals remains fixed or has long half-life, the entire SPL collection, as a whole, self-sustains for a long time. This is a *self-replicating property* for the entire collection. This is different from self-replication of DNA molecules as we involve the entire looped 'collection', not just individual DNA sequences.

One special case of the self-replicating collection is the problem of homochirality in which the collection of chemicals that remain after a long time are D-form sugars and L-form amino acids. I will show in section 66.5 how to achieve an asymmetric distribution starting from a perfectly symmetric (50-50) distribution of L- and D-forms using the chemical looped structure.

3.8 Looping system and environment

In section 2.1, I mentioned that the system and the environment are typically considered as independent of each other. For example, for a chemical reaction where H_2 and O_2 combine to form H_2O, the system is the actual chemicals and the environment is the correct set of pressures, temperatures and volumes under which the chemical reaction is allowed to proceed. We usually assume that the 'environment' creates the right conditions for the 'system' to proceed with its dynamics. For chemical reactions in a laboratory, humans setup these environmental conditions carefully using our intelligence and consciousness. If nature were to setup these conditions without help from conscious beings, the only suggested approach is random chance.

Therefore, a useful question to ask is if the system can create the right environment to run itself. This is a rather unique loop – the environment drives the system and the system rearranges itself or even creates the right environment – thereby coupling the system and environment tightly. I will give two examples of this later. One is during the explanation of the origin of life (see active material-energy looped system of chapter 59). The other is when explaining consciousness (see fundamental pattern of chapter 16). In the latter case, we, the system, learn, practice and alter ourselves to become good at controlling the very environment we live in (like an infant learning to control her hand to put it in her mouth).

In the rest of this book, I will begin to apply these ideas to explain how the brain works, what emotions and consciousness are, and how life itself evolved from inanimate objects.

4

Brain

Human brain is an extremely complex organ. The source of the complexity is clearly within the billions of neurons with trillions of interconnections between them. They form a continuously adapting dynamical network. Until now, there has been no consistent theory proposed to explain this complexity. How does our brain give rise to seemingly perfect and well-coordinated behaviors like memory, intelligence, consciousness, emotions, realizations and others? How can we create synthetic large-scale complex systems that exhibit these behaviors?

In this chapter, I will present issues unique to our brain and leave the detailed analysis of how to address them to the later chapters. The current research on brain is quite extensive to cover in one chapter. I will, therefore, present a short overview of the types of problems researchers have been working on.

The typical approaches taken include (a) studying the neuron at a detailed level and trying to extend the analysis to billions of interconnected neurons, (b) studying the firing patterns of a broad group of neurons using MRI, EEG and other devices and correlating them with the observed behaviors, (c) using psychoanalysis to infer the functional behaviors of the brain, (d) statistical or probabilistic approaches to explain the observed patterns, (e) using computer models to represent critical aspects of a neuron and simulating the behaviors of millions of neurons (cf. neural networks) and others. Unfortunately, none of these approaches have successfully clarified the above behaviors or offered insight into building synthetic conscious systems.

Given such a variety of approaches, one fundamental question to ask is the following. What constitutes a satisfactory answer/theory? Why would a reader, just by reading the theory, feel comfortable that he understands how the brain works? While the above approaches give partial answers to our questions, when will we be truly satisfied? Can building a synthetic seemingly conscious system help? Or does providing a theoretical explanation for the features of our brain but not an actual construction convince us? For large-scale complex systems, having both theoretical explanations and constructive mechanisms are necessary even to claim self-consistency. This is unlike simple systems and those based on fundamental laws. This will be the goal in the following chapters.

4.1 Causality for large-scale structured dynamical systems

When we observe an event that we do not understand, whether it is scientific or not, we tend to look for a cause. Only when we find a cause, we feel a sense of satisfaction, at least at that instant. With scientific events, we have an expectation of repeatability of the causal event so others can verify as well. Causality is the term used to signify this relationship between cause and effect. For simple physical and chemical systems, causality is well understood and accepted. For large-scale complex systems however, there is a common mistake people tend to make. They sometimes regard logical analysis and arguments as *causal* explanations when they are, in fact, just *correlations* i.e., statements that are obeyed with some probability when the event repeats across multiple systems or different times (analogous to statements based on polling). Let me discuss this distinction in further detail now.

Causality is a fundamental assumption in all scientific theories. If a set of well-defined conditions are maintained (like placing a ball on the inclined plane or shooting electrons as in Young's double-slit experiment), another set of well-defined effects are produced (like the ball rolling down the inclined plane or an interference pattern respectively). This lets us identify the effects with the corresponding causes. When multiple components interact in parallel, causality continues to be applicable except that we need to account for all interactions with sufficient accuracy. Such an analysis has practical limitations when we have large number of components. Yet, from a theoretical standpoint, we do not doubt causality itself even for large-scale complex systems like our brain or cells.

As mentioned above, there is one common source of error when identifying causality within large-scale complex systems. This is with correctly identifying when a statement is a causal one versus when it is a correlation. Consider an abstract statement like 'a ball is rolling down an inclined plane'. This is a logical statement *about* causal physical dynamics whereas a real physical ball that is indeed rolling down an inclined plane right at this instant is the raw causal dynamics. Our brain observing this ball produces causal physical and chemical dynamics as neural firing patterns. For a single neuron, these causal processes are already described in most textbooks on biology (see, for example, Purves [122]). Even when a group of neurons fire corresponding to this event, causality continues to hold – earlier neurons supply inputs in parallel to later neurons.

Now consider the following question. Is the period from the causal dynamics occurring within neurons in our brain to the conscious inference of an abstract statement *about* the observation, namely, that 'a ball is rolling down an inclined plane' causal? Or is it just a correlation? Let us say that we performed the ball rolling event a 100 times. A correlation means that our observation produced approximately similar causal dynamics within our brain, making us generate the above abstract statement, say, 95 times. Examples of times when it does not generate the statement is, say, when you are drunk or drowsy during this period or when the lighting is low that you are unable to see the ball rolling down the inclined plane clearly. Interestingly, our state of consciousness plays a role in our inference about an observed dynamics.

If we follow the entire dynamics in a causal manner starting from a ball rolling to light reflecting off the ball, light falling on the retina, neurons firing and propagating within the visual cortex until we end up saying the abstract statement out loud, we notice that somewhere along this path, a switch has happened from raw causal chemical dynamics (with neural firing pathways) to a statement 'about' the dynamics. How did this happen? In fact, how are you sure that this statement is logical and is even related to the observed event? Are our eyes feeding in correct information –

maybe you have eye abnormalities, squinted eyes, color blindness and so on? Are the subsequent causal neural dynamics within your visual cortex related to your abstract statement – maybe you saw and interpreted the observed event differently from me? This is surely common with complex events involving emotions. In fact, even for non-emotional events like, say, reading a book on algebra, your understanding of the same chapter is typically different from mine. Given these differences in our understanding, other sentences you say about other causal events would, in general, be different from mine. Which abstract statements of external events are accurate descriptions of causality and which ones are just logically correct but correlations? Is there even a distinction between the two?

Our brain being a complex dynamical system has become the source of difficulty in answering the above questions. Now, imagine you are proposing a new theory of structured complex systems like living beings or our brain. If your theory is built on causality of neural dynamics, is it more accurate? Or if it is built using logical stateme-nts like above based on correlations (typically using statistical or probabilistic models), is it more convincing? The typical mistake made when studying features of our brain is to treat correlations as a generalization of causal relations instead.

Other examples for which the above questions become blurry is with statements like "when you are hungry (a cause), you eat food (an effect)". This feels like a causal statement analogous to "when you place a ball on an inclined plane (a cause), it rolls down (an effect)". However, both statements are just generalizations of causality mentioned earlier. It is better to say that the former 'if-then' conditional statement is a 'logical' sentence instead of describing causality. The word 'causality' is typically reserved for describing physical and chemical dynamics rather than purely abstract statements obeyed by these dynamical systems.

The former conditional statement (eating when hungry) is falsified more often than the latter causal statement when the respective event repeats several times. In other words, the former logical statement does not obey a 'physical law' unlike the latter statement, which obeys Newton's laws. A human, though composed of only physical and chemical components has free will, making him follow seemingly arbitrary laws. Therefore, statements that involve living beings, in general, like the example above (eating when hungry) are not good generalizations of causality compared to statements involving nonliving beings. They should be treated as correlations (though they may appear logical) because we can only specify probabilities of occurrence of an event for a given cause.

When building a theory describing behaviors like free will and consciousness, if we use logical statements as above that does not link back to the causal neural dynamics, the theory is unsatisfactory. The resulting 'theory' or logical argument appears nothing more than a high probability scenario. The arguments used within such a theory has words like hunger which are related to average chemical dynamics (across thousands of cells). It is not clear when you have linked the raw causal dynamics to the abstract terms and when you did not. Unless this link is explicitly clarified, our belief in the theory drops.

In summary, there is a clear distinction between simple and large-scale structured dynamical systems when we consider repeating the same experiment multiple times. For a simple nonliving system, we can use causality from each of the experimental outcomes to extract a sufficiently accurate physical law (like Newton's laws or Maxwell's laws). On the other hand, for a structured complex system like a human being, we cannot rely on logical correlations built using experimental outcomes to extract accurate physical laws.

Therefore, causal theories on simple systems do appear convincing to a reader whereas correlation theories on structured complex systems do not. The new theory proposed in this book using SPL systems is a *causal theory* of our brain and complex systems, not a logical correlation theory. A direct implication of any causal theory is that we can construct synthetic systems that exhibit each of the features explained. In this sense, we would expect creation of synthetic life and synthetic consciousness to be clarified through well-defined constructions if the theory of large-scale structured dynamical systems proposed were to be causal and be convincing to a reader.

4.2 Systems view

The organization of a human brain is unique compared to the brain of other species. To explain behaviors like consciousness and free will, it is not sufficient to identify the differences in structure and function of our brain relative to the other species. Mere absence of these structures does not imply that these features cannot be exhibited. In fact, our brain is so resilient to damages that some parts of our brain can take over the function of the damaged region. For example, as a child, if there is a major lesion in Wernicke's area of the left hemisphere, needed for speech comprehension, the brain simply switches the entire speech mechanism to the corresponding area on the right hemisphere (see Jaynes [79]). There are other examples of such healing abilities. While there are unique substructures for each of these abilities, other regions of our brain can pick up those traits under dire circumstances, without any noticeable difference to the human. If this were the case, why don't other species develop the same ability using a similar reorganization process within its own brain architecture?

To answer such questions, what we need is a more direct explanation and a way to construct these features beyond analyzing them. An interesting question to answer is if specific molecules, neurons or substructures of our brain are necessary for a given feature. Is there no other way nature or we could have created consciousness and other features? It is unreasonable, for example, to suggest that there is only one way to become conscious and that the natural evolutionary mechanisms did manage to pick it. The current structures of our brain are surely helpful, but are these details necessary? At a coarse level, these structures exist in several other species though their consciousness is not to the same degree. At a detailed level, even two humans do not have the same structures or connections in spite of similar consciousness. Therefore, in between these two extremes, there exists a unique structure and dynamics, which gives rise to consciousness and other features.

The current research on brain is quite extensive to discuss in this book. I refer the reader to textbooks like Purves [122], Kandel et. al [82] and other active journals. For the purpose of this book, I will choose a few topics and represent them in a systems point of view in order to derive a causal theory of the brain. With systems approach, I want to extract just those representations that allow us to construct analogous synthetic systems, with or without neurons.

To start, consider a human born with an initial structure called the primitive brain. There are several innate abilities he is already expert at, the most notable one being the sucking reflex. As he grows and gains new experiences, this structure is modified by continually forming connections with neighboring neurons. From a systems point of view, when we view our brain as a closed loop system, there are at least three functional aspects of the brain:

(a) *External and internal inputs*: External inputs are those that are received through the sensory neurons, say from the eye, ear, nose, tongue and skin. Internal inputs, however, are of two types. One set of internal inputs come from within the internal organs of the body through the brain stem, like, for example, from the heart for regulating heartbeat, lungs, stomach, kidneys and others. The other types of internal inputs are those within our brain like our thoughts, memory, planning and others. All these inputs propagate across the network of neurons through complex connections to various parts of the brain. The details of how an individual sense organ processes the input itself have been very carefully studied (see, for example, Purves [122], Ornstein and Thompson [118]). However, the ability to translate this information to recognition and other higher functions will be the subject of interest in this book.

(b) *Internal processing*: This is the most elaborate and the least understood part of our brain. Examples of these are long-term and short-term memories, language, speech, abstract thinking, emotions, consciousness and others. The substructures for each of these abilities and their roles are well mapped out. For example, cranial nerves emerge directly from the brain to sense and control several organs of our body (see Kandel et. al [82]). Cerebellum is responsible for muscle coordination, reflexes and balance. Cerebrum is used for most of human thought. Frontal lobe is for planning and personality, parietal lobe for touching and moving our limbs, occipital lobe for controlling vision and temporal lobe for processing sound and short-term memory. Thalamus acts as a sensory gateway to cerebral hemispheres. Limbic system comprising of hippocampus is used for processing and storage of short-term memory. Hypothalamus is used for regulation of autonomous nervous system, amygdala for memory of emotions and several others like these. The current research is continually enhancing and refining these areas and their functional roles as more experiments are performed.

(c) *Output generation*: The outputs from the brain either cause a physical motion, say, by contracting a skeletal muscle using motor nerves or produce a sound as speech. In addition, some neurons are also responsible for controlling hormones, the chemical messengers of the cell. These hormones are important for various regulatory mechanisms like for stabilizing body temperature, heart rate and for controlling other internal organs. The underlying mechanisms and structures for these are quite extensively studied as well (see Kandel et. al [82]). However, the details of why and how they are generated to produce a well-defined output in response to a certain input are not well understood.

The above approach of splitting the problem into inputs, internal processing and outputs is quite common in engineering systems. Here, the analogous components that assist with these tasks are sensors, memory based processors and actuators respectively. For example, feedback control systems like thermostats, temperature regulation in refrigerators, cruise control in cars, autopilot in airplanes and others are applications where this systems level approach is widely used in engineering.

4.3   Mechanics versus meaning

Let me now apply the systems approach to a simple example to highlight the gaps in our understanding of features of the human brain. Consider a task of turning our head while we track a moving car with our eyes. Let us split this task and group them into the above three functional areas. We have external inputs like the activation of light sensitive neurons in the eye to see the car and the sound sensitive neurons in the ear to hear the sound of the moving car. There is internal processing in the brain that recognizes the shape, size, edges, color of the car and its parts, detect motion of the car and several sound patterns of the moving car. There is output generation for balancing and for moving our head to track the car, movement of the eyes, focusing of the eye lens and retinal muscles to form a clear image on the retina. In addition, there are feedback mechanisms to correct our tracking motion and to maintain focus on the moving car. All of this information is aggregated in our brain to produce higher-level thoughts and memory.

We are already familiar with several of the detailed mechanisms for subtasks like how retinal neurons are color sensitive, how they transmit along the optic nerve, detailed mechanisms of how our eyes and ears work and others – see Ornstein and Thompson [118]. Yet, our ability to understand the complete task is far from complete because of a number of unexplained 'gaps'. For example, how do we get a coherent picture of the car from information that was spread across millions of neurons in our brain? How does this information converge and how do we recognize the object as a car? How do we correlate the sound and image information to conclude that they are originating from the same object? How do we know that the car is actually moving and that we are tracking it? Why did you decide to track the moving car? When and why do you stop tracking it? What defines, sets and implements these objectives like tracking a moving car in our brain? What mechanisms let us switch between different objectives in a well-controlled manner? There are several other questions related to past memory, experiences and personal meaning that are unexplained as well. We can try this analysis for most of the tasks we perform in our everyday life, like reading a book, watching TV, playing a game and others. If we analyze them in the same way, we will notice that some of the biggest roadblocks are (a) obtaining coherence and continuity within parallel interactions spread across millions of neurons dispersed throughout the brain and (b) explaining how realization, meaning, emotions and consciousness work. These two problems can be classified as mechanics and meaning.

(1) *Mechanics*: These are tasks understood from an input/output or cause/effect point of view to describe the mechanics of the operation. The intention of the analysis is to figure out the underlying mechanisms that can explain how this task works with no emphasis on what it means and why we are doing it. A computer, for example, works in this manner in a completely mechanical way. These are fairly accurate and logical descriptions. They are typically based on the standard physical laws to explain the dynamics of a specific task.

(2) *Meaning*: This is the personal or subjective aspect. Here you want to know what the task means to you personally. This is well defined only when the notion of 'I' exists i.e., when you are conscious. The emphasis here is on why we are doing the task, what it means and what its purpose is from an individual's point of view. Once these are understood, we can start answering other related questions like: what task you want to perform (i.e., choosing your objective), why you want to perform it, how you coordinate a series of steps to achieve your objective, how you switch between objectives and so on. The mechanics, as discussed above, is well-defined *after* an objective is chosen and a plan that involves a series of steps is defined. Meaning, on the other hand, asks about what happens within our brain when we choose an objective as well as how we choose it and plan for the task. Meaning is always relative to the system studying the process. A computer or a robot, for example, has not yet evolved to a state where it can evaluate a task from its own point of view. It only has mechanics whereas humans have both mechanics and meaning. We, conscious humans, define and design the objectives for a robot and the robot performs it. Unlike mechanics, meanings are not necessarily accepted globally across several people. Some of these examples are referred to as *qualia* while the others are subjective topics.

The mechanics and the meaning are usually quite disjoint from one another. For scientific statements, it is apparent that the mechanics is more accurate. However, for complex human experiences, we have seen several examples where different people interpret both the mechanics and the meaning in different ways. This interpretation is significantly different because of the differences in personality and past experiences. As a result, it is unclear whose account of the so-called 'facts' are accurate anymore. However, the interesting question is not whether one person's account is more accurate than the other, but how a system is capable of giving a meaning with some associated feelings in the first place.

Some examples are the differences between measuring time (mechanics) versus sensing the passage of time (meaning) and identifying a shape (mechanics) versus realizing that it is a specific shape (meaning). The same differences exist between mechanics and meaning for memory, emotions, free will, consciousness, self-awareness and others. In the next few chapters, I will analyze each of these higher human abilities by keeping in mind the distinction between mechanics and meaning, as outlined here.

5

Neurons

Having discussed briefly a collection of problems unique to our brain in the previous chapter, let me start by creating a framework to address them starting from this chapter. I will begin by providing a new mathematical model of a single neuron. Neuron is the fundamental unit of our brain that carries useful information. The biochemical mechanisms of how neurons work are already well understood (see Purves [122]). Using these mechanisms, several different models were proposed to highlight different functional aspects of a neuron (see Kandel et. al [82]). Unfortunately, none of these models have shown us how we can combine millions of neurons to explain higher level features like memory, consciousness and others in a convincing way. The model presented in this chapter, on the other hand, will allow us to extract the common principles applicable at all levels i.e., from a single neuron to groups of neurons. Specifically, I will show that these are: (a) properties of stable parallel loops applicable within a neuron (when viewed as a chemical SPL system) and across neurons (when viewed as a physical SPL system) and (b) how a disturbance propagates both within and across interconnected neurons via these two dynamical SPL networks.

The systems approach taken here with the SPL framework for a single neuron has another important consequence. Namely, the exact chemicals and specific designs of a neuron are not critical to derive the higher abilities of our human brain. Rather, the unique SPL dynamical structure constructed using, possibly, synthetic materials would be sufficient for generating each of these features.

5.1 Model of a neuron

A nerve cell has been studied in significant detail and the processes of its operation are outlined in several biology textbooks (see Purves [122], Purves et. al [123], for example). It is the primary component for transmitting information in the brain.

Figure 3.1 shows the basic subcomponents and processes that occur in a nerve cell. A nerve cell has several dendrites extending from the cell body. Two common categories of dendrites are basal and apical dendrites. Basal dendrites are at the base of a pyramidal cell (which is a type of nerve cell in the cerebral cortex, hippocampus and others) while apical dendrites are at the apex. Both types of dendrites receive signals (discussed shortly) from neighboring neurons. This is passed into the cell body where the signals are integrated. Whether the nerve cell now responds or not depends on the amount of signal accumulation thus far. The response from a nerve cell is in the form of a spike of voltage. This voltage spike is triggered at the axon hillock and propagates along the axon to distant regions. The axon is coated with myelin sheath to allow for fast propagation of the voltage spikes, sometimes reaching as high as 100 meters/second. When the electrical signals reach the axon terminal, neurotransmitters, packed in vesicles, are released to the neighboring neurons through the synapses via a small synaptic cleft. At this point, the process described above will now occur within the neighboring neurons. This results in a cascading firing of neurons.

Let us now look at the signals that are transmitted from one neuron to the next. Each nerve cell contains different ion channels on the cellular membrane. These are pores formed by proteins in the lipid bilayer. These pores are usually selective and permit the passage of one type of ion, like for example, potassium, sodium, chloride or calcium. Some of the channels are gated and allow ions to pass only under certain conditions. Most common ones are voltage-gated and ligand-gated channels. Ligand-gated channels are those that are triggered by chemical signals. They are less selective to permeability of ions than the voltage-gated channels. For example, less selective ligand-gated channels gated by cyclic nucleotides convert odor and light to electrical signals and are used in sensory neurons.

When there are no inputs, a nerve cell maintains a resting potential through the diffusion of K^+ ions through open K^+ channels. However, when neurotransmitter inputs fall on the dendrites from neighboring neurons, they open gated Na^+ or K^+ channels that either depolarize or hyperpolarize the cell's membrane potential compared to its resting potential. The sudden influx of ion flow into the cell is termed as the signal received from neighboring neurons. The aggregate effect of all inputs from several neurons produces a net voltage increase as a function of time shown in Figure 5.2(a). Until the threshold voltage is reached, the neuron is essentially inactive. The threshold voltage is analogous to the activation energy in a chemical reaction. It is the point when the sodium current exceeds the potassium current causing further depolarization. It is observed that the threshold potential is different for different nerve cells (see Purves [122] and Purves et. al [123]).

As the number of ions accumulate at the axon hillock, this causes Na^+ and K^+ ion channels at the axon to open and thus producing an action potential. The action potentials are generated every time the total number of ions exceeds a fixed number near the axon hillock. The voltage difference generated from the accumulation of ions on one side of the axon hillock creates a burst of ion flow. These result in time-varying oscillatory bursts. The total number of ions needed to create the burst is approximately the same each time. The initial voltage disturbance that triggered the action potential is amplified by the opening of ion channels at the nodes of Ranvier (Figure 5.1). This disturbance now propagates along the axon.

The series of voltage spikes travel along the axon at extremely high velocities (sometimes reaching almost 100 meters/second) without a drop in voltage. This is primarily possible because of proper shielding by the myelin sheath causing low ion-leakages. The voltage spike propagates like a wave *disturbance* much like water waves and is not through actual ion movement itself.

When the voltage spikes reach the synapse i.e., at end of the axon, it triggers the production of neurotransmitters. The neurotransmitters, for example acetylcholine, are synthesized in the axon terminal and packaged into vesicles. The enzymes needed in this case (or in the case of peptide neurotransmitters) are synthesized in the cell and are transported along the axon through microtubules. As a number of the action potentials reach the synapse, Na+ channels open which depolarizes the axon terminal membrane. This causes voltage gated Ca2+ channels to open and trigger the fusion of vesicles with the pre-synaptic membrane.

This causes the release of neurotransmitters at the synaptic cleft onto the neighboring neurons. Neurotransmitters released at the synaptic cleft through exocytosis are proportional to the rate at which the voltage spikes arrive at the synapse. After a brief period, the released neurotransmitters like acetylcholine are broken down by the corresponding enzymes like acetylcholinesterase. The components are taken back into the pre-synaptic cell to be reused and hence a chemical loop is closed.

Note that all the ion channels that we have described so far make the ions move by diffusion in the direction of concentration gradient. The reverse process that moves ions against the concentration gradient, needed for the physical loop, is accomplished through membrane pumps, most notably Na-K pump. These pumps consume energy so that the process can repeat itself. The operation of the Na-K pump itself is a loop (see Purves et al [123]), where three Na+ ions inside the cell binds to the membrane protein, followed by phosphorylation through the consumption of energy from ATP molecule. This causes a conformational change to the protein on the lipid bilayer, which releases Na+ ions to the outside of the cell and lets two K+ ions to bind. Dephosphorylation releases K+ ions into the cell, thereby completing the loop. The process repeats itself.

Let me now summarize the above processes that occur within a neuron as a few simple steps by taking a systems viewpoint. Figure 5.1 expresses the corresponding processes in a way suitable to develop a mathematical model.

1. Inputs as neurotransmitters reach a given neuron from a number of neighboring neurons. These inputs fall on the dendrites of a given neuron. They trigger the opening of several ion channels. This produces an accumulation of voltage potential that differs from the cells' resting potential (Figure 5.1(a)).
2. When the accumulated voltage potential exceeds a threshold value at the axon hillock, an action potential is generated by this neuron (Figure 5.1(b)).
3. The action potentials of the same magnitude are then transmitted along the axon to the axon terminal (Figure 5.1(c)).
4. The arrival of action potential at the axon terminal triggers the release of neurotransmitters. These are packaged as vesicles and released through the synapses as outputs on neighboring neurons (Figure 5.1(d)).
5. The released neurotransmitters open several chemically gated ion channels on the neighboring neurons. They become the inputs for the post-synaptic cell. The process (step (1) above) now repeats on the post-synaptic cells.

Information passed from neuron to neuron using the above process is very fast, efficient and uses directed pathways (namely, along the axons). This contrasts the transfer of information indirectly in non-nerve cells where diffusion is the primary mechanism, sometimes facilitated through proteins or hormones. Examples of these indirect and distributed transport of chemicals are: from bloodstream to the cell

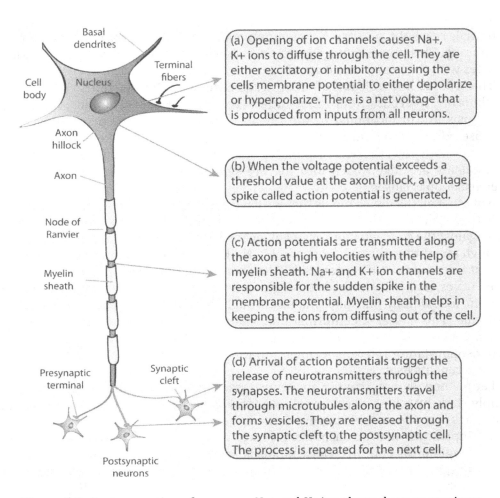

Basal
dendrites

Terminal
fibers

Cell
body

Nucleus

Axon
hillock

Axon

Node of
Ranvier

Myelin
sheath

Presynaptic
terminal

Synaptic
cleft

Postsynaptic
neurons

(a) Opening of ion channels causes Na+, K+ ions to diffuse through the cell. They are either excitatory or inhibitory causing the cells membrane potential to either depolarize or hyperpolarize. There is a net voltage that is produced from inputs from all neurons.

(b) When the voltage potential exceeds a threshold value at the axon hillock, a voltage spike called action potential is generated.

(c) Action potentials are transmitted along the axon at high velocities with the help of myelin sheath. Na+ and K+ ion channels are responsible for the sudden spike in the membrane potential. Myelin sheath helps in keeping the ions from diffusing out of the cell.

(d) Arrival of action potentials trigger the release of neurotransmitters through the synapses. The neurotransmitters travel through microtubules along the axon and forms vesicles. They are released through the synaptic cleft to the postsynaptic cell. The process is repeated for the next cell.

Figure 5.1: Basic operation of a neuron. Na$^+$ and K$^+$ ion channels open causing a voltage potential at the axon hillock. If this potential exceeds a threshold value, the neuron fires an action potential that transmits along the axon. This causes it to release neurotransmitters on the postsynaptic neurons causing this process to continue.

(hormones diffuse and bind to specific receptors to trigger a series of cascading chemical reactions inside the target cell), between cells through the extracellular material or through endo- and exocytosis where specific macromolecules are carried in and out of the cell by forming vesicles.

Let me now suggest a mathematical model of a neuron that aligns with the above steps (1)-(5). The first step results in an accumulation of ions within the cell. The net increase in voltage produces a movement of ions inside the cell. Each neuron is inactive until its voltage exceeds a threshold value. Once the threshold voltage is reached (step (2)), the flow of ions follows Ohm's law to a first-order approximation. This can be expressed as $V(t) - V_{th} = R \ dq/dt$ where $V(t)$ is the net voltage, V_{th} is

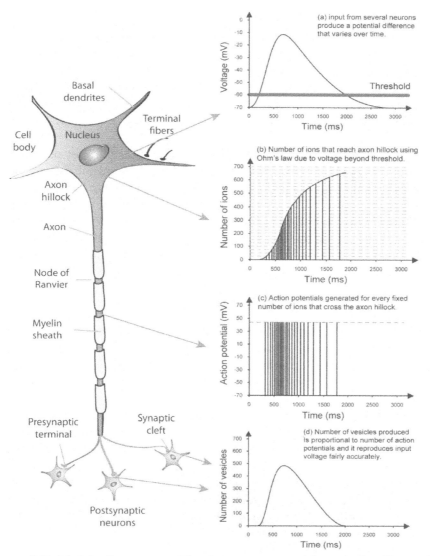

Figure 5.2: Model of a neuron. The data represented here is for illustration purpose only. The input voltage is shown in (a) and its integral in (b). We divide the y-axis in (b) into equal intervals like we do with Lebesgue integral (unlike Riemann integral that divides x-axis into equal intervals). A neuron spike of same magnitude is generated at every instant corresponding to a given y-value as shown in (c). If we count the number of spikes at equal intervals, we see that the original input signal is reproduced accurately in (d).

the threshold, R is the resistance in the flow and $dq(t)/dt$ is the current. Let $u(t)$ be the value of voltage beyond the threshold value i.e., $u(t) = V(t) - V_{th}$. It generates a flow of charges according to $u(t) = R \, dq(t)/dt$ with resistance R. Since the neuron is inactive until the threshold is reached, we will only consider the period when $u(t)$ is non-negative.

The number of ions as a function of time is shown in Figure 5.2(b) and is obtained by integrating Ohm's law for $u(t)$. As the number of ions accumulate at the axon hillock, this causes Na+ and K+ ion channels at the axon to open and thus producing an action potential. The action potentials are generated every time the total number of ions exceeds a fixed number near the axon hillock. The voltage difference generated from the accumulation of ions on one side of the axon hillock creates a burst of current flow.

These result in time-varying oscillatory bursts. The total number of ions needed to create the burst is approximately the same each time. The initial voltage disturbance that triggered the action potential is amplified by the opening of ion channels at the nodes of Ranvier (Figure 5.1). This disturbance propagates along the axon. This is shown in Figure 5.2(b) by taking the time samples corresponding to equal intervals along the y-axis (i.e., along the ion axis). The time instants when the action potentials fires are at unequal times, however. It is computed as follows. Let $y(t)$ be the number of charges flowing at time t and let $x(t) = \int_0^t u(\tau)d\tau$. Then

$$y(t) = \frac{1}{R} \int_0^t u(\tau)d\tau$$

Let Y be this fixed number of ions, which when passed across a point, creates a spike along the axon of the neuron. Hence, every spike of voltage is created along the axon at time t_n as $t_n = x^{-1}(nYR)$ where $n = 1, 2, ...$ Since $u(t)$ is positive in the time period of interest, $x(t)$ is a monotonically increasing function and hence its inverse $x^{-1}(.)$ exists.

The voltage difference corresponding to the action potential is about 110mV from the resting potential. It is approximately the same for each voltage spike at each of these time intervals, as shown in Figure 5.2(c). This is maintained from the opening of ion channels at each nodes of Ranvier. This causes a propagation of disturbance rather than the actual flow of ions along the axon. The series of voltage spikes travel along the axon at extremely high velocities (sometimes reaching almost 100 m/s) without a drop in voltage. This is primarily possible because of proper shielding by the myelin sheath causing low ion-leakages.

When the voltage spikes reach the synapse i.e., at end of the axon, it triggers the production of neurotransmitters. The neurotransmitters, for example acetylcholine, are synthesized in the axon terminal and packaged into vesicles. The enzymes needed in this case (or in the case of peptide neurotransmitters) are synthesized in the cell and are transported along the axon through microtubules.

If the vesicles are accumulated every T units of time, then the output signal $\hat{u}(kT)$ generated at instant kT (where k is a positive integer index) is obtained by counting the number of voltage spikes arrived during the interval $(k-1)T$ to kT as

$$\hat{u}(kT) = \alpha \sum_{(k-1)T \leq t_n < kT} 1$$

for $k = 1, 2, \ldots$. Here α is a constant of proportionality that lets us translate the number of vesicles into a voltage potential produced at the postsynaptic neuron. Simplifying this expression by using $t_n = x^{-1}(nYR)$ mentioned above, we get

$$\hat{u}(kT) = \alpha \sum_{(k-1)T \leq x^{-1}(nYR) < kT} 1 = \alpha \sum_{\frac{x[(k-1)T]}{YR} \leq n < \frac{x[kT]}{YR}} 1$$

$$\hat{u}(kT) = \alpha \left[\frac{x[kT] - x[(k-1)T]}{YR} \right]$$

i.e.,

$$\hat{u}(kT) = \left(\frac{\alpha}{YR} \right) \int_{(k-1)T}^{kT} u(\tau)d\tau$$

If T is sufficiently small, we can approximate this as

$$\hat{u}(kT) \approx \left(\frac{\alpha T}{YR} \right) u[(k-1)T] \tag{5.1}$$

In other words, the output of the neuron is proportional to the aggregate input to the neuron with a short delay T, as shown in Figure 5.2(d). Therefore, we can say that the purpose of the neuron as suggested by this model is to *ensure reliable transmission of the input signal across large distances to the output* (cf. $\hat{u}(kT) \propto u[(k-1)T]$). This happens through the conversion of the input analog signal $u(kT)$ into a digital form of (same strength) voltage spikes.

The interesting point to note here is that nature transmits useful information reliably within a neuron by using the typically undesirable feature, namely, resistance to its advantage. While we humans have the luxury (because of our consciousness) to choose what is useful and what is not, nature cannot. Instead, nature needs to discover designs by taking advantage of whatever is available, even if they seem undesirable like resistance, friction or imperfections. Otherwise, the problem of creating structured complexity will either require a Creator or become a chance event. Later in chapter 6, we will see another example of such a design – imperfections in the dendrite structure gives rise to an ability to create new and useful relationships. Also, we will see that the solution to the problem of origin of life from inanimate objects uses similar designs (see chapter 52 and beyond).

As a number of the action potentials reach the synapse, Na$^+$ channels open which depolarizes the axon terminal membrane. This causes voltage gated Ca^{2+} channels to open and trigger the fusion of vesicles with the pre-synaptic membrane. Neurotransmitters released at the synaptic cleft through exocytosis are proportional to the rate at which the voltage spikes arrive at the synapse. A short while later, neurotransmitters like acetylcholine are broken down by their respective enzymes like acetylcholinesterase. The components are taken back into the pre-synaptic cell to be reused and hence a chemical loop is closed.

The process described above now repeats on the post-synaptic neurons. The signal proportional to $u(kT)$ of each neighboring neuron reaches the post-synaptic neuron via its basal and apical dendrites as shown in Figure 5.1(a). The post-synaptic neuron will then aggregate all of these input signals and transmit the voltages reliably just as the prior neurons. This signal propagates along its axonal pathways to subsequent neurons according to the model described above. The process now continues producing a cascading effect of signal transmission. In this manner, our brain is capable sending messages in the form of electrical signals reliably and repeatedly to distant regions (cf. retinal neural firing patterns triggered by an image of a tree propagating through the visual cortex to other regions within the brain).

5.2 Looped mechanisms

The operation of a neuron and/or a group of neurons contain several looped chemical and physical processes. Yet, the model described in the previous section did not emphasize this aspect as much. Rather, the model described the underlying mechanisms as a linear sequence of steps. In this section and the next, I will highlight the SPL structure of a single neuron and that of the brain network. I will use this structure to show that the linear sequence of steps can be explained equivalently as a disturbance propagation in a SPL network. This alternative form is a more general representation of propagation of dynamics in large-scale structured dynamical systems. It becomes important later in chapter 52 and beyond when we discuss the problems of origin of life and directed evolution of life forms.

A linear sequence of steps in a large-scale structured dynamical system suggests the existence of a 'designer' who planned and setup the system. Consider a few examples of such sequence of steps: (a) the sequence of steps like transcription and translation when creating an enzyme from DNA, (b) the sequence of steps in response to a bacterial infection, (c) the sequence of steps to heal a cut or a wound, (d) the sequence of steps to increase your heart rate when you run faster and so on.

For each such long and well-coordinated chemical sequence of steps, it makes us wonder: (i) how these sequence of steps were created for the first time and memorized subsequently, (ii) who decided that these were a good sequence of steps to make a cell survive better, (iii) why survival of the cell is important that this sequence of steps should be followed i.e., is there a greater purpose to life, (iv) what the source of memory of these sequence of steps is – it is surely not the DNA and (v) where the sequence of steps are stored in an organism like, say, within specific regions in a cell or is it distributed within the entire organism? Our view as a linear sequence of steps is the cause of difficulty in answering these questions.

I will show that the equivalent view as a disturbance propagation within an SPL network answers these questions quite easily – the memory of the sequence of steps is simply in the SPL network itself. To show this, let me first identify several coordinated physical and chemical looped processes occurring within a neuron (see Figure 5.3). Each of these can be viewed as regulation mechanisms of some kind. In the next section, I will use these collection of looped mechanisms to show how a linear sequence of steps can be equivalently represented as a disturbance propagation with the SPL network.

    1. *Generation of resting potential*: The resting potential of a neuron is about -70 mV. For most animal cells, K^+ ions play a critical role in the generation of resting

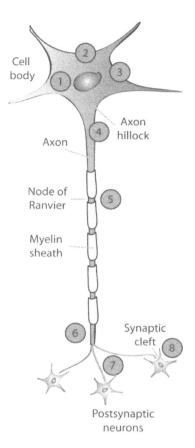

Cell
body

Axon
hillock

Axon

Node of
Ranvier

Myelin
sheath

Synaptic
cleft

Postsynaptic
neurons

Figure 5.3: Looped processes within a neuron. At (1), a nerve cell maintains resting potential of -70mV through a looped process in which a higher concentration of K^+ ions is maintained inside the cell. At (2), Na^+ and K^+ ion concentrations vary as a looped process using Na-K pumps as the ion channels open/close through the release of neurotransmitters. At (3), Na^+/K^+-ATPase protein structure oscillates as Na^+ and K^+ ions are pushed in and out of the cell. At (4), voltage spikes are generated at the axon hillock through a looped process in which voltage-gated ion channels cause a local flow of ions. At (5), the above looped process repeats at each node of Ranvier causing a disturbance to propagate along the axon. At (6), the generation and transport of neurotransmitters occurs through a series of looped processes. At (7), the formation and degradation of vesicles occur through looped process. At (8), the release, breakdown and re-uptake of neurotransmitters on the post synaptic neuron occurs through several looped processes.

potential. The uniporters, cotransporters, pumps and other active transport mechanisms (consuming ATP) ensure that the concentration of K^+ ions are higher inside the cell than outside. At the same time, most cells have potassium-selective ion channels open at all times. As a result, there is a diffusion of K^+ ions through these open K^+ ion channels to the outside following direction of the concentration gradient. This increases the concentration of negative ions inside the cell. This builds up a membrane potential which starts to impede the diffusion of K^+ ions. This balance is termed as equilibrium potential resulting in a net zero transmembrane K^+ ion flux. Similar equilibrium potentials exist for other ions like sodium, chlorine and calcium (Wright [175]). The resting potential is not an equilibrium potential because of the constant consumption of energy. Yet, it is dominated by the equilibrium potential of K^+ ions because they have the greatest conductance across the membrane.

This is a physical loop involving influx and efflux of ions across the membrane to maintain a steady resting potential (Figure 5.3 step (1)). The resting potential is a stable state. If there is a disturbance to the resting potential, say, when neighboring neurons release neurotransmitters or when sensory neurons receive external inputs like light or sound, the neuron responds to this change by producing an action potential. However, subsequently the neuron reverts back to the resting potential after the disturbance propagates and/or dissipates, thereby maintaining stability.

2. *Physical loop of Na^+ and K^+ ion flow*: As discussed in the previous section, the basic operation of a neuron involves generating action potentials that fire and propagate from neuron to neuron. This process relies on the flow of Na^+ and K^+ ions in and out of the cell. When neurotransmitters from neighboring neurons fall on the dendrites of a neuron, they open ion channels causing an influx of Na^+ ions into the cell and K^+ ions out of the cell. This eventually leads to the generation of action potentials at the axon hillock. These fire along the axon releasing neurotransmitters to the post-synaptic neurons and the process continues.

If the Na^+ ions enter the cell to eventually cause the firing of action potentials, what is the process that will push these Na^+ ions out of the cell so the process can repeat at a later time? The Na-K pumps in a nerve cell performs this action so the firing mechanism would not come to a halt. A Na-K pump pushes 3 Na^+ ions outside and 2 K^+ ions inside the cell in one operation. The influx of Na^+ ions and the subsequent pumping of these ions out of the cell using Na-K pumps form a physical loop (see Figure 5.3 step (2)).

3. *Oscillations in Na^+/K^+-ATPase proteins' structure*: The Na-K pump mentioned previously is a Na^+/K^+-ATPase protein located in the plasma membrane of nerve cells. This protein consumes 2/3 of cellular energy in nerve cells and 1/3 of the energy in other cells. The protein's conformational changes oscillate back-and-forth between two different states making it behave like a looped system.

When the ATP binds to the pump, 3 intracellular Na^+ ions bind as well. ATP then hydrolyzes resulting in the phosphorylation at the pump and a conformational change exposing the Na^+ ions to the extracellular space. This change in shape causes Na^+ ions to be released as well. At this stage, 2 K^+ ions from the extracellular space bind to the pump causing dephosphorylation at the pump. This reverts the conformational change and, therefore, transports the K^+ ions into the cell. This original form of the pump has higher affinity to Na^+ ions. This cause the K^+ ions to be released and Na^+ ions to be bound. The looped process now repeats with the binding of ATP (Figure 5.3 step (3)).

4. *Oscillations when generating voltage spikes at the axon hillock*: The axon hillock is the junction at which the firing of a neuron as action potentials are initiated. The ions that have entered the neuron due to the opening of ion channels at various dendrites from a collection of pre-synaptic neurons accumulate within the cell. The accumulation is partly countered by Na-K pumps. Nevertheless, there is a net accumulation of ions at the axon hillock. When the voltage difference at the axon hillock exceeds a threshold value, the neuron fires action potentials.

   Action potentials result from the positive feedback loops triggered by the opening of voltage-gated ion channels. A voltage-gated ion channel is a cluster of proteins located within the plasma membrane that responds to changes in membrane potential. The proteins are capable of more than one conformational change, at least one of which results in higher permeability to a set of ions (like, say, sodium). When the membrane potential exceeds the threshold value, it causes the voltage-gated sodium ion channels (Na_V, V for voltage-gated) to open. The influx of Na^+ ions contribute towards depolarizing the cell (i.e., cell potential becomes higher than the resting potential). Subsequently, potassium ion channels are opened that cause K^+ ions to exit the cell. This has an effect of hyperpolarizing the cell. The Na_V channels close at the peak of the action potential while the K^+ ions continue to exit the cell.

   At small voltages beyond the resting potential, the dominant effect is the efflux of K^+ ions. There is no action potential generated as a result. However, as the voltage potential exceeds a threshold value, the Na^+ ion current dominates. This causes further rise in membrane potential, which in turn open more Na_V channels. This results in a positive feedback loop causing the cell to fire an action potential. During an action potential, the Na_V channels cycle through deactivated → activated → inactivated → deactivated states. This is a looped process that occurs each time an action potential fires (see Purves [122] for further details – Figure 5.3 step (4)).

5. *Disturbance propagation loops along the axon*: After a neuron fires an action potential at the axon hillock, the voltage spike propagates along the axon until it reaches the synapse. The speed of propagation can sometimes reach as high as 100 m/sec. The underlying mechanism for the propagation of voltage spikes is analogous to a travelling water wave. In a water wave, water molecules move locally up-and-down in a loop. Analogously, Na^+ and K^+ ions move in-and-out laterally in a loop orthogonal to the length of the axon at each node of Ranvier. The looped process has two basic steps repeating itself (Figure 5.3 step (5)), one followed by the other: (a) the influx of Na^+ ions and efflux of K^+ ions through the opening of voltage gated sodium and potassium ion channels and

(b) the reverse process of pushing Na$^+$ ions out and the intake of K$^+$ ions using Na-K pump. Both these steps are exactly identical to that described previously in (2)-(4) above.

The process starts with the voltage spike generated at the axon hillock. The excess Na$^+$ ions due to the voltage spike diffuse along the axon to the first node of Ranvier. This triggers the above looped process to initiate at this location by first depolarizing the region. The depolarization opens voltage-gated sodium channels at this location causing a rush of Na$^+$ ions to enter, thus generating an action potential. Now, the excess ions begin to diffuse to the second node of Ranvier causing the voltage-gated ion channels to open there. The process now repeats and continues to the third node of Ranvier and beyond. In this manner, the voltage spike jumps from one node of Ranvier to the next. These jumps creates an appearance of propagation of a voltage spike along the axon as if it were a real flow when in fact it is just a disturbance propagation analogous to a travelling water wave. The presence of myelin sheath covering the axon and hence exposing the nodes of Ranvier is primarily responsible for high speeds of propagation of voltage spikes.

6. *Regulation loop to generate neurotransmitters*: When voltage spikes reach the synapse, neurotransmitters packed inside vesicles are released on post-synaptic neurons. The chemical process of creating neurotransmitters is clearly regulated. We do not want either excess or insufficient neurotransmitters generated within a nerve cell for proper functioning. Most common neurotransmitters are glutamate, GABA, dopamine, acetylcholine and serotonin. They are characterized as either excitatory or inhibitory ones. Excitatory neurotransmitters increase the chances of a post-synaptic neuron to fire action potentials while inhibitory ones decrease the chances. Both types of neurotransmitters are generated within the cell body and are transported via microtubules to the synapses. I refer the reader to textbooks like Purves [122] and Kandel et. al [82] for details on the mechanisms and the sequence of chemical reactions needed to regulate the concentration of neurotransmitters.

7. *Looped process of formation and degradation of vesicles*: The neurotransmitters to be released to the post-synaptic neuron are packaged into vesicles. They are released through the process of exocytosis. The number of vesicles formed are well-regulated at the synapses. The formation of vesicles followed by the breakdown of the vesicles after the neurotransmitters are released forms a looped process. This looped process is initiated when action potentials reach the synapse (Figure 5.3 step (7)). I refer the reader to Purves [122] and Kandel et. al [82] for details on the sequence of chemical reactions needed for this process.

8. *Release, breakdown and re-uptake of neurotransmitters*: When an action potential reaches the synapse, the depolarization at the membrane results in the opening of several voltage-gated calcium ion channels. This causes a sudden influx of calcium ions into the pre-synaptic cell. The high calcium concentration activates calcium-sensitive proteins attached to the

neurotransmitter containing vesicles. The changes in the shape of these proteins make the vesicles fuse to the pre-synaptic membrane and thereby releasing the neurotransmitters in the synaptic cleft. These neurotransmitters diffuse through the cleft and bind to the receptors of the post-synaptic cell. After a short while, the neurotransmitters break loose from the receptors primarily through thermal fluctuations. They are either reabsorbed using special reuptake pumps or are broken down metabolically. This completes the looped process for regulating and recycling the neurotransmitters at the synapse (see Purves [122] and Kandel et. al [82] for further details).

In addition to these looped processes, there are several others that are essential for the proper functioning of any cell. Some of these are: (a) ATP-ADP cycle needed to power a number of chemical reactions in the cell, (b) several looped set of chemical reactions for the generation of ATP using the citric acid cycle (Figure 2.8) and (c) looped set of reactions to create and breakdown a number of proteins starting from the DNA as part of the transcription and translation steps – needed for most chemical reactions since the proteins take the role of enzymes.

As can be seen here, all these loops help in conserving the chemicals while consuming just a minimal amount. The main input consumed in larger quantities is external energy. Another point to note is that each of the looped processes described above is interconnected to one another as well. These interconnections create an SPL network composed of both physical and chemical processes.

5.3   Disturbance propagation in an SPL network

In the previous section, we have discussed several looped processes individually in considerable detail. In this section, I will show that chaining these looped processes will allow us to view the sequence of steps occurring in a neuron (in response to a stimulus) as a disturbance propagation within an SPL network. This viewpoint is important because until now memorizing and executing a long sequence of well-coordinated steps within a cell seems to require a 'designer' to plan, arrange and order these steps (see the discussion at the beginning of the previous section).

Most problems involving creation like the problem of origin of life requires discovering and executing a large number of rather special sequences of steps. This seems like an impossible task without a designer. In the nonliving world, we never see more than 2-3 well-coordinated sequence of steps occurs without the help of an intelligent being. If, on the other hand, it does occur, we are quick to dismiss it as a coincidence (cf. an occasional appearance of a human face in the clouds – who has arranged the clouds in this form? – the only satisfactory answer is just random chance). Existing physical laws do not provide simple answers for the *creation* of such structured and well-coordinated patterns.

In this section, I will show that we do not need to rely on random chance or a designer to setup and arrange a well-coordinated sequence of steps naturally. This result, however, is only true within a special class of dynamical systems, namely, the SPL systems, which we consider in this book. Specifically, we will see that the memory for creating and following a wide range of complex sequences of well-coordinated steps is simply the physical and chemical SPL network itself represented as, say, metabolic networks (see Jeong et al. [80]; Nicholson [114]). Execution of a given sequence of steps can then be seen as the propagation of disturbance within an SPL network. In chapter 58, I will generalize the result as follows:

Claim 5.1: Any linear sequence of steps occurring within living beings, cells or our brain can be represented as disturbance propagation in an appropriate SPL network.

This is a rather deep result because it converts the most difficult and constantly recurring problem of creating and executing a well-coordinated sequence of steps, necessary for the existence of life, into the equivalent problem of designing a suitable SPL network. Later, in chapters 57 - 68, I will show how easy it is to design suitable SPL networks naturally. This would allow us to use the above equivalence to solve difficult problems. Specifically, I will show how to address the problem of origin of life from inanimate objects without requiring a designer even though nature needs to follow a large sequence of steps until, say, the creation of an RNA replicase, at minimum. Note that the occurrence of a precise (or a sufficiently precise) sequence of steps seem to violate the second law of thermodynamics, namely, that we cannot create order from disorder in a closed system. However, I will show that for a special class, namely, SPL systems, there is a way to avoid the above difficulty with the second law of thermodynamics.

If there is a designer or a planner, we need satisfactory answers to questions like: what is the source of memory of the planner – is it in the brain of another conscious being (remember that there is no life yet!) or is it in some other molecular or physical structures as complex as our brain or DNA (neither of which are present yet!)? As of now, whenever we discover a detailed mechanism behind a given feature (like, for example, our discovery of the sequence of steps followed during the transcription and translation of DNA into a protein), we place a lot of emphasis on discovering what these sequence of steps are, but we have little information, if at all, on how these sequence of steps were discovered and memorized for the first time so the cells continue to follow them almost flawlessly each time. The only answer for now is by chance. However, using Claim 5.1, the answer to this question becomes the SPL network itself. Nature only needs to increase the size and complexity of the SPL network over time. When the SPL network reaches a threshold complexity, these and several other sequences of steps would be followed entirely naturally i.e., without the help of any conscious or intelligent designer.

Therefore, with Claim 5.1, the problem of origin of life switches from one of identifying the sequence of steps that led to the creation of the first life forms to one of creating more complex physical and chemical SPL networks. What will come as even more surprising is that the problem of creating a conscious system (even synthetic ones) will turn out to be one of creating more complex physical and chemical SPL networks with special properties. I will discuss these later in chapters 10 - 50. For now, I will specialize the above problem of disturbance propagation to the case of a neuron and consider other generalizations in chapter 58.

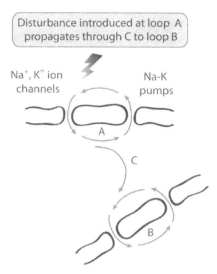

Figure 5.4: Disturbance propagation in a simple 2-looped system. Here loops A and B are in a stable state until a disturbance is introduced. The disturbance at A causes more Na$^+$ ions to flow into the system, which leaks via C into the loop B. In this way, the disturbance propagates to loop B. Energy is consumed by both loops during this process. Loops A and B will eventually return to the steady-state conditions.

Let us first consider a simple two-looped system interconnected as shown in Figure 5.4. Imagine each loop to be a physical loop with molecules moving back-and-forth through a pipe, a conduit or even through two openings – one for intake and one for exit. The looped dynamics itself can be simply viewed as a steady-state condition analogous to a state of dynamic equilibrium with reversible chemical reactions. In the case of a neuron, we can think of this loop to be the flow of Na$^+$ and K$^+$ ions in-and-out of the neuron. On one side, we have Na$^+$ ions entering and K$^+$ ions leaving the neuron either through diffusion or through the opening of Na$^+$, K$^+$ ion channels. On the other side, we have Na-K pumps that push Na$^+$ ions out of the cell and bring K$^+$ ions into the cell. This produces a steady-state condition referred to as the resting potential. Therefore, under normal conditions, the concentrations of chemicals flowing in-and-out as a loop remains stable and steady for both loops A and B. Let us assume that the diffusion of chemicals from A to B via the conduit C shown in Figure 5.4 is negligible under normal conditions.

Let us now examine what happens if this system is subject to a disturbance, namely, a sudden increase in the flow of chemicals, at A as shown in the Figure 5.4. This disturbs the resting state for the loop A. As a result, A has an increased flow of chemicals within the loop. The disturbance now propagates from loop A to loop B via the conduit C connecting the two. This causes loop B to operate beyond its resting state. After a while, the excess molecules leak through the system and each loop returns back to its respective resting states. In this simple example, the disturbance at A has propagated through the loops and the system returns back to the original steady-state condition after a while.

A generalization of this example is one in which we have a large number of loops interconnected linearly as shown in Figure 5.3. A disturbance at the first loop propagates all the way to the last loop linearly. Each of these loops return to their resting states once the disturbance propagates. This generalization can be seen as the mechanism behind the propagation of voltage spikes across the axon of a neuron. The initiation of the disturbance is the excess Na^+ ion accumulation at the axon hillock (i.e., beyond its threshold value). This causes an action potential to be generated at the axon hillock, which is beyond the resting potential. At each node of Ranvier, there is a resting state. These are disturbed serially, one after the other, as the disturbance propagates along the axon. The process is identical to the examples discussed in Figure 5.3 and Figure 5.4.

Figure 5.5 shows another example of an interconnected looped system. Here, each

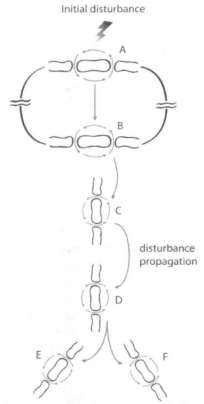

Figure 5.5: Disturbance propagation across a chain of loops. Here, each of the loops are in a steady-state condition initially before a disturbance is applied at loop A. This disturbance will propagate along the chain from A to D and eventually to E and F. After the disturbance propagates, all loops return back to their original resting states. During this process, energy is consumed by all loops.

looped process is regulated locally. The concentration of chemicals in each loop can be considered to be in a state of dynamic equilibrium. Energy is certainly consumed to maintain this steady-state. In most cases, small amounts of input chemicals are also required just to maintain steady-state conditions because of diffusion and other leaks, though these effects are minimal for a well-regulated system. Even though all loops are drawn in Figure 5.5, the dynamics of these loops are not observable as much. Some reasons for this are: (a) the steady-state conditions are not visibly dynamical (for example, water and water vapor in a closed bottle are in a state of dynamic equilibrium with no visible conversion from one form to another) and (b) the steady-state concentrations of chemicals for some of the loops can be quite low. These effects seem to hide the existence of these loops. Nevertheless, it is best to keep track of them at all times. We never know when, where and what types of disturbances are introduced in the system.

If we now introduce a disturbance in this system at A, it is easy to understand how it propagates through the system before the loops return back to steady-state conditions (see Figure 5.5). We can analyze similarly if the disturbance is initiated at another location like C and/or at multiple locations simultaneously. Energy is typically consumed by all loops as the disturbance propagates through the SPL network. From these examples, it should become clear that once we identify and draw the SPL network, it is a simple matter to understand how the disturbance, introduced somewhere within this network, propagates.

As an example, if you cut your finger, you have introduced a disturbance to the looped network represented within your cells and your body. The blood oozing through your finger, the blood clotting mechanism triggered and so on are just the propagation of disturbances. I refer the reader to Garrett and Grisham [48] for details of which chemical reactions occur during the healing of the cut.

The speed of disturbance propagation vary considerably depending on the nature of the SPL networks. In the case of neurons, these speeds sometimes reach as high as 100 m/sec. In most cases, the speed depends on several factors like:

(a) how long it takes to transfer the disturbance from one loop to the next,

(b) the time period of each looped process,

(c) the length of the interconnected looped pathway,

(d) whether the pathways are linearly connected or if there are several parallel pathways as well,

(e) the timely availability of extra energy as each looped process responds to the disturbance and

(f) whether the looped processes occur in an efficient manner (cf. myelin sheath covering the axon to enable fast propagation of action potentials in a neuron) or if they occur openly inside a cell with less assistance from additional structures.

In the previous examples, we considered physical loops with molecules moving within a loop. The disturbance was also a physical one in that it altered the concentration of chemicals. We could have considered either a chemical loop or a disturbance that is based on a chemical reaction. For example, a disturbance could be triggered by a series of chemical reactions, which open ion channels at the dendrites causing a sudden influx of ions. An example of a regulated chemical loop is Kreb's cycle. A disturbance in the form of excess glucose intake causes more ATP to be generated.

The examples discussed until now illustrate the idea of disturbance propagation in a conceptual sense. Let me now apply this to the case of a neuron. I will first list the sequence of steps that occur when a neuron fires and then show how to represent this sequence as a disturbance propagation within the loops identified in the previous section (see Figure 5.6).

1. A stimulus, say, a chemical, pressure, stretch or a membrane potential, triggers several ion channels on the nerve cell membrane to open. This causes a flow of ions into the neuron (see Figure 5.6).

2. Ions accumulate in the cell when several neuron sources open multiple ion channels at about the same time. These ions reach the axon hillock.

3. The density of sodium ion channels is greatest at the axon hillock. If more Na^+ ions enter than K^+ ions leaving, the membrane potential increase and can exceed a threshold value. When this happens, several more Na^+ channels open at the axon hillock that generates an action potential.

4. The jump in the Na^+ ions causes the voltage spike to travel along the axon. At the first node of Ranvier, the membrane potential once again exceeds the threshold value causing a large number of Na^+ channels to open. This produces a voltage spike of the same magnitude.

5. In this manner, a voltage spike appears as a disturbance that propagates along the axon. At each subsequent node of Ranvier, a jump in voltage potential causes Na^+ ion channels to open locally creating a voltage spike of the same magnitude. This propagation continues until the voltage spike disturbance reaches the axon terminal.

6. At the axon terminal, the voltage spike causes several voltage gated Ca^{2+} channels to be opened. This causes a sudden influx of Ca^{2+} ions at the axon terminal.

7. The Ca^{2+} ions bind to the surface proteins of vesicles. They trigger reactions that make the vesicles to open up and release neurotransmitters into the synaptic cleft.

8. The neurotransmitters diffuse through the synaptic gap and bind to the surface proteins of the post-synaptic neuron. This forms a stimulus on the post-synaptic neuron to open a number of ion channels. The above steps now repeat on the post-synaptic neuron.

Let me now outline the same steps above as disturbance propagation in an appropriately chosen network of loops (see Figure 5.6). When choosing the loops, we should ensure that each loop is a stable one. By this, I mean that if we perturb the loop by a small amount, there should be enough mechanisms in place to help the system restore its original state.

1. When the ion channels open up and let Na^+ ions flow into the neuron, this causes a disturbance to the existing stable state of the neuron. The original

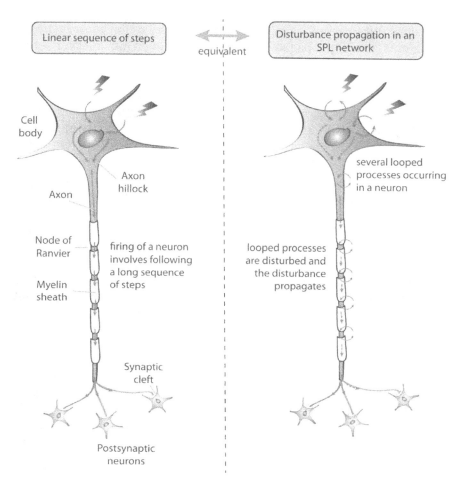

Figure 5.6: A linear sequence of steps for the operation of a neuron represented equivalently as disturbance propagation in an SPL network. On the left, the operation of a neuron is represented in the traditional way as a linear sequence of steps. On the right, we represent them as disturbance propagation within the SPL network of a neuron.

balance will eventually be restored using Na-K pumps in the neuron. This is one of the looped systems. The stimulus mentioned earlier creates an initial disturbance to the stable condition of this looped system (see Figure 5.6).

2. However, before the local equilibrium condition is restored, the disturbance is propagated within the neuron to the axon hillock. A local stable loop here uses the same mechanism of Na-K pumps to maintain a steady-state condition. In this state, the accumulation of the ions at the axon hillock does not result in increasing the membrane potential beyond the threshold value. In this case, the stable state will be eventually restored.

3. Once again, before the locally stable state is restored, the membrane potential

sometimes exceeds a threshold value. The rate of accumulation of ions in this case turns out to be much faster. The disturbance that has reached the axon hillock now gets to propagate beyond, by opening a much larger set of Na^+ ion channels at the axon hillock. This disturbance shows up as an action potential. At the axon hillock, there is a natural stable loop as well. Under normal conditions, the Na^+ ions entering and the K^+ ions leaving the neuron are balanced to maintain a resting potential.

4. At each node of Ranvier, a local loop maintains a stable state of ion flow in and out of the axon. However, before the effect of the disturbance at the axon hillock is nullified, the influx of Na^+ ions propagates along the axon. They reach the first node of Ranvier. This causes a disturbance in the stable loop at the node. This disturbance opens a number of Na^+ ion channels locally and creates a sudden increase in membrane potential. This results as an action potential of the same magnitude at this node.

5. The local disturbance at each node of Ranvier propagates to the next node. The Na^+ and K^+ ions are then restored back to their normal state after the disturbance propagates further along the axon. In this way, there is a local stable loop at each node that helps the disturbance reaching a node to be propagated to the next node. The propagation happens before a given node has a chance to restore to its original stable state. This is analogous to how water wave disturbance propagates. The voltage spikes represent the disturbances that appear to propagate as a wave along the axon until it reaches the axon terminal.

6. At the axon terminal, there exists a local stable loop that maintains a certain concentration of Ca^{2+} ions. However, the disturbance that travelled along the axon creates a higher potential difference relative to the exterior of the cell. This membrane potential triggers new voltage gated Ca^{2+} ion channels to open up. In other words, the Na^+ ion disturbance was converted into a Ca^{2+} ion disturbance at the axon terminal (see Figure 5.6).

7. Before the effect of the Ca^{2+} ion disturbance could be nullified, the Ca^{2+} ions bind to the membrane proteins of the vesicles that contain neurotransmitters. Now, the creation and the maintenance of the vesicles at the axon terminal is itself a stable loop. The neurotransmitters themselves are created within the nerve cell and are transmitted along the axon via microtubules. This process is controlled through a series of stable loops. Once these chemicals reach the axon terminal, they are packaged as vesicles. There is a stable state maintained for this process. However, the disturbance in Ca^{2+} ions causes the vesicles to open up and release the neurotransmitters into the synaptic cleft. In other words, the Ca^{2+} ion disturbance was converted into a disturbance in the vesicle looped system.

8. Several parallel processes occur now from multiple looped systems. One looped system is with the post-synaptic cell where the disturbance from the neurotransmitters opens ion channels as they bind to the surface proteins. This is same as Step (1) above but occurring on the post-synaptic cell. Some of the

neurotransmitters unbind from the post-synaptic cell just from thermal effects. The second looped system is from the disturbance in the re-intake of neurotransmitters by the presynaptic cell. A related looped system involves metabolic breakdown of the neurotransmitters caused from the same disturbance. In this way, we see that one initial disturbance can trigger multiple parallel disturbances as well (see Figure 5.6). The disturbance now propagates through the post-synaptic neurons.

As we can see from this example, we can translate a linear sequence of steps as disturbance propagation within a suitable network of physical and chemical loops. We then showed how an initial disturbance is converted into suitable forms as it propagates within this network in a specific way. We have seen two instances of parallel propagation in this example. In one scenario, the axon can have multiple branches leading to several different neighboring neurons. This natural parallel cascading effect is at a physical level because of the branching of the dendrites. Another scenario we have seen is at a chemical level. The production of vesicles, creation of neurotransmitters, uptake or breakdown of neurotransmitters from the synaptic cleft and the ion channel imbalance in the postsynaptic cell are triggered in parallel from a single disturbance, namely, from the release of neurotransmitters into the synaptic cleft.

I have skipped several chemical reactions that are part of the process for the sake of clarity. This includes the creation of several enzymes and the other necessary intermediate chemicals. However, each of those steps can similarly be mapped as disturbance propagation in an appropriately chosen network of loops. This and the more general case of disturbance propagation in a SPL network will be discussed in chapter 58.

5.4   Consequences of the neuron model

The model of the neuron proposed here has several interesting features. Let me discuss a few of them here. Among these, one of the most interesting aspects is that a neuron uses electrical resistance to its advantage to transmit useful signal across the network. On the contrary, with most man-made designs electrical resistance is regarded as an undesirable property. Considerable power is wasted because of resistance. Unwanted heat is generated for which we need to design additional cooling systems and so on. It is interesting to note that nature has discovered a clever way around these problems. In a sense, discovering such a solution is necessary for otherwise, the heat generated in our brain because of electrical signals transmitted across 100 billion neurons and 100 trillion interconnections would be unimaginably high. This would have rendered the brain useless.

SPL property generalizes from a single neuron to billions of neurons: One of the biggest challenges when proposing a model for a neuron is the following. How can we take the model of a single neuron and use it to explain features exhibited by a system like our brain that contains billions of neurons? The features of a single neuron typically get lost when you assemble billions of neurons. Consciousness, sensing space, passage of time and self-awareness are examples of features exhibited only when we have billions of neurons. An individual neuron does not have these features. There is also no apparent connection between these features like consciousness and the model of a single neuron. How then can we regard the model of a single neuron as useful when studying problems like consciousness? A similar dilemma exists whenever we study any large-scale structured dynamical system. For example, knowing the detailed quantum-mechanical equations of motion of a single electron, proton and other elementary particles do not seem to help us understand how a living cell (composed of billions of these elementary particles) functions.

The model of the neuron presented in this book is unique in this respect because the underlying principles used to explain the function of a single neuron is identical to the structure exhibited by billions of neurons in our brain (as we will see). These common principles are the stable parallel looped dynamics that occur at different scales – at a microscopic scale in a neuron and at a macroscopic scale within the brain network. In the previous sections, I have already emphasized several looped mechanisms that occur in a neuron when proposing the new model. The neuron also exhibits the concept of disturbance propagation within an SPL network. These notions generalize to billions of neurons as well. I will introduce other related concepts like fixed sets and abstract continuity that are unique to SPL systems later when discussing the problem of consciousness. Therefore, we can say that the study of SPL systems is the underlying theme both for studying a single neuron and billions of neurons. What is even more interesting is that SPL systems are critical even when studying the problem of origin of life, as we will see later.

Design and analysis easier with millions of neurons: The above model of neuron shows that the total input is reliably transmitted to the output neurons without distorting any incoming information (see equation (5.1)). Therefore, any interesting feature our brain exhibits depends entirely on the connections instead. Had there been errors or had the neuron modified the input-output behavior, you would have several million parameters to tweak for each feature because of millions of neurons required for any given feature of our brain. This is especially the case as no two neurons are ever the same. These parameters would bring in so much uncertainty or complexity that it would be impossible to predict and control them to achieve any desired behavior. As a result, the problem of designing, controlling and analyzing features of our brain that involves millions of neurons becomes easier with the new model of the neuron, as we will see in greater detail in subsequent chapters.

Unequal sampling for reliable communication: The approach of unequal sampling on the x-axis by taking equal samples on the y-axis has similarities to how Lebesgue integral is computed (versus Riemann integral). One way to transmit a signal $u(t)$ to the destination is to simply sample equally along the x-axis with sampling time T (Riemann approach). We then transmit the sampling time T and the discretized sequence $u(T), u(2T), u(3T), \dots$ to reconstruct the original signal (cf. Figure 5.2(b)). However, this is less reliable because the voltage spikes would need to be of different amplitude, requiring the need to have elaborate mechanisms for their generation. Furthermore, controlling the flow of ions to get these precise voltage amplitudes $(u(T), u(2T), u(3T), \dots)$ is error prone. As a result, the regenerated signal at the destination would end up being considerably different from the signal *u(t)* we were intending to send.

There are several other ways to transmit a signal reliably, which we use in engineering situations (like, say, sending the coefficients of a discrete Fourier transform and others). The approach illustrated here is yet another way, which would avoid most of the above-mentioned issues. The approach is to compute the integral of the signal $u(t)$, discretize along the y-axis equally (instead of the x-axis), sends 1's at each of these unequally sampled times, collect and count the 1's arrived in every time interval T and plot these counts at $T, 2T, 3T, \ldots$ to reconstruct the original signal (see Figure 5.2). This is the Lebesgue integral approach. It is easy to propose several engineering designs to implement this approach, but I will not discuss them here.

Resistance as an advantage: The generation of action potential, in simple terms, uses Ohm's law, i.e., as current flowing through a resistor. While we normally view resistance as undesirable because it results in generation of heat and loss of energy, neuron has used resistance advantageously to achieve reliable communication of its aggregate input voltage to all destination neurons. I have mentioned this previously in some detail. It is interesting to note that nature has discovered a clever way around these problems with heat generation and power loss. In a sense, discovering such a solution is necessary for otherwise, the heat generated in our brain because of electrical signals transmitted across 100 billion neurons and 100 trillion interconnections would be unimaginably high. This would have rendered the brain useless.

Capturing input voltage intensity and its rate of increase: The peak value of $u(t)$ manifests itself within the slope of the straight line in *x(t)*, which in turn determines the length of equally spaced sampling time intervals (see Figure 5.2(b)). Higher peak value implies that the y-axis equal spacing portion of the signal is close together. If $u(t)$ reaches the peak value fast, the sampling points will be dense. Hence, both the intensity of the signal and the speed to reach this intensity are captured with this approach. This information is reliably transmitted through the neuron using the mechanism described here.

Input and output saturation: Saturation at the input ion channels, at the axon hillock, along the axon and at the output synapse are non-ideal factors that play a critical role in the functionality of the neuron. For example, if the neurotransmitters are constantly being delivered to a given neuron from several inputs, the Na^+ channels will stay open for a long time. Since the flow of Na^+ ions is along the concentration gradient, as the cell accumulates enough ions, the concentration inside and outside the cell becomes equal. The cell saturates as a result, especially if Na-K pump is not fast enough to push Na^+ ions out of the cell. Similar situation happens at the output since the number of vesicles at the output synapse is limited. If the rate of action potentials is high for a sustained period, no more vesicles exist to be released until the neurotransmitters are broken down and are brought back into the cell. In other words, whenever there is a physical or chemical loop and a flow in the loop aided by external inputs, there is always saturation if the rate of inputs is significantly higher than the flow inside the loop. This results in either a backpressure at the input or leaks into other loops.

Preventing infinite loops: In a network of neurons (say, a simple example of three neurons connected in a triangular loop where the output of one is the input to another), the presence of a constant external input on one neuron cannot result in infinite loops of firing because of the above physical limitations stemming from energy requirements and loop saturation at the input and at the output as discussed earlier.

Sharing of ions across neurons: The Na$^+$ and K$^+$ ions that flow in and out of the cell are shared across several neighboring neurons. If the ion channels on a number of neighboring neurons are all open during the same period, then the net effect of membrane voltage on each of these cells is lower than when only a few neurons' ion channels are open. This makes it appear as if the threshold on the neurons is higher because it takes longer to reach the threshold voltage. In the case of photoreceptor cells on the retina, this sharing of ions has a positive effect because it helps in amplifying the signal for the neurons corresponding to the edges or boundaries of the image. The neurons in the interior of the image have lower voltage because all neighborhood neurons share the ions. However, the neurons on the boundary of the image has higher voltage because only half of the neighborhood neurons are competing for the ions as light did not fall on the other half of the neurons to excite them (see, for example, Feynman [41]). In section 34.2, I will discuss this once again for detecting edges.

Threshold voltage as activation energy: As mentioned earlier, the role of threshold is similar to that of activation energy in a chemical reaction. If the neuron is already actively firing voltage spikes, all input neurotransmitters are used effectively because the flow of ions contributes directly towards action potential. If, however, the voltage is below the threshold, some of the energy is lost in trying to bring the neuron to the threshold value first. Therefore, if the neuron stops firing, restarting it takes much longer than when it is already firing analogous to how activation energy imposes a similar constraint with a chemical reaction.

Transmission delay: In addition to reliable transmission of information along the complex networks of neurons in the brain, it is also critical to send the information quickly without any transmission delay. This also provides a survival advantage to the organism. The mechanism of firing the same amplitude voltage spikes is significantly better to avoid delays than sampling along the time axis with variable amplitude spikes, which is what we use in several engineering systems (see the discussion on unequal sampling above). Covering the axons with myelin sheath has helped reach transmission velocities of up to 100 m/sec. The presence of active loops is another way to avoid delays because the neurons are already above the threshold value. The disturbance is effectively transmitted along the path without any loss.

Avoid ideal assumptions: The model described here already accounts for non-ideal conditions like electrical resistance, input and output saturation, transmission delays, diffusion and other leaks. The model does not change its qualitative behaviors as a result. This is important because when we combine millions of neurons to explain behaviors like consciousness and others later, the fundamental model of the neuron as an SPL network does not need to be altered. Most other models of neurons have idealized assumptions to some degree that begin to cause difficulties when we combine millions of neurons. As a result, these models start to become probabilistic and correlation-based instead of causal-based ones.

6

SPL Network of Neurons

In the previous chapter, I have proposed a new model for a neuron by identifying the underlying mechanisms as disturbance propagation within an appropriate physical and chemical SPL network. In this chapter, I want to generalize this model of a single neuron to more general SPL network models applicable to very large collection of neurons. Since most features exhibited by our brain like memory and consciousness involve millions of neurons, such generalized models become necessary. Currently, no other existing model of a neuron generalizes in a similar way to all of brain to explain features like consciousness. My objective over the next several chapters is to use the extended framework of SPL networks for understanding various features of our brain.

In this chapter, I want to show that a network of neurons behave like an SPL network as defined in section 2.4. I will pick some simple examples of network of neurons and show how the dynamics across this group of neurons give rise to dynamical loops. I will also discuss the mechanisms behind the creation and deletion of connections between neurons. This allows the network to grow and change over time. As an example, I will examine how a set of loops generate and vary when an image falls on the retina. I will use such examples to identify an important property of SPL networks, namely, that of a *fixed set*. The notion of fixed sets become critical later when we study the problem of consciousness.

While the discussion in this chapter will refer to SPL networks within large collection of neurons, the models and the features themselves do not need to be identified with neurons at all. Neurons and the network of neurons are only used to show the similarity of a feature to a corresponding feature exhibited by the human brain. The main result is the theory of SPL networks itself for which we use neurons and the brain as illustrative examples. In fact, the systems suggested here can be built *synthetically* with other components including electronic or inorganic materials. What are important are the properties they should obey both individually and collectively as SPL networks.

Note that the precise set of chemical reactions and dynamics occurring in individual neurons or in a specific collections of neurons within a given person is not the same compared to another person. Yet, the SPL network and its structural/dynamical properties exhibited by all humans are similar (see chapter 44 which addresses a generalization of this question, namely, how every human brain though different at a structural and dynamical level still experiences happiness, sadness, self-awareness, space, passage of time and consciousness in an identical manner). The objective of this and the next few chapters is to extract just those minimal properties of SPL networks necessary to build features as complex as consciousness, even synthetically. Therefore, deviations when compared to experiments with specific neurons or groups of neurons are to be expected as they may be nonessential for the functionality. In this sense, SPL networks become foundational concepts when studying large-scale structured dynamical systems.

6.1   Structural and dynamical loops of neurons

Consider a typical network of neurons within our brain. In general, such a network will be quite densely interconnected (see Ornstein and Thompson [118] for a sample illustration). In these simple illustrations, we typically show a few neurons but with a large number of dendritic interconnections between then. If we extend such figures to include millions of neurons, it is easy to imagine that the resulting structure gets enormously complex. Yet, to contain the complexity, it is best to start with simpler networks and describe the mechanisms that allow the system to grow the structure gradually. What are the important patterns to look for within this complex network?

One of the important structures visible in Figure 6.1 is the presence of a large number of 'structural' looped pathways between the neurons. Later in this section, I will discuss how to use these structural loops to identify 'dynamical' loops as well (i.e., the dynamics of firing of neurons along the structural loops using the model described in chapter 5). Let me choose a simpler network with fewer number of interconnections to illustrate these concepts more easily. Figure 6.1 is one such network. For the sake of clarity, I have not shown the myelin sheath for all axons in Figure 6.1. An example of a directed structural looped pathway in Figure 6.1 is *ABCA* in which the axon from neuron *A* forms synapses on neuron *B*, the axon from neuron *B* forms synapses on neuron *C* and finally the axon from neuron *C* forms synapses back on neuron *A*. As shown in Figure 6.1, there can be multiple synapses from one neuron to another neuron, in general. This is a helpful feature when the pre-synaptic neuron begins to fire because it increases the probability that the post-synaptic neuron will also fire as it receives inputs from multiple synapses.

The example *ABCA* is a 3-neuron directed looped pathway (see Figure 6.1). More generally, we can have an *n*-neuron directed structural looped pathway $A_1A_2...A_nA_1$ in which the axon from neuron A_i forms synapses on neuron A_{i+1} for i = 1, 2, ..., n-1 and the axon from neuron A_n forms synapses back on neuron A_1 to close the structural loop (see Figure 6.1). In a real network of neurons in our brain, the number of structural loops between a set of *n* neurons will be large. This is because the number of connections between close-by neurons are enormous, at least in an adult. In a newborn, it is observed that the number of interconnections are small and less denser (see Ornstein and Thompson [118]). Over time, the number of connections grow considerably as we have new experiences and as we memorize them within our network. The density of interconnections is a measure of the number of memories an individual has. A newborn has low density compared to an adult.

Therefore, for any collection of *n* close-by neurons, it is quite common to have several *2*-neuron, *3*-neuron, ..., *n*-neuron looped pathways in an adult (see Figure 6.1). As a rough estimate, an adult human has about 100 billion neurons and 100 trillion interconnections. In other words, on average, we have about 1000 connections per neuron to neighboring neurons. For small values of *n* (say, *n* < 10), it is likely that we

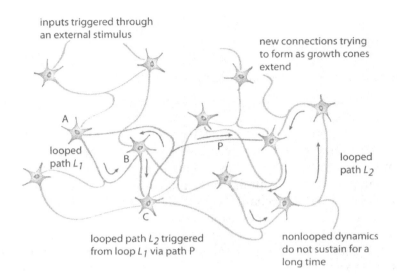

inputs triggered through
an external stimulus

new connections trying
to form as growth cones
extend

A

looped
path L_1

B

P

looped
path L_2

C

looped path L_2 triggered
from loop L_1 via path P

nonlooped dynamics
do not sustain for a
long time

Figure 6.1: A simple network of neurons with fewer interconnections. This is an illustrative network used primarily to introduce concepts like structural and dynamical loops. Two structural loops L_1 and L_2 are interconnected by a pathway P. If neurons A, B and C fire, then the structural loop L_1 is considered as a dynamical loop as well. The firing of neuron C causes action potentials to be transmitted via pathway P. This will cause structural loop L_2 to become a dynamical loop as well.

would find a few regions in our brain that form a fully connected directed graph locally. For these values of n, we will have a large number of structural loops (approximately $2^n - n - 1$). This may not be the case for a newborn since the density of interconnections is quite low. For larger values of n, however, the number of structural loops will be much lower than the above approximate value since it is unlikely to have a fully connected graph. Let us now turn our attention to the more important concept of dynamical loops.

Definition 6.1 (dynamical loop): A structural loop of neurons is called a dynamical loop during a time period of interest if all the neurons within the structural loop are active (i.e., their voltage potentials exceed the threshold voltage) and are firing action potentials.

If at a given time, some or all of these neurons within the structural looped pathway are not firing, then this structural loop is *not* a dynamical loop at that instant. In this book, we are only interested in dynamical loops. In fact, all SPL networks studied in this book only focus on dynamical loops, not static structural loops. Dynamical loops are clearly a much smaller subset of structural loops of neurons. This is because even though our brain has structural loops almost everywhere in our brain, only a small portion of our brain will be actively firing neurons at any given time.

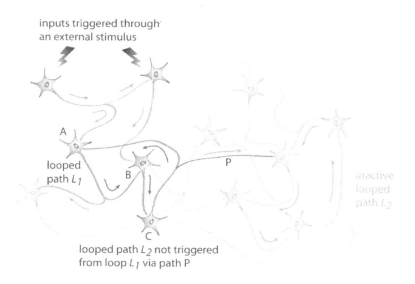

Figure 6.2: A sample network showing the existence of a dynamical loop L_1 while comparing it with a structural loop L_2. An external stimulus triggers a few neurons causing them to fire action potentials. These signals transmit using the disturbance propagation mechanism of chapter 5 from neuron A to B to C to form a dynamical loop $ABCA$. In contrast, the structural loop L_2 is inactive because the signal via P is not sufficient to exceed the threshold voltage of the interconnected neuron. The inactive neurons in this figure are shown in a lighter shade.

These dynamical loops also change over time depending on which neurons are firing at that instant. The cascading firing of neurons will obey the model described in chapter 5. The disturbance propagation that occurs within a single neuron as described in section 5.3 will now proceed across neurons. The interconnected pathway of neurons in the form of a structural loop causes the neural disturbance to go around several times before they eventually die down. Let us look at this in more detail.

Consider the 3-looped *ABCA* neurons of Figure 6.1. What happens when the neurons connected to dendrites at A begin to fire? A number of ion channels open at these firing locations as described in detail in chapter 5. A sudden influx of Na$^+$ ions occurs at neuron A at each of these locations. This is an initial externally-introduced disturbance on the simple 3-looped directed network of neurons. This disturbance propagates within neuron A causing it to fire action potentials (provided the voltage potential exceeds a threshold value) along its axon resulting in the release of neurotransmitters onto neuron B (see Figure 6.2). The underlying mechanisms behind the disturbance propagation was described in section 5.3. Neuron A has effectively transmitted the disturbance to neuron B. The process of disturbance propagation within neuron B is now similar to how it was for neuron A. The disturbance continues to propagate to neuron C next (as long as neuron B's voltage potential exceeds its threshold value). Since we have a structural loop *ABCA*, the disturbance propagation from *C* will now continue back to neuron *A*. Neuron *C* releases neurotransmitters back on neuron *A* causing a new disturbance to be introduced back on *A*. This dynamical loop is unique and interesting because we now have a way to restart the disturbance propagation pathway from *A* to *B* to *C* and then back to *A* even if the initial externally-triggered disturbance at *A* is terminated. Therefore, the dynamical loop of firing of neurons *A*, *B* and *C* continue for a short period (say, for 4-5 cycles) even with no external disturbances after the initial dynamical loop *ABCA* is triggered. Of course, we do need an initial external disturbance to jump-start the dynamical loop. The same mechanism and analysis can now be applied for disturbance propagation within larger loops of neurons as well i.e., for any *n*-neuron loop directed pathways.

The dynamical loop *ABCA* is analogous to a similar ball-rolling structure with inclined planes shown in Figure 2.4 (see section 2.4). One main difference, however, is that in Figure 2.4, each ball physically moves and completes the looped pathway along the inclined planes. On the other hand, what propagates and completes the entire neural loop in Figure 6.1 is a disturbance of ion flow, not the ions themselves. Nevertheless, all mechanisms like merging, linking and chaining of physical SPL systems described in Figure 2.7 continue to be applicable for dynamical loops of neurons as well. In each of these larger looped network of neurons (like Figure 6.1), it is quite convenient that an ion disturbance propagates while the actual ions themselves do not travel large distances.

As we merge, link and chain multiple *n*-neuron loops analogous to the procedure described in section 2.5, we begin to form a complex stable parallel looped network of neurons. Therefore, the resulting system of neurons satisfies all the conditions specified in section 2.4 for an SPL network. Our brain network, therefore, is an example of a SPL dynamical system.

How do we create and grow large dynamical looped networks in our brain? What are the different ways of generating dynamical loops from structural loops? These are the questions we will discuss next.

6.2   Creating and triggering looped network of neurons

Let me briefly discuss the basic mechanisms for forming new connections between neurons in our brain. These mechanisms form the basis for creating larger and complex network of loops over a period of time. I refer the reader to textbooks like Purves [122] and Kandel et. al [82] for further details. A given neuron transmits an ion disturbance along the axon (see Figure 5.1). When this reaches the synapses, neurotransmitters are released from the pre-synaptic neuron to the post-synaptic neuron across a small gap junction. The connections between neurons are formed if the synaptic structures are created from the axon. The developing axon needs to extend through the extracellular space to reach a destination neuron.

The developing axon has actin microfilaments that extend from the plasma membrane and attach to a target neuron (see Purves [122] and Kandel et. al [82]). This expansion can proceed like a sheet. The tip of such a growing axon is called lamellapodium. In addition, there are file extensions called filopodia that rapidly form and disappear around the circumference of the sheet. Both of these are termed as growth cones (see Gordon-Weeks [57]). It is still an active area of research. Laminins of the basal membrane are secreted into extracellular matrices. These interact with the integrins of the growth cone to guide the forward movement of the growth cone. Growth cones sense the guidance molecules such as netrins, slits, ephrins and semaphorins. They either attract or repel a developing axon towards a target neuron. The receptors to these molecules eventually trigger the cytoskeleton of the growth cone to reorganize and make it move either away or towards these guidance cues. As the target neuron is reached, synaptic structures are formed creating a new connection (see Purves [122] and Kandel et. al [82]).

Note that the above description suggests the formation of connections between neurons to be directed locally by the active and repeated firing that occurs because of neural dynamical loops. There is no global coordinator to create and direct specific connections between neurons or for the growth of axons. Instead, they can be explained with local effects using guidance molecules. In chapter 7, I will show how this helps create an associative model for memory – depending only on the relative relationships between loops, not on the exact set of neurons and connections. This new model of memory fundamentally deviates from all existing models of memory proposed in the literature (as we will see in chapter 7). We will see that the model captures several features of memory including associative events like triggering one event while thinking about another event.

Let us now consider the case when our brain SPL network has already grown to have sufficient complexity. We now want to see the possible ways of triggering the dynamical loops in our brain. One typical way is through the excitations triggered from external sensory neurons as the initial disturbance. These are, for example, excitations triggered from the images formed on the retina, sound patterns reaching the ear, taste, odor and touch based inputs. These external signals cause changes in the respective input-sensitive neurons. For example, retinal neurons have photopigments which are receptor proteins that change conformation when photons fall on the cell. They cause the activation of several G-proteins. This in turn causes a series of chemical reactions (see Purves [122] and Kandel et. al [82] for details) that eventually hyperpolarizes the cell. The initial photon disturbance has, therefore, translated into a chemical disturbance in which the retinal cells reduce the amount of neurotransmitters released. A similar disturbance initiation and propagation occurs within the four other sensory regions as well.

The other common source for triggering dynamical loops are purely internal excitations, collectively termed as thoughts. Unlike the previous case where we were able to attribute the root cause to one of the five sensory regions, the precise physical locations of neuronal regions triggering these internal excitations are harder to identify in each case. For each thought, the location of the regions does vary considerably. Yet, we do have an initial disturbance that triggers subsequent neural firing loops initiated internally from other neural firing loops.

The latter case of dynamical loops is more prevalent in an adult than a newborn.

In some cases however, there is a way to identify these internally generated loops as follows. Initially, a set of dynamical loops are triggered by inputs from the sensory neurons like, say, when you look at an object or an event (the initial disturbance). However, subsequently, these externally driven loops become the source of generating several other internal loops resulting in a cascading chain of internally triggered loops (i.e., propagation of disturbance through the SPL network of our brain). For example, when you see a person, you think of when and where you met him, what you were doing then and so on. The loops triggered when you visually see the person are the externally driven loops whereas the loops triggered by your subsequent thoughts are the internally driven loops caused by the externally driven loops.

6.3   Why study just a subset called dynamical loops?

From the discussion in the previous section, you may be wondering why I am identifying loops as important in the above examples/discussion and why I am keeping track of them. Why do I not keep track of the entire collection of neurons that fire during the same period or just a collection of linear pathways of firing patterns? There are several reasons for this.

One reason is that identifying all firing neurons is an over-specification of a given event. An analogy is with curve fitting when you perform an experiment repeatedly with slight variations. You do not want to over-fit a curve to the data because it would then lose predictive ability under new circumstances. New experiments will, most likely, not fit this over-fitted curve rendering it less useful. To see the similarity, consider an event of looking at a table repeatedly. Some of the common variations to this event come from looking at the table up-close, from a distance, at an angle, while walking towards or away, while jumping up and down and so on. Each of these variations initiate different set of firing neurons from the retina. Yet, we still want to identify the table to be the same one from these images.

A second closely-related reason is that we want to satisfy a *continuity property*. Continuity here refers to the fact that small changes in the events produce small changes in the collection of firing neurons. We do not want an approximately similar event to produce large variations in the firing pattern. Unfortunately, if we keep track of all firing neurons for a given event, minor variations of this event does produce large variations in the set of firing neurons. Therefore, it is best to pick only a subset of the entire set of firing neurons thereby avoiding over-fitting as well. Which subset should that be? The answer proposed here is to pick those neurons that form dynamical loops. In fact, we will see in this book that this is the only subset that works correctly for all problems from the origin of life to consciousness in human beings.

In the above example of looking at a table, this event triggers the firing of a large collection of neurons initiated from the retinal neurons. The firing disturbance of the retinal neurons propagate through the optic nerve to the visual cortex and subsequently to further regions within the brain. At this instant, among the entire collection of firing neurons, we will have several dynamical looped pathways, linear non-looped pathways and others (cf. Figure 6.1). We now want to satisfy the following condition. Any subset of firing neurons we want to identify with the event should be such that small variations in the event keeps the chosen subset almost invariant. Such a subset would then satisfy the continuity property.

As another example, if the event of interest is the act of listening to a short story, it is clear that a large number of variations of the same story exists. Your parents, your friends, your teachers and others may have told you the same story in slightly different ways. Each of these events are variations of a base event, namely, that of listening to the same story. We do want identify all variations as the same story. However, the neural firing patterns generated on each account will be considerably different if we look at the entire set of firing neurons each time. This is one reason why we need to identify just a suitable subset common to all of these variations. This subset can then be used to identify the story event. Such a special subset that satisfies the continuity property is a special collection of dynamical loops called a *fixed set*, which I will introduce in the next sections. The detailed properties of fixed sets will be discussed later in chapters 14 and 15. The notion of fixed sets is essential for memory and consciousness.

I have picked looking at a table and listening to a short story as two examples to illustrate the notion of similarity within variations. However, it should be clear that this concept (of identifying similarity within variations) is broadly applicable to *every* event that you do recognize correctly. Other examples are recognizing touch by a specific person, voice/tone of people familiar to you, tastes, concepts, abstractions and, in fact, every object, event or statement you correctly recognize around you at every instant (whenever you are conscious).

In addition to the above two reasons for why we pick loops as important, dynamical loops have several other important properties discussed previously in chapters 2 and 3 that make them ideal for associating approximately repeatable events to invariant subsets in large-scale structured dynamical systems. These include their self-sustaining property, their ability to consume just energy and reusing existing inputs as much as possible, the minimal amount of energy losses, the ability to coordinate and account for coincidences and so on.

Besides, we do not want to make idealized assumptions as is commonly done with most existing theories. Therefore, variability in the event should be accounted by picking a suitable subset instead of an entire set from just one variation of the event. How we capture this variability is the focus of the next couple of sections. Let me now briefly discuss the abstractions behind our sharp central vision. I will use it as an exemplary case to illustrate the notion of similarity and dissimilarity detection.

6.4 Visual signatures

Consider looking at an object like a table. From the optics of our eye lens, the image of the table will be inverted when it falls on the retina. Our retina has two types of cells – rods (about 90 million) and cones (about 4.5 million). Cone cells are capable of distinguishing different frequencies of light (and hence are responsible for color vision) whereas rod cells are responsible for night vision. Cone cells are less sensitive to light than rod cells, which can respond to even a single photon.

Our retina has a small region called fovea, which has the highest concentration of cone cells though it only occupies about 0.01% of our visual field. Fovea is responsible for our sharp central vision. The density of cone cells at the fovea is extremely high. It is estimated that about 50% of the nerve fibers in the optic nerve carry information that originates from the fovea even though fovea occupies only about 1% of the retina. The size of the fovea is about 1.0 mm in diameter compared to 22 mm in diameter (and ¾ of a sphere) for the entire retina.

Whenever we look at objects around us, we turn our eyes and head to focus on the object. When we do this, we are effectively ensuring that the image of the focused region falls right on the fovea. For example, when you are reading a sentence in a textbook, you are scanning the words from left to right, say. While you are scanning and focusing on individual words one-at-a-time, you are turning your eyes to ensure that the specific word looked at that instant falls on the fovea. As you move to the next word, you have turned your eye just enough to ensure that the new word now falls on the fovea. In this manner, as you keep switching the words while reading from left to right, you are ensuring that the current word read always falls on the fovea.

Even though you are focusing on a single word at that instant, it is clear that you can sense the presence of the neighboring words as well. These neighboring words appear fuzzy. However, when you turn your attention to them, you would have effectively refocused their images so they fall back on the fovea. Therefore, unless the word falls directly on your fovea, it will not appear clear and focused. A mere turning of your eyes towards the fuzzy word to make it fall on the fovea (and subsequently readjusting the focal length of your eye lens) makes the image clear once again.

The above situation is not limited to words when reading a sentence. When we are looking at a table and when we quickly scan the edges of the table, we are ensuring that the image of the edge at each instant keeps falling on the fovea. When we look at different objects in the room as we take a quick glance around, we are ensuring that the image of each object during the glance falls on the fovea at each instant.

Given this, consider what happens when we look at a table, say, from different distances, at an angle, while walking towards or away from it and so on. The image that falls on the retina in each of these cases is approximately centered around the fovea. We also do not focus on just one point on the table when we look at it. Rather, we keep scanning the table almost unconsciously from one end to another, across the edges, sometimes in a zigzag way, sometimes at different textures or patterns on the table, actively change the focal length of our eye lens and so on. The net effect of the scanning produces a neural firing pattern in and around the fovea that propagates along the optic nerve to the visual cortex and beyond.

During the period when we are looking and scanning at the table, say for about 1-2 seconds, the union of all neurons firing in the neighborhood of the fovea constitute the 'signature' of the table. This pattern of neurons firing on the retina propagates through the optic nerve to the visual cortex and beyond. It is best to include an appropriate subset of firing neurons from the entire pathway as part of the table's signature, i.e., not just limit it to the neurons firing on the retina. Similarly, if we are looking at a chair at another instant, the firing patterns produced both on the retina and the subsequent pathways within the brain can be used to define a signature by picking an appropriate subset. In other words, for each object that we are capable of recognizing, we can define a signature by picking appropriate subsets of neural firing patterns.

What is this appropriate subset and how do we pick it? The goal of the next few chapters is to address this question. We want to identify just the minimal set of properties that should be obeyed by this subset so it would make the study of features exhibited by large-scale structured dynamical systems easy. These properties, as you may have guessed, are related to stability, parallelness and dynamical loops introduced in this book. I will define additional definitions using these three concepts and apply them to different problems of interest.

The notion of a signature of a recognizable object is not limited to visual images alone. It is applicable to sound patterns, taste, touch, odor and even abstract concepts like numbers and scientific terms. In fact, any object, event or concept that is recognized should have a well-defined notion of a signature. The subset, which is identified as the signature for each of these cases, will be called as a ***fixed set***. In the next few sections, I will begin by briefly discussing two properties that should be satisfied by a fixed set – a similarity condition and a dissimilarity condition. These two conditions will give a motivation for choosing an appropriate subset mentioned above (namely, the fixed set) whenever we talk about our ability to recognize objects, events and others.

6.5   Time decay pattern of looped networks

When our brain receives sensory inputs corresponding to a given object or an event E, it generates a neural firing pattern P that propagates deep into the brain. Such a neural pattern P contains both looped and non-looped firing neurons. Each neuron operates according to the model described in chapter 5. Consider applying external sensory inputs, like visual inputs, corresponding to an object E for a short period T_1. After time T_1, we turn off the external inputs. We want to know what happens to the firing pattern P now. How fast will the rate of decay of the neural firing pattern be? Which neurons will stop firing first and which ones later?

One reason for answering these questions is to identify the important subsets of neurons to keep track of for a given event. Is it a single neuron or a special collection of neurons? Existing approaches to memory try to identify a single neuron or a small collection of neurons for, say, a table, a chair, the letter 'A', a sound for a word and so on. One of the objectives of this book is to show that such models do not generalize well to cover all interesting cases like intelligence and consciousness exhibited by our brain. Instead, I will propose an alternate model of memory (see chapter 7) based on SPL networks to handle such diverse topics.

Let us consider some examples first. If the object E is a table and the sensory input surface is the retina, then we are assuming that the image of a table falls on the retina for a short period T_1. After time T_1, we either look away (and see another object) or close our eyes. Other similar examples are listening to a specific sound for a period T_1 and then not hearing it subsequently, touching a specific patterned object for a period and then moving your hand away, thinking of a specific topic for a period and then thinking of an unrelated topic next and so on. In response to the external inputs falling on the retina (in the form of a table image), a collection of neurons from the retina begin to fire. These neurons propagate and cause several neurons in the visual cortex and other regions of the brain to fire as well. Let us identify all of these neurons together as the neural firing pattern P. Pattern P of neurons along with their interconnected pathways contain both looped and non-looped pathways.

Given that the external stimulus (like, say, the visual sensory inputs) are turned off after time T_1, the collection of neurons within pattern P that continue to fire will slowly start decreasing. Eventually, after time T_2, all of these neurons within P will stop firing. Some of the questions we want to answer are: are there any interesting sets of neurons worth tracking as neurons within pattern P stop firing between time T_1 and T_2? Which neurons stop firing first, which next and which ones last between time T_1 and T_2?

To answer these questions, note that after time T_1, some of the first set of neurons to stop firing are (a) those that are not part of any looped pathways and (b) those that do not receive inputs from looped neurons. This is because, with no external stimulus, these neurons no longer receive neurotransmitters from neighboring neurons. On the other hand, neurons that are part of a loop (like $ABCA$ and L_2 in Figure 6.1) will continue to fire a bit longer even if no 'external' inputs fall on these neurons. This is just the self-sustaining property of a loop, which in the case of neurons, consumes energy in the form of ATP in each cell. This dynamics will continue at least for a few cycles after all external inputs are turned off.

As an example, in Figure 6.1, if the external inputs are turned off, the first set of neurons that stop firing are the ones that receive the external inputs as well as neurons that receive weak signals from these firing neurons. If both loops L_1 and L_2 were firing before T_1, then L_2 will fire for a bit longer than L_1 because L_2 continues to get 'external' inputs via P whereas L_1 does not. As time progresses, at least one of the neuron within L_1 will stop firing. This will cause the entire loop L_1 to stop firing as well almost immediately. This implies, loop L_2 will stop getting inputs from L_1 (see Figure 6.3). Similar to the previous stage, one of the neuron within L_2 will stop firing next, causing the entire loop L_2 to stop firing as well almost immediately. Finally, the non-looped neurons connected from L_2 will stop firing. For this simple example, it was easy to determine the order in which the neurons in pattern P will stop firing (see Figure 6.3).

The above analysis is similar to how we study the propagation of disturbance in most physical systems as well. For example, we can apply a disturbance like a sound or as a water wave for a period T_1, remove the disturbance and observe what happens. If the system in which the disturbance propagates is sufficiently complex, it is likely that the dynamics propagates along some looped pathways as well. The decay of disturbance happens first at the source of excitation and then proceeds outward. Where there are looped pathways, the disturbance continues a bit longer. These typically manifest as resonance or oscillations. Eventually, the entire disturbance will dissipate and die down after time T_2. In simple cases, this appears as the propagation

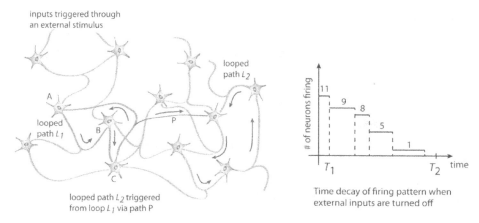

Figure 6.3: Time decay of neural firing pattern. When external inputs are applied to an SPL network for a period T_1 and then turned off, the set of neurons that stop firing proceed in stages. Looped pathways are important set of neurons to track because they self-sustain a bit longer even with no external inputs. On the right we see that the number of neurons that stop firing drop in big steps – whenever an entire loop stops firing.

of a wave (like water and sound waves). For more complex cases and for SPL networks, we can say that the above dynamics is a generalization of wave propagation phenomena in which looped pathways play a critical and an equivalent role of wave oscillations. In SPL networks, the disturbance propagation is not linear unlike sound and water waves. Nevertheless, the interesting subsets to study are loops instead of individual non-looped components.

As we can see, looped pathways are critical for producing a gradual slowdown of neural firing patterns. Also, looped pathways produce large 'step' changes in the way neurons stop firing at least as a reasonable approximation i.e., there is a big jump in the number of neurons that stop firing before a single neuron within a loop stops firing versus afterwards (see Figure 6.3). While it may be true that a single neuron within a given loop may be regarded as the critical one (like, say, the one that stopped firing first within this loop), it is better and more natural to treat the entire loop rather than such individual neurons as the important subsets to study. This is especially true because the neuron that stops firing first within a given loop changes depending on the situation (or, more specifically, depending on the collection of neighboring neurons that are still firing).

For example, in one instance, one part of the loop may receive more inputs compared to the other, while in another instance the opposite happens. This typically manifests as observing the same event with different contexts. If your context is 'watching a movie in a theater' versus 'watching a movie at your home', the thoughts related to 'eating popcorn' will receive signals of different strengths from different regions of your brain. Therefore, which neurons within the 'eating popcorn' loops stop firing may be entirely different in both situations. Nevertheless, studying the entire loop is a better and robust approach.

This aspect of studying loops is a new and significant change in the approach presented in this book compared to existing research (like with neural networks and other branches) where they typically emphasize special critical neurons or arbitrary non-looped networks. Later in chapter 9, I will discuss construction of SPL networks in which we take advantage of looped pathways and their self-sustaining properties.

6.6 Similarity, dissimilarity and fixed sets

In the previous section, we have seen that loops are an important subset to keep track of when looking at the time decay patterns after an external stimulus is turned off. In this section, let us look at additional conditions we want these loops to satisfy when trying to associate external objects and events to internal neural firing patterns. These conditions are termed as similarity and dissimilarity conditions while the resulting subsets of loops are called fixed sets. We will see that fixed sets become important later when studying the problem of consciousness (see chapter 14).

Let us pick a collection C_1 of similar objects or events like similar chairs or similar tables. Consider the corresponding collection $\mathcal{P}_1$ of neural firing patterns using the model of a neuron described in chapter 5, one for each object or event from the collection C_1. Specifically, an object $C_{1i} \in C_1$ generates a neural firing pattern $P_{1i} \in \mathcal{P}_1$ for $i = 1, 2, ..., n_1$. Given this setting, there are at least two desirable features we would want the collection $\mathcal{P}_1$ to satisfy.

(a) We would want the intersection of all elements P_{1i} in $\mathcal{P}_1$, where each element P_{1i} is a set of looped and non-looped firing neurons, to be a nonempty set S_1 i.e., $\bigcap_{i=1}^{n_1} P_{1i} \stackrel{\text{def}}{=} S_1 \neq \emptyset$.

(b) We would prefer S_1 to contain several dynamical loops as well i.e., S_1 is not just *any* arbitrary set of neurons that happen to fire for every object $C_{1i} \in C_1$ falling on the sensory surface. Rather, it contains dynamical loops as well.

In general, it is not possible to satisfy these two conditions for S_1 for all such collections C_1. This is especially true for an infant. For example, at this age, a child's brain does not form enough memories to identify multiple different chairs as a chair, different tables as a table and so on. We do notice this from the number of errors he makes and the number of times you correct him so he learns to associate objects to words correctly. For an adult, such errors are rare for visual objects because we do have well-formed visual networks. Only occasionally we make mistakes when we encounter, say, visual illusions, disguises, cluttered rooms or, more typically, with cluttered photos. The number of times we are exposed to such situations are fairly low. Instead, more common situations when S_1 is empty for an adult are with, say, abstract mathematical concepts like groups, rings and fields rather than physical objects. With abstract mathematical concepts, even though we see several similar examples of a concept (like a ring or a vector space), it requires a great deal of practice to recognize similarities. Each similar mathematical example $C_{1i} \in C_1$ generates different neural firing patterns $P_{1i} \in \mathcal{P}_1$ in our brain will less connections between them. As a result, we initially have $S_1 = \bigcap_{i=1}^{n_1} P_{i1} = \emptyset$. We need sufficient training (in the form of homeworks and exercises) to create new connections in our brain that

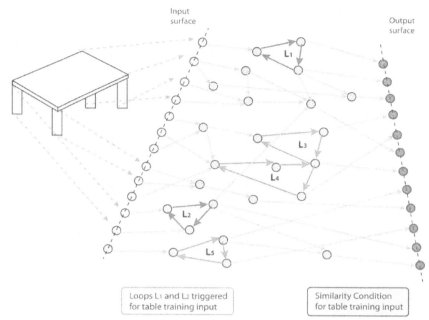

Figure 6.4: A sample SPL network that can be used to identify an image. Here, loops L_1 and L_2 can be used to identify an image of a table. In a more general scenario, we would want to include a collection C_1 of tables that are similar. We then identify 'common' loops analogous to L_1 and L_2 that are triggered for all of these similar tables.

helps us relate these examples, thereby ensuring $S_1 \neq \emptyset$. When $S_1 \neq \emptyset$, thinking of one example allows us to think of other similar examples as well (analogous to: thinking of one table allows us to think of other similar tables as well).

Other examples of collections C_1 that are non-visual are: words and sounds heard from different people with different accents, different food items with approximately similar tastes, some simple analogies in which the objects and events are different but the concepts are the same (like with numbers in which the concept '3' is triggered when working with 3 apples, 3 bananas or 3 chairs), scientific concepts that can be explained through multiple examples (like gravity) and so on.

What happens when the set S_1 for a given object or event is both nonempty and has dynamical loops? To illustrate this using a simple example, consider Figure 6.4 in which we have a table that triggers loops L_1 and L_2. In this case, we could call $S_1 = \{L_1, L_2\}$. Of course, in Figure 6.4 we are assuming that all variations from collection C_1 for a table are indeed going to trigger loops L_1 and L_2. We can use S_1 to uniquely identify the object (i.e., the table) or an event. This identification, in spite of variations within the collection C_1 (as well as C_2 and C_3 discussed next), makes S_1 a robust model for 'recognition'. We can think of S_1 as the *equivalence class* for the object or event. Recall that an equivalence class is a binary relation ($\sim$) that is reflexive ($a \sim a$), symmetric ($a \sim b \Rightarrow b \sim a$) and transitive ($a \sim b$ and $b \sim c \Rightarrow c \sim a$). An equivalence class allows us to define a partition within a collection, which is our goal here in trying to satisfy conditions (a) and (b) above.

We do not want to identify, say, a letter 'A' or an object like a table by a set of neurons or pathways unique to a single instance of 'A' (like a specific font or a specific handwritten letter 'A') or a specific table in your house. Existing models of memory are not explicit about what variations are accepted. There is, however, an implicit assumption, which may or may not include collections C_1 (C_2 and C_3 discussed next). Our goal here is to identify letter 'A' or an object like a table loosely using a notion like an equivalence class (i.e., several similar tables grouped together) with a nonempty set S_1 of neurons (see Figure 6.4). Of course, to clarify this better, we need a way to distinguish different objects from each other as well. This will be discussed shortly as the *dissimilarity condition*. In chapter 7, I will discuss several issues with existing models of memory and then propose a new model of memory using the current discussion on similarity and dissimilarity conditions to avoid all these difficulties.

Another collection C_2 that is important to consider here is with a single object C but with multiple variations. For example, each element C_{2i} for $i = 1, 2, ..., n_2$ in the collection C_2 comprises of the same chair, but when looked from a distance, from up-close, at an angle, while walking towards it or away from it and so on. More generally, a single object or event is subject to a set of small continuous and dynamical transformations (like translations, rotations, motion, shear and others). As before, we form a corresponding collection $\mathcal{P}_2$ of neural firing patterns, one (P_{2i}) for each object or event (C_{2i}) from the collection C_2. Once again:

(c) We want the set S_2, computed as the intersection of all elements of $\mathcal{P}_2$, to be both nonempty i.e., i.e., $\bigcap_{i=1}^{n_2} P_{2i} \overset{\text{def}}{=} S_2 \neq \emptyset$.

(d) We want S_2 to contain several dynamical loops.

Even here, an infant will most likely have an empty set S_2 for several different collections C_2 of visual objects than an adult.

Examples of collections C_2 that are non-visual are: same word or a sentence heard with different variations from a speaker (like with different intensities, tones, pitch, noises and other distortions), same word written with different handwritings or fonts, same taste or odor with different intensities and variations, same abstract concept encountered in different scenarios (like the same concept used in biology, physics, computer science and mathematics) and so on.

A third collection C_3 that is useful to consider here includes either a single object or a collection of similar objects that are partially occluded. As an example, consider a table obstructed by other objects. We still want to recognize it as a table. We, once again, form a corresponding collection $\mathcal{P}_3$ of neural firing patterns, one for each scenario in the collection C_3 (say, a table obstructed in different ways). As before (see (a)-(d) above), it is desirable to have the intersection S_3 of all elements P_{3i} for $i = 1, 2, ..., n_3$ of $\mathcal{P}_3$ to be both nonempty i.e., $\bigcap_{i=1}^{n_3} P_{3i} \overset{\text{def}}{=} S_3 \neq \emptyset$ and S_3 to contain several dynamical loops. This would be common among adults than for infants.

Examples of collections C_3 that are non-visual are: seeing partial words or sentences, hearing multiple voices simultaneously, recognizing partial concepts or ideas and so on.

144

One difference between collection C_3 and the previous ones is that it is not clear how much occlusion is acceptable for recognition. In other words, if only a subset T of S_2 or S_3 is triggered because of the occlusion, is this sufficient to uniquely identify the object in question? The amount of occlusion that lets us recognize an object varies from person to person (though prominent for non-visual objects than visual ones). As a result, it is best to associate the object, event or concept with the subset S_1 or S_2 identified using collections C_1 and C_2 respectively. The collection C_3 will then trigger only a subset T of S_1 or S_2. The question we need to ask then is if the subset T is unique enough to help us identify the object, event or concept unambiguously. If it is not, the system (like ourselves or a general SPL network) would move to a different location to avoid the occlusion, thereby attempting to trigger the full set $S_1 \cap S_2$.

In chapter 7, I will describe mechanisms for ensuring that sets S_1 and S_2 does indeed satisfy both non-emptiness and dynamical loops conditions over time by appropriately evolving our brain network with new memories. We will also see that these mechanisms are generic enough to be applicable to all SPL networks. Therefore, using the mechanisms from sections 7.4 to 7.8, it will be possible to create synthetic SPL systems that satisfy the above conditions for the set $S_1 \cap S_2$. Let me now summarize the current discussion as the definition for similarity condition.

Definition 6.2 (similarity condition): An SPL network is said to satisfy a similarity condition for a given object, event or concept if and only if a collection C_1 of similar but distinct variations of the said object, event or concept and/or a collection C_2 of a small set of continuous and dynamical variations of a single object, event or concept produces dynamical patterns P_1 and P_2 respectively within the network whose intersection S is both nonempty and contains a large collection of dynamical loops.

The idea with satisfying similarity condition is quite simple. We want the SPL network to identify _similar_ objects, events or concepts to be one and the same. Here, I have identified two common ways we can define the notion of similarity – different objects that are similar or the same object subject to small continuous and dynamical variations.

The third common way, namely with occlusion as discussed with collection C_3, is not included in definition 6.2 since it has considerable variability. While it is true that we need to trigger the entire set S to identify the object, event or concept in question, it is likely that a subset T of S is sufficient in most situations. This is common when the context helps in narrowing the possible ambiguity in the choices. In such cases, the analysis with similarity condition should be combined with the _dissimilarity condition_ discussed next. Once we have identified an object with a set S, objects in the collection C_3 can be used to evaluate how much occlusion is tolerable (i.e., how big the subset T should be before we can use it to identify the same object).

What happens when the similarity condition is not satisfied for a given object or event? This means that you have looked at the 'same' or a 'similar' object but have failed to trigger the minimal common subset S. Other scenarios when this could happen include whenever we have empty intersection and when there are a lack of

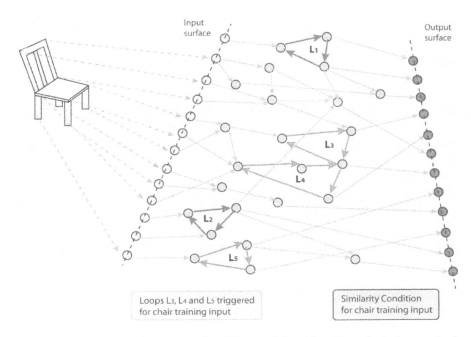

Input
surface

Output
surface

Loops L₃, L₄ and L₅ triggered
for chair training input

Similarity Condition
for chair training input

Figure 6.5: A sample SPL network of Figure 6.4 to identify a chair. Loops L_3, L_4 and L_5 can be part of the chair fixed set assuming these are the common loops triggered for collections C_1 and C_2. In order to say that a chair and table are dissimilar, we would prefer to have uniquely distinguishable set of loops (see Figure 6.4 which identifies L_1 and L_2 for a table). There are common features in a table and a chair like the legs, which are not captured here and in Figure 6.4.

dynamical loops across a reasonable range of variations. In each of these cases, unless we alter the SPL network, the model does not seem to capture reality accurately.

Definition 6.3 (fixed set): Fixed set for an object, event or concept is defined as the nonempty set S made of several dynamical loops that was previously identified using the similarity condition of definition 6.2.

The first point to note in the definition of a fixed set is that if two different objects or events produce the same fixed set, then we cannot use it to distinguish the two. Therefore, if you are able to distinguish a table from a chair unambiguously, this implies that there must exist two distinctly different dynamical loops of neurons, one for each object.

The precise location of the dynamical loops of neurons that constitutes a fixed set is not important as long as they can be used to uniquely identify a given object or event. For example, you can identify a chair by looking straight at the chair (Figure 6.5). Or you can identify the same chair without directly looking at it, but by looking at a neighboring table. Or you can identify the same chair when it is tilted or turned at an angle. Or you can identify the same chair by closing your eyes and imagining the chair (to a degree). Or you can identify the same chair in your dream with your eyes closed. All of these examples imply that the image of the chair does not have to fall at a precise location on the retina if you were to correctly identify it. As long as its fixed set (cf. loops $\{L_3, L_4, L_5\}$ in Figure 6.5), which is typically located deeper in the brain, is triggered, you can still 'see' the chair, even with your eyes closed. This can be seen as a generalization of the similarity condition discussed in definition 6.2. How do we create a fixed set that can be used to identify the object uniquely in all of the above cases? This will be discussed in detail later (see sections 7.8 and chapter 14).

There is another important motivation for defining a fixed set as above. This is directly related to the problem of consciousness as we will see in chapter 14. The choice of the word 'fixed' in fixed set will become clear as well when we relate it to the notion of 'truths' required when analyzing and building conscious systems. Let us now look at how we can distinguish two different objects as different using just the neural firing patterns within our SPL network.

As an example, in Figure 6.4, we have $\{L_1, L_2\}$ as the table fixed set while in Figure 6.5 we have $\{L_3, L_4, L_5\}$ as the chair fixed set. Given that they are disjoint from each other, we can say that a table and a chair are different and hence can be distinguished. However, this example is just for illustration only because it does not capture the fact that a table and a chair have several similar features as well. In the next chapter, we will change this example to show how to express the similar features like shapes, edges and others that are common across multiple objects in addition to the differences.

Definition 6.4 (dissimilarity condition): Two different objects A and B are said to satisfy a dissimilarity condition if and only if the two fixed sets identified for A and B, namely, S_1 and S_2 respectively are disjoint from each other i.e., $S_1 \cap S_2 = \emptyset$.

With most visual objects, it is quite rare that two objects A and B have nothing in common. For example, two objects may have similar colors, similar edges, angles, shapes, textures and so on. Nevertheless, we tend to ignore these common similarities when the objects are considerably different like with a table and a chair, at least, at first glance. Therefore, each concept or component that we can recognize individually would need a fixed set. The legs of a table, the top of the table, the corners and any other explicitly recognizable patterns and components are different fixed sets. These must differ from the table fixed set as well. An infant during his first few years of his life forms these fixed sets from everyday experiences. In the later chapters, I will discuss various mechanisms to create fixed sets for vision, touch and other sensory inputs as well as internal fixed sets related to thoughts, abstract concepts and events, language and others. The collection of all fixed sets will themselves be interconnected to form a 'self-sustaining membrane', which will turn out to be both necessary and sufficient for consciousness (see chapter 19).

In summary, the goal of defining fixed sets and other conditions is to ensure that (a) similar objects are classified as the same and (b) dissimilar objects are distinguishable from each other. Whenever we want to expand the brain network (or general SPL networks), we want to ensure that both similarity and dissimilarity conditions are satisfied for the entire collection of objects, events and concepts we want to model. In the next chapter, I will propose a new memory model that discusses the mechanisms behind the growth and adaptation of the SPL network satisfying both these conditions.

7

Memory

In the previous chapter, we have generalized the SPL model of a single neuron to a very large collection of neurons as well. Our objective over the next several chapters is to use the extended framework of SPL networks to understand various features of our brain. I will begin by showing that the problem of human memory and the extended model using SPL networks are closely related to each other. In the process, I will identify several features of SPL networks including the ability to create new networks, modify existing ones, merge two or more SPL networks and so on. Each of these features do not need to be identified with neurons at all. Rather, the systems suggested here can be built with other components including electronic or inorganic materials as long as they obey a set of properties discussed later both individually and collectively as SPL networks. In this regard, we can view the discussion in this chapter as a new model for synthetic memory using ideas borrowed from the SPL network features of neurons and our brain. In this chapter, I will extract several minimal properties of SPL networks and show how to build several features of our brain synthetically without using neurons directly. Therefore, deviations when compared to experiments with a specific neuron or with groups of neurons are to be expected since the intent is to capture the abstract properties while not making ideal assumptions.

In this chapter, I will build a new model of memory using physical and chemical SPL networks. Memory is a fundamentally important feature of the human brain. Every event we observe and experience is memorized in some form. Without memory, it would be difficult to understand the external world and even become conscious. For example, we would not know if the table (or even our own hand) we saw just a moment ago is the same as the one we are watching now. Of course, we need other features like sensing space and the passage of time as well, but, at minimum, we need memory.

One accepted model for memory represents it within the synaptic strengths between any two neurons and with the growth of new connections between them. For example, if you are memorizing a tree or a letter 'A', specific neurons unique to these visual inputs are first identified. Then, we say that these objects are memorized if we observe formation of new synapses and strengthening of existing synapses for these neurons (see Purves [122] and Kandel et. al [82]). While this model highlights the importance of the underlying chemical processes, identifying the object ultimately to one or more neurons, synapses or their strengths has serious drawbacks. In this chapter, I will mention what these drawbacks are and show how to avoid them by proposing a new model for memory. Comparison of this new model with the traditional model of a neural network and with deep learning will be explored in the future.

7.1   Overview of the new memory model

The core idea of the new memory model is to represent the memory of any object or event as a set of *dynamical loops* of interconnected neurons (i.e., fixed sets – see section 6.6), not as a static set or as a static looped network of neurons. With the same example above of memorizing a tree or the letter 'A', the memory in the new model corresponds to the formation of a specific set of looped dynamical neural firing activity. These dynamical loops of neurons are not fixed for all time. The set of loops triggered for a tree when you were an infant will be different from the set of loops triggered for the same tree when you become an adult. The only requirement when the loops do change is to ensure the relative interconnections between other loops (i.e., between other related objects or events) to be consistently maintained. How this is achieved will be described later in this chapter.

If you are looking at an object like a tree, several neurons fire within the brain starting from the retina and cascading further into the brain. Some of the firing activity forms dynamical loops within the graph-like network of our brain and others do not (see Figure 6.1). If you now do not look at the object anymore and wait for a short while, most non-looped dynamics die down. Among the looped dynamics that does remain and self-sustain, a specific subset (that obey, say, similarity and dissimilarity conditions – see definitions 6.2 and 6.4) is identified as the memory in the new model. Reliability and repeatability of the same looped dynamics for a given input with minor variations becomes possible through similarity condition. In the beginning (like when you were an infant), it is likely that no subset satisfies these special conditions and, hence, you do not form a memory in the traditional sense. However, through repeated observations and other experiences including emotional ones, new connections will form resulting in new looped dynamics. With the expanded set of loops, if all of the conditions in addition to similarity and dissimilarity (specified later) are satisfied, I say that you have successfully created the memory of a tree.

I want to emphasize the importance of dynamics in the new model of memory unlike conventional memory models, which are mostly static (like some sort of shape memory or the ability to maintain a steady voltage). The new representation is not within individual neurons, within simple connections between neurons or even within any long sequence of neural firing paths. There will surely be neural firing paths to reach the loops or to go from one set of loops to another set. Yet, it is important to pick specific dynamical loops itself as the correct representation, not smaller linear paths, non-looped subsets of it or even the entire looped and non-looped paths. For some events, memory will be represented within simple neural firing loops while for others they are represented as a large network of loops. We will see in this chapter that any subset other than dynamical loops is not rich enough to address all the observed issues with human memory.

The mechanisms described for the conventional memory models do apply in the new model. For example, any drugs or chemicals that alter synaptic strengths or a set of connections will affect the memory even within the new memory model. However, they are at a level of detail 'lower' than the neural firing loops. To address higher human abilities including consciousness, emotions and others, we need to go a level of detail 'higher' and pick the dynamical loops as the underlying memory model, as will be seen in later chapters as well.

7.2   The problem of memory in complex systems

Let us first look at a set of questions that we would want to explain clearly when trying to define and understand human memory with a goal of using this understanding to build memory within synthetic large-scale structured dynamical systems.

(a) What set of objects, events or concepts need to be memorized? These can be either static like images or dynamical like videos and sounds.

(b) What is the system that is trying to memorize the above events? This can be a human brain or a synthetic complex system.

(c) What is the mechanism to encode the event and then store it within the system?

(d) What is the mechanism to retrieve and recollect the event at a later time? This recollection need not necessarily be identical to the original event.

(e) What is a mechanism to verify that the event was accurately memorized and retrieved? Which systems are performing this verification?

(f) What relationships exist within the data being memorized and how is this information memorized as well?

(g) If the system has emotions and consciousness, then how can it store and retrieve the personal meaning or feeling associated to this memorized event? This information is typically unique to the system and distinct from the actual event itself. Someone else watching the same event may never be able to guess this information. For example, you liked a scene in a movie because it reminded you of a similar experience from your childhood. Such a connection is unrelated to the event and is either unknown to me or may even be opposite to your experience (i.e., I may dislike the scene).

When discussing memory, it is important to note that the information content in an event is actually endless. It is fundamentally impossible to describe it in a generic language used for communication by humans (using say, images, speech, text and actions). As a result, the actual event and the memory of the event need not be identical.

Most people think of computer memory as an analogy for human memory. Let me first discuss briefly how to address the above items for a computer memory, except for the last item about personal meaning, as it is not relevant to a mechanical computer. For item (a), any event or object we want to memorize is typically in a generic language composed of text, images, video, sound, mathematical equations and others. For item (b), we are using the computer as the system trying to memorize these events. For item (c), the encoding chosen is typically a well-defined algorithm represented in a Boolean language of strings of 0's and 1's that a computer can store easily. For example, some sample encodings are ASCII encoding for English alphabet, GIF for images, MPEG-2 for video and so on. These algorithms do not change over time, especially after the event is stored. The algorithms themselves are also stored using a different encoding, which is sometimes hard-coded in the processor itself (like, say, with ASCII encoding). Each of these encoded sequences of 0's and 1's are stored using electronic circuits based on flip-flops and registers as well as others based on magnetic storage disks and flash memory in a reliable manner.

We store a given event as a file. We choose a filename, identify a specific location on the hard disk and then write the encoded content as strings of 0's and 1's. We associate the filename with a pointer location on the hard disk for easy retrieval and maintain the list of filenames, say in a table at a known location. This addresses the encoding and storing of an event (item (c)).

When retrieving the memory (item (d)), we usually search through the list of filenames as opposed to the content of the memory to identify the correct file. This gives us the exact location on the hard disk to read from. We can then decode it back to the representation of the appropriate language, say, video. The retrieved information is identical to what is stored.

The retrieved information is sent to an output device like a computer screen, printer or speakers that renders the information back into the original form of, say, text, images, videos and sounds. A conscious human can now verify that a computer has stored and retrieved specific information (item (e)). He can validate this even if he does not know the details of the encoding mechanism, where it is stored and how it is stored. Whenever the verifying system is external to the memorizing system, a conscious being takes the role of the verifying system. The information retrieved from the memorizing system should be in a form that can be evaluated or understood by the verifying system. This is true whether you are retrieving memory from a hard disk or from another human brain.

Is there a way to verify the accuracy of the memory by the computer itself without the help of any conscious being? One way to do this is as follows. Imagine a computer storing a short video of the external world. Consider next a video camera that takes another video (a copy) of the video being replayed on the screen (like the photo of a photo). If the new copied file contents and the original file contents are sufficiently similar, we can be sure that the memory storage and retrieval mechanisms of the computer are quite accurate. In other words, we want the original video and video of the original video to be sufficiently similar. We refer this as ensuring self-consistency to emphasize the fact that this is just a mechanical evaluation.

Next, the relationships within the data content like, say, the edges, shapes and object recognition (item (f)) are itself not given any special emphasis when memorizing in a computer. It is expected that conscious humans extract this information subsequently at a future time using special algorithms. Conscious beings are necessary to assist with this task of identifying what the useful information is and how to extract it from the stored data content. This describes all aspects, namely, item (a) – (f) for a computer memory model.

It is quite common to compare human memory with the above model of computer memory. For example, it is suggested that specific neurons along with some connections exist for each event or object memorized (say, neurons that correspond to letter 'A', image of a tree, sound of a turning fan and so on) analogous to the encoding in a computer. It is mentioned that the strengthening of the synapses is a way to store the memory effectively (see Purves [122] and Kandel et. al [82]) analogous to how we store in a computer. Retrieval of the memory is assumed to occur when these specific neurons are excited and when we generate an output like a sound or an action like drawing or writing analogous to how specific bytes are retrieved from the hard disk in a computer followed by displaying them on the screen.

I will show that this model is not adequate when applied to human memory. It does not address the following features properly: (i) relative and associative storage locations, (ii) ability to retrieve these specific neurons, (iii) dynamic and adaptive nature of the neural network like, for example, maintaining our original memory even when changing network structure as when we grow older, (iv) ability to switch from one memorized event to another memorized event, (v) ability to store from short-term to long-term memory locations, (vi) subjective meaning associated with the memory events, (vii) discovering the static and dynamical relationships within the memorized data content and others. These difficulties will become obvious when we discuss important corollaries of our new memory model later in chapter 8 and when studying the problem of consciousness.

In this chapter, I will present a new model of memory and address the mechanics of it. The personal meaning for memory will be discussed later in chapter 22 because it is coupled with realizations (chapter 48), emotions (chapter 49) and consciousness (chapters 10-47). Consciousness is necessary for defining the meaning of a memorized event. For example, if you are unconscious like during deep sleep, in coma, when severely drunk or under general anesthesia, then for most events, the subjective meaning associated with a memory do not exist.

7.3   Memory model – a dynamical SPL network

The new memory model is a special case of the stable parallel looped (SPL) network. The SPL network consists of an input surface, an output surface and an internal network as shown in Figure 7.1. Each node within the network models a neuron as discussed in section 5.1. In Figure 7.1, we have represented a neuron simply as a circle and the directed edges as the set of output connections from a given neuron (cf. Figure 6.1 which shows a similar illustrative network with pictures of neurons). It is easy to translate Figure 6.1 into a picture similar to Figure 7.1 and vice versa. In this section, I will describe the model, its construction and show how it can be trained to exhibit features unique to memory.

Figure 7.1 looks pictorially similar to the traditional neural network discussed in the machine learning community, but with at least three differences: (a) each node is a dynamical system, (b) the model of each node is quite elaborate and is similar to the one discussed in section 5.1 and (c) the network has both structural and dynamical loops.

There are three distinct stages when creating an SPL network. The first one is to create a *structural* SPL network. The second stage is to create a *dynamical* SPL network. For an SPL network, knowing all the structural components and creating static structural loops alone is rarely sufficient for creating memories. The correct set

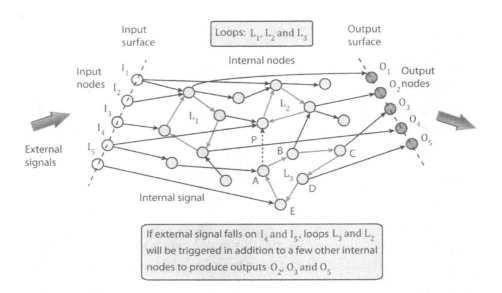

Figure 7.1: A sample SPL network represented as a directed graph. A neuron is represented as a node while the neurotransmitter outputs from one neuron to another is represented by a directed edge. In this figure, we have shown an input surface like the retinal neurons, an output surface like the motor neurons and an internal set of neurons. We have highlighted three loops (L_1, L_2 and L_3) and have suggested how to trigger them using external signals. This can be seen as a neural network with a few differences: (a) each node is a dynamical system and (b) both structural and dynamical loops are present.

of looped dynamics should occur at the correct location and at the correct time. The third stage is to specify how the SPL network *evolves* in a desired manner over time to solve specific problems of interest. Let me discuss each of them now.

Figure 7.1 shows a simple network of neurons, which we will use to illustrate some of the basic features of the new model for memory. When discussing the model below, I will frequently refer to neurons and their functional properties as the exemplary case. However, it should be kept in mind that the description below is at a systems level applicable to even synthetic systems with similar properties. In fact, one of the goals of this chapter is to construct a synthetic SPL system exhibiting memory using the mechanisms described here.

Figure 7.1 shows a few specialized neurons as the input neurons. They can be viewed as sensory neurons. As the inputs fall on these neurons, they translate the input signal into an appropriate neural-firing pattern. These neural spikes are transmitted as a disturbance to neighboring neurons along the connections represented as directed arrows (see discussion of the model of a neuron in section 5.1). A given neuron can take multiple inputs and generate an output that propagates to multiple neighboring neurons. As mentioned in chapter 5, a given neuron fires only when the voltage at the axon hillock exceeds the threshold. The outputs from several neurons propagate deeper into the network causing a cascading effect of neural firing patterns for any given input task. Let us look at the functional features of this network.

Inputs: The external inputs for a given task reach the top nodes using, say, sensory neurons like at the eye or the ear. This produces a cascading neuron firings that travel through a number of connections to deeper internal neurons like *A*, *B*, *C*, *D* and *E* producing several firing patterns.

Internal patterns: Several complex neural firing patterns are formed, in general. Any given neural network has several structural loops of interconnected neurons. Of particular importance are patterns among the structural loops of neurons that are simultaneously excited to form a dynamical loop as well (section 6.1). For example, there is a dynamical loop *ABCDEA* pattern that can form in which the output disturbance of *A* goes to *B*, the output of *B* goes to *C*, and so on until the output of *E* goes back to *A*. If the threshold voltages for these three neurons are low, they can even sustain the loop of neural firings for a while. Another example is one in which a cascading effect of one dynamical loop triggers another dynamical loop. In Figure 7.1, *ABCDEA* loop can trigger another loop (shown in dotted line) via the link *P*. Since *ABCDEA* is a looped network of neurons with internal energy source (ATP within neurons), the firing activity within the loop sustains for a longer time compared to a non-looped pathway. This makes dynamical loops suitable to transmit outermost neural signals to deeper parts of the brain. Non-looped firing patterns, shown in Figure 7.1, will not sustain long enough compared to the looped firing patterns like *ABCDEA*. Since they die down quickly, they do not either propagate deeper into the brain or produce interesting firing patterns.

Create connections: New connections are formed when a developing axon of an actively firing neuron extends its growth cones towards neighboring neurons as described in considerable detail in section 6.2. However, this growth of new connections is typically a slow process. The same excitation must occur repeatedly under different contexts or with different input patterns. A continuous and constant input is not as effective for forming new connections compared to several slightly different input tasks. Pulsating neuron firings produced from voltage spikes generate a force at the tip of the growth cones causing them to extend radially outward. Growth cones then sense guidance molecules and extend towards them to reach a neighboring neuron (see Gordon-Weeks [57]).

Delete connections: Existing connections or synapses that are weak or unused between a source to a destination neuron gets deleted if the firing pattern from the source to destination does not either sustain for long enough (because of the absence of loops) or is not excited for several input tasks or experiences over a period of time. If a given neuron actively fires for a number of inputs, the synapses are strengthened using specialized proteins both at the pre-synaptic and post-synaptic neurons (similar to long-term potentiation mechanisms – see Lynch [96]). The production of these proteins is diminished when the synapses are weak or unused. As a result, the existing connections are deleted.

Outputs: The cascading loops of neurons (see Figure 7.1) can grow and extend into deeper parts of the brain linking with a number of other existing loops that are formed simultaneously when performing entirely different tasks. These extensions eventually link with regions in the brain that produce motor actions or other outputs (like talking to oneself or with others and movement of hands and legs). Recall that neurons are already grouped structurally into regions for controlling hands, legs, processing sounds and others (see cranial nerves, topographic map, sensory-motor regions, peripheral nervous system and other regions of our brain). For example, when an infant makes random sounds, he excites neurons in a specific region responsible for speech thereby producing loops within these regions. If we excite these regions artificially (like, say, by placing electrodes thereby triggering ion channels to open), the infant would produce the same sound. The seemingly random sounds strengthen a specific part of the neural loops. When multiple such disjoint regions are eventually connected, the brain network grows to be capable of performing complex tasks.

For example, at around 6 months of age, you see an infant shaking and kicking his legs even with no apparent external inputs. These actions cause the output generation loops (motor actions in this case) to be trained and created independently. This is similar with sounds as well (at about 1.5 years of age). He now has two distinct regions in his brain – the input regions that have a large number of loops created as he senses objects in the external world and the output regions that also have a large number of loops created through, say, kicking and random sounds. These two regions can now be linked using experiences that use both sets of loops simultaneously. For example, there are loops of neurons that are formed and excited when an infant repeatedly sees a ball every day. These input loops are extended to regions deeper into the brain over a period of, say, a week. When the cascading input loop path for the image of a ball 'intersects' the output sound generation loops that were independently trained, you now have a link that lets an infant generate a specific sound whenever he sees the image of a ball. This is one way he learns to *say* the word 'ball' when he *sees* a ball. The inputs i.e., the image of a ball can be viewed as a cause and the output sound generated can be viewed as the effect. We can map this process in Figure 7.1 with the input surface as the retina and the output surface as the sound generated for a given image.

Basic storage mechanism: One of the crucial steps when storing experiences as memory involves forming new connections between neurons. The underlying mechanisms for this using guidance molecules and growth cones are described in detail in section 6.2.

How do we coordinate these mechanisms to produce a memory for a given task? Let us take a simple example. An image on the retina causes neural firing patterns that extend across several layers of cerebral cortex where special processing like edge recognition, motion detection, complex shape analysis and others are performed (see Ornstein and Thompson [118]). When you are an infant, you have fewer connections within your brain. As a result, the disturbance triggered from the retina does not propagate too far. However, through repeated excitations (like, say, when you are looking at the same object several times in a week), the growth cones for the developing axons extend to form new connections to neighboring neurons. As the pathways extend gradually, the possibility of forming dynamical loops and, hence, fixed sets (section 6.6) increases. Besides, the same image input now proceeds further into the brain. In section 7.8, I will show how this can be used to satisfy both similarity and dissimilarity conditions thereby letting us memorize objects and events.

Clearly, repetition of the tasks is critical to form fixed sets because the formation of new connections through growth cones is a slow process. The structural loops and new connections forming constantly in our brain just from our everyday experiences with or without any consciousness is the main process for storing our experiences in our brain. In the later sections and chapters, I will map several specific events and experiences into the new memory model.

Note that the above description using neural dynamical loops as memory does not depend on the exact set of neurons and connections involved. Rather, it is an associative model for memory – depending only on the relative relationships between loops. Simultaneity of occurrences of events, learning by rote or by practicing frequently are examples of ways neural loops can be linked together. The model captures associative events like triggering one event while thinking about another event. Therefore, in this model, there is no global coordinator needed to direct the growth of axons. Instead, they can be explained with local effects using guidance molecules.

Basic retrieval mechanism: Unlike a computer, retrieval of human memories are not always identical and accurate. It is reliably accurate when you have practiced memorizing the event for a long time. For example, '4+3' is memorized accurately as '7' for an adult and is no longer computed using the rules of addition. Several words, alphabets, their pronunciation and meanings are memorized accurately too. When there are inaccuracies, it is likely that our brain stored or retrieved them incorrectly. For example, when you are in a drowsy or drunk state, it is the retrieval that is inaccurate because what is stored did not change compared to an alert state.

The core requirement for recollecting a given memory in the new model is to excite the specific loop patterns (or fixed sets) that correspond to that memory. Recall that when you memorize an event, the basic storage mechanism discussed above represents the memory as a fixed set (the details to be discussed in the next few subsections). Therefore, in order to recollect the memory, you need to be able to excite the corresponding looped patterns in the brain. If these loops are not excited like during a drunk or drowsy state, you do not recollect that aspect of the memory.

How do you excite each of these loop patterns in your brain? In a normal situation, the process of retrieving a given memory is peppered with a number of conscious decisions during the entire time. For example, if you are recollecting your recent vacation, you need to make conscious choices on what you recollect, in which order and in what detail. You sometimes attempt to recollect the events by consciously thinking about these questions repeatedly. Another way to excite these patterns is by thinking about simultaneous events that occurred during the storage of your memory so you can trigger the same continuity of neural pathways. Sometimes just observing external events triggers a similar pattern from your memory. This then triggers a neural continuity of firing patterns and, hence, several related memories.

The full details of recollection (for both mechanics and meaning) will become apparent only when we study the problem of consciousness later in chapter 10 and beyond. Yet a couple of points to keep in mind are that (a) retrieval is usually not identical among different people even if the memory stored is approximately the same and (b) branching from one memory to another during retrieval is common within a looped network model.

7.4   Structural SPL network

Figure 7.1 shows the structural network or a graph consisting of nodes and interconnections between them. Each node is an abstraction of a neuron as described in section 5.1. One main difference between SPL network of Figure 7.1 and a traditional neural network is the presence of a large number of loops. Let us identify each node in the SPL network with a unique symbol n_i, where $i = 1, 2, ..., N$. Each node n_i has a set $\mathcal{I}_i$ of input connections. Each input connection I_{ij} in $\mathcal{I}_i$ connects node n_j to node n_i for $j = 1, 2, ..., |\mathcal{I}_i|$, where $|\mathcal{I}_i|$ is the cardinality of set $\mathcal{I}_i$. Similarly, each node n_i has a set $\mathcal{O}_i$ of output connections. Each output connection O_{ij} in $\mathcal{O}_i$ connects node n_i to neighboring nodes n_j for $j = 1, 2, ..., |\mathcal{O}_i|$. Let us assume that we have mechanisms to add new connections and delete existing connections between two nodes. Equipped with these mechanisms, we are now in a position to alter the structural network in any manner.

If the SPL network of Figure 7.1 is a synthetic one, we would want a way to represent the structural network so others could recreate the same network at a different place. Towards this end, we have the ability to take a visual representation of a network and create the necessary nodes along with their corresponding input-output connections. We can also store the structural SPL network into a text file and use it to reconstruct the graph at a later time. Both the visual representation and the serialization to a text file become useful later when creating a dynamical SPL network and when training the network to solve specific problems. When we train the network with specific inputs, we will alter the network. It is convenient to save a partially trained network so we could rebuild such partially trained networks directly instead of starting from the original network each time. In this way, we could continue the training from where we left off at a later time. Therefore, such intermediate representations would save us time and effort by allowing us to incrementally train the network. A synthetic structural SPL network can be constructed in one of many ways: using mechanical components, electronic components or as a software program within a computer.

7.5   Dynamical SPL network

Each node within an SPL network performs a sequence of actions repeatedly and independently analogous to a neuron. The basic operation of a node is shown in Figure 7.2. It is an abstraction of the neuron model described in chapter 5. If the node is implemented as a software program in a computer, the operation of a node runs on a single thread. If there are 100 nodes in a graph network, the parent process

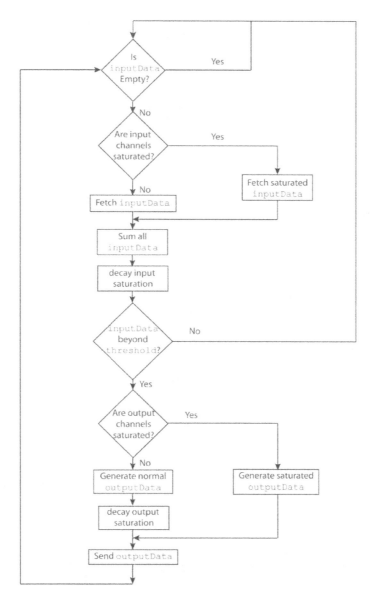

Figure 7.2: Basic operation of a node. The operation of a node described here is an abstraction of the model of a neuron described in chapter 5.

instantiates 100 different threads ideally, one for each node to run its dynamics repeatedly and independently.

Consider one-thread-for-one-node implementation as an example (while recalling the operation of a neuron of section 5.1 for comparison purposes). At a given instant during the thread's execution i.e., within its `run` method, the node n_i first fetches all the data d_{ij} present within each of its input channels I_{ij} (analogous to the accumulation

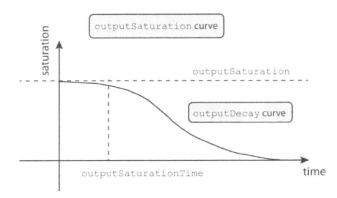

Figure 7.3: Output saturation and decay curve. As a neuron fires for a while, the output neurotransmitters will be saturated. It will take a while before the output generated will be normal again.

of Na⁺ ions from neighboring input neurons). Here, i and j reminds us that the input signal was sent from node n_j to node n_i. The node n_i waits for a short time (input delay), possibly zero indicating no delay. It resets the input channels to zero to indicate that the input signals are consumed (see Figure 7.2). It then computes the sum of all input signals $(\sum_j d_{ij})$. If the sum exceeds a threshold value T_i, the node execution waits for a short time (output delay, possibly zero indicating no delay). After the possible wait, the node n_i checks if the output has exceeded a saturation value V_i.

If the saturation value is exceeded, the node generates an output $P_i = V_i$ through each of the output connection nodes (cf. the analogous conditions when a neuron fires action potentials). The node also increments an internal counter to count the number of times output P_i is set to output saturation V_i. The output P_i is queued to the input data d_{ij} for each output node (cf. neuron n_i sending output to post-synaptic neurons). The internal counter is used to model the decay of output saturation for a neuron. If the counter reaches or exceeds the output saturation time (Figure 7.3), which happens as the thread executes repeatedly, an output decay curve pattern is followed for a specified number of iterations (Figure 7.3). This output saturation curve (Figure 7.3) is a tunable curve. I will discuss how to pick this curve shortly.

If the output data P_i did not exceed the saturation value V_i, the thread iterates through each of the output connection nodes in the output map O_{ij} for $j = 1, 2, ..., |O_i|$ and appends the value equal to an output scaling factor F_i times the data beyond the threshold i.e., $F_i(\sum_j d_{ij} - T_i)$ as the input data within each of these nodes. This analogous to the output neurotransmitters released by a neuron as the disturbance propagates along the axon and reaches the synapses. Each output connected node receiving the input from node n_i checks for an input saturation state (similar to output saturation curve of Figure 7.3). The procedure for deciding whether to accept the input data using the input saturation curve is analogous to the procedure for deciding whether to send the output data using the output saturation curve discussed earlier.

For example, if the input saturation U_i is reached, the node does not accept the output data. Instead, it only accepts input saturation value U_i until an internal input counter exceeds input saturation time. Subsequently, an input decay curve (similar to output decay curve) is used before the node begins to accept the real input data.

One important point to note is that the input decay curve applies to the inputs received even if the node does not exceed the threshold value and even before the input saturates. In other words, a slow decay rate of the input data accumulation applies at all times of the operation of the node (cf. the analogous mechanism of Na-K pump in a neuron).

This describes the operation of a single node on a single thread in a computer. This operation is almost identical to the neuron model discussed in chapter 5. One important observation is that a given node n_i does not generate output data P_i during one iteration of its thread's execution, if the sum of the inputs $\sum_j d_{ij}$ is below the threshold value T_i. When this happens, node n_i is said to be in an **idle state**. If the opposite situation happens and node n_i does generate output data P_i, it is said to be in an **active state**.

Consider a dynamical SPL network with 100 nodes, each running on 100 threads on a computer as described above. The above observation implies that the internal data keeps propagating within the network continuously until (a) all the propagating data decays to zero through the input and output decay curves or (b) the propagating data falls below the threshold values T_i's of the nodes. If external inputs are not constantly supplying data into the network, all data within the network will eventually die down (or drop down to zero). However, not all nodes will stop outputting data simultaneously. Some of the nodes will continue this 'dynamics' (of non-zero input and output data) for a long time, while others decay to zero quickly. This is where dynamical loops within the SPL network become important.

If a node pathway forms a structural loop and if all nodes within the loop are outputting data (i.e., their total input exceeds their respective threshold values), then these set of nodes will continue 'feeding' on each other's input data to self-sustain for a long time. These set of nodes are said to form a dynamical loop. This contrasts the set of non-looped pathways of nodes. Their dynamics die down once the data propagates from one end of the pathway to the other end (see Figure 7.1 for examples of both looped and non-looped pathways).

Therefore, if we look at the operation of the dynamical SPL network as a whole, we observe two distinct qualities among the set of nodes. The first quality is an idle state and the second quality is an active dynamical state. At any given time, some set of nodes are idle. Even though every node runs continuously and independently on its own thread, I still call the state of the nodes that are not outputting any data as idle. The next step is to look for interesting patterns produced within the entire dynamical SPL network. If we have 100's of nodes subject to external input signals, with some nodes idle and others active, the simplest and a unique novel pattern identified here is the *dynamical loop* among the active set of nodes. All other interesting complex patterns within the SPL network are built using this fundamental dynamical loop pattern. In section 7.4, we have seen that the denser the number of interconnections, the more the total number of structural and dynamical loops present in the network.

Non-looped dynamical pathways are considered important only if they help interconnect two distinct dynamical looped pathways. For example, in Figure 7.1, one complex pattern is the set of nodes that are part of L_1, L_2 and the interconnected nodes between them, if they are all simultaneously active. This pattern is clearly broken down into two fundamental loop patterns (namely, L_1 and L_2). In a dynamical SPL network with 100's of nodes, a typical complex pattern is composed of more than 10-20 set of fundamental simple loops. In general, it is possible to perform such a decomposition of a complex pattern into fundamental loops in more than one way. This is because it is likely that several of the loops would overlap giving you multiple ways to label the fundamental loops.

Most algorithms or designs involving a graph consider loops as an undesirable feature. This is because infinite loops tend to make the algorithm run indefinitely causing the computer system to hang (or become unresponsive). Therefore, most graph algorithms detect loops primarily to avoid them. On the contrary, dynamical loops here are the fundamental patterns that will help us solve learning related problems. Instead of avoiding these loops, we actively create more of them. Nevertheless, infinite loops of a special type, namely, those that continue to run for a long time causing large regions in the network active, are undesirable even here. We do not want entire connected parts of the SPL network to enter and remain in an active state. I call such a state as a ***fully active*** network. This is undesirable because the SPL network is no longer capable of distinguishing one pattern from another. For example, consider two different images that fall on the network (cf. retina in the case of our brain). If in both cases the entire network becomes active, there is nothing different in the output to distinguish the two images i.e., the effects being the same cannot be used to differentiate even different causes.

Therefore, we need to avoid the possibility of a fully active network. One way to satisfy this requirement is by ensuring that the dynamical loops do die down fast enough. To do this we need to tune the set of parameters suitably. Recall that the set of tunable parameters for each node are: input delay, output delay, input saturation curve (i.e., input saturation, input saturation time and input decay curve), output saturation curve (similar parameters as input saturation curve), threshold values and output scaling factor.

7.6   Tuning the dynamical SPL network parameters

With multiple parameters available to tune, it is quite easy to satisfy our objective of avoiding a fully active network. I will describe one such approach. First, note that the specific values of these tuning parameters are not universally fixed for all situations. They depend on factors unique to the embodiment of the dynamical SPL network. For example, if the SPL network is implemented on a single CPU computer, the set of factors that affect the tuning are: the clock speed of the computer, whether it uses 32-bit or 64-bit architecture, the sizes of L1 and L2 cache, the number of cores in the CPU, the maximum number of simultaneous threads that are run without compromising the speed, the amount of RAM available, the speed of the RAM, the hard disk access speeds and others.

Yet, a generic tuning algorithm can be stated as follows. Pick a collection of 'test' networks with the total number of nodes ranging from 10 to 100 (Figure 7.1 is one such 'random' network). Choose these random networks such that some have no structural loops (like a tree structure instead of a graph), some others with a few structural loops and others with a large number of structural loops. The denser the connections within the network, the larger will be the number of structural loops in it. Choose the loop lengths also to range from 3-20. Longer loop lengths (i.e., beyond 20) can be broken down into smaller ones under most scenarios.

To each such random network, supply external inputs to trigger one or more dynamical loops. After a few loops have triggered, shut off the external inputs. Measure the amount of time it takes for each loop to continue its dynamics starting from the instant the external inputs are shut off. For non-looped pathways, measure the amount of time it takes for the dynamics to pass from one end of the pathway to the other. Tune the parameters for the nodes such that the ratio of the self-sustaining time of dynamical loops to the time for non-loops is at least a factor of 10. It helps to keep the self-sustaining dynamical loops stay excited for a few seconds, instead of being in the fast microsecond range or slow hundreds of seconds range.

To understand how to tune these parameters, let me list the relationships between the parameters. These relationships offer suitable guidelines.

(a) The higher the threshold, the quicker the loop dies down and vice versa.

(b) The faster the input decay, the quicker the loops die down and vice versa.

(c) The higher the input saturation and output saturation, the longer the loops stays active and vice versa.

(d) The higher the input and output delay, the quicker the loops die down and vice versa.

(e) It is important to keep the output scaling factor to be strictly less than one to ensure stability of the looped dynamics. Otherwise, there will be magnification of the output data, which will eventually push the output data to infinity as the looped dynamics repeat several times.

Initially, tune the parameters for all nodes to be the same. Later, each node picks the parameter values randomly in a small neighborhood of these previously tuned parameter values (say, +/- 5% of the original values). The continuity with dynamics implies that small variations in these parameters will produce small variations in the amount of time a loop sustains. Choose the parameters such that most of the test networks satisfy the objective of avoiding a fully active network.

As the density of the network connections increase, it becomes difficult to avoid a fully active network. Estimating the *critical connection density* (expressed in terms of the minimum and maximum degree of the network) beyond which the objective is no longer be satisfied is a necessary step during the tuning stage. Later, when we use the dynamical SPL network for training and for solving specific problems (like recognition), we should ensure that the network modifications does not increase the density of the network to be beyond this critical density. One way to avoid reaching the critical connection density is by adding more number of nodes. With a large number of nodes, we now have unique pathways and hence a large set of loops while maintaining the density to be quite low. More unique dynamical loops help us solve a wide range of problems, as I will discuss later.

7.7 Construction and operation of an SPL network

Consider a random graph as the initial SPL network. In this section, I will discuss one way to construct the SPL network from this graph within a single CPU computer. We start by creating a `Graph` and a `Node` class. We instantiate as many `Node` objects as there are nodes in the random graph. Each `Node` has a unique `ID` as its member variable. For each node ID, we specify `(ID, Node*)` for each input connection (`inputMap`). Similarly, we specify `(ID, Node*)` for each output connection (`outputMap`). Once we specify `inputMap` and `outputMap` for each node, we have successfully created a valid structural SPL network.

To create a dynamical SPL network, we need to create a thread pool manager, which instantiates as many threads as there are nodes in the structural SPL network. Each thread takes a `Runnable` object with a `run` method that specifies what the thread does when it is in a start state. The run method in the node is an infinite loop that keeps processing input data and generates output according to the algorithm mentioned in Figure 7.2. Next, we start all the threads by calling their `start` methods. The dynamical SPL network is now operational.

Whenever a new node is added to the graph, a new `Node` object is instantiated. The structural input and output connections are setup as above. Finally, a new thread is instantiated and started with this `Node` as its `Runnable` object. When an existing node is deleted from the graph, the corresponding `Node` object is destroyed, the `inputMap`'s and the `outputMap`'s for all nodes to and from this node are updated by removing the entry corresponding to this node. For the dynamical part, the corresponding thread for the deleted node is stopped as well.

When a new connection between two existing nodes is added, the `inputMap` and the `outputMap` are updated with a new entry corresponding to this connection. When an existing connection between two existing nodes is deleted, the `inputMap` and the `outputMap` is updated by deleting the entry corresponding to this connection.

The instantiated dynamical SPL network as described until now does not seem to be doing anything useful. This is because no node receives any external input data. As a result, the `run` method does not output any new data. The method simply returns doing nothing. Therefore, every node in the network is in an idle state i.e., it is waiting for input data to flow into the system. The most common source of input data is through an input surface. Recall that some nodes in the network are specially identified as input nodes with a corresponding input surface and output nodes with a corresponding output surface (see Figure 7.1). Except for these two special cases, all other nodes are treated as internal nodes. A given SPL network has several input surfaces like a transducer for light or image inputs, another for sound inputs and yet another for touch or pressure sensory inputs and so on.

Image input surface: Consider 100x100 pixel-sized images as input to the SPL

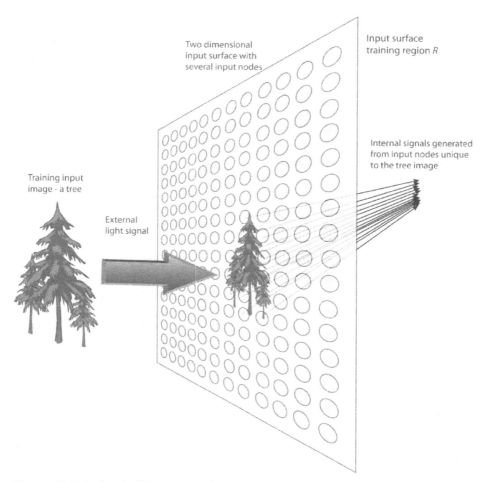

Figure 7.4: A simple illustration of an image input surface. A large grid of nodes are picked as the input surface from which signals are transmitted to the rest of the SPL network. We need a deterministic way of transforming an external image into a firing pattern of the nodes on the input surface.

network. The input transducer translates light information in an image into digital signals that the network can use. The digital information in this case is the RGB color information of an image. When working with such images, we need to use a corresponding 100x100 grid-based input surface of nodes with a one-to-one mapping to the pixels of the image (Figure 7.4). Each pixel has RGB color information unique to the image. For example, if the image is a table, the background is treated as having no RGB information (i.e., zero values for input data) while the foreground sends the numerical values of RGB into the input nodes' input data. The data is sent as bursts of values with a sampling frequency of, say, 10 milliseconds (or a factor of 100-1000 times the rate of non-looped pathway's decay time, which was tuned earlier). This is the flow of input data into the SPL network through those input surface nodes that have the foreground of the image being processed. If the image is a car, a different set of input nodes propagate the input data containing the RGB color information because the foreground covers a different region or shape on the input surface. Since there are three colors, the input surface grid should have 3x100x100 number of nodes.

Given this flow of input data to the input nodes, the threads corresponding to these nodes will no longer be idle. They will process the input data and push outputs to the other internal nodes via the output connections. The flow propagates deeper into the network as long as the external input image continues to remain on the input surface. As a result, some portions of the SPL network become active while others continue to stay idle. If the external images were removed, some of the dynamical patterns within the SPL network would continue to self-sustain because of the presence of dynamical loops within the network. Eventually, the entire SPL network would return to an idle state as before. This is a typical operation of the SPL network when subject to external image inputs.

Sound input surface: If the source of input is a sound, the first step for transducing the sound into a digital signal is to identify and represent the time-varying frequencies of sound. The human audible sound frequency range is 20 Hz to 20,000 Hz. In one example of sound input surface, it contains about 20,000 nodes mapped in a one-to-one manner to the corresponding sound frequencies with, say, 1 Hz resolution. Ideally, we may want to use higher resolutions (like, say, 200,000 nodes with a 0.1 Hz resolution). Furthermore, it is sometimes advantageous to consider, say, 10 nodes for a given frequency (say, for 200.1 Hz) as well. This acts as a source of redundancy. Another representation could be to choose a high resolution of nodes for certain frequency ranges but lower resolutions for other frequency ranges.

The input data into the SPL network now comes via these input nodes corresponding to a given sound input. For example, if the sound input is that of a ball bouncing on the floor for a period of 10 seconds, then the sound frequencies during these 10 seconds are first computed from this sound input. Each of the input nodes corresponding to these frequencies will receive a digital numerical value corresponding to the amplitude of the sound at that frequency. In this manner, a pattern of input nodes receives input data during the 10-second period. This input data propagates into the network as the threads for each of the nodes receiving the data processes and produces output data. As mentioned for the image input surface, even after the 10-second period (i.e., after the external sound input stops), the internal network has self-sustaining dynamical patterns due to the presence of dynamical loops for a brief period. Eventually, the entire SPL network would return to an idle state as before. This is a typical operation of the SPL network when subject to external sound inputs.

Multi-touch input surface: If the input surface is a rectangular area that is capable of detecting touch inputs (like a touch-screen), the input transducer already exists that detects the touch and sends an (x, y) coordinates as digital signal to the computer. In this case, an input node grid is created and mapped in a one-to-one manner to each (x, y) location of the touch screen surface. If the touch screen surfaces are capable of recording the intensity of touch as well, the node at the specific (x, y) location of the touch receives an intensity value. Otherwise, it receives just a value of '1' indicating that the specific location is touched. As you produce a pattern on the touch screen with your finger or a stylus, the intensity information is sent to the corresponding nodes of the input surface. This is the source of input data to the input nodes. They cause a propagation of input data within the SPL network similar to how it did in the image and sound input case. Therefore, several nodes in the network will become active. They self-sustain even after the touch input is removed. Eventually, the entire SPL network would return to an idle state as before. This is a typical operation of the SPL network when subject to external touch inputs.

Using this approach, we now can use input transducers for other types of sensors as well. The only requirement is that the transducer converts an external signal into a digital signal that the SPL network is capable of using. This applies to temperature sensors, chemical sensors and others. The mapping of the sensory information into an input surface nodes is typically straightforward and similar to the above three cases (images or video, sounds and touch).

7.8   Similarity and dissimilarity conditions

Consider an already-tuned initial dynamical SPL network implemented within a single CPU computer. The network has an input surface and an output surface along with a random initial graph of internal nodes (similar to Figure 7.1). Consider a training set $\mathcal{T}$ of N images $I_1, I_2, ..., I_N$ of everyday objects (like tables and chairs). Our objective is to modify the initial SPL network so we can distinguish each of the objects from the training set as well as from a test set that has similar but new never-before-seen objects.

Definition 7.1 (partitioning the training set): A training set $\mathcal{T}$ is said to be partitioned into k distinct subsets $\mathcal{S}_1, \mathcal{S}_2, ..., \mathcal{S}_k$ if and only if they satisfy the following conditions:
  (a) $\mathcal{T} = \mathcal{S}_1 \cup \mathcal{S}_2 \cup ... \cup \mathcal{S}_k$ i.e., all images in the training set are in one of these subsets.
  (b) All images in a specific subset $\mathcal{S}_i$ (for $i \in \{1, 2, ..., k\}$) are visually similar to each other.
  (c) Any two images $I_1 \in \mathcal{S}_i$ and $I_2 \in \mathcal{S}_j$ for $i \neq j$ are required to be considerably dissimilar from each other.

For condition (b) above, we have already described two common ways to group a set of images into a given subset $\mathcal{S}_i$ – see definition 6.2 for similarity condition. They are: (i) by taking different objects that look similar (like distinct but similar chairs) or (ii) by taking the same object but subject to several small continuous and dynamical transformations (like translation, rotation, distortion and others). Such a classification requires another conscious human being in the beginning. However, later we will see that the SPL network itself will become capable of performing this classification.

It is clear that the above definition can be easily generalized to other types of elements like sounds, videos, actions, events and others. The above conditions merely state that partition $\mathcal{T}$ is composed of *dissimilar groups of similar images*. If we pick dissimilar images $I_1 \in S_i$ and $I_2 \in S_j$ for $i \neq j$ from the training set as inputs to the initial SPL network, it is apparent that, in general, there is no obvious pattern produced within the network that could be used to uniquely distinguish them as different objects. The more the number of such images within the training set, the less likely it is for the initial SPL network to distinguish them uniquely. Our objective, therefore, is to modify the initial network so we identify simple and unique patterns for each dissimilar group $S_1, S_2, ..., S_k$.

Recall the discussion in section 6.6 on what similarity and dissimilarity conditions are in relation to an SPL network of neurons. Let me represent these conditions in more precise terms here. Since we have partitioned the training set $\mathcal{T}$ into multiple subsets $S_1, S_2, ..., S_k$, let me identify the images within each subset S_i as $\{I_{ij} \mid j = 1, 2, ..., m_i\}$ for $i = 1, 2, ..., k$. Here m_i represents the number of images within each subset S_i. With this notation, let me represent the set of dynamical loops generated when an image I_{ij} from the training set $\mathcal{T}$ falls on the input surface (like the retina) to be $\mathcal{L}_{ij}$ for $j = 1, 2, ..., m_i$ and $i = 1, 2, ..., k$ (see Figure 7.5).

Definition 7.2 (similarity condition): From definition 6.2, an SPL network is said to satisfy a similarity condition for the training set $\mathcal{T}$ if and only if for each S_i, $i = 1, 2, ..., k$, the intersection of loops $\mathcal{L}_{ij}$ generated for the elements I_{ij} in S_i for $j = 1, 2, ..., m_i$ is nonempty i.e., $\bigcap_{j=1}^{m_i} \mathcal{L}_{ij} \stackrel{\text{def}}{=} \mathcal{L}_i \neq \emptyset$.

For an initial random SPL network, the condition $\mathcal{L}_i \neq \emptyset$ will not be satisfied for some $i = 1, 2, ..., k$. If that were the case, we want to alter the SPL network until $\mathcal{L}_i \neq \emptyset$ does satisfy for all $i = 1, 2, ..., k$. In the next few sections, we will discuss how to do this.

Definition 7.3 (fixed set): The fixed set of an object, event or abstraction identified using a collection S_i of similar objects, events and abstractions respectively is the nonempty common set of dynamical loops $\mathcal{L}_i$ defined in definition 7.2.

The choice of the word 'fixed set' will become clear when we relate it to the notion of 'truths' required when analyzing and building conscious systems later in chapter 14. If the SPL network generates an empty set $\mathcal{L}_i$ for a given object, identified by the collection S_i, then we do not have a fixed set for this object yet. We would want to modify the network until we have a nonempty set $\mathcal{L}_i$. When this happens we identify $\mathcal{L}_i$ as the fixed set for the object in question. With the SPL network of a human brain, we can say that we have memorized an object if and only if its fixed set $\mathcal{L}_i \neq \emptyset$.

For example, if we considered a table and included a collection of similar images of a table in S_i (see Figure 7.5), then if the above set of dynamical loops $\mathcal{L}_i$ is nonempty, then $\mathcal{L}_i$ is the fixed set for a table. We use $\mathcal{L}_i$ within the SPL network as an internal representation to identify an external table. The fixed set $\mathcal{L}_i$ is the memory for the external table. If and only if $\mathcal{L}_i$ is triggered through any means, we say that the SPL

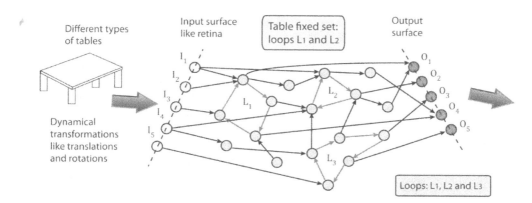

Figure 7.5: Satisfying similarity condition and creating fixed sets. In this example, the training set includes different types of tables and several dynamical transformations of the table all grouped into one partition. If we assume that these variations fall at nodes I_1, I_2 and I_3, then the intersection of all loops are L_1 and L_2. These two loops can be identified as the table fixed set even though this network has an additional loop L_3. In a real SPL network, the input, the internal nodes and the set of loops will be far more in number.

network has 'recollected' the table.

Definition 7.4 (dissimilarity condition): From definition 6.4, an SPL network is said to satisfy a dissimilarity condition for the training set $\mathcal{T}$ if and only if for each pair of collections $\mathcal{S}_i$ and $\mathcal{S}_j$ ($i \neq j$) with fixed sets $\mathcal{L}_i$ and $\mathcal{L}_j$ respectively, there exists two sets of dynamical loops $\mathcal{P}_i \subseteq \mathcal{L}_i$ and $\mathcal{P}_j \subseteq \mathcal{L}_j$ such that $\mathcal{P}_i \cap \mathcal{P}_j = \emptyset$.

For example, consider the fixed set $\mathcal{L}_i$ for a table $\mathcal{S}_i$ and the fixed set $\mathcal{L}_j$ for a chair $\mathcal{S}_j$ (see Figure 7.6). In Figure 7.6, one way to choose these fixed sets are by letting $\mathcal{L}_i = \{L_1, L_2, L_3, L_4\}$ and $\mathcal{L}_j = \{L_1, L_2, L_5, L_6\}$. With such a choice, , $\mathcal{L}_i \cap \mathcal{L}_j = \{L_1, L_2\} \neq \emptyset$. This is because both the table and the chair will trigger several common loops corresponding to primitive features like edges, angles and colors. In fact, the input to the SPL network for almost all images starts from the same input surface (like the retina) and then diverge away in different directions within the network. As a result, several pathways at the initial stage will be common across all of these objects. Since the common features across a table, a chair and all other objects are primitive features like edges and angles, it is safe to associate those initial dynamical loops with these primitive features.

In spite of $\mathcal{L}_i \cap \mathcal{L}_j \neq \emptyset$, we would want to distinguish a table from a chair. One way this becomes easy is if we start looking at subsets of dynamical loops $\mathcal{L}_i$ and $\mathcal{L}_j$ after the pathways have diverged in the longitudinal direction as shown in Figure 7.6. The further along we look at the diverged pathways, the higher the chance of picking two sets of dynamical loops that have nothing in common. In Figure 7.6, these correspond to dynamical loops $\mathcal{P}_i = \{L_3, L_4\}$ and $\mathcal{P}_j = \{L_5, L_6\}$ respectively, rather than the entire looped sets $\mathcal{L}_i$ and $\mathcal{L}_j$. Clearly, $\mathcal{P}_i \cap \mathcal{P}_j = \emptyset$ even though $\mathcal{L}_i \cap \mathcal{L}_j \neq \emptyset$. Therefore, we can

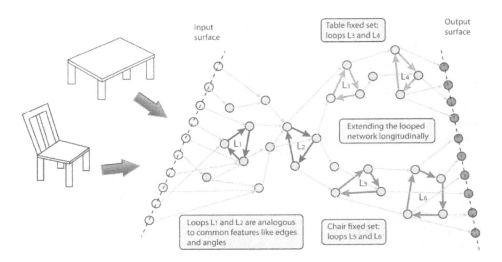

Figure 7.6: Satisfying dissimilarity condition by extending the network longitudinally. In this example, L_3 and L_4 will be identified as a table fixed set even
though L_1 and L_2 are also triggered simultaneously. The reason is that L_1 and L_2
will be triggered even for a chair. Therefore, L_1 and L_2 can only be identified as
the common features of tables and chairs, not a table or a chair individually.
Here, L_5 and L_6 will be identified with a chair. The advantage is that we can
extend the network longitudinally for each dissimilar object from the training
set to satisfy dissimilarity condition.

use $\mathcal{P}_i$ to identify a table and $\mathcal{P}_j$ to identify a chair. Then, $\mathcal{P}_i$ becomes the table fixed
set while $\mathcal{P}_j$ becomes the chair fixed set. The relationship $\mathcal{P}_i \cap \mathcal{P}_j = \emptyset$ implies that a
table is different from a chair.

Let us now discuss how to modify the SPL network to satisfy the above similarity
and dissimilarity conditions for a training set $\mathcal{T}$. By picking suitable training sets like
images, sounds, videos and so on, the SPL network can be made to memorize several
everyday objects and events. If the training set $\mathcal{T}$ is a collection of images, then the
SPL network built addresses the image recognition problem. If $\mathcal{T}$ is a collection of
sounds, the SPL network addresses the sound recognition problem and so on. In
section 7.11, I will discuss how to merge multiple SPL networks that addresses both
image and sound recognition problems. In chapters 10 and beyond, we will use such
modified SPL networks when discussing problems like free-will and consciousness.

7.9 Training and modifications to the SPL network

A synthetic SPL network starts with a small and simple random initial network. The
objective then is to modify and increase the size/complexity of the network by
introducing structure unique to a given problem. The randomness is modified and
replaced with definite structure in two ways. One is with a guided approach using a
training set. The other is from natural modifications within the network from
exposure to everyday external inputs beyond the training inputs. I will first describe

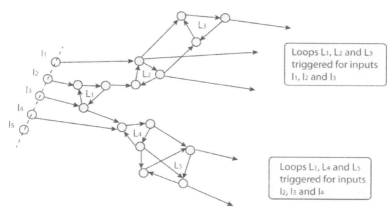

Loops L_1, L_2 and L_3
triggered for inputs
I_1, I_2 and I_3

Loops L_1, L_4 and L_5
triggered for inputs
I_2, I_3 and I_4

Figure 7.7: A sample network to illustrate how different inputs can trigger different loops and how we have the flexibility to grow the network in different directions to enable memorizing distinct objects. In this example, if inputs I_1, I_2 and I_3 fire, loops L_1, L_2 and L_3 are triggered. We are assuming that even though L_1 sends output to L_4, the node in L_4 does not exceed its threshold to fire. Similarly, inputs I_2, I_3 and I_4 cause loops L_1, L_4 and L_5 to fire.

the guided approach using image recognition as a representative example. The other applications will become clear later in the book though I will not discuss them in great detail.

Definition 7.5 (training an SPL network): A dynamical SPL network is said to be trained for a given training set $\mathcal{T}$ composed of images, sounds, videos, events, actions and others if and only if the SPL network satisfies both similarity and dissimilarity conditions for a given partition of $\mathcal{T}$ into subsets S_1, S_2, ..., S_k obeying the conditions of definition 7.1.

For the similarity condition, it is easiest to pick small variations using geometric and dynamical transformations like translation, rotation, stretching, moving closer or farther and others of a single element. To get more diversity, we could also pick different objects of the same type of element. The reference to dynamical patterns of interest is primarily a collection of dynamical loops.

    Similarity condition: Let me now describe how to satisfy the similarity condition for a given image A within the training set. For the similarity condition, we consider several variations of the image A in the form of small translations, rotations, deformations and even dynamical variations like the act of moving the image closer or farther. Let these variations be represented as A_1, A_2, ..., A_n. When the input surface is subject to each of these images for a short while and then removed, the initial random SPL network would potentially generate a set of dynamical loops ($\mathcal{L}_1$, $\mathcal{L}_2$, ..., $\mathcal{L}_n$ respectively). However, the most likely case is that the dynamical loops for all of A_1, A_2, ..., A_n are distinct from each other. This is a problem because all of these images are supposed to represent the same image A and, yet, there are no common set of

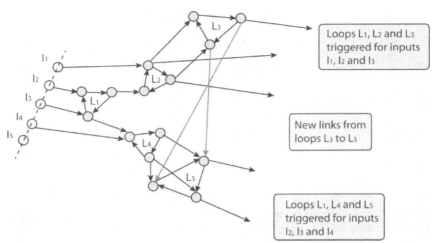

Figure 7.8: Add connections from one set of loops to another set. When modifying the SPL network, it is necessary to form links between different loops to propagate information from one region to the other. In this figure, we are adding new connections to link loops L_3 with L_5.

dynamical loops among them i.e., $\mathcal{L}_1 \cap \mathcal{L}_2 \cap ... \cap \mathcal{L}_n \stackrel{\text{def}}{=} \mathcal{L} = \emptyset$. Therefore, to satisfy the similarity condition, we must alter the SPL network so the set of dynamical loops $\mathcal{L}$, namely, the fixed set for the image A (definition 7.2), becomes nonempty. We will use $\mathcal{L}$ to uniquely identify image A in spite of minor geometric and dynamical variations. The process is as follows. Figure 7.7 shows that small changes in an SPL network can produce large differences in the set of dynamical loops triggered. In Figure 7.7, inputs on nodes I_1, I_2 and I_3 trigger loops L_1, L_2 and L_3 while inputs on nodes I_2, I_3 and I_4 trigger loops L_1, L_4 and L_5. Extending the pathways longitudinally magnify the deviations significantly. This forms the basis for satisfying similarity and dissimilarity conditions.

I will first pick two variations A_1, and A_2 of image A. I will show how to achieve $\mathcal{L}_1 \cap \mathcal{L}_2 \neq \emptyset$. Following this, I will generalize the mechanism to three-image variations A_1, A_2 and A_3 by reducing it to the previous two-image variations problem. Then, we will be in a position to use the equivalent of mathematical induction to generalize the process to n number of image variations as well.

Two-image variations: Consider the list of dynamical loops $\mathcal{L}_i$ triggered for each image variation A_i within the SPL network for $i = 1, 2, ..., n$. Take a pair A_1, A_2 and the corresponding set of loops $\mathcal{L}_1$ and $\mathcal{L}_2$. If $\mathcal{L}_1 \cap \mathcal{L}_2 \neq \emptyset$ in the original SPL network itself, there is no need to modify it. If $\mathcal{L}_1 \cap \mathcal{L}_2 = \emptyset$, we need to modify the SPL network by adding new nodes and/or new connections to ensure it is nonempty. There are multiple ways to achieve this.

1. *Adding connections*: In the first approach, we add new connections. Using the nodes for all loops within $\mathcal{L}_1$ and $\mathcal{L}_2$, let us use the neighborhood information from the geometric layout to add new connections from nodes in $\mathcal{L}_1$ to nodes in $\mathcal{L}_2$ that are close to each other (computed using the distance method). As an

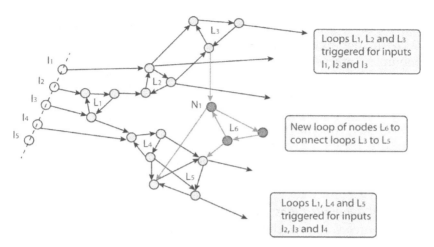

Loops L₁, L₂ and L₃ triggered for inputs I₁, I₂ and I₃

New loop of nodes L₆ to connect loops L₃ to L₅

Loops L₁, L₄ and L₅ triggered for inputs I₂, I₃ and I₄

Figure 7.9: Adding new nodes into an SPL network to connect different loops. We could add a number of nodes or even loops of nodes to link one set of loops with another. We will discuss several applications of these basic operations in order to grow a looped network.

example, consider $\mathcal{L}_1 = \{L_1, L_2, L_3\}$ and $\mathcal{L}_2 = \{L_4, L_5\}$ with no node common between $\mathcal{L}_1$ and $\mathcal{L}_2$ as shown in Figure 7.7. From the list of loops L_1-L_3 and L_4-L_5, pick any two loops with one from each set. The choice is either random (like L_1 and L_5) or more structured, say, by choosing the distance between the pairs to be close enough. Once you picked two loops (say, L_3 and L_5), create a new connection between two of the nodes within the loops, say, from a node in L_3 to a node in L_5 (see Figure 7.8). You may even create two connections in both directions like, say, from L_3 to L_5 and from L_5 to L_3. With such a new connection, you are now guaranteed to have nonempty intersection of loops (i.e., $\mathcal{L}_1 \cap \mathcal{L}_2 \neq \emptyset$). Specifically, after this modification of the SPL network (Figure 7.8), let us re-run the training inputs for images A_1, A_2 and update the list of loops. For the new SPL network, we now have $\mathcal{L}_1 = \{L_1, L_2, L_3, L_5\}$ and $\mathcal{L}_2 = \{L_4, L_5\}$. If we included connection in the reverse direction as well, then $\mathcal{L}_2 = \{L_3, L_4, L_5\}$. Therefore, the $\mathcal{L}_1 \cap \mathcal{L}_2 = \{L_5\} \neq \emptyset$ (or $\{L_3, L_5\} \neq \emptyset$ if we included connections in the reverse direction) in the modified SPL network. The requirement of achieving nonempty intersection is critical for satisfying the similarity condition (see definition 7.2).

2. *Adding nodes*: In the second approach, we add new nodes. As before, consider the example in which $\mathcal{L}_1 = \{L_1, L_2, L_3\}$ and $\mathcal{L}_2 = \{L_4, L_5\}$ with no node common between $\mathcal{L}_1$ and $\mathcal{L}_2$ i.e., $\mathcal{L}_1 \cap \mathcal{L}_2 = \emptyset$ (Figure 7.7). As before, we pick two loops that are close enough (or we may even pick the two loops randomly), say, L_3 and L_5, for the sake of argument. Add a new node N_1 geometrically in between these two loops. Update the geometric layout because of the newly added node N_1. Now, we create at least two new connections, one from loop L_3 to N_1 and another from N_1 to loop L_5. Therefore, node N_1 acts as an intermediate node

connecting two existing loops L_3 and L_5 (Figure 7.9). In addition to adding a single node N_1, we could add an entire loop of nodes like L_6 as shown in Figure 7.9. With such new nodes and connections, we are now guaranteed to have nonempty intersection of loops i.e., $L_1 \cap L_2 \neq \emptyset$. Specifically, after this modification to the SPL network, let us re-run the training inputs for images A_1, A_2 and update the list of loops. In the new SPL network of Figure 7.9, we now have $L_1 = \{L_1, L_2, L_3, L_5, L_6\}$ and $L_2 = \{L_4, L_5\}$. If you decide to add a reverse pathway i.e., from L_5 to N_1 and N_1 to L_3 as well, then $L_2 = \{L_3, L_4, L_5, L_6\}$. Therefore, the $L_1 \cap L_2 = \{L_5\} \neq \emptyset$ (or $\{L_3, L_5, L_6\} \neq \emptyset$ if we had added the reverse pathways) in the modified SPL network. The requirement of achieving nonempty intersection is critical for satisfying the similarity condition (see definition 7.2).

3. *Additional generalizations*: Several generalizations are now possible using the above two core approaches i.e., adding new connections and/or adding new nodes to guarantee the same objective of nonempty intersection of loops. Let me list a few of them now.

    (a) We add multiple connections between multiple loops instead of just one connection between only two loops (like L_3 and L_5 in the above example of Figure 7.8).

    (b) We add multiple nodes N_1, N_2, ..., N_m simultaneously (Figure 7.9). Using the new nodes, there are multiple ways to satisfy nonempty intersection property, some of which are as follows.

    (i) We build new loops among N_1, N_2, ..., N_m and then link them to existing loops L_1-L_5 (Figure 7.9).

    (ii) We build multiple connections between existing loops L_1-L_5 via the new nodes N_1, N_2, ..., N_m (Figure 7.9).

    (iii) We build multiple connections between N_1, N_2, ..., N_m to other existing nodes in the network creating new loops first. Then, we link these loops with existing loops L_1-L_5. These loops now have some new nodes and some existing nodes.

    (iv) We add new loops of nodes rather than non-looped set of new nodes (Figure 7.9).

    (v) We add an entire sub-network into the existing network and interconnect them to the existing loops L_1-L_5.

Therefore, with the above variations to two core mechanisms, we are able to guarantee nonempty intersection for any pairs of transformed images. Each of the large set of possibilities like those described above makes it easy to satisfy the nonempty intersection property.

The above mechanisms may be performed manually using a set of simple tools that let us visualize the network. They may also be performed automatically by writing algorithms for these mechanisms in a computer. The manual visual approach in which a human is involved is useful for some special situations like when we want to train the network for a specific image or a specific concept. Training for a specific concept will become important when we consider other applications like sound and speech recognition, teaching specific words in a language and teaching specific actions. On the other hand, if we want to train the network for a large collection of training inputs, the automatic approach works quicker and easier.

Three-image variations: Given that we are able to guarantee $\mathcal{L}_1 \cap \mathcal{L}_2 \neq \emptyset$ using the above mechanisms, our next step is to guarantee $\mathcal{L}_1 \cap \mathcal{L}_2 \cap \mathcal{L}_3 \neq \emptyset$ when we include A_3 (and its dynamical loops $\mathcal{L}_3$). We solve this problem by reducing the 3-image problem ($\mathcal{L}_1$, $\mathcal{L}_2$ and $\mathcal{L}_3$) into the above 2-image problem ($\mathcal{L}_1$ and $\mathcal{L}_2$) using the following simple idea. Choose $\mathcal{L} \stackrel{\text{def}}{=} \mathcal{L}_1 \cap \mathcal{L}_2 \neq \emptyset$ after modifying the network as above. Now, $\mathcal{L}$ and $\mathcal{L}_3$ are the two new set of dynamical loops we will work with. Our objective can now be restated as $\mathcal{L} \cap \mathcal{L}_3 \neq \emptyset$. However, this is exactly the same problem we just solved if we replace $\mathcal{L}$ by $\mathcal{L}_1$ and $\mathcal{L}_3$ by $\mathcal{L}_2$. Therefore, using any of the above mechanisms of adding new nodes and/or connections, we will be able to guarantee $\mathcal{L} \cap \mathcal{L}_3 \neq \emptyset$.

N-image variations: The above process now becomes iterative, thereby generalizing to any number of similar or transformed images A_1, A_2, ..., A_n. The approach is quite simple (analogous to the steps taken when proving a statement using mathematical induction) and is as follows. First, statement P_1, namely, $\mathcal{L}_1 \cap \mathcal{L}_2 \neq \emptyset$ is guaranteed using the above procedure. Assuming the validity of statement P_k (i.e., $\mathcal{L}_1 \cap \mathcal{L}_2 \cap ... \cap \mathcal{L}_{k+1} \neq \emptyset$), if we can show how to alter the SPL network to satisfy statement P_{k+1} (i.e., $\mathcal{L}_1 \cap \mathcal{L}_2 \cap ... \cap \mathcal{L}_{k+2} \neq \emptyset$), then we are in a position to extend the solution iteratively to P_n (similar to mathematical induction) to conclude that statement P_n (i.e., $\mathcal{L}_1 \cap \mathcal{L}_2 \cap ... \cap \mathcal{L}_{n+1} \neq \emptyset$) can be satisfied for all n. For this, we first assume that our SPL network is altered to satisfy $\mathcal{L}_a \stackrel{\text{def}}{=} \mathcal{L}_1 \cap \mathcal{L}_2 \cap ... \cap \mathcal{L}_{k+1} \neq \emptyset$. Now, using the above mechanisms for two sets of dynamical loops applied to $\mathcal{L}_a$ and $\mathcal{L}_{k+2}$, we can modify the dynamical SPL network to guarantee $\mathcal{L}_a \cap \mathcal{L}_{k+2} \neq \emptyset$, i.e., statement P_{k+1} is satisfied. Hence, P_n can be satisfied for all n.

Therefore, the resulting dynamical SPL network has nonempty collection of dynamical loops common to all variations A_1, A_2, ..., A_n of a given image A from the training set $\mathcal{T}$. This nonempty collection ($\mathcal{L} \stackrel{\text{def}}{=} \mathcal{L}_1 \cap \mathcal{L}_2 \cap ... \cap \mathcal{L}_n \neq \emptyset$) is termed as the **fixed set** for the image A and will now be used to uniquely identify image A.

One important design consideration when modifying the SPL network using the above approaches is to track the density of the network. When we add new connections, the network becomes denser. On the other hand, when we add new nodes, the density does not increase. Since our objective is to never reach the critical density identified as part of tuning the node parameters mentioned earlier (section 7.6), we need to switch the approach from adding connections to adding new nodes if we observe that the density is beginning to increase. If the density is low enough, we continue adding new connections instead of new nodes.

Dissimilarity condition: Having trained (or modified) the network to satisfy the similarity condition for a single image A, our next set of objectives are (a) to train the network for a second image B and (b) to ensure that the dissimilarity condition is satisfied for images A and B. The mechanism to satisfy objective (a), i.e., the similarity condition for a second image B using transformed images B_1, B_2, ..., B_m, is identical to the above mechanism for image A. Assuming we have an SPL network that satisfies similarity condition for both images A and B, we now need to satisfy objective (b).

This is quite easy as well.

The main idea involves extending the dynamical looped pathways for images A and B in different directions. For example, in Figure 7.7, we had previously satisfied the similarity condition for two images A and B individually by defining, say, the fixed sets $\mathcal{L}_A = \{L_1, L_2, L_3\}$ and $\mathcal{L}_B = \{L_1, L_4, L_5\}$. Unfortunately, with these choice of fixed sets for A and B, the dissimilarity condition is not satisfied i.e., $\mathcal{L}_A \cap \mathcal{L}_B = \{L_1\} \neq \emptyset$. Now, to fix this problem, we can first extend the pathways longitudinally (see Figure 7.6). Specifically, when we were trying to modify the SPL network to satisfy the similarity condition for image A, let us assume that we have created new loops by adding new nodes/connections, say, along the top part of the network by extending the network longitudinally (see Figure 7.6 where we deviate the pathways from L_1 and L_2 in different directions, namely, to L_3 and L_4 for the table image longitudinally). Similarly, when satisfying the similarity condition for image B, let us assume that we have created new loops along the bottom part of the network (see Figure 7.6 where we extended L_1 and L_2 to L_5 and L_6 for a chair image longitudinally).

However, this step has not satisfied the dissimilarity condition yet for the current definition of table fixed set as $\mathcal{L}_A = \{L_1, L_2, L_3, L_4\}$ and chair fixed set as $\mathcal{L}_B = \{L_1, L_2, L_5, L_6\}$ since $\mathcal{L}_A \cap \mathcal{L}_B = \{L_1, L_2\} \neq \emptyset$. The fact that both pathways (for a table A and a chair B) started from loops L_1 and L_2 is necessary because the input surface is the same for both A and B. The initial pathways and the sub-network from the input surface will be similar for both A and B. These correspond to similar features within A and B like edges, angles, primitive shapes and others.

Now, to satisfy dissimilarity condition, we can partition the set of loops generated for each image into two groups – the common loops near the input surface and distinct set of loops from the diverged pathways. For image A, $\mathcal{L}_A = \mathcal{L}_c \cup \mathcal{L}_1$ where $\mathcal{L}_c = \{L_1, L_2\}$ and $\mathcal{L}_1 = \{L_3, L_4\}$. For image B, $\mathcal{L}_B = \mathcal{L}_c \cup \mathcal{L}_2$ where $\mathcal{L}_c$ is same as above and $\mathcal{L}_1 = \{L_5, L_6\}$ Here, $\mathcal{L}_c$ represents the common loops near the input surface while $\mathcal{L}_1$ contains dynamical loops on the top part and $\mathcal{L}_2$ on the bottom part (Figure 7.6). With these definitions, clearly $\mathcal{L}_1 \cap \mathcal{L}_2 = \emptyset$.

Therefore, instead of treating $\mathcal{L}_A$ and $\mathcal{L}_B$ as the fixed sets for images A and B respectively, we regard $\mathcal{L}_1$ and $\mathcal{L}_2$ instead as the fixed sets. With this new identification (i.e., $\mathcal{L}_1$ and $\mathcal{L}_2$ as fixed sets for A and B respectively), we satisfy both the similarity condition (since $\mathcal{L}_1 \neq \emptyset$ and $\mathcal{L}_2 \neq \emptyset$) as well as the dissimilarity condition ($\mathcal{L}_1 \cap \mathcal{L}_2 = \emptyset$).

In other words, by choosing different regions within the network to extend the pathways of nodes and dynamical loops as we train for the similarity condition, we will be able to satisfy the dissimilarity condition as well. This procedure can now be generalized for any number of distinct images $A, B, C, ..., Z$ along with their dynamical transformations and/or other similarity variations. The interesting aspect with the resulting modified network is that there is a common region, typically closer to the input surface (like the retina) that is re-used by all of the images. However, as the pathways extend deeper and farther from the input surface, each image proceeds in a uniquely distinguishable direction. It is the latter aspect that lets the SPL network satisfy the dissimilarity condition as well.

Now, if we continue to modify the dynamical SPL network to satisfy both the similarity and the dissimilarity conditions for all images in the training set $\mathcal{T}$, I say that the SPL network is fully trained for the specific training set.

7.10 Connectedness condition

In section 7.9, we have seen how to take an initial random SPL network and modify it to produce a new SPL network that is fully trained for a single training set. The similarity and dissimilarity conditions are satisfied for each element (like an image) in the training set. Now, I will discuss how to train the network for two or more training sets. In general, each training set may or may not use the same input surface. For example, if the two training sets are (a) images of human faces and (b) images of geometric shapes, then the input surface (like the retina) is the same in both cases. On the other hand, if one training set is images of household objects and the other training set is the sounds made by these household objects, then the input surfaces are different (like the retina and the ear respectively).

If the input surface is the same for two training sets, it is better to use the same SPL network to train both of them instead of training them independently using two separate SPL networks and then merging them later. The reason is that the sub-network closer to the input surface should eventually be the same. Therefore, if we started with two different SPL networks with different sub-networks, the merge process will not offer any guarantees regarding similarity near the input surface. The merged SPL network will no longer be successfully trained for both training inputs. The pathways from the input surface in the merged network are not what they were separately trained for.

Therefore, using the same SPL network and training it serially, one training set after the other, is a much simpler solution in this case. The serial training approach is identical to the approach of section 7.9 if we define a new training set as the union of the two training sets.

From this result, it follows that the collection of all training sets $\mathcal{T}$ that rely on multiple input surfaces can be partitioned into m disjoint groups $\mathcal{T}_1, \mathcal{T}_2, ..., \mathcal{T}_m$, with one group per input surface. Examples of training sets that do not obey this partitioning scheme are action-based ones, which do not rely on input surfaces (events like throwing a ball, moving your hand to the left or raising your hand). Excluding these training sets, we now have a unique advantage with SPL network design – create separate and independently trained SPL networks, one for each input surface, and then merge them in a specific manner, to be discussed in section 7.11, to create a new unified SPL network that is fully trained for all training sets.

As an example, consider a trained SPL network $\mathcal{N}_1$ for common everyday images and another trained SPL network $\mathcal{N}_2$ for common everyday sounds (see Figure 7.10). Our objective is to create a single merged SPL network $\mathcal{N}$ that is fully trained for both these training sets. Since the two SPL networks receive inputs from different input surfaces, we have several options for merging them. For example, we could do a simple union of two networks $\mathcal{N}_1 \cup \mathcal{N}_2$ while keeping the two sub-networks disjoint

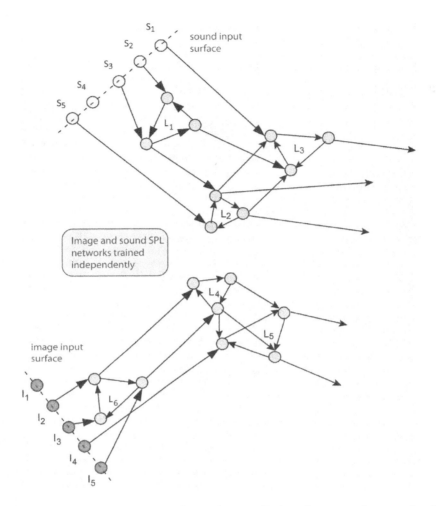

Figure 7.10: Two SPL networks with two distinct input surfaces trained independently. Here, we illustrate two sample SPL networks, one for images and the other for sounds that are trained independently on their respective training sets N_1 and N_2. These two SPL networks are not merged yet into a single SPL network.

from each other (i.e., $N_1 \cap N_2 = \emptyset$) or we could randomly interconnect nodes from each network. However, these two approaches do not capture the external reality. For example, these approaches may allow a link between the dynamical loops (or fixed set) of a car-image from one network N_1 to the dynamical loops (or fixed set) of a ball-bouncing-sound in the other network N_2 even though they are unrelated to each other. This is undesirable and should be eliminated if we want the merged SPL network N to capture external reality, which in this case is that a car-image should only be associated with a car-sound while a ball-image to a ball-bouncing sound.

Our objective when merging two SPL networks N_1 and N_2 is to allow only those interconnections that model external reality accurately like, say, a car-image fixed set connected to car-sound fixed set, but not to ball-bouncing-sound fixed set and so on. The key idea here is that we interconnect only those set of dynamical loops that are excited at about the *same time* even if the loops are physically located very far from each other. Of course, we want to create such interconnections only if the event (of triggering both fixed sets at about the same time) repeats multiple times.

For example, a car-image fixed set, originating from image input surface like the retina, is, in general, physically located quite far in our brain from the car-sound fixed set, originating from sound input surface like the ear. In the external world, it is quite common that when we see a car-image, we do hear a car-sound. The near-simultaneity of occurrence of these two events also repeats multiple times. On the other hand, two events like a car-image and a ball-bouncing sound occur neither simultaneously nor repeatedly. They may be simultaneous once or twice by chance, but not at the frequency of simultaneous repetition of car-image and car-sound.

Near-simultaneity of occurrence of certain pairs of events can be considered as 'truths' in the external world – table-image and table-sound should be associated (i.e., are causally related), fan-image and fan-sound should be associated, fan-image and table-sound should not be associated and so on. We want to make sure our SPL network memorizes such 'truth' relationships (to be discussed more in chapter 14) that exist in the external world. It is even reasonable to call such relationships as 'primitive' laws of nature. The external reality enforces which pairs of events should be associated with each other and which ones should not, using the notion of near-simultaneity of occurrence. We term such an association as satisfying the connectedness condition.

Definition 7.6 (connectedness condition): An SPL network is said to satisfy a connectedness condition for a training set $\mathcal{T}$ if and only if any pair of repeatedly occurring and near-simultaneous (or 'quick' causal) events $E_1 \in \mathcal{T}$ and $E_2 \in \mathcal{T}$ in the external world represented by fixed sets $\mathcal{L}_1$ and $\mathcal{L}_2$ respectively are connected together via one or more, preferably looped pathways in, possibly, both causal directions, namely, from $\mathcal{L}_1$ to $\mathcal{L}_2$ and $\mathcal{L}_2$ to $\mathcal{L}_1$.

Note that pathways in this context refers to not just a single linear set of connected nodes. Rather, it is common to have hundreds of directed pathways from $\mathcal{L}_1$ to $\mathcal{L}_2$, say. Also, it is common to have pathways in both directions i.e., from $\mathcal{L}_1$ to $\mathcal{L}_2$ and $\mathcal{L}_2$ to $\mathcal{L}_1$. This captures the notion that if you are presented with event E_1, you will automatically recollect event E_2 (via the connected pathways) and vice versa. I will use notation $E_1 \leftrightarrow E_2$ to represent this. We can see that there is a causal relation from E_1 to E_2 and vice versa. Here, I want to emphasize near-simultaneity more than just a causal relation i.e., the causal relation should be fast enough to make us memorize the both events as related to each other.

Sometimes, one direction is more prominent than the other. Here, I will use the following notations: $E_1 \rightarrow E_2$ represents the fact that event E_1 lets us recollect event E_2 i.e., E_1 is the cause and E_2 is the effect. The notation $E_1 \leftarrow E_2$ represents the converse –

E_2 is the cause and E_1 is the effect. In both cases, however, I assume that the causes and effects are fast enough to let us memorize the two events as related to each other. For example, hearing a fan-sound (E_2) automatically makes us remember an approximate fan-image (E_1) via these prominent connected pathways (i.e., $E_1 \leftarrow E_2$). On the other hand, seeing a non-rotating fan does not automatically trigger the fan-sound fixed set though a few pathways do exist. With conscious effort, however, it is possible to recollect at least a crude version of the fan-sound using these sparse set of pathways.

7.11 Merging multiple SPL networks

Let us now revisit the problem of training two separate SPL networks $\mathcal{N}_1$ and $\mathcal{N}_2$, one for images (training set $\mathcal{T}_1$) and the other for sounds (training set $\mathcal{T}_2$) and then merging them into a single SPL network $\mathcal{N}$ (see Figure 7.10). When do we say that we have merged the two SPL networks? From the discussion in section 7.10, we need to follow these steps below:

(a) Train an SPL network $\mathcal{N}_1$ to satisfy similarity and dissimilarity conditions for training set $\mathcal{T}_1$ (see Figure 7.10).

(b) Train another SPL network $\mathcal{N}_2$ to satisfy similarity and dissimilarity conditions for training set $\mathcal{T}_2$ (see Figure 7.10).

(c) Create a new training set $\mathcal{T}$ containing directed pairs of events $(I_i \rightarrow S_i) \in \mathcal{T}$ for $i = 1, 2, ..., n$ such that $I_i \in \mathcal{T}_1$ and $S_i \in \mathcal{T}_2$ satisfying the following condition. An element $(I_i \rightarrow S_i)$ belongs to the training set $\mathcal{T}$ if and only if I_i and S_i occurs repeatedly and almost-simultaneously (or is 'quickly' causal versus causal event extended over a long period) in the external world for $i = 1, 2, ..., n$. Specifically, if I_i occurs, then S_i occurs at about the same time in the external world (i.e., I_i is the cause and S_i is the effect – though the assumption is that the time interval between cause and effect is short). An element $(I_i \leftarrow S_i)$, on the other hand, implies that if S_i occurs then I_i occurs at about the same time in the external world. In some cases, it is likely that occurrence of I_i implies the occurrence of S_i and vice versa. In such cases, we use the following simpler notation $(I_i \leftrightarrow S_i)$ to represent it.

As an example, let I_i represent, say, an image and S_i represent its corresponding sound. Some examples of these are car-image and car-sound, ball-image and a ball-bouncing sound, a fan-image and the fan-sound and so on. We do not ever want to include $(I_i \rightarrow S_j)$ for $i \neq j$ into the training set $\mathcal{T}$ if such a directed pair cannot occur near-simultaneously and repeatedly in the real world (like a fan-image I_i and car-sound S_j).

(d) Form new connections between SPL network $\mathcal{N}_1$ and SPL network $\mathcal{N}_2$ to create a merged SPL network $\mathcal{N}$ such that $\mathcal{N}$ satisfies similarity and dissimilarity conditions for training sets $\mathcal{T}_1$ and $\mathcal{T}_2$ as well as the connectedness condition for training set $\mathcal{T}$ (see Figure 7.11). For the connectedness condition, if $(I_i \rightarrow S_i) \in \mathcal{T}$, we take the fixed set $\mathcal{L}_{Ii}$ of I_i and create directed pathways from $\mathcal{L}_{Ii}$ to the fixed set $\mathcal{L}_{Si}$ of S_i (see Figure 7.11). Similarly, if $(I_i \leftarrow S_i) \in \mathcal{T}$, we take the fixed set $\mathcal{L}_{Si}$ of S_i and create directed pathways from $\mathcal{L}_{Si}$ to the fixed set $\mathcal{L}_{Ii}$ of I_i. If $(I_i$

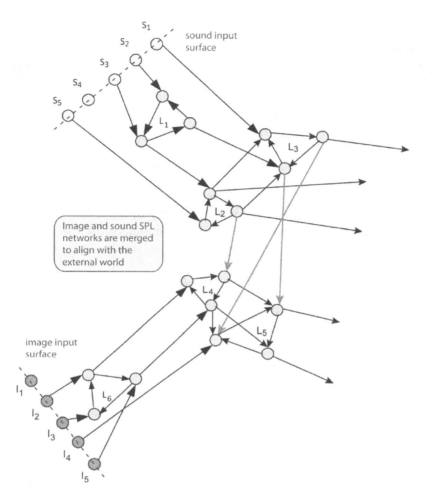

Figure 7.11: Merging two individually trained SPL networks to satisfy connectedness condition. In this example, we have two distinct input surfaces, one
for images and the other for sounds. We have two training sets $\mathcal{N}_1$ and $\mathcal{N}_2$ to
train each of these inputs independently to satisfy similarity and dissimilarity
conditions in two separate SPL networks. Next, we create a new training set $\mathcal{T}$
that expresses the directed relationship between images and sounds. The goal
of the new training set $\mathcal{T}$ is to satisfy the connectedness condition.

$\leftrightarrow S_i) \in \mathcal{T}$, we create directed pathways in both directions, namely, from $\mathcal{L}_{Ii}$ to
$\mathcal{L}_{Si}$ and vice versa. The mechanisms for creating these directed pathways are
those described in section 7.9.
Steps (a) and (b) above are already discussed in considerable detail in section 7.8.
Since our main objective is to merge SPL networks $\mathcal{N}_1$ and $\mathcal{N}_2$, we do not want to
interconnect them in an arbitrary way. We want to interconnect them only in a
specific way that captures the external reality. For this, we first create a new training
set $\mathcal{T}$ as described in step (c) with pairs of events occurring repeatedly and near-

simultaneously (quickly causal) in the external world. Then, we create connected pathways from the fixed set $\mathcal{L}_{Ii}$ of I_i to the fixed set $\mathcal{L}_{Si}$ of S_i, say, as described in step (d) for at least three reasons.

Firstly, the training set $\mathcal{T}$ captures external reality and not any arbitrary set of relationships that can theoretically be modeled like, say, in a computer. Secondly, causality in the external world is represented naturally by having directed pathways from one fixed set to another. Thirdly, connectedness condition is required both for the storage and retrieval of memorized events (i.e., fixed sets).

Recall that the set of dynamical loops in each SPL network satisfying the similarity and dissimilarity conditions are such that they are always structurally and dynamically connected. This is obvious because the only source of input data is from the input surface and each triggered node must necessarily have a connected pathway from the input surface. If there is no such pathway to a specific node, that specific node will remain idle. Connectivity of the dynamical loops $\mathcal{L}$ for a given element in the training set is guaranteed when training one external input at a time.

However, such connectivity of dynamical loops is no longer satisfied when we train each training sets independently on a separate SPL network and later plan on merging the networks together (see Figure 7.10). For example, the set of dynamical loops $\mathcal{L}_{car\_image}$ for an image of a car and $\mathcal{L}_{car\_sound}$ for the sound from a car are individually connected within their sub-networks. However, as yet, they are disconnected from each other and, in fact, physically located far from each other (see Figure 7.10). This is true even if both inputs are triggered simultaneously, as is the case in the external world.

Satisfying the connectedness condition is quite easy once we have identified which pairs of events should be connected, namely, those from the new training set $\mathcal{T}$. The underlying mechanisms are precisely those described in section 7.9.

Similarly, an algorithm to detect disconnectedness of dynamical loops is straightforward as well. We first list the complete set of loops triggered for $(I_i \rightarrow S_i) \in \mathcal{T}$ for a given $i \in \{1, 2, ..., n\}$. By simply traversing the nodes within these dynamical loops, we should be able to detect whether two sets of loops are connected or disconnected. We do this for every pair of loops. It is also possible to use one of the many graph connectedness algorithms from computer science literature to detect disconnectedness of dynamical loops.

If we detect that two or more set of loops are disconnected, we add new nodes and/or new connections between the nodes using any of the mechanisms mentioned in section 7.10 to guarantee connectedness. We continue this process for each element in the new training set $\mathcal{T}$. Once this is completed, the resulting modified SPL network is called the merged SPL network. The merged SPL network is no longer arbitrary. It captures a subset of the true external reality.

Generalizing the process of merging $p > 2$ independently trained SPL networks $\mathcal{N}_1$, $\mathcal{N}_2$ ..., $\mathcal{N}_p$ is now an iterative process. At each step of the iteration, we take two SPL networks and merge them into one. This decreases the total number of SPL networks to be merged by one. Therefore, after $(p - 1)$ iterations, we will have a single merged SPL network.

While the process of merging multiple SPL networks is simple, the nontrivial aspect is the creation of the correct set of training inputs in each iteration that does capture near-simultaneity of pairs of events from the external world. It would be better if this task does not require a conscious being to set up training sets each time. Let us see how the external world can address this issue automatically.

7.12 Perpetual source of associative inputs

The unique feature of the above connectedness condition for merging two or more networks is that the external world acts as a perpetual source of training inputs so the SPL network can learn automatically. We no longer need the help of any conscious being to prepare the training set. For example, just exposure to everyday external inputs like sounds and images automatically provide near-simultaneous set of inputs to the SPL network. These sources naturally create a disconnectedness situation in the set of dynamical loops triggered. Our SPL network would then be modified to ensure that the connectedness condition is guaranteed as long as the near-simultaneous inputs repeat multiple times.

Algorithmically, we maintain the collection of disconnected set of loops under a given external scenario. For each pair of dynamical loops, we count the number of times they occur simultaneously while still remaining as a disconnected set of dynamical loops. If this count is larger than, say, 5-10 times, we add new nodes and/or connections to ensure that the connectedness condition is satisfied for this pair of dynamical loops. In other words, we do not attempt to satisfy connectedness condition when two inputs occur simultaneously just once or a small number of times. This captures the learning notion that if an event repeats several times, we memorize it eventually.

This is an important application – capturing the true external reality within a dynamical SPL network entirely autonomously. A human is no longer needed to teach that a car image and car sound should go together or that the word 'car' and an image of a car should be associated and so on. Any associative inputs that exist in the external world are automatically mapped into the SPL network just using the mechanisms described until now. Triggering the loops or fixed set corresponding to a car sound will automatically trigger car-image fixed set and not a ball-image fixed set. Such an autonomous learning feature is unique with SPL networks unlike any other existing model or theory.

This notion of associative inputs is not limited to two disjoint input surfaces like an image input surface and a sound input surface. They can work even with a single input surface. For example, we almost always see chairs whenever we see a table. Therefore, seeing a table reminds us of chairs as well. This implies triggering a table fixed set should trigger a chair fixed set (either a chair image-fixed set or a chair-word fixed set). For an infant, this association is not formed initially. His SPL network is disconnected with respect to these two fixed sets. However, with repeated exposure to tables and chairs, he will eventually satisfy a connectedness condition (see Figure 7.12). Note that he does not satisfy a connectedness condition between a table fixed set with a car fixed set.

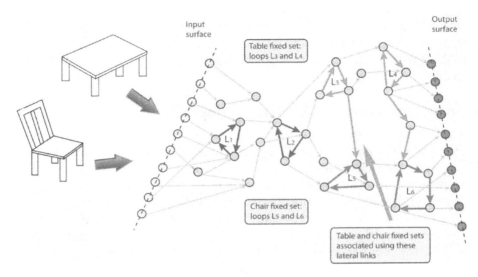

Figure 7.12: Connectedness condition captures the notion of general associative relationships. In this example, we want to capture the fact that seeing a table will remind us of chairs. Therefore, when we satisfy the connectedness condition, it would allow the SPL network to trigger a chair fixed set (either image or a word) whenever a table fixed set is triggered. Here lateral links connect loops L_3 and L_4 (table fixed set) with loops L_5 and L_6 (chair fixed set).

I say that the dynamical SPL network has now transitioned from a learning-through-training to a self-learning state. The SPL network now uses events from the external world to become quite complex thereby capturing several facts. At all stages, the connectivity condition ensures that we have large self-sustaining interconnected dynamical loops in response to external events. A self-learning SPL network is now capable of behaving similar to a self-learning human.

8
Features of the Memory Model

In the previous chapter, we proposed a new model for memory. In this new model, memory is directly related to the formation of a path consisting of both structurally and dynamically interconnected loops. Information propagates along these paths in a cascading manner from an externally observable cause to an externally observable effect. Non-loops are not given special emphasis in this model except when they help in transitioning from one loop to another. In this chapter, I will highlight some of the key aspects of the new memory model.

8.1   Loops, non-loops or the entire path

If the looped paths are not yet formed for a given event and instead you only have non-looped paths, you do not memorize the corresponding parts of the event. It is tempting to generalize and say that the entire path, not just the looped portion of the path, is responsible for memory. The difficulty is that such a representation becomes over-specified, rigid and lacks robustness. It would not fit with the adaptive nature of our memories. For example, as you learn new skills, your brain network changes. The entire path would necessarily change as a result. If the entire path were used as memory, we would not have a well-defined notion of memory. It would be difficult to keep track of each of these changes as you grow and experience new events. Furthermore, deleting connections either naturally or due to accidental brain damage would be catastrophic if you use the entire path for the definition of your memory. However, our personal experiences show that this is not the case. Our memories are robust to a large number of natural changes. Therefore, the entire path is not a good choice for memory.

Similarly, any other special structure within the paths other than loops is not a good choice as well because it would not satisfy all of the features mentioned in chapter 3 for loops. Some of these are: an ability to sustain dynamics for a long time, consumption of less energy, ability to coordinate and an ability to make parallel processing problem tractable. Memory, as we are aware of with the human brain (as opposed to a computer memory), already exhibits each of these features as well as several others that will be described in chapter 9. The new memory model does capture them correctly.

One simple qualitative scenario in which dynamical looped structure appears to be the correct choice is with how sometimes our sensory inputs as well as our thoughts are 'locked' into specific memories. For example, with vision, examples are images like a cube that appear to either go inside the paper or project out of the paper, or images of bumps or dents that appear to lock our brain's state into one way or the other and, more generally, with other optical illusions. After locking our view into a particular pattern for these illusions, it is hard for us to unlock and view it in the other way with ease. Neural loops, as described with SPL networks, compared to non-looped models address these examples more accurately.

In the case of sounds, for example, consider the ticking sound of a clock that was once inaudible because of noise in the room. When there is a brief moment of silence, let us say that we hear the sound distinctly. After that instant, even for considerably higher noise level, we continue to hear the ticking sound. We have essentially locked onto that ticking sound after we have recognized it. Once again, neural loops, unlike non-looped pathways, is a better model to explain how even a small sound will continue to self-sustain easily.

For the case of smell or taste, a mere statement that a particular food you are about to eat is spoiled is enough for our sense organs to be locked into this state (via neural loops versus non-looped pathways). We will sense a similar malodor whether it is truly present or not.

With touch, for example, we feel ticklish when a person merely pretends to tickle us without actually touching. An initial tickle makes the corresponding neural loops to be triggered. They will self-sustain even with a small subsequent touch inputs unlike non-looped representations of the initial tickle.

In addition to these examples, we have already discussed several others that fit well with the new model of memory in the previous chapter. In other words, we can say that looped dynamics is the only structure that fits naturally within a structured complex dynamical network like our brain for the definition of memory.

8.2   Dynamics versus statics

Dynamical firing pattern is critical for memory, not just the static structural loops or any other static structure. There are clearly several structural loops within the complex network of neurons in our brain. However, only a specific looped set of neurons that are simultaneously excited to form dynamical loops, for a given event, are responsible for its memory. The structural network of neurons is re-used for a large number of events. For example, a given event can excite 2 loops with, say, 10 neurons in a given structural network while a different event can excite 3 different loops with, say, 7 of the same neurons as before but an additional 10 new neurons within the same network. The requirement of dynamics for memory makes the structural network highly efficient and smaller while still having a large capacity for storage of events. Ignoring dynamics would not answer other features of our brain like our ability to perceive the passage of time, which will be discussed in chapter 40.

More fundamentally, without dynamics, the structural network is uniform and cannot capture the notion of meta-information (namely, *about* the memory) in an autonomous way. We would need meta-information to be stored elsewhere. We would also need another subsystem to keep track of it. For example, information like where is a given memory stored, what does the memory mean and how is it related to other events that either occurred at about the same time when this memory was stored or occurred at other times but related to this memory (cf. meta-information like the file allocation table in a computer) should be maintained and updated regularly. Keeping track of this meta-information for our adaptive brain network would be impossible if we did not have any dynamics within the definition of memory.

Another scenario in which the dynamics in the loops, the timing and the speed of firing pattern are important for memory is in a sleepy or a drunk state when your reflexes are slow (i.e., neuron firing rate is slower). Your recollection of an event is not good, even though the structural network of neurons, the connections and the static structural loops themselves did not change compared to an alert state. The timing and order of dynamical excitation is necessary in this case as is captured by the new SPL-based memory model. This contrasts with computer memory. With computer RAM or hard disks, the notion of storing or retrieving is static i.e., if the memory is created, then until it is overwritten, we consider it as memorized. We do not view the computer as retrieving an incorrect memory because of the time of the day or for some other time related events.

One may argue that this is a drawback with human memory. Can we not fix this issue, especially if we are designing synthetic memory? What is the advantage of making mistakes when storing and/or retrieving memory that we should include it as a feature in the new model as well? The point is that as the complexity grows, there will be valid situations when we do want to limit the amount of information that is processed. This is especially true if we want the synthetic system to exhibit a feature called free will. Part of the advantage of free will is to have choices and an ability to switch between these limited choices, as we will see later in chapter 28. This involves retrieving memories based on time and other states. The distinction between static network of connections and dynamical excitations are critical to maintain in a large-scale structured SPL system.

8.3   Robustness to variations

The SPL network model for memory implies that the exact neurons, the connections or the loops themselves need not be fixed for a long time. They can change over time as long as the cause and effect are unchanged. This gives rise to robustness in our memory from changes in internal neuron structure due to constant exposure to new learning experiences or due to damage either from accidents or diseases in the brain. The damages may cause you to forget the memory of certain events because of broken connections and discontinuities in the paths. Once you re-experience those events over a sustained period, the paths are re-routed at just the few broken locations to form continuous and connected looped paths once again. You can now recollect the entire past event effectively. There have been extreme scenarios where patients suffering from amnesia or patients in coma have had sudden recovery in some cases. These scenarios do appear to fit well within this memory model.

8.4   Increasing complexity generating simplicity

As the SPL network is fully trained with multiple training sets, the complexity in terms of the total number of nodes and the number of interconnections increases considerably. Yet, the complexity of recognizing a given image does not change because the dynamical loops are still maintained. In fact, several of the nodes generate output data quickly because they receive input data, which already exceeds the threshold value (see section 5.1). There is no need for these nodes to wait until sufficient input data is received. If the complexity of the SPL network is higher, the number of input connections for a given node is higher as well. As a result, a given node exceeds the threshold value quickly and, hence, causes faster propagation of input and output data within the SPL network. Problems such as image recognition becomes easier compared to when the network is less complex (see section 9.2).

Additionally, the accuracy of the image recognition solution, for example, increases with the complexity of SPL network. This is because we can ensure that (a) the set of dynamical loops are distinct for a large number of distinct images, thereby making it easy to recognize a large number of objects in a new image (dissimilarity condition) and (b) the set of dynamical loops have nonempty intersection for a large set of dynamical transformations of images, thereby making it easy to recognize the same image even under unexpected and complex but realistic variations (similarity condition). Satisfying both of these conditions only happens by increasing the set of dynamical loops. This surely increases the complexity of the SPL network.

Therefore, increasing the complexity of the SPL network makes image recognition (section 9.2), sound recognition (section 9.4), pressure or touch detection and other input detection problems easier. These other problems will not be discussed in this book though they are very similar to the image and sound recognition problems discussed in this chapter. This is a novel feature of the new memory model, unlike other existing models. With most existing models, the corresponding problems become difficult as the internal complexity increases.

9

Applications of Memory Model

The new memory model discussed in chapter 7 had several advantages over the conventional model for memory. We have seen several of these advantages in chapter 8 and in the previous chapter. In this chapter, I will mention a few other applications briefly to highlight the broad applicability of the new memory model. The concepts introduced like fixed sets and others will also become critical later when we study the problem of consciousness.

From a purely computational point of view, it is interesting to note that the problem of parallel computations become tractable when we use loops as a way to synchronize events. Otherwise, keeping track of every pathway becomes too difficult to manage. Instead, we need to just detect loops and study how to hop from one loop to another while building relationships between external events with an internal representation. This will start becoming clearer as we look at more applications in this chapter.

9.1   Distinguishing foreground from background

One important task for image recognition is to distinguish a foreground object from the background. I will now describe briefly how to use a trained SPL network of chapter 7 for this task (see, for example, Figure 7.12). This task is also important later when sensing three-dimensional space, which I will discuss in chapter 35.

1.   Consider an initial random SPL network tuned with the node parameters as specified in section 7.6.
2.   Pick a given training set of images $\mathcal{T} = \{A_1, A_2, ..., A_n\}$ like, say, the common household objects (tables, chairs and others).
3.   As explained in section 7.9, train the initial SPL network to satisfy similarity and dissimilarity conditions for each image from training set $\mathcal{T} = \{A_1, A_2, ..., A_n\}$, where A_i are distinct images for $i = 1, 2, ..., n$. Let the set of dynamical loops triggered by the trained SPL network be $\mathcal{L}_1, \mathcal{L}_2, ..., \mathcal{L}_n$ for each respective image $A_1, A_2, ..., A_n$.
4.   During the training phase, each image is individually presented onto the input surface within a specific region R (say, within a rectangular region in the center of the sensory surface like the fovea on the retina – see Figure 7.4). After training, if a new image A falls within the same region R, the trained SPL network would trigger a set of dynamical loops $\mathcal{L}$. We can compare $\mathcal{L}$ with $\mathcal{L}_1$, $\mathcal{L}_2, ..., \mathcal{L}_n$ corresponding to the images from the training set. If, however, the new image A falls partially or completely outside region R, the set of dynamical loops $\mathcal{L}$ triggered should not be compared with $\{\mathcal{L}_1, \mathcal{L}_2, ..., \mathcal{L}_n\}$. We need to move and transform the image A so it centers within the region R.
5.   Several approaches can be used to move and transform the image A. The most

common ones, in which we do not factor the specific characteristics of the image A, are either random movements or choosing specific curved paths to move and transform the image A. Other classes of approaches are those in which we use the characteristics of the image A itself. For example, we can trace along an edge that exists within the image A (provided the SPL network detects edges, which happens whenever an edge is within the training set). Another example is to trace across contrasting colors (after the SPL network is trained to detect contrasting colors). Each of these methods is easy to represent algorithmically.

6. The criterion for detecting a foreground object from the background can now be expressed as follows.

    a. Compute the cardinality $|\mathcal{L} \cap \mathcal{L}_i|$ for each of the sets $\mathcal{L} \cap \mathcal{L}_i$ for $i = 1, 2,$ ..., n as the image A moves and transforms according to any of the approaches specified in step 5.

    b. Somewhere along the path, the cardinality $|\mathcal{L} \cap \mathcal{L}_i|$ is maximum for each $i = 1, 2, ..., n$. This maximal location along the path is typically different for each image $A_1, A_2, ..., A_n$.

    c. If the maximal cardinality of $\mathcal{L} \cap \mathcal{L}_i$ when using a given image A_i is zero, the new image A under consideration is not similar to A_i. On the other hand, if, at any given time, the moved/transformed image A falls within the region R and is similar to one of the image A_i in the training set, then it is guaranteed that $|\mathcal{L} \cap \mathcal{L}_i|$ is maximal. This follows directly from the similarity condition for the trained SPL network.

    d. Compute the maximum value of the maximal cardinality for $i = 1, 2, ...,$ n, namely, $|\mathcal{L}_{\hat{\imath}}| = \max\limits_{i=1,2,...,n} |\mathcal{L} \cap \mathcal{L}_i|$ for some $\hat{\imath} \in \{1, 2, ..., n\}$. If this maximum value $|\mathcal{L}_{\hat{\imath}}|$ is zero, then the new image A is not within the training set. In this case, we treat the image A as the background even though it may contain an object O that we humans can identify. In other words, from the trained SPL network's point of view, this object O is undetectable and is, therefore, a background instead of a foreground object. If, on the other hand, the maximum value $|\mathcal{L}_{\hat{\imath}}|$ is not zero, then the new image A is a foreground object. Which object $A_1, A_2, ..., A_n$ within the training set is most similar to image A in this case? Clearly, the answer is $A_{\hat{\imath}}$ where $\hat{\imath} = \underset{i=1,2,...,n}{\operatorname{argmax}} |\mathcal{L} \cap \mathcal{L}_i|$, namely, the $\hat{\imath}$ for which the maximum above is attained. Similarly, the location along the path and the corresponding transformation for which the cardinality $|\mathcal{L} \cap \mathcal{L}_i|$ was maximum can also be identified.

7. We now have distinguished a foreground object from the background by moving and transforming the object while comparing the resulting dynamical loops with the fixed sets $\mathcal{L}_1, \mathcal{L}_2, ..., \mathcal{L}_n$ for the objects $A_1, A_2, ..., A_n$ in the training set $\mathcal{T}$.

The above algorithm has given us a way to refocus the new image onto the region R even if it was initially at a different location. This is analogous to turning our head or eyes to refocus the image to fall on the fovea (the central sharp vision region) of our retina i.e., we turn our eyes and/or our head to look 'directly' at the image.

This algorithm is especially useful for the general image recognition problem when we have a new image with multiple different objects in it. We would be able to detect each object within the image, one-at-a-time, analogous to how we scan different objects in an image one-at-a-time to recognize all of them eventually. I will discuss the general image recognition problem next.

9.2 Image recognition

In this section, I will discuss how to recognize images using an SPL network. The algorithm presented here can be generalized to address sound recognition, pressure or touch detection, detection of chemicals and other inputs using appropriate sensors as well, some of which I will discuss in the next few sections.

The problem of image recognition can be simply stated as follows. Given any image containing several objects, a trained SPL network is said to solve the image recognition problem if it is capable of correctly identifying each of the objects in the image. I will now describe how an SPL network solves this problem. Since SPL network is a learning system, it may not recognize all the objects correctly from a random image right away. As it is subject to different training inputs or normal everyday external inputs and as it continuously modifies its network in a specific way (as described in section 7.9), it starts to recognize more and more accurately. The steps to perform image recognition are as follows.

1. Pick a random SPL network as the initial network with one implementation, say, in a computer. Assume that the node parameters are tuned as specified in section 7.6 to avoid a fully active network.

2. Build a training set of images that we want our SPL network to recognize. For example, if we want to recognize everyday household objects, then we pick a training set of such images. If we want to recognize human faces, the training set will correspond to only a large collection of human faces. If it is just fingerprints, the training set will correspondingly include a large collection of them and so on. Other exemplary training sets are geometric shapes, different handwritten text, different symbols in a language (like Cantonese, Russian and others), different mathematical symbols, different components of a car, different electronic circuit symbols, sheet music symbols and so on.

3. As explained in section 7.9, train the initial SPL network to satisfy the similarity and the dissimilarity conditions for each image from our specific training set $\mathcal{T}$ = $\{A_1, A_2, ..., A_n\}$, where A_p are distinct images for $p = 1, 2, ..., n$. Figure 6.4 and Figure 6.5 picks common household images as an example of the training set to illustrate how image recognition works.

    When training the SPL network (Figure 9.1), we pick each image A_p ($p = 1, 2, ..., n$) individually to train it. First, we partition the training set $\mathcal{T}$ into k distinct subsets $S_1, S_2, ..., S_k$ as described in section 7.8. Each image A_p is in one of the subset S_i. Each subset S_i contains m_i similar images I_{ij} ($j = 1, 2, ..., m_i$) and two distinct subsets S_p and S_q for $p \neq q$ contains images that look considerably

191

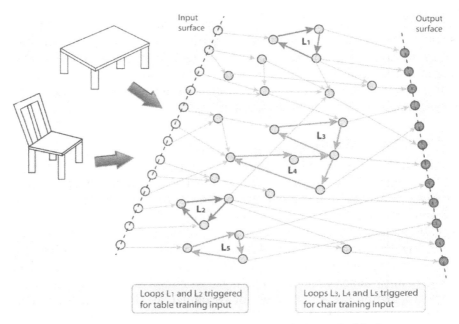

Input surface

Output surface

L_1

L_3

L_4

L_2

L_5

| Loops L_1 and L_2 triggered for table training input | Loops L_3, L_4 and L_5 triggered for chair training input |

Figure 9.1: An SPL network trained with simple household objects, as an example. At the end of training, we will have an SPL network that satisfies similarity and dissimilarity conditions for the training set $\mathcal{T}$.

different from each other. The trained SPL network has a collection of dynamical loops $\mathcal{L}_{ij}$ for each image I_{ij}, where $i = 1, 2, ..., k$ and $j = 1, 2, ..., m_i$. They are guaranteed to satisfy the following conditions:

    a. $\mathcal{L}_{ij} \neq \emptyset$ for $i = 1, 2, ..., k$ and $j = 1, 2, ..., m_i$. Note that there are a total of $m_1 + m_2 + ... + m_k$ nonempty conditions.

    b. $\bigcap_{j=1}^{m_i} \mathcal{L}_{ij} \overset{\text{def}}{=} \mathcal{L}_i \neq \emptyset$ for $i = 1, 2, ..., k$. This is the similarity condition (see section 7.8). $\mathcal{L}_i$ is termed as the fixed set for similar images represented by subset $\mathcal{S}_i$.

    c.   For each pair of collections $\mathcal{S}_i$ and $\mathcal{S}_j$ $(i \neq j)$ with fixed sets $\mathcal{L}_i$ and $\mathcal{L}_j$ respectively, there exists two sets of dynamical loops $\mathcal{P}_i \subseteq \mathcal{L}_i$ and $\mathcal{P}_j \subseteq \mathcal{L}_j$ such that $\mathcal{P}_i \cap \mathcal{P}_j = \emptyset$. This is the dissimilarity condition (see section 7.8 and Figure 7.6).

4.   When the trained SPL network satisfies all of the above conditions (Figure 9.1), it implies that (a) any two dissimilar images in the training set belonging to $\mathcal{S}_i$ and $\mathcal{S}_j$ $(i \neq j)$ are distinguishable from each other using the corresponding fixed sets $\mathcal{L}_i$ and $\mathcal{L}_j$ and (b) slight dynamical variations or transformations to each image belonging to a given subset $\mathcal{S}_i$ are recognized as the same image. In the example of Figure 9.1, this implies that a table and a chair are surely distinguished as distinct images based on the set of dynamical loops triggered. Specifically, dynamical loops L_1 and L_2 are used to identify a table while dynamical loops L_3, L_4 and L_5 are used to identify a chair. At the same time, a

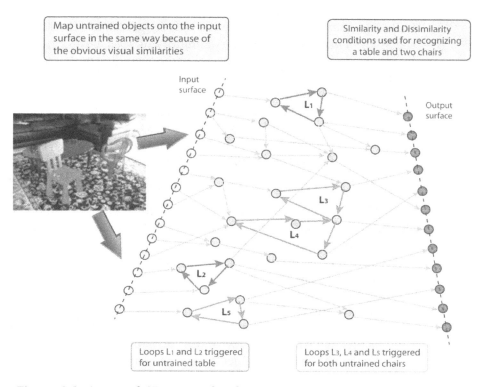

Figure 9.2: A trained SPL network subject to new input images. When a new image falls on the input surface, a set of loops will be triggered some of which correspond to the fixed sets from the training set. Using these loops, we can identify a fixed set from the training set that is most similar to the new image.

translated, rotated or zoomed table still produces the same set of dynamical loops (L_1 and L_2 in this case – Figure 6.4). This completes the training process.

5. Consider a new image composed of two or more images from the training set. Figure 9.2 shows such an example in which we have a table, two different chairs, a sofa and others. This new image is clearly not from the training set. Furthermore, the tables and the chairs in the new image, even if taken separately, are not from the training set as well. The SPL network in step (4) was trained for some representative tables and chairs, not the specific tables and chairs of the new image.

Therefore, the trained SPL network is said to have solved the image recognition problem in the limited sense (limited by the collection of training sets) if it can recognize the objects from the new image presented here (i.e., the table and the two distinctly different chairs in Figure 9.2).

6. When a new image falls on the input surface of the trained SPL network, the following steps occur for image recognition.
    a. The set of foreground objects are identified and distinguished from background objects, as detailed in section 9.1. In Figure 9.2, the set of

foreground images that the SPL network identifies correspond to the table and the two chairs. Since the sofa is not part of the training set, the SPL network is not capable of identifying it as a foreground object. It just treats it as a background object instead. For now, I will call the three foreground objects the SPL network has identified as objects O_a, O_b and O_c. The objective of the image recognition problem is to identify object O_a with a table and objects O_b and O_c to a chair.

b.   The dissimilarity condition satisfied for the SPL network for objects O_a and O_b (as well as object O_a and O_c) implies that both objects generate disjoint set of dynamical loops. Specifically, $\mathcal{P}_a \subseteq \mathcal{L}_a$, $\mathcal{P}_b \subseteq \mathcal{L}_b$ and $\mathcal{P}_c \subseteq \mathcal{L}_c$ correspond to dynamical loops for O_a, O_b and O_c respectively satisfying $\mathcal{P}_a \cap \mathcal{P}_b = \emptyset$ and $\mathcal{P}_a \cap \mathcal{P}_c = \emptyset$. To verify this, we first obtain the set of dynamical loops for each of the three foreground objects O_a, O_b and O_c (i.e., $\mathcal{L}_a$, $\mathcal{L}_b$ and $\mathcal{L}_c$ respectively). A simple set-comparison of $\mathcal{L}_a$, $\mathcal{L}_b$ and $\mathcal{L}_c$ implies that the objects $O_a \neq O_b$, objects $O_a \neq O_c$ and objects $O_b = O_c$. This is useful information already – we know that two objects (O_b and O_c) are the same and both of these are distinct from the third object (O_a).

c.   Next, we use the similarity condition to identify and associate each of these objects O_a, O_b and O_c to corresponding objects from the training set. For this, we compare $\mathcal{P}_a$ with the corresponding fixed sets defined in step 3(b) above for the objects from the training set i.e., with $\{\mathcal{L}_1, \mathcal{L}_2, ..., \mathcal{L}_k\}$. The set-comparison will show that $\mathcal{P}_a$ is a subset of $\mathcal{L}_i$ for some $i \in \{1, 2, ..., k\}$. In the more general case, $\mathcal{L}_a$ may be compared directly to $\mathcal{L}_i$. Here, $\mathcal{L}_a \cap \mathcal{L}_i$ will be a maximal nonempty set for some $i \in \{1, 2, ..., k\}$. The maximality can be computed in any number of ways. The simplest approach is to find the largest value for the total number of elements in the set. Therefore, we identify $\mathcal{L}_a$ (or $\mathcal{P}_a$) with $\mathcal{L}_i$, $\mathcal{L}_b$ (or $\mathcal{P}_b$) with $\mathcal{L}_j$ and $\mathcal{L}_c$ (or $\mathcal{P}_c$) with $\mathcal{L}_k$ for some $i, j, k \in \{1, 2, ..., k\}$.

d.   Since each fixed set $\{\mathcal{L}_1, \mathcal{L}_2, ..., \mathcal{L}_k\}$ from the training set is already identified with a real physical object like tables, chairs and others, we now know that $\mathcal{L}_a$, $\mathcal{L}_b$ and $\mathcal{L}_c$ (or $\mathcal{L}_i$, $\mathcal{L}_j$ and $\mathcal{L}_k$, respectively) are a table and two chairs respectively.

7.   The process of recognizing objects from any other new image is identical to the process described here.

Therefore, the above approach solves the image recognition problem for any new image. It is possible to pick several training sets when using images from our everyday objects. A given SPL network may be trained with more than one training set as well.

9.3   On dynamical transformations for recognition

In section 7.8, we mentioned the similarity condition involving dynamical transformations when training an SPL network. In the previous sections too, we have noticed the importance of dynamical transformations for image recognition problems as well. Here, I will give additional examples to suggest that invariance under a set of dynamical transformations is, in fact, common for most learning tasks. In fact, in chapter 14, we will use them once again (namely, as fixed sets) for creating a framework to study consciousness itself.

When you look at an image like a table or an alphabet like 'A', we recognize it even if we rotate, reflect, stretch and distort, draw or write in a reasonably messy way (maybe with poor drawing or handwriting skills). The set of symmetries that lets us recognize an image is broad. This is true even with sounds that have a correspondingly different set of symmetries. Different accents from non-native speakers constitute some examples of such variations. It is interesting to note that we do have other examples with touch as well. For example, if someone writes alphabets on the back of your body, you can recognize them with sufficient accuracy. These symmetries are translations, rotations and several spatial areas on your body.

Looped dynamics with our new memory model offers a unique perspective into why this is possible unlike non-looped and non-dynamical based memory models. The simplest explanation is that each of these dynamical transformations triggers the same set of loops. These loops are reached, typically, from multiple pathways. The similarity condition encapsulates this idea of adapting an SPL network to implement this feature. Besides, looped dynamics self-sustain longer than non-looped dynamical excitations of a neuron or at a synapse as claimed by the other models.

Once we have trained our brain (or an SPL network) to satisfy similarity condition, we are capable of detecting the object to be the same in spite of a large set of dynamical transformations. On the other hand, if you pick an image, for which the similarity condition is not satisfied yet, like, for example, an image you have never encountered before, you would not be able to identify that the transformed image is the same as a distorted version of the original image. We can verify this with examples like a modern day painting, ancient hieroglyphs, Cantonese alphabets to a non-Cantonese speaker, different terrains on different planets, certain cellular substructures pictured using electron microscope and others. If you see any of these images at one instant and then at a later instant show you a slightly transformed version of the same image, you would not be able to guess that both are the same, as you would with 'A' or a table.

In addition to spatial symmetries, our brain (or an SPL network) can be trained to satisfy similarity condition for time-based symmetries as well. For example, the sound input or the touch input on the back of your body can take, say, about a second. If the neurons that are excited as you trace an alphabet on your back does not stay excited as dynamical loops, the chances of recognizing the letter is diminished. This is inherently captured in our model as dynamics of sustained loops of excitation. Models that do not take dynamics into account (specifically as loops) cannot explain such temporal symmetries.

During early childhood, we satisfy similarity condition for a small set of transformations for each object or event we recognize. As we grow older, we continue to extend this set to a wider range of transformations. In chapters 13-14, I require the set of dynamical transformations as a fundamental feature when defining truths and fixed sets needed for understanding consciousness.

9.4 Sound and speech recognition

The problem of sound and speech recognition is almost identical to the image recognition problem described in the previous section if we use the SPL network model of chapter 7. This is unlike the case with traditional neural network or other machine learning algorithms. Let me outline these steps now.

1. Pick a random SPL network as the initial network with one implementation, say, in a computer. Assume that the node parameters are tuned as specified in section 7.6 to avoid a fully active network.

2. Build a training set of sounds and speech that we want our SPL network to recognize. Some examples of such training sets are sounds of words in a language (including from several different languages) and sounds of everyday objects.

3. As explained in section 7.9, train the initial SPL network to satisfy the similarity and the dissimilarity conditions for each sound from our specific training set T = $\{A_1, A_2, ..., A_n\}$, where A_p are distinct sounds for $p = 1, 2, ..., n$. The input surface for sound was described previously in section 7.7. As with image recognition, we could use multiple sound/speech training sets and train different specialized SPL networks.

    When training the SPL network, we pick each sound A_p ($p = 1, 2, ..., n$) individually to train it. First, we partition the training set T into k distinct subsets $S_1, S_2, ..., S_k$ as described in section 7.8. Each sound A_p is in one of the subset S_i. Each subset S_i contains m_i similar sounds I_{ij} ($j = 1, 2, ..., m_i$). These correspond to dynamical transformations like, say, different accents and voices (for example, differences in male, female and children voices) analogous to geometric transformations like translations and rotations for the case of images. Two distinct subsets S_p and S_q for $p \neq q$ contains sounds that sound considerably different from each other. The trained SPL network has a collection of dynamical loops $\mathcal{L}_{ij}$ for each sound I_{ij}, where $i = 1, 2, ..., k$ and $j = 1, 2, ..., m_i$. They are guaranteed to satisfy the following conditions:

    a. $\mathcal{L}_{ij} \neq \emptyset$ for $i = 1, 2, ..., k$ and $j = 1, 2, ..., m_i$. Note that there are a total of m_1 + m_2 + ... + m_k nonempty conditions.

    b. $\bigcap_{j=1}^{m_i} \mathcal{L}_{ij} \overset{\text{def}}{=} \mathcal{L}_i \neq \emptyset$ for $i = 1, 2, ..., k$. This is the similarity condition (see section 7.8). $\mathcal{L}_i$ is termed as the fixed set for similar sounds represented by subset S_i.

    c. For each pair of collections S_i and S_j ($i \neq j$) with fixed sets $\mathcal{L}_i$ and $\mathcal{L}_j$ respectively, there exists two sets of dynamical loops $\mathcal{P}_i \subseteq \mathcal{L}_i$ and $\mathcal{P}_j \subseteq \mathcal{L}_j$ such that $\mathcal{P}_i \cap \mathcal{P}_j = \emptyset$. This is the dissimilarity condition (see section 7.8).

4. When the trained SPL network satisfies all of the above conditions, it implies that (a) any two dissimilar sounds in the training set belonging to S_i and S_j ($i \neq j$) are distinguishable from each other using the corresponding fixed sets $\mathcal{L}_i$ and $\mathcal{L}_j$ and (b) slight dynamical variations or transformations to each sound

belonging to a given subset S_i are recognized as the same sound. This completes the training process.

5. Consider a new sound pattern composed of two or more sounds from the training set. An example is the speech pattern for a new sentence containing a few words from the training set. The trained SPL network is said to have solved the sound or speech recognition problem in the limited sense (limited by the collection of training sets) if it can recognize the sounds of words from the new sentence presented here.

6. When a new sound pattern reaches the sound input surface of the trained SPL network, the following steps occur for sound or speech recognition.

    a. Each sound or word in the sound pattern generates a set of dynamical loops in the trained SPL network. The set of similar sounds previously trained by the network are analogous to the foreground objects of image recognition. The remaining sounds are analogous to the background objects which cannot be recognized by the SPL network.

    b. The dissimilarity condition satisfied for the SPL network allows us to recognize each different sound or word in the sound pattern. For example, for distinct sounds O_a and O_b, the network generates different sets of dynamical loops $\mathcal{L}_a$ and $\mathcal{L}_b$ respectively. We can pick $\mathcal{P}_a \subseteq \mathcal{L}_a$ and $\mathcal{P}_b \subseteq \mathcal{L}_b$ such that $\mathcal{P}_a \cap \mathcal{P}_b = \emptyset$. A simple set-comparison of $\mathcal{L}_a$ and $\mathcal{L}_b$ implies that sounds $O_a \neq O_b$.

    c. Next, we use the similarity condition to identify and associate each of these sounds O_a and O_b to corresponding sounds from the training set. For this, we compare $\mathcal{P}_a$ with the corresponding fixed sets defined in step 3(b) above for the objects from the training set i.e., with $\{\mathcal{L}_1, \mathcal{L}_2, ..., \mathcal{L}_k\}$. The set-comparison will show that $\mathcal{P}_a$ is a subset of $\mathcal{L}_i$ for some $i \in \{1, 2, ..., k\}$. In the more general case, $\mathcal{L}_a$ may be compared directly to $\mathcal{L}_i$. Here, $\mathcal{L}_a \cap \mathcal{L}_i$ will be a maximal nonempty set for some $i \in \{1, 2, ..., k\}$. The maximality can be computed in any number of ways. The simplest approach is to find the largest value for the total number of elements in the set. Therefore, we identify $\mathcal{L}_a$ (or $\mathcal{P}_a$) with $\mathcal{L}_i$ and $\mathcal{L}_b$ (or $\mathcal{P}_b$) with $\mathcal{L}_j$ for some $i, j \in \{1, 2, ..., k\}$.

7. The process of recognizing sounds or words from any other new sentence is identical to the process described here.

Therefore, the above approach solves the sound recognition problem for any new sound or sentence. It is possible to pick several training sets when using sounds from our everyday objects. A given SPL network may be trained with more than one training set as well.

9.5   Associating images, sounds and words

In the previous sections, we have seen how a trained SPL network is capable of recognizing several images and sounds. We created a separate SPL network for each task. However, ideally, we would like to merge both SPL networks to create a single SPL network that performs both tasks. In section 7.10, we have already seen how to do this, namely, by satisfying the connectedness condition. In this section, let us look at how we can form other types of associations both within and between related images, sounds and words from a language.

Associating images to sounds and vice versa: Consider a new training set T with pairs of events of the types $(I_i \to S_i)$, $(I_i \leftarrow S_i)$ and $(I_i \leftrightarrow S_i)$ as described in section 7.11. Here $I_i \in T_1$, a training set of images and $S_i \in T_2$, a training set of sounds for $i = 1, 2, ..., n$. The pairs above like $(I_i \to S_i)$ is assumed to occur repeatedly and near-simultaneously in the external world. The directed arrow in the pair $(I_i \to S_i)$ suggests if image I_i occurs in the external world, then sound Si occurs within a short period.

Consider a trained SPL network N_1 for the training set T_1 as described in section 9.2 and another trained SPL network N_2 for the training set T_2 as described in section 9.4. It is now possible to follow the steps outlined in section 7.11 to merge both SPL networks into a new SPL network N that satisfies the connectedness condition for the training set T. The resulting SPL network N now associates several images to sounds and vice versa. For example, a car image would trigger the fixed sets corresponding to a car sound, a fan sound triggers fixed set for a fan image and so on. By creating a large training set T of near-simultaneous pairs of events, it is possible to capture true relationships between sounds and images that exist in the real world.

Associating images or sounds to words and vice versa: A word 'car' that refers to a visual image of a car has two representations. One is the image of the word car, say, in English (unrelated to the picture of a car) and the other is the sound of the word 'car'. The three representations (picture of a car, picture of the word 'car' and the sound of the word 'car') have different fixed sets, in general. They are also stored in different regions in the brain or in different regions within a trained SPL network.

Using the mechanisms described in sections 9.2 and 9.4, we can train three separate SPL networks, one for each representation. Furthermore, using the steps outlined in section 7.11, we can merge these SPL networks iteratively into a single SPL network that satisfies connectedness condition for three new training sets each emphasizing these linked pairs. In this specific case, we can train for all six directions of linked pairs – since triggering any one of car-image, car-word-image or car-word-sound fixed set would trigger the other two fixed sets.

Generating outputs from fixed sets: We can add a layer of 'output' nodes to make it easier for the SPL network to communicate with a conscious human being. Until now, the only way for us to know if the SPL network is correctly trained is to manually look at the excited nodes and dynamical loops in the network and perform a manual comparison with a previously identified list of dynamical loops (or fixed sets) for each trained element. This is tedious as the number of trained elements increase to cover several everyday objects and events. To automate the process, we can add 'output' nodes that when triggered converts a digital signal into an output signal understood by devices like, a computer monitor or a speaker. Such output transducers already exist. We just need a way to connect the output from our SPL network to these output transducers.

This is easy as well. Firstly, we add a new set of nodes as loops that, if triggered, generates a specific output. For example, we add new loops that if manually triggered generates a word 'car' on the computer screen. We add another set of loops that if manually triggered generates a 'car' sound on the speaker. For each trained object from our training sets, we could create new sets of loops and the corresponding set of output words or sounds on the monitor or speaker respectively. Secondly, we add pathways from the fixed sets for each of car-image, car-word-image and car-word-sound to the above two sets of output loops (i.e., monitor output nodes and speaker output nodes). The mechanisms to add these pathways are the same as those described in section 7.9.

After finishing these steps, when an the SPL network receives (a) a new sentence as input to the sound input surface or (b) a new picture to the image input surface, it outputs a corresponding sound or text on a monitor for each identifiable or trained component.

Associating images to other images, sounds to other sounds and words to other words: The connectedness condition can now be generalized to any pair of events that occur in the external world both repeatedly and near-simultaneously. It is no longer restricted to events from two different input surfaces (image and sound). For example, tables and chairs are almost always together. This is a relationship between two images that can be memorized by an SPL network by satisfying the connectedness condition between a table fixed set and a chair fixed set. Other examples include similar objects you find across most living rooms (or kitchens). With images, the number of such examples are enormous – everything we see around of us like inside the house, out on the street, in shops, in malls and so on. The connectedness condition when applied to such repeated and near-simultaneous occurrence of images lets us memorize several new relationships that exist in the external world. These relationships are not universally true. Yet, an SPL network can memorize them temporarily and later alter them if the configuration/relationships change considerably in the future.

The situation is same with pairs of sounds or pairs of words. Some sounds and/or topics occur together several times though they are not as common as images occurring together. The easiest examples are those that let us predict one sound or word when we hear another sound or word. The predictive ability is the result of memorizing the repeated occurrence of a set of events by satisfying the connectedness condition. The connectedness condition manifests itself as capturing a causal relation.

Memorizing long chain of events: Consider a trained SPL network that realizes image recognition, sound/speech recognition, associating images, sounds and words as above. It is now easy to modify the SPL network to memorize a long chain of events. We do this iteratively by satisfying the connectedness condition pair-wise until we form a long chain corresponding to an event composed of images, sounds and words.

9.6 Performing actions

One powerful application of the SPL network design presented here is that it can be used to perform actions in response to image, sound/speech or a textual input. Consider an SPL network trained to recognize images, sounds/speech and with associations to one another aligning them to the external reality (sections 9.1-9.5). If such an SPL network is capable of performing actions, it would exhibit features more powerful than an robot, as we will see. For a system to perform actions, we first need several actuators. These are, for example, motors connected via rigid bodies to produce motion. Humans have already discovered several devices, which we use them every day, capable of generating force and motion.

Let us pick one such device (like, say, a robot). A few exemplary actions include moving left, right, top, bottom, turning and others that are feasible with most mechanical systems. Other actions include an ability to draw on a screen using a set of motion patterns in a plane. Let us now create a new training set T based on the above collection of actions. Each element in the training set is meant to associate images, sounds, speech, words and long chain of events to the above set of actions.

Our objective is to pick output nodes with commands that make the mechanical device perform actions. To do this, the first step is to identify a set of digital commands that cause the machine to perform actions. For example, a microcontroller can be programmed to generate voltage signals in certain patterns. These patterns can make machines like robots, say, in an assembly line perform assembly of specific components of a car. A collection of such software programs can be made accessible by the SPL network, if the implementation of the SPL network is within a computer, say. We create special nodes as a representative implementations of these software programs.

The next step is to create a set of dynamical loops in our SPL network that can be designated as the fixed set for each action (like turning the device left or right) from the training set T. Of course, if we trigger these dynamical loops, the device does not yet perform any actions. We need to link the action fixed sets to the special nodes mentioned above designated for performing actions. For example, turning-left fixed set is linked to the special node that makes the device turn left and so on. Thus, if the output loops becomes active, the output signal from the linked special nodes is translated into a signal that triggers the digital command to produce a leftward movement.

The final step is to link the fixed sets for images, words and/or sounds to the action fixed sets using the mechanisms described in section 7.9. For example, if the input is a word called 'turn left', the output should be an action performing a left turn by the mechanical device. For this, when the fixed set for 'turn left' is triggered, we link it to the 'turn left' action based fixed set, which in turn triggers the special node to make the device turn left. Similarly, if the input is a sound called 'turn left', the output is once again the action of turning left. The same applies with all other action based elements from the training set T. In this manner, the SPL network can be trained to perform actions in response to sounds, images and words.

9.7 Self-learning – choices

Once the SPL network is trained to perform actions (section 9.6), it becomes possible for the system to choose which inputs (e.g., image input surface, sound input surface and others) to process rather than simply processing any input that falls on the input surface. This gives rise to the notion of self-learning and choices for the SPL network. Consider an SPL network to have subsystems capable of performing actions i.e., has motors and arms. Let us train this SPL network to recognize images, sounds, speech, with associations to one another aligned with the external reality (sections 9.1-9.5) and to perform actions (section 9.6).

Examples of actions that the SPL network can perform are turning left, right, top and bottom, moving forward and backward, push, pull and others. In section 9.6, we have seen how to associate each of these actions to a sound or words. For example, if the image is a table with food on it, let us assume that we have trained the SPL network to move forward. If the image is a wall, assume that the SPL network is trained to move backward. If the sound corresponds to 'turn left', the SPL network is trained to turn left and so on. We can now communicate with this SPL network using words and/or images and the system responds by performing the corresponding action. In other words, humans can command the system to do tasks. However, equipped with this ability, the SPL network can also perform these actions even if a conscious human does not ask it to do the task.

For example, if the system turns left, the inputs that fall on the image input surface (like a retina) is now different from what it was before. This makes the system process new images. If the processing triggers fixed sets for a wall, say, the SPL system will move backward since the system was previously trained to do so. When the SPL system moves backward, once again new images fall on the image input surface. If, for example, the new images contain a table with food on it, the SPL system would move towards that direction since it was previously trained to do so.

In this manner, the SPL system is capable of performing new actions as it processes new images caused by an action, which in turn makes it to process new images, perform new actions and so on iteratively. In general, the system processes them in the order of associative relationships memorized within the SPL network. This feedback between sensing and actuation allows the system to learn and memorize new facts or relationships that exist in the external world. For example, in the previous case, when the system identifies food on the table, it is likely that there is a drink present almost always. This correspondence between food and drink is probably not previously trained, in which case the SPL network has created a new association between food and drink fixed sets by satisfying connectedness condition. This is an example of self-learning by the SPL system. The sensory-actuator feedback loop has allowed the SPL system to learn several such relationships without the help of any conscious human.

As the number of such relationships stored in the SPL system increase tremendously, a notion of choice forms naturally within the system. At any given instant, the SPL system is now capable of detecting several images, sounds and words. Which of these should the SPL system process first and in what order? Each of these tasks appear as choices, though predictable and deterministic to a human initially. It is only when the number of detectable objects and the corresponding set of memorized relationships increase, it starts to appear as though the SPL system is deciding between one of many choices. This is a uniquely new feature that has emerged in an SPL system.

9.8 Predicting future events and actions

Consider an SPL system trained with several features as described in section 9.7. Such a system is now capable of predicting future events and actions in response to image, sound/speech or a textual input using the associative relationships stored as links between dynamical loops for a sequence of steps. These relationships align with the external events as discussed in section 9.6.

As the trained SPL network receives external inputs, the set of dynamical loops corresponding to the object in question is initially triggered. However, the existence of links between dynamical loops cause the triggering of several new dynamical loops automatically. These subsequent loops predict what the next likely object or event might be. The predictions may or may not be correct. If the actual external inputs do confirm the predictive loops, the corresponding nodes are already beyond the threshold state. This makes the excitation of subsequent predictions easier and faster.

In another example, sensory loops may be combined with action loops to produce interesting predictive features. For example, when a table image fixed set is triggered, the relationship links that was memorized through repetitive past scenarios will trigger action loops to make the input surface move (section 9.6) to where the chairs might be. The prediction need not necessarily be correct, but the system can move the input surface autonomously in a natural direction aligned with the external world. Later, in chapter 20, I will discuss the importance of predictions to the problem of attaining a feeling of knowing.

9.9 Creating a geometric map within our brain

Near-simultaneity of two events that occurs several times produces excitations in the brain that will be stored and linked to one another, according to our model (cf. connectedness condition of section 7.10). This is helped by brain regions like reticular activation system where sensory information from both eye and ear converges. This can be generalized as an ability to link one topic to another. You can learn new things through near-simultaneous occurrence of events. For example, ball can be associated easily to the fact that it rolls because this is how it is commonly observed in real life. Even seemingly disjoint topics can be related to one another because of the way they commonly occur in everyday life (cf. connectedness condition of section 7.10). Since the threshold voltages for two simultaneous tasks are already exceeded (as the corresponding neurons are already firing), further sustained excitations propagate the neuron firing patterns deeper into our brain. This will eventually converge to link the two repeated tasks.

The same connectedness criteria is applicable to sound generation and hearing as well. Different languages have unique sound patterns that are easily recognizable if you learn the language. For example, languages like Cantonese, German, Spanish and some Indian languages have unique pronunciation patterns. This gives rise to different accents when native speaker switch their language and talk in, say, English instead. Some sounds are difficult to generate while others are difficult to understand when heard by non-native speakers of that language even though it seems quite easy for the native speakers. The simultaneous occurrence of these sound patterns creates the necessary neural loops. They are sufficiently etched in your brain through repeated usage over most of your childhood that it is quite a challenge to change this structure (analogous to habitual patterns).

Furthermore, the feeling of continuity that we get in our everyday life relates to the creation of a conceptual "geometric map" of events or objects of things that occur in common scenarios. The notion of a distance between objects is initially established through near-simultaneity of occurrence of objects in nature. This happens at an early childhood just as edges and angles begin to be distinguished in our brain at this age (see Ornstein and Thompson [118]). In chapters 30-34, I will begin by showing how to create neural loops for simple geometric features. I will build on these neural loops and connectedness criteria through near-simultaneity to show how we are capable of sensing a three-dimensional world around. This space-time simultaneity creates the geometry that lets us quickly identify in-place versus out-of-place objects.

For example, an infant or a toddler starts to recognize the living room in his house and the relative location of objects like furniture (i.e., the geometric map) quite easily. On the other hand, it is not easy for him to recognize a similar living room in his parents' or a friends' house. This geometric map for an infant is not as evolved compared to an adult. We, as adults, do not notice the subtle differences in the arrangement of similar furniture and instead recognize the room as a living room.

9.10 Detecting similarity using storage proximity

The conventional models of memory do not address the following questions about memory. Is the memory for the alphabet 'A' closely located to 'B' and/or if it is farther from 'Z'? Is the lower-case alphabet 'l' stored close to number '1'? For example, the ASCII code in a computer for '1' is 49 and 'l' is 108. There is clearly no inference about the visual (geometric) similarity between these two symbols just by looking at their internal ASCII representation. Such a representation is, therefore, unsuitable for our brain. With a computer, there is a human to interpret that '1' and 'l' is indeed similar even though their corresponding ASCII codes lack this information. There is no one to interpret such similarity within our brain. The same applies with similar sounding words like 'right' and 'write', similar tasting foods, similar thoughts, similar concepts like in mathematics and others. As parents, we do not have to teach our children these similarities. They do learn this by themselves. Their brain has the correct architecture to discern this information. It would be serious defect if the memory model does not capture this.

Similarity and differences between objects is related to the proximity of storage representations. Otherwise, there is no easy way to identify the similarity by the system itself. The similarity information is always present within the external objects themselves. However, this information will be lost if it is not represented as memory within our brain network. A computer, for example, has lost the similarity information when it represented '1' and '1' as 49 and 108, respectively. In addition, we need to ensure that this relative geometry (of storage regions) is generated when we are, in fact, learning them serially. For example, let us assume that an infant has memorized and stored in his brain the number '1' in a reliably recognizable way after sufficient practice. If he starts to learn and memorize the lower-case alphabet 'l' next, how can we ensure that it will necessarily be stored physically close to the number '1', given that our brain is extremely complex?

In our new memory model, the neural loops (or fixed sets) generated for similar objects like '1' and 'l' are quite similar. The mechanism to achieve this in an SPL network was described in considerable detail as the similarity condition of section 7.8. Conceptually, we took advantage of the fact that the images on the retina are, in fact, almost identical resulting in a large common subset of neurons to be excited. Therefore, closer to the eye, neural firing pathways are similar. In section 7.8, I showed how to take this initial similarity to generate common neural loops for both symbols.

In reality, we do notice the errors from ambiguity in an infant, at least, initially. For example, he would say the sound for the number '1' instead of the letter 'l'. Over a period, these fixed sets deviate to produce different sounds for '1' and 'l' (see section 7.8 on dissimilarity condition). In this way, the memory regions for '1' and 'l' are physically located closely within our brain even though you learnt them at entirely different times in your life. The similarity is inherently detected by our brain, but the differences must either be taught by someone else or self-realized through additional context and experiences.

Here is another nontrivial generalization of how our memory model captures similarity and dissimilarity conditions of section 7.8. Consider the task of talking non-grammatically correct sentences while using valid words. If the storage representations were not close together and linked according to the connectedness criteria, it would be easy to do this. However, humans cannot talk such non-grammatical sentences easily compare to a computer. If you try to create a sentence using words from a few of the previous sentences, but placed in a random order to sound incoherent, you will observe that it is quite difficult to talk such sentences aloud. The storage representation, the structure of commonly occurring sentences, similarity of objects and events are closely coupled, making it difficult to change the order randomly.

9.11 Explosions

Let me briefly discuss a specific situation that can occur when we have multiple interconnected loops. The possibility is one in which our brain can generate an explosive set of excitations. Here, I am using the term 'explosion' to refer to a state in which small inputs produce large excitations simultaneously across multiple regions of our brain. Examples of such explosions are emotional experiences and realizations. I will discuss them in further detail in chapters 48 and 49.

Let us assume that we have encountered several disjoint topics in the past that have produced different pathways of loops. As an example, let us say that we have loops A_1, A_2, ..., A_n interconnected when memorizing topic 1, loops B_1, B_2, ..., B_k interconnected for topic 2 and loops C_1, C_2, ..., C_m interconnected for topic 3. The three topics could be, say, from geometry, algebra and calculus respectively. When working on a problem within each topic individually, only the corresponding subset of neural loops will fire. We may not be in a position to see the similarity between these seemingly disjoint topics. There are, however, special problems that do highlight the relationship between these topics. When we solve such special problems, it is possible that a few connections between A_1, A_2, ..., A_n and B_1, B_2, ..., B_k are formed.

A unique situation occurs when we add, say, three connections – one from A_n to B_1, B_k to C_1 and C_m to A_1 when solving a new problem. These three special connections result in a sudden explosion of excitations i.e., all loops A_1-A_n, B_1-B_k and C_1-C_m fire simultaneously even though we were thinking of just one topic, say, topic 1. This is a sudden sense of realization that geometry, algebra and calculus are related to one another. This is a unique state in which a large number of new loops are formed and/or excited with just a few number of new connections. In general, explosions are characterized in this manner – a small number of inputs, causing small structural changes, result in an enormously large set of sustained loops. Not only does this connect several disjoints topics together, it does so by producing an explosion. The energy of this explosion and the number of new neural loops that fire, as a result, are high enough to produce excitations at a global scale in the brain.

This explosion has a cascading effect that creates new connections, thereby linking new causes to effects. While memorizing a topic typically requires you to practice repeatedly, in this case explosions, the collection of excitations is so strong that you form new memories with just one experience alone. It is already well known that emotions and realizations let you memorize topics very easily even if you experienced it only once. In chapters 48 and 49, I will show how looped explosions form a basis for the 'mechanics' of human realizations and emotions. In contrast, consciousness does *not* occur through such looped explosions (see chapter 23).

10
Consciousness

Until now, our study of large-scale structured dynamical systems focused directly on SPL systems. We have seen how to create an infinite family of SPL systems and have seen how to use them to propose a new model for memory. As part of this, we have seen clear advantages of the new memory model over existing approaches. Starting from this chapter, the discussion will be of a slightly different flavor. We will still use SPL systems and many of the concepts introduced previously, but our eventual goal is to build a synthetic conscious system instead. This is not a simple task for several reasons. One of the main difficulties comes from a lack of formalizations to study the topic of consciousness rigorously and constructively. In comparison, for memory, taking a constructive approach was a bit easier, though we did take the extra time to include several features and applications of human memory into a new SPL systems-based model. The problem with consciousness is that the discussions very often gravitate towards philosophical implications with little, if any, constructive solutions. Specifically, we cannot use philosophical analysis to build a machine that can become self-aware, that exhibits free will, sense space or the passage of time.

Therefore, I will spend a significant portion of time just to extract concrete concepts from what seems like vague and philosophical topics, using examples from our everyday experiences. This extra step of translating from philosophical to concrete concepts was more or less absent in the previous chapters when we were discussing memory and other topics. As a result, you may find the discussion in the next few chapters to have a wider range of detail – from simple to interesting. It is important to be aware of this dichotomy as you read through the next several chapters. We will revisit some of the most fundamental concepts like truths, sensors, predictions, causality and others that we have already taken for granted and will start analyzing them from a constructive angle. But keep in mind that our goal is to extract rigorous concepts common across several examples so we can make progress in our understanding of such diverse topics like self-awareness, free will and others. The translation to rigorous concepts will involve, as we will see, SPL networks once again. The hope is that, at the end of this journey, we will have a deeper understanding of the detailed structure of consciousness and other related topics with an ability to build synthetic conscious systems as well.

Let me begin by listing several common problems related to consciousness that we want to address. Why do we even want to study the problem of consciousness? The existence of consciousness in human beings is the only reason why it is reasonable to study the very problem of consciousness. In fact, if not for consciousness, we would not even know that we are living in this world. We would not recognize a single event or object that exists around us, even when we are interacting with the external world every second of our lives. This seems a bit strange to visualize. Without consciousness, we are completely unaware of any object around us even though our eyes may be wide open and our other sense organs may be working perfectly. How does it feel to live and yet not know that you are alive? It is as if you are sleepwalking your entire life or you are under general anesthesia that is simply not wearing off.

This is not as fictional as it sounds. More than 3.9 billion years have passed since the existence of life on earth. Millions of complex life forms have existed and lived during this period. Yet, none of them were able to realize what nature had created the same way we do. Over this period, several events in the living and the nonliving world have repeated. Yet, generation after generation, the organisms that lived did not even realize that they were doing the same tasks repeatedly like eating, sleeping, hunting, walking around and others (at least the same way we realize). Until consciousness was discovered in human beings, the beauty of nature went completely unnoticed. If humans become extinct due to some unfortunate accident in the future, then life would once again continue its course, the mechanical way. No one would be present once again to understand and realize what humanity had created. The houses, the bridges, the roads, buildings, computers, machines and others that humans once created become lifeless obstacles to other species. How can we comprehend this?

10.1 The problem of consciousness

Consciousness is one of the most important features that nature had ever discovered on earth. Everything else that humans created is dwarfed relative to the discovery of consciousness. Yet, we take consciousness for granted. If someone says that '*there is nothing in this universe*', a conscious person would immediately recognize this statement as a contradiction. On the other hand, how would an unconscious person evaluate if the above statement is incorrect? How could he prove or disprove it? Consciousness is the only way to disprove this statement because we do see a world around us. It is not an illusion. It is as real as our very own existence. An unconscious person like a person sleepwalking or a person in coma does not know that you are standing next to him waiting for him to wake up. What is obvious to you that you exist, that the world exists, is not for him. In other words, it is both necessary and sufficient for him to become conscious once again in order to know that the world exists.

Consciousness has also turned out to be the most difficult feature to explain. Several scientists and philosophers have attempted to explain this over hundreds of years unsuccessfully. How can a dynamical system develop consciousness by itself? What types of dynamical systems can exhibit consciousness? Why do only certain types of dynamical systems exhibit consciousness but not others? How did it originate in humanity? Or, consider even a single human being. How did consciousness emerge in this person, especially since he was not conscious as a fertilized egg cell, during the development of a newborn or after birth (for about a year)? A single cell or its components are just made of molecules, which by themselves do not have the property of consciousness. Yet, as the human fertilized egg cell grows (consuming only physical molecules once again) to become an adult, he starts to exhibit consciousness. The transition from an unconscious fertilized human egg cell to a conscious adult over the years involved only specific arrangement of physical molecules. What is unique about this organization, both at a cellular level and within a human brain? Would constructing such an organized structure and dynamics in a synthetic system using electronic components, say, give rise to the same property of consciousness? Do we even need to build such a system before we can answer this question? Or can we analyze and answer such questions theoretically?

When we encounter such difficult problems, we usually look for analogies or examples to help us find a pattern. For consciousness, the only example we have is a human being. Unfortunately, our brain also happens to be the most complex organ we have ever known. It comes equipped with several additional features like intelligence, creativity, emotions and everything else we know about our brain. All these complex features cloud the problem of consciousness. Perhaps, complexity of a specific type is essential for consciousness. Nevertheless, we wish there is a simpler conscious system without some or all of these additional features to make it easier to study this problem. Unfortunately, we are not aware of any such systems. Is it even possible to find such 'simple' and yet conscious systems? In the next few chapters, I will clarify each of these issues.

When discussing consciousness, the problem of *qualia*, i.e., the subjective quality of experience, is generally referred to as the 'hard' problem. Emotions are well-known examples of *qualia*. I want to briefly highlight one subtle difference between the problems of consciousness and emotions here. In section 9.11, I suggested that the mechanics of emotions are 'explosions' within multiple interconnected loops (more details in chapter 49). This implies that most emotions are quick and intense when triggered and the intensity dies down just as quickly. Consciousness, on the other hand, is a very involved and gradual process that takes almost a year after birth to develop. After we become conscious, it stays intact for long periods (say, about 15 hours a day when you are awake) for the rest of our lives, unlike emotions, which die down quickly. Any attempt to explain consciousness will face a few challenges because of this slow path towards becoming conscious. During this year or more, which set of experiences are important and which are not, will become nontrivial to justify. In the next few chapters, I will show how these collection of experiences create a specific structure within our brain necessary and sufficient for consciousness.

Given the above issues (including subjectivity originating from *qualia*), what is an unambiguous definition for consciousness? Unfortunately, people refer to different aspects when discussing consciousness. Some of these are self-awareness, free will, ability to know the world around you, sensing three-dimensional space and the passage of time. In this chapter, I will briefly discuss the common issues with each of these concepts. I will address them in detail in later chapters.

The approach I will take to define consciousness and subsequently to propose a consistent theory is as follows. In chapter 11, I will first consider humans as the exemplary case for consciousness. I will identify those features within the broad topic of human consciousness that allow us to propose necessary and sufficient conditions. I call this subset as the **minimally conscious** state. The necessary and sufficient conditions that we propose will allow us to build a formal theory around the concept of a minimally conscious state. My next objective is to generalize this 'guaranteed' notion (because we have both necessary and sufficient conditions) of minimally conscious state to all systems, including other living beings and even nonliving systems. Using the theory built around minimally conscious state, I will then extend it to cover all other features commonly attributed to human consciousness like self-awareness, free will, sensing space, the passage of time and others.

To define 'minimality' with consciousness, we first need to agree that the concept of consciousness has a gradational representation. This can be seen by comparing the degree of consciousness across different species. For example, human consciousness is relatively higher than most other animal consciousness. Among animals, we notice that some species exhibit memory of a few local events and some others seem to understand even death. From these observations, most people tend to accept that consciousness, emotions and other unique human features are present in other species as well, but perhaps to a much lesser degree. Such a gradational representation of feelings and consciousness is considered the least contradictory from a logical point of view, even though there is only a limited, if at all, scientific verification.

Given such gradations, it is logical to suggest that the concept of a minimally conscious state is well defined. Minimally conscious state can be thought of as the lowest degree of consciousness below which the concept has no relation to consciousness. In fact, this is the reason why it is even possible to derive a formal theory for minimal consciousness. After expanding on the necessary and sufficient conditions for minimal consciousness, we will see in the later chapters that the resulting theory clarifies and lets us evaluate the degree of consciousness in rigorous terms for other living beings and, potentially, for synthetic systems as well.

Therefore, we can say that the above process has allowed us to pick any reasonable definition for consciousness (not just minimal consciousness considered here) and eventually translate it to the theory proposed here. Let me now begin by discussing each of the topics and issues commonly associated with consciousness.

10.2 Self-awareness

Self-awareness is a common feature associated with consciousness. It relates to a system's ability to know about its own existence. All known nonliving objects do not seem to have this property. From a synthesis perspective, imagine trying to create a system capable of knowing itself. If you attach a camera to a computer and let the camera look at itself, will the computer eventually know itself? Unfortunately, it is not that simple. For example, vision is not important for self-awareness (cf. the existence of conscious people who lack vision from birth). Our self-awareness does not seem to stem only from our sensors looking at ourselves externally. Our internal sensors and actions from within our own body seem critical for self-awareness.

We can analyze and identify all subsystems within our body that are critical for self-awareness. However, can this analysis answer how to build a synthetic system guaranteed to be self-aware? We should be able to start with a simple dynamical system and start adding components/features in an attempt to make it self-aware. Yet, this is not obvious even if we have identified the critical subsystems. We need answers to some fundamental questions. For example, would the front part of the system know that the back part or any attached limbs to be part of its own body? Does a single physically connected cell, which does survive and behave as a whole entity, know parts of its own cell? How about a colony of bacterial cells, a collection of cells in your brain, an entire plant, a dog, a human or a non-living system like a robot? How could some collections become self-aware but not others? If you were to assign awareness to any of these systems (including a human brain), to which part would you do so – its front, its back, its middle or some combination of them.

Answers to these questions are critical before we can build synthetic conscious systems. Theorems 23.1 and 23.2 in chapter 23 provides answers to these questions in precise terms for any system – living and non-living.

10.3 Ability to know

We use the word 'know' in a number of situations. We use it when talking about tangible objects (like "do you know this person or this object?"), about abstractions (like "do you know algebra?") or about actions (like "do you know how to play a piano?"). What does *knowing* something mean? When can you say that you know what you were observing, learning or doing? Can a synthetic system know anything?

There are times when you clearly feel you know that you do not know. For example, sometimes you see a person on the street and you think you know him, but you just cannot remember his name. As you are struggling to remember the context, how can you *know* that you do not know? Later, when you do remember his name as well as the entire context, you have a 'feeling of knowing' for who he is and also a feeling that you know that you know him. How did recollecting a few events produce such a drastic change? Similarly, is there a case when you do not know that you know? For example, sometimes when we do a task, we are surprised at how well we did it, especially because we feel we have never learnt it consciously.

In this book, I consider 'knowing' as the most fundamental aspect of consciousness. In fact, I treat knowing to be part of the *minimal consciousness* notion introduced earlier. This is the minimal feature that should be present in any system if we want to claim it as a conscious system. Theorem 22.1 of chapter 22 presents necessary and sufficient conditions for attaining a feeling of knowing.

It is easy to see that knowing is a more general concept encapsulating consciousness itself. For example, knowing yourself is self-awareness. Knowing that you have a choice and picking one while obeying seemingly arbitrary laws is free will. In addition, we know the existence of an external world around us. We know that there is space and we sense the passage of time. We know a number of geometric properties like angles, edges, shapes, sizes and distances. We perceive the world in a cohesive manner. We know dynamical properties like motion, force and mass. In general, for any feature related to consciousness, there is an aspect of attaining a feeling of knowing for the feature beyond understanding the mechanics of operation. If we exclude the feeling of knowing for the above features, the system behaves no different from a nonliving system (or a machine).

10.4 Free will

When a human is presented with a number of choices, say, for food, we know that an external observer cannot predict which item he is going to pick first to eat, which one next and so on. Taking a probabilistic approach, say, by conducting a survey across several people does not let us predict what a specific person would do. Can any other model or a set of equations capture his choices, which are not by themselves random? Or, is there an inherent uncertainty at an atomic or molecular level that results in a net unpredictability? What appears strange is that there is a 100% certainty for the decision maker himself. He also seems to have full control over his decisions.

What is the cause of this inconsistency in predictions between the two people involved? Is it because the observer does not know what is happening in this person's brain? If so, what information should we provide to the observer? Should we provide all details at the lowest level like the states of individual neurons and all of their interconnections? Or, should we just provide details at a higher level like the list of all past decisions and events for a finite period, say, the past 10 years? However, this person himself does not know details at both these levels. Yet, he is not only picking the item he wants with 100% certainty, but he also knows that he is making a choice. This unique ability that humans have acquired is free will.

How can we build a system in which individual components obey the standard laws of nature but the system as a whole gets to obey seemingly arbitrary laws (at least, to a degree)? In chapters 26-28, I will present the root cause of free will.

10.5 Sensing space and the passage of time

We rarely consider the question of existence of space and the passage of time when working with everyday problems even though we fundamentally rely on them. Are these notions real or are they just a representation in our brain? To compare, let us first look at a few similar examples. We know that color, for example, is not a property of the object. A green leaf does not have the 'appearance' of green in it. What feels green to you like the brightness, the intensity of the color, the contrast and so on – it is even difficult to express this feeling in words even though you and I know what we are talking about – is it the same for both you and me? Clearly, this is not the same for you and a color blind person. Similarly, when you clap your hands, no 'sound' is produced. Rather, air molecules just compress and expand at certain frequencies, which we 'perceive' as sound. The same applies to taste, odor and touch. All of these are just representations within our brain. When I have internal thoughts within myself, even though I am not making any audible sound, I still do 'hear' my own voice, which feels identical to an external sound.

Is this the same situation with space and time as well? How does a given system know that there is a world outside itself? What representation of memory and internal brain structure allows this? What internal structure is needed to capture the geometry of space, its emptiness and then let the system know about the existence of this geometry? Emptiness in space itself does not trigger any signals on the sensory organs (both eye and skin). Then, how does our brain detect empty space in the *XY-*

plane, not just depth, but between objects even when we close one eye? The information that the 2-D retina surface receives at any given instant does not have empty gaps in the *XY*-plane. This is because the background objects behind the emptiness would indeed send light signals to excite 'all' neurons on the entire 2-D retina surface (albeit in a less focused manner). Even if you have one eye or if you are completely blind at birth, our brain is still capable of creating a consistent view with spatial emptiness. More importantly, how are you able to know about this emptiness when you are conscious? With unconscious or less conscious states like sleepwalking or severely drunk, you do not 'know' about the emptiness even though you can move around avoiding obstacles. A robot can be programmed to avoid obstacles in a similar way. However, would it 'know' about the emptiness or is it analogous to an unconscious sleepwalking state?

When you close one eye or when you look at distant stars (or distant airplanes), the perception becomes two-dimensional. We never sense the third dimension, even though the depth is quite large (i.e., the depth distance between stars is too high and yet we do not perceive this depth at all). Similarly, there are instances when you lose your sense of direction of, say, left or right analogous to loss of depth perception – like in an ocean with no visible land in any direction until we pick a reference point. Can a synthetic system create a three-dimensional spatial representation too?

With time, even when we are sitting idle, we do sense the passage of time. If we do the same task repeatedly, we know that we are repeating the task, unlike, say, a computer. There are other situations when we feel time passing by either slowly or very quickly. For example, when you are waiting to meet someone, time seems to pass slowly. On the other hand, when you are enjoying, say, a party, time seems to pass quickly. What is the source of these differences? How should we best represent time in our brain to explain each of these scenarios?

More generally, if a dynamical system obeys and follows the laws of nature, how can it look *outside* of its raw dynamics and know *about* its dynamics? For example, how do we know *about* time and the passage of time just from observing and interacting with time-varying dynamics? A machine simply follows the laws of physics. How can it know about the laws of physics, including knowing about the passage of time? It is as if the system is observing the very flow of dynamics by stepping out of the external world. Since not all collections or arrangements of dynamical systems allow such a possibility, what is unique about our brain's internal interconnected structure to make this happen?

10.6 Cohesive perception

The world around us appears cohesive and continuous whenever we are conscious. There are no holes or patches even as we close our eyes intermittently, blink or turn around. How does consciousness alone create such a cohesive perception for a given system, synthetic or natural? Consider a few related questions. The distribution of red, green and blue sensitive photo-pigments among cone cells is uneven – 75% red, 20% green and 5% blue (see Roorda and Williams [131]) – and non-uniform (in a single eye and both eyes). Yet, red, green and blue rectangles have identical edges and appearance (cf. colored rectangles on a computer screen if it has a similar non-uniform distribution of RGB). How does our brain manage to create such an identical visual perception?

How does a 1-inch sized retina correctly represent objects of different sizes (even 10 feet tall) when the focal length of our eyes are constantly changing? What subsequent chemical reactions and neural firing patterns starting from the 1-inch retina are capable of estimating and projecting an image that feels like 10 feet tall, which we can verify by taking measurements as well? To compare with a synthetic system, consider one with a camera taking pictures of sizes much larger that the size of its film. How does such a system know the difference in size relative to its internal representation, which is a maximum of a few inches at all times?

As another example, when we change our prescription glasses, our perception of sizes and distances is altered. Temporarily, we feel that objects appear farther than they really are when our short-sighted power increases and we wear new prescription glasses. How does our brain recalibrate itself within a few days? As an adaptive system, our brain continuously rewires its connections. How does it recalibrate to keep memories and experiences consistent and identical? For example, the perception of a car remains the same even after 10 years? If a sensor like a thermometer were constantly changing its internal structure like our brain, how would it recalibrate to continue measuring the same temperature? What mechanisms make this possible when you have millions of neurons rewiring through new experiences?

When interacting with external world, how do we move our hand precisely to the left or to the right without ever knowing the correct set of motor neurons and the correct order to activate them? Until our death, we are completely unaware of and have no access to the internal chemical sequence of steps to perform any action, even though we are capable of performing action itself with high precision. In comparison, a robot requires detailed internal knowledge to control its movements. Our control mechanisms seem to be from 'outside' the system (say, by evaluating our position and momentum relative to the external world) even though the underlying power sources and actuators are 'inside' our body. What internal structure is needed to make this happen? I will discuss each of these in the later chapters 29-41.

10.7 Uniqueness of consciousness across all humans

Several billion humans lived or are currently living on earth. It is impossible to have two humans with the same internal brain structure for any finite period. There is also no known substructure within our brain which, if present, guarantees consciousness. Instead, there are only structures, if absent, cannot give rise to consciousness. Yet, the property of consciousness is identical in all human beings. This is the same with most feelings and emotions as well (not what caused the emotion, but the emotion itself). How can we explain this interesting fact? Surely, there are different degrees of consciousness like with hypnosis, sleepwalking, schizophrenia and others. Yet, irrespective of their internal brain structure, how do all humans achieve the same conscious state? Notice that raw complexity of a brain is not enough to explain this.

This is because a human and an animal, say, do not have the same degree of consciousness and other feelings. We need to be explicitly clear on what type of structure within a complex system leads to consciousness. Moreover, we need to show how every human being, but not other species, creates this structure. Then, we should show that this abstract structure is 'identical', in some sense, among all human beings, even though the physical (both static and dynamical) structure, as a network of neurons, is widely different.

If we were to build a synthetic system with the same abstract structure, will it also attain this unique property called consciousness? Will it be identical to human consciousness if we build the system using inorganic chemicals?

10.8 Using consciousness to evaluate consciousness

The more we know about consciousness, the more we can use it to eliminate or analyze whether another human being or a synthetic system is conscious or not. For example, our free will can be used to force a relationship between two seemingly independent events. Let me discuss this briefly here.

One of the most important ways to use our own consciousness to evaluate consciousness in another system (living or synthetic) is to use our free will in an effective way. Let us say that we are communicating with a system externally and using it to determine if this system is conscious or not. Since I am conscious and I am setting up conditions at-will, only I know ahead of time whether the answers to my questions are true or false. It is possible to pose a large set of questions with no *apriori* answers. For example, some of the questions for which only I know the answers are pointing at my hand and asking how many fingers I raised, making my hand move left or right and so on. I can choose at-will whether to raise two, three or more fingers. Similarly, I can ask whether I am touching an object or not, but choose at-will whether to actually touch it.

In the above type of examples, the external system is forced to use multiple sources of information (like touch, vision and hearing) simultaneously to guess the answer. What I show, what I say, what the system senses (by seeing or hearing), what it responds back as an output and my ability to evaluate its answer are all linked together because of my free will. I can build a large set of interlinked questions, facts, actions and events for which only I know the answers (true or false) ahead of time because of my free will. If the answers from the system reveal that any of these links are broken, I will have a way to detect them.

This structure within the truths is abstract and invisible. Our universe has a considerable amount of additional such structure as well. We constantly keep mapping it within our brain as we grow and gain new experiences. The solution to the problem of consciousness that will be discussed in this chapter will outline such detailed structure to help with this problem. Even though this approach is not sufficient, the structure among the truths imposes severe restrictions on systems that claim to be conscious. We can, at least, use this approach to quickly eliminate rather than confirm whether an external system is conscious or not.

10.9 The problem of *qualia*

In the previous section, I have suggested one approach to evaluate whether an external system (living or synthetic) is conscious or not, namely, by elimination rather than by confirmation. Using the above approach, we, effectively, show that evaluating consciousness in a synthetic system is as hard as evaluating it for another human being. Let us now address the same issues in a more direct way, i.e., if I have created a system and claimed that it is conscious, how can we evaluate if my claim is correct or not? In order to answer this question directly, we need a definition of equality of 'features' of complex systems or, in general, equality of two complex systems. For example, is the consciousness in a system *equal* to the consciousness in a human being?

One simple definition of equality for complex systems is point-wise equality. Here, we use one-one correspondence between molecules for both systems. We start with identical physical arrangements at a given instant and setup identical external conditions as well. This will kick off similar dynamics in both systems. With this definition of equality, if two people are equal and one person is conscious, then we can conclude (or, more precisely, assume) that the other person will be conscious as well. The point-wise equality, unfortunately, is not a useful definition. It is too idealized, impractical to use and is even theoretically impossible to validate. Besides, any two existing normal human beings who are already conscious do not obey point-wise equality. In fact, this is true with not just consciousness, but with every higher human abilities like emotions, visual perception and other feelings. No two human beings ever have identical brain structures. Therefore, we need an alternate definition of equality with, at least, the following two properties.

(i) *Equal under comparison*: If two normal human beings already exhibit a given feature, then our definition of equality should show that the features are equal for them as well.

(ii) *Unequal under comparison*: If a human being and another system like an animal differ considerably with respect to exhibiting a given feature, then our definition of equality should show that the features are not equal for them as well.

If we have such a definition of equality, then is it reasonable to use it to compare, say, the consciousness in a human being and consciousness in a synthetic system? We can use the equality to identify (a) when the consciousness in two different normal human beings is similar, (b) when the consciousness in two human beings in different conscious states (like dreaming or coma) is different and (c) when the consciousness in a human being and another animal is different. If this definition of equality is satisfactory, we can use it for every other feature like emotions, appearance of colors and other feelings as well.

In general, there are several possible ways to define such an equality to meet these conditions. In section 44.2, I will describe one such definition. Let us classify the set of all features for which we want to apply the above equality into two broad groups. The first set is the physically measurable or detectable features (like mass, charge, shapes or forces). The second set is purely subjective features, commonly referred to as *qualia*. Examples of this are emotions, consciousness and other feelings, in general. The above equality is particularly difficult to evaluate for the latter subjective set i.e., for *qualia*. For example, it is easier to show that two systems have equal shapes than equal feelings like fear.

Let me discuss this problem of *qualia* by picking one example here – the visual appearance of color. When I look at a red or a blue colored circle, it appears a certain way to me. How can I know if these colors appear the same way to you as well? In fact, how can I even describe to you how a red color appears to me? The first step is for both of us to agree on the names of the different colors, distinguish two different colors as different, identify similar colors as same and specify which new color is produced when different colors are mixed together. The information obtained through these experiments are, what I call, purely relative information. On the other hand, the precise appearance of a given color to a given individual is, what I call, the absolute information or the *qualia* of the color. Therefore, in spite of these agreements, how do I know if you and I perceive the *qualia* of a given color the same way? If, at any instant, we do not agree on a color for a given object, we try to first identify a 'physical' cause for this. The assumption is that there should be something physically different to produce different qualia. This is termed as *physicalism*. The problem, however, is that no matter how many discrepancies we try to address in this way, it never gives any information about the *qualia* of color itself.

Secondly, the *qualia* of color is not a property of the object. It is a property emerging from entirely within an individual's brain. In the future, it may be possible to identify specific regions in the brain that is responsible for the *qualia* of color. Unfortunately, such a physical identification does not help me know how a red color appears to you. How can I describe to you how a red color appears to me? Whenever I try to describe the appearance of color (like it is bright and has a certain shade), I am forced to use physical phenomena. When I do this, I am only identifying red color relative to other colors. I am not describing the absolute quality or *qualia* of red itself.

For example, there are tetrachromats, especially among women, who have four-color pigments in their eye instead of three like most of us. If I take a group of objects and group them based on same colors according to me, the tetrachromats will, most likely, not agree that all the objects in, say, one group are of the same color. While the disagreement can be identified to a physical cause within their brain, is their appearance of red color for them different from mine? Even if it is, how can we ever know what that difference is? The only information acquired from the above experiments is relative similarities and differences, which provides no information about the *qualia* itself.

This problem with color is applicable to any *qualia* including emotions, consciousness and other feelings. This is a serious limitation with *qualia*. It is not merely a practical limitation. The root cause is in the very definition of *qualia* itself. *Qualia* is, by definition, associated to a quality that is not directly related to any physical phenomena. Given this theoretical limitation, it is necessary to have a suitable definition of 'equality' (as discussed above) that can still be used to determine when the *qualia* in two different systems are the same. This will be discussed in chapter 44.

With a new definition of equality, we are not avoiding the problem of *qualia*. Rather, we are separating the problem of absolute information present in *qualia* from the relative information.

10.10 Predicting future decisions

If we know how our brain and our consciousness works, can we use this knowledge to predict future actions and decisions of a given person? Generalizing this, can I know your future decisions as well as the entire society's future? Excluding practical or computational limitations, are there any theoretical limitations? Is it possible to improve the accuracy of predictions over time to any desired level? Since conscious beings have free will, we believe that there is a way to alter our future and falsify such predictions. We think that we can use our free will to formulate and control our choices. Besides, we feel that these choices themselves are entirely private to us i.e., are completely unobservable to an external system. We want to know if this is truly possible or if such control and unobservability with free will is just an illusion. In section 44.5, I will discuss these topics in some detail.

10.11 Is consciousness genetic or learnt?

When we raise a child, we teach him several facts and qualities from science, art, music and others. However, we never teach them self-awareness, free will, sensing space and passage of time. We do not have to teach these qualities even though they are not born with consciousness (cf. a fertilized egg and a growing fetus not having consciousness yet). They seem to become conscious automatically. We only explain the existence of such concepts, what they mean and the words used to describe them by linking them through analogies so we can all agree on the topics and discuss them in a meaningful way. With examples like how to play a piano, we are afraid that a child might never learn it if we do not teach him. Are we not afraid that he may never become conscious if we do not teach him?

Philosophers have created imaginary systems called philosophical zombies. This zombie is assumed to do everything identical to a human except exhibit consciousness. Are we not worried about generalizing this notion to a human child? His brain is a complex system with no consciousness at birth. Is it possible that he grows imitating everything a normal adult does, except exhibiting consciousness? If not, what is unique about his brain architecture that avoids this case?

The other possibility is that we consider consciousness to be genetically inherited. In this book, we are not only interested in addressing the problem of consciousness, but we want to explain the origin and evolution of life as well. Therefore, even if the creation of human consciousness now has a genetic component, there should be a period when consciousness did not exist in early humans. How did the evolutionary transition from no consciousness to consciousness occur thousands of years ago? The transition cannot be gradual or incremental.

For example, how can we have a stage for consciousness of a system that exhibits only 40% of self-awareness, half passage of time or only 'partial' free will? What does 40% self-aware even mean? What does making improvements to this 40% self-aware system mean so it improves to get 50%, 60% all the way to 100% self-aware? Does it even make sense to have incremental improvements for such concepts? For problems like finding a cure for cancer, an incremental approach is indeed natural. Over time, we keep improving our techniques to eventually find the cure for cancer. The existence of *qualia* for consciousness, however, makes the notion of a quantitative error, and, hence, an incremental approach, unreasonable.

Our objective in the later chapters is to discover the structure and the process of becoming conscious both now and from an evolutionary point of view.

10.12 Transition to consciousness

Consider a human fertilized egg cell or a growing embryo. While there is some debate on whether to call it as alive, it is certainly unreasonable to suggest that a growing embryo is conscious to a degree comparable to adult consciousness. As a matter of fact, it is generally accepted that a human becomes conscious to a degree comparable to an adult (with features like self-awareness, free will, sensing space, sensing passage of time and others mentioned previously) only after about a year after birth. Therefore, when viewed as a physical system, a human has transitioned from a non-conscious fertilized egg stage to a fully conscious adult human. During this transition, only physical and chemical processes have occurred like food consumption, growth and replication i.e., nothing meta-physical was introduced.

If we compare the structure of a human brain before and after the above mentioned transition, we notice that there is an increase in structured complexity of the brain network. This is easy to understand if we compare the number of memories of objects and events stored in an adult relative to a newborn. This increase in structured complexity of our brain occurs throughout our lifetime. What is unique is that the improvement occurs even for an infant when the initial structure is the least structured. Each of the improvements require considerable amount of energy used in a directed manner over several months. What are the mechanisms that ensure a continuous improvement in our brain structure over long periods of time?

Why would a system (nonliving or living) adapt by itself to improve its internal structure? We rarely, if at all, see nonliving systems that create order from disorder. While this is a violation of the second law of thermodynamics only for a closed system, an open system that has an internal source of energy allows such a possibility. Yet, we have never seen nonliving systems evolve in a direction of creation of order in which the complexity is at least as high as that of a single cellular organism even if it has access to large amounts of external energy.

The problem of transitioning to consciousness for the first time for each human being requires the creation of an enormous amount of order within his brain network (cf. the problem of origin of life in chapter 52 that creates an enormous amount of order within a single cell). For example, when you are born, you do not have an ability to focus your eye lens so the image falls correctly on the retina instead of in front or behind. Over a period, you automatically learn this ability. We are never afraid that our newborn son or daughter will not learn such abilities and will not become conscious even though we are not helping directly.

One of the primary objectives of the new theory of consciousness is to explain how the above-mentioned transition has occurred and how it is occurring in every individual even today. I will provide mechanisms that allow a system to transition to consciousness by itself i.e., without any help from other conscious beings.

10.13 Mechanisms with versus without consciousness

When humans want to build any system, we first use our consciousness to help decide 'what' to build. Once we define this objective, we look for a solution. The final system built would not have an ability to know what it is doing. Only we conscious humans know what the system was meant to do. The system itself simply obeys the physical laws and performs what we designed it for. For example, once we feel the need to record family events, say, we think of various ways to build a 'camera'. The camera we build simply takes pictures because of its design and dynamics. The camera does not have any internal mechanisms or abilities that allows it to know that it is taking pictures – only we conscious humans know this. Besides, the camera does not take pictures on its own – only we do. Furthermore, in this case, we have created order (a camera) from disorder (basic raw materials scattered around). The direction of creating order was guided and maintained for a considerable amount of time by our consciousness, which is clearly external to the system being created (the camera).

The designs and mechanisms for almost all man-made systems like computers, fans, televisions, telephones and others follow the same pattern. Let me highlight a few issues from this discussion: (a) choosing and setting the objective happens outside the system achieving the objective, (b) the mechanisms needed to build the system, after we find a solution to the objective picked earlier, require help from external conscious humans, (c) the system needs help from conscious beings to operate after it is built and does not operate on its own and (d) the creation of order from disorder happens over a long period of time with help from conscious humans.

The mechanisms needed to build almost all man-made systems require conscious beings. If conscious beings disappear, so will the construction of all man-made systems, even if we have already setup an assembly line. The assembly line and the machines involved do not have mechanisms to find the raw materials they need to build themselves. No new machines or designs will be invented while building existing machines will cease. In other words, the solutions (i.e., designs and mechanisms) proposed until now for various man-made systems rely heavily on conscious beings. This is a problem when building a conscious system for the first time (as well as when creating life for the first time i.e., the problem of origin of life, which I will discuss starting from chapter 52).

If the construction of almost any synthetic system require consciousness, how can a conscious system be created for the first time? To build a conscious system for the first time, we can only rely on mechanisms that do *not* require consciousness. What are these mechanisms? Each of the issues (a)-(d) mentioned above becomes challenging. What to build i.e., the eyes, the ears, how they are interconnected and processed within the brain and other structures within the brain, all of which seem to have a definite purpose needs to be encoded within the system. If someone else is responsible for encoding this, who creates him? A mere survival advantage cannot explain the creation of features for a single individual – like, for example, an infant learns to auto-focus his eye lens to make images fall precisely on the retina, he learns to track a moving object, he learns to sit, stand and walk, he learns to turn towards a source of sound and others. He was not able to perform any of these tasks at birth.

The problem of constructing a system meeting a certain objective has multiple solutions, designs or mechanisms, in general. For most man-made systems, we choose only those solutions that makes the system efficient, reliable, durable or robust. These solutions almost always require consciousness since we need to pick the correct materials and process them in a correct manner to achieve the desired properties and efficiencies (cf. the processes followed when creating a transistor). Unfortunately, for the problem of creating a conscious system for the first time by nature itself, the same criteria of efficiency and others cannot be applied. We have to sacrifice several features simply to avoid *any* dependency on other conscious beings.

For example, if nature wants to build something similar to a camera with an auto-focus mechanism (i.e., an eye), it needs a design as well as an ability to specify and work towards this objective without any consciousness. None of our existing man-made designs for a camera would work because they require consciousness. A newborn does not have an ability to auto-focus his eye lens so objects at any distance would form images precisely on the retina. Rather, he learns this ability over a period of several months. This and any other learning abilities are rarely considered genetic. Since he is not conscious yet (to the same degree as an adult), he has no awareness of why it is important or worthwhile to learn this ability by himself. Yet, almost all humans (and even all other animals with eyes) learn to auto-focus quite naturally. What are the mechanisms that let them move in the direction of increasing order and specifically to auto-focus their eye lens?

Note that not all learning abilities for a human require us to specify mechanisms without consciousness. For example, our ability to learn mathematics is a directed task and creates structured complexity within our brain. However, explaining this structure by using consciousness to setup the objective and to try for it is an acceptable solution. It is only tasks that a newborn learns within the first 6-12 months when he does not have an adult-level of consciousness that should be explained without requiring consciousness. These examples are, say, an ability to auto-focus eye lens, an ability to track a moving object and others mentioned earlier (which even animals are capable of learning), not, say, the ability to learn mathematics, build a computer, a television and others.

In this book, I will propose several mechanisms (see, say, chapter 16) for SPL systems that let them discover such objectives naturally and let them adapt to find a solution naturally, namely, without any consciousness. Any theory on consciousness should distinguish mechanisms with versus without consciousness and address issues (a)-(d) mentioned above. I will show that an SPL framework is a unique architecture that allows even a non-conscious system to automatically define 'natural' objectives and find corresponding solutions by itself, thereby addressing issues (a)-(d) mentioned above. It does this by expending energy in a directed manner even with no immediate benefit (see chapter 16).

10.14 The seat of consciousness

Some people have an intuition that makes them question if the problem of consciousness can ever be solved. The intuition appears logical when they try to 'personify' consciousness. They refer to a 'mind' that directs the 'body' to perform actions. When we try to give mind a physical representation, we have an infinite regress. Who tells the mind, to tell the body, what to do? At the next level, who tells the component that tells the mind what to do and so on, *ad infinitum*? We have infinite levels in the hierarchy of mind-body problem that poses a fundamental problem even

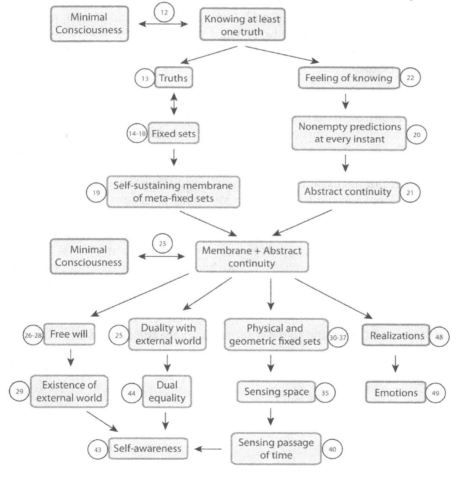

Figure 10.1: Outline of the chapters on consciousness. The chapter numbers are indicated next to the topic in the figure. The reader can refer back to this figure, as needed, to understand the relationship between the various topics.

from a conceptual point of view. Whether to build a synthetic conscious system or to study the problem of consciousness in a philosophical sense, we need a way to avoid the infinite regress. A lifeless system, obeying the laws of nature, should evolve to understand the laws, interpret them and use them to its advantage by sensing and controlling itself. How do we address this issue? In the later chapters, I will show ways to reduce the infinite layers of hierarchy into a finite number of them (or, specifically, into just two layers - chapter 19).

One of the central results is to show that consciousness is related to achieving a unique dynamical state called the self-sustaining membrane of meta-fixed sets for a certain special subset of SPL systems (see chapter 19). This is a unique state of synchronization of dynamical SPL systems in which the system senses what it controls and controls what it senses in a self-sustaining looped manner. Such a self-sustaining dynamical membrane, as we will see later, resolves this issue of the seat of consciousness both constructively and conceptually. We no longer need to worry about who tells the mind and/or body what to do. Instead, the system as a whole achieves autonomity by entering an unique state of synchronization through the self-sustaining membrane.

With the above introduction to various topics related to consciousness, let me briefly highlight the organization of the next several chapters (see Figure 10.1). I will begin by presenting several well-accepted examples of conscious and unconscious states (chapter 11). The purpose is to use these examples to study consciousness both from a positive and negative point of views. Next, in chapter 12, I will present the first and the simplest form of necessary and sufficient conditions for consciousness. The next several chapters is to exploit these necessary and sufficient conditions to create a dynamical theory of consciousness. Once we have a dynamical theory (see chapter 23), I will explore other topics commonly attributed to consciousness like free will, sensing space, sensing the passage of time, self-awareness and others.

Since the topic of consciousness is too broad, it helps to understand how the discussion in the next several chapters is structured. At a conceptual level, I have highlighted this relationship between topics in Figure 10.1. The reader can refer back to Figure 10.1 as needed when reading a given chapter both to recollect the previously covered topics and to get a sense of where the discussion is heading.

11
Conscious and Unconscious States

In order to study the problem of consciousness formally, we first need several examples of altered states of consciousness ranging from fully conscious to unconscious states. These examples, as we will see, will provide several insights during our analysis later. Most discussions on consciousness tend to gravitate towards philosophical arguments with little details on how to design experiments validating the argument. The examples of conscious and unconscious states discussed in this chapter will provide a way to limit the number of such untestable hypotheses.

Another way to limit the number of theoretical hypotheses is to ask how to construct a synthetic system using the explanation provided for a feature. These examples of conscious and unconscious states once again will guide us when trying to construct synthetic features. Throughout the book, I will emphasize and require such constructive explanations, both natural and synthetic, for each of these topics. Let me now discuss yet another reason why examples of conscious and unconscious states are important – they are for proving logical 'theorems' on consciousness.

11.1 On proving logical statements on consciousness

In the later chapters, you will see that I will present 'theorems' stated in terms of necessary and sufficient conditions for features like consciousness and others. This may seem counterintuitive because these statements are not as rigorous as mathematical statements and yet I am calling them as theorems (like, for example, a statement like humans are conscious if and only if they know at least one truth – Theorem 12.1). You may be wondering how we can 'prove' such non-mathematical statements. What does proof even mean? Mathematical theorems and proofs are statements that we are absolutely sure of within an appropriate axiomatic system. How can we be absolutely sure of non-mathematical statements such as the one above?

The answer is that these statements should be seen as purely logical statements. To evaluate if a statement on consciousness (i.e., a claim like above) is logically correct, we first need all positive and all negative examples of consciousness to test against. Our next task is to unambiguously verify if the statement is true for all positive examples and false for all negative examples of consciousness. If we succeed at showing this, then from a logical point of view, we have 'proven' the statement. This is the best we can do, unlike in mathematics, given a lack of a rigorous axiomatic system. This is not a drawback of the theory presented here.

Logical correctness can be made as rigorous as possible whenever we have a large collection of positive and negative examples. In this chapter, I will present such a list of positive and negative examples of consciousness. I will keep referring back to these examples whenever we want to 'prove' any such logical 'theorem'. Over time, the hope

Scenarios with no or low degree of consciousness	Scenarios comparable to normal consciousness
Newborn	Dreaming
Sleepwalking	Meditation
Deep sleep	Mental disabilities
General anesthesia	
Completely drunk	
Hypnosis	
Knocking a person unconscious	
Entering a state of coma	

Figure 11.1: Different scenarios with low degree of consciousness and with normal degree of consciousness. Here, we itemize a number of common scenarios that can be used as exemplary cases when studying the problem of consciousness more formally. These scenarios with serve as counterexamples when analyzing different situations later in the book.

is that this list of positive and negative examples will grow and be refined for precision. Our confidence in the 'proof' improves as long as the verification of the statement against positive and negative examples is unambiguous and convincing beyond any reasonable doubt. For this reason, I will explore each such statement from several points of view and by using multiple examples.

11.2 Scenarios with no or low level of consciousness

Let me list some situations when we are not fully conscious. I will keep referring to these examples throughout the book whenever I want to show necessity and sufficiency of an argument for a given feature related to consciousness (see Figure 11.1).

1. *Newborn*: An infant who is just a few weeks old is an example of a 'normal' human being who is not conscious yet to the same degree as an adult human. This is a useful example because there are no artificial chemicals or abnormalities that cause this state unlike the other examples below. Every part of an infant's brain and body works correctly, as it should. The only difference is that his brain does not yet have the necessary structure and interconnections comparable to a conscious adult.

2. *Sleepwalking or deep sleep*: When you are sleepwalking, you are in a state of deep sleep. Therefore, you are not conscious. Contrary to what most people believe, you are not dreaming or acting out your dreams in this state. You tend to act in a mechanical way using your hard-coded memories. Even in the case

of normal deep sleep, you are not conscious. You do not know how much time you were in this state. It is only after you come out of this state, you can evaluate how long you have slept.

3. *General anesthesia or completely drunk*: When you are completely drunk or when you were given general anesthesia, there are specific chemicals that block the neurotransmitters from binding at the synapse to open the Na-K channels. This hinders the neuron's ability to fire in your brain. As a result, the overall brain activity is lowered significantly. However, these chemicals eventually wear off. When this happens, you transition out of this unconscious state. You do feel a little dazed for a while. Yet, as with any other unconscious state, you do not remember or recollect events well enough while in this state. In the two examples discussed here, the unconscious state is induced externally by chemicals, unlike the previous examples.

4. *Hypnosis*: Hypnosis is a rather interesting discovery of a lowered state of consciousness. While the research is still debatable, it has received quite rigorous scientific treatment in recent years. The people who can indeed be hypnotized do not remember all the details later. They seem to be able to comprehend and respond appropriately during hypnosis. However, they act in a way that is somewhat detached from current reality. They behave partially consistent with a different time period, like re-experiencing a past memorized event. Their actions do show that their behavior is guided by memory (typically in response to a hypnotist's question) instead of reality. It is actually quite surprising that a hypnotist can *synchronize* your neuron firing patterns to align with his voice commands. It is nontrivial to attain such perfect synchronization for any large-scale complex dynamical system, if not for stable parallel loop architecture.

   External chemicals like with general anesthesia or severely drunk cases do not induce this state. However, it is still externally induced through sounds. You need to be able to bring your brain waves to sufficiently low frequency oscillations that you can still understand language. You can say that the equivalent self-induced synchronized state of a low level of consciousness is dreaming. This description will become clear after we understand how brain waves are created (see sections 15.1 and 15.2).

5. *Knocking a person unconscious or entering a state of coma*: One way to knock a person unconscious is by giving a strong blow to his head or by briefly cutting off blood supply to his brain. This highlights the fragile state of consciousness. Our consciousness is quite robust to gradual changes, but is intolerant to shocks in the brain. Just a moment before becoming unconscious, all parts of your body are well-coordinated and quite firm. However, when you are unconscious, you are numb and floppy as a lifeless person. If the damage to your brain is not severe, you recover to a conscious state fairly quickly. If, on the other hand, the brain damage is severe, you can enter into a state of coma. Recovery from coma is not easy unless the damaged regions can fundamentally be re-routed. We have limited techniques to achieve this in a consistent and reliable way.

In summary, the following are the list of less conscious states discussed here: a normal infant, sleepwalking, deep sleep, under general anesthesia, severely drunk, hypnosis, knocked unconscious and coma.

11.3 Scenarios comparable to normal consciousness

Let us now look at situations when you are conscious but to a somewhat lesser degree than a fully awake state (see Figure 11.1).

1. *Dreaming*: One common example of a less conscious state is during sleep, specifically when you are dreaming (i.e., during rapid eye movement (REM) sleep). You may or may not remember your dream in detail when you wake up. The events themselves may not appear logical compared to the external world. However, you do realize and feel that you are fully aware of yourself while you are dreaming.

2. *Meditation*: Another useful state of consciousness occurs during meditation. Believers of meditation sometimes even view this to be a 'higher' state of consciousness than the normal state. This is certainly a highly controlled state of synchronization within our brain. This is because you are able to achieve and maintain a normal level of consciousness even when the brain activity is significantly low (with delta or theta brain waves), i.e., almost comparable to that of deep sleep. Recall that during deep sleep you are not conscious whereas in this state, you are fully conscious. With this synchronization, you are quite effective at controlling your thoughts as well. The backlash with the topic of meditation is that the language people use to describe it sounds non-scientific and spiritual, which is probably the reason why this was not analyzed in detail (similar to how hypnosis is regarded).

    It is important to note that achieving such a state is not an easy task. You need a significant amount of practice. There is still some disagreement on whether people can achieve such a higher state of consciousness. Nevertheless, it is quite clear that such a theoretical possibility of synchronization exists for any complex system, living or nonliving. I will show in section 50.4 that this is a well-defined and an important state for a complex SPL system along with a clear representation within our brain architecture.

3. *Mental disabilities*: People with mental disabilities like dementia, schizophrenia and autism do become conscious even if takes a bit longer to communicate with them effectively compared to a normal person. Since the brain architecture for these people is different from a normal human being, the ability to learn skills that require voluntary action is impaired. For example, achieving normal levels of speech, logic, planning and intelligence are delayed. This delay will only partially slow down their ability to become conscious. While a normal child takes about a year to become conscious (to a degree comparable to an adult), people with these disabilities would sometimes take longer. Nevertheless, all of them will become conscious. People suffering from schizophrenia, for example, may hear other voices, but they would still know themselves distinctly from these voices at a given instant.

It is important to highlight that all of the above examples have been studied in great scientific detail. The clinical research is extensive for each of these states. I will refer to these examples throughout this book and present more details, as needed.

In summary, the following are the list of conscious states discussed here with a degree comparable to a fully awake state: dreaming, meditation and people with mental disabilities like dementia, schizophrenia and autism. I will eventually merge these and the states discussed in the previous section into a unified theory of consciousness.

12
Necessary and Sufficient Conditions for Consciousness

There have been several attempts at providing new frameworks for consciousness (Baars [6]; Dennett [32]; Chalmers[19]; Crick and Koch [28]; Tononi [154]). Each of these attempts recognize individual aspects within the complexity of our brain network as the source of consciousness. A few examples of these approaches are: identifying the neural correlates for consciousness (Koch [86]), taking an information theoretic approach to consciousness by studying the connectivity in graphs (Tononi [154]), looking at the origin of consciousness within humanity (Jaynes [79]), sensorimotor account for visual consciousness (O'Regan and Noë [117]), quantum mechanical view for consciousness (Hameroff [62]) and evolutionary views on consciousness (Edelman [34]; Dennett [32]). While it is difficult to do justice in reviewing these works in any detail, it is nevertheless known that none of them have specified both *necessary and sufficient* conditions for consciousness in a satisfactory way. In fact, for this reason, it is well-recognized that consciousness is still an unsolved problem.

In the previous chapters, I have presented a new model for neurons, memory and other features like image recognition, sound recognition and self-learning using SPL networks. However, the problem of consciousness is considerably different from each of these problems. The main difference is that consciousness, self-awareness, free will, sensing space, passage of time and others are topics that discuss *about* the systems. For example, we want a synthetic system that exists in space and changes in time to know *about* these facts as well as about itself or its components. Using only dynamics occurring in the external world, how can a system know *about* the dynamics?

The language used to describe about the dynamics is itself not connected to dynamics. It is, say, in English sentences which is purely abstract and unrelated to any physical phenomena like mechanics, electromagnetics, gravity and others. Obviously, entities responsible for describing about the system do not even have physical or chemical properties. For example, the laws of physics obeyed by the system are not obeyed by sentences in English language. How do we bridge these two disconnected spaces – English language space (used by our thoughts and awareness) and physical or chemical dynamical space (used by our brain network)? Without a way to close this gap, there is less hope in addressing the problem of talking *about* the system and hence solving the problem of consciousness.

Therefore, the main objective now and in the next several chapters is to introduce concepts and designs that let us move seamlessly from one space to another i.e., from abstract language space to dynamical space and vice versa. While this is a difficult task in general, the only way to achieve this objective is if we work with necessary and sufficient conditions whenever we translate concepts from one space to another.

For example, the central result of this chapter is to show that it is both necessary and sufficient to *know* at least one *truth* for minimal consciousness. Note that this 'if and only if' statement is entirely in the abstract language space. Yet, it gives us an equivalent way to study consciousness by understanding what truths are. In the later chapters, I will translate the notion of *truths* (a concept in abstract language space) into the notion of *fixed sets* introduced earlier (in dynamical space – see section 6.6) using, once again, equivalent conditions. Subsequently, I will translate *knowing* (a concept in abstract language space) into a notion of *abstract continuity* (in dynamical space) using equivalent conditions. Figure 12.1 summarizes these steps and suggests an alternate way to approach the problem of consciousness. It suggests that the difficulties with studying consciousness and its interactions with other topics like free will, sensing space and time can be simplified using the necessary and sufficient conditions proposed in this book.

In other words, one of the central themes for solving the problem of consciousness is to use equivalent conditions to move from an abstract language space to dynamical space and vice versa thereby giving us a way to talk *about* the system using dynamics occurring within the system.

12.1 Why necessary and sufficient conditions?

How can we gain confidence in a theory on consciousness unless we work with necessary and sufficient conditions? For example, Tononi [154] has suggested that "consciousness is integrated information". Necessity of integrated information i.e., that which is beyond the sum of its parts is well justified. However, why is integrated information sufficient to create a conscious system? What is the nature of this integrated information that would make it sufficient for consciousness? When we build a synthetic system (like a machine) in which the information is well integrated, what are we integrating and why would we regard such an integrated system as conscious? How much information should be integrated before we can view the system as conscious? Besides, the integrated information in terms of purely static structural properties of the brain network (for example, static structural regions like fusiform gyrus in area V8 identified as neural correlate of color – Crick and Koch [28]) cannot be sufficient. This is because a static network and regions such as fusiform gyrus are structurally the same, even if manually stimulated, both when we are conscious and when we are not (like during deep sleep and under general anesthesia). Therefore, static structures cannot be the complete set of equivalent conditions. Dynamical properties must necessarily be included for consciousness. Yet, there is less discussion on which physical and chemical dynamical features are necessary and sufficient.

As another example, O'Regan and Noë [117] have mentioned that visual consciousness is the outcome of sensorimotor contingencies. Here again, necessity of actions and sensing is well justified. However, why and what sensorimotor interactions would make them sufficient for consciousness? If we build a machine, how many sensorimotor contingencies should be included and in what order? When

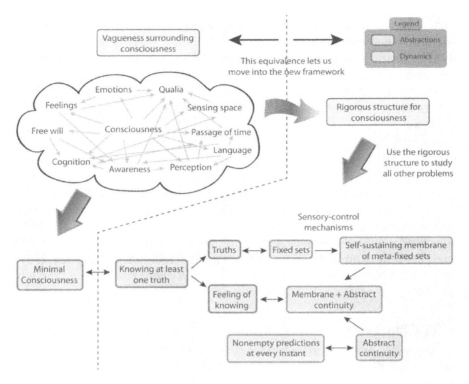

Figure 12.1: The central result is that minimal consciousness is equivalent to knowing at least one truth. This result lets us move away from the vagueness surrounding consciousness to the study of truths, the structure and relationships between them and the feeling of knowing using a dynamical framework of stable parallel loops. The detailed analysis, at times, appears to be unrelated to consciousness. However, the resulting structure built using these details offer necessary and sufficient conditions. Specifically, this approach reveals that minimal consciousness is equivalent to the existence of a self-sustaining dynamical membrane of meta-fixed sets made of abstract continuous paths.

such a machine becomes complex, what is the nature of this complexity and what resulting structural and dynamical properties make it sufficient for consciousness?

We can now evaluate each of the proposed frameworks in the existing literature on consciousness in a similar way. They all fall short with either necessity or sufficiency. This is one reason why we are not yet convinced with any existing theory on consciousness. Guaranteeing sufficiency is, in general, quite hard than necessity. As another example of necessity, researchers are trying to identify certain regions in our brain, which if absent, cannot give rise to consciousness. The only good example of sufficiency of consciousness is the existence of a normal human brain. This, unfortunately, is not a helpful statement to develop a formal theory of consciousness.

Another requirement for a theory of consciousness is that it should explain how to *build* synthetic conscious systems in addition to *analyzing* how our brain gives rise to consciousness. As an analogy, to fully understand what a computer is and how it works, we should also understand how to build it, at least, in theoretical terms. All existing models and discussions on consciousness either ignore or postpone this aspect. This makes them necessarily incomplete.

Towards satisfying this requirement, specifying necessary and sufficient conditions become critical once again. Without these conditions, there is no sure way to evaluate theoretical or experimental correctness. For example, how can we answer if neurons or a brain is necessary for consciousness? How can we be sure? Existence of humans only says that they are sufficient. It is unreasonable to suggest that there is only one approach to become conscious and that the natural evolutionary mechanisms did manage to pick it using random mutations. If we are attempting to build a synthetic conscious system, when will we stop our pursuit and say that we have reached our goal if we do not have these necessary and sufficient conditions?

Finding necessary and sufficient conditions for consciousness also avoids the difficulties involved in discovering the theory incrementally. This is critical because there is no clear notion of error to help us make incremental improvements to the theory on consciousness. For example, there is no incremental stage for a theory of consciousness that explains, say, only 40% of self-awareness, half passage of time or 'partial' free will. What does making improvements to such a theory mean? For problems like finding a cure for cancer, an incremental approach is natural in which we keep improving our methods over time. However, for consciousness, the existence of *qualia* (the subjective quality of experience) makes the notion of a quantitative error, and, hence, an incremental approach, unreasonable.

In this book, I depart from all existing approaches by providing both necessary and sufficient conditions for each concept introduced, even if it appears trivial at first glance. The theory will be extended using only necessary and sufficient conditions linking abstract concepts to dynamical notions and vice versa. For example, the central result of this chapter is the following – *minimal consciousness is equivalent to knowing at least one truth* (Theorem 12.1 – recall the discussion in section 11.1 on what 'proving' a theorem is in the context of logical statements such as these, as opposed to proving mathematical statements). At first glance, this equivalence sounds either trivial or that we have pushed the problem of consciousness into the problem of knowing. The objective of the later chapters is to clarify and show how to build a dynamical systems framework for both truths and knowing. Here, stable parallel looped (SPL) dynamical systems discussed in the previous chapters will provide the underlying framework. They help generate an infinite family of increasingly complex and highly interacting stable systems. Such large-scale structured complex systems are necessary since simple systems are rarely conscious. Using this family, I show that *a system is minimally conscious if and only if it has a threshold self-sustaining dynamical membrane of meta-fixed sets* (Theorem 19.2). If the membrane exists, you are minimally conscious and if it 'tears down', you are not conscious. Furthermore, Theorem 19.2, unlike Theorem 12.1, now allows us to (a) build synthetic conscious systems and (b) study human consciousness empirically as well.

12.2 Why pick knowing for minimal consciousness?

In the above discussion, I have picked 'knowing truths' as the minimal set for consciousness. Why should we pick them? One reason is because 'knowing' is so fundamental that most of us treat it as synonymous to consciousness. For example, knowing yourself is self-awareness. Knowing the existence of the external world involves knowing the existence of space, knowing the passage of time, knowing the external objects and having a cohesive perception. Knowing your interactions with the external world includes your free will. Knowing your feelings is related to knowing your subjective quality of your experiences (*qualia*). Therefore, if we can formalize the concept of knowing (see Figure 12.1), it appears that we can extend it to the above specific forms of knowing quite naturally.

Conversely, if we do not include knowing into the minimal set, the resulting definition does not appear to have any relationship to the concept of consciousness. For example, if you indeed do not know anything including yourself, why would you consider yourself conscious? This situation does occur in each of the seven less conscious states mentioned in chapter 11 (like deep sleep, coma and under general anesthesia – see Figure 11.1).

If someone says that *'nothing exists in this universe'*, then even though the statement appears obviously incorrect, we can guarantee that it is false *if and only if* we are conscious. To convince ourselves that this is true, consider one of the less conscious states mentioned in chapter 11, like, for example, when we are in deep sleep (see Figure 11.1). Even though you and the rest of the universe continue to exist physically, you, as a system in deep sleep, are incapable of 'knowing' your existence or the existence of the universe. This is true with any fact for which an adult conscious human can evaluate truth, including knowing parts of your own body. Even if you are repeatedly performing the same task, you do not know or realize what you were doing, why you were doing and that you were even doing it, to the same degree as adult conscious humans. This statement is at the heart of every problem related to knowing anything in the universe. This is the reason why I use it to define minimal consciousness.

12.3 Minimal consciousness = knowing at least one truth

In this section, I will present one of the central results on consciousness. These are the necessary and sufficient conditions for minimal consciousness expressed in terms of knowing a fact. Recall that the term 'minimal' is used here to identify those aspects of the human brain that are commonly attributed to consciousness by most people (section 10.1). Therefore, 'minimal consciousness' refers to those minimal set of features which if not present is not related to human consciousness. In the previous section we have seen that 'knowing' is one such minimal notion. Theorem 12.1 formally relates this minimal notion of consciousness to knowing (recall the discussion in section 11.1 on what 'proving' a theorem is in the context of logical statements such as these, as opposed to proving mathematical statements).

When stating Theorem 12.1, I only use human consciousness as an example even though other species may have consciousness to some degree. This is because, from a scientific study, humans are the only examples of consciousness who have been extensively studied. For all other species, the research has only been through way of generalization. For example, only after identifying specific neural correlates for consciousness in the human brain, researchers tried to identify similar regions in the brain of other species. If these regions were present, they generalize and state that this organism also exhibits consciousness to a degree, possibly lower than human consciousness. Definition 12.1 presented later in this chapter is one such generalization applied to all systems, living and nonliving. While Theorem 12.1 is a result applicable to human beings referring to 'human' consciousness, definition 12.1 generalizes to all dynamical systems referring to 'minimal' consciousness instead.

Let me begin by stating Theorem 12.1. I will discuss the proof towards the end of the section.

Theorem 12.1: A human is minimally conscious if and only if he knows at least one truth.

First, let us look at the intuition behind this result. I have mentioned earlier that if someone says *there is nothing in this universe*, a conscious person would immediately find it to be false whereas an unconscious person would not know the difference. In fact, if we say just about anything about our universe, an unconscious person cannot evaluate if it is true or false. I can tell him that he has 10 fingers on his right hand, and in another 10 seconds say that he has 3 fingers, then 7 fingers and so on. I can completely contradict myself with every sentence or action. Yet, he would not realize that I am contradicting myself when he is not fully conscious. Every sentence or event feels like a new event to him. He either responds to it or not, in a completely mechanical way.

As another example, imagine that you are under general anesthesia (say, for some dental surgery). After the surgery, your anesthesia begins to wear off. However, there is a brief period, say, about 30 minutes when you are a little disoriented, even though you can talk and walk, at least with some help. This is different from the situation when your anesthesia is strong enough to cut off your motor abilities and perception entirely. During this period, you are not conscious of all the events that occur. You can walk, look around, talk to your spouse and answer questions that she might ask. You are popping in and out of your consciousness during this period. You do not know the amount of time you were in this state, i.e., you do not realize the passage of time. After several hours, when your anesthesia completely wears off, you only remember few events. These are a subset of events that occurred when you popped into a conscious state. During most of this period, you were acting in a mechanical way. This is generally the same situation with other unconscious states (section 11.2) like when you are fully drunk, during deep sleep, sleepwalking or when you are knocked unconscious. In some cases, you react mechanically to questions posed to you. In fact, if someone takes a video of you or tells you later how you acted during this period, you would be surprised (and possibly embarrassed).

What can you infer when you look at such a person in this partially unconscious state? What is he looking at? Even if he is smiling at you, waving his hands, what does it mean to *him*? Since he is only responding in a mechanical way using his past memories, you can lie about the events. He is not processing this information well enough to verify independently if your statements are true or false.

However, are there some true statements that you can never lie to him whether he is conscious or not? Could they be something very personal to him like about his own body, his own emotions or pain? More generally, are there regions in his brain that always evaluate the truth of, at least, a few statements or events with 100% accuracy, independent of his state of consciousness?

My claim is that there is no such region in a human brain and no such truth statements. The intuitive reason is that truth is not present in the raw dynamics of a physical or chemical system (like within the neuronal dynamics of our brain). It is an abstract notion, namely, something that states *about* a dynamical event. Truth is not part of the event nor is it inherently present in the event itself. Consciousness is necessary to claim if an event is true or not.

Consider, for example, any truth related to physical phenomena, i.e., physical laws like the existence of gravity. Gravity existed and will continue to exist irrespective of whether there are conscious beings or not. However, nothing about gravity stands out as inherently 'true', different from non-true physical phenomena (like, say, 'humans can fly'). In other words, if a physical phenomenon does indeed occur, we may as well regard it as true. The objective with science then is to identify just those minimal set of true phenomena called laws, using which we can (a) derive and hence discover other true physical phenomena and (b) identify and eliminate non-true phenomena to a degree. The main difficulty with this approach is to decide which set of physical phenomena did occur and which ones did not. This verification step is when we require consciousness. Without consciousness, we cannot partition true from non-true physical phenomena. In fact, we keep refining our true set of laws as we improve our verification step with new accurate machines and with repeated experiments. Therefore, the notion of true physical phenomena is relative to consciousness, which for now is human consciousness, as it is the only highly evolved form that makes verification reliable and repeatable.

Next, consider truths related to non-physical phenomena, i.e., abstractions like logical and mathematical statements. Here, a statement like '2 + 2 = 4' is true whether there are conscious beings or, in the extreme case according to some, whether the physical universe exists. However, if we look closely, such mathematical statements and theorems are true only if a set of axioms are regarded as true. The objective with mathematics is once again similar to science. Identify just those minimal set of axioms using which we can (a) derive other true theorems and (b) eliminate non-true mathematical statements. Gödel (see Hofstadter [71]) has shown that there are limits to such axiomatic approaches when we try to satisfy both completeness (related to (a) – discovering all truths) and consistency (related to (b) – not having any contradictions). Specifically, he has shown that both (a) and (b) cannot be satisfied simultaneously if the axiomatic system does encompass arithmetic.

At this point, the mathematical approach is no different from the previously discussed scientific approach. Namely, the main difficulty is to decide which set of axioms are acceptable as true and to decide what inconsistency means. These two verification steps require consciousness. Without consciousness, the number of axiomatic systems will be enormous. There will be no way of knowing which systems are useful and which ones aren't, especially since we cannot eliminate one from the other. Recall the big philosophical shift that occurred in mathematics with the discovery of non-Euclidean geometry (see Courant and Robbins [25]). It was shown that mathematics cannot decide whether the geometry of the physical universe is Euclidean or non-Euclidean. Therefore, as with physical phenomena, the notion of true abstract phenomena is relative to our consciousness.

In other words, there is no region in our brain that will help us evaluate at least a small subset of statements or events to be true irrespective of our state of consciousness. If you truly know that I am lying to you, then you are already conscious. To be able to say that you know something implies that I have to attribute consciousness to you first. A machine cannot know anything because it is not conscious. You do not need anyone's help to evaluate the truth when you are conscious (at least for several simple events). You just know it. The statement that you are evaluating does not have to be universally true so everybody agrees on. It could be unique to you. Similarly, you simply do not know the truth or falsity, when you are not conscious just as a table does not *know* that it is on the ground.

From the above discussion, we see that there is a link between consciousness and truths. Let me briefly explore the representation of truths (more details in chapter 14). We are used to describing truth statements as sentences in a language. We do need to 'define' the words used in these sentences though. This is not an easy task when you consider teaching the meaning of words (like, say, the concept of the word 'sharing') to a 2-3 year old. When we try to explain the meaning of a given word using other sentences, we realize that there are additional new words within them that are undefined for him. We need to define these as well, before we can define the initial word. This process can iterate several times. For most of the simple words, we can get around this inherent recursion soon enough. However, for some complex words (like responsibility, jealousy and others), this gets quite unwieldy that we simply give up (explaining to a 2-3 year old). We instead say that 'he will understand this word when he gets older'. This is an inherent problem with any abstract language.

Let us eliminate three sources of misunderstandings with abstractions in language to help us understand the link between truths and consciousness better. Firstly, as mentioned earlier, there are undefined words and relationships between these words that are unclear. For example, when you say to a 2-3 year old that he has 10 fingers on his right hand, he does not understand several words in this sentence. He, probably, does not have memory for some words, like fingers, hands and the concept of 10 or numbers, in general. As a result, he might believe your sentence when you say it. If you subsequently say that he has 3 fingers on his right hand, he still does not know the clear contradiction that an adult identifies. This is because he does not know the definition of '10', that '3' is different from '10', that '3 < 10', that '3' is not a synonym of '10', that a set cannot simultaneously have both '3' and '10' elements and several others. Since these are not yet defined for him, he simply accepts them as definitions. He begins to form both these sentences (with 10 fingers and with 3 fingers) as a memory without realizing that both sentences are talking about the same hand. He does not know or care that both these sentences are related and are contradictory.

Secondly, there are synonyms in our language that cause additional redundancy. For example, if you show a cat and say that it is a 'dog', he does not care because the word 'dog' is not defined to him yet. He simply starts to associate the word 'dog' to the animal cat. He may even form a memory if you repeatedly make this (incorrect) association. You can later show the same animal cat and call it a 'sheep' as well. Even if he has memorized and associated it previously with the word 'dog', he will not feel that this is a contradiction. This is because he does not know that there should be a unique word for a given animal. In fact, we do have synonyms for words. Why is this situation any different? He treats 'sheep' as a synonym for 'dog' for the animal cat. As a result, these words and sentences, even if you say them one after the other, are not contradicting at all. They are simply memorized as alternate ways to describe the animal cat.

Thirdly, memory is not fixed. It changes over time as we gain more experiences. This alters our notion of truth, as we grow older. For example, we hear a number of fictional stories, say, about Santa Claus, when we were a child. We believe them as true. However, as we grow older and gain a lot more experiences, we are able to identify the contradictions and inconsistencies. We fix these inconsistencies to form a consistent picture with all the facts that we have experienced until then. We continue this process throughout our life.

Now, a 2-3 year old does not fully understand truth from falseness. The above three issues with language only make matters worse. Nevertheless, he is conscious. This is because he knows *himself*. He knows his hands, his legs, his emotions, his pain and others. You cannot lie to him about his own emotions or pain. The only lie, if at all, is the word you used to describe it, but not, say, the pain sensation itself. He knows, for sure, if he got hurt irrespective of what word you used to describe it. He may not know the truths about the external world or abstract statements that rely on natural language, but he definitely knows himself.

An unconscious person, on the other hand, does not know himself. Let us consider hypnosis as an example of a low degree of consciousness to evaluate this (see section 11.2). It has been shown that hypnosis is quite effective to manage pain and pain perception. It is also possible to trigger and alter the experience of the emotion to a certain degree. For example, the hypnotist can ask a person to eat hot and spicy peppers while saying that they are sweet instead. When the hypnotized person eats it, he does say that it is sweet, contrary to what he would have said if he was fully conscious. He is not merely expressing his taste in words incorrectly. Hopefully, he is not pretending or lying (at least with some cases). This is because sweating is partially an involuntary response within the body. In fact, he will show all expressions related to happiness and joy as if he were eating a sweet delicious item. Of course, this is not true with every hypnotized person.

Hypnosis is not yet well accepted within the scientific community. Yet, the important point is that in these lower conscious states, you do not know your own emotions, pain and perception of truth. I can lie about your very own emotions and pain provided you are capable of understanding a language to respond. This is true with cases like hypnosis, completely drunk or other lower conscious states. You would not be able to evaluate the truth of my statements. Therefore, my claim is that there is no region within our brain that evaluates the truth of a few events with 100% accuracy irrespective of our state of consciousness.

We tend to think that emotions and pain are unique to us. If someone tries to lie about them, we think we can catch their lie. If someone hits you hard and instead say that he kissed your hand, would you not know the difference? It turns out you would not know it, when you are not conscious. A 2-3 year old can clearly know the difference, unlike an unconscious person. An unconscious person may not cry or retreat his hand in response. When he does seem to retreat his hand, it is either due to a reflex action or because of a mechanical response from past memory.

The process to change the truths for a conscious person is also a slow process. For example, an emotion called acceptance is one way to change your truths. This emotion is learnt over years of repetitive contradictions within your own personal emotional truths itself. Let me now summarize the above discussion formally as proof of Theorem 12.1.

Proof of Theorem 12.1: The statement of the theorem is that a person is minimally conscious if and only if he knows at least one truth. To prove this result, I will consider each direction individually.

Minimally conscious ⇒ *knowing at least one truth*: If I state several obviously false statements (as evaluated by an adult conscious human), like that I have seven fingers on my hand or that I pinched you but said that I patted you instead and so on, you, as a conscious person, know that I am not telling the truth. Similarly, I cannot lie to you about the basic existence of objects, shapes, sounds and others. In other words, you already know several truths whenever you are conscious.

In addition, from the above discussion, we have seen that the contra-positive is also true, namely, that if you do not know even a single truth, you are not conscious.

Knowing at least one truth ⇒ *minimally conscious*: Here, I will show the contrapositive, namely, that if you are not conscious, you do not know even a single truth, by considering all known cases when we are indeed not conscious or even partially conscious. Examples of these are during deep sleep, under general anesthesia, when knocked unconscious, during coma, severely drunk, when sleepwalking and even in a newborn baby (Figure 11.1). In these cases, you do not *know* the existence of objects around you, even though you may avoid bumping into them when walking. Your ability to know basic logical facts is lost. In fact, as discussed above, you do not even know your own emotions, feelings, hands, legs and other parts of your body. When I pinch you, you will not know unless you wake up i.e., become conscious.

We can analyze thoroughly each of the above seven unconscious or less conscious cases individually to convince ourselves that you do not know even a single truth. The analysis for each of these cases is almost identical that I will only discuss one scenario here. For example, under general anesthesia like, say, during a surgery, we do not know any truths including our body, our feelings and pain. Only afterwards, but not during the surgery, it is sometimes possible to remember parts of the events which were memorized mechanically. The analysis is similar with the other six cases above (like when knocked unconscious, during deep sleep, severely drunk, coma and others).

In partially conscious states like when the anesthesia is wearing off, when you are about to fall asleep or when you are drunk, you are switching between a conscious and an unconscious state quite randomly. Therefore, your ability to know even a single truth also switches accordingly. Q.E.D

This is the central theorem for consciousness. The advantage with Theorem 12.1 is that we can now study the theory of consciousness as a theory of truths and their relationships, a possibility we never had before. This is the first result presenting necessary and sufficient conditions for consciousness, unlike any other theory. The rest of the book explores the equivalent approach taking both an analytical and synthetical perspective. Theorem 12.1 shows that we should address two aspects clearly, before we can gain a deeper understanding:

(a) what is a truth and

(b) how would we be able to *know* even a single truth (or any fact, in general)?

For the former topic, we need a formal definition of truth. We need to express it in a way that does not require consciousness. I will discuss this in the next chapter. For the latter topic, we need a formal definition of 'knowing'. When we say that we *know* some event or statement, we want to understand what that means, when and how it happens in our brain. I will discuss this in chapters 20 - 22. I will state necessary and sufficient conditions for attaining a feeling of knowing in chapter 22. The concept of knowing is by far the most important aspect of Theorem 12.1. However, to be able to address it clearly, we first need to explore truths and the structure of interconnected truths in greater detail.

Once these two aspects of Theorem 12.1 are clarified, I will present a large repository of truths starting from chapter 16. As part of this, I will discuss extensions to the new theory on consciousness beyond minimality to include other features of human consciousness like free will, sensing space, sensing the passage of time and self-awareness. Therefore, we can view Theorem 12.1 to be applicable to 'all' dynamical systems including living and nonliving, whereas the extended theory applies to other aspects of human consciousness. Let me state this generalization formally as the following definition (or axiom).

Definition 12.1 (minimally conscious): A dynamical system is minimally conscious if and only if it knows at least one truth.

This definition is formally valid for human beings as we have seen with Theorem 12.1, i.e., the set of dynamical systems for which definition 12.1 is applicable is non-empty. The definition is also reasonable and well defined for all other living beings that appear to have at least some degree of consciousness. Therefore, the statement is valid for every known example of a system with some degree of consciousness. This allows us to generalize consciousness to all dynamical systems, living and nonliving, treating the above statement as a definition of a *minimally conscious* state.

When working with truths, it is interesting to note that in reality we will know more than one truth at any given time. Our ability to learn truths is not a serial process. Our brain does not develop and grow by knowing exactly one truth, then two truths, three truths and so on. The process of knowing truths is not discrete either. We will know many truths simultaneously or none at all because of the inherent nature of the structure of truths (see Corollary 23.1 of chapter 23).

Even though Theorem 12.1 relates consciousness to truths, it does not say how to make the set of truths nonempty. We know that the set of truths are empty during an embryonic stage. We are definitely not conscious as the embryo develops and grows during the nine months of pregnancy. This is partly because we do not have a fully formed brain yet. The set of truths remain empty even at birth and, possibly, several months after birth.

However, as an adult we are absolutely sure (at least, to ourselves) that we are conscious. As a result, from Theorem 12.1, the set of truths is nonempty for an adult. How did this transition occur from an empty set, during the fertilized egg state and before birth, to a nonempty set for an adult? What is the mechanism that ensures truths to be continuously formed for a given person? Besides, how do we know that truths are being formed for a person other than you. This is especially needed when you are, say, building a synthetic conscious system. The above theorem does not seem to directly assist us with these questions. Therefore, in the later chapters, I will develop an elaborate theory that will help answer these and several other such questions in greater detail. Here, definition 12.1 as a generalization applicable to all dynamical systems becomes necessary.

13
Truths

Theorem 12.1 has given us a way to move away from the vagueness surrounding the problem of consciousness. Recall the discussion in section 11.1 on what 'proving' a theorem is in the context of statements such as Theorem 12.1, as opposed to proving mathematical statements. We have seen that Theorem 12.1 required us to focus on two topics – truths and knowing. The only reason why studying these two topics is equivalent to studying minimal consciousness is because Theorem 12.1 is both necessary and sufficient. On the contrary, all existing theories have only stated either the necessary or the sufficient conditions, but not both. This made them necessarily incomplete.

In this chapter, I will begin the alternate and yet equivalent study of consciousness by exploring the nature of truth. Later, from chapter 20, I will explore the topic of knowing. As mentioned briefly in section 12.3, truth statements are pure abstractions, clearly distinguishing themselves from real physical and chemical dynamics. How does our brain represent them within our chemical neuronal architecture? How do we take an abstract notion like truth, described in a natural language like English, and represent it by a dynamical notion, described in terms of physical and chemical reactions? In the next chapter, I will discuss these representational questions. In this chapter, I will focus on defining what truths are.

13.1 Linking abstractions to dynamics

Some examples of truths are personal truths like 'my hand' or 'my emotions' and physical truths like causality, which states that cause always precedes an effect. When we were still developing from an embryo, we are not conscious. Our set of truths is empty. However, when we become an adult, we do become conscious. During this transition, the universe around us existed and proceeded in very much the same way. Causality is, indeed, obeyed whether we are conscious or not. In fact, other conscious people, unlike an unconscious person, existed to assert to this fact. Since we fundamentally believe in the continuity of events as a truth, we find no contradiction after we are conscious, when someone says that causality is true even while we were not conscious.

Causality is a statement about physical dynamics. We can determine the truth about causality only by looking at the dynamics. However, since the dynamics is the same before and after you are conscious, we conclude that causality is not inherent in the dynamics itself. It can, therefore, be termed as an abstraction. In other words, no matter how long and how many dynamical systems we observe, causality is not present *within* the dynamics. It is an abstraction *about* the dynamics.

This is true with any truth statement. For example, I know that 'this is my hand' is a truth. Others can look at my hand; study its shape, its internal dynamics and so on.

Yet, they cannot feel and conclude the truth about this statement the same way I feel and conclude. They can, however, feel and state a similar statement about their own hand. Any statement about my hand is only a generalization through analogy. Therefore, such statements are abstractions.

Mathematical concepts and theorems are other examples of abstractions. For example, '3' is an abstraction that refers to 'any' three objects. When we have any three objects, there is no concept of '3' within them. You simply define this concept as an abstraction. You create relationships between such concepts and start to prove theorems using them. We want to include definitions like '3' as truths as well. In mathematics and other fields, we started studying abstractions in their own right since they outline important relationships between seemingly unrelated concepts. The meaning that the truth refers to does not give rise to the abstraction or truth itself.

It is important to relate abstractions with dynamics. This is because every statement or thoughts that a conscious person has are abstractions. However, the world around us, including us, is dynamical in nature. We have been creating abstractions, like mathematics and language, and are studying them independent of the dynamics. When we study algebra, we do not think in terms of physical objects that are rolling or bouncing. Similarly, when we form sentences in a language, the grammatical rules have no relation to the physical objects.

This lack of connection between abstraction and dynamics turns out to be problem if we want to study consciousness. For example, Newton's laws of motion, Maxwell's laws of electromagnetism and others are abstractions describing real physical dynamics. These abstractions are represented in a form that is completely disjoint from and devoid of any real physical dynamics itself. For example, these representations can 'live' inside books as printed matter, inside CD's, DVD's, hard disks, photos and others, without any dynamics even though they talk about dynamics. Until now, we did not view this as a problem. However, if our objective is to create a synthetic system that is attempting to become conscious using the above set of representations, we do have a nontrivial problem to address.

I have briefly mentioned one such issue in section 9.10. For example, I have mentioned that the ASCII code for '1' (the number) and 'l' (the letter) in a computer are unrelated. They are 31 (for number '1') and 6C (for the letter 'l') in hexadecimal, respectively. These, inherently visual objects, were represented in an abstract way with no relationship to the dynamics triggered by our visual sensory system. However, these two symbols look alike. This information is lost because of our poor abstract ASCII representation. Our brain, on the other hand, maintains the similarity relationship using the dynamical loop representation of neuron firing patterns as discussed in chapter 7. We do not have to 'teach' our brain that '1' and 'l' look similar, unlike a computer model. This necessity to teach similarity and dissimilarity becomes pronounced for more complex patterns like paintings, movies or concepts.

Similarly, we do not have to teach our brain that 'write' and 'right' sound similar, unlike a computer. The sound wave dynamics automatically capture this information, which is represented in a dynamical form within our brain. In fact, we have a large number of such relationships between abstractions that do not have to be taught because our brain represents them as dynamical loops, unlike a computer. Our new SPL memory model satisfying similarity and dissimilarity conditions (chapter 7) is a dynamical model that already captures this aspect.

In the later chapters, we will see more applications of a dynamical model that existing models cannot capture. Specifically, in chapter 46, I will show that *all models that attempts to create conscious systems must necessarily converge to a dynamical SPL network model*.

One issue raised here when the abstractions are represented independent from dynamics is that such systems are incapable of capturing relationships by themselves. An external conscious being should assist such systems to teach similarities as well as differences. For example, if we have a brown square and a green rectangle, a computer would need to be taught a property called color and say that both shapes have different colors. We need to teach 'sides' and say that both have same number of them. We need to teach 'angles' and say that both have four right angles in them. On the other hand, since these abstractions like color, edges and angles are represented in a dynamical fashion in our brain (or within an SPL network – chapter 34), we can recognize them naturally in all other objects as well. A child recognizes these visual similarities in color, edges and angles without any effort, though he does need to learn the abstract concepts and/or their names.

Furthermore, conscious humans can even create new properties that were not defined yet, like specific wallpaper patterns or new shapes. We can find new relationships between them as well (see section 9.7 for how to do this with SPL systems). For example, if the brown square is a bread slice, then we can identify a new property 'food'. This can trigger a memory of other food items that you eat with bread like jam and butter and so on. All of these related events are triggered naturally within our brain's dynamical representation. A computer, on the other hand, would not be able to create these abstract relationships by itself without any such links with dynamics.

Additionally, abstractions with no relation to dynamics allow the possibility of creating complete gibberish structures as well. For example, if we consider letters in an alphabet as an abstraction, represented with no relation to dynamics, then it is perfectly valid to create absurd words and meaningless sentences. Even with geometric shapes, any random combination of shapes is a valid design. Similarly, we can generate utterly incomprehensible sounds while speaking. However, there is no value in allowing abstractions to be combined in any possible way as above. This is another problem with computers making them difficult to capture the real physical world because they allow such generality with abstractions. With humans, such gibberish is not common and requires effort.

We need to constrain and restrict abstractions to only a useful set. For example, with language, we want to allow only grammatically correct sentences. With sounds, we want to allow only those that we agree upon for a given language. If the abstractions do not have a corresponding dynamical representation, the rules for restrictions are difficult to learn and discover automatically. Humans, on the other hand, do not depend on the rules to avoid talking gibberish. In fact, you have to put additional effort to create a meaningless (or grammatically incorrect) sentence using meaningful words. The dynamical representation in our brain naturally forces us to create almost correct sentences or generate correct sounds for words (after we have learnt them). The memory model proposed in chapters 7-9 captures this aspect unlike computers or other existing theories.

The link is sometimes referred to as the emergence of abstractions from dynamics. Temperature is one example of an emergent property. It is only defined in a statistical sense for a large collection of molecules. No smaller subset of just a few particles has a notion of temperature. The property, however, is derived from within physical dynamics. While there are several emergent properties within, say, uncoordinated molecular motions (like states of matter, malleability, ductility and so on), our interest in this book are with emergent properties derived from within the SPL dynamical structure of an adult brain.

13.2 Definition of truth

From Theorem 12.1, we need a formal definition of truth for the problem of consciousness. The usual definition of truth is a statement that is a fact or can be verified as a fact. However, what is a fact and how can you verify it? Theorem 12.1 states that an unconscious person (say, under general anesthesia) does not know any facts, nor can he verify one. Therefore, if we use this standard definition, it does not apply for an unconscious person. Furthermore, this definition expects a truth to be universally true. There is no emphasis on the validity of a truth over time and other dynamical variations. What is an alternate definition of truth that avoids these issues?

Let me begin by analyzing the problem of how to verify a fact, before proposing a new definition. Consider a truth statement like, say, 'sun rises in the east' (ignore the inaccuracy of this statement closer to north and south poles for now). How can we use the standard definition of truth to verify that this is a fact? Let me assume that I am already conscious. This is a nontrivial assumption as it implicitly assumes self-awareness, free will, existence of an external world, sensing three dimensional space and time.

In addition to consciousness, we need to make several assumptions specific to the statement we are trying to verify. Let me list a few of them here to highlight the existence of several implicit assumptions we seem to take for granted.

(a) We have to accept a definition of east, say, by drawing perpendicular arrows on the ground for east, west, north and south.

(b) We have to wake up in the morning to check that sun is *above* the horizon every day to claim that it rises (as opposed to sun setting in the evening).

(c) We have to assume the definition of a straight line. We have to assume that it can be extended infinitely in either direction without self-intersecting. This is purely an imaginary extension in our brain. This imagination clearly requires consciousness, specifically, existence of an external world, space and time. We can then use our imagination to extend the eastern arrow towards the sun and feel that it will eventually intersect the sun when it rises above the horizon.

(d) We assume that we can generalize the validity for all time simply by repeating the same experiment just a few times. If the same result occurs every time we repeat, then we assume that this result will continue to occur in the future as well. This ability to generalize is valid because of our free will assumption. We will see several examples later in which I show how to use free will to prove nontrivial statements like, say, proving the existence of an external world – see chapter 29. In the current example, using free will, any conscious being can perform the same experiment at any time or place, completely at-will and get the same results. Therefore, the result itself is independent of our observation and is, hence, generalizable.

These are just a sample set of assumptions. If you analyze the problem more carefully, you can easily list several other assumptions. If I accept all of these assumptions and the reasoning that uses these assumptions, does this mean that I have verified the original fact? As a conscious person, I surely feel this.

The above list of assumptions (a) – (d) can still be considered as high-level assumptions. It is possible to further breakdown these assumptions into simpler ones. Let me list a few examples, once again to highlight the hidden assumptions.

(e) The assumption (a) has implicitly assumed that we understand the notion of perpendicular lines, their similarity properties and others. We also know how to draw them.

(f) The assumption (a) also assumed that we are capable of understanding any grammatically correct sentence even if we have never heard a given sentence before in our life. For example, you have never read or heard several of the sentences in this book before in your life. Yet, you can understand them. This is surprisingly nontrivial and leads to an important notion of abstract continuity (see chapter 21).

(g) The assumption (b) has implicitly assumed that we are looking at the same sun every single day. This is not as trivial as it sounds. We can take multiple approaches to convince ourselves about this. The easiest and the most common one is that we generalize based on other dynamical phenomena we observe every day. This is a direct consequence of our consciousness, specifically, from our ability to judge that an external world exists (see section 29).

(h) The assumption (b) also assumed that we, our consciousness and the world, like, say, the arrows we have drawn on the ground have not changed when we sleep and wake up to perform the experiment the next day. This is similar to the reasoning given in (g) above.

(i) The assumption (c) implicitly assumed that when we are indeed looking in the direction of the sun, not in any other direction. This is again not obvious. Note that the images that fall on the retina are actually upside down. Yet, we feel that we are viewing the world in an upright direction. Can we imagine the world to be upside down (and the physical laws suitably modified)? After understanding the complete solution to consciousness, you will see that an upside down world is not self-consistent. There are other reasons why this assumption is nontrivial, as can be understood if we consider the following common variations with our eyes: squinted eyes, only one eyed and other eye abnormalities. Are we still facing the correct direction of the sun in each of these cases?

It is possible to further breakdown assumptions (e) – (i) into even simpler ones. You can continue this iterative process several times. I will not be doing this now because I will be doing this in the later sections, under different contexts. Nevertheless, the steps I am performing should be quite clear.

In other words, even for a simple well-accepted truth like 'sun rises in the east', there are a very large number of underlying assumptions or truths. It is interesting to note that the situation is the same for almost all truths. Given this, what can we conclude from the above analysis and how can we use it to propose a new definition of truth? Towards this end, let me list a few points below that are common to most statements.

1. *All convincing assumptions are truths*: How do you characterize each of these assumptions? They are statements that a majority of people accept as true because they appear convincing to us. This is what a truth is according to the standard definition. Therefore, it is reasonable to include all convincing assumptions as truths.

2. *Lower-level assumptions are closer to the problem of consciousness*: How would you classify the lowest-level assumptions? It is already becoming clear with assumptions (e) – (i) that they are becoming generic, namely, they are applicable to other scenarios as well. In addition, these lowest-level assumptions are closer to the problem of consciousness. For example, that the sun is the same every day, ability to understand language, facing directly at the sun, ability to generalize through repetition are directly related to existence of an external world, space and time as well as free will. The lower we go with the assumptions, the more closer we get to the problem of consciousness.

3. *Assumptions and the analysis are generically applicable to all scenarios*: Is this example unique or does the description given here extend to other scenarios? For example, let us say that we want to verify as a fact whether 'earth goes around the sun'. This example is certainly nontrivial because until Copernicus made this claim, people believed that 'sun goes around the earth'. Let me briefly explain one way to verify this fact. The intention is to show that the analysis and the assumptions are similar to the previous example.

   To verify that the earth goes around the sun, you would need to add a vertical axis at the location where you have drawn east, west, north and south. You now have a three-dimensional coordinate system. Relative to this system, you can measure the various angles corresponding to the locations of various stars at a fixed time during the night. You tabulate this data in a book and perform the same measurements every day. You will observe that the angles seem to precess indicating that the stars are not moving around the earth. They seem to translate. After measuring for a year, you will notice that the stars have returned to the same locations (when you compare with the tabulated data).

The tabulated data now repeats itself. You now generalize that even if you continue these measurements into the future, they repeat every year (valid because of your free will – see (d) above). This experiment can be performed with the sun as well. You can then conclude that the earth is going around the sun. In this analysis too, we have assumed (a) – (i) as well. There are new additional assumptions about space and time that are more explicitly required in this case. Nevertheless, the basic approach and assumptions are equally applicable.

You can see that this approach and a similar set of assumptions are applicable even for, say, Darwin's theory of evolution. Darwin and Mendel, for example, measured properties of plants and animals in detail across several generations. They formulated laws generalized to all past and future situations using free will. This is not to say that such statements are 'universally' true. For example, 'sun goes around the earth' though believed for a long time turned out to be false.

4. *Dynamical network of truths*: What I have shown here is that a statement you want to verify as a fact can be broken down into a large number of assumptions that can be arranged and organized into components tied together with processes. To a conscious person, all of these lowest-level assumptions and the processes are very convincing. He will, therefore, conclude that the original statement is equally convincing and is, hence, a valid fact.

Any given statement is expressed as a complex dynamical network of lowest-level assumptions interconnected in a specific way. A suitable path in this network can be thought of as a process that should be followed to create a convincing argument. This network of truths becomes highly interconnected as we include a number of everyday experiences. I will discuss this in more detail in future sections.

5. *Structure of assumptions for an unconscious person*: What happens when I do not assume consciousness in the above analysis? The dynamical network of truths mentioned above can still be applicable to an unconscious person. However, the difference is that he does not know that each of these elements in the network is a truth. This is a direct consequence of Theorem 12.1. Therefore, the entire structure you have created is an unverifiable network of statements.

This is not a good situation. To understand this, look at what will happen when I replace a given statement in the network with its negation. The unconscious person would not discover any contradictions. He does not know or care which of, the statement or its negation, is 'more' accurate. In fact, he does not even care if both of them are present simultaneously. This is not acceptable in a mathematical axiomatic system. However, this is exactly what happens in a computer. Imagine a computer that generates mathematical theorems algorithmically starting from a set of axioms. If I include a new axiom that contradicts one of the theorems it generates, will the computer stop and claim that this is not allowed? We know that it will simply keep generating more and more absurd statements as if they were all theorems (unless you specifically program to let it look at the entire list, detect errors and throw exceptions).

A conscious person, on the other hand, immediately notices the contradiction (unless they are involved statements that requires deep analysis like 'earth goes around the sun'). I will show how this is possible for a conscious person in section 50.5. Therefore, the standard definition of truth does not help in ensuring that we create the correct network of assumptions instead of an absurd network.

From the above discussion, we can say that one of the most important steps towards creating a correct theory of consciousness is to ensure that every statement, event or assumption is brought to the same footing, using a formal definition of truth. We can further classify them as axioms, theorems, lemmas and corollaries, but the basic notion of truth should be uniform across all events, objects and statements.

Here is a new definition of truth that avoids the above-mentioned problems. The intuition I use comes from the commonly accepted notion that *if an event keeps repeating the same way even under different variations, we tend to accept it as a truth (or reality)*. Conversely, if the event 'changes' under approximately repetitive experiences, we do not accept it as a truth. Therefore, 'repetitions' and 'variations' are both necessary to define truth. Excluding either one would alter the very meaning of truth significantly.

Definition 13.1 (truth): A truth is any representation like an object, event or a statement that does not change significantly (a) over a finite period or a finite repetitive scenarios and (b) under a sufficiently large set of dynamical transformations.

In this definition, the dynamical transformations capture the notion of variations while a large set of variations over a finite period captures the repetitions. The word 'dynamical transformations' refers to the fact that the variations do not have to be limited to static rearrangements (transformations) of the system in space and time. There could be a series of dynamical variations that transform the system over a period of time. As an example, a balloon is a system with several dynamical variations as it moves around in space and time for a finite amount of time before it pops.

Consider any statement or event that we are evaluating if it is a truth or not. The above definition says that we should do the following. We should identify a set of dynamical transformations that are natural for the event. For example, these could be different locations, times and other variations when the event can occur. We then check if the event is occurring approximately the same way in each of these varied scenarios. We then repeat this verification process over a finite period to check if the event still behaves the same way. If it does, we accept it as a truth. If not, we say that it is not a truth.

This definition agrees with our intuition that if an event occurs the same way over and over, we believe it as a truth. If not, it is not a truth. The definition simply tries to make these intuitive notions precise so we can connect it to dynamics. Namely, repeating over and over are identified as a set of dynamical transformations because in each repetition the system is transformed a bit. Also, we capture the realistic case that the number of repetitions do not have to be infinite.

It is clear that mathematical abstractions like the concept of numbers, your emotions, feelings, pain, space and time does obey this definition. A ghostly image does not satisfy this definition because it fails to repeat under several variations. Most lies (non-truths) also do not satisfy definition 13.1 because variations in each repetition produce considerably different results each time. When working with physical systems, evaluation of truth obeying definition 13.1 is easier because experiments can be performed repeatedly and under variations. With abstract mathematical and computational systems, repeated computations and several collections of examples can be used as a way to evaluate the truth of a statement.

If the person observing the event is conscious, he believes the event to be true if the above conditions are satisfied. If the person is not conscious, it is still better to memorize mainly those events that are fixed compared to the surrounding events that do change. Memory can simply be stored in a mechanical way. Notice that the above definition of truth is more generic and it de-emphasizes the role of consciousness in its definition. There is an indirect reference to consciousness through verification of the statement or event.

It is instructive to look at each of the assumptions (a) – (i) above to see if the above definition 13.1 is applicable for all of these cases. For example, the definition of east, sun rising above the horizon, definition of a straight line and others are statements that are fixed over long periods of time. These are repeatable events. They remain fixed because either you are using these terms (east, straight line) consistently every time or the dynamics repeats the same way every time (sun rising above the horizon). The dynamical transformations are, at minimum, space-time transformations (say, space-time locations where you did this experiment repeatedly). The terms like east or the dynamical events like sun rising, do not depend on the location in space and time. Note that 'sun rising in the east' is not accurate closer to the north and south poles. We do understand this case and exclude such corner cases.

Let me now explain several of the terms introduced in the new definition 13.1 in more detail. We will also study various examples and properties of truths.

13.3 Types of truths

I want to emphasize that truth is not just defined with respect to a statement. It is generalized to include events and the existence of objects as well. The existence of a table in front of you is a truth. These other generalizations may seem unnecessary for a conscious being. However, before the system has become conscious, it is important to use real physical events and objects as valid truths to memorize them.

Among the set of truths, some can be called as *universal* truths. If a statement, event or object does not change over time and is the same for *every* conscious being (i.e., we include all 'conscious beings' itself within the set of dynamical transformations of interest), we define it as a universal truth. Some examples of universal truths are the concept of numbers, mass of an object, charge of an electron or proton, continuity, causality within dynamical systems, mathematical logic, emotions, feelings and consciousness.

A given truth does not have to be universally true for all conscious beings. This is useful when you include unconscious beings or any of the previously mentioned altered states of consciousness. For example, a drunk driver could swear that he did not see certain obstacles that are otherwise visible (i.e., true) for a normal conscious being. Therefore, this specific truth is different for him than for a conscious being.

The truths that are not universal are statements that do not change for a finite period of time for a specific conscious being. This distinction is useful to allow a number of other examples of truths that change over time or that change across different conscious beings (like a tooth fairy or Santa Claus). One simple example to highlight the difference between the above two types of truths is, say, the statement about the existence of God. It is not an universal truth. Other examples are those that involve emotions and *qualia*.

13.4 Truths fixed only for a finite period

We defined truth by requiring that it does not change for a finite period of time. We did not require it to be fixed forever. This is because what appears to be true now can later turn out to be false. Existence of Santa Claus is a truth for several years for a child. Similarly, sun going around the earth or that the earth is flat has been a truth for thousands of years until more scientific facts were discovered that led to inconsistencies with other existing facts. These should have been considered as valid truths because they were fixed and verifiable by an individual, at least until a contradiction was discovered. Including these examples as valid truths in the new definition is extremely important for building a realistic and rigorous theory of truths.

13.5 Dynamical transformations

I have required that truth be fixed with respect to a large set of dynamical transformations. The motivation for transformations for evaluating the truth should actually be obvious. When we say that something is not changing or is fixed for a period of time, then inherently the system should have been subject to some changes by the environment. Truth is, essentially, defined as those aspects of this system that does not change in spite of changes in the environment. The changes made by the environment are precisely the dynamical transformations. For example, let us pick one of the events as the average scenario ($\bar{E}$). Then, all other variations are expressed as the set $\{\Delta E\}$ around the average. When we say ice is a solid, there are, clearly, several dynamical transformations (different temperatures and pressures at different times of day, say) that preserves the solid state of ice before it melts into water.

The general set of dynamical transformations were left unspecified. For a given situation, these transformations will become quite clear. In any case, I will discuss a generic set of transformations that are applicable in most common scenarios.

1. *Human brain itself as a set of transformations*: One particularly important set of

dynamical transformations are those that occur right within our brain. Human brain is a dynamical system that is constantly changing. New connections between neurons are formed every day as we grow and gain experiences or memories. The transformations that occur within our brain, as a result of these dynamics, produce changes to the very event, object or statement we are trying to evaluate the truth of. In spite of these transformations within our brain, the concept of a table, chair, numbers, feelings, color and others, for example, do not change. These are valid truths for a given human being even after years of new experiences. Similarly, truth is not the same for every human being. The transformations in your brain may alter the truth or validity of an event or statement compared to my brain.

2. *Space-time transformations*: One common set of transformations that are widely used are geometric transformations like translations, rotations, reflections and stretching. For example, the concept of table that is formed in your brain is the same whether you move towards or away from the table (translation), turn your head around while looking at the table (rotation) or you look at the table through a straight or distorted mirror (reflection or stretching). Another common set of transformations are time-based dynamical transformations. Physical dynamics are examples of such transformations. For example, a basketball is well-defined whether the ball rolls or bounces (dynamics that transforms the basketball). As you can see, the transformations can begin to get more abstract when they are expressed conceptually in terms of these basic ones. For example, changes in functional features, changes in colors, changes in compositions or subcomponents and changes in terminology or definitions (say, in mathematics) are all useful transformations.

If the set of transformations gets larger or a given statement remains fixed for a long time, the statement begins to get closer to a universal truth.

13.6 Examples of truths

I have already given several examples for evaluating the truth of a statement. I will list additional examples briefly and evaluate them with the new definition. For example, when we say that the temperature is 75 °F, there are clearly a number of transformations with particle space-time locations that do not change the average kinetic energy and, hence, the temperature. The same holds with the notion of solids, liquids and gases. A statement like 'this is my hand' is true even when I move, twist and turn my hand in any direction.

In mathematics, we can make a distinction between universal truths and *derived* truths. Axioms are considered as universal truths while the theorems are derived truths. For example, '1' is a single item. The set of transformations are '1' of anything whether it is one candy, one balloon or one of any other tangible object. Note that we only consider tangible objects, for which the truth is already established, when defining '1'. After we define '1', we can treat it abstractly without any relationship to tangible objects. A statement like '3 < 10' is a truth because of the truth about the numbers and the corresponding definitions. Theorems that are more complicated can be derived from definitions (such as '1') and an accepted set of axioms.

Summarizing this, we can say that the definition of truth 13.1 captures the link between abstractions and dynamics. It captures facts that are closer to reality, the time varying nature of truths, universal and other types of truths as well. This definition is both generic and does not emphasize consciousness.

We can use the new definition of truth to create truths. We also use truths to form new relationships between truths and hence create new truths again. Our brain is simply a dynamical network of truths. Truths can change over time. In the next chapter, I will show that truths are represented as *fixed sets* (see definition 6.3 in chapter 6) in our brain. Therefore, we will see that our brain network of truths is equivalent to a network of fixed sets.

14
Truths as Fixed Sets

The main objective of this chapter is to explain how we can represent truths within complex dynamical systems. Note that since truths are abstractions, there are two aspects to truths – the mechanics of storing and retrieving as well as the meaning (or knowing) associated with a truth. For example, if you see a table in front of you, its existence is a truth. You can both (a) detect it mechanically and (b) know about its existence. In this chapter, I do *not* address (b) at all, namely, how you know about truths. I will discuss this topic (b) starting from chapter 20. For now, I only discuss the best way to represent truths in a mechanical sense (i.e., (a) above) that allows features like self-learning, associative memory, capturing all types of relationships between truths including hidden ones, ability to discover new truths and so on. In the previous chapter, we have seen several of these relationships, though in an abstract sense using logical reasoning. Such a reasoning had no connection to the raw physical and chemical dynamics that occurs within our brain. In this chapter, I want to express these same relationships in a dynamical sense using a suitable representation of truths.

There are multiple ways to represent truths that satisfies definition 13.1. I will briefly discuss them first. I will argue that most of these representations do not give rise to features like self-learning and others mentioned above. I will then show that *fixed sets*, which are generalization of loops (see section 14.2), satisfies definition 13.1 and is a better representation of truths. Such a representation is applicable to all complex SPL systems including our brain. They also model emergent properties within SPL systems. This will give us a way to start from dynamics, obtain fixed sets from dynamics and, eventually, lead to the abstract notion of truths.

Therefore, truths and fixed sets will be treated as equivalent. This equivalence suggests that we can move from abstractions to dynamics and vice versa seamlessly in, at least, this one important case (with truths). The surprising and interesting aspect, to be discussed in further detail in subsequent chapters, is that most statements or events in the external world like the existence of external world, space, passage of time, existence of objects, their physical and chemical properties and so on are already truths or interconnections between truths (cf. examples of section 13.6). Therefore, this equivalence lets us move between abstractions and dynamics seamlessly for almost all common experiences related to consciousness. This reduces the well-known knowledge gap between our thoughts, expressed abstractly, and the corresponding physical and chemical reactions, expressed as dynamics, within our brain. Closing this knowledge gap has been one of the biggest challenges with the problem of consciousness until now.

14.1 Existing approaches for representing truths

There are multiple ways to represent truths that satisfy definition 13.1. A common representation is a digital one similar to how we store facts in a computer. Consider, for example, the truth about the existence of a table in front of you. A computer with a video camera attached to it represents such a truth as follows. The camera starts to stream image frames at a given rate (say, 24 frames per second) to the computer. Each frame has a certain resolution like 1080p. These are just numbers corresponding to the RGB representation at each pixel of a rectangular area corresponding to a single frame. An encoding (like MPEG) is applied to all of these numbers for each frame as well as across multiple frames to compress the amount of data needed to store. The resulting encoded data is the representation for any given object. The stored data for a table will be different from that of a chair. The encoding is different for other types of truths like sounds, languages and mathematical concepts as well.

Of course, this representation, by itself, does not satisfy definition 13.1 for truth because representations of different variations are not similar to each other. Let us try to see if we can alter and/or add new information to the above representation to satisfy definition 13.1. First, we should extract just the subset in the image frame corresponding to the truth we want to represent. In the above case of a table, we need to identify just the table from the series of movie frames using some image recognition algorithm. Such a process or an algorithm has its own representation in the computer like, say, as Java source code files compiled into a 'jar' file that can be run in a computer. The input movie frames will be processed by this 'jar' file to extract just the regions in the image frame corresponding to the table, say. So, we already have two different representations to start addressing a specific truth – one is the data representation i.e., of the raw data being processed and the other is the algorithm representation i.e., of the algorithms that process this raw data.

Second, we need a collection of dynamical transformations that act as different variations of the event being observed (existence of a table, in this case). Some examples of these are: looking at a table from a distance, from nearby, at various angles and so on. In each of these variations, the image recognition algorithm needs to be constantly improved to (a) recognize the table as an object and (b) to identify this object as the same table among the image variations. Among the two representations, the raw data representation of the table digitally is considerably different for each variation, even though the encoding has not changed. This is simply because the table is at, say, a different angle in one image compared to another image. So, our effort goes towards improving our algorithms rather than trying to change the data representation. A typical algorithm identifies a number of features from an image frame for each variation and represents them as, say, a vector of numbers. It then compares this feature vector with other feature vectors previously identified for each object using some similarity metric. The object is classified as a table if it is most similar to the feature vector of a table. Several such classification algorithms exist in the literature (see Hastie et. al [67]).

Initially, a conscious being is required to improve the algorithms so the image recognition algorithms become better. Conscious humans also change the data representation for reasons like reduced storage space and others. However, we can assume a time when the algorithms have sufficient accuracy that the presence of conscious humans is not a requirement for detecting and representing truths.

Given this, what is the final representation of a truth in a computer for, say, the existence of a table? When the computer receives images from its video camera, the raw data representation has considerable variability to be useful across multiple dynamical transformations. If we have to look for components that do not change under all these variations, we will not be able to identify the table to be the same just from the raw data in the image frames. It is, therefore, reasonable to compute the average of all feature vectors mentioned earlier and pick it as the representation for the table. The algorithms themselves should not be part of the representation of a truth because it is common across entirely different truths related to visual images like tables, chairs and so on.

One of the main difficulty with such a representation is clearly a lack of self-learning ability for the system. There is no notion of learning by exploration by the system itself. Conscious humans are needed to guide the system. The second difficulty is that the system finds it nontrivial to capture the relationships between the truths. Recall the discussion in section 13.2 where I have identified a large number of relations that exist between truths. Such relationships exist everywhere around us. The representation of truths as a vector of numbers is a poor one to help compare and relate seemingly unrelated truths like, for example, relating an image, its sound and a corresponding emotional experience. Other examples are natural relationships between abstract concepts present in a long event like all the events that occurred in your recent vacation. The third difficulty is that the representation is a static one, not a dynamical one. There are several examples of truths that appear to be the same externally but provide different meanings based on context like time of day and other internal states of the system. If we use the current raw data representation, we will have difficulty capturing such dynamical notions. Besides, a number of features satisfied by the new memory model described in chapter 7 are not captured by the current digital representation of truths.

Other representations that have similar issues are identifying truths by simply referring to a collection of unstructured set of neurons that are excited when you are thinking about an abstract concept. For example, when you are thinking about a 'table' or when you are presented a set of images/sounds, we try to locate the regions or the set of neurons excited within your brain using an instrument like MRI. We repeat this experiment several times. This lets us identify a set of excited neurons common across all trials. This common set is then identified as the dynamical representation of the abstract concept of 'table'. No structure within this common set is identified. This makes it very difficult to compare the common sets for widely different abstract concepts like for tables, chairs, mathematical concepts, a pizza, sound of an airplane, gravity, your happiness and so on. All other issues mentioned previously are also not addressed satisfactorily using such representations. The new memory model using loops and fixed sets avoids most of these difficulties, which I will discuss next.

14.2 Representing truths as fixed sets in SPL systems

What is unique about complex dynamical systems like our brain? Any complex interconnected network of neurons is bound to have several structural and dynamical loops as discussed in chapter 6. To understand the role of loops, let us first look at the general properties of dynamical systems.

There are several ways to view a dynamical system. It is common to use phase-space (or state-space) representation when studying them mathematically. Here, we first represent the full state of a system in terms of n independent generalized position variables $x_1(t), ..., x_n(t)$ and a corresponding set of n independent generalized momentum variables $p_1(t), ..., p_n(t)$. Such a system is said to have n degrees of freedom. Each of these variables are a function of a single time variable t. For example, a simple pendulum of length l has a single degree of freedom $\theta(t)$ and a generalized momentum $p(t) = l\dot{\theta}(t)$ that specifies its location at any given time.

The dynamics of a system is now expressed as a set of $2n$ first-order ordinary differential equations. If $X(t)$ is a n-dimensional vector of position variables $(x_1(t), ..., x_n(t))$ and $P(t)$ is the corresponding n-dimensional vector of momentum variables $(p_1(t), ..., p_n(t))$, we can represent the dynamical equations of motion as follows: $\dot{X}(t) = F(X(t), P(t), t)$ and $\dot{P}(t) = G(X(t), P(t), t)$ where $F(.)$ and $G(.)$ are two functions. For example, for a simple pendulum, $\dot{\theta}(t) = \frac{1}{l}p(t)$ and $\dot{p}(t) = -g\sin\theta(t)$ where g is the acceleration due to gravity.

If the initial conditions $X(0)$ and $P(0)$ are known, we can solve these first-order differential equations for all time $t > 0$. The solutions $(X(t), P(t))$ for $t \geq 0$ can be plotted on a $2n$-dimensional plane as a directed pathway starting from $(X(0), P(0))$. For each point $(\bar{X}, \bar{P})$ in the $2n$-dimensional plane, we can compute directed pathways by solving differential equations with initial conditions $X(0) = \bar{X}$ and $P(0) = \bar{P}$. This $2n$-dimensional plane is termed as the phase-space representation of the dynamical system. For example, Figure 14.1 shows the phase-space representation of a simple pendulum. This system has a single stable point at the origin. If the initial condition is at the origin, the system continues to stay at this location. For small initial conditions (within a region called a separatrix), the pendulum oscillates back-and-forth represented as a pseudo-elliptical looped pathway in the phase-space. If the initial conditions are outside the separatrix, then the pendulum can make full turns and move away from the origin (in phase-space). For other examples and more formal study of linear and nonlinear dynamical systems, I refer the reader to textbooks like Khalil [84] and Isidori [76].

Another way to view a dynamical system is as a continuous 'movie' with sequential frames of events along the time axis. With this representation, we see that every frame is a continuous transformation in space parameterized using a time variable.

I will use both representations during the current discussion to highlight important concepts qualitatively. For example, since the definition of truth has transformations in them, we will use the movie representation to clarify this aspect. The frames i.e., the transformations deform continuously along the time axis from one instant to the next. Secondly, the definition of truth also talks about events to be

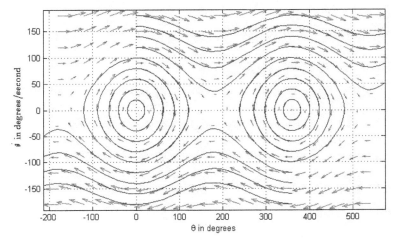

Figure 14.1: Phase space representation of a simple pendulum. If the initial condition is at the origin, then the system stays at this location. If it has small amplitude, then the pendulum keeps oscillating back-and-forth represented by a pseudo-elliptical looped pathway in phase space. If the initial velocity is fast enough, then the pendulum makes a full turn and moves away from the origin. These are represented by lines at the top and the bottom that go to 360°, 720° degrees and so on.

unchanging over time. It is easier to see this aspect as a 'fixed' state in the phase-space representation. Therefore, both viewpoints offer insights into the properties of dynamical systems needed for the current discussion.

Let me now relate the definition of truth expressed in abstract terms in chapter 13 using the dynamical systems concepts discussed above. The definition of truth has two aspects: (a) a set of dynamical transformations (i.e., the variations) and (b) the ability for an event to remain unchanged in spite of these variations (see definition 13.1).

To see equivalent concepts in dynamical systems, note first that continuous transformations have a notion of fixed points. For example, Brouwer's fixed point theorem states that any continuous transformation from a closed unit ball to itself has at least one fixed point. There are several interpretations of this theorem. One such interpretation is as follows. Let us take water in a cup and stir it for a while. The water comes to a halt displacing molecules from the initial location to some other locations. The water molecules before and after the stirring have moved within the cup in a continuous way. Brouwer's fixed point theorem says that if we perform point-wise comparison of points before and after the stirring, there is at least one point that did not move its position. This point that did not move is called a fixed point. The existence of this fixed point under the above continuous transformation, is a nontrivial result. There have been several proofs, generalizations and even far reaching applications of Brouwer's fixed point theorem (see Dugundji and Granas [32]).

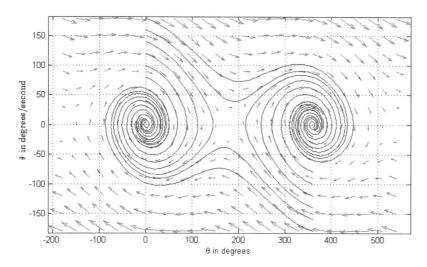

Figure 14.2: Phase-space representation of a damped pendulum. This system has several asymptotically stable points at multiples of 360°. For every initial position and momentum of this pendulum, the system eventually converges to one of these stable points.

The notion of fixed points appear close to our definition of truth. A fixed point is one that does not change under a single continuous transformation. A truth is quite similar – it is a statement or event that does not change over a period of time under a sufficiently large set of dynamical transformations. We now need an extension of the notion of fixed points to cover beyond a single continuous transformation to include a set of dynamical transformations. For this, we will look at the notion of limit cycles in nonlinear dynamical systems.

In a dynamical system, fixed points are analogous to stable attractors. Figure 14.2 shows the phase-space representation of a damped pendulum (cf. an undamped pendulum in Figure 14.1). The equations of motion are slightly altered: $\dot{\theta}(t) = \frac{1}{l}p(t)$ and $\dot{p}(t) = -g\sin\theta(t) - \alpha p(t)$. This system has an asymptotically stable attractor at the origin (and at every multiple of 360°). Namely, it is the rest point when the pendulum comes to a halt. If the system starts its dynamics at any point other than these asymptotically stable attractors, it eventually converges to one of these stable points. Physically, all of these stable points correspond to the same rest position (vertically down with no oscillations) of a pendulum. In phase-space, this single rest position corresponds to multiple points corresponding to multiple full rotations. This convergence is represented by a spiral pathway to the stable points (Figure 14.2).

In addition to asymptotically stable points, some nonlinear dynamical systems have stable limit cycles as well (see Khalil [84]). Figure 14.3 shows the phase-space representation of a simple nonlinear dynamical system that has a limit cycle. The

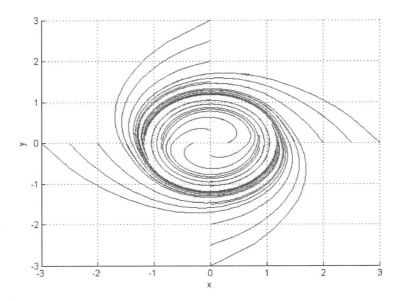

Figure 14.3: A simple example of a nonlinear dynamical system that has a limit cycle. The unit circle in the phase-space is a limit cycle for this system. All initial conditions inside the unit circle spiral outward to converge to this limit cycle while all points outside the unit circle spiral inward to the limit cycle. The points on the unit circle move around but stay on the limit cycle. This system also has an asymptotically 'unstable' point at the origin.

dynamical equations of this system, represented in the polar coordinates, are: $\dot{r} = r^2(1 - r^3)$ and $\dot{\theta} = r^2\{(1 - r^2) + \sin^2 \theta\}$. If the initial conditions in the phase-space are outside (or inside) the limit cycle, the evolution of dynamics causes the pathways to asymptotically approach the limit cycle. If, on the other hand, the initial conditions are on the limit cycle itself, they continue to stay on the limit cycle. I will call these limit cycles as "fixed sets" instead of fixed points. If the system enters a stable limit cycle, it stays on this limit cycle. The entire set of points on the limit cycle can be viewed as "fixed" even though no single point on it remains fixed. If the system is sufficiently close to the limit cycle, but not on the limit cycle, then it proceeds towards the limit cycle because of stability.

The term 'invariant sets' is sometimes used to refer to the notion of fixed sets introduced here. Clearly, the entire phase-space is a trivial invariant set. Our goal is to identify nontrivial subsets of the phase-space that are invariant for a nonlinear dynamical system. Not all systems have limit cycles, but when they do, these systems sustain their dynamics for sufficiently long time when the initial conditions let them enter the limit cycle. Fixed points (i.e., stable points), on the other hand, typically are points of rest and are not as interesting as limit cycles.

For structured complex SPL systems, we have seen the pervasive nature of loops already. Even in the inorganic chemical world, limit cycles are shown to exist (see section 2.6 on chemical oscillations and Field and Noyes [43]). If the system has sufficient complexity and a high degree of interconnections, it is easy to form a large number of dynamical loops. In fact, it is impossible not to have loops within an adult human brain network. For example, even a simple connected directed graph with hundreds of nodes will have several loops. Our brain, on the other hand, has billions of nodes and trillions of interconnections. Of course, a newborn's brain does not have as many interconnections and loops compared to an adult human.

If an external input corresponding to an object or event triggers parallel excitations from the sensory surface like the retina, these propagate into the brain to form a number of dynamical loops that sustain the firing patterns for sufficiently long time (see chapter 6). As the inputs are constantly supplied, the loops continue their dynamics for a while and are, therefore, stable. Each such loop triggered in response to an external input can be regarded as a limit cycle. The set of neurons that are part of the loop are together fixed with respect to the overall dynamics even though no individual neuron within the loop is static. Loops, as discussed in this book, are an important generalization of invariant or fixed sets. Therefore, we can say that *fixed sets are good candidates for truths because they do not change for a period of time under a set of dynamical transformations.* Truths are 'abstractions', like statements and events, that do not change over a period of time in spite of several variations. Similarly, fixed sets are 'dynamical' representations that do not change under a set of variations in initial conditions.

Let us now look at the loops of neurons that fire in response to an external signal and compare them across several different variations of the external signal. Consider an SPL dynamical system like our brain. Let us pick the stable looped firing path of neurons when you are, say, looking at a tree. As you now walk towards, away or in any other direction from the tree while still looking at the tree, the dynamics and the corresponding set of loops within the brain keeps altering. These set of loops are quite different for an infant compared to an adult because of the low degree of connectivity in an infant.

If we look at the loops of neurons before and after your motion (i.e., before versus after walking or turning), the situation is quite analogous to looking at the water in the cup mentioned earlier, before and after the stirring. Just as the water in the cup before and after is subject to a continuous transformation, your motion (as walking while still looking at the object) also produces several continuous transformations in the neural firing patterns of your brain before and after your motion. For the water cup case, Brouwer's fixed point theorem states that there is at least one point that did not move its location. The expectation for the dynamics in your brain subject to your motion is that we have a similar notion of points, not one but several, that are fixed and also not individually like the water cup case, but together as a set of points, namely, the fixed set. Each of the points (or molecules) within the fixed set would move their locations, but when viewed over a period, this set as a whole stays in the same region. In other words, these dynamical transformations, namely, walking or turning, have produced "fixed" sets that do not change (as a set – even though individually, they do change). This is precisely the definition of truth 13.1 of chapter

13.

Our goal is to identify proper subsets that are fixed over a small enough time period. We are not interested in the entire firing set of neurons because they trivially satisfy the invariance condition. For every continuous transformation of interest, there will always be an invariant set of neural loops for a short period as long as our brain network is sufficiently dense. The question is: does one invariant set A_1 for one continuous transformation T_1 have anything in common with another invariant set A_2 for another continuous transformation T_2 i.e., is $A_1 \cap A_2 \neq \emptyset$? For example, consider transformation T_1 to be walking towards a table and T_2 to be tilting your head while still looking at the table. Our brain generates two invariant sets (i.e., a collection of loops) A_1 and A_2 respectively for each of these two transformations. If $A_1 \cap A_2 = \emptyset$, then our brain says that whatever we saw when walking towards the table is different from what we saw when tilting our head. This is not a good situation because we want to say that we saw the same table in both cases. This situation can happen for a newborn baby because of his sparse network of connections in his brain. The truth statement we want to capture in this case is that 'what we are seeing is the same table'. In a colloquial sense, this is a truth statement because the table being the same does not depend on who is watching it and how. Therefore, we want to satisfy $A_1 \cap A_2 \neq \emptyset$ in this case.

In general, there will be a much larger set of transformations $T_1, T_2, ..., T_n$ for every such 'truth' statement that can be considered as valid or reasonable set of variations (like walking towards, away, turning, tilting, jumping, running, seeing different colored cars and so on). We want our brain to generate invariant sets (loops) $A_1, A_2, ..., A_n$ respectively that eventually satisfies $A \stackrel{\text{def}}{=} A_1 \cap A_2 \cap ... \cap A_n \neq \emptyset$. I say 'eventually' because it is likely that our brain starts satisfying the above nonempty condition for a smaller set of transformations initially (say, for a newborn). Over time, he would learn and form enough interconnections to realize the truth under much larger set of variations. The nonempty intersection A is what we call as the *fixed set*. Notice that this is precisely the similarity condition and fixed set defined in chapter 6.

It is important to realize that the nonempty intersection condition can satisfy even for adults, not just for newborn babies. In the case of adult humans, these typically correspond to events for which we have not yet formed memories. For example, for abstract mathematical concepts like groups, rings or fields in algebra, it is likely that we do not have fixed sets that let us identify as groups or rings just by looking at a set of examples even though we have seen these examples in the past. In such cases, we have not yet formed 'fixed sets' for the concept of groups or rings relative to the set of transformations represented by these examples. When you are learning to play piano, you do not have fixed sets for, say, major and minor chords and other patterns, as you look at the sheet music. However, with enough practice, you do recognize common patterns as fixed sets. A doctor with less experience does not have fixed sets if we include a large number of variations of sicknesses compared to a more experienced doctor. He will only have $A \stackrel{\text{def}}{=} A_1 \cap A_2 \cap ... \cap A_n \neq \emptyset$ for small n i.e., for small number of variations. However, with experience, he will create nonempty intersections A for a large number of variants $N > n$ of a given sickness. He can make use of this nonempty intersection to predict your sickness even under new situations.

In the case of infants, the lack of fixed sets is more prominent even for several common visual and auditory inputs. For example, an infant when seeing the same person or object at different angles, distances or in different rooms (i.e., a typical set of dynamical transformations) does not yet have a common set of neural loops triggered. In other words, he does not have a fixed set for this object or a person yet. Hence, the same object or person does not yet feel the same to him. It is not a truth because it is not fixed over time. This may feel strange to an adult – how is the same table looked under different contexts not the same? But for an infant, with his current network of neurons, all these variations may indeed produce different sets of firing neurons. The existence of a table is not a truth yet for an infant if there is no common set of firing neurons across all of these different contexts. In fact, how can an infant say that he is looking at the same table under different contexts if the resulting firing neurons are different each time i.e., when $A = \emptyset$? In the current literature as well, researchers do look for a common set of neurons across different variations as a way to identify the object. However, in this book, we are not looking for 'any' common set of neurons, but only those neurons that form stable parallel looped dynamics (for reasons covered in previous chapters and sections).

This invariant set is precisely the minimum set of loops that did not change their dynamics as a whole over a finite period and those that are common before and after your motion i.e., they are fixed as a set, not individually. They may not exist yet, but if they do, they form a natural converging fixed set for this transformation. This fixed set (as the minimum set of common loops) is precisely the reason why a tree looks and feels like a tree no matter what you do, like say, walking, running, jumping, turning your head at an angle and others. These different types of motions like walking, jumping and others are the set of transformations needed in the definition of truth.

Recall the definition of truth in section 13.2 and see the similarity with a fixed set: truth is an event that does not change for a finite period under a sufficiently large set of variations – a fixed set is a collection of dynamical loops that continue to fire for a finite period under a sufficiently large set of dynamical transformations. While truth was a statement expressed abstractly in a language like English that only a conscious human can understand, fixed set is a statement expressed entirely in terms of dynamics that occurs within an SPL system. The latter representation, namely, the dynamics described by a fixed set can occur in a system irrespective of the state of consciousness of that system. Since there is a one-to-one correspondence between truths and fixed sets, I will treat them to be equivalent and interchangeable concepts. *We, therefore, have a way to move from statements expressed in an abstract language (like English) to an equivalent physical and chemical dynamics occurring within SPL systems and vice versa.*

In fact, everything we see around us should be represented as a fixed set. When you see a table, a chair, a tree or a car, you know them correctly even though that specific car or that specific tree is something you have never seen before. A child can identify this easily as well even though his set of experiences are limited. The process of creating fixed sets may take time, but they do happen. For most sensory related fixed sets, this happens in the early childhood. However, for more abstract concepts, the creation of fixed sets happens even when we are adults.

14.3 Location of fixed sets in an SPL system

Where are the fixed sets physically located within our brain (or, in general, in an SPL system)? The answer is simply that they can be anywhere within the brain (or within an SPL system). Some fixed sets corresponding to edge detection, angle detection, motion detection and other primitive visual properties are closer to the sensory regions (like the primary visual cortex). The same applies with primitive sensory properties for sounds, tastes, odor and touch. Other fixed sets corresponding to concepts, objects and words in a language can be considerably deeper within the brain.

Wherever they are, it is important to realize that fixed sets are those dynamical loops that can be used to uniquely identify a given object or event. For example, you can identify a chair by looking straight at the chair. Or you can identify the same chair without directly looking at it, but by looking at a neighboring table. Or you can identify the same chair when it is tilted or turned at an angle. Or you can identify the same chair by closing your eyes and imagining the chair (to a degree). Or you can identify the same chair in your dream with your eyes closed. All of these examples imply that the image of the chair does not have to fall at a precise location on the retina if you were to identify it. As long as its fixed set, which is typically located deeper in the brain, is triggered, you can still 'see' the chair, even with your eyes closed. This is a generalization of the similarity condition discussed in definition 6.2 (see section 6.6). Section 7.8 discussed how to create fixed sets satisfying the similarity and dissimilarity conditions in SPL systems.

We can use the same notion of fixed set even for output generation just as we use it for inputs. For example, when you want to move your hands, legs or eyes, you need to trigger a set of motor neurons. The motor neurons release neurotransmitters, typically, acetylcholine onto the muscle fiber. These bind to the postsynaptic receptors causing the muscles to contract. Now, there are multiple set of neurons within the brain that can trigger a given output response. These 'loops' of neurons can be identified as the fixed sets for each type of control action. We will soon see that it is advantageous to identify the fixed set within the brain as the representation of the object rather than just the unique pattern at the sensory surface (see how we tried to satisfy similarity and dissimilarity conditions in section 7.8). In the same way, it is useful to identify a fixed set within the central nervous system for a given control action rather just the set of neurons that are directly connected to the muscle fibers. This lets us study sensory and control problems in a similar way. For example, Georgopoulos et. al [52] (and their subsequent generalizations to three-dimensional movements) have introduced the notion of preferred directions in the motor cortex. They have seen that a population of neurons encodes direction of movement.

How do we create fixed sets in a generic SPL system? In chapters 7 and 9, we have already seen how to create a large collection of fixed sets using suitable training sets for images, sounds, words in a language and so on. We have also seen how to let the system create more fixed sets automatically through self-learning. For this, we have seen the need to satisfy similarity, dissimilarity and connectedness conditions. The above mechanisms and patterns for creating specific fixed sets are for generic SPL systems. Yet, we can easily extend them to apply to SPL network of neurons as well. Even though I will present these mechanisms in the later chapters, most of the analysis needed for consciousness only requires the existence of these specific fixed sets. The mechanisms for both creating and linking the fixed sets are particularly important when we want to construct synthetic conscious systems.

As a human is exposed to new experiences, he learns and forms several fixed sets in his brain. These are well-defined irrespective of whether we become conscious or not. Two different people need not have loops in similar regions of the brain for the same object, like say, a tree. Both do need to form fixed sets and relate different fixed sets together. However, each person would create a fixed set for the same object at different locations and will relate to different fixed sets in different ways. These manifest as different experiences and different memories recollected by different people even when observing the same object or event. The existence of fixed sets in our brain is the cause of recognition of the object itself. The concept of the object does not change even though the dynamics within our brain is subject to changes continuously. We will revisit this topic once again in chapter 24. The existence of fixed sets in a sufficiently large-scale structured and interconnected SPL system like our brain is a fundamentally important property.

What is the nature of a fixed set in our brain? When everything else is continuously changing and moving around, the fixed set does not change. Fixed set is, therefore, an emergent property from within a dynamical system. It is an abstraction converged from the internal physical dynamics within the brain. Every fixed set can be viewed as an abstraction that may or may not be associated to, say, a word in a language like English. It is possible to start creating 'shortcuts' like words in English to relate to a fixed set and then manipulate the abstractions (or words) directly. This allows us to identify and generate richer patterns, which manifests as intelligence and ingenuity. We will revisit each of these advantages of fixed sets in the later chapters.

How are different fixed sets related? What happens when several of the fixed sets are excited almost simultaneously? Can these interrelated fixed sets self-sustain and stay excited for a long time? How do they give rise to features like free will, sensing space and passage of time? I will discuss each of these questions in the subsequent sections. Specifically, I will show how a sufficient number of interrelated fixed sets in a large-scale structured SPL system can give rise to consciousness (see chapter 23). For now, we can say that *every abstraction, including truths, is represented in our brain as a fixed set*. Each of the relationships between abstractions is purely relative, expressed as links between fixed sets. The mapping between our external experiences and the internal representation as relationships between the corresponding fixed sets will be carefully explored in the rest of this book.

We can certainly identify where each of these fixed sets are physically located in

Figure 14.4: Meta-fixed sets are fixed sets created from other fixed sets instead of directly from external inputs. They allow us to talk *about* the system. An example of a fixed set is a specific table. A meta-fixed set, on the other hand, represents a generic table instead. We can similarly create meta-meta-fixed sets and so on. In this figure, even though we have shown the arrows in only one direction (from fixed sets to meta-fixed sets and so on), there will be pathways in the reverse direction as well.

the brain by performing experiments. For example, if you are looking at a table and have excited the corresponding fixed set, we can locate the specific region for this fixed set using brain-scanning techniques like PET, MRI or MEG. However, what is sufficient for our discussion is just the existence of each of these fixed sets, which can be guaranteed whenever there is a sufficient complexity of our brain network.

14.4 Meta-fixed sets – about the system

Given that all abstractions are represented as fixed sets in our brain, it is useful to distinguish abstractions representing a given object from abstractions that talk about a collection of objects. I use the term fixed sets for the former abstractions while I use meta-fixed sets (namely, fixed sets of fixed sets) for the latter abstractions. It is the meta-fixed sets that provide our brain or general SPL systems an ability to talk *about* the system. For consciousness, it is not sufficient to describe facts or truths within a stable parallel looped framework, but it is also important to talk *about* the facts themselves and represent it back into the same stable parallel looped framework.

If you look at any object in your room, say a table, there is a fixed set formed in your brain, according to definition 7.3. This corresponds to loops in your brain that remain excited under several geometric transformations like when you walk towards the table or when you turn your head while still looking at the table. These dynamical loops are clearly specific to the table you are looking at. We treat the representation of this 'specific' table as a fixed set instead of a meta-fixed set. Fixed sets can, therefore, be thought of as an internal representation of external phenomena.

There are, however, additional fixed sets in your brain. For example, the word 'table', the notion of a generic table (not this specific one), the properties of a generic table, the shape and the materials that a table is made of and other functional properties have corresponding fixed sets as well. I want to call these as meta-fixed sets instead of merely fixed sets even though the previous definition of fixed set is equally valid for these cases. Meta-fixed sets, unlike fixed sets, do not have a direct association to external phenomena. Nevertheless, meta-fixed sets can both excite and be excited by fixed sets directly (see Figure 14.4).

The reason why we call them as meta-fixed sets is that they no longer correspond to a specific table even though they have several invariant dynamical transformations. If you see any other table, the same meta-fixed set in addition to the original fixed set mentioned earlier, will be excited. These meta-fixed sets are generalized from a 'base' fixed set, which is a specific table. As you gain experiences and see multiple different tables in your life, you form the above generalized meta-fixed set. Each example of fixed set created from multiple instances of similar objects (like a table or a chair) is truly a fixed set of fixed sets. I, therefore, call them as meta-fixed sets.

Meta-fixed sets are, as a result, new abstractions resulting from several fixed sets. Fixed sets are related to external objects directly whereas meta-fixed sets are related to external objects indirectly. This abstraction lets you talk about the fixed sets themselves (see Figure 14.4). You can form sentences about tables. For example, you can say 'I placed my computer on a table', without referring to any specific table. With descriptions such as these, we are using meta-fixed sets as abstractions.

In Figure 14.4, we have shown the direction of pathways from fixed sets to meta-fixed sets and so on. This is typically the case when a child is learning new concepts. He would start with simple fixed sets that are specific examples he encounters (like a specific basketball, a specific set of numbers and so on). Over time, he will either encounter or begin to use these specific fixed sets in a number of different situations. As a result, he will be forced to create fixed sets from other fixed sets instead of directly from the external inputs. These become the meta-fixed sets as shown in Figure 14.4. In the process, there will be connections from meta-fixed sets back to fixed sets as well (i.e., he can recollect a specific table – a fixed set – when he thinks of a table, in general – a meta-fixed set). In this way, there can be loops between fixed and meta-fixed sets as well. With more experiences, he can even form meta-meta-fixed sets and so on.

The unique aspect of the SPL network and the concepts of fixed sets is that the representation of meta-fixed sets is once again the same form as fixed sets, namely, a collection of nonempty intersection of loops generated from multiple variations of an object, concept or an event. Even though Figure 14.4 appears as a hierarchy, the presence of loops between meta-fixed sets and fixed sets makes them more general than the conventional hierarchical structures. There is no strict notion of 'levels' in a hierarchy with the SPL representation of Figure 14.4. In a standard hierarchical descriptions, we also have different representations at different levels. This difficulty is completely eliminated here because everything is eventually represented back as a collection of loops.

Let me briefly mention another interesting example of a meta-fixed set, which we will discuss more later as well. This is a special meta-fixed set associated with consciousness, namely, the notion of 'I'. When you move your hands, legs, fingers and other parts of your body, you have fixed sets that correspond to each of these parts of your body. However, you also need to have a new meta-fixed set that corresponds to 'I'. This meta-fixed set will be triggered each time any part of your body sends a message to your brain. The word 'I', the associated feeling and other aggregated properties applicable to the entire person can similarly be called as meta-fixed sets.

The formation of these meta-fixed sets is a much slower process than the formation of the corresponding body-part-fixed-sets. This slow process suggests that a human being can know and use different parts of his body while still not knowing himself or not becoming self-aware (see section 43).

In fact, there are so many meta-fixed sets formed in your brain that it is valuable to study the structure of the interwoven meta-fixed sets directly. In chapter 19, we will see that it is possible to create an entire self-sustaining 'membrane' of meta-fixed sets, provided the brain has a certain asymmetric structure.

15
Properties of Fixed Sets

In the previous chapter, we have seen that each truth is represented internally as a fixed set. Let us now try to understand the intricacies behind simultaneous excitation of fixed sets as well as the interrelationship between several fixed sets.

15.1 Simultaneous excitation of fixed sets

In this section, I want to show that our consciousness imposes limits on (a) the time duration required to excite a given fixed set and (b) the time interval for how long the fixed set is capable of self-sustaining in the absence of inputs that triggered it. In the next section, I will show how these limits translate to generation of brain waves of certain frequencies in order to be conscious. Let me work with an example to highlight this.

Let us say that you are in the kitchen and you are looking at the dining table. As you look at the table, a number of neurons from the retina are excited. They propagate within the brain to produce a set of loops of neuron excitations. Now, let us say you move around the kitchen while still looking at the table. You walk towards it, away from it, bend down and look at it. Maybe someone else is even moving the table as you observe it. During all of these continuous changes, as you are still looking at the table, there are new excitations in the brain that will intersect with some of the previous excitations. We can identify the common set of neural loops as the table fixed set as discussed in the previous chapter. We can guarantee the existence of such fixed sets during these types of continuous motions whenever there is sufficient complexity and a high degree of connectivity in our brain network (cf. how we trained an SPL network to satisfy similarity and dissimilarity conditions in chapter 7).

Under repeated observations, the same table fixed set will be excited each time. This is possible as soon as the synaptic strength of the interconnections between the corresponding neurons becomes high enough. We can, therefore, say that we have recognized the table object whenever we trigger this table fixed set, at least, in a mechanical sense and vice versa. Let me generalize this statement as follows: *the mechanics of recognition involves identifying the fixed set for any given object, event or experience.* I will discuss the feelings associated with recognition later in section 22.

Now, if you are no longer looking at the table, you still have excitations in your brain that corresponds to the table fixed set. This is primarily because the table fixed set consists of several loops. We have seen that loops have a property that the dynamics will sustain naturally for a sufficiently long time. This sustained motion produces an equivalent effect of having a thought related to the table even while you are not looking at it anymore. This would have not been possible if not for the notion of loops introduced in this book. In this sense, a theory of consciousness relies on the notion of loops in a fundamental way.

Next, you turn you attention to the chair. Again, as you are moving around the kitchen, let us say that you are looking at the chair. This creates and excites a chair fixed set with respect to a similar set of continuous transformations (like moving closer, turning and others). While you have switched your focus to the chair, we will also assume that you look at the table occasionally. This occasional look at the table excites the table fixed set of neurons and keeps them in an excited state (i.e., the voltages for these neurons are beyond the threshold values – see chapter 5 that describes the model of a neuron). As you look at both the table and the chair, your brain has two simultaneously excited fixed sets, corresponding to the table and the chair.

Let us now assume that there are four identical looking chairs around the dining table. As you are walking and looking at the other chairs as well, the fixed sets corresponding to the remaining chairs are excited. However, they are pretty much the same as the previous chairs' fixed set (because they look similar). In addition to the chair fixed set, you mechanically start to count the chairs. You create a fixed set corresponding to '4' (number of chairs) as well. This is purely an abstract fixed set not related to the tangible objects. Here, we are assuming that you have already learnt basic mathematical abstractions like numbers and counting. You now have, at least, three fixed sets in your brain, one for the table, another for the chair and another for number '4'. Note that I am simplifying the argument for the sake of clarity. Strictly speaking, several other fixed sets representing the geometry and relative arrangements will also be excited (like edges, angles, textures and others). You can continue this analysis with, say, a number of other objects on the table and other memories that trigger from these. Each time, you create more and more fixed sets that are simultaneously excited in your brain. All of these fixed sets are typically in different regions in your brain because they have unique distinguishable signatures (see chapter 7 on the new memory model).

Let us now examine the amount of time needed to excite any given fixed set and see how it relates to the simultaneous excitation of fixed sets. What would happen if it takes a long time to excite a given fixed set like, for example, let us say that a table fixed set takes about 500 milliseconds to get triggered? This could happen if the neurons along the pathway from the retina to the table fixed set are still below the threshold voltage preventing the generation of action potentials (i.e., they do not fire). In this case, we would have to wait for all these intermediate neurons to fire before we could reach the table fixed set. Two scenarios when this is quite common are: (a) for an infant and (b) for an adult trying to recognize patterns that he has never learnt before, like, say Chinese characters. In both these cases, you do not have a dense interconnected subset in the regions of interest. In general, your low memory is because of a sparser network, which results in slow excitation of fixed sets.

There is another reason for slow identification of a fixed set. This happens when you are not trained with the set of continuous transformations needed for the fixed set. Your walking, jumping and others vary the dynamics significantly that your brain is not able to create enough number of loops that can be grouped as a common fixed set. For example, for a child who just learnt how to walk, a number of new continuous transformations are introduced due to his 'wobbly' walking. Prior to this, the only transformations for him were with him lying on the floor, rolling over and crawling. When he looks at the table while walking unsteadily, sometimes losing balance and falling down, the internal dynamics corresponding to these continuous transformations do not necessarily produce a common core fixed set compared to, say, when he is sitting. He needs to create new memories corresponding to all of these new continuous transformations. This would improve the interconnections and the common regions of excitations within his brain structure.

While the above situations can occur under normal circumstances, partially unconscious scenarios also result in a slow identification of a fixed set. For example, when you are drunk, feeling sleepy and when you are under mild general anesthesia, specific neurotransmitters explicitly block the neurons from firing, causing no cascading firing patterns as well. As a result, the loops and, hence, the fixed sets are triggered slowly, if at all.

Let us now assume that a table and a chair each take 500 milliseconds to excite its respective fixed set. Let us say that the concept of '4' takes, say, 800 milliseconds and so on. This time duration is also an indication of how long neural loops can sustain the excitation when you are not looking at the object anymore. If you have a good memory of these fixed sets, the recollection should be faster as you see the object more often. Let us assume that this is not the case, for now.

Now, if you are looking at the table for 500 milliseconds and then you switch to look at the chair for another 500 milliseconds, the fixed set (i.e., the neural loop excitations) corresponding to the table that you are no longer looking at has pretty much died down. Only the chair fixed set will continue its excitations. Next, when you start counting the number of chairs (for 800 milliseconds), the excitations corresponding to the table and the chair fixed sets have disappeared as well. This slow pace of reaching the excitations for each of these fixed sets is beginning to cause a serious issue. Only one fixed set is staying excited at any given instant of time. As a result, you are no longer able to form continuity across several fixed sets.

In effect, you have excited a table fixed set, followed by a chair fixed set in a sequential order instead of simultaneously. This sequence erases the previously excited table fixed set from your brain. This means that the voltage for most of the neurons, which are part of the table fixed set have dropped below the threshold value (see chapter 5 that describes a model of a neuron). When you now switch your attention back to the table, there are no remnants of excitations corresponding to the table fixed set from the previous instance. As a result, you start to form the table fixed set once again 'from scratch'. It is as if another new table appeared in front of you. You do not have any continuity in the neural loop excitations in your brain to claim that this is the same table as the previous one. From your brain's perspective and its internal dynamics, this is a new table.

As a result, every object you see around you is a new object due to these slow recognition mechanics of fixed sets. You have lost the truth with the fixed sets. You cannot identify any contradictions. This is similar to the example mentioned earlier in section 12.3 about the number of fingers in your hand. I had mentioned that any incorrect statement about the number of fingers in your right hand does not feel contradictory when you are not conscious. Each sentence would appear as a new one for you if the corresponding fixed sets are excited in the slow, say, 500-millisecond range. The situation was similar when you were in an unconscious state like general anesthesia, sleepwalking or completely drunk as mentioned in section 11.2. Even if you attempt to verify the statement by trying to count your fingers, your fixed sets corresponding to the incorrect sentence "you have 10 fingers on your right hand" have disappeared. You may indeed discover that there are only 5 fingers on your right hand. However, you no longer remember that I had said you have 10 fingers instead. Similarly, when you were sleepwalking, completely drunk or in general anesthesia, your ability to form fixed sets quickly is severely impaired. Special chemicals block your ability to form loops, similar to the situation when you have a sparse network of neurons in your brain (as an infant).

Therefore, if we have slow excitations of fixed sets, we can no longer be called as conscious. We cannot sense the passage of time. Every single event is a new event. We are not bored of doing the same task over and over, like a machine. We have no idea that the task we are doing now is the same as the one we did just a while ago. For example, we have seen people sometimes repeating the same sentence over and over when they just begin to wake up from general anesthesia or when they are considerably drunk. They do not even realize this. In fact, they are surprised when you later say that they have been repeating the same sentence (after they have become fully conscious). This suggest that we can become conscious only if the rate of excitation of fixed sets is faster, say, in the millisecond range.

The slow rate of exciting a fixed set has created an inherent discreteness in the flow of events. Two consecutive events are no longer linked together in a continuous way. Even though there is continuity in the physical world, the slow processing within your brain is forcing a discontinuity. If you imagine looking at events only every 500 milliseconds, the world around you will appear discontinuous. You cannot understand causality because you probably never remembered the cause for a given effect, due to these large gaps and quick dissipation of excitations.

15.2 Frequency of fixed sets – origin of brain waves

From the example in the previous section, it became clear that a number of fixed sets should be quickly and simultaneously excited in our brain in order to have a sense of continuity. This is necessary for consciousness i.e., for sensing the passage of time and to sense dynamics itself. In this section, we will see how different brain wave patterns are produced because of the different rates of excitations of multiple fixed sets.

Imagine you are looking around scanning objects in your living room. Since you are already conscious, every object you see like tables, chairs, books and others can be recognized in a matter of milliseconds. There is hardly any object in your living room where you have to focus and think in order to recognize it. Let us call each object, subcomponent, attributes like its color, shape and size, and in fact, any feature you recognize around you, as a 'recognition tick'. Imagine that we have plotted the number of recognition ticks on the y-axis and time on the x-axis. We would get a dense set of data points corresponding to a single recognition tick for most of the timeline along the x-axis. However, occasionally, we recognize multiple objects at about the same instant. We will see these few scattered data points as well.

In order to plot a figure that shows the recognition ticks when we are fully awake and conscious, we will have to perform some experiments on a few test subjects. Let us perform a thought experiment instead for now. How does this plot look when we lie down to sleep and until the time, we start dreaming? During this period, we are not fully conscious, at least, after we relax and do fall asleep. This is one scenario, namely, when you are trying to fall asleep, where it is easy to study the transition from conscious to unconscious state, unlike other unconscious scenarios mentioned earlier in section 11.2 (like coma, sleepwalking and hypnosis). Let us try to understand how our perception and awareness of the world changes during this period until we fall asleep.

When you close your eyes, you are still able to recognize, say, the sound of the fan, feel the surface of the bed and have many thoughts about events that occurred during the day. You can also have several other random thoughts. Initially, your recognition ticks are no different from when you are wide-awake. However, as you are slowly relaxing and entering a sleep state, the number of items you recognize start to decrease and your thoughts become disorganized. You probably stopped recognizing the sound of the fan. The recognition ticks become widely spaced in time. Occasionally, you wake up because of a sharp sound, an itch on your hands, or even the thought of performing this experiment. When you wake up briefly, the number of recognition ticks increase sharply. You will re-recognize, say, the sound of the fan that was previously lost. However, provided the disturbances are not prominent, you return to a drowsy state. Eventually, you enter a deep sleep state.

If you plot the recognition ticks during this entire period, you will notice that they will start initially with dense ticks similar to an awake state. They will become sparser as time progresses. Whenever you wake up due to minor disturbances, you will have dense spikes. As you continue plotting these ticks, they will eventually drop to a few ticks per second.

Let us roughly estimate the density of recognition ticks. When we scan, say, our living room, we can recognize, at least, 20-25 facts in one second. For example, there are many colors, shapes, edges, angles, motion, small features of objects, large objects and so on that are excited simultaneously in our brain. Of course, we cannot list them or express all of these facts using words in a language like English, during this one second. We, however, do not feel any discreteness in our thoughts. Let us say that, on average, we have 20 or more recognition ticks in one second when we are fully awake. This means that it takes roughly 50 milliseconds or less on average to recognize any single feature if done serially. In reality, it is much faster for some facts (and hence the corresponding fixed sets) and slower for others. For example, some primitive facts (or fixed sets) corresponding to edges, angles and motion are already in an excited state because they are present in almost all objects around us. Now, as you start to fall asleep, the recognition ticks start to drop to low levels. As the ticks drop to 1-2 per second, the time it takes to recognize a feature increases to about 500-1000 milliseconds. Therefore, the average time for each recognition tick, as you begin to fall asleep, increases from about 50 milliseconds to almost as high as 1 or 2 seconds.

As mentioned with the example in the previous section, if the recognition ticks drops to low values, you are no longer conscious. Notice, however, that there is no difference in external dynamics, say, with the sound of the fan or people walking around during this period, when compared to the case before you started falling asleep. The source of the difference exists only inside your brain. There are explicit mechanisms in our brain to either cut off or release inhibitory neurotransmitters during this period. This causes fewer loops and excitations within our brain. The main difference is that the exact same external dynamics (like the sound of a fan) simply goes unnoticed by you as you become drowsy and fall asleep.

Let us now translate time durations for the average recognition ticks into frequencies. For frequency, we want to count the number of recognizable objects in one second. If an object takes 500 milliseconds to recognize, this corresponds to 2 Hz (i.e., 2 recognizable objects in one second). Similarly, if the objects take between 33 to 40 milliseconds to recognize, they correspond to a frequency range of 25 Hz to 30 Hz.

From our previous estimates, we can say that when we are awake, the frequency of recognizable objects is 20 Hz or higher. When we fall asleep and enter a deep sleep state, the corresponding frequency falls as low as 0.5 Hz. These frequencies seem to correspond to what scientists have been measuring as brain wave activity under various consciousness states. The electrical activity produced by the brain has been measured by placing electrodes on the scalp. This measurement process is called electroencephalography (EEG). The brain activity measured is plotted as a graph. It comprised of multiple oscillations and is termed as brain waves.

The study of brain waves is quite extensive. In particular, this has been used to understand sleep cycles. They are broadly classified into five frequency regions. These are delta waves (0.5 – 4 Hz), theta waves (4 – 8 Hz), alpha waves (8 – 13 Hz), beta waves (14 – 30 Hz) and gamma waves (> 30 Hz). Now, our above analysis suggests that the frequency, measured as the number of recognizable objects per second, starts in the 20 Hz and even higher i.e., gamma range. As you start to fall asleep, the number of recognizable objects per second keep dropping to very low levels. You start in gamma range (> 30 Hz) corresponding to less than 33 millisecond recognition ticks and beta waves (14 – 30 Hz) corresponding to 33 - 70 millisecond recognition ticks. In these states, you are either conscious or highly alert, as brain wave research suggests. You then begin to relax and enter the alpha range (8 – 13 Hz) corresponding to 75 – 125 millisecond recognition ticks. Then, you start to get drowsy and enter the first stages of sleep. This is the theta range (4 – 8 Hz) corresponding to 125 - 250 millisecond ticks. Finally, you do enter a deep sleep state in the delta range (0.5 – 4 Hz) corresponding to 250 – 2000 millisecond ticks. This is the stage where you do not remember anything, even the passage of time itself. You do not remember how long you were in this frequency range.

Gamma waves are thought to be associated with perception and consciousness (see Buzsáki [17]). Gamma waves were also shown to be present during the process of awakening and during active rapid eye movement (REM) sleep. As we can see from our analysis, during these states, we do have a large number of recognition ticks (about 30 per second). The fact that gamma waves are generated during REM sleep also suggests that we remember our dreams during REM sleep. This is also related to the fact that during REM sleep you have about 30 recognition ticks per second similar to an awake state. Infants and children 2-5 years old only have delta and theta waves even while they are awake, unlike adults. This correlates with the number of objects and events infants recognize (say, less than about 8 per second).

This is not merely a mathematical coincidence. The physical mechanics for recognition ticks were already mentioned earlier as the identification of fixed sets. Fixed sets are represented as loops. The brain wave patterns produced in our brain are the result of electrical activity of sustained excitation of loops from each of the neurons belonging to the fixed sets. Non-looped electrical activity do not sustain long enough to be registered by EEG devices as significant and repetitive activity. They are ephemeral and show up as noise. The rhythmic pattern measured by EEG correspond to repeatable and sustained electrical activity, namely, the loops. This is one reason why loops are quite critical in the study of large-scale complex systems like our brain. The fixed sets triggered here correspond to recognition ticks. The frequency of recognition ticks, as a result, correlates well with the brain wave frequencies. Fixed sets slowly decay as you enter deep sleep. When the number of fixed sets decreases, so does the brain wave frequency. This can be identified as the root cause of the origin of brain waves. The mechanisms are, fundamentally, loops (of electrical activity of neurons) and fixed sets for each of the objects.

Let us now briefly discuss two additional scenarios that show additional correlation between fixed sets (or recognition ticks) and brain waves. These are (a) brain wave patterns for infants who are awake and (b) brain wave patterns during REM sleep.

It is known that infants generate delta and theta wave patterns even when they are awake unlike adults who have beta and gamma wave patterns instead. Adults generate theta and delta waves only when they are drowsy or in deep sleep. Clearly, infants are not drowsy or in deep sleep while they are awake and are, on the contrary, quite active. How do you explain this difference in brain wave patterns? As mentioned earlier, an infant recognizes fewer objects compared to adults because of his limited experience. He has a sparser brain network structure, especially in the forebrain, giving rise to a fewer number of fixed sets. The recognition ticks are typically less than 5-10 per second. He is, therefore, in the theta and delta ranges owing to the current density of his brain network.

REM sleep is known to produce dreams in adult humans. Of all stages of sleep, this is the only stage where you have vivid memory when you wake up. You do not necessarily remember every single dream. You also tend to forget your dream quickly. Yet, we generalize and say that every time we enter REM sleep, we do experience events that appear real. This period happens to correlate well with rapid eye movement in humans and hence the name REM. Why do we remember the events vividly only during this period and not during any other stages of sleep? It has also been already shown that REM-type electrical waves are produced in all adult mammals from their forebrains while infant mammals, including human babies, generate from brainstem until the baby develops its sensory systems (Siegel [139] and Solms [143]). Besides, it was shown that during REM sleep the brain wave pattern is similar to when you are wide awake i.e., beta and gamma waves. In other words, using the above equivalence, the frequency of recognition ticks is in the same range as when you are awake. A large number of fixed sets become synchronized even when you are asleep (during REM sleep) purely from the existing memorized SPL structure of the brain. This makes you conscious during this period though you are still sleeping. I will discuss more on this later in section 19.5.

It turns out that measuring brain waves is one way to determine if you are conscious or not. This is already well-known (see Buzsáki [17]). From the current discussion, we now have a new equivalence: *a high number of recognition ticks produce a gamma wave pattern and vice versa*. While gamma waves are known to be necessary for consciousness, the above equivalent description now states that high number of recognition ticks are necessary for consciousness. This is a new result derived from the theory of SPL systems introduced here. It provides a clearer understanding of the origin of brain waves by relating them to the number of fixed sets or loops triggered per second. In chapter 35, I will use simultaneous excitation of fixed sets to describe the structure of space-time and other interesting scenarios for a conscious being.

The unique aspect of the above equivalence is that it is applicable for all SPL systems, not just our brain. The number of fixed sets triggered simultaneously as discussed here is a new form of synchronization in large-scale structured SPL dynamical systems. Our goal is to propose more such equivalences for each concept related to human consciousness. This is critical because, together, all of these equivalences will help us choose alternate materials and designs when constructing features related to consciousness (i.e., synthetic consciousness). Until now, we have assumed that neurons, our brain network and regions like hippocampus and others, along with specific neurotransmitters and other organic chemicals present in a human brain, say, are critical for consciousness. These equivalences suggest that this is not the case. We can start to build synthetic conscious systems using these equivalence relations.

16
Fundamental Pattern

So far, we have seen how truths were critical for consciousness (Theorem 12.1), how truths can be represented as fixed sets within SPL systems and how the relationships between truths can be expressed as appropriate links between fixed sets. We then started looking at the restrictions imposed on conscious systems when multiple fixed sets are triggered simultaneously (chapter 15). In this regard, we have seen that a number of fixed sets should be quickly and simultaneously excited in our brain in order to have a feeling of continuity. This allowed us to explain the origin of brain waves. In particular, we saw that gamma wave patterns were necessary for consciousness. Other brain wave patterns resulting from different rates of excitation of linked fixed sets as we recognize different features highlight different states of consciousness (like alert and deep sleep).

Let us now look at how we can create fixed sets naturally. In chapter 7, we have already seen how to create fixed sets synthetically in an SPL system with help from conscious beings. However, systems like our brain do form several fixed sets even before we become conscious. How do they do this? What is the source of drive to create new fixed sets, especially for a newborn who is not yet conscious? How does a system continue to maintain this drive as if it has acquired an awareness to learn or a determination to complete a task? Can a non-conscious or a nonliving system create or choose a purpose on its own, start the task, monitor its progress, stop the task, work towards a goal and redo some or all the above steps at a later time, if necessary, while expending its available energy in a useful manner? Each of these steps are highly ordered with a seemingly well laid out plan to achieve a desired future state.

Indeed, we have never seen nonliving systems perform such ordered sequence of steps by themselves. For most autonomous control systems like autopilot in an airplane, cruise control in a car, temperature regulation in HVAC systems and others, conscious humans do the initial planning/ design to achieve specific objectives like minimum error when tracking a path or when regulating around a value. Only after we estimate the parameters and choose the correct control parameters do we let machines operate by themselves (while continuously supplying energy).

Even if a nonliving system has an internal energy source, it is counterintuitive to imagine that it will use its energy appropriately to perform a well laid out plan. In this regard, while the second law of thermodynamics is not violated (because of the internal energy source), it is still hard to accept the existence of such nonliving systems unless someone presents clear mechanisms and designs for such physical systems. Mere non-violation of the second law of thermodynamics is not a satisfactory explanation for the existence of such systems, though such explanations are common (cf. arguments for creating life for the first time).

Consider a simple task of looking at objects around us. Our eyes autofocus our eye lens continuously so the images of objects fall correctly on the retina. As we turn and look at nearby objects instead of faraway objects (and vice versa), our eyes consume energy and refocus accordingly with high precision. If the images fall either in front (resulting in near-sightedness) or behind the retina (resulting in far-sightedness), we would see the images as blurred. A newborn does not have this autofocus ability. It is only after several months of practice does he learn this ability (i.e., not an inborn ability). Yet, almost all humans and several animal species do acquire this ability without any help from adults. This contrasts with other learning tasks where parents do help like with learning to stand or walk.

We can offer several mechanisms for abilities such as this. However, it is important to distinguish mechanisms that require external conscious beings versus those that do not. An analogous man-made system, namely, a camera that can also auto-focus its lens requires external conscious beings to build (i.e., when choosing the correct materials, when assembling the camera correctly and others). Such mechanisms cannot be used for our eye lens. While requiring consciousness for skills like learning mathematics and others are reasonable – since you are already conscious – tasks like auto-focusing eye lens, tracking a moving object, turning your head towards a source of sound and others should not depend on external conscious beings. We learn them at an infant age when we are not yet conscious without help from other conscious beings.

What are examples of such mechanisms? A survival advantage is not a valid mechanism since newborns spend considerable amount of energy and for several months i.e., it is not a pre-programmed task. In general, learning using our brain is not considered pre-programmed. For example, mathematics, that everybody learns, is not encoded back in the DNA. Since not-yet-conscious infants and other animals do not plan for a future benefit when learning, there should be a non-conscious mechanism behind this. This is what I call as the fundamental pattern. In the next chapter, I will discuss several examples relying on the same underlying mechanism.

16.1 Auto-focusing eye lens

Before I describe the fundamental pattern in more general terms, let me take a specific example. This is our ability to auto-focus our eye lens so images form correctly on the retina instead of in front or behind. I will propose a mechanism that does not require consciousness, which I will later generalize as the fundamental pattern. While presenting the mechanism, I will make assumptions several of which are obeyed by our brain and our eye lens. Furthermore, these assumptions will be simple enough that other SPL systems can be built to obey these conditions and, hence, the fundamental pattern naturally.

When an image does not fall on the retina, it will appear blurred to us. There is an area on our retina called fovea, which is responsible for our sharp central vision. This has an unusually high density of neurons. We use this region to obtain detailed visual information. Nearsightedness (myopia) and farsightedness (hyperopia) are examples of eye defects in which an image falls in front or behind the retina, respectively (see Figure 16.1). Since the distance from the eye lens to the retina is more or less constant,

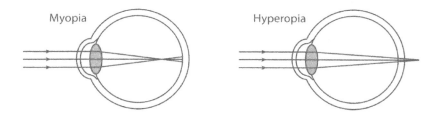

Figure 16.1: Common eye defects in a human. Nearsightedness occurs when the image falls in front of the retina while farsightedness happens when the image falls behind the retina. In both of these cases, the image will appear blurred to us. Appropriate corrective lenses can be used to fix these problems by ensuring that the image falls on the retina.

the only way for images of different distances to continue to fall at the same distance i.e., on the retina is if the focal length of our eyes change to compensate for different distances. If this change is not accurate enough, we will have the above eye defects. Of course, we are aware that different objects are at different distances only after we become conscious. Therefore, it is critical for our eyes to re-adjust the focal length of our eye lens constantly using a well-defined control mechanism.

Our eyes are capable of altering the focal length of our eye lens by changing the tension of the zonular fibers by either contracting or relaxing the ciliary muscle. This changes our eye lens from a flat to a spherical shape. The control of these muscles happens through a parasympathetic signal from the cranial nerve *III*, which is part of the oculomotor nerve. These signals release neurotransmitters, through the appropriate synapses, onto the ciliary ganglion. This will eventually control the contraction of the ciliary muscle. None of the organisms with eyes are born with an auto-focus mechanism. Instead, they need to practice and perfect this process. If an organism does not learn to auto-focus, it has a disadvantage with basic survival (as it cannot react quick enough when there is inaccuracy in predicting distance of a predator or a prey).

Figure 16.2 describes the problem of auto-focus in simple terms. An external object at a given distance from the eye lens forms an inverted image on the retina. Depending on the focal length of the eye lens, the inverted image can fall either in front (*A*), behind (*C*) or on the retina (*B*). If the image falls anywhere in the range *A-C* except at *B*, the image will appear blurry. Our eye needs to constantly readjust its focal length so the image falls on the retina at *B*. If the external object moves either closer or farther, the image will no longer fall at *B*. If we turn our head to look at a different object either closer or farther from the previous object, the new image will no longer fall at *B*. If an object is moving (like a car on a road) and we are tracking its motion, the distance of the object from your eyes change. As a result, the image does not fall on the retina at *B*. Our brain needs to readjust the focal length very quickly. If you blink or close your eyes briefly and look at a different object, the image will not fall at *B*. All of these cases are quite common in our everyday experiences. It is, therefore,

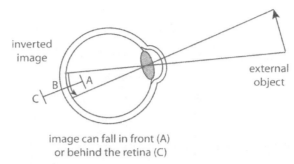

inverted
image

external
object

B

A

C

image can fall in front (A)
or behind the retina (C)

Figure 16.2: The problem of auto-focusing an image on the retina. The image of an external object is inverted as it falls on the retina. Depending on the focal length, the image can fall in front (A), behind (C) or precisely on the retina (B). The task is to re-adjust the focal length of the eye lens so the image always falls at B. This should happen both for the first time for an infant even for stationary objects (when his brain did not learn to refocus yet) and subsequently for an adult as the external objects move or we look at objects at different distances.

necessary for our brain to develop auto-focus ability.

How can an organism practice and learn this skill, directing its time and energy towards this task, while it is still not conscious? If this learning ability requires dedicated and directed effort from an organism, there is a chance that the organism can give up and not learn this skill. I will show here that this situation never happens in most cases. The organisms' brain and eye architecture is such that there is a natural convergence to the auto-focus solution.

The problem of auto-focus in engineering systems is not entirely new. We have built cameras and camcorders that do auto-focus fairly accurately in spite of disturbances and measurement errors, using various types of control algorithms. Linear and nonlinear control systems is an active area of research. However, it is important to note that these control algorithms are not 'natural', in the sense that they cannot be discovered by non-conscious systems automatically. A nonliving system like a camera has no natural notion of adapting itself to converge to an auto-focus solution that we have discovered. Instead, conscious beings need to first define an objective like minimizing an error in regulation or tracking. Then, we need to place actuators and sensors in a specific way to ensure we can satisfy this objective. If not for conscious beings, a nonliving system has no reason to correct its mistakes primarily because there is no notion of a mistake from a nonliving systems' point of view. If a camera does not auto-focus correctly, it is still a perfectly valid dynamical system.

Therefore, these nonliving systems and the corresponding engineering control algorithms require an external conscious being to evaluate the quality and to subsequently alter their designs. Such a strong dependency on an external conscious being makes these algorithms unsuitable for problems like auto-focus, tracking a moving object, turning our head towards a source of sound and others mentioned

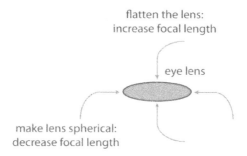

Figure 16.3: Simple model to change the focal length of the eye lens. One signal flattens the lens thereby increasing the focal length while another signal causes the lens to become spherical thereby decreasing the focal length. Ciliary muscles and the zonular fibers are examples of mechanisms that allow such a control.

earlier. Our goal here is to discover a much more 'fundamental' algorithm that is natural within the architecture of a human brain (i.e., the underlying assumptions should fit with our brain's internal SPL structure). External conscious beings do not teach an infant how to auto-focus his eye lens, track a moving object and others, even though he is not born with these skills. Over a period of time, his brain naturally converges to an appropriate control algorithm. In this chapter, I will discuss one such algorithm. This algorithm is so generic and has a wide range of applicability that I call it as the 'fundamental pattern'. I will discuss several applications of the fundamental pattern in the next chapter. Let me begin by listing a few assumptions below, stated in general terms, even though the main example of interest is our brain.

Assumption 16.1.1: There exists components like the ciliary muscle and zonular fibers that allow the focal length of eye lens to change.

This assumption establishes a causal connection between the eye lens and a set of components connected to it that directly cause the change in focal length. In Figure 16.3, we show a simple way to achieve this using two types of control signals. One signal tries to flatten the lens thereby increasing the focal length while the other signal tries to make the lens spherical thereby decreasing the focal length.

Assumption 16.1.2: There exists regions like the cranial nerve *III* that can be used to control components like the above-mentioned ciliary muscle and zonular fibers responsible for changing the focal length of the eye lens, say, by generating a parasympathetic signal, albeit randomly initially.

This assumption provides a means to change (or not change) the focal length by generating (or not) a signal at a distant region. As an example, in Figure 16.3, we are suggesting that the two control signals shown are actually triggered somewhere deeper in the brain (like from the cranial nerve *III*). This is shown as region *X* in Figure 16.4.

Assumption 16.1.3: The regions that receive sensory inputs from the retina like the cranial nerve *II* and visual cortex are not too far from the regions that control the focal length of the eye lens like the cranial nerve *III*.

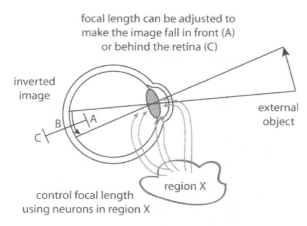

focal length can be adjusted to
make the image fall in front (A)
or behind the retina (C)

inverted
image

external
object

region X

control focal length
using neurons in region X

Figure 16.4: Control region X for adjusting the focal length of our eyes. The region X of neurons can be quite far from the eye. Using these neurons, it should be possible to make the image fall anywhere in the range A-C around the retina. The goal is to become better at adjusting the focal length so the image falls at B at all times.

Assumptions 16.1.1 and 16.1.2 together suggest that we can trigger a signal anywhere in between the causal pathway from regions like the cranial nerve *III* to the eye lens to change the focal length of eye lens (depicted as region X in Figure 16.4). These assumptions, however, only suggest that we can change the focal length randomly initially, say, at birth. Assumption 16.1.3 states that the sensory pathway and the initiation of the control signals (say, at the cranial nerve *III*) are not too far from each other.

Our task now is to present a mechanism that takes our unstructured random abilities and convert them into an ordered sequence of steps. During the process, the system would need to connect the sensory regions to the control regions to create a well-defined order. These connections should be directed so the system learns to auto-adjust the focal length in proportion to the distance of objects being viewed even when the system is not conscious (or alive – with synthetic systems) and when there is no help from other conscious beings.

Figure 16.8 shows one model for a system that performs this task naturally. In Figure 16.2, we have the sensor (like an eye) which receives light signals for objects at different distances. A convex lens converges the light rays and forms an inverted image on a sensory surface (like the retina). Since the system does not yet have an auto-focus mechanism, the image can form either in front (A) or behind (C) the sensory surface resulting in a blurry image. A region X shown in Figure 16.4 represents one way to control the focal length of the eye lens (like the cranial nerve *III*). The initial structure is not well-connected to help with auto-focus. The change in focal length is initially random (assumptions 16.1.1 and 16.1.2) and the sensory inputs propagate and disperse to different regions. Yet, there are a few observable patterns.

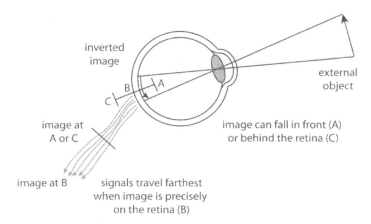

Figure 16.5: Signal propagation relative to where the image falls on the retina. If the image falls either in front (A) or behind (C) of the retina, the image looks blurry. In contrast, if the image falls on the retina at B, the image will be sharp. As a result, the signal propagates farthest if the image falls at B. The farther the image is from the retina on either direction (i.e., towards A or C), the shallower the signal propagates.

Firstly, new connections tend to form naturally from repeated sensory inputs as described in chapter 7, causing the input signal to propagate deeper into the brain over time (see Figure 16.5). This results in a gradual increase in the density of neurons, as can be seen in a newborn. Secondly, let us examine the difference in firing patterns produced for the two cases: when the image is on the retina (at B in Figure 16.5) versus when it is in front (A) or behind (C) the retina. In the latter case (i.e., at A or C), the image does not have any sharp edges and will appear blurred. Recall that an edge is characterized as a boundary with two distinct neural firing patterns on either side (cf. edge of a table – see Figure 16.6). When there is a sharp edge i.e., with image precisely on the retina, one side of the edge has several actively firing neurons (see neighborhood Q in Figure 16.6), corresponding to the image, while the other side has no firing neurons, corresponding to a lack of an image. In comparison, neighborhood P in Figure 16.6 is entirely inside the image, causing it to generate action potentials for all neurons in P. For neighborhood R, on the other hand, there will be no neurons that will fire as it is completely outside the image. This situation is, however, different when the image is blurred, i.e., with image falling in front or behind the retina. There will be no sharp edge in these cases. Instead, a small fuzzy band will be present outside an edge, which causes the edge to appear fuzzy.

Let us now see how a sharp image on the retina causes the neural firing pattern to propagate farthest as shown in Figure 16.5. In chapter 5, we have seen that in order for a neuron to fire, it needs to open several Na^+ and K^+ ion channels. A sudden influx of Na^+ ions into the cell causes depolarization of the cell. When the net voltage exceeds a threshold value, it will eventually result in the generation of an action potential (see section 5.1). Therefore, the amount of available Na^+ ions in the extracellular space

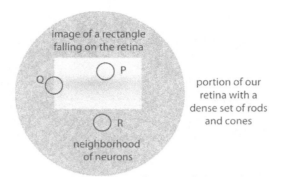

Figure 16.6: Detecting edges when the image is sharp versus when it is fuzzy. In this example, a rectangular image falls on the retina. The difference between a neighborhood of edge neurons Q versus interior neighborhood P is that half of the neurons in Q do not fire whereas all neurons fire in P. Na$^+$ ions from the extracellular space enter neurons in order to fire. These ions are shared across all neurons in P versus only half of the neurons in Q. As a result, the firing signal is stronger in Q compared to P.

contribute towards a neuron's firing ability – the more the number of neurons firing simultaneously in a small neighborhood, the more the sharing of available Na$^+$ ions happens between them and the less Na$^+$ ions each neuron gets, resulting in less rate of firing and less total duration of firing. The converse situation happens when there are less number of neurons firing simultaneously in a small neighborhood.

The above analysis is particularly suitable for neurons on the retina, specifically on the fovea, the region responsible for sharp central vision. This region has a very high concentration of rods and cones relative to the rest of the retina. We can use the above discussion to help us distinguish signals from 'edge' neurons (neighborhood Q in Figure 16.6) compared to 'interior' neurons (neighborhood P) corresponding to an image. I will discuss edge detection in further detail in chapter 34. As shown in Figure 16.6, for neurons like Q on the edge, only half of the neighborhood neurons fire whereas for interior neurons like P, all neighborhood neurons fire. Therefore, the signal from edge neuron Q will be stronger than the signal from interior neuron P (from the above-mentioned relative sharing of Na$^+$ ions from the extracellular space) provided the image falls precisely on the retina, not in front or behind the retina.

This translates to either a deeper or a shallower penetration of the firing pattern within the brain (see Figure 16.5). The cascading propagation of neural firing pathways is maximized when the image falls precisely on the retina. Generalizing this to other SPL systems requires the following additional assumption.

Assumption 16.1.4: The mechanism of signal propagation from a given node in an SPL system is such that the strength of the signal is inversely proportional to number of simultaneously active nodes within a small neighborhood.

This assumption can be easily satisfied by incorporating an analogous notion of

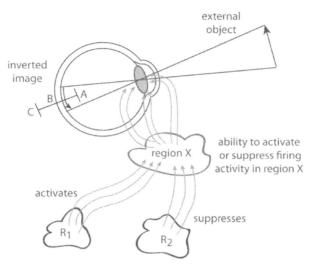

Figure 16.7: Activating or suppressing firing activity through two regions R_1 and R_2. Region R_1 behaves analogous to a positive emotion. The more you do (using region X) what you like, the more you continue to do (using region R_1's activation). Similarly, region R_2 behaves analogous to a negative emotion. If you dislike what you do (using region X), you either stop doing it or revert your action (using region R_2's suppression). In the case of controlling focal length of our eye lens, regions R_1 and R_2 are not emotions though the analogy is reasonable.

sharing of Na^+ ions across all simultaneously active nodes within a given neighborhood of nodes (see Figure 16.6). This is natural in active physical systems i.e., those that generate useful energy, whenever a group of subsystems compete for shared resources and/or energy.

With this additional assumption 16.1.4, the signal from the retina does not propagate deeper in the brain when the image is in front or behind the retina whereas it does when it falls on the retina. Initially, this latter situation happens only by chance. Yet, when it does happen, new connections are formed and existing neural pathways extend gradually deeper into the brain over a period (see Figure 16.5). In other words, the directional extension of existing pathways is natural, though slow. Note that this does not explain how our eyes learn to auto-focus.

Let us now look at the brain structures formed on the control side i.e., those responsible for changing the focal length. From assumption 16.1.2, we have regions like the cranial nerve *III* that send signals to change the focal length. There should be at least two looped regions R_1 and R_2 that change the tension of the zonular fibers by either contracting or relaxing the ciliary muscle (see Figure 16.7). Let us identify looped region R_1 as the one that sends signals to produce an excitatory response on the neurons controlling the ciliary muscle in region X while looped region R_2 sends signals to produce an inhibitory response on the same neurons. These signals, therefore, produce positive and negative response respectively. For example, if the

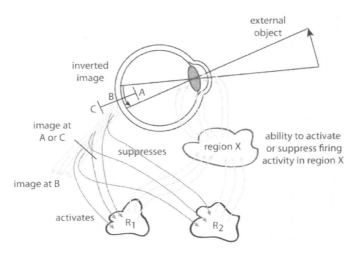

Figure 16.8: Auto-focus mechanism. The sensory region connects to the control regions R_1 and R_2 at different points along the sensory pathway. The signals reaching R_1 amplifies activity in R_1 whereas those reaching region R_2 suppresses the firing activity. The signal strength in R_1 and R_2 is an integral of the sensory signal along the pathway. We now have a closed-loop control system capable of auto-focussing our eye lens.

current signal is increasing the focal length of the lens (by making the lens flatter), then triggering region R_1 will enhance the signal and make if further increase the focal length. Similarly, if the current signal is decreasing the focal length of the lens (by making the lens spherical), then exciting region R_1 will make if further decrease the focal length. When triggering region R_2, the effect is the opposite i.e., if the current signal increases the focal length, exciting region R_2 will make it decrease the focal length and vice versa. When both regions R_1 and R_2 are excited, the relative strength of both signals determine whether the net effect will enhance or diminish the current change in focal length.

To make it easy to remember the effect of regions R_1 and R_2, we can use the following analogy: region R_1 is analogous to producing a positive emotion while region R_2 is analogous to producing a negative emotion – if what you currently did produced a positive emotion, you will continue to do it for some more time while if it produced a negative emotion, you will try to revert what you were doing. Keep in mind that there is no emotion involved in the case of changing the focal length. For now, it is just an analogy although we will see later that the same idea (of having two regions – a positive control region R_1 and a negative control region R_2) generalizes to other examples in which appropriate positive and negative emotions will indeed be generated as well.

The next task is to understand how to control triggering of regions R_1 and R_2. But, before we do that, let us summarize the above discussion as a refinement to assumption 16.1.2.

Assumption 16.1.5: There exists two control regions R_1 and R_2 in an SPL network, which produces positive and negative signals respectively causing it to either enhance the original effect (of changing the focal length of the lens) or diminish it further.

How do we control triggering of regions R_1 and R_2? For this, we need a way to connect the sensory regions to the control regions. Since the sensory and control regions are close enough, from assumption 16.1.3, there is a natural tendency for the neural pathways from the sensory regions to extend towards the control regions R_1 and R_2 over a period of time. We now have a round trip pathways from the sensory regions to the control regions and back as shown in Figure 16.8. Notice that as the sensory signals have an ability to transmit deeper depending on where the image falls (i.e., in front (A), behind (C) or on (B) the retina cause it to generate a sharp or a fuzzy image – see above when discussing assumption 16.1.4), we will, in general, have multiple pathways from different points along the length of this pathway towards regions R_1 and R_2 (see Figure 16.8).

These multiple pathways allows us to perform an analog of mathematical integration of the original signal quite naturally. As the sensory signal from the retina propagates deeper towards the control regions, the total input signal reaching regions R_1 and R_2 accumulates from all neurons along the pathway i.e., an integral along the pathway (see Figure 16.8). However, the output from regions R_1 and R_2 are considerably different from each other according to assumption 16.1.5. Let us summarize the above discussion as an additional assumption here.

__Assumption 16.1.6:__ The inputs that fall on region R_1 in an SPL network increases the chances of firing action potentials (say, by depolarization in the case of neurons) while the inputs falling on region R_2 decreases the chances of firing action potentials (say, by hyper-polarization in the case of neurons).

Assumptions 16.1.5 and 16.1.6 are quite easy to satisfy with neurons because of the existence of neurotransmitters and the corresponding receptors that can either depolarize or hyperpolarize a neuron. For example, GABA and glycine are examples of neurotransmitters in the retina that act as inhibitory molecules, the release of which decreases the probability of firing an action potential by the postsynaptic neuron. Similarly, glutamate and acetylcholine act as excitatory molecules in several cases causing an increase in the above probability of firing. In addition, the same neurotransmitter can act as both excitatory and inhibitory molecules depending on the receptors present on the postsynaptic neuron. Given these possibilities, assumption 16.1.6 requires us to find two regions R_1 and R_2, where R_1 reacts in an excitatory manner while R_2 in an inhibitory manner when both of them receive the same input signals from anywhere along the pathway corresponding to an image. The same applies to the outputs from regions R_1 and R_2 falling on region X (i.e., assumption 16.1.5).

Therefore, each of the assumptions 16.1.1-16.1.6 can be satisfied using entirely natural mechanisms with considerably high probability when the underlying architecture is neuronal-based SPL system as described in chapter 7. With synthetic SPL networks, satisfying these assumptions is even simpler, since a conscious human can assist.

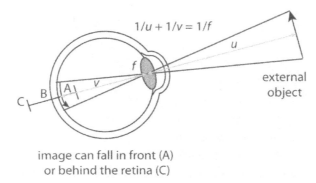
image can fall in front (A)
or behind the retina (C)

Figure 16.9: Basic optics of a convex lens. If u is the distance of the external object from the lens, v the distance of the image from the lens and f the focal length, then a convex lens satisfies the relation $1/u + 1/v = 1/f$. We can use this relation to determine how one or more variable changes in response to a change in another variable.

Let me now show how our eyes naturally develop the ability to auto-focus an image on the retina using these assumptions. If an image falls in front or behind the retina (Figure 16.8), since the edges are not sharp, the signals transmit only short distances. Over a period of several days (or months), multiple things happen: (i) new connections are formed that extend inward gradually and randomly, (ii) the focal length is changed randomly and is controlled through regions R_1, R_2 and X and (iii) new connections extend randomly towards regions R_1 and R_2 because of assumption 16.1.3. Occasionally, the change in focal length causes the image to fall precisely on the retina. This creates an image with sharp edges causing a deeper penetration of signals because of assumption 16.1.4 (see Figure 16.6). Over a period, the network structure naturally modifies to the one shown in Figure 16.8.

From the basic optics of a convex lens, we can easily show the following relation to be satisfied: $\frac{1}{u} + \frac{1}{v} = \frac{1}{f}$ where u is the distance of the external object to the lens, v the distance of the image from the lens and f the focal length of the lens (see Figure 16.9). Using this relation, it is clear that if the object moves closer to our eye (i.e., u decreases), then the image moves further from the lens (i.e., v increases) provided that the focal length of our eye lens is unchanged. Similarly, if we increase the focal length f, the image distance v increases if the external object does not move.

The problem of auto-focus is to show that wherever the image falls between A-C, the system alters the focal length f to converge to B. In other words, we want to show that B is the stable equilibrium point for the dynamics of the combined sensory-control system of Figure 16.8. For this, we would need to analyze the following four distinct cases. I will provide the details for the sake of completeness.

1. Image falls in front of the retina (between A-B) and the focal length f increases initially:

   a. As f increases, v increases since u is unchanged. This implies that the image moves in the direction from A to B.

    b. In this case, we would want to continue moving the image from A to B in order to focus the image. In other words, we want to continue increasing the focal length f.

    c. Since the image is beginning to become sharper, the signals along the sensory pathway starts to propagate deeper as shown in Figure 16.8 (assumption 16.1.4).

    d. This causes an increase in the strength of the signal falling on the control regions R_1 and R_2.

    e. From assumption 16.1.6, this causes region R_1 to be excited more while region R_2 to be suppressed.

    f. The input on region X from R_1 will be more while the input from R_2 onto region X will be less (from assumption 16.1.5). This implies that the amplification from R_1 is higher while the suppression from R_2 is lower. As a result, region X will amplify the initial effect of increase in focal length from both an increase in R_1 and a decrease in R_2. This will make the image continue its motion in the direction from A to B causing the image to become even sharper (see Figure 16.8).

2. Image falls in front of the retina (between A-B) and the focal length f decreases initially:

    a. As f decreases, v decreases since u is unchanged. This implies that the image moves in the direction from B to A.

    b. In this case, we would want to revert the direction (i.e., from A to B) in order to focus the image. In other words, we want to increase the focal length f.

    c. Since the image is beginning to become fuzzy, the signals along the sensory pathway start to retreat as shown in Figure 16.8 (assumption 16.1.4).

    d. This causes an decrease in the strength of the signal falling on the control regions R_1 and R_2.

    e. From assumption 16.1.6, this causes region R_1 to be excited less while region R_2 to be suppressed less.

    f. The input on region X from R_1 will be less while the input from R_2 onto region X will be higher (from assumption 16.1.5). This implies that the amplification from R_1 is lower while the suppression from R_2 is higher. As a result, region X will suppress and eventually revert the initial effect of decrease in focal length from both an decrease in R_1 and an increase in R_2. This will make the image revert its motion in the direction from A to B causing the image to become even sharper (see Figure 16.8).

3. Image falls behind the retina (between B-C) and the focal length f decreases initially:

    a. As f decreases, v decreases since u is unchanged. This implies that the image moves in the direction from C to B.

    b. In this case, we would want to continue moving the image from C to B

in order to focus the image. In other words, we want to continue decreasing the focal length f.

c. Since the image is beginning to become sharper, the signals along the sensory pathway starts to propagate deeper as shown in Figure 16.8 (assumption 16.1.4).

d. This causes an increase in the strength of the signal falling on the control regions R_1 and R_2.

e. From assumption 16.1.6, this causes region R_1 to be excited more while region R_2 to be suppressed.

f. The input on region X from R_1 will be more while the input from R_2 onto region X will be less (from assumption 16.1.5). This implies that the amplification from R_1 is higher while the suppression from R_2 is lower. As a result, region X will amplify the initial effect of increase in focal length from both an increase in R_1 and a decrease in R_2. This will make the image continue its motion in the direction from C to B causing the image to become even sharper (see Figure 16.8).

4. Image falls behind the retina (between B-C) and the focal length f increases initially:

a. As f increases, v increases since u is unchanged. This implies that the image moves in the direction from B to C.

b. In this case, we would want to revert the direction (i.e., from C to B) in order to focus the image. In other words, we want to decrease the focal length f.

c. Since the image is beginning to become fuzzy, the signals along the sensory pathway start to retreat as shown in Figure 16.8 (assumption 16.1.4).

d. This causes an decrease in the strength of the signal falling on the control regions R_1 and R_2.

e. From assumption 16.1.6, this causes region R_1 to be excited less while region R_2 to be suppressed less.

f. The input on region X from R_1 will be less while the input from R_2 onto region X will be higher (from assumption 16.1.5). This implies that the amplification from R_1 is lower while the suppression from R_2 is higher. As a result, region X will suppress and eventually revert the initial effect of decrease in focal length from both an decrease in R_1 and an increase in R_2. This will make the image revert its motion in the direction from C to B causing the image to become even sharper (see Figure 16.8).

From the above analysis, we see that for any initial position of the image anywhere in the range A-C and for any initial change in focal length, the new change in focal length will proceed in a direction that causes the image to converge to B. In all these cases, the image gets sharper making B a stable equilibrium point for all dynamical changes. If the initial control signal makes the image sharper, then the new control signals from regions R_1 and R_2 will cause the image to become even more sharper while if the initial signal makes the image fuzzy, the new signals from R_1 and R_2 will revert the effect to make the image sharper.

When the image is already precisely on the retina at B, there is no reason to change the focal length, but if it did change (either because you turned to look at objects at different distances or if the objects moved away), then the above analysis suggests that the new control signals generated will cause it to refocus the image to fall precisely on the retina at B once again. This can be seen as a regulation of the image to make it fall precisely on the retina, not in front or behind. The position B is a stable equilibrium point – perturbations to this state brings it back to the same state.

In summary, the mechanism described here can occur naturally since it is possible to satisfy assumptions 16.1.1-16.1.6 quite easily with finite probability. The resulting control mechanism (Figure 16.8) allows our eye lens to automatically focus an image on the retina.

It is important to emphasize that several control algorithms can be designed for the problem of auto-focus of our eye lens. However, each of these algorithms should be carefully evaluated to see if they can be discovered automatically by a synthetic system without help from conscious beings. For example, the auto-focus mechanisms in a camera or a camcorder does not fall in this category. They, most definitely, require conscious beings to design them initially. Only after they are designed, can they operate with little help from conscious beings. Our goal here was to design an algorithm that an infant can self-discover and converge to a solution with no help from his parents or other conscious beings.

16.2 Auto-focusing moving objects

Given the auto-focus mechanism described in the previous section, let us now consider a few variations to this problem. What happens when an image that was previously auto-focused moves away or moves closer? What happens when you switch from looking at the current object (which was already auto-focused) to look at a closer or a farther object instead? In both these cases, we would like our eye lens to refocus quickly and correctly.

When the object moved away, using the current focal length, the image will fall in front of the retina (at A). When this happens, assumption 16.1.4 implies that the firing signal no longer propagates as deep inside the brain. The pathways retreat naturally. This effect is not caused by any internal change in the focal length. Rather it is an unintentional effect triggered by the external objects' dynamics. Yet, the 'integral' signal generated from this pathway to regions R_1 and R_2 decrease. Our brains' intentional response of changing the focal length, though initially random, will quickly readjust to make the image fall precisely on the retina as described in the previous section (steps (1) and (2)). For example, if the initial random change caused the signal to move deeper (because the image is sharper), then the total signal on regions R_1 and R_2 increases. This causes the initial random change to be amplified until the image falls precisely on the retina (see step (1) of the previous section). Similarly, if the initial random change caused the signal to move even shallower (because the image become even more fuzzy), then the total signal on regions R_1 and R_2 further decreases.

This opposes the initial random change causing it to revert the direction of the change in focal length and hence make the image sharper (see step (2) of the previous section). In effect, the image will be refocused to form precisely on the retina.

The analysis is identical when the object moves closer. Our eye lens will quickly refocus to form the image on the retina. If we turned our head to look at a distant (or closer) object instead, the analysis is once again similar to the case when the object moved away (or moved closer). Our eyes will refocus to form the image precisely on the retina.

We can say that our brain architecture has adapted itself to have a stable equilibrium state for the problem of focusing the image on the retina. Even though the initial architecture was sufficiently random, just a set of simple and natural assumptions 16.1.1-16.1.6 have allowed the system to alter itself to create a unique architecture capable of solving the problem of auto-focusing the eye lens without the help of any external conscious beings. This is critical because we want all species with eyes to acquire this ability entirely naturally irrespective of whether they are conscious. In fact, we want an infant human to acquire this ability without the help of a parent. For some other tasks like learning to stand or walk, we may believe that, as parents, we are helping them. But for tasks like learning to auto-focus eye lens, tracking a moving object, turning towards a source of sound and others, we do not help at all. Infants learn entirely by themselves even though they were not born with any of these skills (except in the cases when there are obvious abnormalities in the most critical organs involved for the task i.e., say, the eye itself – abnormalities in the brain rarely prevent them from learning these tasks).

16.3 Creating adaptive self-regulating systems

The control mechanism proposed in the previous sections for the problem of auto-focusing eye lens can be easily generalized to a number of other cases. I call this generalization as the fundamental pattern of physical SPL systems. It is important to note that the discovery or creation of this mechanism does not require consciousness or help from other conscious beings. An SPL system is capable of adapting itself provided it satisfies a small set of mild assumptions. I will summarize these assumptions here and discuss several applications in the next chapter. In each of these applications, we have, at minimum, two sets of components – one for sensing a change and the other for controlling it. The objective of the SPL system then is to adapt or alter itself so it learns to self-regulate itself to a naturally desirable state. The definition of a desirable state in each case, as we will see, will not depend significantly on who is evaluating the system. This is particularly important when the system is not conscious because there is no concept of a desirable state or an ability to evaluate a state among nonconscious systems.

__Assumption 16.3.1__: There exists a causal connection between the component that senses a signal (or a pattern) and the component that controls or changes this signal (or the pattern).

For the problem of learning to turn your head towards a source of sound, the sensory components are located near the ear and the control components should allow you to turn your head. We want, at minimum, a causal connection between these two regions. This causal connection will not, in general, be producing the desired effect just yet. With other examples, sometimes the components that cause the signal to change need not be part of the brain. They could be your neck, hands or legs.

Assumption 16.3.2: There exists regions within the SPL network that directly interfaces with the above-mentioned sensory and control components, allowing the SPL network to sense the changes and control them.

This assumption allows us to use the SPL network to sense and control the entire system, analogous to how our brain does to control the body. Once again, these regions do not yet sense and control the system in the correct direction to produce desirable changes. The next assumption states that the two regions – sensory and control – are close enough that they have a finite probability of interconnecting.

Assumption 16.3.3: The regions that receive sensory inputs within the SPL network are not too far from the corresponding regions that control them.

Assumption 16.3.4: The mechanism of signal propagation from a given node in an SPL system is such that the strength of the signal is inversely proportional to number of simultaneously active nodes within a small neighborhood.

This assumption states that in any small neighborhood of a given node, the node has to compete with the neighboring nodes for a shared set of inputs (like Na^+ and K^+ ions in the case of neurons) before it can fire. The more the number of simultaneously active nodes, the less will be the share of available inputs for a given node and vice versa. This relationship does not extend well to large neighborhoods of a given node. If the signal generated here is large or strong, I sometimes refer to it as an *explosion*. When only a few nodes are active in a small neighborhood, the possibility of generating an explosive output signal is high and vice versa.

In the case of auto-focusing our eye lens, an explosion is created when the image is precisely on the retina i.e., when the image is sharpest. For all other cases, the image is fuzzy and we do not have an explosion in the strength of the signal that propagates along the sensory pathways. This explosion can also be understood from Figure 16.6 where we have seen that the signal strength is highest for the neighborhood Q.

Assumption 16.3.5: There exists two looped regions R_1 and R_2 in an SPL network, which produces positive and negative output signals respectively, resulting in either enhancing the original control signal or diminishing it further.

This is a nontrivial assumption compared to the previous ones. Yet, conceptually, it is a natural assumption. It states that there are two types of drives for any living being – a positive drive and a negative drive. The positive drive enhances the original action while the negative drive suppresses it. For living beings, these are innate features. We are, therefore, generalizing this notion to all SPL systems by treating it as an assumption. The two regions R_1 and R_2 would, in general, be at different locations for different tasks. In some cases, it is possible to simplify this assumption to show how regions R_1 and R_2 can be created naturally as well (see section 63 which considers the problem of designing a simple SPL system that moves towards a direction containing more food when presented with two sources of food).

In some other cases, we will see that region R_1 behaves as a source of positive emotion (likes) and region R_2 as a source of negative emotion (dislikes). For example, consider an infant who is learning to put his hand in his mouth. This is a task almost all infants learn by themselves. In fact, we notice that there is a period in their life when they start putting almost everything they reach (like toys and others) into their mouth. Of course, they stop doing this after several months. For this task, initially we do observe two distinct emotions guiding them – a positive emotion like, say, some degree of happiness when he succeeds and a negative emotion like, say, some degree of frustration, resulting in crying too, when he fails (like, say, when you pull his hand out of his mouth or when his hand slips out of his mouth accidentally).

Assumption 16.3.6: The inputs that fall on looped region R_1 in an SPL network increases the chances of generating output signals (like firing action potentials, say, by depolarization in the case of neurons) while the inputs falling on looped region R_2 decreases the chances of generating output signals (say, by hyper-polarization in the case of neurons).

This can be seen as a refinement to assumption 16.3.5 where we qualify what the inputs do to regions R_1 and R_2. While the outputs are positive and negative signals, the inputs that fall on these regions determine whether there will be a higher or a lower chance of generating the respective output drives.

One implicit assumption is that the SPL system does not have critical structural, sensory-control related or other defects that prevents the system from functioning correctly. If an SPL system now satisfies assumptions 16.3.1-16.3.6 for a specific learning task, it follows that the resulting system adapts itself to converge to a stable equilibrium state with no help from other conscious beings. In most cases, identifying and characterizing the stable equilibrium state will be quite easy as we will see in chapter 17 for a number of examples. Furthermore, the analysis of the stability of the equilibrium state is similar to the one presented in section 16.1 for the problem of auto-focusing our eye lens.

Let us see how an SPL system can self-adapt to discover a new stable equilibrium state. One of the first important questions to answer when any system tries to self-adapt is: what is it adapting to? What is the state it is trying to converge to over a long period of time? If the system is self-aware, this state can be chosen by the system itself like, say, with learning music or mathematics. However, for several problems like auto-focusing eye lens and others discussed in chapter 17, we would not want to wait until the system becomes self-aware so it can choose an appropriate state to adapt to.

What properties should be exhibited by this state so the system can self-discover it? One such property can be characterized as follows: the equilibrium state should help satisfy assumption 16.3.4 naturally i.e., there should be a high probability of forming an *explosion*. For example, for the problem of auto-focusing eye lens, consider the difference between the two states – the image falling precisely on the retina versus falling in front or behind the retina. Between these two states, we note that the set of neurons firing at the edges are fewer (and hence resulting in a sharper boundary) for the former case while they are more in number (and hence resulting in a fuzzier boundary) for the latter case. This shows that the SPL system can produce an explosion when the image falls precisely on the retina and not otherwise. As a result, assumption 16.3.4 can be used directly to distinguish the two cases. The situation is the same for other examples like tracking a moving object, turning your eyes towards a source of light and others that we will discuss in chapter 17. In each of these examples, we will be able to *find a state that is both natural and intuitively reasonable for which an explosion is possible.*

Let me now generalize the mechanism described in the previous section and state it briefly in a form that is applicable more broadly in several situations. This mechanism is what we call as the fundamental pattern. It helps create adaptive self-regulating SPL systems. Consider an application in which we want an SPL system to learn a new task without the help of any conscious being. Let me briefly outline the steps here (for comparison purposes, see the steps described in section 16.1 for the problem of auto-focusing eye lens).

1. Let us assume that the sensory and control regions of an SPL system satisfy assumptions 16.3.1-16.3.3. These are the minimal set of conditions needed so an SPL system can discover an appropriate sensory-control mechanism for a given task.

2. Among the possible states the system can be in for the task in question, identify a single sensory state S that satisfies assumption 16.3.4 i.e., the state that can potentially produce an explosion. The goal is to show how the SPL system can adapt itself to converge to state S. In the process, we want to show that S is a stable equilibrium state i.e., small variations around state S will bring the system back to this state.

3. As the system is subject to several sensory inputs over time, new pathways are established that extend and connect the sensory and control regions albeit randomly initially. These new pathways also connect to regions R_1 and R_2 of assumption 16.3.5.

4. When the sensory system is in state S above, the pathways extend farthest. This causes a large amount of inputs to accumulate in both regions R_1 and R_2. As a result, R_1 is activated more while R_2 is suppressed more from assumption 16.3.6. This causes the control signals to be amplified in the positive sense – both increased R_1 output and decreased R_2 output amplifies the signal because of the special nature of these two regions specified in assumption 16.3.5. This attempts to keep the system in state S.

5. If the sensory system moves out of state S above, the sensory signals start to contract. This causes a decrease in the amount of inputs that fall on regions R_1 and R_2. As a result, R_1 is activated less while R_2 is activated more from assumption 16.3.6. This causes the control signals to be amplified in the negative sense – both decreased R_1 output and increased R_2 output causes the control signal to reverse because of the special nature of these two regions specified in assumption 16.3.5. This attempts to bring the system back to state S.

6. The above steps implies that state S is a stable equilibrium state – any change (internal like, say, the system turning away or external like, say, the external

object moving away) that causes the system to come out of state S will result in producing control signals that tries to bring the system back to the state S. In other words, the system converges by adapting itself to self-regulate around the state S.

The above steps constitute the mechanism we call as the fundamental pattern. In the next chapter, I will apply these steps for a number of examples to show how an SPL system can self-learn and adapt itself without help from other conscious beings. The above steps, though described as a serial process, are, in fact, iterative in nature. Different regions in the SPL system need to evolve through multiple exposure to sensory inputs over a long period. Each of these regions (like the sensory regions, control regions, regions R_1 and R_2 and others in between) are activated in different proportions under different conditions, sometimes even when it is not related to the task as well. For example, the regions controlling your ability to turn your head towards a source of sound is activated not just for this task, but even for other situations like when looking at or interacting with toys. As a result, some regions will become more evolved than the other. Yet, eventually when the development of each of the critical regions and their interconnections for a given task reach a threshold state, the fundamental pattern will become viable.

17

Applications of the Fundamental Pattern

In this chapter, I will present a few applications of the fundamental pattern. We will need these mechanisms later when we show the existence of an external world, space and the passage of time. From the previous section, it was clear that for the fundamental pattern to be applicable, we need to choose an objective that produces a suitable explosion within an SPL system. After this choice, the rest of the sensory-control mechanism is a direct mapping of the algorithm presented in section 16.3. This algorithm is fairly generic for most tasks.

Let me first list a few questions whose answers will turn out to be direct applications of the fundamental pattern. How do you turn yourself towards the correct source of light when a light beam flashes? When we hear a specific sound, how are we able to turn our head and body towards the correct source of sound? How does our brain learn to converge both of our eyes on one object, instead of one eye looking at one object and the other at a different object? How can we track a moving object accurately with our eyes? After we detect the odor of, say, a tasty food, how do we move towards its source correctly? How do you learn to move your hand to pick an object with high precision? Let me begin by addressing these questions next.

17.1 Turning your eye towards a source of light

Our eyes have six orbital muscles. Four of these muscles control the movement of the eye, one for going up (superior rectus), one for moving down (inferior rectus), one for moving to the side towards your nose (superior oblique) and the last for moving to the side away from your nose (inferior oblique). The other two muscles (lateral and medial rectus) control the movement of the eye as the head moves. Each of these muscles are controlled by several cranial nerves in the brain. Specifically, these nerves are cranial nerve *IV* for superior oblique, cranial nerve *VI* for lateral rectus and cranial nerve *III* for all other muscles. The retina of our eye also has a region called fovea with a high density of photoreceptors (cones), about 50 cones per 100 microns. It is estimated that the optic nerve carries about 50% of nerve fibers just from the fovea, which occupies only 1% of the retina (see Montgomery [110]).

Clearly, when light falls on the fovea, as opposed to the other regions on the retina, it will create an explosion of excitation (in relative terms), which propagates through the brain. If there is a bright light source in one part of your visual field but not on the fovea, then by turning your eye towards this source of light so it falls directly on the fovea, your eye suddenly generates an explosion of firing pattern (because of the high density of photoreceptors in the fovea). We now need to learn how to turn our eye towards the light source using these orbital muscles to generate explosions. We use the fundamental pattern. As mentioned in step (2) of section 16.3, to use the fundamental pattern, we need to look for a natural sensory state S that produces an explosion i.e., a state satisfying assumption 16.3.4 compared to other related states. We have already identified this, namely, to turn our eyes towards the source of light so that the image directly falls on the fovea. Of all the possible movements of the eye, only these set of control actions, generated from the cranial nerves, will produce an explosion.

The remaining steps (3)-(6) of the fundamental pattern can now be applied similar to auto-focusing task discussed in section 16.1-16.2. Over time, our brain will create two regions R_1 and R_2 with properties obeying assumptions 16.3.5 and 16.3.6. They ensure that state S is a stable equilibrium state. If either the source of light moves away from the fovea or if you turn your head so the image moves away from the fovea, the control mechanism through regions R_1 and R_2 will output signals on the appropriate muscles causing our eyes to turn back towards the source of light.

The key point to note is that the learning mechanism that turns our eyes towards the source of light is a natural one – not because it provides us with a survival advantage (cf. an organism that is incapable of learning this skill – it can die relatively easily in the presence of other predators) or that it is a desirable state to achieve, as evaluated by us. Rather, among all available sensory states with the possibility of the image falling anywhere on the retina, the one state that is uniquely different and the one that produces an explosion of neural firing patterns is when the image falls on the fovea. The SPL architecture of our brain now allows us to alter its internal structure to take advantage of this unique state and convert it into an stable equilibrium state.

Most textbooks mention incorrectly that our brain learns these skills because of their survival advantages. When a sensory-control algorithm is desired, they provide an engineering solution like the design used in a camera or a camcorder. However, these algorithms are not valid because a non-conscious system cannot learn them by themselves. This is a necessary requirement when studying the problem of consciousness.

Even with the other examples discussed in this chapter, we will see that one of the critical step is to identify such an unique and natural state S first i.e., one that produces an explosion of neural firing patterns compared to the other related states. Applying the fundamental pattern now becomes straightforward and analogous either to the current example or to the problem of auto-focusing eye lens discussed in sections 16.1 and 16.2.

Notice that an infant who has just learnt to turn his eyes towards a source of light uses this new skill rather involuntarily. However, as he grows older and learns free will (see chapter 28), he will have a choice whether to turn to the direction of light or not.

The important outcome of learning this new skill is that it correlates our movement of eyes to the 'correct' direction of the source of light. What we see with our eyes is correlated to what is 'really' in the external world. We are not looking at some random direction and claiming that we can see an object or that it is indeed in front of us. This may seem obvious to conscious beings, but we need to be sure that we can distinguish reality from imagination. The underlying mechanisms should work even when we are not conscious. Such precise correlations with the reality, not some artificial (or abstract) information memorized and stored like, say, in a computer without any connection to reality, will eventually help us give rise to our own consciousness. Many complex situations occur in the external world. It is nontrivial to claim that the external world is not just a dream. This example gives us, at least, one simple guarantee: the direction in which you are looking and the direction where the object is located are indeed correlated and almost identical. Each of the examples discussed in this chapter do make similar guarantees.

17.2 Converging both eyes onto the same object

For a human, the optic nerves, optic chiasma and optic tracts are connected to reach the lateral geniculate nucleus from both of our eyes. The temporal retinal fibers from the left eye (to the left of your retina) and the nasal fibers from the right eye (to the left of your retina) are part of the left optic tract. These fibers correspond to the right visual field. The structural arrangement ensures that the information from the right visual field from both eyes is processed in the left visual system in the brain (see Purves [122] and Kandel et. al [82]). The same applies with the left visual field from both eyes i.e., it is processed by the right visual system. The interesting point is that approximately the same information from both eyes reach the same region in the brain. The convergence of visual information from both eyes due to this unique structural arrangement introduces the possibility of explosions in a natural way. In addition, the control mechanism for moving both eyes is partially synchronized even at birth. They do not move independent from each other.

Let us now see how we can apply the fundamental pattern. As discussed in section 16.3, the first step is to identify a natural sensory state S that can produce an explosion compared to other related states (assumption 16.3.4). If both eyes are looking at different objects, the information sent from each eye, though converging to the same location (say, the left or the right visual system), are fairly disjoint from one another. The input Na^+ and K^+ ions needed for neurons to fire action potentials at these regions are now shared across most of the neurons. The pattern of neurons corresponding to different images on different eyes – like a table and a chair – ends up triggering much broader set of neurons compared to the case when the images are the same. As a result, these sensory states do not produce a strong signal at the left or the right visual system. On the other hand, if both eyes are looking at the same object, this unique sensory state S now sends approximately the same information to left and right visual system. Therefore, we now have a strong signal (i.e., an explosion).

Another way to explain why there will be explosions (in relative terms) when the images are the same on both eyes is by decomposing the input patterns from both eyes as (a) input from one eye and (b) the differences or delta's from both eyes. As an example, (a) corresponds to common sub-patterns between a table and a chair, say, when you overlay a table image on top of a chair image and (b) corresponds to the differences between them. With this decomposition, we notice that the contribution from (b) is small when both eyes look at the same object compared to the case when both eyes look at different objects. As a result, we will have an explosion for the former case.

The remaining steps of the fundamental pattern (section 16.3) can now be applied just as we did with the previous examples. Over time, our brain will create two regions R_1 and R_2 with properties obeying assumptions 16.3.5 and 16.3.6. They ensure that state S is a stable equilibrium state. Small perturbations to this state will cause the control mechanism through regions R_1 and R_2 to output signals on the appropriate muscles causing both of our eyes to re-converge back onto the same object. I refer the reader to sections 16.1-16.3 for further details since the underlying mechanisms are quite analogous.

17.3 Tracking a moving object with your eyes

When we become conscious, we have a choice whether to track a moving object or not. In section 28.3, I will discuss how we know that we have choices, how we learn to choose one from a set of choices and how we know that we are making the choice.

When we do decide to track a moving object, our sensory-control mechanism follows the object quite accurately. Let us briefly examine what happens on the retina when we are tracking a moving object. The image on the retina moves away from the fovea (responsible for sharp central vision) as the object moves. When you try to track it accurately, you are turning your eyes so that the image falls back on the fovea at an approximately original location. In this manner, we are constantly attempting to make the image stay on the fovea by turning our eyes or our head, if necessary, whenever we are tracking a moving object. The sensory state S corresponding to the image staying on the fovea acts like a stable equilibrium state.

How does our brain learn such a mechanism when an infant is born without this ability? We will once again use the fundamental pattern to show that consciousness is not necessary to learn this skill. As discussed in the previous sections, the first step is to identify a natural sensory state S that is capable of producing an explosion compared to other related states. Among all positional variations of an image in a short duration (as the object moves or you move your eyes or head), the one state that produces the strongest signal is when the image returns back to the original state. Note that the original state has a certain strength of the signal. We can even assume that we have auto-focused our eye lens to form a sharp image, resulting in an explosion. When the image moves away and we are not tracking it, the deviation of the image position on the retina starts to get higher. The set of actively firing neurons become large (some neurons from the original region and some from the new region). The Na^+ and K^+ ions in the extracellular space are shared across these large number of neurons and hence will not result in an explosion. On the other hand, the set of firing neurons are small when the image returns back to the original region, thereby producing an explosion.

We can now apply the remaining steps of the fundamental pattern. These steps for self-learning are nearly identical to the previous cases. Once our brain learns and adapts itself for this task, the tracking sensory state behaves as a stable equilibrium state. We can summarize the stability aspect as follows: (a) we have an initial explosion from auto-focus, (b) the image moves away, (c) this weakens the signal strength, (d) you generate a suitable control signal from regions R_1 and R_2 to track the image and (e) the explosion returns. In this manner, small perturbations to the original state are restored.

17.4 Tracking a curved boundary with your eyes

Whenever we look at a large-sized object, we have a natural tendency to trace its boundary with our eyes. For example, if we look at a table, a door or a car, we naturally scan the boundary of the object quickly and without much difficulty. The boundary does not have to be a straight line and yet we trace it with ease. In this section, we will see that the underlying mechanism for learning this skill is the fundamental pattern. As with the previous examples, consciousness will not be necessary to learn this skill. This task is closely related to the previous task (section 17.3) of tracking a moving object.

As discussed in the previous sections, the first step is to identify a natural sensory state S that is capable of producing an explosion compared to other related states. Here, we have an image of a curved boundary on our fovea at a given instant. As we now turn our eyes or our head by a small amount, the curved boundary stops falling on our fovea for most of our eye/head movements. There are, however, two directions of motion of our eye/head that makes a curved boundary fall back on the fovea. These correspond to either a forward or a backward motion along the curved boundary. Only for these two directions of motion, our brain continues to produce the strongest signal. For all other states, there is no explosion.

Let us assume that we have auto-focused our eye lens to form a sharp image. When we turn our eyes/head away and we are not tracking the curved boundary, the deviation of the curved boundary on the retina starts to increase i.e., the set of actively firing neurons begin to change considerably. Specifically, when we have an image of a boundary on our retina, we would have two distinctly different neural firing patterns, one on either side of the boundary. This results in a strong signal from neurons on the boundary (see section 34.2). However, when there is no boundary, all neurons in the region share the extracellular Na^+ and K^+ ions evenly, resulting in a relatively weaker signal. The absence of a strong signal except for the two states S (forward and backward direction of motion along the curved boundary) can be used by the fundamental pattern to make our self-control mechanisms converge to these two states S.

We can now apply the remaining steps of the fundamental pattern. These steps for self-learning are nearly identical to the previous cases. Once our brain learns and adapts itself for this task, the tracking sensory state behaves as a stable equilibrium state. We can summarize the stability aspect as follows: (a) we have an initial explosion from auto-focus, (b) the image moves away as you turn your eyes or head, (c) this weakens the signal strength, (d) you generate a suitable control signal from regions R_1 and R_2 to track the curved boundary and (e) the explosion returns. In this manner, small perturbations to the original state are restored. This results in tracking the curved boundary.

17.5 Turning your head towards a source of sound

When you hear a sound, the excitations produced from both of your ears reach superior olivary nucleus via the cochlear nuclei. Two specialized nuclei called the medial and lateral superior olives receive information from both ears. The signals from each ear has a slight differential in most cases. These time and intensity differences are believed to be captured by these two superior olives (see Purves [122] and Kandel et. al [82]).

When an infant hears a sound, he usually does not know the source and direction of the sound. However, after several months, he is capable of turning his head in the correct direction. How is he capable of learning this skill even before he is conscious (to the same degree as an adult)? It is worth highlighting that several animals (bats, dolphins, toothed whales and others) have even more acute abilities when it comes to detecting the direction of a source of sound than humans. To learn this skill, we can once again use the fundamental pattern because there exists a natural sensory state S that produces an explosion compared to other related states. As you turn your head around while listening to a sound, the excitations received from both ears (at the medial and lateral superior olives) have different strengths of signals. For example, the ear closer to the sound has a stronger signal than the ear away from the sound. However, there is a particular position of your head for which the source of the sound is equidistant from both your ears. For this position of your head, we have a stronger cumulative signal and hence a possibility of an explosion. The fundamental pattern can, therefore, be used to train yourself to identify this location (position and direction) for different types of sounds.

As discussed in the previous sections, we can now naturally apply the remaining steps of the fundamental pattern. Our brain uses appropriate regions R_1 and R_2 to make the above sensory state S (one that produced an explosion) as a stable equilibrium state.

Note that humans after they become older are additionally capable of locating the source of sound even without turning their head. This, however, requires memories to be stored to provide directional information i.e., specific meta-fixed sets, one for each direction (or angles around you). This is analogous to creating specific meta-fixed sets, one for each angle of an edge (or line segment) used for identifying curves with your eyes (see section 17.4). We cannot apply the fundamental pattern directly for these new cases because there is no explosion produced when learning each angle individually. Instead, we need to learn to trigger a unique meta-fixed set for a specific angle. I will discuss this later in section 34.4 for the case of visual detection of angles in edges.

17.6 Moving towards a source of odor

Humans and several other animals, particularly dogs, have a good sense of smell. They use this ability to turn or move towards the source of smell (coming from edible food, bad odors and others). We do not use this ability as commonly as we use our eyes and ears. Yet, it is an important skill necessary for our basic survival. Newborns, while they seem to have an ability to detect a wide range of odors, do not have an ability to identify the direction of the source of odor and correctly 'turn' towards that direction. Over time, they do learn to identify this direction with sufficient accuracy. How do we learn this skill even without a high degree of consciousness? In this section, I will show that the fundamental pattern can once again be applied to this case.

The olfactory sensory neurons in the olfactory epithelium detect odors in most vertebrates. These sensory neurons extend into the brain to the cranial nerve I (olfactory nerve). The sensory neurons are activated when the odorant molecules that dissolve in the mucus lining (as they pass through the superior nasal concha) bind to the receptor proteins. A cascade of chemical reactions occur that eventually opens cyclic nucleotide-gated ion channels (CNG) causing an influx of Ca^{2+} ions into the cell. This causes the neurons to fire action potentials. These signals propagate to the olfactory cortex and other areas including amygdala, responsible for emotions, via the olfactory bulb converging on special structures like glomeruli to give rise to an overall perception of odor (see Morris and Schaeffer [112] for further details).

For the problem of learning to turn towards the source of odor, we first need to identify a sensory state S that generates the strongest signal (i.e., an explosion) among all possible directions. Clearly, the direction facing the source of odor naturally satisfies this condition. Also, this state exhibits the properties of a stable equilibrium state – any perturbation that turns us away from this direction generates a weaker sensory signal. As discussed with all the previous examples, the fundamental pattern can use this property to learn to control our ability to turn towards the source of odor. I will omit the details of how to use the fundamental pattern now having identified the state S that generates the explosion. This is because the remaining steps are identical to those described in section 16.3 (and for the previous cases).

17.7 An infant learning to put his hand in his mouth

Most infants go through a period when they seem to put almost everything they can reach into their mouth. They do grow out of this ability several months later in some cases. In other cases, it may turn into a habit like sucking the thumb. However, it is interesting to see how they learn this skill when they were not born with this ability. In fact, you can clearly see a period of, say, about a few weeks when they are struggling, but trying hard to put their hand in their mouth. You do see signs of happiness when they succeed and signs of frustration (resulting in crying sometimes) when a parent removes his hand or a toy from his mouth. In this section, I want to show that learning this skill is a direct application of the fundamental pattern.

To use the fundamental pattern, we first need a sensory state S that is capable of producing an strong explosive signal relative to other sensory states. In this case, the sensory states of interest can be a combination of visual and taste inputs. If the infant relies on visual inputs as well, then several of the previously discussed tasks should first be learnt prior to learning this skill. These are, for example, the ability to auto-focus eye lens, the ability to move towards a source of light, the ability to track a moving object and the ability to converge both eyes onto the same object.

In addition to these, one inborn skill necessary here is the sucking reflex – an instinctive reaction to suck anything that touches the palate (roof of his mouth). We can already say that a strong signal is generated from the taste buds when the sucking reflex is triggered. Therefore, this is the special sensory state S needed for the fundamental pattern to be applicable. We can now apply the remaining steps of the fundamental pattern to turn state S into a stable equilibrium state. I will omit the details here as they are quite analogous to the previous cases. One subtle difference compared to the previous cases is that regions R_1 and R_2 do correspond to a positive and a negative emotion respectively.

18
Common Fixed Sets

In the previous two chapters, we discussed a new mechanism for SPL systems called the fundamental pattern. It is a sensory-control mechanism that the system uses to self-learn a task with no help from conscious beings. This was an important mechanism because it showed that a number of features common across several species can be learnt using just the SPL structure. In this chapter, I will present more mechanisms for creating such common features necessary during our (and other species') everyday interactions with the external world. They implicitly depend on terms like motion, direction, action, neighbors and size. In this chapter, I will express these terms in terms of the internal dynamics that they refer to so we can generalize them to all species.

The objective is that we want to avoid using the above terms in such a way that only a conscious being understands. Even though these terms are intuitively obvious, the internal dynamics is far from simple and is generally different across different people. For example, to identify a term, we typically require several thousands of neurons to fire. However, we aggregate all of this firing activity into a single term when we refer to, say, an angle, an edge, and others. Each such term is a compact representation of the more complicated internal dynamics that occurs within the brain. Later, we develop new terms for complex events and objects like trees and cars that rely on just these simple terms. Once again, these new terms inherently deal with the firing activity from tens of thousands of neurons.

The problem with using these compact terms (i.e., abstractions referring to dynamics) is that the natural relationships that exist between these terms are lost. We have to recreate them manually. For example, words like 'basketball' and 'marble' have nothing in common. However, in reality, there are several hidden relationships like, say, their spherical shape, their differences in sizes and others. Such information is lost when used as words. The hidden relationships can only be inferred through semantics, which in turn requires consciousness. Had we used the internal neural firing activity, this information would (and, in fact, should) be present.

The difficulty with capturing hidden relationships can be seen when we try to represent them in a computer. A human would first need to manually encode features like shape and others as additional information when trying to identify these objects. Machine learning techniques then uses these features to discover similarities and dissimilarities between different objects. The involvement of a human is unreasonable if a system is attempting to become conscious and wants to maintain the internal dynamics and their relationships consistently. For example, let us say that a ball is rolling to your left on the ground. It appears as though this statement is so obvious that we do not need any further explanation. However, let me list a number of implicit assumptions we have made here because of our consciousness. These assumptions become questionable if the system is not yet conscious.

1. We assumed that the ball and the ground are external objects, which is not valid yet.
2. We assumed that the ball is on the ground. This is not clear because we do not know if the ball is touching the ground or if it is floating and rolling in air, with the ground being a background image.
3. We assumed that the ball is moving. Instead, you may be the one moving or turning while the ball is at rest.
4. As the ball is moving, we are assuming that it is the same ball for the entire time, even at the new location. We cannot know this. Maybe it is similar to a pattern seen in casinos where we see different light bulbs flash in a linear sequence creating the impression of motion.
5. As the ball is moving, you are not in a standstill position. How can we assume that your own motion is perfectly nullified to create an impression that you are watching only the ball move?
6. We are also assuming space and time in the entire description. This is not obvious before you are conscious.

If we analyze more carefully, I am sure we will discover a few more assumptions. Even for such a simple dynamical event, which takes a mere 5 seconds to complete, we have made several assumptions. The point to note, however, is that a number of these assumptions (like with the existence of objects, space, time and motion) are common across most events.

Therefore, the first important step is to represent the above terms using dynamics and without requiring consciousness. It should come as no surprise that we will use specific fixed sets to identify with each of these terms. We have already seen that the correct way to represent abstractions and truths are as fixed sets. In this chapter, I will consider a system with sensory-control mechanisms like the fundamental pattern observing the external world using one-dimensional (1-D) or two-dimensional (2-D) sensory surface. Using the common patterns that occur in our external world, I will discuss the types of fixed sets we can form and study what they mean.

18.1 Sensory surfaces – problem formulation

Consider an SPL system that has a 1-D sensory surface AB and other features described until now (like memory model features of chapter 8, fundamental pattern and others). You can imagine such a surface to be the cell wall of a single cellular organism or, say, the sensory surface that detects pheromones (like, say, in an ant). In contrast, examples of 2-D sensory surfaces are like the surface of the retina or the surface of our skin. I will discuss 2-D surfaces later.

The region from A to B has a finite number of components placed along the surface that detect, say, pressure from touching or disturbances in light or sound. Let us assume that these components are arranged in the following sequence from A to B: A, $A + \Delta x$, $A + 2\,\Delta x$, $A + 3\,\Delta x$, ..., $B - 2\,\Delta x$, $B - \Delta x$, B. Here Δx is the distance from one component to the next. Strictly speaking, Δx need not be the same across all components from A to B, as stated here. We assume this for simplicity. The reasoning below does not change when we use Δx_1, Δx_2, Δx_3 and so on.

The sensory signals from the surface propagate deeper into the SPL system, which can be as complex as a human brain. In this chapter, I will explain a few mechanisms to create fixed sets when inputs fall on a 1-D sensory surface. These basic mechanisms are generic enough to extend to 2-D case as well. It is important to emphasize that most of the discussion here is about identifying patterns on the sensory surface, not of the external world. Only in the case of direct-contact based sensors like touch, you can infer useful information about the external world as well. For other non-contact based sensors like eyes and ears, the patterns like sizes, shapes and angles on the retina or in the inner ear could be completely different from the corresponding properties of the external objects.

18.2 Memories of patterns on the sensory surface

Let me reiterate the mechanisms previously described in chapter 7 as a way to create memories for patterns formed on the 1-D sensory surface. We have previously described that the SPL system can form fixed sets for each such pattern using basic abilities like forming new connections, modifying and deleting existing connections (see section 7.3 describing the new memory model). For a 2-D sensory surface, the type, the number of patterns and the fixed sets formed are more elaborate. These include, for example, fixed sets for different shapes, textures and colors of patterns.

18.3 Detecting time-variations at a given region

How can our brain differentiate whether the inputs falling on a specific region of the sensory surface are time-varying or static? What are the fixed sets for detecting these time-variations? Consider an input that falls at A and moves to $A + 2\,\Delta x$ via component $A + \Delta x$. This is an example of a time-varying pattern that covers a region from A to $A + 2\,\Delta x$. On the other hand, a static pattern within the same region i.e., from A to $A + 2\,\Delta x$, is one where an input falls on the entire region at the same instant, instead of progressively moving from one end to the other, say. We want our brain to detect this time-varying aspect of the inputs. It is obvious that when our brain does detect this difference eventually, it will be represented as an appropriate fixed set. This fixed set and the underlying mechanism is critical for a number of features (involving motion and tracking), as we will see later in this section. Let me now discuss how to create this fixed set.

Since there is a time lag Δt_1 between the excitation of neurons at A and $A + \Delta x$ (likewise Δt_2 between $A + \Delta x$ and $A + 2\,\Delta x$), the neurotransmitters released on target neurons from A and $A + \Delta x$ have this time lag. These neurotransmitters open the gated Na^+ and K^+ channels with a similar time lag Δt_1. When the channels are open, it generates a voltage difference (see section 5.1). These target neurons fire when the voltage exceeds the threshold value.

Consider the variation of voltage over time for a single neuron that is subject to constant input. If there are interconnections between neighboring neurons, then the target neuron at A would receive inputs from $A + \Delta x$ and, possibly, from $A + 2\,\Delta x$ as well. In this situation, the target neuron at A will have a small spike of voltage at about Δt_1 and $\Delta t_1 + \Delta t_2$ after the first input from A. This is because of the interconnections from the neighboring neurons (from target neurons at $A + \Delta x$ and $A + 2\,\Delta x$). These inputs open gated channels at the target neuron at A at Δt_1 and $\Delta t_1 + \Delta t_2$ respectively, causing a small spike in voltage. The time-variation of voltage at target neuron A would have two spikes at Δt_1 and $\Delta t_1 + \Delta t_2$.

Let us consider the static situation where the pattern does not 'move' from A to $A + 2\,\Delta x$, but instead is simultaneously excited at all three locations (A, $A + \Delta x$ and $A + 2\,\Delta x$). The time-variation of voltage at target neuron A would be flat or constantly increasing. There would be no spikes in the voltages compared to the previous case when the pattern moves. This is because there are no time lags to generate the spikes when all inputs are simultaneous. This is an important distinction in the measurable behavior of the voltage signal at the target neurons. This can be used to distinguish patterns that vary with time (due to motion or dynamics, typically) versus static patterns on the sensory surface.

How do the spikes in voltages in time-varying case (rather than the static situation) help with creating or exciting unique fixed sets? There are two ways it will help: (i) they will create a surge of voltage that helps in exceeding the threshold voltage of either the current neuron or a subsequently connected neuron and (ii) they can help in creating new connections (synapses) with neighboring neurons. A small burst of voltage helps in extending the growth cones of an axon compared to a gradual increase in voltage. A burst, similarly, helps in exceeding the threshold quickly compared to a gradual increase in voltage.

Hence, time-varying patterns, that produce these bursts in voltage, help in exciting existing neurons and their interconnections easily as well as in creating new connections between neighboring neurons. This is one way to explain why watching a movie helps in memorizing events more easily that looking at static pictures from this movie. The time-varying dynamics in the movie produces surges in voltages that help in memorizing the event more effectively than when there are no surges while watching a static image.

It is important to note that these fixed sets do not help in sensing the passage of time (though this fixed set seems to identify the time-varying aspect). The fact that all dynamical systems are implicitly expressed in terms of time makes it a little harder to isolate this concept and identify unique fixed sets. I will discuss sensing the passage of time in more detail in chapter 40.

18.4 Identifying a pattern even as it moves away

When a pattern moves from one region P to a neighboring one Q (assuming P and Q are disjoint from one another) on the sensory surface, how does the system identify it as if it were the same pattern? This is not obvious for the following reason. The complete information our brain uses to help with this problem has two parts: inputs directly from the sensory surface (like, say, from P) and neural pathways that propagate deep into the brain (like say, from the retina extending along the optic nerve into the brain). The former inputs are disjoint from our assumption for P and Q. The latter inputs, i.e., the neural pathways originating from P extending into our brain are also disjoint from the corresponding set of neural pathways for Q. The complexity of our brain suggests that the pathways disperse along different directions. Therefore, there is no reasonable way to compare the two patterns at P and Q, let alone say that they are the same.

The key idea needed to solve this problem requires using our self-control mechanisms (like the fundamental pattern). When the pattern moved from P to Q on the sensory surface, you move yourself (or your sensory surface) in the opposite direction using these self-control mechanisms so that the pattern falls back onto the location P. Now, you can compare the original sensory surface inputs, the deeper neural pathways to the newly returned sensory surface inputs and the new neural pathways deep in the brain. This comparison is meaningful unlike the previous case (one at P and the other at Q) because the regions on the sensory surface are approximately the same (both near P). Once again, the actual comparison is performed by triggering an appropriate fixed set. I will discuss the details next. For now, I will only consider lateral motion of the pattern, not changes resulting from motion of the external object along the depth direction.

There are two cases to consider here when we try to identify the patterns as the same one: (i) we analyze the entire motion from P to Q and (ii) we only analyze the end locations P and Q. The former case is quite common for a conscious being as he focuses on the object (spending real physical energy) during the entire motion from P to Q. The latter case happens in situations when we are distracted and are thinking of something else, not observing the entire motion and, instead, only looking when the pattern is at location P and then, subsequently, only after it has reached Q.

In either case, the system's self-control actions are important to simplify the problem. As mentioned earlier, we use self-control actions to move the pattern at Q in the opposite direction back to the same location at P in order to compare the dynamical effect at both the sensory surface and deep inside the system (using, say, the fundamental pattern). For example, if a ball moved from left to right, the image on the retina moves from right to left and, in turn, you move your eye from left to right to ensure that the image falls back on the fovea. In essence, you are tracking the moving ball using the fundamental pattern (section 17.3).

Let me now describe the mechanism (using detection of time-varying inputs in section 18.3 above) in terms of neurons and their interconnections within our brain. When the pattern falls on the retina at P at time instant t_1, it excites a set of neurons that propagate within the brain to form fixed sets specific to the location at P. As the pattern moves towards Q at time instant t_2, there will be bursts of voltage spikes at P (and the corresponding fixed sets) through the interconnections as explained in section 18.3 above. We can now group the neurons into three sets at time instant t_2: (1) neurons at P that are not excited, (2) neurons that are excited the same way as at time instant t_1 and (3) set of neurons that are newly excited.

If the system uses its self-control mechanisms to move the eye in the opposite direction (to track the moving pattern) at time instant t_3, the role of set (1) neurons gets interchanged to set (3) neurons and vice versa, while set (2) neurons stay the same. This oscillation produces new bursts of voltage spikes near P once again (using section 18.3 above). Now, when the system moves in the opposite direction, it does not necessarily stop at P. It will overshoot to the 'left' of P. At time instant t_4, the pattern moves again towards Q. This is counteracted by the system's opposite motion (possibly, imperfectly again) at time instant t_5. This oscillatory motion between the pattern's and systems self-motion continues for a while until it stabilizes to an equilibrium position at P.

As the pattern moves and the system tracks it, most of the neurons near P experience these oscillatory bursts of voltage spikes (even with imperfect tracking). This variation of voltage spikes is unique to tracking a moving sensory pattern from P to Q. This produces a new fixed set according to the memory model of chapter 7 (similar to section 18.3 above). Our brain can use this to infer that the pattern is the same even as it moves from P to Q.

Now, for both cases (i) and (ii) mentioned in the beginning, we can use this unique fixed set to say that the pattern is the same as it moves from P to Q. The above explanation directly refers to case (i). For case (ii), we are not looking at the pattern's intermediate motion from P to Q. Let the pattern form at P at time instant t_1. After a long period, say at time instant t_{10}, the pattern forms at Q. If the system moves in the opposite direction at t_{10} so that the pattern falls back at P, this will trigger the above unique fixed set, as long as it was already created through case (i). Typically, as an infant you would have tracked a moving pattern a number of times (i.e., case (i)) that you do not need to track it anymore and instead focus on something else during this intermediate period (i.e., case (ii)). Therefore, triggering this fixed set suggests that the pattern is the same even in case (ii).

I have implicitly assumed that the system has linked the sensory patterns to the control actions. This is a reasonable assumption. A number of loop pattern mechanisms can help with this assumption. Furthermore, the system has an ability to create self-control actions reliably and accurately in response to the motion of a pattern from P to Q i.e., for tracking a moving object. We have seen one way to make this happen, namely, using the fundamental pattern (see section 17.3 for vision).

18.5 Fixed sets for neighboring subsequences

Let us say that a pattern is produced at some point on the surface AB and moves towards another point. From the system's point of view, this can happen in a few different ways: the system turns around, or it stays at the same place while the external objects turn around, or it stays at the same place but different static external objects touch it in a sequential and rhythmic way (like light bulbs flashing in a sequence to produce a feeling of moving lights).

The sequence of triggering of inputs is always a subsequence in either A, $A + \Delta x$, $A + 2\ \Delta x$, $A + 3\ \Delta x$, ..., $B - 2\ \Delta x$, $B - \Delta x$, B or the opposite sequence from B to A. The dynamical events that occur in our universe have a notion of continuity that ensures that the patterns cannot be random. The events triggered from the external world onto the sensory surface never allows the possibility of a random sequence like $A + 5$

Δx, $B - 6\,\Delta x$, $A + 3\,\Delta x$, $A + 7\,\Delta x$, $B - 2\,\Delta x$, $B - 5\,\Delta x$ and so on, as the pattern moves. Only if we view time-varying patterns as theoretical abstractions, such random sequences are possible. However, the underlying continuity of physical systems prevents this possibility in real situations. This continuity naturally defines a notion of neighboring chemicals. Only contiguous subsequences from A to B or from B to A are memorized as fixed sets from everyday events. The formation of a fixed set for a repeating pattern is a direct consequence of our new memory model (see chapter 7) and the time-varying detection mechanism of section 18.3 above.

18.6 Fixed sets for sizes on the sensory surface

The fixed sets considered here are for identifying different sizes of patterns on the sensory surface, not different sizes of the external objects (which will be discussed later in section 18.10). We have seen in section 18.5 above that, the patterns formed on a sensory surface can only be specific contiguous subsets of chemicals. They create fixed sets corresponding to neighboring subsequences. If the sensory surface has two patterns formed at a given location, then it is possible to distinguish them using this fixed set if and only the two patterns have different sizes. If two patterns of the same size (say, $5\,\Delta x$) are at two different locations on the sensory surface, then we can use mechanism of section 18.4 above to identify if both patterns are of the same size (i.e., by moving it back to the same location for comparison purposes using the fundamental pattern).

To be able to create a fixed set unique to a size, the system needs to take the movement of a given pattern across the sensory surface into account. The conventional definition of size already assumes this movement of the pattern along the sensory surface. We can now use the notion of self-control (using, say, the fundamental pattern) as described in section 18.4 above to form a unique fixed set corresponding to size. The movement of a pattern by a conscious being can be either by one's own motion or by the pattern's own motion (treating the pattern as triggered by an external object). When a pattern of fixed size moves from location P to Q, self-motion moves it back from Q to P (say, by turning the head to track the moving object). This creates a repeatable loop that can be used to verify if the size of the pattern is the same when it falls back at location P as explained in section 18.4 above. Once again, without this comparison back at P, it would not be possible to establish if the size is the same or not. Returning to P accidentally is less likely to occur than through self-control mechanisms like the fundamental pattern.

Here, I have not talked about the 'true' size of the pattern (like, say, of a table). Rather, I have explained the fixed sets corresponding to the size on the sensory surface (of, say, the same table). How can we infer this true size of each of the external objects? I will discuss this later in chapter 36.

18.7 Motion detection at a given location

If a pattern moves from A to $A + 2\,\Delta x$, how can the system detect that there was a motion in this region on the sensory surface? The adjacent subsequence fixed set mentioned in section 18.6 is triggered both when there is motion or when a static pattern of 'size' *2 Δx* occupies these three chemicals simultaneously. We need to use a different fixed set to distinguish these two cases uniquely. This is a special case of the time-varying detection mechanism of section 18.3. We use it to create a new fixed set using the bursts of voltage spikes at *A*, only when the pattern moves from *A* to *A + 2 Δx*. This new fixed set is triggered only when the bursts are produced at *A*. This happens only when there is a motion away from *A*. This is the motion detection fixed set at this location. There are several such fixed sets created based on motion of patterns at different regions on the sensory surface.

18.8 Detecting single dimension

In order to detect a single dimension, we should create fixed sets that allow us to detect forward and backward directions first. We have shown that the subsequences memorized as fixed sets are from *A* to *B* or in the reverse direction in section 18.5 above. This gives rise to a notion of direction: a forward (from, say, *A* to *B*) and a reverse (from *B* to *A*). With the previous motion detection fixed sets, the system was only able to know if the pattern had moved. Our objective now is to know the direction in which the pattern has moved. We want a fixed set that distinguishes specifically if a pattern has moved from *A* to *A + 2 Δx* or if it has moved from *A + 2 Δx* to *A*. This becomes easy using the above mechanisms (like section 18.3), as we will see now.

The first thing to note is that, using mechanism of section 18.3 above, when a pattern moves from *A* to *A + 2 Δx*, there are three bursts of voltage spikes created at neurons downstream from *A*. However, when the pattern moves from *A + 2 Δx* to *A*, the bursts are created downstream from *A + 2 Δx* instead. We can use this important difference to create new fixed sets to distinguish the forward from the reverse motion.

However, these fixed sets are unique to the position of the chemicals. If the forward motion in the direction of *A* to *B* occurs at, say, *A + 10 Δx* instead of *A* or *A + 2 Δx*, the corresponding fixed sets will be different from the above ones (i.e., the ones at either *A* or *A + 2 Δx*). Therefore, we cannot call this the forward sequence fixed set. How do we get a unique directional fixed set across the entire 1-D sensory surface?

Let me call the above fixed sets created from the bursts using mechanism of section 18.3 as 'first-level' fixed sets. Now, there can be a 'second-level' downstream fixed sets (i.e., a meta-fixed set) created using several of the first-level fixed sets to keep track of motion beyond *A + 2 Δx*. For example, the bursts at *A*, *A + Δx* and *A + 2 Δx* triggers three new fixed sets as the pattern moves beyond *A + 2 Δx*. These three new fixed sets have time lags *Δt*'s. These time lags will now let us use the exact same mechanism of section 18.3 above, recursively. It helps us detect time-varying inputs except the difference now is that the inputs are fixed sets themselves instead of inputs from the sensory surface. These create new bursts at second-level and, therefore, generate new second-level fixed sets.

As you continue this across a few levels i.e., deeper into the brain, you eventually have a fixed set that is triggered for *any* forward motion anywhere on the 1-D sensory surface. The cascading interconnections will eventually trigger this unique forward-direction fixed set deeper in the brain whenever there is a forward motion. The same applies with the backward motion as well. We can now call them as the forward and backward motion fixed sets.

These two directions represent the notion of a single dimension. When the sensory surface is two-dimensional like, say, with our retina, we can use a similar process to detect the existence of two independent dimensions. In this case, we can independently produce and detect the motion of patterns in the sideways and up/down directions. We will have similar unique directional fixed sets to help us conclude the existence of two dimensions. Furthermore, the generalization extends to three dimensions as well using our internal body inputs. Our hands, for example, can move independently in each of the three dimensions, say, about the shoulder and the elbow joints (which act as 3-D sensory surfaces). The inputs from these joints are transmitted through the brain stem in a similar way to create necessary fixed sets for each of these directions. We, therefore, know and detect whether our hand moved sideways, up and down, forward and backward or in any other direction.

Now, we have ignored one important scenario with these fixed sets. This is the self-motion from the system itself. It is possible that the system can rotate in a way that the pattern can actually move in the opposite direction. For example, let the pattern move from A to B while the system tracks the moving pattern. If the system rotates faster than the pattern's independent motion, the resulting pattern on the sensory surface will actually reverse direction. The pattern on the sensory surface moves from B to A. Yet, the system needs to conclude correctly that the object is moving from A to B. The system should detect the change in *relative motion*, relative to itself.

This implies the system should have a different fixed set to detect and 'subtract' its own self-motion. When we discuss body fixed sets later, I will show how to create these new relative-motion fixed sets. Motion relative to the system's own body will be possible to detect once system's own body motion can be detected using appropriate fixed sets (discussed later in section 18.13).

18.9 Comparing two or more patterns

When we have two patterns triggered on the sensory surface, how can the system identify both of them as similar or dissimilar? Similarity of simple patterns like intervals, circles and squares are easy using just the definition of the fixed set (see section 7.8 on similarity and dissimilarity conditions). Compared to 1-D sensory surfaces, 2-D sensory surfaces like our retina generate many more complex patterns like, say, the shape of a car, a tree, a table and a chair. Analyzing similarity and dissimilarity among them is nontrivial and requires much more than basic memory. We have already discussed this in considerable detail in section 7.8. We want to now generalize these mechanisms to cover more cases.

For each of the above complex patterns, our brain triggers multiple fixed sets simultaneously, one for each subpattern within a pattern. For example, if you are looking at two cars, you will initially trigger several common fixed sets like color of the cars, shapes of the wheels and other geometric arrangements. It may be possible to detect dissimilarity quickly (using comparison mechanism of section 18.4 above) if some of these common fixed sets are different. However, the difficulty is when most of these are approximately the same. One way to handle this case is to say that if after a while we do not trigger any disjoint set of fixed sets, we conclude that the two patterns are similar. Typically, when you identify differences, you also memorize where to look to find these differences using the fundamental pattern (as long as the patterns are repeatable enough to create a new fixed set). This would, for example, allow us to quickly turn our attention to these subtle differences when we look at the same two cars a few days later.

There is one important difficulty that still needs to be resolved. This is with how we associate the same fixed sets triggered in parallel, once for each pattern, to the correct pattern. For example, the wheels, the doors and the windows are similar for each car though the colors may be different. Yet, our brain needs to associate the correct color (and other differences) to the correct car. It cannot mix these fixed sets across the patterns given that they are triggered in parallel and at different regions in our brain.

To address this correctly, we need a self-sustaining membrane of meta-fixed sets and abstract continuous pathways. These two new concepts will not only address this issue, but will turn out to be critical for explaining how we attain a feeling of knowing and how we become conscious. They will be discussed in detail in chapters 22 and 23 respectively. The basic idea is that memories, though represented as fixed sets, are not disjoint and discrete set of items. Instead, they are well-connected both spatially and temporally and they have rather special continuity properties as well.

18.10 Comparing patterns not on sensory surface

When we want to compare two patterns A and B generated at different locations of the sensory surface, we use self-control mechanisms like the fundamental pattern (see also section 18.4) to make the second pattern fall on the region of the first pattern or vice versa. With both patterns now falling on the same region, it becomes possible to compare them using the fixed sets triggered by both of them. This approach is reasonable when we are comparing patterns that fall directly on the sensory surface. However, if the two patterns A and B are like thoughts that are internal to our brain, how can we compare them? In a few simple cases, we may have a way to control the cause of these thoughts by changing the pattern forming on the sensory surface. In such cases, we can continue using the above approach.

To cover more cases, let us propose a generic mechanism. We imagine creating a path from each pattern towards a common region C, which is physically close to both A and B. It is quite reasonable to let the common region C be either the first (A) or the second fixed set (B) itself, though this is a degenerate case. If our brain succeeds in creating such a common region C for patterns A and B (through everyday experiences), we can say that the two patterns are similar in some sense. If there is no such common region C, we can say that the two patterns A and B are disjoint or unrelated.

The intuition behind this is that if there is a common region C, we can use it to move the firing pattern from A to B via C or vice versa. This suggests that thinking about A can lead you to think about B within a short period i.e., there is some association between the two patterns. The assumption that A, B and C are close enough that the neural firing patterns can reach all regions without dissipating is important. If there is no common region C, then it is not typical that we think about pattern B when thinking about pattern A. For example, thinking about pizza can remind you of a funny experience you had, say, during your vacation in some restaurant whereas it might not remind you of an unrelated hiking trip.

In the above discussion, we can only say that the two events are related or unrelated, not that they are similar or dissimilar. Yet, we can attempt to generalize the same approach to latter cases as well when we use self-control mechanisms like the fundamental pattern. The approach is that we first trigger pattern A. Next, using the fundamental pattern, we trigger several regions C_i's that originate from A. For some fixed sets triggered from regions C_i, it is possible to trigger several patterns B_i, some of which self-sustain longer than the others. The pattern B_i that wins (i.e., the one that self-sustains the longest) can now be compared to the original pattern A. These comparisons can be characterized sometimes as analogies. In some cases, two seemingly disjoint topics may turn out to be analogous. Note, however, that evaluating the degree of similarity of the two patterns is beyond the scope of the current discussion as we need a certain degree of consciousness. We require a self-sustaining membrane of meta-fixed sets and abstract continuous paths mentioned briefly in section 18.9 (and which will be discussed in greater detail in chapters 19 and 21).

18.11 Detecting geometric relations between patterns

When multiple objects form patterns on the sensory surface, we can group them into subpatterns, one for each object. We can use the mechanisms explained previously to excite fixed sets for each of these subpatterns. We can now view the entire sensory surface as a collection of, potentially, disjoint subpatterns. With a 1-D surface, they correspond to disjoint intervals with a fixed set generated for each interval. There is a clear notion of sizes of intervals and relative distances between them. With a 2-D sensory surface like our retina, the subpatterns correspond to disjoint regions. We would have geometric properties like (a) sizes, (b) shapes, (c) relative distances and (d) angles between two areas.

Let me address each of these geometric properties now. Firstly, for sizes, we have already discussed how to form fixed sets, one for each size, in section 18.6. We use this mechanism for each of the multiple objects corresponding to the disjoint patterns simultaneously. Secondly, the shape of a simple pattern (like a square, circle or a car) is directly related to the memory of the pattern i.e., we use mechanism discussed in section 18.2. Additionally, we can use the similarity mechanism (see section 18.9) to identify a set of fixed sets. We can distinguish each different shape with a different set of fixed sets.

Thirdly, the notion of distances is not too different from the notion of sizes on the sensory surface. Let us assume that two objects have already been detected using the corresponding fixed sets. The relative distance between two patterns is a new fictitious 'background' object of size equal to the distance between them. Note that these are not true sizes and distances of external objects (which I will discuss later in section 34.7 and chapter 36). They are sizes and distances of patterns 'on' the sensory surface.

In the case of a 2-D surface, the relative distances between two patterns is approximately the 1-D size of an imaginary line connecting the two patterns. We, in fact, use this approach in a dark room if we were to rely only on touch and not on vision to estimate sizes and distances between two objects. Our self-control mechanisms like the fundamental pattern provide yet another way to estimate relative distances. It lets us focus on each of the pattern separately when we scan, say, by turning our eyes or head from one object to another. Therefore, in addition to triggering the fixed sets for each of the patterns, we also trigger the fixed set that is a representative of the amount of turning of our eyes or head, i.e., an estimate of the relative distance between the objects.

Lastly, we can estimate the angles between two objects using the above approach with an imaginary 1-D line. We form fixed sets for the common directions like up, down, left, right, 45° angle and others. We create these fixed sets by linking them to the amount and direction of turning we have to do with our eyes or our head relative at a given object. It is also known that there are specific neurons in our primary visual cortex that are sensitive to a particular angle formed on the retina (see Ornstein and Thompson [118]). This specialization for angles is shown to be the result of early visual experience and is not present at birth (see Tovée [155]). These are the fixed sets formed for a line at a given angle on the retina. We use them primarily to detect angles that correspond to the curved boundary of an object. The ability to track a curved boundary is a consequence of the fundamental pattern (see section 17.4).

I will discuss more mechanisms for sensing other geometric properties later in chapter 34. As you may have noticed, the mechanisms involving complex patterns start to rely more on the membrane of meta-fixed sets and abstract continuity.

18.12 Motion detection via changes in distances and angles

In section 18.7, we have discussed one way to detect motion within a 1-D pattern. This involved comparing the foreground with the background directly. It extracts the minor time-variations in the firing patterns when an object moves and uses them to create the necessary fixed sets. Let me now discuss yet another way to detect motion by observing changes in relative distances and angles between patterns. We can consider the former approach as a low-level primitive mechanism is always at work. The latter approach, on the other hand, works only after you have a membrane of meta-fixed sets and several other meta-fixed sets for relative distances and angles, even though the underlying mechanism is still based on the one described in section 18.7.

Motion detection is such a critical aspect for consciousness that there are multiple mechanisms to ensure a high level of accuracy. For example, with eyes, it has already been shown (see Ornstein and Thompson [118] and Tovée [155]) that a certain specific set of neurons in the primary visual cortex fire only in response to the motion of an object. If the objects in the visual field are stationary, these neurons do not fire.

With ears, we can detect motion by sensing the variation in the pitch as the object that generates sound moves towards or away from you. With touch, we can detect motion either through direct contact or indirectly, say, when a sudden gust of wind rubs against your skin due to the object's motion (like, say, you sense that someone moved or has opened the door). The alternate mechanisms are especially prominent in humans who are unable to see with their eyes.

The new relative motion based mechanism works the following way. Imagine looking at objects in a room where most of them are static relative to each other. When you move your eyes, turn your head or refocus objects at different distances from you, the entire pattern on your retina moves as a whole. There is no *relative* motion between different object patterns on your retina. There is translation and rotation of the entire pattern instead. On the other hand, if the objects in the room moved around, then the relative distances and angles between objects change. The new pattern produced on the retina is no longer a simple translation or a rotation. The new pattern can, nevertheless, be expressed as a continuous transformation of the previous pattern. Your eye sensor using, once again, the mechanism of section 18.3 above, detects this relative motion between objects. This is an important way to distinguish whether you have moved or if the objects have moved.

Our ability to detect changes in relative distances and angles is easier when the objects are closer together and when they are in our field of view. In this case, even when you are focusing on a specific object, we can detect the motion of a nearby object as the distance between the objects change. Asymmetry in the geometric arrangement of objects, which is quite common in our external world, is particularly helpful to detect changes in relative positions. For example, when an object moves closer to another object on the right, it also moves farther from the objects on the left. A single motion (say, to the right) produces a number of contrasting changes in relative positions within neighboring objects.

The fixed sets mentioned in section 18.11 correspond to different relative distances and angles between objects. When one of the object moves relative to the others, as explained in section 18.3, time-variations in the firing pattern is produced that triggers the neighboring fixed sets. Depending on whether the relative distances decrease or increase, the correct neighboring fixed set is excited using the mechanism described in section 18.3.

Note that the variations in the relative distances and angles are continuous. Therefore, the corresponding fixed sets created from everyday experiences should be close together as well. This is identical to the notion of neighboring subsequences as described in section 18.5. Arbitrary jumps in the relative distances and angles do not occur in the natural world. For example, the relative distances can only vary as, say, 5 cm, 5.1 cm, 5.2 cm, 5.3 cm and so on, in a linear way, either in the forward or the backward directions (analogous to mechanism in section 18.8).

This is the same situation with relative angles as well. The fixed sets for neighboring relative angles are closer together as well. As you learn to track a moving object, the reference point you use to measure the relative angles and distances changes (like when you trace a curved edge of an object using the fundamental pattern – see section 17.4). This lets you identify the pattern as the same even as it moves around as explained in section 18.4. To know the cumulative change in distance or angle, you need to track the object fast enough that all fixed sets corresponding to the entire time period stays excited. You can then use your memory of the beginning event to compare with the end location (see section 36.3 for further details). For complex or longer paths, this cumulative estimation requires higher human abilities.

18.13 Detecting changes with or without self-motion

Our objective here is to identify unique features that allow us to distinguish if patterns produced on the sensory surface come from the system's own motion or if they are produced without the system's involvement. For example, we want a way to distinguish if I moved a ball or if the ball was moved without my involvement, say, by someone else. When we look at the way a sensory-control loop is formed, we see that a sensory pattern on the surface is eventually linked to an action via an internal path of excitations. Some sensory patterns have no associated actions, while others do have a subsequent step of triggering a control action. In the former case, the system is simply observing the event whereas in the latter case, it is actively involved in moving the object.

One of the basic mechanisms for forming and linking fixed sets is through near-simultaneity of events. Most of our everyday memories that link seemingly disjoint topics are based on this mechanism. In general, there are multiple control actions like turning your head or eyes, moving your hands or legs, twisting your body and bending down along with several sensory patterns. The near-simultaneity of sensory information and control action is itself stored internally as new fixed sets. These fixed sets are triggered if and only if the sensory pattern and the associated control action are almost simultaneously excited. There are multiple such fixed sets corresponding to different tasks. They are used to identify near-simultaneity of sensory pattern and the control action uniquely. Hence, the system has a way to distinguish if a change in sensory pattern was triggered through its own action via this fixed set.

Even though action fixed sets are derived from patterns on the sensory surface initially, the independence of the patterns after sufficient practice makes this link less important. The system can generate an action even without a corresponding sensory input. We can both identify and trigger a given fixed set through other fixed sets, not just from the external sensory inputs.

18.14 Self-control mechanisms to nullify an action

With these above mechanisms, we now have several ways to create different types of useful fixed sets. These fixed sets help us analyze and understand complex everyday scenarios. Other examples that follow easily from the previous mechanisms are, for example:

   (a) how a pattern's size can grow on both sides – this is equivalent to an object getting closer,

(b) how a pattern's size can shrink on both sides – this is equivalent to an object moving further away, and

(c) how a pattern's size can grow and move to the left.

The fixed sets for such scenarios can form only when we consider system's self-control. The way to do this is as follows. First, observe that in the above examples, the patterns generated on the sensory surface change in a well-defined manner. Given this, if the system responds by moving in an opposite direction, the net change in the pattern can be nullified.

This is a generic looped process that helps detect change by making sure that an 'error' signal is close to zero. The system can use self-control mechanisms like the fundamental pattern to learn this. If you recall from step 2 of section 16.3, in order to use the fundamental pattern, we need to look at a natural sensory state S that produces an explosion i.e., a state satisfying assumption 16.3.4 compared to other related states. We have already identified this, namely, to turn the sensory surfaces such that they generate signals that are opposite to the original patterns. Of all the possible movements of the sensory surface, only these set of control actions will produce an explosion.

As an example, for (a) and (b), a sensory pattern moving closer (or farther) can be detected when the system learns to move in the opposite direction i.e., backwards (or forwards). For (c), the sensory pattern moving from left to right is detected when the system moves from right to left. These control actions are the unique sensory states S for which the system produces explosions.

Each of these fixed sets are also independent of the objects that generate the pattern. The system, if sufficiently complex, can form more relationships between the above fixed sets through the commonly observed patterns. These relationships between fixed sets need not always result in truth (cf. fixed sets based on physical laws, which are always true). If the events do repeat sufficient number of times, the system either forms a new fixed set or relates existing fixed sets.

18.15 Patterns and fixed sets on a 2-D sensory surface

Even though we focused primarily on 1-D sensory surfaces, most of the mechanisms discussed above naturally extends to 2-D sensory surfaces as well. I will discuss this briefly.

Consider a 2-D sensory surface like the surface of a retina. Such a system can form several different shapes and sizes. The system stores fixed sets to identify these patterns, as discussed in section 18.2. The movement of the patterns can be in any 1-D line on the 2-D surface (i.e., can have different subsequences along different directions) using mechanisms discussed in sections 18.3 – 18.7. The notion of left/right direction for 2-D pattern is again relative to a given object or to you. It can be distinguished using different fixed sets, similar to the 1-D case discussed earlier, namely, using a forward direction – from A to B and a reverse direction – from B to A (as discussed in section 18.7 above).

Since several distinct patterns are possible on a 2-D surface, the complexity formed through the interrelationships between these patterns can be very elaborate. We can create more links between the sensory-control mechanisms based on the different types of patterns sensed via their unique fixed sets. The system can fine-tune the relationships between fixed sets (both at sensory and control regions) for each pattern on a 2-D surface by identifying, say, the gradients and internal textures within a given shape.

18.16 Fixed sets for other organisms

Let me briefly discuss the degree and the extent to which the above fixed sets are created in various other organisms.

Firstly, there are organisms that cannot sense light and sound as effectively as an eye or an ear and there are others that have a small brain with very few neurons. Some examples are hydra, jellyfish and earthworms. The patterns detected by these organisms using chemicals or touch are quite limited compared to the patterns that are possible when they are capable of detecting light. Furthermore, the set of fixed sets that can be formed are very limited because of the small size of their brain. Their sensory surfaces and their internal brain have abilities to create fixed sets from patterns that are detected only locally. The number of memories and interconnections are fewer and are constantly overwritten with new experiences. As a result, these organisms do not have a threshold network of "abstract continuous paths" (discussed later in chapter 21) that let them distinguish and identify themselves from the external world. They appear to be behaving in a mechanical way.

Secondly, there are organisms like ants or bugs that have eyes. Yet, they distinguish fewer patterns than a mammal. Their sensory surfaces are small and have lower resolutions. If we compare this to a human, it is analogous to looking at objects from a very close distance. At those distances or resolutions, most objects appear the same. They cannot really notice that these objects have different textures, shapes or sizes. Nevertheless, these organisms can store enough fixed sets to be capable of determining the existence of an external world. However, they do not have a large enough brain. This limits the total number of fixed sets and the breadth of local geometric information (like 3-D space) they can store. As a result, their dynamical membrane (discussed later in chapter 19) is discrete and is not rich enough to 'know' the existence of the external world. We can explain the seemingly complex abilities they exhibit using the interlinking of these fixed sets and mechanisms like the fundamental pattern.

Thirdly, we have mammals like dogs or cats that have quite highly evolved network of fixed sets. These mammals do have abilities to sense the existence of an external world. They have the ability to form most of the fixed sets mentioned above. They even have a rich dynamical membrane of meta-fixed sets. However, the ability to 'know' themselves is not easy even if they do have an ability to sense external world. The set of abstractions and the interconnections between them (like we have with a language) are the limiting factors with these organisms. With limited abstractions, the ability to ask questions about their own existence is not easy. A 3-4 year old child is also in a similar situation. In general, even if you are 'minimally conscious', knowing that you are conscious is a nontrivial step. I will discuss this in chapter 43.

19
Dynamical Membrane of Meta-fixed Sets

In the previous chapters, we have seen why knowing truths was necessary and sufficient for consciousness, how truths can be represented as fixed sets within SPL systems and how the relationships between truths can be expressed as appropriate links between fixed sets. We also saw how brain waves were generated when a number of fixed sets are simultaneously excited (chapter 15). Given the critical role played by fixed sets for the problem of consciousness, we continued our discussion by proposing several mechanisms for creating different types of fixed sets just from everyday interactions with the external world.

Let us now look at additional structure that exists when we have a large number of fixed sets. When several fixed sets interlink over time, the SPL network structure starts to grow from a locally dense structure to a globally dense one. This creates the possibility of generating new fixed sets that spans across wide regions within our brain, linking multiple regions and multiple input sources. This is a natural extension of structured complexity that allows us to solve new and difficult problems. This suggests that there is a need to study the properties of interwoven structure of fixed and meta-fixed sets.

In this chapter, I will show that there exists an unique threshold state of interwoven structure for most large-scale structured SPL systems called the *dynamical membrane* of meta-fixed sets. The central result is that this dynamical membrane self-sustains for a sufficiently long time under mild conditions. We will see that this self-sustaining property of the membrane plays a critical role in exhibiting features like free will, controlling what you sense, switching what you control based on what you sensed and so on. Also, since fixed sets are candidates for truths, the existence of a dynamical membrane makes it ideal for our ability to 'know' the truths. In chapter 22, I will explain how we 'know' truths by presenting both necessary and sufficient conditions. These conditions, as we will see, requires the dynamical membrane in an essential way.

I will begin by discussing how our brain interconnects sensory-control regions across distant regions through everyday experiences. This will follow naturally using the mechanisms described in the previous chapters. Additional questions that will be addressed in this chapter are: what is the dynamical membrane of meta-fixed sets? How do we create it for the first time? How can the membrane be broken and re-created again? Can the membrane move, shrink or expand? How does this membrane self-sustain for a long time without dying down? Is there a way to measure changes in the membrane? How does the membrane attain a self-sustaining property? It is worth emphasizing that the dynamical membrane of meta-fixed sets is a property of SPL systems, not just our brain.

One of the interesting outcomes of the self-sustaining dynamical membrane is that it addresses the problem of the seat of consciousness discussed in section 10.14. We no longer need an external agent or even a concept of mind to control the body. The question of who tells the mind to tell the body to perform a task is automatically resolved with the unique state of synchronization achieved by the self-sustaining dynamical membrane. With this new model, the self-sustaining membrane itself senses and controls its own dynamics.

19.1 Collection of fixed and meta-fixed sets

Let us first begin by looking at some examples of meta-fixed sets. Recall that meta-fixed sets are fixed sets created from other fixed sets instead of creating from external inputs directly. For example, the word 'table' is a meta-fixed set because it is not directly associated to a given external table. Rather it is associated to a 'class' of external tables. Meta-fixed sets are critical when you are trying to identify a new object to something you have seen in the past. A never-before-seen object can still be identified as a 'table' if it has similar body, legs and shape to that of a table you did see before. This is because the fixed sets for the body, legs and shape will eventually trigger the 'table' meta-fixed set.

Meta-fixed sets are the most common types of truths we use in our natural language and other abstract representations. However, the dynamical transformations necessary to create meta-fixed sets are less common than for creating fixed sets because of fewer repetitive scenarios. For example, to create 'table' meta-fixed set, a child would need to encounter different tables under different scenarios at different times. Other examples are words in a language and mathematical abstractions (like numbers and concepts). These can take years to form. To form an interwoven network, we do need a large number of fixed and meta-fixed sets (or truths). A given fixed set can either be true or false. It can be regarded as a truth only when the fixed set repeats several times and stays excited continuously in our brain.

Some examples of 'external' truths (fixed or meta-fixed sets) that we have already encountered are causality, physical laws of our universe, mathematical logic and space-time geometry (like lines, curves and time durations). Some of the 'internal' truths are colors, emotions, pain and neural excitations from our own body parts (like from our heart, hands, legs, eyes and neck).

External truths other than the primitive ones are not entirely reliable because we need to keep observing the event using our sense organs. Internal truths like explosions resulting from emotions and pain are also not as frequent because we are not subject to these feelings often. They are more valuable after you are conscious, similar to external truths. When we try to perform tasks by controlling our own motion or parts of our body like hands, legs, neck and eye movements, the inputs are present at all times. The peripheral nervous system informs every change in our body to the central nervous system whenever we are awake. These are, therefore, the best set of truths that continue to stay excited for the longest periods of time.

Generalizing this, all basic sensory signals present at all times are good candidates for truths. For example, with eyes, even though we see different objects each time, there are always fixed sets corresponding to edges, colors, angles, motion and intensity of light among different objects. With ears, we hear background noise with variations in pitch, tone and intensity most of the time. With touch, though the fixed sets triggered are less common compared to eyes and ears, they are not negligible because we are in constant contact with other objects. This is especially true for an infant, say, when lying down or being held by parents, whether standing, walking, sitting or holding objects. These yield meta-fixed sets from different regions of the body through the peripheral nervous system.

With odor and taste, the meta-fixed sets occur for comparatively much smaller time periods. For example, an infant clearly has fixed sets with the taste of the milk he drinks. This affinity to a 'fixed' taste is obvious when we switch to bottle-feeding or when we pick a different formula for the milk. With respect to odor, as parents hold him most of the time, he senses their unique odor and generates corresponding fixed and meta-fixed sets. This is apparent, say, when a crying infant calms down when his mom or dad holds him (involves both touch and odor), as opposed to some other person holding him.

I will use several of these examples of fixed and meta-fixed sets when discussing the structure of the dynamical membrane. The main point we want to highlight here is that the total number of fixed sets starts to grow and become interconnected throughout our life. These interconnections are not random. Rather, they align with our experiences. It is this entire collection of interconnected fixed sets that play an important role for the formation of the dynamical membrane.

19.2 How to self-sustain dynamics for a long time

The importance of dynamical loops (not static structural loops of neurons like *A-B-C-A*, which exist for most dense graphs) and fixed sets are that they are the simplest, nontrivial and pervasive patterns. The firing pattern for such dynamical loops already self-sustain for a short while, even when the external inputs are turned off. If we now have a collection of dynamical loops, how can we increase the total amount of self-sustaining time? Here, the total self-sustaining time refers to the amount of time it takes for almost all loop dynamics to die down. This is measured after all external inputs are turned off. We do not expect 'all' looped dynamics to be active when computing the total self-sustaining time. Rather, we want enough number of loops to be active.

For example, it is possible that at time t_1, loops on the left side are active. Then, a few seconds later, the interconnections between them cause the loops on the right side to become active while the left side loops die down. Subsequently, the active loops move to other regions. This process of interchanging of firing neurons from left to right and vice versa can repeat several times. In fact, more complicated firing patterns can occur between different regions causing the overall self-sustaining time to become large compared to when the regions are sufficiently disjoint from each other.

To see why this is important for consciousness, consider the time it takes to complete a single dynamical loop. If it is fast, our brain would be able to trigger many parallel as well as serial cascading sequences of fixed sets in one second. This gives us an ability to 'know' several truths (discussed later in chapter 22) simultaneously at any given instant. This is required to provide a cohesive view, as opposed to a discrete view, of the external world. A cohesive view is necessary for consciousness. Therefore, the higher the number of self-sustaining active fixed sets, the greater our chance to become conscious and vice versa.

A direct way to estimate the total number of simultaneously excited fixed sets is by using brain waves (see chapter 15). Under normal conditions, the electrical oscillations recorded by electroencephalogram (EEG) correspond to the sustained loops of firing activity in a given region. Nonlooped dynamics are ephemeral and appear as noise. Therefore, higher brain wave frequency corresponds to faster repeatability of looped dynamics. This is possible when we have a high number of fixed sets excited at a given instant and vice versa (empirical study is a future area of work). Gamma brain waves (> 30 Hz) correspond to a conscious state (Buzsáki 2006). As our brain wave frequencies become lower, i.e., from beta 12-30 Hz to alpha 8-12 Hz to theta 4-7 Hz and to delta < 4 Hz, our conscious state lowers from alert to relaxed, to drowsy and to deep sleep, respectively. This corresponds to a decrease in the number of fixed sets. Indeed, when we approach an unconscious deep sleep state, we do notice that we know less number of truths (or, equivalently, fixed sets).

Therefore, to maintain gamma wave patterns, how does our brain create and excite a large number of self-sustaining fixed sets at every instant? The following claim specifies one way to do this.

Theorem 19.1: Consider an active SPL system containing two loops. The dynamics of the entire system will self-sustain longer if the loops are positively linked together than if they were isolated, assuming the external inputs have been turned off.

What is the difference between positive and negative linking of two loops? In the case of physical SPL systems, we note that looped dynamics has an inherent direction of motion. When we have two loops that interact, we do not want the two directions to collide and hinder the overall motion. For example, consider physical SPL systems like those considered in chapter 2 (see Figure 2.4). If multiple such SPL systems interact, negative linking of the two loops refers to the arrangement that causes the overall motion to die down i.e., one loops forward motion links to another loops' reverse motion. Positive linking, on the other hand, causes one loop to aid the other loop to boost the overall motion i.e., one loops forward motion links to another loops' forward motion as well.

In the case of chemical SPL systems like our brain, positive linking refers to the fact that the one loop releases excitatory neurotransmitters on the other loop boosting the dynamics in the second loop. Negative linking refers to the fact that the one loop releases inhibitory neurotransmitters on the other loop, causing it to suppress the dynamics in the second loop.

To see why the above theorem is true, note first that each neuron and, hence, the loops of neurons have an *active* energy source (as ATP in cells). This energy source is used to regenerate chemicals like the neurotransmitters and to drive the ions through the corresponding Na-K and Ca^{2+} ion pumps in order to self-sustain the dynamics. Secondly, in the positively linked case (i.e., neural synapses linked through excitatory neurotransmitters instead of inhibitory ones), the outputs from the first loop provide inputs to the second loop and vice versa. Therefore, the additional neurotransmitter inputs from the second loop open new ion-channels on the first loop using the *active* (i.e., with an energy source) chemical mechanisms (Kandel et al. [82]). This increases the total self-sustaining time for the linked-loops case. The result generalizes to all active SPL systems including synthetic ones.

If we take hundreds of loops, then from the above claim the total self-sustaining time for the entire system will be much higher if they are interconnected in a positive manner than if they are disjoint or sparsely connected. Furthermore, the stability of the entire interconnected looped system is guaranteed because of the directional nature of the input-output dynamics within the loops (see chapter 2).

19.3 Sensory-control loops across distant regions

We have seen that our memories of simple and primitive events are stored as fixed sets according to the new memory model. Each of these primitive events can come from different input sources like from vision, hearing, taste, odor and touch. While simple events are well-represented as fixed sets, what is a good representation for memories of complex events like your recent vacation, details of your house, its layout and others?

In the previous section, we have seen that interconnecting the fixed sets is one way to increase the amount of time the looped dynamics can self-sustain. Therefore, one obvious model for memorizing complex events is to connect each fixed set, corresponding to the smaller primitive events, along a linear sequence in the same order as the causal flow of events. For example, you can list the sequence of smaller events that occurred during your recent vacation one after the other along the natural causal timeline. Each of these smaller events is represented, as before, as fixed sets. The linear causal sequence then becomes a linearly linked list of fixed sets.

However, given that each fixed set for the smaller events is not necessarily close to one another to be able to interconnect in a linear manner, a better representation would be a graphical or network model of fixed sets instead. In the network model, we still need pathways that allow us to traverse from one fixed set to the next, but not necessarily for the entire causal timeline, rather just parts of it. We can verify this easily when we try to recollect events from our recent vacation. We do not necessarily recollect them in the same causal timeline except for shorter scenes. Our recollection hops between different scenes in a non-causal order, some of which are even in the reverse direction of causality.

The other advantage with the network model is that different fixed sets can be reused across different events. In fact, this gives rise to the associative learning ability that we are all well aware of – starting from one memory, we switch to a completely different memory through common associations. The SPL memory model described in chapter 7 is indeed a network model unlike a computer memory model. The computer memory is closer to a linear model as can be seen with, say, events stored in a linear sequence of movie frames along a causal timeline.

We have already seen in the previous chapters that looped models were a necessary replacement to avoid the difficulties observed with linear causal representations of simple memories. Similarly, we now see that loops of fixed sets (i.e., loops of loops) are natural representations of complex memories for the same reasons. As the density and complexity of the network increases, it is only natural to form loops of fixed sets even across distant regions in our brain. This is a slow process, however.

There are two qualitatively different ways to link fixed sets. The first one can be seen as extending the existing set of loops *longitudinally* for a given abstraction (Figure 7.6), while the second one links two or more abstractions *laterally* (Figure 7.12).

The former link helps a given fixed set grow and spread out deeper into the brain (or in a synthetic SPL system). This is necessary if we want to keep increasing the number of distinct objects we want to distinguish. Examples of these are fixed sets for edges extended longitudinally to angles, which then extend all the way to shapes of objects. The cellular mechanisms for these links are based on extensions to growth cones at the end of a developing axon known as filopodia and lamellipodia using guidance molecules like netrin, slit, semaphorins and ephrins as discussed in chapter 6.

The latter link helps us associate two different events based on the simultaneity and repetitiveness of occurrences. For example, you may associate eating a particular food with visiting your grandmother's home because she always makes that special item for you. Such associative memories (lateral links) are more personal than the longitudinal links. The cellular mechanism identified here is long-term potentiation (LTP) (Kandel et al. [82]; Graham [58]). This mechanism has been shown to associate two or more distinct memories in the hippocampus, cerebral cortex, cerebellum and others (see associative learning or Hebb's postulate summarized as 'cells that fire together, wire together' – Kandel et al. [82]; Gazzaniga et al. [49]; Graham [58]).

During an infant age, the fixed sets are isolated and sparse. As an infant interacts with the external world, he maps the repeated experiences by creating several fixed sets as discussed in chapters 7 and 18. The above two linking mechanisms are now combined to create a dense 'mesh' of fixed sets, both longitudinally (Figure 7.6) and laterally (Figure 7.12). The links between loops are not arbitrary. For example, an infant links object-fixed-sets with shapes, boundaries, colors, textures and pattern fixed sets. Boundary fixed sets are linked to edge and angle fixed sets. Sound fixed sets are linked to image and word fixed sets (of a language). The motion fixed sets for your hands and legs are linked based on the common repeatable actions and so on. If a table fixed set is triggered, it causes a chair fixed set to be triggered automatically using these links and vice versa. You do not create links between a table fixed set and a balloon fixed set, for example, i.e., arbitrary links between fixed sets are not formed. They depend on your experiences. In general, the dynamical relationships between fixed sets are directly related to the chain of events occurring at that instant. With a synthetic SPL system, we can easily control the creation of such specific links.

In this manner, an infant builds a large repository of fixed sets during the first few years of his life and links them together according to his personal experiences. As the mesh of fixed sets become dense, it is possible to create new loops across distant regions. Specifically, the dense mesh of fixed sets with multiple intersecting pathways allows the possibility of forming larger loops with fixed sets themselves (i.e., loops of loops). Theorem 19.1 of the previous section now says that such a positively connected network of loops self-sustains for a long time in which the neighboring loops help each other to continue the dynamics.

In order to create loops of fixed sets, sensory-control mechanisms like the fundamental pattern are important. Recall that during the sensory-part in a sensory-control mechanism, external inputs fall on the sensory surface. The firing patterns from the sensory surface propagate deeper into the brain until they excite a collection of fixed sets. These fixed sets are internal representations for external objects. We have discussed a large number of mechanisms in the previous chapters that highlights how these fixed sets are formed and excited.

Now, to complete the sensory-control loop, we need the control-part of the mechanism. For this, we first need a repository of control fixed sets (similar to the sensory fixed sets). The same mechanisms described in the previous chapters for creating sensory fixed sets works equally well for creating action fixed sets as well. For example, we would create fixed set R for moving our hand to the right, L to move to the left, T to move to the top, B to the bottom and so on for other directions. The role of fixed set R is that if we trigger it, our hand would move to the right. The same applies with all other action fixed sets. There will, in general, be many different ways to trigger the fixed set R, analogous to multiple different ways to trigger a sensory fixed set. The difference, however, is that sensory fixed sets are typically triggered when external objects fall on the sensory surface whereas control fixed sets are triggered internally through other fixed sets.

This difference between sensory and control fixed sets makes the control fixed sets somewhat inaccessible unless we link the two of them together in the following way. The fixed sets triggered during the sensory-part extend longitudinally and laterally. They eventually link to the control fixed set causing our brain to generate a control action. The control action typically results in moving the sensory surface in different directions. This causes new external signals to fall on the sensory surface producing new fixed sets, which in turn trigger new control fixed sets and the process continues (see a similar example discussed earlier on how the fundamental pattern works in Figure 16.8). These are what I call as the sensory-control loops that span across distant regions. They form in most SPL systems. I will discuss additional sensory-control mechanisms (beyond the ones discussed in chapters 16-18) that help in creating more such distant loops in subsequent chapters.

The sensory-control mechanisms like those based on the fundamental pattern allows the system to learn new events automatically without the need to know the internal chemical details of our brain. Furthermore, our brain can work entirely at an abstract layer bypassing the knowledge of all internal dynamics. For example, consider the following problem. How can we perform a given task with high precision without directly controlling or even knowing the full set of required internal neurons? As an example, let us say that you want to move your right hand to the left by, say, one foot (see section 17.7 for a similar example of an infant learning to put his hand in his mouth). How can we perform this action with sufficient accuracy without ever knowing any details of which neurons to fire? Indeed, conscious humans perform this task accurately every day without ever knowing and controlling the correct motor neurons directly. Keep in mind that this task cannot be performed unless the correct set of motor neurons does fire for a correct period in a correct order (modulo a few variations).

The situation is the same even for non-action based tasks like controlling our thoughts that span across multiple topics. Here too, we need to ensure that the correct set of neurons fire for the correct period in the correct order within the correct set of regions in our brain (modulo a few variations). How do we do this so easily every day without ever knowing where these neurons are and how to control firing of these neurons for each given task?

Generalizing this, we see that after we become conscious, every action, thought, memory and event is entirely expressed in an abstract language unrelated to the dynamics that occurs within our brain. When you are driving a car, planning an event, preparing dinner or playing a sport, every input through your sense organs and every action you perform with your body are aggregated into abstract representations (say, using your native language or your emotional states). You, as a conscious individual, only understand this aggregate view of the external world. At no point would you even be thinking about the internal brain dynamics like which neurons are firing, which ones should fire, how do you make them fire, where in the brain are these neurons located and so on. In fact, any attempt at understanding these details interferes with your ability to perform the task at that instant. You can surely ponder on these questions later, but not while performing the task. Even if you do think about these questions, you can only think in the same (aggregated) abstract language.

The creation of sensory-control loops described earlier answers the above questions. They allow us to control the formation of a global set of neural loops using local set of fixed sets. We create complicated loop structures using smaller looped or fixed sets structures rather than working directly with low-level interconnections between individual neurons at each time.

The sensory-control loops can be thought of as an interconnection of two separately trained and learnt pathways: (a) the pathways from the sensory surface to the fixed sets and (b) the pathways from the control fixed sets to the muscles that generate an action. The missing link between the two pathways is a new pathway from the sensory fixed sets to the control fixed sets. When this missing link is formed, we are capable of moving our hand precisely using an 'indirect' sensory input rather than directly triggering the control fixed set. With more such sensory-control loops, we no longer control our hand motion by directly exciting the control fixed sets. This is the case with other thoughts and actions too – we only control them indirectly through other sensory fixed sets after the sensory fixed sets link to the control fixed

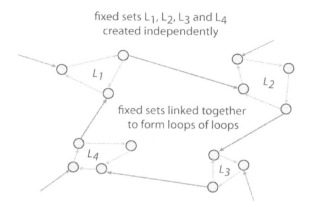

Figure 19.1: Linking multiple fixed sets to form larger loops spanning distant regions. In this example, loops L_1, L_2, L_3 and L_4 are shown to be formed independently through distinct everyday experiences. These loops are also at distant regions in the brain. Some of them could be sensory loops and others control loops as well. Over time, it is possible to link these distant loops together to form a larger loop, provided the corresponding external abstractions do occur near-simultaneously.

sets. Clearly, these mechanisms generalize to all SPL systems.

There are other ways to generalize the above sensory-control loops as well. One way is to have multiple pathways connected together to form a larger loop spanning distant regions in our brain instead of just two pathways (i.e., sensory and control) mentioned earlier (see Figure 19.1). An example of this is *realization* that will be discussed in detail in chapter 48. Let us say that you learnt or experienced different parts of a topic or an event at different times (like algebra and geometry in mathematics). Each of these parts is represented as a partial pathway in our brain, which when combined later will become a larger loop. Once each part is memorized reliably, it will so happen that one day you will experience the last piece of the puzzle (like realizing that some parts of algebra is just same as geometry and vice versa). At this time, your brain successfully completes the larger loop creating an 'explosion' of neural excitations. When this happens, you will have a feeling of realization. Each of the smaller seemingly unrelated topics or events becomes related to one another.

As shown in Figure 19.1, multiple smaller pathways of fixed sets are learnt independently and together they combine to form a larger loop. The formation of a larger loop using more than two pathways is not reliable. The outer larger loop is not closed unless each of the multiple inputs trigger in a sequential order producing a cascading effect. It is nontrivial to create such a larger loop because it spans across large and distant regions of our brain – the possibility of repetitions is low. Additionally, we need to coordinate the timings of each of the parts of the outer loop. If the timing between two consecutive inputs is not correct, there is a good chance it will not result in triggering this larger loop. This is one reason why realizations are less common. Yet, it is important enough to be noticeable because of the explosive

327

effect it generates.

Other generalizations are: (a) using multiple sensory inputs, their respective fixed sets and partial pathways to form the larger loop instead of just from a single sensory input pathway – examples are when you use both your eyes and ear (or your nose and your taste buds) to perform an action and (b) using multiple control fixed sets and pathways to form the larger loop – examples are when you use both your hands and legs at the same time.

The sensory-control loops and their generalizations appear to be slow and inefficient. They are prone to several errors. Yet, with proper training, this becomes one of the most powerful approaches to perform complex tasks. Here are a few other advantages:

- When we have minor damages to our brain, the distributed nature of these loops makes it possible to re-route the connections at only select regions and the net effect is that you retrieve your 'lost' memories.
- If we were to rely on accessing the neurons directly to perform a task, this would pose an insurmountable problem for just every day activities. The sensory-control loops, though tedious to form for the first time, avoids this difficulty by using the outer abstract loops rather than the inner neural loops.
- When we have a large network of neurons firing in parallel, it is usually difficult to coordinate them unless we have a way to serialize them. The larger sensory-control loops described here are directional in nature. This makes the neural pathways proceed in a serial manner even as they propagate across distant regions.

In the next section, we will see how the network of fixed sets converges to a special state called the dynamical membrane of fixed and meta-fixed sets. This, as we will see later, is a necessary structure for creating consciousness (including a synthetic conscious system).

19.4 Formation of dynamical membrane of meta-fixed sets

Using the mechanisms discussed in the previous sections and chapters, our brain begins to create a large network of fixed sets. The network starts to grow denser over time. This is especially prominent in the first few years of our lives. Theorem 19.1 can be applied to the network iteratively to show that the entire network self-sustains for a much longer time than if the network were sparser. The fundamental pattern (see chapter 16) can be applied to create denser network connecting sensory and control regions. The other mechanisms described in chapter 18 can also be used to create additional interconnections between fixed sets. These connections align with the repeatable sequence of events that occur in the external world. The relationships in the external world begin to be accurately represented within our brain via interconnections between fixed sets.

The dense network will soon reach a threshold state of connectedness. This threshold-connected state is unique and well defined for a given individual. When a portion of this threshold network is excited, the connectedness property helps spread the excitations across significant portion of the network. This allows the possibility that most of the network can stay excited for a long time. The sensory-control loops aid this process. This results in a unique dynamical state as well. I call this unique static and dynamical network as the *self-sustaining membrane of meta-fixed sets*. The self-sustaining membrane is a property of SPL dynamical systems. It can be characterized as the threshold network of fixed and meta-fixed sets that is capable of self-sustaining its dynamics for a sufficiently long time. I use the term 'membrane' to highlight the dynamical and oscillatory nature of the network of fixed sets.

The membrane receives sensory information from an input surface causing it to trigger control actions on an output surface. This in turn supplies new sensory information triggering additional control actions, thereby producing a stable self-sustaining looped process. During this stable dynamical state, the membrane shrinks in some regions and expands in other regions depending on the external signals or the internal inputs (like thoughts) at that instant. The static network of fixed sets should have a threshold density in order to have a self-sustaining dynamical membrane. Let me state the existence of this unique state formally as the following theorem. In the rest of the chapter, I will discuss various properties of the self-sustaining membrane.

Theorem 19.2: There exist active SPL dynamical systems that self-sustain as a whole for a sufficiently long time, in which sensory information triggers control actions and control actions triggers what the system senses, continuously in a stable looped manner. This dynamical state is termed as the self-sustaining threshold membrane of fixed and meta-fixed sets.

Why do we have an interconnected network of fixed and meta-fixed sets in our brain? Notice first that when we perform tasks, the excitations in our brain correlate with the motion of hands, legs and the rest of the body. By correlations, we mean that there are regions in our brain that can be identified uniquely as a source of motion of hands and legs. Why is this correlation important? If they were not correlated, it is difficult to have repeating loops and fixed sets. Furthermore, the dynamics in our brain will have no connection to reality.

Aligning with the real world is an unavoidable requirement when you use dynamical systems as the framework for complex systems. Without such correlations, we cannot say that our senses are indeed detecting, say, the motion of our hand. Two dynamical systems are inherently connected, by definition, if and only if there is a unique correlation between the two dynamics. Abstract systems with no connection to dynamics, as with our existing computers, exist and operate in a virtual world. Our brain does not use this unreal or abstract framework. This forces appropriate interconnections between fixed and meta-fixed sets. For example, causality should be embedded into our brain. Otherwise, we will quickly be out-of-sync with reality. If I move my hand, our brain should produce an event unique to my hand motion. This event should not be mixed with a different event like moving my leg. They can be linked, but there has to be an independent excitation to each of them. The correlation of the brain with the body is critical.

To understand how and why the interwoven structure of meta-fixed sets forms, let us consider a simple event. Let us say that a toddler has a red ball in his room he likes to play with regularly. He looks at the ball, picks it up and throws it around. Now, looking at the ball is quite independent from picking up the ball, i.e., he does not have to pick up the ball when he looks at it. Similarly, picking up the ball is independent from throwing the ball. Each of these objects and actions are memorized in his brain as fixed and meta-fixed sets. If he practices the entire task several times, the interconnections between these sets are memorized. What I mean by these interconnections is that seeing a red ball will trigger him, via the interconnections, to pick up the ball and throw it.

As an analogy, in Figure 19.1, we can consider fixed set L_1 as seeing a red ball while fixed set L_2 as picking up the ball and throwing it. Now, let us say that he has learnt to say the word 'ball' through some other experience unrelated to the above scenario. He can now link this meta-fixed set (saying the word 'ball') when he picks up the ball to throw. Continuing the analogy, in Figure 19.1, we can say that fixed set L_3 is like saying the word 'ball'. When he learns to laugh (independent of this task), he can link this meta-fixed set as well when he throws the ball (laughing can be considered as fixed set L_4 in Figure 19.1 and so on). He can memorize the room where he plays and can associate it with this event. Continuing this process, we see that the structure of the fixed and meta-fixed sets start to become more dense and interwoven simply through everyday repeatable experiences.

We can test that his brain indeed formed these interconnections between the meta-fixed sets with a simple experiment. For example, we can read him a book that has the picture of a similar red ball. When we do this, we will notice that he not only recognizes the red ball in the book, but it will also trigger his memory of playing with the ball via the interconnections of all the above meta-fixed sets (from the analogy mentioned above with links between fixed sets similar to the ones discussed in Figure 19.1). He usually expresses this by pointing you towards the room where he plays and wants you to take him there.

We sometimes simplify this and say that he is just recollecting his memory. While this is true, what I want to point out additionally is that our brain forms a specific interconnected structure within what we vaguely refer to as memory. This structure within the fixed and meta-fixed sets is not random. Our experiences create and enhance them daily. In the later chapters, I will describe several patterns to explain additional ways to generate complex interwoven structure of meta-fixed sets. In the current discussion, I have used fixed sets and meta-fixed sets interchangeably. This is because humans (unlike other animals) do not just form a fixed set and leave it as is, disjoint from others. They link it with several other fixed and meta-fixed sets to form a complex interconnected pattern of meta-fixed sets. I call this structure the *self-sustaining dynamical membrane spanned by meta-fixed sets*. This structure stays excited for long periods even though parts of it might expand, shrink or move around.

Becoming conscious for the first time is a slow and gradual process because we need a large number of fixed and meta-fixed sets. We need to build the dynamical membrane using them. It takes several months to a year. However, after we become fully conscious, we can transition in and out of consciousness easily. For example, we do this every night when we sleep. This is because these fixed sets are stored structurally within our brain architecture as memory. We can then use this structure easily to control the dynamics necessary for memory.

There is a considerable difference between the observable states of a person when the membrane exists versus when it does not exist. The existence of the membrane, for example, has a profound effect on emotions, realizations and feelings. This will be discussed in chapters 49, 48 and 22. For now, I will consider how the membrane shrinks during dreams.

19.5 Shrinking the membrane during dreams

When we dream, we are conscious. There is a clear sense of myself watching and responding to events. I feel my own existence. However, the set of truths we know, including about our own body, are considerably different compared to when we are awake. For example, we do not usually see our own hands, legs and most of our body in our dream as often as we would when we were awake. We do, however, 'feel' and sense their existence clearly. In contrast, all external objects and events appear quite normal in our dreams, though the physical laws they seem to obey are quite different.

To understand why the set of truths and, hence, the membrane has shrunk, first note that the neuron inputs that reach the brain from the rest of our body when we were awake are quite different compared to when we were dreaming. When we are awake, our peripheral nervous system constantly sends neural signals to our brain about every change that occurs in our body. This information reaches the brain stem through the spinal cord. Our brain is able to process this to determine positional information of parts of our body, touch sensations and information from internal organs. However, during active asleep, it is well known that there is inhibition of all motor activity itself. We are unable to move our hands and legs, say, because of the neuronal mechanisms of postsynaptic inhibition of motoneurons. Motor neurons in the brain stem and the spinal cord are strongly hyperpolarized. This causes muscle atonia (Morales et al [100], Takakusaki et al [150]) with glycine as the neurotransmitter (Chase et al [19]) during REM sleep.

In other words, the truths corresponding to our own body are cutoff in our brain at the brain stem and spinal cord during sleep. As a result, their existence is not directly known. However, we can know them indirectly by stimulating directly the same fixed sets in the brain, skipping the pathway from peripheral to the central nervous system (cf. phantom sensation in section 50.2). We can sense this in one of two ways: (a) directly when there is a leakage of neuronal inputs into your brain to excite the appropriate set of truths, or (b) indirectly via the links between meta-fixed sets from scenarios that require your hands or legs (like say, when you need to run or catch a ball in your dream). This indirect mechanism suggests the possibility of exciting the body fixed sets through the interconnections of the other fixed sets. A generalization of this notion is called dual space representation, which will be discussed later in chapter 25.

Therefore, when the body inputs are cutoff during sleep, it shrinks the membrane of meta-fixed sets. Sometimes, we even sense an inconsistency resulting from the body inputs, as they are cutoff in our dreams. For example, if in our dream we are in a hurry while driving a car, running away from someone or pedaling a bike, we find it hard to do these tasks. We may be pressing the gas pedal or the bike pedal hard, but the car or the bike does not seem to move as fast.

Other meta-fixed sets not excited during our dreams, causing our membrane to shrink further, are the physical laws of our universe. These meta-fixed sets are not excited through your external sense organs because of the above-mentioned inhibitory mechanisms. The implication is that you do not necessarily obey physical laws in your dreams. For example, you sometimes fly in your dreams or perform other physically unreal things. The physical truths now are only based on memory and thoughts instead of reality.

Given that so many fixed sets are not excited during our dreams, our dynamical membrane shrinks considerably though it maintains the self-sustaining nature. What will happen if the membrane shrinks further down? We can reach a state when the membrane becomes so small that it 'tears down'. The self-sustaining ability in this state is considerably lost. This is indeed what happens when we are in a state of deep sleep i.e., when we are no longer dreaming. We do not have a membrane. As a result, we are not even conscious and we do not remember anything during these periods (see the next section 19.6 for further discussion).

19.6 Membrane variation for each brain wave pattern

There is a good correlation between membrane variation and the corresponding set of brain waves produced. Let us examine this briefly. The membrane (which can be viewed as a threshold connected network of actively firing fixed sets) keeps varying – shrinking sometimes, expanding other times, stretching to different regions of the brain and so on. In section 15.2, I have mentioned that brain waves are generated because of the simultaneous excitations of fixed sets. When we are awake or we are in REM sleep, we have beta or gamma wave patterns generated in our brain. When we are in deep sleep, they are either delta or theta waves.

In section 15.2, I associated brain waves with the frequency of simultaneously excited fixed sets. We can now additionally qualify them as those arising from oscillations in the dynamical membrane. Clearly, the dynamical membrane itself is composed of simultaneous excitation of a large number of meta-fixed sets (arranged as a network) for a sufficiently long period. Some of these simultaneous meta-fixed sets are from external sensory inputs, internal inputs or our thoughts. If the membrane exists, we observe beta and gamma wave patterns (i.e., loops with frequencies greater than 20 Hz). At low frequencies like delta or theta waves, the dynamical membrane does not form. At these frequencies, we do not have enough number of fixed and meta-fixed sets to synchronize and form a membrane.

Therefore, we see that when we have gamma waves, we have a large number of interconnected fixed sets that stay excited, giving rise to the formation of a self-sustaining dynamical membrane. Conversely, when we have a self-sustaining dynamical membrane, our brain produces gamma waves. Equivalently, we say that our brain generates gamma waves if and only if the dynamical membrane exists. Since the existence of gamma waves is known to be associated with consciousness, this result shows that the existence of a dynamical membrane will give rise to consciousness as well. I will discuss the necessary and sufficient condition for consciousness expressed in terms of the dynamical membrane later in chapter 23.

On the other hand, when we have delta or theta waves, the number of simultaneously excited fixed sets are sufficiently low that you cannot form a self-sustaining dynamical membrane. If our brain activity does fall to abnormally low levels, then in this case, we will not be conscious. This corresponds to generation of delta or theta brain waves indicative of deep sleep.

Therefore, larger self-sustaining dynamical membrane spanning across distant regions of our brain is an important state that is closely related to consciousness. In the next section, I will compare all other altered states of consciousness discussed in chapter 11 like meditation, hypnosis, completely drunk state and others to the existence, 'size' of the dynamical membrane and its self-sustaining ability. The general pattern we observe is that low level of consciousness has smaller and possibly minimal self-sustaining ability of the membrane. Sometimes the dynamical membrane may even pop in-and-out of existence, which translates to popping in-and-out of consciousness for the human.

Since the existence of a self-sustaining dynamical membrane is a property of stable parallel looped dynamical systems, the above discussion and, hence, its relation to consciousness can be extended to all synthetic SPL systems as well. The precise necessary and sufficient conditions expressed in terms of the dynamical membrane for all SPL systems will be discussed later in chapter 23.

19.7 Membrane variation as state of consciousness vary

Let us now look at the list of altered states of consciousness mentioned in chapter 11 and see how the dynamical membrane is affected in each of these scenarios. I will argue that the dynamical membrane does not exist in any of the unconscious states. Examples of scenarios when we are not fully conscious are (see Figure 11.1): with infants, sleepwalking, deep sleep, severely drunk, general anesthesia, coma, knocked unconscious and hypnotism. In all of these states, it is clear that the number of meta-fixed sets are either severely decreased or occur at a much slower pace. Let us look at each of these cases below.

The situations with infants, sleepwalking and deep sleep were already discussed in the previous sections 19.5 and 19.6 as well as while discussing the origin of brain waves. In these cases, the number of fixed sets that are simultaneously excited are quite low giving rise to delta and theta waves. Therefore, the membrane cannot form as there is not enough self-sustaining activity to maintain the membrane in an excited state for a long time.

With severely drunk states and general anesthesia, the formation of the loops (i.e., fixed sets) is chemically blocked. The Na^+ and K^+ ion channels are blocked chemically effectively preventing the neurons from firing action potentials. The number of these fixed sets drops significantly. Our rate of recognition of events (as fixed sets) decreases considerably. As a result, the membrane as a whole cannot self-sustain. It breaks down.

In the case of coma, for example, most of the structure of the brain could be intact. However, the critical junctions that trigger most meta-fixed sets are severed. As a result, the larger sensory-control loops (section 19.3) needed for the dynamical membrane to form are difficult to reach and excite. This causes a severe drop in the number of meta-fixed sets triggered and maintained, causing a loss of consciousness. Therefore, there is no membrane formed.

This situation is approximately true even when you are knocked unconscious. The difference is that there is no permanent damage to these critical junctions. The blood supply is cutoff temporarily causing a sudden drop in the number of loops and the corresponding meta-fixed sets. You lose your body stiffness because several of the sensory-control pathways triggered via these critical junctions cannot be excited anymore. The control mechanisms needed to maintain this stiffness is broken. Hence, there is no membrane.

The brain waves generated during hypnosis is theta waves. During sleepwalking as well, theta and delta waves are generated. You have control of your movements in both situations. With hypnosis, you are able to communicate seemingly effectively while you are still partially unaware of your own self. This is a rather important stable and synchronized state in our brain. Specific regions of our memory are capable of being excited, since our communication abilities are intact. It appears that our membrane is completely passive (cf. having an active membrane while we are fully conscious) and is controlled by an external person. This is a situation where only the self-sustaining nature of our membrane is broken, not the membrane itself (see section 19.2). When the external environment triggers the variations in this membrane, they die down quickly because of its passive nature. The ability to self-sustain the membrane is maintained externally by the hypnotizer. As a result, this is a rather unusual and interesting state of synchronization.

In other words, for all of the above less-conscious scenarios, we have seen that the dynamical membrane is not excited similar to an awake state. The interesting aspect, however, is that we do form memories (of several meta-fixed sets) within our brain even with the above less conscious experiences. The reason is that we can excite the regions corresponding to these meta-fixed sets directly within our brain without the need to go through the external sensory surfaces or through our body. For example, to trigger a chair meta-fixed set, we do not need to 'see' a chair using our eyes. Instead, we could trigger the meta-fixed set directly through other interconnected fixed sets like how we do in our dreams. Similarly, to trigger the meta-fixed set corresponding to our hand motion, we do not actually need to move our hand. Instead, we trigger the meta-fixed set in our brain directly using the interconnections with the other fixed and meta-fixed sets (like it happens in our dreams). The generalization of these scenarios is called the dual-space representation, which I will discuss in chapter 25.

We have already seen this ability to synchronize, namely, by directly triggering fixed sets in our brain, in the case of dreams. During dreams, we were able to become conscious through self-excitation within our brain even though the inputs from our body are cutoff at the brain stem. However, this forced synchronization (and control) of meta-fixed sets during dreams do not typically sustain long enough. You do need your physical body inputs to trigger these regions consistently and repeatedly. As a result, you do not stay naturally in a dream state for a long period.

Let us now turn our attention to one of the most important properties of the dynamical membrane – its ability to self-sustain. Once we clarify this ability, we will be able to generalize the above self-excitation situation with dreams to other cases as well: like with meditation, out-of-body experiences, virtual reality and re-living an experience simply by thinking or imagining an event. Each of these cases does need a focused effort, training and a bit of concentration.

19.8 Sensing and controlling the dynamical membrane

In order to understand how the membrane can self-sustain, we need to have mechanisms for sensing and controlling the membrane as a whole. The terms sensing and controlling have an analogous meaning to the standard definition even in this context, though they are subtly different. I will discuss this briefly now. When we use the word 'sensing', we usually have two separate aspects we refer to: (i) the varying phenomena and (ii) the system that detects this variation, namely, the sensor. For example, in the case of eyes, a variation in the light patterns is the phenomena and the retina is the subsystem that detects this variation. The subtle difference in the case of the dynamical membrane is that both of these refer to the same component. The variation in the membrane is the phenomena and the membrane is the subsystem that detects its own variation.

There is actually a third aspect too, namely, the causal chain of subsystems that produce the variations themselves (for example, a causally linked chain of external objects that produces the variations in the light patterns, which is eventually detected by our eyes). Such a similar causal chain exists in the case of dynamical membrane as well. The typical components that are part of this causal chain are external objects, internally generated motor actions (since humans have an active internal energy source), emotions and other thoughts.

The situation is similar when we use the word 'control'. There are three aspects too: (a) an object A that we want to control, (b) another object B that controls it and (c) the causal chain of variations that make B perform the action. In the case of dynamical membrane, both (a) and (b) refer to the same object. The dynamical membrane controls its own variations. The causal chain, as with the sensory case, can be both external and internal components.

Given this, we can state the sensory-control problem as one in which the membrane senses its own variations and controls itself so it can self-sustain as a whole for a long time. In the case of feedback control systems that self-regulate like thermostat in a refrigerator, cruise control in a car and autopilot in an airplane, we solve a similar problem that achieves a specific objective. These objectives are regulating a specific temperature, maintaining specific speed or tracking a specific trajectory respectively. For our case, a similar objective is self-sustaining the dynamics (or keeping the membrane stable) for the longest time. Therefore, we see that the dynamical membrane' self-sustaining ability is indeed a sensory-actuator feedback problem.

Since the membrane is interwoven, if there are small variations at one part of the membrane, the interconnections can magnify this effect and produce considerably large variations throughout the membrane. Typically, these large variations correspond to the excitation of a collection of fixed and meta-fixed sets. The small variations correspond to, say, the raw sensory inputs. When the magnification of small variations happen through the interconnected network, we say that the membrane has sensed the original small variations. Otherwise, the membrane did not sense these variations. The process of sensing is continuous because there is a continuous supply of both external and internal inputs (like memories) as long as you are awake and conscious.

Controlling the membrane is similar too because of the interwoven nature of the membrane. The control mechanisms are linked to the sensory mechanisms as described previously (see, for example, the fundamental pattern). Therefore, small variations at one part of the membrane will be sensed first by the membrane. This causes the control regions to be activated as well, via the sensory-control loops (like with the fundamental pattern). The control mechanisms then make the system move. As the system moves, new inputs fall on the membrane, which causes the membrane to vary once again. In other words, the membrane eventually controls the initial variations.

One approximate analogy to help visualize this process is with a large soap bubble. Small variations in the form of wind disturbances change the shape of the soap bubble. These small variations eventually produce large variations in the soap film because of its connectedness, causing the shape to wobble considerably. As the soap bubble wobbles, the internal pressure in some regions become high (or low) relative to other regions. This generates a force that changes the shape once again. In this manner, the wobbling process is stable and self-sustains for a while.

Let me list a few examples for which either the sensing or the controlling of the membrane is prominent. Consider standing on the edge of a rock, closing your eyes and spreading your hands as if you are flying. If there is a gust of wind to help your imagination, you realize that you are able to excite the fixed sets corresponding to a real flying scenario like on a glider. In this example, you are not actively controlling your thoughts. Instead, you are letting your thoughts be guided freely by the subtle sensory input variations (like wind gusts and slight imbalances as you are on the edge of a rock) and your past memories. This is an example of sensing the rhythms and self-oscillations of the membrane.

Similarly, if you are sitting in your car (or on a train) and, say, the neighboring car moves backwards, there are times when you suddenly feel that your car moved. You may even react by pressing the brake pedal harder. Even though the principle of relativity implies that (a) your car is static while the neighboring car moves and (b) your car moves while the neighboring car is static are equivalent, you still feel that your car moves and not the other way around. This is because your meta-fixed sets are stored and interlinked as memory in your brain by watching surrounding objects move while you drive or sit in a moving car. You do not have as many memories of watching other cars move past you, while you stay still at one place. Even with other similar illusions involving relative motion, in spite of the equivalence with the principle of relativity, our experience takes a form that is consistent with our past memory.

As another example, if you play a video game for a long time that involves either flying or racing, your brain creates several meta-fixed sets from this experience. This is linked with the rest of your membrane that even if you close your eyes, you feel as if you are driving on a racetrack. This is another example where you sense the variations of the membrane excitations and its self-sustaining ability. A similar example if with a long road trip. If you were driving continuously for like 10-15 hours on a road trip, even in the night when sleeping, you would feel as if objects are moving past you similar to your driving experience. Another similar example is with seeing a phantom image or a ghost in a dark room, especially after watching, say, a horror movie.

With the above examples, the membrane shrinks or extends 'inadvertently' using the links between the new and existing meta-fixed sets. Let us now see examples for which we can control this process deliberately.

Let us imagine visualizing an image of a hand holding a bat and hitting a ball, or an image of a hand holding a pen and writing on a piece of paper. To make this thought experiment work better, it is better to close our eyes as we imagine these events. During your imagination, you form some partial images of a bat, a ball or a hand holding a pen and so on. These images are not as vivid as seeing a real bat, ball or a pen. Yet, you can visualize quite reasonable images if you try harder.

However, there is an important difference. With each of these images in your brain (like a bat hitting a ball or a hand writing on a piece of paper with a pen), you have absolutely no sensations or feelings associated with the corresponding motion. For example, you do not feel that it is you who is hitting the ball with a bat. If feels as if someone else is swinging the bat. Similarly, you do not feel as if it is your own hand writing on a piece of paper. Rather, it is just a fictitious hand that appears as if it is moving. It feels no different from watching someone else's hand move in a dream. The image and the corresponding motion (of the event) you are visualizing does not feel part of your own experience. The interesting question is: can you now focus your attention and control your thoughts so that you feel this visualized image to be your own hand?

The answer is that it is possible to control your thoughts and simulate the above scenario (to sense and feel the fictitious image to be as real as your own hand) for a short period. One way to do this is by first exciting the fixed sets related to your real hand. You can do this by gently imitating these same tasks. For example, you can position your hand on a flat surface, curl your fingers as if you were holding a pen and wiggle your hand by a tiny amount as if you were writing on a paper (pretend as if you were writing a long sentence). Of course, you are still closing your eyes and imagining the scenario but with weak movements. This weak stimulation will indeed excite the correct fixed sets in your brain needed for writing on a piece of paper (see

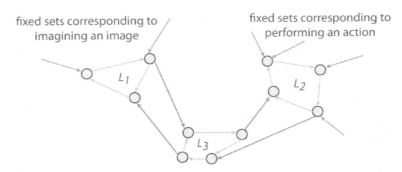

fixed sets corresponding to imagining an image

fixed sets corresponding to performing an action

L_1

L_2

L_3

synchronize L_1 and L_2 fixed sets by
expanding the dynamical membrane through L_3

Figure 19.2: Expanding the dynamical membrane to link distant fixed sets. In this example, we have a fixed set L_1 triggered through our imagination (like imagining writing on a piece of paper). Another fixed set L_2 is triggered through the actual writing on a piece of paper. Our task is to synchronize both these fixed sets by expanding the dynamical membrane via other fixed sets like L_3 so we can feel that our imaginary hand feels the same as our real hand.

Figure 19.2).

What you have successfully excited so far in your brain are: (1) a fixed set corresponding to a fictitious image of a hand writing on a piece of paper and (2) a fixed set corresponding to your real hand motion. These two fixed sets are potentially at different regions of your brain. What we need is a way to extend the excitations and link them together in order to make us feel that both represent the same hand. With some focused effort, the regions in your brain associated with your initial imaginary hand can be made to synchronize and converge with your stimulated meta-fixed sets of your real hand motion (at least for a short while), including the related pathways (see Figure 19.2). When this happens, the imaginary hand will begin to 'feel' as if it were your real hand (remember to close your eyes during this simulation). The interwoven links of meta-fixed sets formed as memory through past experiences have helped you achieve this synchronization.

Meditation is another example to gain control over your thoughts in a similar way, if trained consistently. I will discuss meditation in some detail, later in section 50.4.

We can generalize and say that focusing on any topic (say, a mathematical problem, planning an event, playing a musical instrument and others), in general, is an example of our ability to alter the membrane to create peaks in appropriate regions. You can synchronize and excite all the necessary links as a path from your current state to a desired destination state. You start from your initial state and trigger intermediate fixed sets that are relevant to this problem. As you trigger enough number of fixed sets, they begin to form a continuous connected path, which we can eventually control and extend it to the destination (see Figure 19.2).

In fact, recollection of a memory by following such continuous connected paths from a cause to an effect, as mentioned in chapter 7, is a special case of traversing along these paths of abstract fixed and meta-fixed sets. The representation of the new memory model with a geometric map, landmarks and turning points along the path, is one such example.

We can actively create memories using standard well-known techniques like analogies, trying multiple examples, rote, triggering appropriate emotional explosions and others. When you create the appropriate memories in a specific order, you direct the formation of the membrane structure in your brain in a way unique and personal to you. From these personal experiences, you form a membrane structure giving you abilities quite different from others. These abilities can be with intelligence, engineering skills, artistic skills, sports and games, musical prowess and with other faculties humans have created.

19.9 Self-sustaining the dynamical membrane

The sensing and control mechanisms discussed above gives us an ability to create a self-sustaining membrane. There are at least two constant sources of excitations. One is from motion or variation in the external objects like say, birds flying, cars moving, people walking and others. The other is from our own self-motion like say, walking, turning our eyes or head, doing some task with our hands or legs and thinking about things we did or will do. These excitations reach the membrane and constantly provide energy to keep it excited.

If the external variations around you, your internal thoughts or your self-motion settle down to a low degree that the membrane is no longer capable of self-sustaining, the natural state we enter is a resting or a sleeping state. On the other hand, when we are conscious, the dynamical membrane always exists. The membrane is not a static network. It moves across different regions of the brain depending on the types of variations you experience. It shrinks and expands as necessary. When you focus on a topic, the membrane expands to that region to include those meta-fixed sets as well. It later contracts back to your previously excited internal meta-fixed sets. This happens while maintaining the continuity between the transitions (see Figure 19.2).

When we relax or fall asleep, the membrane first disintegrates resulting in a loss of consciousness (during deep sleep state). It, subsequently, recreates through self-synchronization. There are minor variations that can be sensed through various sense organs, from residual thoughts, recent memorized activities and the existing fixed and meta-fixed set structure. These variations accumulate over a period in a coherent way because of the directional nature of stable parallel loops. This gives rise to a self-synchronized membrane. This state of self-synchronization while asleep is experienced as dreams.

When we are awake, the membrane is self-sustaining because of the continuous supply of inputs from different internal organs, external sense organs, emotions, logical thoughts mediated through an external language and actions performed using motor neurons. Humans are never static when they are awake. Their head, eyes and body moves around resulting in a constant source of external inputs. The sensory-control loops discussed previously are used to produce the self-excitation and control of the fixed sets. Our control mechanisms generate new sensory inputs while the sensory inputs generate new control mechanisms. This process repeats for a while in a stable manner maintaining the self-sustaining oscillations of the membrane. The self-sustaining membrane also aids in creating new fixed sets and new links between them.

It is important to keep in mind that the self-sustaining dynamical membrane is a property of complex SPL systems just as concepts like fixed sets, sensory-control loops, fundamental pattern, explosions, memory model and others discussed previously. They, along with a dynamical membrane, can exist if there is sufficient asymmetry in the complex system. The asymmetric structure is critical, for otherwise, the loops are locked into symmetric patterns. Unfortunately, symmetry does not yield sufficient complexity. The specific fixed and meta-fixed sets that are part of the dynamical membrane (like say, body related fixed sets, space, time, causality and other external dynamical phenomena) is what determines whether the system can become conscious or not. We will look at the relationship between the dynamical membrane and consciousness in further detail in chapter 23.

Recall from Theorem 12.1 of section 12.3 that in order to become conscious, we needed to know at least one truth. However, we now saw that truths (i.e., fixed and meta-fixed sets) do not exist in discrete numbers. Rather, they form an interwoven structure, which we called as the dynamical membrane. Therefore, knowing one truth became equivalent to knowing the entire membrane of truths (or meta-fixed sets). This may come as a bit of a surprise. All we wanted to look for is a single truth to explain consciousness. However, we now see that we either know an enormous number of truths (i.e., the entire membrane) or none at all. It is now easy to appreciate how intricate the problem of consciousness is – a detailed and a rather special set of structures are needed before we can address the problem of consciousness.

20
Predictions

Until now, we have focused on one important aspect of consciousness as stated in Theorem 12.1, namely, truths. These discussions have culminated in identifying an important structure in our brain, namely, the self-sustaining dynamical membrane of meta-fixed sets. Even though the brain is so complex with billions of neurons and trillions of interconnections, its stable parallel looped structure has provided an elegant underlying structure spanned by the fixed and meta-fixed sets. We will now begin to analyze the other important aspect of consciousness stated in Theorem 12.1, i.e., our ability to *know* something. The basic question we want to address in the next several chapters is: how are we capable of knowing anything – from a simple object, event, word, picture, our hands, legs and so on, to pretty much anything we experience at every instant when we are conscious? It turns out that to understand how we know truths, we need to look at how predictions and continuity within abstractions work. In the next several chapters, I will discuss these topics and see how they relate to the dynamical membrane of meta-fixed sets.

Let me begin by focusing on the problem of predictions first. Recall that when we think of predictions, we typically think of predicting events at discrete time instants instead of at every instant. For example, when we look at a trajectory of a ball, we try to predict where the ball will reach when it lands. We do not seem to care about predicting its location along the entire trajectory at every instant. Similarly, when we are reading a novel, we think that we are not predicting anything, at least most of the time. Rather, we seem to be only reading and understanding the plot. Only at specific time instants, we seem to try to predict what will happen next.

This, as I will show later, is not a correct or a complete view. Both when looking at a ball's trajectory and when reading a novel, I will show that our brain is predicting 'something' at every instant. As an example, to understand even a single sentence requires a large number of predictions, as we will see later. This notion of prediction as our ability to guess at every time instant is generic enough that it extends to all cases. Therefore, I will incorporate it in the definition of prediction itself.

When defining predictions, one other point to keep in mind is that, we should not rely on help from conscious beings. To achieve this, we need to work with several repeating events. If an event is non-repeating, it is hard to predict what will happen next because this is the first and the only time the event occurs. We will not have an opportunity to form a memory (i.e., a fixed set). On the other hand, when an event repeats several times, we will memorize parts of the event. We will then be in a position to guess what will happen next when the same event begins to occur once again. Therefore, let me start by discussing the importance of repetition and by identifying the set of events that repeat in our everyday life.

20.1 Repetitions and their properties

When an event repeats several times, it is easier to form memories, at least for parts of the event. The repetitions help create fixed sets that satisfy both the similarity and dissimilarity conditions of chapter 7, a necessary set of conditions to satisfy for forming memories. We may be under the impression that it is still possible to create a memory even if the event does not repeat. This is not true in general except when there is an emotional explosion or when the event is a long and complex one with interactions with several familiar objects. Consider a simple event like looking at a specific object that we have never seen before. If we stare at this object just once, but long enough, would we be able to create a memory? Unfortunately, this is less likely within the SPL structure of our brain because of input saturation.

On input saturation: If we keep staring at an object A, the input ion flow via the open Na-K ion channels among these firing neurons will eventually saturate and stop. This is a problem because we would then not know if the object A exists anymore. When the sensory neurons that were once firing saturate and stop firing now, we have no signal to indicate that the object is still present. Of course, this cannot happen easily under normal circumstances for a few reasons: (a) humans are rarely static and we rarely stare at a single object long enough that all relevant sensory neurons saturate, (b) the network is densely connected that it is possible to excite different neurons at different times to avoid saturation of all relevant sensory neurons and (c) we have a looped architecture that allows the neurons to self-sustain the dynamics via multiple pathways for a long time. Our ability to control our own motion produces new input patterns on our sensory regions, which makes saturation less likely.

There are, however, cases when this does happen. An infant with limited control mechanisms (as he does not move much) can experience saturation and, therefore, fall asleep. In the case of adults, one common scenario where we experience saturation of external inputs is when we lie down to sleep. When this happens, we do not hear, say, the sound of the fan, feel the surface of the bed, hear low volume television sounds and others in spite of the inputs falling on our ears. It is as if these objects do not exist anymore when we are asleep even though the inputs from these objects do reach our sensory organs in exactly the same way as when we are awake. Such saturation and loss of perception of objects is less likely to happen when we are awake.

On identifying an object as the same when an event repeats: While repetition is useful, how do we detect an object that reappeared as the same or if it is a new one? For example, in section 15.2, we tried to quantify the rate of excitation of multiple fixed sets. We looked at the fixed sets for tables, chairs and others. We wanted these fixed sets to be excited in about 50 milliseconds in order to produce beta or gamma waves. We saw that if we do not have a strong memory formed for table and chairs, then it is important to have the table and chair fixed sets to stay excited long enough to avoid being forgotten. On the other hand, if we do have memory of these fixed sets, even if they are not excited now, they can be quickly excited. However, how do we know if it is the same table, we saw yesterday? The answer is that two objects are the same if the corresponding fixed sets are the same. To be precise, the following three sets of fixed sets should be the same: (a) the fixed sets for the objects themselves, (b) the relative relationships between the fixed sets within the objects and (c) the relative relationships between fixed sets of neighboring objects.

On detecting repeating events: In the previous discussion, we looked at whether an object in an event is same or not. Let us generalize this and ask if an entire event is approximately the same or not. Specifically, each time an event repeats, we do not want to say that it is a new event. To do this, we need to create, first, a collection of meta-fixed sets $\mathcal{P}$ in our brain corresponding to the repeating event E. When an event E repeats, the same collection of meta-fixed sets $\mathcal{P}$ will be re-triggered. The next step is to create a new meta-fixed set R linked from the above collection of meta-fixed sets whose sole purpose is to detect the specific repeating event E. In other words, this meta-fixed set R is triggered if and only if most of the event E repeats, not just a small part of the event E. If there is no such fixed set R, we would not be able to distinguish this event E to some other collection of events that together trigger the same collection of meta-fixed sets $\mathcal{P}$. Note that how we created a meta-fixed set R is analogous to how we created any specific fixed set in the first place from a collection of dynamical transformations (see section 14.2).

The idea with having a new meta-fixed set R is two-fold. Firstly, we want to distinguish two events that use most of the same collection of meta-fixed sets, but occurring in sufficiently different timeline order, as different events. Secondly, we want to capture only a very small subset of permutations among the collection of meta-fixed sets as repeatable events. We want to reject all permutations that violate the known physical laws of nature though they may be theoretically possible.

Let us make these a bit more precise. Consider a collection of meta-fixed sets A_1, A_2, ..., A_n triggered in that timeline-order for a given event E. As an example, a simple event E can correspond to dropping a ball and catching it. Let us say, for simplicity, that we breakdown the event E into smaller sub-events, which trigger the following meta-fixed sets: A_1 = the act of releasing the ball, A_2 = the ball falling down, A_3 = the ball hitting the ground, A_4 = the ball bouncing, A_5 = the ball moving upwards and A_6 = the act of catching the ball. Repetition of event E corresponds to triggering the same sequence of sub-events $A_1, A_2, ..., A_6$ once again in that timeline-order.

The first above-mentioned reason for creating a new meta-fixed set R is that a considerably different timeline-ordered permutation $A_{p1}, A_{p2}, ..., A_{pn}$ of meta-fixed sets should be treated as a different event F. The second point says that among all theoretical permutations of $A_1, A_2, ..., A_n$ (i.e., among $n!$ permutations), we only want to consider a very small number of them as valid repetitions for which we create new meta-fixed sets like R. These valid repetitions are only those that are physically possible in the external world. We do not care to detect repetitions among the physically impossible permutations. In the above example, almost all permutations of $A_1, A_2, ..., A_6$ are invalid in the physical world. For example, $A_1, A_3, A_6, A_2, A_5, A_4$ is physically impossible (ball hitting the ground A_3 followed by ball catching A_6 followed by ball falling down A_2 is not physically possible).

Unlike this example, it is possible that in a more complex example we can have several valid permutations or valid permutations of subsets of $A_1, A_2, ..., A_n$. Let us call the collection of all these valid permutations (including those for subsets) as $\mathcal{P}$. We partition the collection $\mathcal{P}$ into disjoints classes $\mathcal{P}_1$, $\mathcal{P}_2$, ..., $\mathcal{P}_k$ in which each class is an equivalence class i.e., permutations within a given class are considered to originate from similar events while permutations across two different classes are viewed as generated from considerably different events. Let us now create a new meta-fixed set R_1, R_2, ..., R_k, one for each equivalence class $\mathcal{P}_1$, $\mathcal{P}_2$, ..., $\mathcal{P}_k$. An event corresponding to an equivalence class $\mathcal{P}_i$ is considered as repeating if and only if the corresponding meta-fixed set R_i is triggered. Note that the discussion above is analogous to how we needed to satisfy similarity and dissimilarity conditions when identifying a fixed set (see section 14.2), except that it is generalized to a larger event that includes multiple meta-fixed sets.

Humans have a natural ability to detect repetitions from a large class of events like with a collection of questions, sounds, images, time sequence of events (say, in a movie) and others. There is no special training needed to do this as well. Even a young child is capable of detecting repetitions without any help from parents. A computer, on the other hand, seems to detect repetitions only if pre-programmed. Humans detect repetitions even if the event is not exactly identical, but has just a few key components (making it fall into an appropriate equivalent class $\mathcal{P}$).

For example, if we have the same colored object, like a red ball, red balloon, red car that are visible, the correct repeating pattern we should identify is the color 'red' in spite of the obvious object differences. If there are different colored balloons, we identify the correct repeating shape pattern, namely, the 'balloon' (and not the colors in this case). If several objects come from a distant location towards us, we identify the repeating pattern as movement towards us (of any object). If a number of different objects are dropped and they all fall down due to gravity, we identify the repeating pattern as falling down irrespective of the specific object that is currently falling down at that instant.

A computer capable of detecting several such repetitions is not easy unless it has an SPL architecture that uses fixed and meta-fixed sets as its internal representation. Humans and several animals can perform such pattern recognitions quite easily because of its brains' SPL architecture.

On predicting repetitions at every instant and before the event occurs: If we or an SPL system takes hours to analyze in order to detect a repeatable pattern, we only behave as a passive system. As the external events are constantly changing, we need to keep up with these changes and detect *before* the repeating event finishes and not afterwards. At minimum, we want to perform pattern recognition and repetition detection in real-time.

Predictions in unconscious systems are quite simplistic compared to conscious human beings. Let me briefly discuss one aspect of it. If you do tasks in such a way that your guesses are always behind the actual occurrence of events, then you are simply obeying the laws of physics. This is also related to the fact that as we detect repetitions in real-time, we want to avoid waiting for the entire event to be finished before we detect it. When you hit a ball, we are not merely asking you to predict where the ball will land. This is just predicting the 'end result' of an action. Instead, we want you to keep predicting what happens almost every second of the event.

If your predictions are behind the actual event every second (i.e., the actual event happens before you were able to predict), then you are simply following causality. It turns out that in this case, you can never sense the passage of time (see chapter 40). Unconscious systems, typically, behave in this way. At about 6 months to one year of age, a child can already make predictions at every instant. For example, just from looking at a feeding bottle he can predict a sequence of events that he is used to doing. He may not predict all of these events at once. However, as you start repeating these tasks in the correct sequence, he will be able to predict each sub-event, one after the other, every second along the way before it occurs.

As another example, if a question is being asked that was asked before, we analyze it pro-actively and 'predict' what the question itself might be in addition to proposing a collection of possible answers. We have anticipation as a feature and are *actively* trying to guess the question as well as the answers. Later in chapters 21 - 22, I will discuss how prediction plays a critical role to attain a feeling of knowing. This will even help devise better predictive designs for unconscious systems.

20.2 Repeating events in our nature

Given that repetitions are important for predictions, let us try to identify as many repeating events as we can. We want only those repetitions that we can detect with our sense organs i.e., we exclude repetitions at, say, a subatomic level. Our first intuition is that most repetitive events are external dynamical phenomena. For example, day and night cycle, seasonal cycle, oscillating pendulums or swings and water waves in a beach. However, these naturally occurring repeating events seem so few in number that they hardly matter in creating meta-fixed sets and, hence, help with predictions. Are there more repetitive events than these?

A much larger class of repeating events comes from considering simply *static objects*. The key point to realize is that *we* are not static even though the object itself is static. We can move around as we observe the objects. This introduces variation and repeatability.

For example, you are in your living room that has tables, chairs and other furniture. As you re-enter the room each time, you see the same objects repeatedly. The neural inputs from your eyes, ears and other sense organs are similar each time you view the same static object. This is the same situation when you look outside at cars, trees, leaves, houses and others. Your motion like turning your head, walking, coming back to a similar location causes even static objects to create repeatable dynamics within your brain. The number of these static events is enormously high when you include primitive features like colors, sounds, shapes, edges, angles, motion, smell, tastes and others. These primitive events are the most common ones that even an infant is capable of detecting initially. When we consider this large class of events and combine it with our free will, we now have an almost infinite number of repetitive events.

Among the set of static objects we considered above, one important subclass are our very own body parts. This is especially important for an infant. His hands, legs and other externally visible parts of his body are always there with him even as he moves them. As he is observing the external world, if he looks back at his own hand, the inputs to his brain, repeats. His free will combined with his own body parts creates an enormous number of repetitive events and provides many opportunities to sense dynamics, space and the passage of time, as we will see later. We will now see how the problem of prediction becomes easier when you have such a large number of repeating events.

20.3 Definition of prediction

When we think of predictions, we implicitly assume that we have asked a question. We are now looking to answer the question before someone else does or before the event finishes. As we start searching for answers, we can say that we are making predictions. The questions, in general, can be posed naturally using 'who', 'what', 'when', 'where', 'why' and 'how'. In fact, if we are working on a new topic or observing a new event, we consciously ask ourselves these questions to get a better understanding of it. Therefore, making predictions and attaining a feeling of knowing are related.

As with any abstraction mentioned earlier, a 'question' is learnt by associating it with dynamics. We know that a question is defined as a statement for which an action is expected. A collection of actions are already well-defined for an infant. They are directly related to dynamics. We, therefore, create the association between a question statement and the action dynamics in a mechanical way using imitation. Once you have an ability to ask questions by yourself, we can define what prediction means.

Informally, a prediction is the creation of a statement, called an answer, for a question about an event that has not yet occurred, either as fast as or before the event has occurred completely. Our goal is to take this informal definition and express it in a dynamical system form suitable for all stable parallel looped systems.

Predictions have significant value only if you are attempting to answer at every instant as the event unfolds. It is not useful to simply wait for the entire question to be asked. If you take sufficiently long time to discover an answer during which time the event has already occurred, then you are not really predicting anymore. Humans are fast at predicting that we not only guess the answer before the event occurs, but we also predict *before we even ask the question*. When I look at a table, I already know that it is a table even before I have a chance to ask myself what it is. This is true with most of the *static* objects we see around us. We do not even realize that we have asked a question. Before we asked the question, we already know the answer. As a result, we do not feel the need to complete the question statement.

We can rephrase the above informal definition for prediction as follows: a system is predictive with respect to a given dynamical event if and only if it performs an action that can be directly correlated to the future state of the original dynamical event. Otherwise, I say that the system is simply obeying the causal laws of nature with respect to this dynamical event.

Just to clarify, the dynamical event is the one that is occurring at this instant. Here, we can view a typical dynamical event as a 'chained' sequence of causal events with a cause leading to an effect and an effect producing the next cause and so on. We are attempting to predict what is going to happen next for this dynamical event (i.e., its future state F). What the above statement means is the following: a system predicting the future state F of the event should perform an action that is directly correlated to this future state F before the event reaches state F. The action then can be identified as the system's response as the event is unfolding. For example, the action can be: talking out loud telling what is going to happen next or extending your hand in a direction to catch a ball before the ball reaches you. These actions indicate that you have predicted the future state (of, say, the ball) correctly because you reacted to it before it occurs.

Let me now restate the above definition of prediction differently. This time, I want to state it using the language of fixed sets so it is suitable for stable parallel looped systems.

Definition 20.1 (prediction): A stable parallel looped system is said to predict a dynamical event if and only if it excites a sufficiently large subset of the meta-fixed sets that corresponds to the future state of the event.

In other words, if and when the future state of an event occurs, it is naturally going to trigger a series of meta-fixed sets in your brain. We say that you have predicted this event, before it has occurred, if and only if you trigger the very same set or a sufficiently large subset of these meta-fixed sets right at this instant itself. In this way, the stable parallel looped system is a bit ahead of the real event at every instant.

In the definition of prediction, I want to ensure that the system continues to trigger a subset of the future meta-fixed sets, as the event unfolds, i.e., at every instant, not just once every minute or at the end of the event. In this sense, prediction provides a way to have a continuity of thoughts and hence knowing itself (as we will see in chapter 22). It will also play an important role for sensing the passage of time (see chapter 40).

To be clear, both predictive and non-predictive systems with respect to a given dynamical event do obey the causal laws of nature. However, a non-predictive system does not seem to keep track of the chained causes and effects, as a memory, to do something useful with it. For example, a later cause-effect pair within this chain does not have any memory of what the earlier cause-effect pair did. As soon as the cause-effect pair advances within its chain, the system simply forgets the link. A predictive system, on the other hand, will store these links for a given dynamical event, say, on a hard disk for a computer or as a stable parallel looped network of interconnected neurons for our brain. When the dynamical event repeats, it can use this memory to make useful predictions about the future state of the event.

In general, nonliving systems and machines are some examples of non-predictive systems. The sensors for these systems do receive inputs from the external event. However, a vast majority of these external inputs are ignored. Most existing computers are behind the real event, essentially post-analyzing the event. Some machines (like autopilot in an airplane, missile guidance systems, temperature control in a refrigerator and other feedback control systems) do have predictive ability but in an extremely narrow sense. Some of their deficiencies are (a) they predict a narrow set of dynamical events i.e., they do not identify or even look for complex patterns, and (b) their predictions do not occur at every instant.

A non-predictive system simply reacts and responds to dynamical changes. When an event repeats, the system starts from the beginning once again with little help, if at all, from the previous iteration. The effort and energy consumption is more or less the same as before. A lack of a sustained looped dynamics is, in general, a reason for this problem.

A predictive system, on the other hand, is proactive. It is capable of performing its own internal dynamics fast enough that it has some *free time* available to perform other actions. If a ball is rolling towards you, one action you can perform with your available free time (because you were fast enough to predict what is going to happen next) is to extend your hand to reach for the ball before it reaches you. You are, in this case, waiting for the ball to reach your hand (i.e., waiting corresponds to your free time). This waiting (or the free time) even allows you to sense the very passage of time, as we will see later in chapter 40.

In general, the SPL system waits for the external dynamical event to proceed towards what it already predicted as a future state. This is possible because of the sustained looped dynamics or the memory (fixed sets) stored within its network.

20.4 An example of predicting a repeating event

To understand the issues with prediction within our brain, let us now try to analyze what happens continuously within the system and the environment as we watch an event. Consider the following simple repetitive experiment. A person takes a ball in his right hand from a bucket of balls. He stretches his hand horizontal to the ground and drops the ball. The ball falls down and bounces a few times before it comes to a stop. He then picks up another ball from the bucket and continues the cycle several times. You are an observer watching this entire event repeat itself, from a distance. What happens in your brain at every instant as you watch this event? Your brain continuously analyzes every instant and compares it with your existing memory. We do not wait for the hand to take the ball completely out of the bucket before we start analyzing. Even as the hand moves to reach for the ball, the continuous time-varying input from our eye feeds into our brain. In fact, our brain processing is actually ahead of the real event itself.

For example, after the first few repetitions, your brain forms fixed sets for several sub-events in this loop. Some of these are for, say, the ball bouncing several times, the person's hand releasing the ball, taking a ball from the bucket and sound of the bouncing balls. These fixed sets are linked to form a continuous related event. When the person now starts to pick up a ball from the bucket, your fixed sets are triggered sequentially and almost instantly via the active existing links. For example, the fixed set for picking a ball is quickly excited. The links with neighboring neurons cause a cascading effect of excitation to trigger fixed sets of the next sub-event. This manifests as a prediction that the person is going to raise his hand and then drop the ball. When the person does, in fact, bring his hand to a horizontal position, the new input patterns strengthen the excitations in the corresponding paths of neurons in your brain. This reinforces neural excitations to produce a further cascading effect of firing pattern. You eventually predict that the person is going to drop the ball and, subsequently, that the ball is going to bounce a few times. As the repeating event starts to unfold, you are constantly predicting and strengthening your memory links about what will happen next.

After a large number of repetitions, your memory becomes sufficiently rich. Each repetition lets you predict further into the future because most of the meta-fixed sets are triggered almost instantaneously. You even predict the entire event just as it starts. Of course, our predictions need not be correct at all times. What is important is the fact that our brain's cascading firing of fixed sets are well ahead of the real event giving us multiple choices as possible predictions at all times.

You also start to predict *about* the event itself. For example, you will guess that the person is going to repeat the task once again even before he starts to do the task. This abstraction is beyond predicting the details of the event. It is a statement about the event itself. This contrasts with the previous predictions, which were not about the event. Rather, they were predictions of the actual event. The repetitive nature of the event creates higher-order loops or meta-meta-fixed sets as well.

Let us now briefly contrast this to scenarios when we are not conscious or when we are not perceiving the passage of time. When we are not conscious, our predictive abilities are very low, if at all. In these cases, we are always behind the real event and are simply obeying the direction of causality. Without predictive abilities, it would be impossible to detect if the repetitive event occurred a short while ago or just now. It becomes difficult to distinguish between new versus old events even when the events repeat (see section 15.2 on the origin of brain waves). For example, under general anesthesia, severely drunk and other cases, we see people repeating the same words, sentences or events several times without realizing that they have already mentioned it.

20.5 Dynamical structure during prediction

Let us now outline the structure of predictive systems (for both conscious and unconscious beings) in more detail. Figure 20.1 shows how the brain of a conscious person works when performing a task that takes a finite amount of time. Imagine an event that starts at time T_0 and proceeds until time T_f. This event, for example, could be a ball falling down as discussed previously or someone asking you a question like 'what is your name'. We can generalize the previous discussion about repeating events to the case when the event is not repeating as well in the following way. At time T_0, the input from the event reaches your sensory organ, say, your eye or your ears. The parallel excitation of the neurons on your eye from this image produces a cascading sequence of firing patterns in your brain. The meta-fixed sets are excited quickly and your brain begins to excite the possible options at time T_1 using past memory. If there are multiple choices, all of these possibilities are, at least, partially excited. Additional external inputs like other objects in the room or presence of sunlight provide real-time data that cause some of the choices to be excited more than

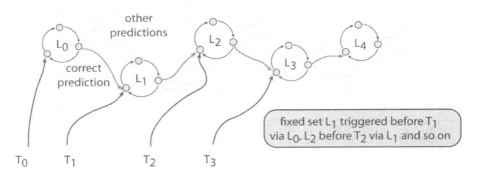

predictive fixed sets triggered in the same sequence
before the external inputs trigger them

other
predictions

L_0

correct
prediction

L_1

L_2

L_3

L_4

fixed set L_1 triggered before T_1
via L_0, L_2 before T_2 via L_1 and so on

T_0 T_1 T_2 T_3

external inputs trigger fixed sets at times T_0, T_1, T_2, ...

Figure 20.1: Predicting external event by a conscious being. In this example, external inputs trigger fixed sets L_0 at time T_0, L_1 at time T_1, L_2 at time T_2 and so on. However, as fixed set L_0 is triggered, several predictive meta-fixed sets are triggered automatically based on your past memory. One of this prediction L_1 is the correct one while the others are valid but incorrect w.r.t the external event occurring at this instant. Fixed set L_1 could be triggered before T_1, i.e., before the true external event triggers the same fixed set. In this case, we can say that we have predicted L_1 before the event has occurred. Similarly, we could predict L_2, L_3 and so on before T_2 and T_3.

average while curtailing excitations for some other choices. This narrows down the choices considerably.

In Figure 20.1, we have represented other available choices in dotted lines. This is to emphasize that we are only recollecting them partially. When the real event occurs at T_1, it provides the complete details making some of these dotted lines into dark lines (representing more complete excitations). The specific sequence of events causes the fixed and meta-fixed sets to be linked together. If the same sequence of events repeats, these links will be further strengthened. Otherwise, new links are formed from this junction to other fixed sets causing your network of fixed sets to grow through experiences.

If the number of choices is very few, your memory produces a quick recollection of the entire path of fixed sets until the final time T_f. This is not as rare as it sounds. For example, if a ball is falling down between time T_0 to T_f, there are hardly any choices at all. When a conscious person does not have any memory of the current event, the neuron architecture forms the necessary links from the new neuron loops (fixed sets) to the existing memory automatically through simultaneity of occurrence of events as described in detail in chapter 7. In this manner, with more repetitions, our brain's excitations are capable of being ahead of the real event itself.

Now, it should not come as a surprise that a system is capable of being ahead of the

predictive fixed sets are rarely triggered before the external event

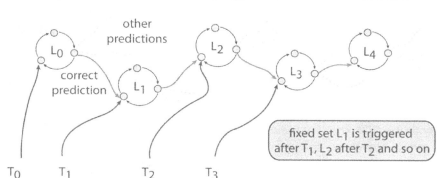

external inputs trigger fixed sets at times $T_0, T_1, T_2, ...$

Figure 20.2: Predictions for an unconscious system. Here, among all possible predictions after fixed set L0 is triggered, picking the correct prediction is passive and slow. It takes longer than T1 to trigger fixed set L1. The real external event has already occurred before the unconscious system could predict L1. The same continues with L2, L3 and so on. The internal predictions are too slow compared to the external event unlike for a conscious being.

real event itself. To see this, note that whenever we have an adaptive system that tries to synchronize with another system (or the environment), one of the system leads and the other system follows. When humans are viewed as an adaptive system, our brain is the system trying to synchronize events with the external environment. Initially, our brain lags for any new event because of a lack of memory. The external events occur faster than our ability to analyze and predict. When the event repeats multiple times and our brain adapts to form memories, we begin to lead compared to the external event. The neuron loops help us excite the necessary regions in our brain faster.

It is difficult for both the system and the environment to be in perfect sync. One possible reason is that our brain is adaptive while the external system simply follows the laws of nature with less need to coordinate. There, however, are other examples for which partial synchronization is possible. For example, consider a competitive environment in which both systems are adaptive, like, say, two humans competing in a game or multiple species fighting for the same food resources. Here, situations like stalemates (giving rise to cooperation) as well as clear winners (giving rise to predator-prey models) are possible. In such cases, we can say that the leader is better at predicting than the follower, at least temporarily.

Let us now consider how Figure 20.1 changes for an unconscious being. Figure 20.2 shows the equivalent representation when an unconscious person performs a task. While the fixed set at T_0 is being discovered, the real time has already advanced to T_1. The same situation happens when processing the event at T_1. The processing is slow enough that the real time advanced to T_2. The unconscious being is always lagging the real external event. T In addition, the set of additional predictive choices are quite limited as well. The speed of computation for an unconscious person is slow because of the inability to form a large number of loops. This is true for unconscious states like anesthesia, sleepwalking, deep sleep, severely drunk, hypnosis and knocked unconscious. For these case, the loops are curtailed using externally supplied or internally generated chemicals preventing cascading excitations of neighboring fixed sets.

Yet, as mentioned earlier, there are missile guidance systems and, in fact, all control systems that are capable of predicting the events. These systems do lead the external event. However, the class of repeating events they detect are inferior. As a result, the predictive ability is quite limited. In the next section, I will show how to relate repeating events and predictions to our dynamical membrane of meta-fixed sets. This relationship is critical for the problem of consciousness.

20.6 Fixed sets for predicting repeated events

We have already seen in chapter 19 that the dynamical membrane of meta-fixed sets is an important structure for consciousness formed within our brain through our experiences during the first few years of life. In this section, I will show that repeating events, by virtue of inherent loops of repetition of dynamics, can be viewed as generating a special fixed set as well.

When an event repeats, it usually does not repeat in an exactly identical way. This is because of the disturbances, imperfections and interactions with neighboring systems. We cannot avoid these unless you completely isolate the systems. These non-ideal situations introduce variations, which can be grouped as a set of dynamical transformations. For example, if dropping the ball (example from the previous section) repeats a hundred times, we can abstractly represent this scenario as a *single* repeating event along with a hundred small variations around this single event. The single repeating event represents the *average* scenario. The hundred small variations are perturbations from the average scenario. These variations define the set of dynamical transformations under which the average scenario does not deviate significantly.

When we start to view this way, we can say that the average scenario is 'the' single fixed set under all of these dynamical transformations (cf. the definition of truth in section 13.2). Of course, we do have the standard space-time dynamical transformations to make the average scenario into a fixed set as well. For example, you can watch the repeating event at a different time, you can walk towards the objects, away from the objects, tilt your head while watching them and other such space-time transformations. The entire repeating event, viewed as the average scenario, is still a fixed set within our neural loops.

The neuron architecture of our brain that observes these repeating events will eventually create and store this fixed set (i.e., the average scenario). An important consequence of this fixed set is that if the entire event does not repeat itself, then this fixed set is not excited. If the ball dropping experiment deviates considerably (like, say, the person kicks and plays with the ball before completing the event), you may not be able to detect this event as a repetition of a previous scenario.

Note the previous fixed set (average scenario) may initially be excited even under a new modified scenario because of short-term memory. However, as the event stops repeating, the short-term memory is altered (through the interconnections of neurons) that the old fixed set is no longer excited for the new modified scenario. Therefore, this fixed set is excited if and only if the above event repeats. From this, we can conclude that the average scenario fixed set in our brain uniquely identifies with the original repeating event.

Analogies can be viewed as generalizations of complex repeating events. Two widely disjoint topics can have similarities as analogies, which can be detected if properly trained. These are possible only when you represent meta-meta-fixed sets back into the same form of stable parallel loops as meta-fixed sets. Another nontrivial repetition involves creating a pattern by generalizing several examples. For example, if we drop multiple objects that have seemingly nothing in common, the repeating aspect of this event is an abstract fact that objects are being dropped. We can detect such abstract analogies or generalizations by identifying the corresponding unique fixed set (i.e., the average scenario).

20.7 Predictions viewed from the dynamical membrane

Now that we have seen how predictions are related to fixed sets, let us see how to express what happens in our brain as we predict, in terms of the dynamical membrane. After our dynamical membrane is formed, let us imagine that we are watching a repeating event. Our membrane, as mentioned in section 19.4, is self-sustaining with help from several existing thoughts and external inputs.

(a) At time instant T_1, an external input specific to the repeating event falls on our retina. These external inputs propagate through our brain as cascading set of neural loops of excitations until a fixed set specific to the object A we are observing is excited (see Figure 20.1).

(b) When the fixed sets for object A are excited, further links lead to exciting a meta-fixed set of object A that lets us identify object A as an abstraction representing similar objects. Now, the dynamical membrane extends to include this meta-fixed set (because the membrane is a connected component in the graph) and creates new excitations that self-sustain neural links surrounding the meta-fixed set. If the neural inputs from this meta-fixed set have sufficient magnitude of excitation, say, from a dense structure of memorized interconnections, then a 'local peak' in the membrane is generated. This peak is directly a result of the external inputs (see Figure 20.1). I call this a 'real' meta-fixed set (as opposed to a 'predicted' meta-fixed set). This occurs at time instant $\hat{T}_1$.

(c) The real meta-fixed set is linked to a number of other meta-fixed sets, which are part of the membrane. These links were created through our experiences and memory. The connected nature of the membrane allows the newly formed peak to propagate to other regions on the membrane thereby creating additional peaks. We can view these new peaks, informally, as choices resulting from our thoughts triggered from the original input event. They are generated at time instant $\hat{T}_2$. I call these as predictive meta-fixed sets primarily because they are not generated from external inputs directly (see Figure 20.1). Rather, they take current external inputs and past memory to create new outcomes via the continuity of interconnections within meta-fixed sets.

(d) So far we have seen that the external inputs at time instant T_1 produced a real meta-fixed set on the membrane at time $\hat{T}_1$. This resulted in generating predictive meta-fixed sets on the membrane at time $\hat{T}_2$. The external event continues its dynamics and produces the next output at time instant T_2. This output will, most likely, correlate with one of the predictive meta-fixed sets. If the external input matches with one of the predictive meta-fixed set, then our prediction is fairly accurate (see Figure 20.1). Otherwise, our guesses were incorrect. The time interval $T_2 - T_1$ is usually small. We can, therefore, call these as instantaneous predictions, as opposed to predictions when the event has nearly completed. The process continues with the real event proceeding according to its natural dynamics. The inputs continue to fall on the retina at time instants T_3, T_4 and so on. The real fixed sets are generated at T_3, T_4, ... while the predictive meta-fixed sets are generated at $\hat{T}_4$, $\hat{T}_5$ and so on respectively.

(e) The time instant T_2 can be either before or after $\hat{T}_2$. If $T_2 < \hat{T}_2$ (or more precisely, $T_2 - T_1 < \hat{T}_2 - \hat{T}_1$), this says that the real external event has already occurred while our brain is still trying to predict the possible outcomes. In this case, it is taking much longer for us to predict the outcomes. As a result, our brain in behind reality and we are simply obeying the laws of our nature. This is not a good situation for consciousness or for detecting the passage of time, as noted earlier (see Figure 20.2).

If $T_2 - T_1 \geq \hat{T}_2 - \hat{T}_1$, this says that the time taken to predict is faster than the time taken for the external event to proceed to the next step. Our predictions are ahead of the event. It is better to call this situation as a true prediction because our guesses were generated *before* the event occurred, unlike the previous scenario (see Figure 20.1).

If we call $\Delta T_{21} = T_2 - T_1$ and $\Delta \hat{T}_{21} = \hat{T}_2 - \hat{T}_1$, we will have $\Delta T_{21} \geq \Delta \hat{T}_{21}$ when our guesses occur before the actual event. In this case, we have an additional amount of 'free' time available, namely, $\Delta T_{21} - \Delta \hat{T}_{21} \geq 0$. During this period, we think we know what the next step of the event will be. It is perfectly reasonable to have incorrect predictions as long as they are not empty. This non-emptiness is needed later in chapter 21 when we discuss abstract continuity needed for attaining a feeling of knowing. We predicted quickly enough that we have a sufficient amount of free time available before the real event actually occurs. What can we do during this period? In chapter 40, I will show how this free time can lead us to detect the very passage of time, provided we have several *instantaneous* predictions, not merely end result predictions. If $\Delta T_{21} - \Delta \hat{T}_{21} < 0$, then, intuitively speaking, we do not have any free time available to detect the passage of time (see section 40.4 for more details).

We can see that the above process has a nice continuity with the flow of events when the predictive meta-fixed sets are nonempty at every time instant. In this nonempty case, we are able to predict at every instant. Quite obviously, this is not the case most of the times. In fact, it is true that we can only predict after we have memorized the repeating event or after we have understood the logic to apply it to a new scenario. When we have not yet mastered these techniques, our predictions will be empty. It is, therefore, a prerequisite to memorize several events through an iterative process before we can apply the above outlined steps. Let me briefly describe this.

Firstly, we will need to memorize and form meta-fixed sets for each of the objects we view during the event. This typically happens already during our very early childhood. We need a dynamical membrane that contains recognizable objects from our everyday life like, say, tables, chairs, books, balls, houses and others, as well-connected meta-fixed sets. Secondly, we need to form sufficient links between these objects based on near-simultaneity of occurrence of objects like, say, chairs are always seen with tables, trees are always present along the roads and so on. Once we have a core set of these links as well, we will be in a position to form new relationships through the event you are currently observing.

If this event does not yet have any predictions, at least the meta-fixed sets for this event are triggered on the membrane. As there are no predictive meta-fixed sets at $\hat{T}_2$, the real event at T_2 will produce a peak on the membrane that corresponds to the real meta-fixed set. This will give you an opportunity to form new links that connect meta-fixed sets at $\hat{T}_1$ to meta-fixed sets at $\hat{T}_2$. The two peaks at both time instants are linked through the self-oscillations of membrane and the *instantaneous* nature of the time intervals between T_1 and T_2. A notion of continuity within the meta-fixed sets on the membrane between $\hat{T}_1$ and $\hat{T}_2$ is established. The process continues at time instants $\hat{T}_3$, $\hat{T}_4$ and so on. If you look at Figure 20.1, the external event sequence along the timeline T_1, T_2, T_3, and so on is mapped in our brain in a linearly connected way onto the dynamical membrane along the peaks at $\hat{T}_1$, $\hat{T}_2$, $\hat{T}_3$, and so on. We will study this useful mapping in detail in the next chapter.

When there are no predictions, you form new continuous links between the meta-fixed sets across the membrane to align with the real external event. We do not memorize every minor detail of the external event through this process though. This is when repetition becomes critical. As you watch a repeating event often, your memory becomes stronger resulting in the creation of new links (see chapter 7). Additionally, variations in a repeating event (i.e., the dynamical transformations) are also critical to establish a *robust* set of links between meta-fixed sets.

After practicing or watching the event for a while, we would have created an interconnected structure on the membrane corresponding to this event. When the same event repeats later, we will have valid nonempty instantaneous predictive meta-fixed sets. The process outlined in Figure 20.1 would now yield a smooth flow of events because we will have nonempty predictions every step along the way. In this manner, we can combine the above prerequisite memory-formation iterative steps with the above steps (a)-(e) for exciting predictive meta-fixed sets on the membrane for almost all events we experience in our everyday life.

21
Abstract Continuity

In the previous chapter, we have seen how the progress of an external event along a linear time variable is mapped onto the dynamical membrane as a linearly connected and 'continuous' path of meta-fixed sets. This mapping allowed us to predict future states of an unfolding event before the event finishes. As we experience a large number of events every day, we create a highly networked but linearly connected continuous paths of meta-fixed sets through our memories.

Since meta-fixed sets are associated with abstractions, this suggests that we can define a notion of continuity among abstractions via the continuity of the underlying dynamics. However, we will see that this is still a weak form, not sufficient for our purposes. I will, therefore, discuss a stronger notion called abstract continuity in this chapter. We will also see why this concept of continuity within abstractions is important to study in the first place. In chapter 22, I will show that these notions help attain a feeling of knowing. This, as you recall, is both necessary and sufficient for becoming conscious (according to Theorem 12.1).

21.1 Generalizing continuity to abstractions

The notion of continuity is well studied for dynamical systems. Is there a similar notion when working with abstractions, like with sentences in a language, paintings with images, speech with sounds or with our thoughts? I want to show that our approach of representing abstractions using dynamics within an SPL system helps in achieving this generalization fairly easily.

What does defining continuity for abstractions mean? In mathematics, we define continuity of functions in multiple equivalent ways. For example, if the inverse of any open set is open, then the function is continuous. Intuitively, this says that small causes will produce small effects. The dynamical systems we observe in nature are all continuous. However, humans have discovered abstractions like language and mathematics that appear discrete instead. How can we generalize continuity to these situations? The answer lies with using dynamical systems as a way to represent abstractions with the help of loops and fixed sets described in chapter 14. But before we do this, let us first understand the motivation for generalizing the definition of continuity to abstractions.

Consider simple sentences in a language like 'I like to eat pizza tonight'. Compare it with another sentence like 'pizza to like I tonight eat', which is a simple rearrangement of words from the previous sentence. Clearly, we understand the first sentence but not the second, even though the second sentence is a simple rearrangement of words from the first sentence. We could have made the rearranged sentence difficult to understand in a number of ways, but we did not. For example, we could have rearranged the letters in the words (like 'zipaz' instead of 'pizza') or scrambled up all the letters in the sentence, not just within words. Instead, we simply rearranged the words, each of which we do understand very well. Such a simple change was enough to make you not understand the sentence.

You may argue that we do understand the second sentence if we just try a little longer. Yet, my point is that there is genuine question to answer here. It may be that a short sentence like the one considered here has so few words that the meaning is well determined even if we rearrange them arbitrarily. If we take a longer sentence, may be even two sentences or better yet an entire paragraph (say, the previous paragraph) and rearranged the words, there is almost no hope of understanding the resulting paragraph. This is even without rearranging letters in the words. Therefore, for the sake of simplicity, I will consider the above two simple sentences as illustrative examples in the following discussion.

So, why do we understand the first sentence and why do we not understand the second sentence? How did a simple rearrangement of the original sentence destroy our feeling of understanding (or knowing)? At the outset, the second sentence seems like a simple rearrangement, but apparently, this has resulted in a drastic change within our brain. The feeling of understanding that is unique to a conscious being disappeared by a trivial rearrangement of words in a sentence. The feeling reappears as we gradually rearrange the sentence, one word at a time, in the correct order.

Clearly, the first sentence is grammatically correct while the second one is not. We say that the simple rearrangement turned the original sentence into a grammatically incorrect sentence. However, why should conscious humans (even a 2-3 year old) understand grammatically correct sentences? There is nothing sacred about grammatically correct sentences with respect to generation of a feeling of knowing. In fact, for an infant who does not yet know any language, both the above sentences, grammatically correct or not, are equally valid (or equally invalid). As we grow older, it is quite amazing that we preferentially understand only grammatically correct sentences, not any others even if we have never heard them before. For example, several sentences in the book are entirely new to you. You have never heard them before and yet you understand them simply because they are grammatically correct. Why or how? The situation is true with partial sentences if they are arranged in an approximately grammatically correct way (like say, 'want pizza tonight').

If we look closely, the notion of grammar is not fixed over time. As a result, an adult may find it difficult to understand a grammatically correct sentence, just like an infant, under some scenarios. For example, the grammar and sentences used during Shakespeare's time is not necessarily understandable now. Another example is with people speaking different dialects of the same language. Some dialects have slightly different grammar as well. The sentences in each of these dialects are either partially or completely incomprehensible. Yet another example is when you are just beginning to learn a new language. Your spoken language feels 'broken' to me and creates similar incomprehensible situations as above.

From these special examples, we see that our inability to understand or attain a feeling of knowing cannot be attributed just to grammatically incorrect sentences. We can understand all grammatically correct sentences. However, not all sentences we understand are necessarily grammatically correct.

Another reason for not attributing grammatical incorrectness to our inability to attain a feeling of knowing is because the situation discussed here is not unique to sentences in a language. In fact, we can generalize and ask the following question – why do we understand anything at all? For example, if we take a familiar image like, say, a chair, sounds in a language like from a recording or even more complex events, we do understand them or, more generally, we have a feeling of knowing. These examples are not too different from grammatically correct sentences discussed earlier. To see the similarity, let us convert the problem of understanding these cases to the problem of not understanding them using a simple rearrangement (similar to how we converted an understandable grammatically correct sentence into a non-understandable sentence through a simple rearrangement of words).

If we take a chair and rearrange its subcomponents (like the base, legs, headrest and others) *spatially* like in a jigsaw puzzle, you lose the feeling associated with recognition of chair completely. Here, the rearrangement is in two or three dimensions compared to the previous case of sentences where the rearrangement was in one dimension. If you take a recording of your sound and rearrange them along the timeline, you encounter a similar loss of feeling of understanding. In each of these examples, we did not even try hard to rearrange them in complex ways (like say, with scrambling the words from 'pizza' to 'zipaz').

Let us now see how the transition from not having a feeling of knowing to having one happens. Consider a jigsaw puzzle. The incorrect arrangement of the pieces, analogous to incorrect arrangement of words in a sentence, makes us not understand the picture. As we start arranging them in the correct order, we begin to attain a feeling of knowing of the entire picture (analogous to how it happens with a grammatically correct rearrangement).

Similarly, consider an event in which we are trying to recollect a person whom we just forget, but know that we have seen him before. In this case, we can say that the thoughts and events are incorrectly 'arranged' that we are not able to attain a feeling of knowing of that person. As we start to arrange them in a correct order, namely in the order of how we met him and other memorable events, we suddenly attain the feeling of knowing. I will discuss several more examples in this and the next chapter.

If we want to generalize the notion of grammatically correct sentences to images and sound as well, there is no equivalent term. I, therefore, call all of these scenarios, with sentences, images sounds and events, as *abstract continuity*. In the case of sentences, having abstract continuity and satisfying grammatical correctness are equivalent. In the above examples, I say that we have broken the 'continuity' by rearranging subcomponents. Our objective is to understand what the necessary and sufficient conditions are to attain a feeling of knowing for all cases like correct arrangement in sentences, correct arrangement of jigsaw-like images, correct arrangement of sound variations, correct arrangement of events and so on.

At first, it appears strange to call the above grammatically correct sentence ('I like to eat pizza tonight'), that merely has 6 words, as continuous. Let me justify it now using a few continuity-related properties exhibited by such grammatically correct sentences. Firstly, notice that there is an important source of variation in this sentence. For example, we can replace 'eat' with, say, 'cook', 'make', 'bake' and 'order', 'pizza' with 'pasta', 'bread', 'salad' and 'rice', 'like' with 'don't like', 'love', 'want' and 'don't want' to create new sentences that are all equally understandable. These variations are not infinite in number, but we still do have enough possibilities. With images and sounds, we can imagine morphing them so that they do not alter the content drastically. Some of these variations include space and time-based dynamical transformations. The resulting morphed images and sounds will still be understandable. These variations suggest that the set of abstractions we want to call as continuous are 'close' together. This is, however, not enough to generalize continuity to abstractions. Our process of identifying these benign small variations appear quite adhoc to be applicable to all abstraction-based scenarios (like with images and sounds).

Let us fix this by converting the adhoc process into a systematic procedure instead. Recall the discussion in chapter 14 where we have created abstractions within our brain using dynamics, namely, as meta-fixed sets. The advantage with this representation is that SPL dynamical systems are inherently continuous. The loops (or meta-fixed sets) for each of these closely related abstractions are physically linked together using paths of neuron connections. We can use this fact to make the above process less adhoc as follows. Consider the two previous sentences once again:

(i) I like to eat pizza tonight.
(ii) pizza to like I tonight eat.

What happens in our brain as we read each of these sentences? When someone reads each of these sentences slowly word-after-word, our brain is trying to predict which word is going to come next. For example, when we hear 'I' of the first sentence, in a quick fraction of a second, we are able to predict, 'I want...', 'I like...', 'I am...' and others as immediate possibilities. We may not say it aloud, but our brain is already trying to predict after we hear the very first word. Next, when we hear 'I like', we can predict 'I like music...', 'I like to...' and others as possibilities. We may not have definite answers or correct choices with our predictions. Nevertheless, we do predict a few valid choices. This is exactly analogous to the discussion in section 20.5 (see Figure 20.1). For 'I like to', the nonempty choices are 'I like to do...', 'I like to play...' and others. For 'I like to eat', the nonempty predictions are 'I like to eat candies...', 'I like to eat ice cream...' and so on. Similarly, we have nonempty predictions for the rest of the sentence as well. The choices need not be correct at every instant relative to the sentence we are hearing. Just being nonempty is sufficient.

On the other hand, consider the second sentence 'pizza to like I tonight eat'. When we hear 'pizza', we do have nonempty predictions like 'pizza is...', 'pizza has...' and others. However, the moment we hear the next word 'pizza to', we have lost all our predictions. There is no meaningful word that comes after 'pizza to'. We have an empty set of predictions. The situation does not improve when we hear 'pizza to like' or 'pizza to like I' and so on. Therefore, as we start hearing word-after-word in a linear order, our predictions drops to zero very soon and stays there unlike the first sentence.

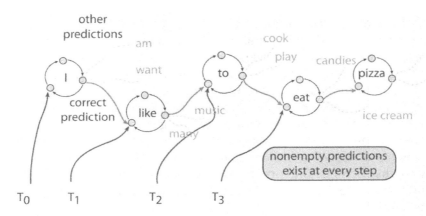

spoken sentence triggers fixed sets at times $T_0, T_1, T_2, ...$

Figure 21.1: Abstract continuity of a grammatically correct sentence when represented internally in an SPL system. Each word in the sentence is represented by its meta-fixed set. For every word in the sentence, we have nonempty predictive meta-fixed sets. The predictive meta-fixed sets for each word is triggered before the correct word is spoken.

This is the fundamental difference between any grammatically correct sentence and an ill-formed sentence for an adult conscious being. All sentences we do understand have nonempty predictions if we break them word-for-word in a linear order like above. This condition is, however, not sufficient for understanding a sentence (see chapter 22 that gives both necessary and sufficient conditions). Note that such a difference exists even for all new grammatically correct sentences that you have never heard before (like sentences in this book) and ill-formed sentences.

Let us see how the neural loop excitations within our brain behave for both sentences. In Figure 21.1, I have shown how this predictive process works for the first sentence by linking neural loops in a well-defined path from the beginning of the sentence to the end. As we reach a meta-fixed set for a given word (say, 'I'), the neural excitations extend outward as well. These extensions would trigger additional meta-fixed sets which we called as predictions (like 'I like...', 'I want...' and 'I am...') *before* we hear the next word. These additional meta-fixed sets will become part of the membrane i.e., the existing membrane extends to include these predictive meta-fixed sets as well. These predictive meta-fixed sets may not be correct, but they do exist i.e., they are nonempty.

When we hear the correct next word of the sentence ('like'), we will include it into our existing structure of excitations within our brain. We continue this process where we trigger additional nonempty predictive meta-fixed sets like 'I like music...', 'I like doing...' and others. The critical point is that after we include the correct data, if the new linked excitations in our brain do not destroy our future predictive abilities, it is perfectly valid to have incorrect predictions previously. For example, we may never have predicted that 'to' will be the next word after listening to 'I like...'. Instead, our predictive meta-fixed sets were 'music' and 'doing', for example. Nevertheless, when we incorporate the actual word 'to' and correct our predicted sentence to 'I like to...', we have not destroyed our future predictive abilities. For example, our future nonempty predictions are, say, 'I like to eat...', 'I like to run...' and others.

In other words, we expect our guesses to be wrong most of the time that we do not even attempt to be exhaustive with our predictions. Rather, we just look for one or two predictions. These nonempty choices exist and intersect the meta-fixed set membrane (or rather the membrane extends to include these predictive meta-fixed sets). With nonempty choices, we can guarantee continuity in the dynamical sense within SPL systems and, hence, the terminology 'continuity within abstractions'.

We can compare this with what happens for the second sentence. The extensions to the meta-fixed sets as predictions are empty for 'pizza to...' or 'pizza to like...' and so on. The membrane does not extend as we hear more words from this sentence. Using this analysis, we can now define abstract continuity as follows.

Definition 21.1 (abstract continuity): Abstractions are said to be continuous if their dynamical representations produce nonempty, and possibly incorrect, predictive meta-fixed sets along most linear subsequences. These subsequences are either implicitly present (like with linear sentences in a language) or explicitly created (like with linear paths when scanning a 2D image with our eyes) within the abstraction.

I will use the term abstract continuity to refer to the above definition. Notice that this definition generalizes to jigsaw-like images mentioned earlier and, in fact, to all abstractions. Predictive ability is a universal notion applicable to all events and experiences. Let me now clarify a few subtle points about the above definition.

- The term 'predictive' in the above definition should be understood in terms of the definition 20.1 of section 20.3. Prediction does not simply mean that we are guessing what the final outcome might be while watching an event. Instead, it is our ability to guess at every instant as the event unfolds.
- Abstract continuity is defined whether there is a dynamical membrane or not. This allows us to include animals as systems that have abstract continuity even though they may only include primitive meta-fixed sets involving basic images, sounds and simple abstractions related to dynamics. Furthermore, we want systems that have not yet created a dynamical membrane, like an infant, to be able to have abstract continuity in a local sense.
- The term 'along every linear subsequence' refers to the above discussion on prediction were we want to constantly keep guessing as the linear subsequence grows from the beginning to the end of the sentence. For example, with a sentence, we want to start predicting at every letter and word along the sentence, not merely analyze after you finish reading the entire sentence.
- The notion of linear sequence typically aligns with the standard timeline in most cases. However, when there is a parallel excitation, like with an image falling on a broad region on the retina, there is no natural notion of timeline.

This is because different paths of neuron excitations from the retina reach different regions of the brain in parallel. We need to distinguish this case carefully. In case of abstractions like sentences, speech, thoughts expressed in a natural language, music or mathematical structures, the linear sequence is already implicit. For example, it is the linear order of the sentence structure itself that you read along the standard timeline. In case of abstractions like two-dimensional or three-dimensional images, the linear sequence is explicitly created using (a) the implicit questions we ask ourselves in a natural language or (b) by scanning the image with eyes in a sequence of our choice. For example, for some images we are constantly talking to ourselves asking questions like 'what is this shape doing here?', 'why is this shape tilted like this?' and so on to gain predictive links. In addition, for images, we scan them in various directions like, say, from left to right, top to bottom, along the edge boundaries, zigzag paths to other neighboring objects and so on. All of these one-dimensional directions create a natural linear sequence as well.

- Abstract continuity and standard continuity are qualitatively similar. However, there are a couple of important differences. Firstly, if you draw a one-dimensional line on a piece of paper without any breaks, then it is a continuous path. If this line represents an event, say, like the road you use to drive from your home to a shop, then this is not an abstractly continuous path. It is not sufficient to look straight ahead along the path to have abstract continuity. You need to look laterally along the road like the shops, houses and gas stations as well. These surrounding objects should form memories along with the path itself.

  Abstract continuity forces us to move along a linear path in at least two dimensions. It is like a 'wide-band' of finite thickness. The analogy should work even within the stable parallel loop architecture of the brain. If an image falls on your retina, following the path of excitation of a single neuron deep into the brain, while valid for continuity, is simply a degenerate case to be of any value. We, therefore, exclude this case for abstract continuity. The parallel excitation resulting from an entire image on a finite region of the retina produces a wide enough 'band' of neuron firing patterns that propagate through the optic nerve to the deeper regions of your brain. The pathways will then be along fixed sets on the membrane.

  Secondly, in the definition of continuity with dynamical systems, we can make the neighborhood 'epsilon' (in, say, the epsilon-delta definition of continuity) as close to zero as possible. However, with respect to the definition of abstract continuity, it is not well-defined below finite time duration. This time interval corresponds to gamma wave frequency of brain waves i.e., greater than 30 Hz to about 200 Hz (5 to 33 milliseconds). This is because it takes, as high as, 30 milliseconds to excite a meta-fixed set corresponding to an external or internal input. Even in the best case scenario, any event that occurs faster than 5 milliseconds is not detectable by your conscious self. Our thoughts also occur in this time range. This does not mean that we cannot study external phenomena that are faster than 5 milliseconds. The trick we use here is to record and run the phenomena in slow motion. The recording device does indeed have much faster clock frequencies. But the subsequent play back will be in slow motion at 5 millisecond range. This is analogous to a magnifier to 'look' at distances in the nanometer range because our eyes cannot detect this range directly. However, while we can have magnifiers for space and time to observe external phenomena, we inherently cannot have magnifiers for our own conscious brain, at least as yet. Therefore, for abstract continuity in our brain, time transitions below, say, 5 milliseconds are not relevant.

The above definition, expressed in terms of dynamical representations with predictive abilities, is valid for all types of abstractions we represent in our brain including, say, sounds, images and sentences (see section 21.4 for more examples).

Predictions within a sentence described here with the above two examples is identical with predictions for any dynamical situation like the scenario of section 20.5 (see Figure 20.1 and Figure 21.1) and others. Abstract 'discontinuity' is analogous to watching unrealistic dynamical phenomena, like science fictional oddities or unrecognizable physical phenomena. In these cases, we lose the continuity with our predictive abilities.

What happens when we reach the end of a grammatically correct sentence? The predictive meta-fixed sets become empty just as they do in the middle of a grammatically incorrect sentence. Yet, we do not feel that we do not understand the sentence. Why? To see this, first note that abstract continuity as defined above does not require an infinite sequence nor does it need to continue for an infinite amount of time. Secondly, the beginning and ending of a sentence is not known apriori (cf. with images and events for which there is no clear beginning and ending). An infant would need to learn these notions during his childhood. In fact, if you recall, a 2-3 year old hardly forms complete sentences and, yet, both we and himself understands it well enough. Also, sometimes we do not use full sentences when talking to him. At this age, his abstract continuous paths are quite limited. They are mostly based on actions, tastes, vision, hearing and others basic sensory inputs. He needs these primitive fixed sets to create new abstract continuous paths to sentences in a language (see section 21.3 for additional details on how to create abstract continuity). Specifically, he will need to create abstract continuity for the very notion of beginning and ending of a sentence as well. In this manner, a child learns when to start and stop processing a sentence without disturbing the abstract continuity accumulated so far.

21.2 Verifying abstract continuity

To check abstract continuity for a discrete abstract sentence, we first translate it into a dynamical representation of linked fixed sets within a given stable parallel looped system. Using standard 'dynamical' continuity of neural firing patterns, we verify if there are nonempty predictive meta-fixed sets along the entire pathway at *every* fixed set and most subsequences locally, as we did with the examples considered previously.

I say that abstract continuity is broken if and only if the above dynamical continuity is broken. As mentioned in the definition, it is important to have nonempty predictions for 2-word (like 'like to', 'to eat' and 'eat pizza' as opposed to 'like I' and 'I tonight'), 3-word (like 'like to eat' and 'to eat pizza' as opposed to 'like I tonight') and other subsequences as well, not just when appending 1-word to a growing sequence considered in the previous discussion.

21.3 Creating abstract continuity

How does our brain create abstract continuity for a given event in the first place? If you are learning to catch a ball through repeated scenarios, we identify an average scenario and several variations around it. For example, the ball reaching you correctly is an average scenario while other mistakes like reaching too close or too far from you are common variations. Through each of these experiences, our brain stores nonempty predictive meta-fixed sets for the ball-catching event. These predictions let us reach for the ball quite accurately even in a new situation. Similarly, when you are learning a new topic in abstract mathematics, you create fixed sets for each of the primitive concepts first (like for groups, rings and fields). Through several examples (like real numbers and matrices), you then create *predictive* meta-fixed sets around each fixed set. Now, when you encounter a new theorem, each of these nonempty predictions will give rise to abstract continuity.

The situation is similar when a child learns a language. All sentences for an infant are equally invalid or discontinuous initially. For example, both sentences (i) and (ii) above, do not give any predictive choices as he hears each word i.e., he does not understand either sentence. As he grows older, how does he develop abstract continuity with just sentence (i), but not with sentence (ii)? Sentence (ii) remains in a discontinuous state.

The answer is simple when we start including memories of each of the sentences you hear in your everyday life. As an infant, you hear grammatically correct sentences used by your parents, friends and others at all times. You may not have memories for every word in their sentences. Nevertheless, you begin to form memories gradually within the neural looped structure of your brain. You will memorize (a) the words for each of the object, (b) the geometric shapes of the objects, (c) how it feels when you touch them, (d) how it tastes when you put it in your mouth, (e) how it bounces, rolls, breaks or splashes as you throw it around, (f) the sounds the object makes or (g) how it smells. Each of these meta-fixed sets (vision, touch, taste, sounds and others) are formed according to the memory model described in chapter 7. As several memories are formed and interlinked using near-simultaneity of occurrence of events, your internal brain structure becomes specialized. The links that keep forming through your experiences and those that are maintained through constant usage, lets you form continuity, as defined above, between several abstractions.

If we take grammatically correct sentence (i) above, an infant cannot still attach a meaning to it. He needs to become conscious before he can start understanding new sentences. His consciousness implies that he can understand several words, expressions and actions already at this age. The new sentence (i) would need to be elaborated using actions and words he can already understand. This will let him create links with his existing dynamical membrane between visual/tactile abstract continuous paths with words in a language. For example, a new word 'eat' can be associated with the action of eating, which he already understands. In this way, he is able to expand his dynamical membrane. He does not need to understand every single word in sentence (i). We simply need to use the same sentence, expression or action consistently until he forms the new neuron links between the old and new meta-fixed sets.

Over months of experiences, he slowly transitions to abstract continuity within language. When formulating new sentences on his own, he does not need to create and follow grammatical rules. Instead, he can use the looped network created by the abstract continuous paths. Therefore, hearing sentences in a language that merely 'describes' an event is as real as 'observing' the physical event itself because of these abstract continuous paths.

It is important to note that whether we understand what an infant says or not, he always does understand himself. The sentences he formulates do not necessarily obey our grammatical rules making it difficult for us to understand. However, he has clear meaning about what it means. He always has abstract continuity with his own thoughts. In the beginning, the sentence itself is not conveying any more information to him than what he already knows. The sentence acts as a shortcut to express a complicated scenario in simple terms.

The approach an infant takes is as follows. He first creates abstract continuity for a dynamical event. He expresses it in a language, which, at least, he understands. We correct it to follow our grammatical rules. He fixes his sentences by creating new abstract continuous paths with language abstractions and linking it to the event abstract continuous paths that he already has. In this way, he is able to expand his links between meta-fixed sets from dynamical events to abstract language-based sentences too. The set of abstract continuous paths keep growing, linking more and more events to form a complex looped network or an ever-expanding dynamical membrane.

Creating abstract continuity in each of these cases is a slow process. The abstract continuous pathway for a given event stored within our brain is not like a thin trajectory, but a 'wide band' around the average pathway. Generalizing from these examples, we now say that our brain creates abstract continuous paths for memorizing physical events, understanding mathematical statements, learning sheet music, a new language, designing engineering structures and other common everyday tasks using sensory-control mechanisms and repetitive experiences.

In the case of a synthetic SPL system, creating abstract continuous paths is a manual process of linking fixed sets based on different variations of training examples. For sentences in a language, this involves linking different word fixed sets in such a way that the system has only those nonempty pathways at a given fixed set that correspond to grammatically correct sentences. This involves picking a training set of grammatically correct sentences, say, from the story books a child reads at a young age. For grammatically incorrect sentences, the nonexistence of predictive meta-fixed sets results in a discontinuity.

One interesting case to notice is that a child sometimes forms grammatically incorrect sentences even though he is only exposed to grammatically correct sentences. This is because the evolving cumulative network automatically has extraneous links between meta-fixed sets even though he has never heard these sentences. As a result, he believes that it is a meaningful sentence. This is usually corrected by a teacher or a parent. This teaching process *deletes* some of the existing links thereby pruning the network to stay as close to grammatical correctness as possible. This pruning step can be performed manually even in a synthetic SPL system.

21.4 Examples of abstract continuity

In the next several sections, I will discuss a few sample scenarios to give you a better feeling of what abstract continuity means and how to apply it in new situations. This is a fundamentally important concept for attaining a *feeling of knowing*, as will be seen in chapter 22. It is, therefore, critical to get a feeling of knowing for this very concept itself.

There are a few important points to remember when trying to get an intuition for this concept. Firstly, abstract continuity captures the notion of 'practice makes you perfect'. The mistakes are as important as the correct answers. Each time you practice and make mistakes, you learn and form new predictive meta-fixed sets that link the correct solution to these errors. You can predict what can happen in a new unknown situation because of your memories of past mistakes, which become your predictive meta-fixed sets. Hence, you get a chance to correct them before it becomes too late.

Secondly, abstract continuity gives you an easy way to learn new topics by linking a number of previously learnt topics, i.e., it makes the coordination of a number of complex tasks an easy problem. You do not learn each of the past topics by itself. You create abstract continuity around each topic through practice. This makes the ability to link and create a new path *across* several topics, an easy matter. Had you not created abstract continuity around each topic, you would have needed an extraordinary amount of precision while coordinating all of these complex tasks. Let me discuss several examples to highlight these points.

21.5 Linking multiple sensory events

Consider a ball rolling towards you and touching your hand. As you observe the ball rolling towards your hand, the size of the image on your retina keeps increasing gradually. While looking at the ball, you are also looking at your hand. The size of your hand on the retina, however, stays the same. The image of the ball and the image of your hand can intersect, for example, when you bring your hand in front of the ball. This does not mean that you have touched the ball – the visual input alone does not address this case. We need to combine it with our tactile inputs. The moment the ball touches your hand, both of the images on your retina intersect. At the very same instant, sensory inputs from your skin (touch) are also sent to your brain through the peripheral nervous system. This is a rather special instant when you have two different sensory inputs simultaneously reaching your brain.

The behavior before versus after the ball touches your hand are very different. For example, the touch input is not present before. Therefore, this special instant feels like a discrete event where the ball either touches your hand or not. There is no natural notion of continuity. Before the ball reached your hand, the visual inputs seem to have a natural notion of continuity. After the ball touched your hand, once again there is a natural notion of continuity with both visual and tactile inputs. The transition point between these two cases (when the ball just touches your hand) seems like a step change. Neurons from your skin either fire or do not fire accordingly. How then can we define a notion of abstract continuity in this case?

As mentioned in section 21.3, our brain needs to link multiple meta-fixed sets to create abstract continuity. For this, we need multiple variations of the same event. Imagine repeating the same experiment, namely, the ball touching your hand, multiple times. We observe different variations in the outcomes. These variations will be linked to the original event as nonempty predictive meta-fixed sets. For example, the possible set of variations are: (a) the ball touches at different regions on the hand, (b) you perform self-control actions to move your hand to touch the ball, (c) the ball hits your hand with different forces causing your hand to move in response, (d) the ball just barely touches your hand, (e) the ball misses your hand occasionally, (f) you hold the ball after it touches and so on. Note that each of these outcomes does not have low probabilities. They do occur several times when you play with the ball. Each of these repetitions helps form multiple neighboring meta-fixed sets.

Therefore, the seemingly special discrete event is actually related, in an abstract continuous way, to a number of other possibilities. The ball-touching event is no longer a discrete event. Rather, the step jump before versus after is internally linked to several meta-fixed sets corresponding to each of the above variations in a smooth or continuous manner. The next time the event starts to repeat, these memories from some of the above variations let us generate nonempty predictive meta-fixed sets. These predictions are linked to control actions that prepare you for the immediate future. For example, if the expected outcome in the immediate future is to catch the ball, the control actions are triggered ahead of time to make you extend your hand to reach the ball. Your abstract continuity has helped you prepare for the future tasks. Discrete events are, therefore, converted to abstract continuous events.

21.6 Doing well in uncertain situations

Generalizing the above example, we can say that abstract continuity helps us perform well in most unplanned situations. Figure 21.2 describes this process in an abstract sense. Consider an event E_1 that we repetitively practice over several days. As we practice event E_1, I have represented minor variations around this event as points inside a small circular neighborhood. In the previous section, I have described these variations for the ball-catching event (like, for example, the ball moving to the left, to the right by different amounts and so on). Other examples that have several variations when repeated are: when practicing to play piano (the various mistakes you make), when practicing to play tennis, when practicing mathematics (say, algebra), when

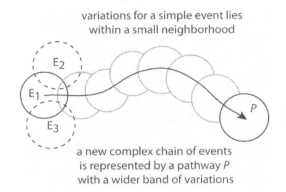

variations for a simple event lies
within a small neighborhood

E_2

E_1

E_3

P

a new complex chain of events
is represented by a pathway P
with a wider band of variations

Figure 21.2: Wider band of variations for a causal chain of events to help create abstract continuity. A given event E_1 memorized in an SPL system has several variations, which represents the mistakes you make while learning. Each event E_1, E_2 and E_3 though similar are learnt at different times. They form overlapping neighborhoods as shown here. When you are learning a new causal chain of events, shown as pathway P, these neighborhoods give rise to predictive meta-fixed sets. They make it easier to memorize the entire pathway P using the wider band of variations.

practicing to repair cars (like, say, a mechanic would do) and so on.

Each of these variations (i.e., the points inside the circle in Figure 21.2) act as predictive meta-fixed sets for the event E_1. Our brain forms memories for each of these points (variations). A given task contains a large number of such variations like E_1, E_2, E_3 and so on. Over a period of time, our brain creates a number of fixed sets for each such event and their variations as shown as additional circles in Figure 21.2. When you now try to perform a task that overlaps with several of these events, it becomes easy to create a path P joining a number of points within these circles (see Figure 21.2). Even though there are uncertainties when performing a given task, both the existence and the ability to find a good path becomes easy because of the nonempty predictive meta-fixed sets at each event E_1, E_2 and so on.

We now say that our brain has created abstract continuity for event E_1 as well as several variations of it. In general, the overlapping circular balls shown in Figure 21.2 is how we should visualize an abstract continuous path P. An abstract continuous path P should always be viewed as a 'wide band' spanned by these circular regions instead of a thin line corresponding to the actual path P itself. Once such a wide band is formed, abstract continuity helps us in performing well in uncertain situations.

21.7 Learning two languages

We used the features of a language as an example to define abstract continuity in section 21.1. Let us now look at abstract continuity when learning two languages. If you speak two languages at home, then as you talk to your child, you occasionally use words from both languages to describe a situation. Your child, who is just beginning to learn to speak, does not know if the words you used are from two different languages.

You may have used words from both languages in a single sentence. You probably used different synonyms at different times when describing the same situation. The synonyms can be from both languages as well. The emphasis on the words in different languages may be different. Each of these situations can be seen as different variations of the current event. Your child is simply associating sounds of words to objects he sees or objects he interacts with. These events do repeat a number of times under different situations.

Each word is memorized as fixed sets. They are also linked together as variations to each scenario and are associated to the physical object. These variations are analogous to the ones discussed with the previous examples. Here again, the variations and the links between the memories help in creating abstract continuous paths P, as shown in Figure 21.2. The abstract continuous paths in the case of language correspond to forming grammatically correct sentences.

A child can learn two languages easily as a result. An older adult, on the hand, finds it difficult to learn a new language because the fixed sets corresponding to the new language cannot be stored closer to the fixed sets of the first language and to the fixed sets for the representation of the physical object as seen through your eyes or sensed through touch. At a young age, these fixed sets are closer because of your limited set of memories and from the near-simultaneity of occurrence of words or events. For an adult, it is more difficult to link these widely distant regions to create abstract continuous paths for the new language. As a result, generating nonempty predictive meta-fixed sets spanned across these distant regions gets difficult when an adult hears a sentence from a new language. The net effect is that an adult finds it difficult to talk and understand a new language. Considerably more effort and practice is needed for an adult than for an infant.

21.8 Learning through mistakes

Abstract continuity captures our intuition that you learn through your mistakes as well. Some examples of this situation are when learning to play a game, to draw or paint, to play a musical instrument, to learn mathematics, to repair cars or even basic skills like learning to read, write, walk, learn or speak a language. In each of these examples, if you learn well within the first few attempts, you are not exposed to several common variations, which are critical for creating a wide band shown in Figure 21.2. Therefore, you will not have abstract continuity for these events.

You will not be able to generate predictive meta-fixed sets when you experience a new variation of the event. For example, when shooting baskets in basketball, it is initially better to make mistakes and correct yourself under common scenarios than to get perfect shots. Making minor mistakes builds abstract continuity, which is necessary to do well in new situations.

21.9 Learning by doing

As another example of abstract continuity, consider the problem of learning by doing a task. We feel that we understand a topic better when we actually perform the task rather than just read about it. For example, reading a subject like mathematics, an engineering field or the theory of music only produces a limited amount of variations. They are not rich enough to provide a wide range of predictions. Of course, this also depends on how much effort we put into thinking about examples, corner cases, failure modes and other types of deviations from the standard textbook details. Since we do not know how much to analyze for each of these cases, our analysis and, hence, our predictions will usually be quite shallow.

However, when we actually perform a task multiple times, we encounter a larger set of variations that forces us to think and do entirely unexpected and new things, most of which are difficult to expect otherwise. For example, when we are playing a piano or a guitar, several errors occur because of the complexity of the music and the coordination (or lack thereof) of our fingers. These are difficult to anticipate. Similarly, when reading a book on how to repair a car, the book only covers a few possibilities than if we actually start fixing the car. The amount of information that can be memorized as fixed sets each time we try to repair a car is not only diverse but spans across different skill sets. This breadth that comes through experience cannot be described in a manual. They give rise to a large number of predictive meta-fixed sets. As another example, reading and understanding a theorem in mathematics only outlines the 'correct' solution. The author does not highlight all the failed attempts to discover the theorem, different examples for which the theorem is true or false, the importance of the assumptions stated in the theorem, the other ways to alter the statement of the theorem and so on. It is only when we try to prove the theorem ourselves, try several examples and exercises do we store several predictive meta-fixed sets that help us relate the subtle connections implied by these theorems.

In general, our actions naturally guide us towards different directions and encourage us to analyze each of these different possibilities quite deeply. These simple to complex set of possibilities are difficult to guess just by thinking about them. As a result, our SPL network grows dense as we perform several tasks rather than just read about them. The abstract continuous paths are, similarly, much larger than when you just read the topics.

21.10 Learning and obeying rules

Memory, as fixed sets, can be thought of as storing and recollecting simple facts. For example, triggering fixed sets for a table, a chair, a ball, a balloon and other objects allows us to uniquely identify external objects. However, in addition to memorizing simple facts, we memorize rules as well. Rules are a series of steps that we consistently follow to perform a given task. When a child tries to add two small numbers, for example, he counts using his fingers in a specific order. He remembers this rule because he applies the same procedure for several sets of numbers.

Generalizing this, we can say that planning of any event involves obeying a set of rules. Memorizing a large number of such rules is equally common in our everyday life (compared to memorizing facts). Rules have the property that they are applicable to a wide variety of objects, events or facts. For example, addition rule is applicable to all numbers. The same applies to most mathematical rules like multiplication, subtraction and more generic algorithms. Grammatical rules are other examples we have discussed in greater detail in this chapter. Most disciplines like music, art, language, engineering and others have several rules that people memorize and follow.

How do we represent rules and the ability to memorize them within the SPL framework? The answer is by using abstract continuous paths. We have already seen this quite extensively with memorizing grammatical rules in this chapter. Rules indirectly refer to predictions. This is similar to grammatical rules – as you complete the sentence, you should have nonempty predictive meta-fixed sets. Your memory of addition rule works by predicting what to do next as you start applying the steps (like when a child starts counting with his fingers). If you have empty predictions along the way, you have not successfully memorized the steps. If you are solving a mathematical problem by applying a series of steps, your predictive meta-fixed sets that you have built through practice drive you to follow the steps correctly. Your ability to finish the problem correctly depends on how well your predictive meta-fixed sets guide you. Another way to state this is that you have memorized the rules correctly.

The process is the same with any rules and with any discipline. Your ability to play and improvise music depends on your predictive meta-fixed sets that you have built through years of practice. The rules in mathematics, algorithms in computer science, patterns in medical diagnosis, music, art and others are all represented as abstract continuous paths in an SPL framework. I will discuss more about this later in chapter 22 when I show how we attain a feeling of knowing of an event using abstract continuity.

21.11 Action-based abstract continuity

Most of the examples we have discussed so far dealt with abstract continuity of sensory fixed sets. Let me now discuss what abstract continuity means when referring to control actions within motor control regions of our brain. The predictive meta-fixed sets for this case would be action-based fixed sets. For example, if you are moving your hand to the left, the predictive meta-fixed sets at that instant are the possible future actions that you can do. Some of these action fixed sets are to continue moving left, change direction to move to the right, up, down or any other direction around that point. As we practice these tasks, we create links between, say, the left motion fixed set to each of the above control meta-fixed sets. Similarly, we have a set of locally linked control meta-fixed sets for moving our head, eyes and other parts of our body. On the other hand, you do not have local links between the fixed sets for your eye movements to the corresponding fixed sets to, say, your knee movement. This implies that turning your eye will not trigger knee movement as a prediction.

Therefore, when we are performing an action-based task, like moving our hand, we have nonempty predictive meta-fixed sets at every instant just as we did for a sensing-based task. This lets us plan our future actions in a meaningful way. We create most of these action-based abstract continuous pathways in our early childhood when we learn to, say, roll over, sit, crawl, stand, walk, run, jump, hop and so on. As children, we try a large set of variations for each of these tasks for several years.

Predictions and abstract continuity in the case of actions allow us to 'know' that you are performing an action. You do not need to move your hand to know what, where, when and how you are going to move. You also do not need to finish the entire action yet. You are capable of combining abstract continuity from actions along with abstract continuity of sensory information or even abstract thoughts to imagine and understand complex scenarios.

The action-based abstract continuous paths are representative of our common everyday tasks. If you learnt gymnastics in your childhood, you have additional set of abstract continuous paths based on these experiences. Similarly, if you learn to play piano, guitar, tennis, dancing or other activities involving physical movements, you create additional unique action-based abstract continuous paths from these experiences as well. These are similar to non-action-based abstract continuous paths that you build, say, from learning mathematics, a new language and others.

Each of the examples discussed in the previous sections should give us an idea on how to analyze abstract continuity in a new situation. We will see in the next chapter 22 that you will not attain a *feeling* of knowing if you do not create abstract continuity via the variations around the desired outcome. When learning a new subject, if you do not have abstract continuity for several topics, you will end up not getting a feeling of knowing for the entire subject itself. This is true with each of the examples mentioned earlier. Abstract continuity is the reason why it is important to practice many different types of examples, perform the tasks and make mistakes too.

22
Feeling of Knowing

In section 21.1 of chapter 21, we discussed two sentences – one that was grammatically correct and the other that was grammatically incorrect. We were able to understand the first sentence but not the second one. We used this fact to define continuity for abstractions. However, we did not address the question of how we understand these sentences. In this chapter, I will discuss what *knowing* means and the conditions under which we can know anything, including these sentences. This is useful for consciousness. In fact, according to Theorem 12.1, consciousness is equivalent to an ability to *know* at least one truth. We have already discussed truths in detail and have identified an unique dynamical structure, namely, a self-sustaining membrane of meta-fixed sets. In this chapter, I will focus on the other important aspect of Theorem 12.1 i.e., knowing. I will show that the existence and the excitation of abstract continuous paths on the dynamical membrane for a given event is equivalent to knowing that event.

Let me note an important point about these necessary and sufficient conditions, in general and for attaining a feeling of knowing, in particular. When we discussed a similar set of necessary and sufficient conditions for consciousness in Theorem 12.1, we did not address 'why' we become conscious when these conditions are satisfied. We simply said that we do. In a similar sense, we will not address 'why' we get a feeling of knowing. Rather, we will just say that if the necessary and sufficient conditions are satisfied, they give rise to a feeling of knowing and vice versa.

This is not unusual with most scientific topics. For example, when you have mass, you have gravity. Why you have gravity and not some other type of force is no more clarified than why you have a feeling of knowing. Yet, we can understand a great deal about the structure of gravity in terms of the curvature of space-time and beyond. Similarly, we will be able to understand the structure of feeling of knowing in terms of abstract continuity and the dynamical membrane even though we do not address the question of 'why'.

Indeed, for the problem of consciousness, we have already seen a rather unique structure, namely, the self-sustaining membrane of meta-fixed sets. We never knew this structure before. We discovered it just by exploring the necessary and sufficient conditions of Theorem 12.1 even though we never addressed the 'why' question. In a similar sense, we have identified abstract continuity as a rather unique structure just by exploring the problem of how we understand grammatically correct sentences but not grammatically incorrect sentences.

In chapter 44, I will revisit this topic of 'why' we become conscious or 'why' we attain a feeling of knowing. This problem of 'why' is related to the problem of understanding *qualia*, the subjective quality of feeling (see section 10.9). The core problem to address is our ability to compare *qualia* across two or more conscious systems. In chapter 44, I will discuss this and highlight some fundamental limitations.

22.1 Knowing = abstract continuity on the membrane

Consider the following three situations: (a) meeting a person you do not know, like a stranger, (b) meeting a person you know very well, like your mother and (c) meeting a person you vaguely recognize but are struggling to recollect the context including his name, like a person you met during a birthday party a year ago but you cannot quite remember. Let us see if these three situations are similar to the two sentences discussed in chapter 21 with respect to the relationship between the problem of knowing and abstract continuity.

What happens in our brain during these situations? We associate a feeling of knowing only to case (b). In case (c), as you begin to recollect more details about this person, you slowly start feeling a sense of familiarity. Eventually, when you do remember his name, you suddenly experience a feeling of knowing. This also triggers a sudden explosion of memories of other related facts about the person and the context when you met him. During this initial explosion, the feeling of knowing is particularly strong. However, over time, it settles down to a normal feeling of knowing similar to case (b) of a familiar person. For case (a), you only know the basic features like, say, that he is a human being with a certain set of physical characteristics like the gender, height and others. Most questions related to knowing him personally yield empty predictive meta-fixed sets.

In order to compare this example with sentences in a language, consider the following three types of sentences respectively: (1) sentences that we cannot understand at all, like a grammatically incorrect sentence (ii) of section 21.1, (2) sentences that we clearly understand like a grammatically correct sentence (i) of section 21.1, and (3) sentences that we partially understand, like a poorly structured sentence formulated by a novice speaker of a language.

With cases (a) and (1), we only have basic meta-fixed sets like physical characteristics of a human being. Similarly, we only know individual letters or words, but not the sentence. With cases (b) and (2), almost any reasonable question you ask about the person like the context when you met him and about any other interaction you had with him have clear answers. These are the nonempty predictive meta-fixed sets needed for abstract continuity. They are analogous to predictive meta-fixed sets for a grammatically correct sentence. With case (c) and (3), some questions about the context do have nonempty predictive meta-fixed sets, but others do not. You start to fill in the empty holes by trying hard to recollect – was it at a birthday party, at a friend's place, at a school and so on? This process is analogous to trying to rearrange a partially grammatically correct sentence to understand it better, by asking a nonnative speaker of the language if he meant this versus that. You are trying to discover an abstract continuous path for both remembering the person and for understanding the sentence. Eventually, you will get a similar feeling of knowing.

There are two interesting questions to answer with the example discussed here. Firstly, for case (c), how do you know that you knew the person once, but not now? Knowing a person is not a single feeling. You can have a feeling of knowing of the meta-information i.e., 'about' the problem, namely, that you knew him before, but not for the actual information, namely, the person itself. I will discuss this in section 43.6. Secondly, what gives rise to a feeling of knowing when all you did was recollect the context, which sometimes is as simple as knowing his name? This is the primary question we will focus on answering in this chapter (see Theorem 22.1).

Notice that among these three situations, attaining a feeling of knowing is most difficult with a complete stranger or with a scrambled sentence. Simple rearrangement is not enough to know what the sentence means. Similarly, you need considerable effort, memories and experiences to get a sense of familiarity with a stranger. The process is, however, similar – you are trying to build a set of nonempty predictive meta-fixed sets for a broad set of questions that can give rise to abstract continuity. The set of predictive meta-fixed sets possible by rearranging a sentence are far less compared to those needed to become familiar with a human being.

From the above discussion, it became clear that not only are these two examples analogous, but the underlying processes used to explain them are similar, namely, the feeling of knowing is equivalent to identifying the specific scenarios that yield abstract continuity. I will continue analyzing this equivalence through the rest of this chapter, but let me formally state it as theorem that generalizes to all cases.

Theorem 22.1: Knowing an abstraction like, say, an event, object or a sentence is equivalent to the existence of a self-sustaining membrane of meta-fixed sets and the triggering of an abstract continuous path from the meta-fixed set representation of the abstraction to the dynamical membrane.

I will informally represent this theorem as 'knowing = membrane + abstract continuity'. We can generalize the above theorem and say that any given feeling is equivalent to the existence of a membrane and abstract continuous paths. I will discuss two important examples of feelings later – realizations in chapter 48 and emotions in chapter 49. It is important to keep in mind that the above theorem only gives necessary and sufficient conditions for the feeling of knowing. It does not say why you have feelings and what they are in a physical sense. In section 44.2, I will introduce the notion of 'dual equality', which lets us compare feelings of, say, two different people. It is not the same as the 'true equality' of feelings that we would want to accept. However, we will see that dual equality is the best we can do to compare feelings of two different complex systems anyway.

It is interesting to note that we do use this theorem either knowingly or unknowingly in our everyday life. For example, we use the above theorem either to convey our ideas effectively to others or to understand other people's ideas. In the former case, we do this by guessing the possible sources of abstract discontinuities for our audience. We try to eliminate them by filling in pertinent details, which will act as nonempty predictions for our audience. In the latter case, we pose a correct set of questions if we perceive abstract discontinuities when trying to understand another persons' point of view. These will fill the empty predictions with some additional predictive meta-fixed sets. In this sense, we can say that the above theorem has turned our intuitive approach into a definite one in a way that is broadly applicable to all cases of knowing.

As another example, consider looking at a picture when we are conscious versus when we are not (like when the general anesthesia is wearing off or when severely drunk). Most of the times when we are conscious, we instantaneously attain a feeling of knowing of the picture. However, there are times when we are distracted with other thoughts that we do not 'know' what we were looking at. Only when we turn our attention back at what we were looking previously, the feeling of knowing is attained. To relate this example to Theorem 22.1, I say that, firstly, our distracted conscious thoughts make up our initial dynamical membrane. As we look at the image, light input through our eyes trigger a collection of meta-fixed sets. If the intersection between these fixed sets and the dynamical membrane is empty, we do not attain a feeling of knowing. If, however, we turn our attention back and actively think about the picture, I claim that we are expanding the neural firing regions of the dynamical membrane to overlap and intersect the corresponding fixed sets initiated from the retinal images. If and only when this happens, I will show that we attain a feeling of knowing.

The theorem also highlights the importance of physical dynamics here. The static structure corresponding to the abstract continuous paths are stored within our brain network (like on a hard disk in a computer). However, this is not sufficient. They should be dynamically excited through the self-sustaining membrane. For example, if you are focused on a different topic, you may not hear or understand what the person is saying. They do not dynamically trigger the abstract continuous paths even though they do exist within our brain's static network structure. One trick we use to avoid this is to excite the internal dynamics by repeating the sentence or thoughts silently within our mind. Even though we may be distracted, by repeating the sentences within our mind, we force the relevant regions to be excited so we can trigger the predictive meta-fixed sets as well.

Let us now better understand the relationship between knowing and abstract continuity. We need to prove both directions of the equivalence of Theorem 22.1. For example, we should think of situations where we 'knew' something (like understanding a sentence, recognizing a person, an image or a sound and others), but did not have abstract continuity i.e., we had an empty set of predictions after we already had a membrane. Conversely, we should think of situations where we did have abstract continuity for an event (i.e., had nonempty predictions) when already having a membrane, but did not attain a feeling of knowing. In order to prove Theorem 22.1, I want to show that both these cases are not possible.

22.2 Knowing $\implies$ membrane + abstract continuity

Let me assume that we are observing an event and it produced a feeling of knowing. We want to show that there exists a dynamical membrane and an abstract continuous path for this event as well. Firstly, it is clear that you should have a membrane of meta-fixed sets. For otherwise, you are not even conscious. Recall that the existence of a membrane is necessary for consciousness. If you are not conscious, you cannot know any truths, including this very event, from Theorem 12.1. This violates the assumption that you already know the event.

Next, we want to show that we should have abstract continuous paths for this event. I will show this by contradiction. For this, I will assume that there is no abstract continuity for this event i.e., there are no nonempty predictive meta-fixed sets at every instant along any linear sequence. Then, to get a contradiction, I will show that such a state is not possible.

One of the first results to show using these assumptions is that you do not have any memory of the current observed event. The intuition behind this result is that memory is never isolated and always creates abstract continuous paths. For example, a few related memories stored for any event are: (a) when you observed the event, (b) what you were observing before you started watching this event, (c) did the event make you happy or sad, (d) did you do any task while observing the event, (e) were there other people watching the event with you and so on. We form these memories without any significant effort. The self-exciting nature of the membrane automatically lets you link the currently observed event with all of these facts naturally (which are answers to these questions). When you observe the event once again, you will have definite answers to several of these questions. This gives rise to valid nonempty predictions, violating our assumption.

Now, these questions are asked automatically because of the self-excitation of your membrane. The nonempty predictions give rise to abstract continuity. However, we just assumed that there is no abstract continuity for this event. Therefore, we can conclude that this event behaves like a completely new event with no past memories even though you may have experienced this event before. This conclusion is valid even if we only consider events like internal thoughts or dreams. Even these artificial memories have a set of nonempty predictive meta-fixed sets and, hence, abstract continuity, which violates our assumption.

Hence, we conclude that the system is observing a new event with no past memories. You may feel that this is a contradiction in itself – how can you feel that you 'know' an event when you have never seen it before? However, this is not a contradiction yet. For example, you can imagine reading a sentence you never saw or heard before in your life, like say, some of the sentences in this book. Yet, you understand them quite well. This is true with several other events in our day-to-day life as well.

To get a contradiction, we should further breakdown the original event into subcomponents (or words) along linear sequences (like a sentence). From the previous section, recall that we get a linear sequence directly from the timeline, like with sounds and sentences, or indirectly, like with scanning 2-D or 3-D object with our eyes. In general, we can do this in multiple different ways quite naturally. For example, an image has edges connected to form boundaries, a sound or a sentence has individual words as subcomponents and so on. If you have predictive meta-fixed sets at each subcomponent, you have abstract continuity, by definition, which will violate our assumption. This case is analogous to sentence (i) of section 21.1. Therefore, to satisfy the above result, we can only have one of two possibilities: (a) some of the subcomponents are new ones with no past memories or (b) all of the subcomponents have past memories but the linear sequence itself is entirely new.

The case (b) is analogous to sentence (ii) of section 21.1. For such linear sequences, to say that you 'know' the sequence is illogical. Therefore, this case is not possible.

For case (a), the only way we can avoid the new subcomponents of the sequence is if they are not critical for the feeling of knowing. For example, a 3 year old understands a sentence even if he does not understand every word in the sentence. For this special case, if we ignore these new subcomponents, the remaining subsequence of subcomponents will have valid memories. As a result, they have abstract continuity, by definition. This violates our assumption. Therefore, we cannot ignore these new subcomponents with no memories. For example, just because a 3 year old does not understand several words like prepositions, conjunctions and others, you cannot create a sentence without those words and expect him to understand the shortened sentence.

You can now continue breaking down each new subcomponent into smaller subcomponents until there is no value in doing so. We can apply the above analysis to end up, finally, in a situation where there is a sequence of new and old subcomponents with the new subcomponents being critical enough that we cannot ignore them. Let us regroup the subcomponents into categories such that each category includes at least one new critical subcomponent. Now, we cannot know each of these categories because it contains at least one unknown critical subcomponent. This eventually leads to a situation where we have a linear sequence of categories each of which is unknown while the entire sequence itself is known. This is also a contradiction. As a result, case (a) is also not possible.

Next, we can analyze each of the multiple such linear sequences of subcomponents in a similar way leading to the same contradiction. We, therefore, conclude that knowing without abstract continuity leads to either contradictions or illogical situations. This shows the result in one direction, i.e., that knowing implies the existence of a membrane and abstract continuous path. Let us now show the converse.

22.3 Membrane + abstract continuity ⟹ knowing

Let us assume that the system already has a membrane and has abstract continuity for an event. I want to show that this gives rise to a feeling of knowing of this event. Let me first provide an intuition for this statement before addressing details. The main difficulty with this result is to show that we achieve a feeling of knowing for *every* event we ever experience whenever we have a dynamical membrane and abstract continuous paths for the event.

1. The first step is to show the existence of, at least, a large class of events for which the above conditions give rise to a feeling of knowing. I have already shown this with grammatically correct sentences in a language in chapter 21.

2. Our next step is to generalize this to other classes of events as well. For this, I will consider two important family of events – dynamical and abstract events. Examples of these are: experiments, illustrative examples in an abstract

theory, historical facts, mathematical facts as well as everyday discussions and arguments. I will show how to decompose events within these families into primitive parts, which I call as components and processes (which are linearly connected sequence of events). This decomposition is analogous to words and sentences in a language respectively. In the case of mathematical facts, the components and processes correspond to primitive abstractions and rules respectively. Therefore, we can now generalize the result shown for grammatical sentences, namely that we attain a feeling of knowing under the stated conditions, to all of these cases as well.

3. The next step is to generalize the above decomposition approach of components/processes or primitive abstractions/rules to all cases. The intuition here is that all events fall into the following types: abstract, dynamical or a combination of both. However, each of these types of events already allowed for such a decomposition. Now, we show that the primitive components become part of the membrane while the linear sequence of events become part of the abstract continuous paths. Therefore, together they give rise to a feeling of knowing for all cases.

4. Finally, we need to address the feeling of knowing for the primitive components and processes directly. Until now, I had expressed the feeling of knowing relative to these primitive features. For the last part, I will show that the true representation of primitive components are fixed sets that are part of the dynamical membrane while the representation for processes and linear sequence of events are abstract continuous paths. The membrane combined with abstract continuity, therefore, represents the totality of your experiences, excluding nothing else. Together, they covers all situations.

If we combine all of the above statements, it implies that the existence of the membrane together with abstract continuous paths gives rise to a feeling of knowing for all events. Let me now discuss each of the above statements in more detail.

For step (1) above, we have already looked at a couple of examples that addresses how we get a feeling of knowing – understanding a grammatically correct sentence and recognizing a person whom we have seen in the past (see chapter 21). With sentences in a language, knowing the simplest words is equivalent to representing them as fixed sets as well as creating abstract continuous paths with real physical objects and actions. These 'primitive' words are recursively linked together to form a complex network, which becomes a part of the dynamical membrane, while knowing the sentence composed of these words requires abstract continuity.

How can we now generalize these examples to cover every scenario that occurs in our everyday life? In the previous section, we have already seen the similarity with other examples like jigsaw images, odd arrangements of objects like tables and with recollection of events.

Let us classify all events that occur around us every day into two types – dynamical events and abstract events. Dynamical events involve real tangible objects. For example, objects moving around, physical systems obeying the laws of nature, people performing tasks while interacting with these tangible objects, people communicating and cooperating on tasks and so on. Abstract events, on the other hand, typically correspond to the description of dynamical events. For example, you hear different statements, sentences and directions (to follow) from other people. You engage in discussions, arguments and debates with other people on a variety of topics.

In both of these families of events, several of them are completely new events. Yet, we feel that we can understand what is happening around us while we observe or perform these tasks. How is this feeling possible when we have never experienced these events before?

For abstract events, several of the discussions, say, are on new topics. You present your own opinion as you hear other peoples' opinions. In spite of these discussions being entirely new, you actually understand the arguments. How do you get the feeling that you understand all types of opinions whether you agree or not? Remember that there are other examples of arguments which make you feel that something is missing in a persons' reasoning even though he has used grammatically correct sentences. What is the cause of differences between these two cases? The sentences themselves do not capture this difference. For example, the degree of understanding of this book by two different people is different even though the set of sentences read are the same. Clearly, the brain structures formed from past experiences and knowledge of relevant prerequisites are different in both people. It seems like these structures are fundamentally important. Yet, for simpler abstract topics, it is indeed possible for two people to attain a similar state of knowing through similar training.

One point to keep in mind is that we should not consider the feeling of knowing to be the same either for different events or at different times for the same event. An argument may appear convincing now, but can change later as you experience new events.

For all events, i.e., dynamical, abstract or a combination of both, it is possible to start from a state when you have no feeling of knowing and transition to a state when you do get the feeling. The generic approach we can take here is to familiarize ourselves with various aspects of the event, through examples and through repeated practice (cf. how you learn and understand, say, mathematics or music). These steps will give rise to abstract continuity, which I want to show is sufficient for attaining a feeling of knowing, when we combine it with the dynamical membrane. We should note that for several events, the transitions that give rise to a feeling of knowing have already happened a while back (say, through normal everyday experiences or through formal education during our childhood). The new events will depend on these past events to make the problem easier.

Given this, let us work through examples of these two families of events and show that they are analogous to sentences in a language for which we have already established the sufficient condition for the feeling of knowing.

For dynamical events, consider building physical dynamical systems. During our childhood, we learn to create abstract continuity for the basic operations that make up any given machine or experiment. These are, say, the set of dynamical operations like assembling parts, gluing, bending, stacking, breaking, moving and applying forces as well as familiarity with a set of materials like plastics, paper, wood, metals, bolts, resistors, capacitors, chemicals, cement, sand and others. Therefore, by using these primitives, we have a feeling of knowing for the construction of other machines involving similar components and operations (cf. do-it-yourself kits). In fact, even reading the setup of an experiment is sufficient to gain abstract continuity and a feeling of knowing without having to build it.

For abstract events like learning and understanding a topic from, say, a research paper, generating nonempty predictive meta-fixed sets is critical to have no gaps in the reasoning. There are several ways to do this. For example, one way would be to derive some of the results discussed in the paper ourselves (after reading once, say). This would immediately highlight the places where we do not have abstract continuity. These are typically the places where we are stuck and are unable to complete the derivation ourselves. We will have to look back at the research paper to see how the author has proceeded. Another way to do this is by repeating the experiment or implementing the algorithm, described in the paper, ourselves. This will also force us identify gaps in our understanding. The advantage with performing experiments is that there is no contamination of results from our physical observation (as we perform the experiment) to our consciousness. We utilize the abstract continuity of basic events (like assembling parts, gluing, bending and others) and abstract continuity of basic materials (like plastics, paper, wood and others) to our advantage as we perform and observe the experiment. In fact, even reading the setup of an experiment is sufficient to gain abstract continuity. There is no need to actually build it. We already have nonempty predictive meta-fixed sets if we simply imagine following the instructions in our head or look at the schematics.

Mathematical statements are another example of abstract events for which we use a set of examples (like integers, functions and matrices) and exercises at the end of a chapter for building abstract continuity. They play the equivalent role of experiments in dynamical systems. In the case of discussions, debates and arguments, we use analogies with topics for which we already have abstract continuity, relevant historical facts and cases (as doctors and psychiatrists use) and precedents (as lawyers use).

We can now generalize this notion of abstract continuity to all dynamical events. We can imagine every dynamical event to be composed of two parts: components and processes. The components refer to real physical objects of our universe. We can see, hear and touch them. Their existence cannot be denied because we have multiple sense organs that can detect them consistently and reliably beyond any doubt. The question of their existence is no different from the question of our own existence. As mentioned earlier, the components are reused in multiple designs.

The processes are sequence of operations that we apply on the components. These processes obey the laws of nature. We apply the processes in a specific order either serially or in parallel to construct the final design. As with the components, these processes are reused in other designs.

Next, we can imagine breaking down the components and the processes into smaller and more primitive subcomponents and sub-processes. In some cases the primitive components and processes become obvious only when arranged in a haphazard way and then rearranged correctly later (cf. the jigsaw-like puzzles).When we encounter some of the same components or processes in a different situation, we create new links specific to this scenario. As the set of scenarios vary over a period of time, a given component or process will end up having a dense neighborhood of meta-fixed sets. This gives rise to nonempty predictive meta-fixed sets or abstract continuous paths for a new design that uses previously familiar components or processes.

The approach of breaking down into components and processes and arranging them in a specific order is identical to how we did with a grammatically correct sentence. The components are equivalent to words in a sentence that are reused in multiple different sentences. Components that cannot physically exist in the real world are analogous to gibberish words (like 'oyu' instead of 'you'), which we clearly eliminate. The processes are equivalent to grammatical rules of a language. In dynamical events, we link the processes by ensuring that they obey the laws of nature. Analogously, grammatically incorrect sentences are not valid.

Now, for a grammatically correct sentence, we have already shown that abstract continuity is sufficient to get a feeling of knowing. From the above analogy, a dynamical event, like an experiment or a design, is similar to a grammatically correct sentence. The abstract continuity in both cases is similar. Therefore, by generalization, we conclude that abstract continuity gives rise to a feeling of knowing for such dynamical events. We can now extend this analysis to all types of dynamical events like recollecting people, personal experiences, recognizing objects, dynamics and others. We can decompose all of these events into components and processes in a similar way.

In fact, such decomposition is equally applicable to abstract events like with abstract mathematics. In fact, mathematics itself is a language similar to natural language, but one, which is more structured. The rules of mathematics are like grammatical rules in a language. Abstract continuity as described with sentences in section 21.1 is more directly applicable to mathematics when we use mathematical axioms instead. Therefore, the above sufficient condition for a feeling of knowing applies here as well. This analysis applies to other abstract events like debates, discussions and arguments mentioned earlier as well. Either each abstraction is a primitive one, like an axiom, or we can define it in terms of other primitive abstractions. For example, all words in a language (including mathematics) are defined in terms of other primitive words. The primitive abstractions themselves need to be self-evident by associating them with dynamical events. We can then express more involved discussions and arguments, ultimately, in terms of these primitive abstractions using a collection of self-evident rules. The primitive abstractions correspond to components and the self-evident rules correspond to processes. Therefore, the generalization is now applicable to these cases as well.

Since the desired result is already true for grammatically correct sentences, I say that the result remains true, via generalization, for both dynamical and abstract events. As all events occurring around us every day can be classified as dynamical events, abstract events or a combination of both, the above generalization now applies to all events. Therefore, in summary, we can say that abstract continuity is sufficient for the feeling of knowing for (a) all grammatically correct sentences in a language (including mathematics), (b) all dynamical events and (c) all abstract events.

Now the last step is to see how to represent the primitive components and processes themselves. If we look at a language, all words are defined in terms of other words and a set of primitive words. The primitive words themselves are associated directly with dynamical objects and events that most conscious beings agree as self-evident.

For example, a straight line is self-evident because we can draw it on a piece of paper to verify its definition. The sound for the letter 'A' (or any word) is self-evident because we can generate a sound and hear it back to confirm that it is the same sound. A shape of the letter 'A' (or any word) is self-evident because we can write the shape on a piece of paper and look at it to confirm again. Similarly, colors, edges, shapes, tastes, odors, words used to describe emotions and definitions or axioms in mathematics are basic abstractions chosen to be self-evident. They are expressed directly in terms of tangible objects and physical dynamics that our sensory organs can detect reliably, repeatedly and at will. Additionally, you have self-evident rules that describe how to combine these primitive abstractions. These rules are obvious in mathematics (like axiomatic principles) and science (like the laws of nature). However, with social discussions, the rules (or laws) are chosen, say, by the society as a whole. These rules do change over time just as grammatical rules change over time.

What makes a primitive component or process so obvious that you simply take it for granted? The primitive components are like colors, edges, angles, sounds or the basic sensory inputs. The primitive processes are like lifting your hand, turning your eyes or head or the basic control actions. These are the most repeatable phenomena in our everyday life. Each of these primitive components is represented and stored in our brain as fixed sets.

Knowing a primitive feature for the first time is equivalent to knowing a feature that you have never seen before. For example, if you are learning a new language like Mandarin and you come across a new character, what should you do to memorize it and recognize it later? Or, if you are learning a new theory like general theory of relativity and you come across the term 'general covariance', what will make you feel that you know the term well? Or, as a toddler, when you are learning English, how do you recognize the letters of the alphabet? In most cases, we use abstract continuity to link with other existing facts (see the previous examples).

The key point to note is that we memorize the primitive components as fixed sets (like, say, for Mandarin or English characters). Fixed sets are those looped neural-firing patterns formed in our brain that are triggered even under several dynamical transformations. Fixed sets are used to *define* new terms and concepts. Therefore, the primitive components and the processes are stored in our brain as fixed sets. They are networked together based on simultaneity of events, creating abstract continuous paths. These paths become the primitive processes. This network evolves to become a membrane of meta-fixed sets (see chapter 19). Once our brain creates a threshold membrane, it expresses every other event and newly learnt components or processes in terms of these primitive components and processes in an abstract continuous way. This happens as the terms used in the new definitions themselves are memorized as meta-fixed sets. In fact, in section 23.1, I will show that the membrane of meta-fixed sets itself is constructed through early childhood as a network of abstract continuous paths.

The self-sustaining membrane presents now a view of the whole entity (like a 'picture') instead of just a collection of seemingly disjoint individual subcomponents ('thousand words' i.e., the expression that 'a picture is worth more than a thousand words'). Together with abstract continuity, we see that we attain a feeling of knowing. This completes the proof.

23

Consciousness = Membrane + Abstract Continuity

In the previous chapter, we have seen one of the most important results for consciousness, namely, that a feeling of knowing is equivalent to abstract continuity on a membrane of meta-fixed sets. From Theorem 22.1, this implies that the consciousness is equivalent to having abstract continuity on a membrane of meta-fixed sets. In this chapter, I want to show that we can view the membrane itself as a network of abstract continuous paths. This implies that each of our daily experiences already create abstract continuous paths whether we are conscious or not. As the abstract continuous paths are stored in our brain, they begin to self-intersect forming complex networks. A network of abstract continuous paths reaching a threshold state that includes space, time and internal body inputs, is an important resulting structure. We identify this threshold structure as the dynamical membrane of meta-fixed sets. Therefore, this is an alternate view of the dynamical membrane. The first view was to see the membrane as a threshold network of meta-fixed sets and the second new view is to see it as a threshold network of abstract continuous paths.

23.1 Membrane = Abstract continuous path network

In chapter 7, we described the memory model simply in terms of linked loops and fixed sets that obey similarity, dissimilarity and connectedness conditions. As we store several memories, we will have paths of linked loops that intersect to form a complex network. Figure 7.12 and others showed such networks as looped paths with turning points at critical junctions. After introducing the notion of fixed and meta-fixed sets, we have seen that there is an equivalent and, actually, a more refined view to the network of Figure 7.12. In chapter 19, we called this as the self-sustaining dynamical membrane of meta-fixed sets.

As we started looking at this structure of the membrane more closely, we realized that we could represent the notion of predictions within the membrane (see Figure 20.1 of section 20.5). Using predictions, we were able to define abstract continuity. The notion of abstract continuity provided a further refinement to our view of the membrane of meta-fixed sets. In this section, I will show that we can view the membrane as a network of abstract continuous paths that intersect one another rather than just viewed as a network of meta-fixed sets. When a threshold set of meta-fixed sets that include space, time and internal body inputs interlink as abstract continuous paths, we create the same dynamical membrane. Let me now state the main result of this section as the following theorem.

Theorem 23.1: The self-sustaining membrane of meta-fixed sets is equivalent to a dynamical network of abstract continuous paths.

To prove this theorem, note that in section 19.4 we have seen how the membrane is formed for the first time using meta-fixed sets as the primitive elements. Let me now discuss how the same membrane is created for the first time using abstract continuous paths as the primitive elements. An infant is born with a primitive brain and without a self-sustaining membrane. He uses fundamental pattern of complex SPL systems (chapter 16), common loop patterns (chapter 18) and others to create meta-fixed sets and link them together. The links are strengthened through repetition and near-simultaneity of occurrence.

As the links are stored in the brain as stable parallel loops, they give rise to predictions. A given fixed set is triggered under a large number of scenarios (like, say, playing with the same ball at different locations, times and situations). Therefore, the same fixed set is linked to different fixed sets using near-simultaneity, with each one corresponding to a new scenario. Therefore, when you trigger this fixed set for a new scenario, you will trigger all of these neighboring fixed sets as well. These are precisely your predictions based on past memories. Even though the set of memories are small for an infant, this gives rise to abstract continuity (i.e., nonempty predictive meta-fixed sets) within a limited context. In fact, this is how a child transitions from a state of no abstract continuity for a given event to a state with abstract continuity. His abstract continuity will appear to trigger responses that are in a sense mechanical, but with sophisticated predictive abilities. As a parent, we are sometimes surprised when a child makes several such nontrivial predictions.

Yet, there are discontinuities with predictions because of a lack of rich set of memories. In other words, an infant does not have abstract continuity at every instant. They are present at brief intervals and at specific locations (like in his own bedroom and other parts of his own house). He loses abstract continuity if he is in a new place, partly because he does not have a sufficient number of meta-fixed sets memorized. Even though he seems to be looking at the world around himself in a new place, he does not have either a fully-formed membrane or abstract continuity (i.e., he has so few predictions).

As the interconnections grow to include a core set (space, time and internal body meta-fixed sets), he will eventually have a self-sustaining set of these abstractly continuous interconnected meta-fixed sets. Until then, his brain is actively involved in creating meta-fixed sets (like colors, edges, shapes, sounds and tastes), linking them based on near-simultaneity of occurrence (like common arrangement of objects in a room, common patterns of sounds for your native language, common letters and words) and performing actions using the fundamental pattern and other loop patterns. These activities help link all the meta-fixed sets along abstract continuous paths.

His brain slowly partitions into a collection of almost disjoint, but abstractly continuous regions with dense local connectivity. Each locally connected region is a small network of abstractly continuous paths. These locally connected regions are, typically, arranged spatially far from each other with sparse links between them. Each abstractly continuous path in a given locally connected region corresponds to an event he has experienced at some point of his life. For example, he has some language-specific regions and other regions for images, sounds, tastes, odors, touch, feelings and actions. These regions start as sparsely interconnected regions. However, as he experiences more events that actively use multiple disconnected regions (like, say, when learning to play a sport like basketball), the interconnections between them start to grow and become denser. It is as if these disconnected regions are beginning to coalesce to form larger and well-connected regions. This coalescing happens over a period of several months.

While this coalescing is occurring in different parts of his brain, we, as parents, observe the infant exhibiting predictive abilities like with the examples highlighted earlier. A larger coalesced region corresponds to a larger network of abstractly continuous paths. However, both the smaller sparsely connected networks and the larger coalesced networks cannot self-sustain the neuron excitations until a certain threshold state is reached. The network corresponding to the threshold size that can self-sustain is what we have termed as the dynamical membrane of meta-fixed sets. This threshold state occurs when his brain network has coalesced enough to include space, time and internal body meta-fixed sets.

There are two parts to this threshold network: (a) a static physical network of interconnections between the neurons and (b) a dynamical self-excitation of a sufficiently large region of the brain. If either state (the static or the dynamic network) does not exist, I say that this threshold network is broken. Therefore, we can say that the membrane of meta-fixed sets is equivalent to the threshold network of abstractly continuous paths.

When the threshold state is reached, the infant will have a self-sustaining membrane that makes his entire brain act as a whole entity. The threshold network is now densely interlinked that viewing the brain as a sum of its parts is no longer correct. His brain senses and controls his whole body, not just the hands, legs and other parts individually. His brain predicts the relative locations of his entire body both in space and in time while performing or observing an event. The abstract continuity for all parts of his body (i.e., nonempty predictive body meta-fixed sets) ensures that it no longer makes decisions as union of parts of his body, but as a whole instead.

A system acting as a whole is truly a feature of complex system based on stable parallel looped architecture. This includes all living organisms whether they are conscious or not. In fact, the self-sustaining membrane, the fixed sets, abstract continuity, fundamental pattern and others are all properties of stable parallel looped systems. Such a whole system can actually 'know' itself.

23.2 Becoming conscious – emergent properties

From Theorem 12.1, we know that it is necessary and sufficient to know at least one truth in order to become conscious. Two aspects needed explanation here. One was what truth meant and the other was how to get a feeling of knowing of this truth. We now have answers to both these topics. Therefore, conceptually, we can combine these results to show how we can become conscious. Let me discuss this from a theoretical point of view for now. In the remaining chapters, I will discuss (a) how to create an important set of meta-fixed sets like space, time and other features from the external world and (b) how to create features unique to consciousness like self-awareness, free will and perceiving the world in a cohesive manner.

We have seen in chapter 14 that truth is best represented as special dynamical loops of neuron firing patterns within our brain, which we called as fixed sets. We saw the ability and the mechanisms to interrelate fixed sets to form a complex network of fixed sets. We called this as the dynamical membrane of meta-fixed sets. This membrane had special properties, the most critical one being its ability to self-sustain for a long time.

In chapter 22, we saw the necessary and sufficient conditions to attain a feeling of knowing. This required the formation of abstract continuous paths on the membrane of meta-fixed sets for a given event. Later, in section 23.1, we saw that the membrane of meta-fixed sets itself can be viewed, equivalently, as a threshold network of abstract continuous paths.

Therefore, to know at least one truth, we needed one abstract continuous path corresponding to this truth on the membrane of meta-fixed sets. However, the membrane itself is a threshold network of abstract continuous paths from Theorem 23.1. What happens when the desired abstract continuous path for this specific truth lies within the threshold network of abstract continuous paths i.e., the membrane? The self-sustaining membrane of meta-fixed sets will excite both the desired abstract continuous path as well as the entire threshold network of other interconnected abstract continuous paths. Hence, *self-sustaining the membrane is both necessary and sufficient to know at least one truth and is, therefore, equivalent to becoming conscious,* according to Theorem 12.1. Let me state this formally as the following theorem.

Theorem 23.2: A system is minimally conscious if and only if it has a self-sustaining threshold membrane of meta-fixed sets.

This is an important theoretical result. Let us look at an interesting corollary. Since the membrane is made up of a large number of abstract continuous paths, each such path gives rise to a feeling of knowing of a corresponding truth. Therefore, we 'know' each of these truths, all at once. In other words, we automatically know a large number of truths even though we started this exploration of just wanting to know one truth.

It is surprising that it is illogical to ask how you know a single truth (like how you know that this is a tree or some other single fact and nothing else). In fact, it is quite common to hear people ask how you know a single fact like your own hand, this table, a chair and so on when discussing the problem of consciousness. They pose questions like 'how do you know this is a chair while a computer does not', almost always emphasizing either a single fact or a small number of facts. Their expectation with such questions is that a person proposing a new theory of consciousness should be able to explain clearly how our brain is capable of knowing a single or a small collection of such facts.

However, we now see that it is simply impossible to know one and only one truth (or a small number of them). A complex SPL system can only be in a state where you are forced to know a large number (specifically, an entire membrane) of truths simultaneously or none at all. Let me state this formally as:

Corollary 23.1: Knowing a single truth is equivalent to knowing an entire membrane of truths simultaneously.

In this sense, consciousness can be seen analogous to a phase transition (like changing a solid to a liquid or a liquid to a gas). When the threshold state of the membrane is indeed reached, you automatically know a large number of truths. When the threshold state is not reached, you simply know nothing. This is what happens during unconscious states like deep sleep, general anesthesia, coma, knocked unconscious and others. What we consider as a conscious state does not form gradually. For example, you never remember _entering_ into a dreaming state in a slow and gradual manner. The same applies with completely drunk, knocked unconscious, coma, under general anesthesia and other unconscious states.

The transition from conscious to unconscious and vice versa is sudden, not a gradual process. It is a discrete switch in state, as perceived by your own consciousness. Note that the internal dynamics itself is smooth and continuous i.e., there is no discreteness. Your conscious perception is what makes it discrete. Exciting the underlying threshold membrane of meta-fixed sets in a dynamical sense during a dream or otherwise is itself a gradual process, just as with a phase transition. We call properties that show such a discrete transition as _emergent properties_. Some examples of these are feelings like the feeling of knowing, realizations and emotions.

There is yet another important subtle point to keep in mind here. Knowing that you are indeed conscious is not the same thing as becoming conscious. The threshold membrane, while it is made up of a large number of abstract continuous paths, may not contain an abstract continuous path to a special truth, namely, for knowing your own consciousness itself. In this case, you do not know that you are conscious even though you are conscious!

For example, a toddler does not know that he is conscious even though he is conscious. Generally speaking, it is true that we do not know a large number of truths. For example, even though you are going to your office every day, you may not know that there is a gymnasium right next to the cafeteria, say, even though you pass by it every day. It is only when you plan on using the facility, you realize that it was located next to the cafeteria. Another example is: until you plan on buying a house, you do not realize that there are 'open house' signs on several streets. It is only when you are actively looking to buy a house, you seem to find these 'open house' signs almost everywhere. While not knowing such facts seem reasonable, it sounds quite strange that you cannot know your very own consciousness. I will discuss this in more detail in chapter 43 on self-awareness. In addition, I will also discuss notions like free will, knowing about the existence of an external world, sensing three-dimensional space and passage of time, detecting geometric properties like sizes, distances and shapes of objects in the subsequent chapters.

24
Sensors

In the previous chapter in Theorem 23.2, we have seen a new alternative set of necessary and sufficient conditions for becoming conscious, namely, to create a self-sustaining dynamical membrane of meta-fixed sets. This result was still partially at a theoretical level because it did not specify what to include in the dynamical membrane. So the goal in the remaining chapters is to provide additional mechanisms to help us create a set of specific fixed sets critical enough that we should include them within the threshold membrane. These are space, time, geometric, physical, dynamical features of external objects, our body fixed sets as well as language.

In order to build these features, one of the first steps for any SPL system is to sense several external inputs. This happens through a collection of sensors. But what is a sensor? The notion of a sensor is subtle when we do not factor consciousness into the system. A conscious human evaluating if an external or an internal component is a sensor or not is considerably different from a not-yet conscious system, evaluating and trying to build a sensor by itself (i.e., without help from any conscious being).

As an example, consider an adaptive unconscious system i.e., one that is changing its internal structure like, say, an infant or a newborn animal. As it changes its internal structure, it would need to recalibrate its sensors to measure the same quantity reliably. For example, if a thermometer that is measuring temperature is such an adaptive system, then as it modifies its internal components, a given temperature measurement like, say, 70 °F should still be measured as 70 °F after its internal modifications. How do systems ensure that?

If an infant recognizes a car when he is 3 months old, how does he continue to recognize it even when he is 10 years old even though his internal brain structure has changed considerably? Indeed, a child does forget a large number of memories as he grows older, but not something like a car. How is he recalibrating his internal structure to ensure he can measure the same event in the same way for several years? In this chapter, I will discuss several of these issues that stem from subtle assumptions made in the definition of a sensor.

24.1 Ambiguity in sensing inside, outside and boundary

Consider the problem of building a sensor and including it as part of a system either by nature or by ourselves. If our objective is to let the system eventually know what it senses, we have the following problem. When we have a sensor like a thermometer, there is a sensory surface, which we can view as the boundary of the sensor. This, with thermometer, is that surface, which should be in contact with the heated surface that we are trying to measure. Relative to this boundary, there is a well-defined outside and an inside as well. For example, if the heated surface is air in the room, then the outside is the air or the room, the inside is the inside of the thermometer bulb and the boundary is the surface of the thermometer bulb. In chapter 29, I will discuss how an unconscious system can show the existence of a well-defined outside i.e., an external world by itself (cf. conscious systems like humans who simply take this fact for granted). Excluding this point for now, I will discuss what the issues are in attempting to distinguish the inside, the outside and the boundary of a sensory surface.

When we view the sensor as a dynamical system, there is a possibility that distinguishing the inside, outside and the boundary is ambiguous. For example, we humans sense the world in three dimensions using our five sensors. We see this world every instant of our life provided, of course, we are conscious. Therefore, if there is any ambiguity with our sensors, it should be resolved to get a consistent view of our world and even to become conscious. Do we really think the world around us looks exactly the way it appears to us?

24.2 Reality observed through our senses

Consider color as an example. We know that it is not an inherent property of the object. This is clear the moment we talk to a colorblind person. Similarly, there are no sounds inherently present in our world. Instead, there are only patterns of compressions and rarefactions of air molecules. The same situation applies with smell and taste. Even with touch, when we stretch our hand to reach a book, we truly believe that we are moving our hand through empty space and are touching the book. Our eyes appears to provide information to our conscious self, which 'confirms' this fact. However, when we are in a less conscious state (like severely drunk or under general anesthesia) and we stretch our hand to reach a book, we do not know that we are moving our hand in space, even though other conscious people can confirm this. What conclusion can you make about the true nature of the world around us then?

Our sensory organs are responsible for providing a view of the world. The world itself is not the way it appears to us. For example, if our sensory organs have much higher sensitivity levels or ranges, then the world will appear completely different from what we see now. If our eyes are capable of generating nerve impulses even for microscopic dust particles, or if our ears can hear ultrasounds like bats, or if our eyes can see a wider range of electromagnetic spectrum like birds, the world around us will appear with different colors and will be too crowded. In these cases, I claim that our brain's architecture should be fundamentally different if we were to still become conscious. This will become clear later (say, after reaching chapter 43). For example, sensing three-dimensional space and the passage of time becomes much harder in this case.

Even though the world is not the way it appears to us, we know that there should be some verifiable facts. For example, we know that a table exists and that it is different from a chair. If we become conscious once and establish this fact, it stays the same until we become conscious again to verify the same fact. This is because the switch from a conscious to an unconscious state and vice versa, is entirely within our brain. Similarly, the color, the texture, the surface roughness and other features of a table may not be the same as it appears. Nevertheless, that the table has four legs, that they are attached at the four corners, that it has the approximate shape of a rectangle with a certain size, that the table is located in front of you are all facts that do not depend on you and your consciousness.

I call these examples as describing the relationships between objects. If you close one eye, the world already looks different. However, the existence and the relationships between objects do not change. Rather than trying to understand how the world appears, we want to understand the relationships within our world, even indirect ones, relative to our own consciousness. In the next few sections, I will focus on understanding these relationships. The framework we have developed so far with fixed sets, dynamical membrane and abstract continuity will turn out to be useful in analyzing this problem. Let us now see how sensors help us understand these relationships.

24.3 Inside, outside and the boundary of a sensor

Consider any of our five sensors (or, in fact, any other man-made sensor). Let me, temporarily, assume that there is an external world and that the sensor has a non-zero finite size. This implies that I have assumed the existence of an inside, outside and the boundary of a sensor. Let us study the issues caused by this assumption, if any.

When the external input you are trying to measure falls on a sensor, it produces a dynamical effect only *after* the input falls on the sensor. While the external input is still on its way towards the sensor, the sensor is completely unaware of this input. For example, if the light reflected from an object does not yet reach the eye, you do not know that the object exists (cf. light from stars that have not yet reached us). Similarly, if the compressions and rarefactions of the air particles do not reach your ears yet, you do not know that the sound exists. The same situation applies with odor, taste, touch and, in fact, with all sensory systems including man-made sensors.

This raises an inherent issue. How can a sensor differentiate if the input is reaching it from the outside, inside or if it triggered just at the boundary? Now, the input itself does not carry information about its own origin with it. You have to infer this positional and directional information through other means. Additionally, the sensor does not know about the existence of this input until after the input has already reached it. The instantaneous effect produced on the sensory system is identical for all the above three scenarios. For example, the change in the height of the mercury level in a thermometer is the same whether you supply the heat internally (if there is an energy source inside the thermometer), externally (through air temperature) or at the boundary (by touching a hot object directly). In all three scenarios, there is a region which we identify as the true 'sensory surface' on the thermometer that acts as the source of change in the mercury level. In the case of eye, this region is the retina. For taste, this region is the surface of the tongue and so on. In other words, we need dynamical changes on the sensory surface to produce subsequent effects on the sensor. We cannot infer the existence of an outside and an inside directly by looking at the output.

Let us see examples of excitations specific to human sensors that occur on the inside, outside and the boundary. When an image of an object falls on the retina, a neuron-firing pattern is generated that propagates deep into the brain. The retina acts as the boundary of our eye sensor. An example of excitations that occur right on the boundary is, say, when you have a fixed black spot in your vision. You know that the spot itself is neither inside your brain nor outside, but it is at the boundary. Similarly, several other such eye conditions exist that affect your vision because of abnormalities directly on the retina.

For excitations inside the sensory surface of our eyes, we can identify subsequent neural-firing pathways deeper in the brain, starting from a given pattern of excitations on the retina. When you are dreaming or when you close your eyes and picture a scene, you are exciting the inside of the sensory surface directly, i.e., there are no excitations on the sensory surface. Yet, you feel the same sensation of seeing an image as if it were external to you. Of course, you do know that the image is inside your brain and not outside by using additional information like, say, sensing the action of closing your eyes or sensing the poor quality of the image, which is common only when you close your eyes.

The outside of the sensory surface (with our eyes) is the external world. Most of the objects we observe daily, when we open our eyes, correspond to such examples. Even with these three distinct regions of excitations – an inside, an outside and the boundary – we see that the end effects are identical.

This situation applies equally well with other human sensors. For example, when you talk to yourself silently (like your thoughts), you clearly hear a sound even though there are no compressions and rarefactions in the air molecules. The sound is identical in every respect to an external sound that you hear through your ears or right at the boundary of the inner ear. Similarly, you can close your eyes and imagine eating your favorite ice cream or imagine the smell of a freshly baked pizza and produce internal excitations that either make your mouth water or smell the odor. This is purely an imagination but has an identical effect to real taste or odor (although considerably diluted in signal strength – this is not so in our dreams, however). You can also similarly imagine someone touching your body when in reality there is no one touching you. The sensation is identical between the inside, outside and boundary scenarios in all of our sensors.

How do we distinguish these three cases for a sensor? An important source of difficulty comes from the fact that the sensor is a dynamical system. In section 24.2, I mentioned about chaining sensors together. For example, we can view the light reflected from an object, reaching the retina, exciting appropriate chemicals, generating neuron spikes and producing cascading firing patterns that continue deep inside the brain, as one long chain of linked dynamical events. Therefore, the outside, boundary and the inside appears to be an artificial classification. Fortunately, there is a separation in the composition of materials like with outside air versus our skin surface versus inside cellular chemicals to help us group these regions together and call them as an outside, boundary and an inside respectively. If the structure, composition and the complex interactions are quite intricate, we no longer make these clear separations. For example, we do not know the boundary for a given thought, a given emotion and others, even though we do sense them. To a conscious being, this separation appears quite natural in some cases. However, when a system is attempting to become conscious, realizing this separation is an important nontrivial task.

24.4 Passive time-varying patterns on a sensory surface

Let us state the above ambiguity in precise terms. If an object moves closer to you in the external space, the excitations produced on your retina can be explained equivalently without introducing a notion of an external world. For example, we can say that the size of the region of excitation on your retina increases with time, using possibly an unknown mechanism. Similarly, if the object moves away in the external world, then it is equivalent to decreasing the size of the region of excitations without assuming the existence of external space. If one of the objects breaks into two pieces when it collides with another object in the external world, we can equivalently explain it from a sensory boundary point of view. For example, we can say that it is a time-varying excitation of the corresponding images directly on the retina without again assuming the existence of an external world. This equivalent representation appears analogous to suggesting that the event occurs on a television screen.

Every event that occurs in the external world can be expressed in terms of time-varying patterns generated directly on all sensory surfaces. What, then, is the need for a notion of the external world? When an infant is about a month old, the notion of external world appears unnecessary when forming memories. The equivalent representation as a time-varying pattern on the retina (or other sense organs) seems sufficient. As he experiences new events, he forms fixed sets and links them together, as outlined in the previous sections. When does this representation as time-varying patterns on sensory surfaces breakdown for him? In chapter 29, I will discuss how an infant resolves this seemingly correct but incorrect equivalence. I will show how his brain is forced to create the notion of an external world.

With a representation involving time-varying patterns, *any* pattern is allowed to form on the retina. You can imagine 'painting' any random picture on the retina. This is allowed initially, say, at or before the birth of a human. This is the starting point. Even though the newborn does not yet know the existence of time, the patterns on the retina can be imagined (by a conscious adult) to be sequentially ordered along the timeline. Not only is each pattern at every instant random across the surface of the retina, the time sequence of patterns are also randomly arranged. There is apparently no connection between the patterns at one instant with the one at the next instant.

However, after the human is born, the set of patterns are severely constrained to allow only patterns that exist in our external universe. Random images are no longer allowed. There is sufficient structure both within a pattern and among a sequence of patterns on the retina. They are grouped into specific well-defined regions on the surface of the retina corresponding to objects visible to us. These patterns repeat themselves quite often as you observe the same object several times. When all random patterns were allowed, it appeared unlikely that two identical patterns can occur in a short period of time. However, repetition of identical patterns is now extremely common.

The patterns are also causally linked in a very specific way. They correspond to a time sequence of events that we observe (even though he may not yet know about space and time). There is a perception of continuity among patterns for most of the time. However, the resolution of our sensors may introduce apparent discontinuities in the sequence of patterns. For example, the switch from one pattern to another may appear discontinuous. The patterns are also linked across the surface of the retina. When an object moves, the pattern moves on the surface of the retina. These patterns are memorized as fixed sets in our brain. The near-simultaneity of occurrence of events and the dynamical constraints of our nature are stored in our brain's stable parallel loop architecture. We even memorize rules that seem to be always obeyed among variations within the patterns (see chapter 18 for a few specific mechanisms). For example, some of the most common rules are based on causality. If a certain set of conditions are satisfied by the patterns, i.e., the cause, then a certain different set of patterns are produced, i.e., the effect.

24.5 Active control of time-varying patterns

The set of allowed time-varying patterns on a sensory surface, even though quite constrained, are viewed as passive patterns. You do not seem to have any control on either the time sequence of patterns or the structure of patterns across the sensory surface at a given time instant. They vary according to the external laws obeyed by other objects (even though you may not yet know about their existence). How can we generate time-varying patterns actively? We can do this if we have an ability to control the formation of these patterns. This is possible by what are called as the body fixed sets.

Among the fixed sets formed in an infant, some need special emphasis. These are the ones generated from one's own internal body inputs. For example, there are fixed sets that correspond to moving one's own hands, legs, head and other body parts. We can classify these inputs into two types: (a) fixed sets generated internally through the peripheral nervous system like, say, hunger, breathing or bowel movements and (b) fixed sets generated by external sensors like, say, by observing patterns from his own hands using his eyes, or by hearing sound patterns from clapping his hands, or by generating taste patterns when placing his hand in his mouth.

If we consider body fixed sets of the second type (i.e., generated by, say, observing your body with your eyes), at first glance, they do not appear any different from fixed sets corresponding to external objects. Both body inputs and inputs from external objects produce dynamical patterns on your retina that are memorized in a similar way. However, there is one specific advantage with body fixed sets. This has to do with the *control* of your hands and legs. The fundamental pattern of complex systems discussed in chapter 16 provides mechanisms where your action and observation forms a closed loop. While there is no notion of 'self' yet, the presence of a number of such action-sensing loops creates self-regulation, at least in a mechanical way. For example, an infant can use these mechanisms to control the motion of his hand to put it into his mouth (see section 17.7). In chapter 17, I have discussed a few other similar examples that uses fundamental pattern like, say, how to control your eyes to focus or to track a given object.

Let us see how the ability to control yourself (even in a mechanical way) affects the time-varying patterns on the retina. When you learn to control your own body fixed sets (using the above mechanisms), you introduce additional structure and constraints within the patterns that form on your retina, say. Your self-control mechanisms (after sufficient practice) provide new ways to produce patterns of your choice. You can control the formation of a specific pattern and a specific variation in the pattern independently that involves using your body, with the above mechanisms.

Several of the rules (like the causality-based rules) that you may have memorized with variations in patterns on your retina can be potentially violated because of your self-control mechanisms. For example, without your involvement, if a ball always rolls a certain way and you memorize it as a rule, then you can fundamentally alter the path of the ball through your self-control mechanisms and break this rule. The generalization of this involves using your free will, which I will discuss in chapter 28. However, even without free will, using just the fundamental pattern, you can still achieve self-control. To highlight this difference, I am only using a narrower set of mechanisms represented by the word 'self-control' instead of the broader 'free will' mechanisms. The only rules that we can call as universally valid, when you include your own self-control mechanisms, are the physical laws of our universe. You need to be able to know what the laws are with or without your involvement. You can only know this distinction after you become conscious.

Therefore, the patterns that can form on the sensory surface are not just passive, but can be actively changed using your self-control mechanisms. In section 29, I will show how this can help in showing that an external world exists.

24.6 Definition of a sensor

Given a number of issues discussed above with sensors, let us now try to redefine what a sensor is more formally with a hope of avoiding these issues. When we want to understand what a sensor is and how it works, we should clarify the following points.

(a) the inputs which the sensor is trying to detect or measure – there is an equivalence class of inputs with transitions between them,

(b) a dynamical system that takes the above inputs and converts them into outputs,

(c) the outputs itself generated by the sensor,

(d) an ability to detect or measure the outputs – conscious human plays an important role to observe the output,

(e) an ability to cope with changes within the dynamical system – this needs recalibration of the sensor.

Definition 24.1: A dynamical system can be considered as a sensor that detects a specific phenomenon if it is capable of producing an effect unique to the observed dynamical phenomena in a reliable and repeatable way.

The reliability and repeatability in the above definition is quite critical. If we were to exclude it, the above definition would not be useful because any dynamical system would then obey this definition. By imposing reliability and repeatability, we narrow the choices of dynamical systems for sensors considerably.

Man-made sensors are reliable and repeatable because we build unique designs and choose correct materials that exhibit the most suitable properties necessary for the function. However, these choices most definitely require consciousness. The materials are so rare and the designs are so intricate that they are almost impossible to form naturally.

On the other hand, sensors in living beings, though equally intricate and use special materials, are built without the help of conscious beings. This is primarily possible because the underlying designs use SPL systems. Looped dynamics in SPL systems provides a natural way to guarantee repeatability and also reliability. For example, if a given pattern on your retina produces a given fixed set (loops of excitations) at a given instant, then the same pattern will trigger the same fixed set at a later time as well as with other dynamical transformations. This was the definition of a fixed set. If instead of a fixed set, we used a specific set of neurons to identify the input pattern, then the reliability and repeatability is drastically reduced. Furthermore, as your internal brain structure changes over time, we just need to maintain the relative relationships between the fixed sets as opposed to making sure that the specific neurons are still triggered.

When a sensor detects inputs, the sensor does not know that it is detecting the input, except when the sensor is part of a conscious being. A thermometer does not know that it is measuring temperature when there is a change in the mercury level. Conscious beings do have an ability to know if a sensor is detecting an input or not. Other systems do not need to 'know' that they are detecting an input. They are simply interfaced with a set of sensors in a mechanical way like the sensors in your car or a thermostat.

Let us begin by looking at several examples of sensors. Human beings have five sensory organs that detect external inputs: eyes (detecting light), ears (detecting sound), nose (detecting gases as odor), tongue (detecting solids and liquids as taste) and skin (detecting pressure from touch). Humans have discovered a number of other sensors like electromechanical, thermal or chemical ones that help in understanding the world around us. Some of the examples are thermometer (measuring temperature), clock (measuring time), galvanometer (measuring voltage), tachometer (measuring speed), gyroscope (measuring orientation) and Geiger counter (measuring nuclear radiation).

24.7 Sensor as a dynamical system

Consider any physical dynamical system. If this system is subject to suitable external inputs, they produce changes within the system. This is equivalent to saying that a cause (external inputs) produced an effect within the system. If we can use the effect to identify the cause reliably, we can say that the dynamical system acts as a sensor that detects the specific external input. The quality of a sensor depends on how well, how reliably and how repeatedly you can identify the cause by looking at the effect. This is the basic principle used by any sensor that attempts to detect a physical quantity. More precisely, every such sensor attempts to detect or measure *changes* in a physical quantity. This implies all sensors are dynamical systems as is expected from (a), (b) and (c) above of section 24.6. For example, a thermometer takes the energy from random collision of molecules (the cause or the source of temperature) and produces a change in the mercury level (the effect).

If there is a change in the cause, you want to see a change in the effect as well provided you want to call the dynamical system as a good sensor. If a change in cause does not produce a change in effect, then just by looking at the effect (i.e., no change), you are unable to conclude if the cause exists or not. An underlying assumption in the above argument is that zero cause produces zero effect (except when there is an offset). This is equivalent to saying that you cannot create something from nothing. This is true even with thoughts. For example, even sentences in a book comes from the author's self-sustaining membrane of meta-fixed sets.

With most dynamical systems, multiple different causes contribute towards the same effect with comparable magnitude and quality. Furthermore, when a cause changes over time, the effect produced within the sensor changes over time as well. If you now couple both these situations, it becomes quite difficult to identify the source of the cause just from looking at the effects. It is considerable additional work to distinguish these two cases. As a result, not all dynamical systems are effective sensors.

24.8 Equivalence classes and transitions between them

We need to be clear about the specific cause that a sensor is trying to measure. Contrary to what we might think, the inputs are abstract entities that usually belong to an equivalence class. An equivalence class is like a grouping of similar items. Elements from two different equivalence classes are considered different from each other, while elements within the same equivalence class are consider similar. Mathematically, it is a reflexive, symmetric and transitive binary relation.

In most cases, the equivalence classes correspond to the resolution of the sensor. For example, if a digital thermometer has a single decimal resolution, then all values from 75.3 °F to 75.4 °F belong to an equivalence class. Here, the output of the thermometer does not vary even though the input varies by 0.1 °F.

To highlight the subtle point about the inputs being abstract entities, consider a reliable sensor. It is possible to have quite detailed dynamical changes within the external system to produce a cause, while the sensor itself does not produce any significant observable effects. Such a situation can happen when there is a dynamical equilibrium between the external system and the sensor. The dynamical equilibrium state is an example of an equivalence class of inputs.

For example, the mercury level does not change in a thermometer even when there are large fluctuations in individual molecular motions outside. In this case, we have a thermal equilibrium between external system and the thermometer. Therefore, to be precise on what a thermometer measures, you can only say that it measures changes in temperature, not the detailed dynamics behind the cause of the temperature (i.e., random molecular motions). When we look at the mercury level, we cannot conclude that there are no dynamical changes happening outside. An abstraction, i.e., a single number that we have created to represent the equilibrium state of molecular motion, namely temperature, itself is not changing, however.

An equivalent way to say this is that a sensor measures a cause, modulo dynamical equilibrium (or an equivalence class). Modulo here refers to the collapsing of all equivalent dynamical states into a single class. When the external system is in an equilibrium state with the sensor, there are no observable effects. The sensor only detects *transitions* to the equilibrium state (or equivalence class). For example, in the case of a thermometer, 'modulo thermal equilibrium' is equivalent to measuring temperature. The thermometer reading is expected to stabilize and come to an equilibrium state before we read the temperature (cf. measuring body temperature during fever). In practical situations, the change in temperature is fast enough that the thermometer does not have sufficient time to reach equilibrium states at every instant with the external system. Nevertheless, we assume, in an abstract sense, that a notion of temperature exists during this period as well and that the thermometer detects all these transitions quickly. We, therefore, believe in the continuity of variation of abstract concept of temperature. Strictly speaking, the continuity of variation is in the height of the mercury level. However, with a linear temperature scale next to the mercury level, we simply attribute the continuity to temperature (as real numbers) as well. This example can be generalized to all sensors that have equilibrium scenarios like galvanometer, gyroscope, tachometer and several others.

The property of continuity while detecting transitions between equivalence classes is a central one to have for any sensor. When there are no transitions (i.e., during an equilibrium state), we say that the sensor is not detecting anything. Instead, the actual detection did already happen in the past when the transition occurred. The current equilibrium state merely refers to a 'memory' of a past transition. For example, a steady mercury level at 75 °F refers to the fact that this temperature was attained at some time in the past. It is currently maintained by the external system and is indicated in the sensor by virtue of thermal equilibrium. From the sensor's point of view, nothing significant that causes a change in temperature is happening outside at this instant, to be able to detect.

It does not matter whether we think in terms of a sensor's point of view or not for a thermometer. However, it does matter when we analyze sensory organs for a human being. Sensors for conscious beings are unlike man-made sensors like thermometers. Our consciousness lets us evaluate the external world from our point of view. We know that something is happening outside only if our sensors detect the above type of transitions. If the external system is in dynamic equilibrium with our sensor, then there are no corresponding neuron excitations. There would, however, be neuron-firing patterns when the system initially transitioned to this dynamical equilibrium state. This firing pattern sometimes has a short-term memory associated with it (as looped excitations). It is this memory that lets us know that we entered a dynamical equilibrium state and, possibly, stayed in this state. There is, however, no direct way to know the existence of external dynamics even though significant energy is spent outside to maintain the equilibrium state.

For example, there is air around us that is in a state of dynamic equilibrium with our skin. Our skin does not detect it unless there is wind that causes a transition out of this dynamic equilibrium state (or we have a past memory). Sometimes, saturation of the neuron signals is responsible for a lack of neuron excitations, but the most common source is dynamic equilibrium. If the location of the wind on your body does not change while its intensity keeps changing, we will produce neuron-firing patterns. If there had been saturation instead, these neurons would not have fired. We have other similar examples with our sensors. If you eat the same food in large quantities for a long time, our brain reaches an equilibrium state with this taste. The taste diminishes significantly after a while. If we hear a sound like white noise (like from a fan) continuously, we feel it as a background noise, which achieves a partial equilibrium state. We can detect any other sound over this white noise quite easily.

One particularly interesting case is with the detection of passage of time itself. What are the equivalence classes while detecting the passage of time? What are the corresponding transitions between these equivalence classes? We want to detect the variation of 'time' along timeline instead of variation of, say, temperature along timeline. This is a rather bizarre and nontrivial question. This is because I am asking how we detect the very flow of time, along which all dynamics are fundamentally based on, as if the flow of time itself were a dynamical event. I will discuss this in chapter 40.

24.9 On creating a sensor that knows what it senses

All man-made sensors detect a specific input. However, they do not know that they are measuring a given input. For example, a thermometer does not know that it is measuring temperature. It simply produces a change in mercury level for a change in temperature. Humans, on the other hand, do know that a thermometer measures temperature or that an ammeter measures electric current. How can we start with such a sensor (like a thermometer) and evolve to a stage where it begins to know what it is actually measuring (like a human)? This is yet another way to look at the problem of consciousness. In the next few sections, I will study this problem and list the important features of such a sensor. For example, I will note that the ability to adapt the internal structure of a sensor is itself critical for this task. However, this raises another problem. If the structure of the sensor itself changes over time, how do you ensure that the sensor is correctly recalibrated? The changed sensor should measure the same input with a high accuracy and reliability. It is like having a thermometer that constantly keeps changing its internal components and, yet, reads correctly the same 75 °F as the outside temperature.

24.10 Chaining sensors

Another way to view a sensor is that it is a signal transducer. It takes an input signal

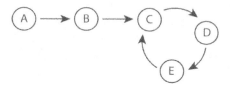

chain of events sometimes are linear like
ABC and sometimes form loops like CDE

Figure 24.1: Causal chain of events. Sometimes the chain of events that occur are linear represented here by *ABC* while at other times they can form a loop represented by *CDE*.

as cause and translates it into a, possibly, different output signal as effect. The qualitative nature of the input and output signals could be completely different. For example, with thermometer, the input signal is temperature as random oscillations of molecules and the output signal is the height of the mercury level. With galvanometer, an electrical voltage difference is transduced into a rotation of a needle.

Since any dynamical system can be viewed as a primitive sensor, consider a system that takes cause *A* and produces an effect *B*. Let us call this as sensor *I*. Similarly, consider a dynamical system that takes cause *B* and produces an effect *C*. Let us call this as sensor *II*. Note that I have taken the effect *B* from sensor *I* to be the same as cause *B* for sensor *II*. With this choice of causes and effects, we can link both sensors *I* and *II* together (see Figure 24.1). This effectively produces a new dynamical system that takes cause *A* and produces an effect *C*. Let us call this as sensor *III*. As a simple example, a change in temperature (cause *A*) produces an expansion or contraction (effect *B*) of several materials. An expansion or contraction (cause *B*) of some alloys like constantan, which itself does not vary much with temperature changes, produces a change in electrical resistance (effect *C*). Combining both these features by, say, embedding constantan alloy within another material that expands or contracts can result in a new system that produces changes in electrical resistance (effect *C*) in response to changes in temperature (cause *A*). This new sensor detects changes in temperature just like a standard thermometer.

We can use the above transitive relationship to chain multiple dynamical subsystems together. This is an effective procedure to produce high quality man-made sensors. You can pick the best materials or designs for each individual subsystem within the chain. The chain of subsystems usually is linearly connected, but it is not unusual to have looped chains. The existence of such looped systems is quite typical in biological systems than with man-made systems.

When the chain is linear, we can observe the final effect and conclude that this system is detecting the initial cause. However, when the chain forms a loop, what is the initial cause and what is the final effect? In this case, we can view any cause of any subsystem (of the chain) as being detected by this system. In fact, for this, we can observe any effect from any subsystem within the loop as well (see Figure 24.1). In the case of a linear chain, this is not true. The cause that we view as being detected should always be *before* the effect observed in the linear path (see Figure 24.1).

Let us look at a special subsystem, namely, a human being, who is part of a chain of sensory subsystems. Strictly speaking, all examples of sensors we have been talking until now already included a conscious human being as part of a chain. For example, when we said that a thermometer measures temperature by producing a change in mercury level, we implicitly assumed that a conscious being exists to observe the change in mercury level.

All man-made sensors can be truly called sensors only because the final effect produced by any of these sensors, for a given cause, is actually an input cause that a conscious human being is capable of processing. In other words, every man-made sensor should produce an effect that one of our sensory organs (i.e., eyes, ears, nose, tongue or skin) can detect and produce an effect within our brain. Humans, therefore, play the role of measuring the output as required by (d) of section 24.6.

Each effect of a sensor should produce neuron excitations at one or more of our sense organs. If any given man-made sensor does not produce an output that one or more of our five sense organs can detect, then a conscious human would not know that the sensor is actually detecting a cause. In this sense, the man-made system is not really a sensor. For example, our nose can detect a gas leak provided certain pungent smelling chemicals are added to natural gas. We detect alpha- or beta-waves (from nuclear radiation) through a Geiger counter because it produces an output that, say, makes a lamp glow, turn a needle or generate audible clicks. Our eyes or ears can process all of these outputs.

24.11 Role of a human in sensory systems

The involvement of a conscious human being raises two important questions. Can we call the dynamical system a sensor when a conscious human is not present? Equivalently, when can we say that a system is detecting an input by itself when there are no conscious beings around to assist? What makes the sensory organs (and the brain) of a conscious human special that it lets us conclude that the external dynamical system is truly detecting a cause by observing its effect?

A human observing the system does not influence most sensory dynamical systems. The dynamical systems behave more or less the same way whether a human is observing the system or not. However, this is not true when the systems are at an atomic scale (i.e., a quantum mechanical system). Each human sensory organ requires sufficient amount of energy as input to produce a unique pattern of neuronal excitations. These inputs with sufficient energy (like a light or a sound pattern) should come undisturbed from the source until they reach our sense organ. They should come directly from the quantum mechanical system (for the sake accuracy and reliability). This implies that the quantum mechanical system itself was supplied with sufficient energy to generate a unique light or sound pattern that our sense organs can detect. This energy disturbs a quantum mechanical system in a fundamental way, as highlighted by Heisenberg's uncertainty principle.

A human does not have to observe the system directly. Instead, he can 'record' the event to view it later. However, the energy supplied so the recording instrument can register a pattern, will disturb the quantum mechanical system in the same fundamental way. In this sense, we cannot avoid Heisenberg's uncertainty principle because conscious human beings are themselves dynamical systems that need energy to experience even feelings including one's own consciousness. It is impossible to have feelings, thoughts or consciousness without consuming any energy.

If we now look at any one of the five sensors in a human, they take an external input (like light, sound) as cause and produces neuron firing patterns. This effect acts as a cause for interlinked neurons and produces subsequent firing patterns leading eventually to the triggering of a fixed set. This is equivalent to the transitive chaining of dynamical subsystems mentioned earlier. One advantage here is that, except for the initial subsystem, all other subsystems take neurotransmitters as inputs and produce neurotransmitters as output. However, the number of subsystems in the chain could easily reach millions even for a simple detection of, say, an image of a tree. This poses a challenge especially since the millions of subsystems are not static. They keep changing by adding new interconnections, deleting several existing ones and changing others through new experiences. This is the problem of recalibration, which I will discuss next. This is the last issue (e) highlighted in section 24.6, namely, how to cope with changes that occur within a sensory system.

24.12 Recalibrating a sensor

With a sensor, can we distinguish the following two cases just by measuring the outputs: did the external inputs change or did the sensor itself change internally? The change registered by the sensor could have been caused by either the change in external inputs are a change within the sensor itself. When a sensor changes internally, we need to recalibrate it to make accurate measurements. Let us state the condition for recalibration clearly.

Inputs to most sensors have a well-defined notion of an increase and decrease in the input. This corresponds to an increase or decrease in the energy (or mass) of the input. For such a sensor to be reliable, we need it to satisfy the following condition at every stable state of the sensor: *a zero net change (an increase followed by a decrease by the same amount) in the input should produce an output that is identical to when there is no change in the input.* If this condition is not satisfied, we say that the sensor needs recalibration. For example, when the temperature is increased from 70 °F to 80 °F and back to 70 °F, we expect the mercury level to return to its original position. Otherwise, the thermometer needs recalibration. If the sensor requires constant recalibration, it is not a reliable sensor.

Why does a sensor need recalibration? This could be simply due to normal wear and tear. The dynamical system that makes up the sensor has moving parts, which wear out on repeated use. The internal structure of a sensor can change over a period of time due to stress, heat, resistance, frictional and other forces. If we do not recalibrate the sensor, then the error in detection produces unexpected results, especially when we use the sensor to perform a task. For example, if you were to take a medication only when your temperature is higher than 103 °F, then a thermometer that is not calibrated accurately can result in your intake of medication incorrectly, causing harmful side effects to your body.

In reality, it is difficult to find a sensor that never requires recalibration. There, however, does exist special materials or designs for some sensors (like mercury in a thermometer), which do satisfy the above zero net change condition for sufficiently long periods of time. Even when you cycle the sensor's inputs, back and forth several times, the output does return to its original state. Discovering such sensors requiring special materials with favorable properties certainly requires human consciousness and intelligence. It is unrealistic to assume that naturally occurring materials and designs make up good sensors that do not require recalibration for long periods of time. Therefore, the problem is interesting when one wants to design a sensor using just naturally occurring materials and still satisfy the zero net change condition.

Now, the traditional way to recalibrate a sensor requires a conscious being. A conscious human performs standard tests and compares the results with expected outputs to recalibrate. Is there a different way to recalibrate a sensor when there are no conscious beings? For example, how can a newborn human, who is not yet conscious, recalibrate his own five sense organs? To understand what this means, consider a human looking at a red ball and recognizing it. We have already mentioned earlier that the inputs from our eyes propagate deeper into the brain by chaining a number of subsystems through the neuron interconnections. This chaining happens even within a digital thermometer, in an analogous way, where a temperature input finally produces a decimal number on a digital display. When we see a red ball, the neuron inputs excited on the retina propagate through the optic nerve to deeper regions of your brain.

Let us use the set of neurons and the paths taken by them at this instant to identify the red ball. Over a period of a month, our brain's internal structure changes significantly because of a number of new experiences. When you now look at the same red ball, the set of neurons and the paths taken then are, most definitely, different from the ones a month earlier. Yet, you have no problem identifying the object as a red ball. Even though your brain's internal structure has changed significantly, your brain has recalibrated in such a way that you still detect the object as the same red ball. This is quite nontrivial. As mentioned earlier, it is like having a thermometer correctly read 70 °F, i.e., the true outside temperature, even when its internal design is considerably altered.

Our brain stores memories from new experiences quite carefully as we grow from a child to become an adult. If our brain does not recalibrate as it adapts, the world around us will look completely different a month later. The colors, shapes, sizes, sounds and tastes will all appear unrecognizable every day. While recalibrating a man-made sensor is already quite nontrivial, it is quite surprising that our brain (and even individual cells) are able to recalibrate our view of the world in a remarkably precise way. This is why the notion of fixed sets (satisfying similarity, dissimilarity and connectedness conditions – see chapter 7) and the SPL architecture of our brain is fundamentally important.

Of course, the error in precision is so minor because our brain constantly recalibrates. Yet, there are situations when the errors do accumulate to produce a different view of the world around us. One example where we notice this clearly is, say, after more than 10 years of new experiences. For example, consider revisiting a place that you have lived about 10-15 years ago like, say, a town you lived for several years when you were a child. You will be surprised to notice that the town, the homes, the roads and other objects will appear smaller after you revisit 10-15 years later. This is because your childhood memories were formed relative to your own 'size' (of, say, your hands or legs). These memories are also formed relative to your measurement of 'time' as well. When you were small, you take longer to walk a given fixed distance (say, 10 feet) compared to when you are taller and older. All of your current memories 10-15 years later when you revisit are being calibrated using your adult sizes and times. Therefore, the comparison now will make the town appear smaller.

Your growth from a child to an adult is large enough to have a noticeable effect in the space-time geometry as memorized by your brain. On the other hand, if you were growing up in the same town, the error in your view of this town would be quite gradual and continuous. You would not detect this minor error at that time. However, if you have not visited the town for 10-15 years, the situation is different. Your brain's internal structure did change enormously with new experiences during this period. The gradual change and continuity that you would have had was now broken because you have not visited the place for 10-15 years. Therefore, the net accumulated error when compared to your past memory of the childhood town appears quite significant.

The best example where recalibration is obvious is with an infant. At that age, an infant's brain is changing so significantly with new connections forming continuously with new experiences. It is so hard to ensure an infant predicts correctly even for the same inputs. If he memorizes and identifies a balloon correctly, it is not necessarily true that he will be able to identify a balloon two days later. If you go on a vacation for a month and return back to your own home, he would have most likely forgotten several aspects of your home even though he has lived there, say, for six months. His brain essentially needs to recalibrate once again and memorize events from your home. The changes in his brain that forces recalibration seems to happen, at least considerably, for almost 2 years of his age.

Our brain is both a complex adaptive system and a sensory system. The adaptive nature of our brain interferes with the sensory aspects of it. Given the above examples and the discussion about the need to recalibrate, how do we ensure our brain maintains a high precision of detection in spite of these issues? How do we recalibrate our brain regularly in a continuous way? This will be the focus of discussion in the next chapter.

25

Duality with the External World

In the previous chapter, we have identified an issue with the adaptive nature of our brain. How do we know if an observed change is from within the system or from outside the system? In this section, I will show that this problem has a clear solution by introducing the notion of a dual space representation of the external world. In this dual space representation, an object expressed in an input/output way is equivalently expressed as a converging set of multiple interlinked object relationships. Viewing an object equivalently in terms of relationship to other objects is termed as the dual space representation of the object.

Therefore, the primal view of an object is the identification of the fixed set for that object. In the primal view, the external object is directly mapped internally to the unique fixed set that it excites. The dual view of the object is the identification of the neighboring objects that are typically associated with this object. In the dual view, we indirectly infer the external object via the neighboring fixed sets. To a conscious human, both these views are equivalent. This equivalence will allow us to address the problem of recalibration of an adaptive sensor discussed in section 24.12 in a clean and natural way. Let us now look closely at these notions.

25.1 Primal versus dual space representation of events

The standard representation of external events we have discussed so far involved using neurons and their connections that form stable parallel loops i.e., fixed sets. This is termed as the primal representation. Primal representation is *directly* correlated to the external object. Every property of the external object that is sufficiently important and detectable contributes towards the primal representation. In chapter 7, we have described a new memory model that stored information as neural paths of linked dynamical loops i.e., as fixed sets from a definite cause to an effect. These fixed sets obeyed the similarity, dissimilarity and connectedness conditions. With this representation, we identify the external object or event using fixed sets.

In this primal form, each object would have a fixed set associated with it. Multiple fixed sets are linked together based on near-simultaneity of occurrence of events or, more generally, using abstract continuity. Even though the links between fixed and meta-fixed sets form regularly with each new experience, we only saw that all fixed sets together create a dynamical membrane that can self-sustain for a sufficiently long time. We did not attempt to identify the fixed set solely from the network of fixed sets itself. This is an alternate way to identify a fixed set. We use the underlying interconnected loops instead of the firing pattern from the external object. Therefore, this is an *indirect* way to describe the same fixed set. Identifying a fixed set entirely from the network of other interconnected fixed sets is what we call as the dual space representation for the fixed set.

This is a reasonable way to identify the same fixed set for the following reason. We can trigger a given fixed set corresponding to an external object either (a) directly from the sensory surface by following through the cascading neural pathways or (b) indirectly through neighboring connections and fixed sets. For example, if you see a ball directly with our eyes, I call that its primal representation. If you close your eyes and imagine the ball by recollecting situations when you played with the same ball, I call that its dual representation. From your point of view, both versions of the ball fixed set are identical and represent the same external physical ball. This is the same with every other physical object, thoughts and everything else. The indirect triggering of a given fixed set (dual representation) is identical to direct triggering of the same fixed set (primal representation). You hardly care how a fixed set is triggered as long as it is triggered.

The primal representation is quite natural and can be viewed as the dynamical representation of the external objects and events. The dual space representation, instead, can be viewed as an abstract representation of the external objects. In the previous section, we have seen that the primal representation has an inherent difficulty with recalibrating an adaptive system. I will show how the dual space representation resolves this problem quite easily.

The dual space representation should not come as a surprise to us. With the example of revisiting our childhood town after 10-15 years (see section 24.12), we have seen that the structure of space and time itself is expressed relative to our own size and speed. It suggested a possibility that the structure of space and time has a dual space representation in addition to a primal representation involving dynamical loops.

There are several other examples of dual space representation. I will discuss some of them in the later chapters. But for now, let me briefly mention a few of them here. For example, hearing a sound through our ears (primal view) is equivalent to triggering the corresponding regions in the brain directly, say, by talking to ourselves (dual view). Tasting a specific item through our tongue (primal view) is equivalent to imagining the item and feeling the same mouthwatering taste without external stimuli though to a lesser degree (dual view). Sensing a physical touch (primal view) is equivalent to imagining and triggering the same sensation of a touch without any external stimuli (dual view). Seeing a physical object directly with our eyes (primal view) is equivalent to attaining a similar feeling of knowing and sensing the same physical object just by imagining it (dual view). Other examples include emotions and pain as well without having a direct external stimuli (primal view) and instead by triggering directly in the brain through other thoughts (dual view) – like by imagining a sad or a horror story. Phantom limbs is another example of dual space representation.

25.2 Membrane = dual space of the external world

Let us now see why the membrane itself is equivalent to the dual space representation of the external world. This will follow directly from the equivalence of the membrane to the network of abstract continuous paths as shown in chapter 23. Recall that an abstract continuous path represents a memory of an event combined with a set of nonempty predictive meta-fixed sets along the path itself. The path is ordered using a timeline that your consciousness senses. Once a threshold network of abstract continuous paths is formed and we have abstract continuity for the specific event, we experience a feeling of knowing for the event (see chapter 22).

Now, any given object itself is represented as a fixed or a meta-fixed set. Therefore, we can use Theorem 22.1 once again to say that knowing an object is equivalent to triggering a suitable set of abstract continuous paths. This set of abstract continuous paths is expressed as a network of fixed and meta-fixed sets. Generalizing this for all objects, we can conclude that any given object can be equivalently expressed in terms of the above network of fixed and meta-fixed sets. This is precisely the dual space representation for that object. Using this result, since every meta-fixed set on the membrane has a dual space representation, the entire membrane itself contains the complete dual space representation of your experiences with the external world.

Figure 25.1 shows how to visualize the primal and dual representations within our brain. Imagine partitioning the brain (shown as a closed circular region) into two parts – an inner and an outer part – separated by the membrane. We view the outer part, i.e., above the membrane, as a region close to the external world. The sensors like the eyes, ears and others directly reach this outer region. The inner region corresponds to meta-fixed sets that are structurally created through a number of past experiences. They are not excited now and are not part of the membrane. This is equivalent to saying that you do not remember all of your past memories at this instant. You can remember as needed by thinking about an event. As you recollect a new event, you extend the membrane to this region as well.

The outer region corresponds to the structure needed for the primal representation. Similarly, the inner region corresponds to the structure needed for dual space representation. If we now take any given meta-fixed set A on the membrane, there are two equivalent ways to define it uniquely. The meta-fixed set

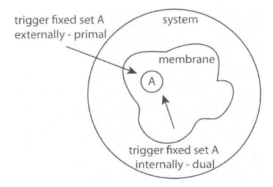

Figure 25.1: Primal and dual representations of a fixed set. A given fixed set A can be triggered in two equivalent ways. In the primal way, the external inputs are the cause whereas in the dual way, the internal inputs trigger A. The dynamical membrane expands as needed to include fixed set A.

can be excited purely from the external inputs that pass through the outer region (see Figure 25.1). This path of linked loops is precisely the primal representation of this specific meta-fixed set A. An alternate equivalent representation comes from reaching the same meta-fixed set on the membrane through the inner region (see Figure 25.1). These paths through the inner region correspond to abstract continuous paths, recollecting events from your past experiences. This alternate way to identify the same meta-fixed set is precisely its dual space representation.

In reality, you do not stay entirely in the inner or in the outer region when uniquely identifying a meta-fixed set. The external inputs generate part of the path in the outer region when you take a quick glance at the object. This excites a number of meta-fixed sets on the membrane, at least partially (see Figure 25.1). They correspond to the most likely representations for the external object. This partial information, by itself, is not sufficient to identify the object uniquely. Instead, these meta-fixed sets produce neighboring oscillations in the dynamical membrane using the interconnections in the inner region. These interconnections are from past memory expressed as abstract continuous paths among the meta-fixed sets. The abstract continuity helps in narrowing the previous choices to just a few. This is the act of thinking about the object. You iterate between glancing the external object (producing detailed excitations in the outer region) and thinking about the choices (producing excitations in the inner region) to help you recognize more aspects of the object. The iterative approach produces convergence from both the inner and outer regions until finally you identify the correct meta-fixed set. The ratio of time spent on thinking compared to glancing depends on the object you are observing and on your past memory stored as abstract continuous paths.

For example, when it is a familiar object, a child will trigger its fixed set through primal pathways as well as the abstract continuous paths through dual pathways quickly enough to get a feeling of knowing. However, if it were a less familiar object to the child, he would take more time with the dual pathways to trigger the abstract continuous paths (i.e., via the neighboring fixed sets) even though exciting the fixed set itself through primal pathways is quite fast. Therefore, he will take quite a while to attain a feeling of knowing (see Figure 25.1).

In the next section, I will discuss a couple of sample scenarios that rely on the dual space representation quite explicitly. Later in section 44.2, I will define a notion of 'dual equality' that helps us evaluate when and how a given feature is the same in two complex systems. This also relies on the dual space representation in a fundamental way.

25.3 Sample scenarios for dual space representation

In the next couple of sections, I will discuss examples that use dual space representation. One is with how the images and our thoughts are triggered during our dreams. The other is how to answer all the questions raised in section 24.12 with respect to recalibration. The answer to the former problem will be from a direct consequence of the definition of dual space representation. The answer to the latter problem lies in understanding how to consistently maintain the relative relationships between all objects in space and time as our brain grows and adapts. The dual space representation will then guarantee that the view of any given object is not distorted because of changes in internal neuron interconnections.

25.4 Dual space representation during dreams

There are a few cases where the above discussion (with Figure 25.1) is different. This is when you get dreams or even during meditation. During dreams, as mentioned earlier in section 19.5, most of the external inputs (including your own internal body inputs) are cut off using special neurotransmitters. Very little inputs, if at all, arrive at the membrane from the outer region (see Figure 25.1). In other words, the primal pathways are essentially cut off. You only have a dual space representation to identify each meta-fixed set.

If a given meta-fixed set is excited on the membrane, ideally the equivalence between primal and dual space representations implies that the brain cannot distinguish which of these representations caused the excitation. However, in reality, it is simply impossible to have this indistinguishability when you are conscious. The brain will definitely know which representation caused the excitation of the meta-fixed set. This is because it is impossible to excite a single meta-fixed set by itself when you are already conscious. The existence of a network of abstractly continuous paths and the membrane itself makes it impossible to reach the meta-fixed set using a path that completely avoids all other meta-fixed sets.

It is, however, possible when you are not conscious and you do not have a membrane of meta-fixed sets. When you are entering an unconscious state naturally, like when you are about to fall asleep, the membrane of meta-fixed sets is so 'thin' (see section 19.7). It is now possible to reach a meta-fixed set using paths that avoid other meta-fixed sets because there are so few of them in an excited state. You probably have experienced this situation already at times. When you are in a drowsy state and you wake up for a brief moment, you feel disoriented sometimes. You are not sure, if what you remember was something that just happened or if it was a dream. This is prominent if you do not wake up fully afterwards and, instead, fall back to sleep immediately.

When you do enter a dream state, the objects, the people, space, time and the events you observe are entirely identified using the network of meta-fixed sets using dual pathways alone. On the other hand, when you were awake, the inputs from your eyes and other sensors create the dual space representation of space and time. The static structure of the membrane helps in creating abstract continuity in your dreams that is somewhat similar to the real world. Entering a dream state, therefore, requires exciting the threshold network to form a dual space representation. This is one prominent example to understand the importance of dual space representation.

25.5 Recalibration with dual space representation

In the beginning of this section, I suggested that it is easy to recalibrate a sensor when you have a dual space instead of the primal representation. Given that the meta-fixed set is uniquely identified with a primal or a dual space representation, our new experiences regularly keeps updating the primal-based signature with the dual-based network. The paths from the primal representation can change as long as the same meta-fixed set is excited. The network of interrelationships formed in the dual space is quite elaborate. It is based on abstract continuity and near-simultaneity of occurrence of events.

When you are looking at the system as a sensor, then detecting the external object is equivalent to triggering its corresponding fixed set using the primal pathways. As the internal structure of your brain changes from new experiences, there is a chance that this primal representation fixed set can change partially. Fortunately, you have an equivalent dual space representation as well for the same fixed set. Changing these neighboring fixed sets as well amounts to forgetting most of your past memories that involves this object. This is highly unusual. You need to change a fairly large network of interlinked meta-fixed sets (or abstract continuous paths) if you have to change the representation of just one object. Even if some of them are broken down, even an occasional recollection of the past event reinforces the links and repairs the minor interconnections. It may be harder to recollect all of the details, but you can at least hope to find a few dual pathways to reach the same fixed set and, hence, both detect it and attain a feeling of knowing. As a result, the sensor is always detecting the same object in spite of elaborate changes in the internal structure of the brain. For example, the meta-fixed set corresponding to a table cannot be broken easily because there is a network of interrelated facts memorized when you think of a table. You would need to break all of these relationships. This is true even with basic features like colors, shapes, edges and others.

This does not mean that you cannot change the network and have a different view of an object now versus a few months later. What I mean is that the dual space representation makes such changes in the network a gradual and abstractly continuous process. For example, you have a fairly detailed network of abstract continuous paths for events related to, say, Santa Claus when you were a child. As you grow and experience new events, you start to break several parts of this network gradually. Initially, these regions of neurons are inaccessible. But eventually they are changed to produce different meta-fixed sets that is abstractly continuous with your new experiences. You begin to forget several details of Santa Claus. You are even quite surprised that you ever remembered some of these facts, say, when your parents show videos of your childhood.

Your brain is rewiring these interconnections between fixed sets through new experiences. It can eventually alter your network drastically, but this happens in an abstractly continuous way. This guarantees that there is no abruptness to your memories, except during critical brain injuries causing amnesia or from gradual degeneration of brain tissue leading to Alzheimer's disease. In these cases, you have an inability to *access* these memory regions even though they may be intact. You maintain a continuity of your thoughts even though the final transformed state is completely different compared to the initial state. This is the adaptive learning feature of our brain.

In fact, your ability to forget (in normal scenarios) is because of reusing several of these meta-fixed sets and from their inaccessibility. At the same time, your ability to recollect is because of the existence of regions of weakly linked past meta-fixed sets. They are not completely inaccessible or reused for other purposes. Nevertheless, they are hard to recollect. You may need to re-experience the event (or revisit the same place, say) if you have to discover an abstractly continuous path to these past memories. The dual space representation can, therefore, be used to explain our ability to both recollect as well as forget past events.

Recalibration, in the case of humans, has a new meaning with its adaptive structure. Namely, when your brain recalibrates, it stores new memories and links them with past memories in an abstractly continuous way while forgetting some of the past events.

The dual space representation makes the entire external universe completely relative to your own consciousness. Consciousness is solely responsible for creating the feeling of absoluteness for a few topics. These so-called absolute feelings are entirely hidden inside yourself and your own consciousness. Consciousness, in fact, makes it even possible to have these absolutes and be aware of them using the dual space representation. If you are not conscious, the notion of these absolutes do not even make sense. They are entirely within each conscious individual and are properties of complex SPL systems. This is an important consequence of our consciousness.

26

Dynamics of Serial and Parallel Systems

Our objective in the next few chapters is to describe the mechanism for free will. Free will is not only important for conscious beings, but it is also critical for proving a number of other features like the existence of an external world, sensing the three dimensional space and the passage of time, as we will see. In this chapter, I will focus on what makes the problem of free will, a difficult one. I will discuss why we do not see free will in nonliving systems or what is necessary within a system to make it exhibit a feature comparable to human free will.

The core aspect of the problem of free will is to understand the difference between serial and parallel systems. Free will arises from the unique properties of parallel systems and the unique way we serialize parallel systems. It does not arise from a serial view of the external world, which is what we employ in most of our analysis. In this chapter, I will discuss what this difference is between serial and parallel systems, give several examples of both types of systems and, finally, specify a notion of *grouping* components as a way to serialize large-scale parallel systems. In the later chapters, I will use different types of groupings to describe the root cause of free will. I will also show how to build large-scale SPL systems that exhibit the property of free will.

26.1 On choice within real physical dynamical systems

Free will, by definition, lets a conscious being identify a set of choices and pick any one of them freely at any given instant without violating any laws of nature. We should assume that the set of choices are reasonable and physically feasible. Free will does not have to work in every single scenario and for every set of choices. Such idealizations are unnecessary and are not relevant to the problem of free will. The important problems that need an answer are how you attain (a) an ability to know a set of choices and (b) an ability to choose one of them. For example, I can move my hand either to the left or to the right, whenever I want with a good enough accuracy. A ball rolling down an inclined plane never turns left or right. It does not realize that it has been going straight all this while, and that it can choose a different direction from the other two valid choices. What is the source of our ability to make a choice?

If we look at the air molecules in a room, each of them has well-defined trajectories. As they are moving along these trajectories, they may hit the walls, other molecules or other objects in the room. They bounce off in a predictable way. We have a good enough description of the laws obeyed by each of these molecules. These laws never let one or more molecules decide to turn left or right just because they are valid choices (if there are no obstacles). Quantum mechanics assigns a probability that a molecule turns left or right. Yet, we would not attribute free will to the molecule and say that the molecule decided to turn left or right if the observation did reveal that the molecule had turned left.

Rather than attribute such complex structures like awareness, choices and free will to a single molecule, our objective is to find alternate explanations for these features. If we did attribute these features to a single molecule, it would imply that all objects in this universe have consciousness and free will with varying degrees. Such a hypothesis, unfortunately, does not shed any light on these features and has little practical use, if at all. In this book, our objective is to identify the root cause of free will that is constructible and less philosophical.

Going back to the motion of molecules in the room, there may be minor forces and disturbances that we may have ignored, which seem to give certain randomness to these trajectories. But they are never of the free will type. If none of the molecules makes decisions of the above type, how can a specific grouping of these molecules i.e., all molecules arranged in the form of a human being, suddenly start making decisions to choose a left or a right motion at-will?

If we look at the trajectories of every molecule in our body, will they not follow well-defined laws just like the air molecules? Are there some neurons or neurotransmitters that decide to choose and operate (or fire) at-will resulting in an aggregate observable choice of left or right direction of motion? It is clear that not all molecules need to make such choices, if at all, because of the cascading set of connections. They can produce an amplification of the signal to result in an observable action. There is no need to make a decision every step along the way. Yet, it is unreasonable to assign the source of free will to specific neurons or neurotransmitters.

Therefore, the laws of microscopic molecules when applied to macroscopic objects, especially human beings, seem to result in behaviors (like free will) that are not compatible with each other. Strangely, macroscopic nonliving systems and even several other living beings do seem to obey the same set of microscopic laws. What then is special with macroscopic living beings like conscious humans?

While most laws are deterministic, quantum mechanical laws bring in inherent randomness at an atomic scale. An electron, for example, has a nonzero probability of existence in several regions within an atom simultaneously. In spite of this randomness, continuity is still guaranteed. An electron would not jump, when observed, from one location to another discontinuously simply because it had a nonzero probability in these regions. How do these probabilities at a microscopic level translate into well-defined decision-based control mechanisms at a macroscopic level? Even if we provide an answer to this question, we still need to answer who is evaluating and making a decision here and how that subsystem got this ability.

We should note that free will is quite different from randomness manifested through quantum mechanics. To an external person, it may appear that you are making choices quite randomly. Specifically, over a period of time, you have a nonzero probability to be anywhere within a given region. However, to yourself, you know what choices you are picking and where you are moving. You know that you are changing your choices, but not randomly. They are not accidental or occurring by chance. You know where you will be after a period of time even though others cannot predict your location.

In the next few chapters, I will show how free will occurs from parallel dynamics and loops. Let me start by discussing the difference between serial and parallel systems next.

26.2 Serial versus parallel view of the external world

When we want to analyze a physical system, we tend to apply the fundamental laws in a serial way. Most nonliving systems do not have sufficient complexity to warrant a parallel analysis. Keeping track of detailed parallel dynamics of subcomponents of a system seems unnecessary and even computationally intractable. Yet, for living beings, it is sometimes necessary to study the parallel dynamics directly. Free will is one such feature where I will show that serial analysis is simply insufficient. In the next few sections, I will state our intuition behind the distinction between serial and parallel systems in precise terms. I will present a few nonstandard examples to highlight the differences between these systems.

When a conscious human observes a dynamical system, he implicitly views it in a serial way along a unique timeline (see section 2.2 for more details). This timeline is well-defined by his ability to sense the passage of time. We can call this as the observer's global timeline. As a result, we can naturally arrange the evolution of a dynamical system sequentially along his perceived timeline. This representation captures the event as a whole, without looking at other internal dynamics within this system in detail, for too long. For example, if you are observing a tree swaying due to a wind gust, you perceive the entire tree along your unique timeline, not the detailed dynamics of every leaf or branch individually, except may be temporarily.

In contrast, a parallel way of observing the same event requires you to first split the event into a large number of interacting subcomponents like leaves and branches. Each subcomponent has its own natural or preferred timeline as well when viewed as a 'flow' event (see Figure 2.3 and section 2.2 for more details). For example, as the tree sways, each leaf that falls down has its own trajectory in its flow-based timeline. The timescales are also different – branches fall faster while leaves fall slowly. We could analyze this at one more level lower – we can draw spatial trajectories of cells within a leaf or even molecules within cells as it falls down (see section 2.2 for detailed examples of parallel analysis).

Some sub-events occur too fast to understand if perceived using a global timeline i.e., the observer's timeline. Keeping track of all the detailed trajectories of each of these subcomponents along their own flow-based timelines is what I call as a parallel analysis (see section 2.2 for more details). For most systems, we rarely study them as parallel dynamics. There are clearly practical limitations, computational challenges and even very little benefit in performing such a parallel analysis. Therefore, we *group* them into larger components (like a leaf or a branch) and study them as a whole instead. Depending on what we want to analyze, we pick different grouping of components at different resolutions. Serial analysis is one that has a small number of groups while parallel analysis has much larger number of groups.

Let me define a few terms and use them to identify serial versus parallel systems.

When we model a given dynamical system using differential equations, we write the equations of motion (like Newton's laws) of several important subcomponents and their interactions. The set of independent variables used to describe them are referred to as the *degrees of freedom*. These are, typically, the generalized positions and momentums of the individual subcomponents. We call the total number of independent variables as the *order of the system*. Even though a real physical system has millions of molecules and, hence, millions of degrees of freedom, we approximate (or group) it to a much smaller order system using rigid body assumptions and others. This is usually sufficient for understanding specific phenomena.

If the order of the system is small, say, less than 100, we can say that we are analyzing the system using a serial model. If, on the other hand, the system has, say, several hundred or more degrees of freedom, I will treat the model as a parallel representation. For example, we view weather systems or fluidic dynamical systems as parallel while rigid body systems are considered as serial systems. We tend to think that classifying systems as serial or parallel is only artificial. There is nothing fundamentally different between them. It appears that as long as we discretize the timeline sufficiently narrowly, say, by choosing the time interval between two measurements to be very small, we can convert a parallel system into a serial one. In this sense, we do not seem to need parallel analysis.

However, in chapter 28, I will show that free will is possible only if we view our brain in a parallel way. Serial analysis of the processes of our brain will never explain free will. In fact, several other human abilities are fundamentally parallel phenomena as well, as discussed in this book. Yet, in each of these cases, loops help in simplifying the complexity within parallel systems.

26.3 Examples of serial and parallel systems

Let me now discuss a few simple examples to illustrate the differences between serial and parallel analysis and to see why this difference is important for consciousness. Since the main difference between parallel and serial systems is in the order of the system, we usually employ a special way to group components so we can approximate a higher order system into a lower order one. Each of the examples below highlight different ways of grouping that we commonly use. In the first two examples, we choose a rigid body grouping as a natural way to reduce the order of the system. This is a generic grouping applicable to most nonliving things (and some aspects of living beings as well). In the subsequent examples, there is no obvious way to group them. In these cases, we rely on our consciousness and our brain's internal structure to help with the grouping.

(i) Balls rolling down an inclined plane: Consider an inclined plane and let us imagine rolling 5 balls down the plane. How would we study this system, namely, how do we answer questions like where the balls will end up, which one will reach the bottom of the inclined plane first and so on? When analyzing this system, one main assumption we make is what is called a rigid body grouping. We treat each ball and the inclined plane as rigid bodies and then apply Newton's laws. We study the trajectories of the center of mass of the balls instead of every single molecule within each ball. The

center of mass moves in a straight line down the inclined plane in a natural timeline. The other molecules in a ball have a cycloid motion, going around and down.

The rigid body assumption implies that the relative distances between the molecules do not change. We also assume that the inclined plane is smooth. The subtle imperfections on the surface of the inclined plane gives rise to a frictional force which we approximate as well. These assumptions together allow us to study the system as a serial system. The parallel motion of all molecules are not important to study for the set of problems we are interested in. Instead, we abstract all the molecules and represent them as a single entity (like the center of mass) and study its motion along a single flow-based timeline.

Furthermore, each ball is studied independently from the other. The equations of motion for a given ball has no interacting terms from the other balls. This assumption adds another simplification to the problem. We can solve each ball's equation of motion separately and plot/observe its trajectory like a flow based pathway independent of the other balls. This is convenient because different balls move at different speeds.

We have two ways to view the motion of these balls. One is to view each ball in its own independent pathway as a flow event. This can be called as a view *along* the direction of motion. Here, we only keep track of a single ball's motion at a time. The patterns we can recognize with this view are if a ball is moving in a straight line, if it is bouncing up and down, if it is turning, if it is hitting some walls or other obstacles and so on.

The other is to view different balls simultaneously. Here, we compare how each ball is moving relative to one other. This can be called as a view *across* the direction of motion. In this view, we are taking time slices and then checking where each of the five balls are at that time instant. In this view, the patterns we can recognize are which ball is ahead, which one is behind, which ball started slowly but overtook others, which ball started fast but fell behind and so on.

I call the first view as the serial view of the system while the second one as the parallel view of the system. Causality applies to the first view in a more direct manner than for the second view. Clearly, each ball's trajectory obeys the causal Newton's laws of motion in a direct way. However, when we look at the time sliced patterns of multiple balls at once, the resulting spatial patterns become somewhat arbitrary, though they indirectly obey the causal laws via the individual ball's motion (see Figure 2.1).

By 'somewhat arbitrary' what I mean is that if we perform the same experiment 100 times and look at three consecutive time slices T_1, T_2 and T_3, which includes all five balls' relative positions, they will be considerably different each time. If instead of five balls, we have 1000 balls rolling down each time and we look at the same time slices, the spatial pattern of the 1000 balls in each of the 100 experimental trials will be quite arbitrary. This is a very important difference that will manifest later as the root cause for free will.

Recall that in section 2.2, we have seen another similar example highlighting this difference i.e., a television tube with thousands of electron beams hitting the screen.

These electron beams are analogous to the thousands of balls discussed here. The spatial patterns produced on the television screen are analogous to the spatial patterns of the 1000 balls at different time slices. With a television, we know that the spatial patterns can be made completely arbitrary i.e., we can project any movie. This implies that the spatial patterns on the television screen do not seem to obey any set of physical laws even though each electron beam does obey the standard laws. Analogously, the 'spatial patterns' of the 1000 balls can be made to obey quite arbitrary laws even though each ball obeys standard Newton's laws. This is a unique property that will later manifest as the root cause of free will provided we sense and control the spatial patterns. In chapter 28, I will show how we can exploit this central difference to create a free will system.

Notice one interesting thing that happens when we take a single ball in the first serialized view and break it down into the trajectories of individual molecules within the ball. Each individual molecule has a cycloidal motion. Interestingly, these motions become the serial view while the entire ball's motion, which we previously called as the serial view, now changes into a parallel view.

Therefore, the notion of a serial view versus a parallel view is some-what relative. It depends on how well a given abstraction like a rigid body grouping obeys the standard laws. If the grouping does not obey the laws well enough, we may need to break it up into different, possibly, smaller groupings thereby converting what we initially had as a serial view into a parallel one. Most of the times, we do not break up the grouping all the way to an atomic scale and start with quantum mechanical laws. The features we want to study partially determines the scale of the components we want to use in the grouping.

(ii) Athletes in a race: The example discussed previously with the balls rolling down an inclined plane is quite generic. Several everyday events fall into the same category and, hence, uses a similar type of grouping like the rigid body grouping. For example, if you are watching athletes in a race like in Olympic games, we have two types of views too: (a) a serial view along the direction of motion tracking a single athlete at a time and (b) a parallel view across the direction of motion comparing multiple athletes simultaneously. Some of these examples are with track and field, swimming, boat racing, cycling, horse racing, car racing like NASCAR and others.

In these cases, we usually do not break up the grouping of, say, a single athlete into components like his hands, legs and body and track the spatial trajectory of these components individually. Instead, we look at the grouping of a single human even though the human is not a rigid body. A human has several degrees of freedom. Yet, our brain naturally groups them as a whole and tracks the entire group as an aggregated pattern. Strictly speaking, our brain keeps track of far more groups of patterns like the surroundings and the spectators as well.

There are two natural groupings here. The first one is a rigid body grouping that is present in, say, the components of a car, a bicycle or even a human (at least in an approximate sense). This grouping correlates well with the external objects' properties. Here, we typically treat a single entity as the grouping instead of breaking it into subcomponents. The second grouping, on the other hand, is unique to our own consciousness. Here, we group multiple components like multiple cars and multiple people as we track the dynamics of an entire 'scene'. Different conscious people watching the same event would have the same first grouping, but considerably different second grouping. This is because different conscious people will focus on different aspects of the scene. The neural-firing patterns within their brains will then proceed in different pathways.

The second grouping unique to our consciousness is not a convenience, but is necessary to become conscious itself. You necessarily have to ignore the detailed parallel dynamics across multiple components and instead group it appropriately to result in a serialized single dimensional timeline, if we want to become conscious. In spite of this serialization at an abstract level, the internal dynamics of neural-firing patterns themselves would work in parallel. There is still the issue that our brain needs to coordinate the neural firing patterns of hands and legs of the athletes. Otherwise, the hands and legs will not appear attached to their body even though it is obvious that they are indeed attached. This coordination will be guaranteed through abstract continuous paths, giving rise to a feeling of knowing and observing a complete human body (see chapters 22 and 23).

(iii) Fluid flow: Consider the flow of water in a river. We want to see how to study this system. Unlike mechanical systems where you have a notion of rigid body as a natural way to group molecules, fluid systems like water or airflow does not have such a natural grouping. Keeping track of every molecule' motion as a parallel view is far too tedious to study such phenomena. Fluid mechanics, therefore, uses a notion of control volume (as opposed to material volume) to study the dynamics. Control volume is an imaginary grouping that consists of a time-varying region through which fluid can flow in and out. By applying Newton's laws on a control volume, we get the familiar Navier-Stokes equation for fluid flow. This is a new way to serialize the parallel view.

Our approach here is quite similar to the previous examples, except that we are using a new *time-varying grouping*. The rigid body grouping in the previous examples did not vary with time. Yet, as before, we have avoided the study of individual molecular motion. We study aggregated patterns of the fluid flow using a control volume grouping. This example shows a new way to serialize a seemingly complicated parallel dynamical system. Note that we did not create new laws to study them. We simply used the same Newton's laws, but reformulated them in a form suitable to control volumes.

This is interesting because while rigid body is an abstraction at one extreme where the relative molecular distances do not change at all, the examples here with fluidic systems are at other extreme. They have no natural substructures moving together as a whole. The structure within a living cell can be seen as something in-between these two extremes. You have a partially rigid structure made of actin filaments and microtubules of the cytoskeleton and a partially fluidic structure with chemicals flowing in and out of the nucleus or in the cytoplasm. What then is a natural grouping for cells? In this book, we have seen that physical and chemical loops form a natural grouping for studying large-scale dynamical systems.

(iv) Reading a sentence: Let me now consider an example of reading a simple sentence to highlight the dynamics that occur within our brain. When we look at a sentence, the image of the letters and words fall on the retina and proceed along the optic nerve to the primary visual cortex. These parallel set of inputs disperse within the brain. What is a natural neural grouping to consider when trying to explain how we recognize words and understand sentences? There is a one-one correspondence with the set of neurons on the retina or along the optic nerve to the shape of the letters and words. However, as we look beyond the primary visual cortex, we lose this one-one mapping because of the complexity of interconnections within our brain network. This complication implies that the neural grouping at the retina has limited use. We have seen that fixed sets for the letters and words are the correct groupings to identify and to keep track of.

Now, as we read the sentence, we need to ensure the time evolution of the parallel dynamics proceeds in the correct order of the sentence sequence. This is possible through abstract continuity as discussed earlier (see chapter 21). The fixed sets are linked together to have nonempty predictive meta-fixed sets. This gives rise to a natural grouping and serialization, resulting in a feeling of knowing as well. It is important to keep the entire abstract continuous path excited as you read a sentence. Otherwise, you will not know or understand the sentence. We can clearly see this difference in a 4-5 year old. A 4-5 year old reads so slowly that he forgets the first word when he reaches, say, the third word unlike an adult. He, typically, is unable to pursue the parallel dynamics fast enough to understand it. Yet, when an adult reads the same sentence quickly, we see that he does indeed understand it.

In general, our consciousness itself chooses a serial abstract continuous path along our unique timeline even though our brain processes parallel excitations for most abstractions. Our focus mechanism (using the fundamental pattern of chapter 16) indeed directs the energy along these paths to both sense and control the direction. This aspect of our consciousness makes it look as if most of our abstractions like pictures, shapes, colors (i.e., appropriate groupings) and thoughts are serial. Therefore, even in this example, we needed to choose an appropriate grouping (fixed sets and abstract continuous paths) to serialize and, hence, make it easy to analyze a complex parallel dynamical system.

(v) Water waves and other examples of disturbance propagation like sound waves: So far we have seen three types of groupings with the previous examples – rigid body grouping, time-varying control volume grouping and fixed sets grouping. In this example with water waves, I want to show that the natural grouping varies the subcomponents themselves considerably in addition to making the grouping vary with time. Consider ripples of water waves in a pond or waves reaching the shore at a beach. When you are observing these waves, you clearly see and feel a water wave pattern approaching you. But, what exactly is moving towards you as the water wave reaches you?

To compare, consider a ball rolling towards you. When the ball is approaching you, the entire set of molecules of the ball do indeed move towards you, though not necessarily in a straight line (i.e., they take a cycloidal path except for the center of mass). However, with water waves at the beach, the most interesting thing is that there is no equivalent 'real' physical motion approaching you from far away. The perception of motion of the water wave is abstract and is entirely within our brain. In fact, only a disturbance (an approximate up and down motion) propagates. Yet, we see and feel as if it were a real motion (like ripples approaching or receding away).

This is true even with sound waves and vibrational waves in solids. The air molecules only move back and forth with a minor displacement. But the speed and distance propagated by the disturbance itself is very large. For example, sound waves travel at about 330 m/sec in air and about 4000 m/sec in glass. They travel several hundred meters quite easily whereas the air molecules themselves move only a few centimeters. Even with nerve cells, the voltage spikes propagate as a disturbance sometimes at 100 m/sec even though the Na^+ and K^+ ions themselves travel only locally (a few centimeters at most). In fact, this is true with all types of waves that have real physical internal dynamics.

Let us now look closely at the crests and troughs of water waves. A set of molecules A form crests at time instant T_1. At the next time instant T_2, a different neighboring set of molecules B form the crest while the previous set of molecules A dip a bit. At time instant T_3, a new set of molecules C form crest and the set of molecules A and B at T_1 and T_2 respectively have dipped lower. This process continues over time.

Now, as the wave propagates, it is the crest that moves towards us. Let us look at this motion at a molecular level. Conceptually, our eyes need to look at one set of molecules A at T_1, then switch to a different set of molecules B at T_2, then switch again to another set of molecules C at T_3 and so on, each time picking only the molecules corresponding to the crest (or a slightly wider region that includes the trough as well). This is because as we observe a wave propagate, we typically focus on the crest and track its position as it moves. Our brain maps these images and the motion as if we were observing the same crest. Therefore, we now have a new abstract grouping that allows us to visualize a moving crest. Had we not kept switching the molecules at each instant and instead observed the same set of molecules during the entire time period, we would have simply seen the molecules move up and down with a slight lateral displacement.

The visual switching of molecules within our brain happens automatically to create the feeling of a wave motion. In fact, it is quite difficult for our brain to see just the non-switching scenario (with up and down motion). How does our brain perform the switching automatically? When the image of the wave pattern (a region of about one wavelength) falls on our retina, our brain takes the 2-D pattern and excites the corresponding *fixed set* at time instant T_1. When the wave propagates by a small amount, the 2-D wave pattern formed by the new set of molecules at T_2 does not differ significantly from the previous 2-D pattern at T_1. We treat this small variation as part of the dynamical transformations included in the very definition of truths and fixed sets (see definition 13.1).

Therefore, even though the set of molecules making up the 2-D pattern have changed completely, the new pattern on the retina excites the same fixed set. Our brain thinks that it is the same 'object' because of the same fixed set, when in reality there is no such object for an extended period of time. Our brain detects the displacement of this imaginary object as another fixed set as described in section 18.3. Therefore, our brain tracks this as a slightly time-varying pattern (fixed set) moving towards us.

The main difference between the natural grouping of this example and the previous examples is that we constantly change the set of molecules we want to track as a group at every time instant.

What set of laws do the wave patterns obey? Even though the individual molecules obey the standard physical laws, it is possible to create wave patterns that obey sufficiently arbitrary laws. We can do this using, say, a more viscous liquid than water and by combining it with an internal energy generation mechanisms like a simple propeller. This example can be generalized to cells since they have internal energy generation mechanisms (mitochondria), better sensing and control mechanisms to guide the movement of molecules and other features. We will discuss this in greater detail starting from chapter 59.

The unique feature of switching the set of molecules in the grouping with this example has one interesting advantage. Since the molecules M_1 at time T_1 are different from the molecules M_2 at time T_2, the forces acting on these molecules M_1 at time T_1 are different from those acting on molecules M_2 at time T_2. Therefore, causality is not strictly maintained. Recall that to maintain causality, we have to track a molecule' effect (i.e., its motion) in response to a cause (i.e., a force acting on it). But after applying forces to molecules M_1 at time T_1, we are no longer interested in tracking M_1 molecules' effect at time T_2. Instead, we are only interested in a different set of molecules M_2 at time T_2. Therefore, before we want to study the causality of molecules M_1, we already switched our attention to molecules M_2. The switching process continues at every instant. This is particularly advantageous because our goal with creating a free will system is to create patterns (like water waves) that do 'not' obey standard causal laws of nature.

Of course, strictly speaking, there are relationships between these forces from, say, the continuity of motion and incompressibility of the liquid. Nevertheless, we can change water to some other liquid with special viscous properties mentioned earlier to lessen the causal interaction-based effects across molecules. With such viscous liquids, we could imagine introducing considerable arbitrariness to the set of laws obeyed by the grouping as a whole (at least, as perceived by you). For example, we can setup conditions to create complex wavy patterns making it seem that they are not obeying any fixed laws of nature.

Another way to lessen the above causal effects arising from interactions across components is by increasing the structured complexity of the system. Water and other viscous liquids still have very simple internal structure. Therefore, the easiest way to lessen the interaction-based causal effects with these liquids is by looking at larger distances i.e., after the wave has progressed several meters. The cumulative causal interaction-based effects drop considerably a few meters away. On the other hand, if the internal structured complexity is very high like within a cell or within our brain, then we do not have to wait until the disturbance propagates to large distances for the causal effects to diminish. The neural firing patterns become intertwined very quickly that large subsets of neurons no longer obey any meaningful causal laws. I will discuss this in greater detail when we study free will in chapter 28.

If, instead of *us* setting up conditions, had the pattern or the system as a whole sets them up by itself, modifies and controls these very conditions, how would it feel to an observer like us? This is not a fictitious question. One example of such system is a human being. The resulting feature is what I call as free will (i.e., I am not considering the philosophical interpretations that people assign to the problem of free will). We will study this in more detail in chapter 28.

26.4 Grouping components – serializing a parallel system

The examples discussed in the previous section have several common features among them when viewed as both serial and parallel systems. In fact, we should emphasize that we get a more accurate description of the external world when we view it as a parallel system. The serial view that we usually take is a simplification. Our consciousness, in a sense, hides the parallelness. We naturally create groups that when tracked gives us an impression of serialness along our perceived unique timeline. We can now generalize this notion of grouping components as a way to serialize any parallel system. Let me summarize this here.

A parallel system containing thousands of components is quite difficult to analyze. The individual paths and the number of interactions between these components soon grow to levels making it computationally intractable. It is, therefore, useful to group a large number of components together as an entity and study its motion as a whole. We can view this process of grouping components as serializing a parallel system. In the previous section, we have seen how to do this with several examples.

From the examples of previous section, it became apparent that there is a hard parallel computation problem that our brain solves using fixed sets and abstract continuous paths. Specifically, the problem is as follows. If you have independent streams of data generated and received through a set of parallel wires, how would you take the correct cross-sections across the streams and track its time evolution correctly? You can imagine these streams to be firing patterns from retina along the optic nerve. If you, say, blink your eye, the streams will become out-of-sync. Also, the streams will travel in different pathways if the network is considerably interconnected. This increases the chances of the streams to go out-of-sync. How would you re-adjust so that you can start tracking the correct cross-sections once again? The solution to free will and, in general, to the problem of consciousness provides a definite approach to solve this problem.

When we start looking at complex systems like our brain, single or multi-cellular organisms, the presence of parallel dynamics becomes quite apparent. What is the best grouping for such complex physical and chemical systems? For most solids, we have seen that rigid body groups are the typical ones we use. For fluids, the typical ones we use are the groups based on control volume. In situations when the fluid dynamics is more energetic or random like air molecules in a room, we use a statistical approach instead. Once we choose an appropriate grouping, we can study their interactions by keeping track of forces and energy transfer between them. This approach is quite reasonable for simple low order systems. However, for complex systems the above groups are not useful.

For example, if we are studying chemical cellular systems, the trajectory of each individual molecule is not important (cf. fluid mechanical systems). Yet, we do need to keep track of different chemicals created by exchanging atoms between molecules (i.e., through chemical reactions). The properties of resulting chemicals created are considerably different from the reactants. For example, H_2O is considerably different from H_2 and O_2. More commonly, our body creates different enzymes using the same set of amino acids. Studying how different components merge or break down, both physically and chemically, is necessary for understanding each of the behaviors exhibited by these complex systems. Either a statistical approach or the detailed tracking of each component is not helpful here.

Instead, we look for external conditions creating a group of components that form a stable pattern for a finite period as the natural group to study. Loops are best examples of such natural groups to study for our brain and for chemical systems. With brain, we have already seen the use of loops as fixed sets in the previous chapters. We do not track the exact set of chemicals and their trajectories. Instead, we only study how these chemicals move in and out to form stable dynamical loops.

When we have multiple parallel loops, we study the interconnections between them, though not just any such interconnections. Abstract continuous paths are the natural set of interconnected pathways to keep track of. In the beginning, we will not have abstract continuous paths between the meta-fixed sets. However, with new experiences and memories, the set of interconnections between the fixed sets start to become denser. A given pathway between linked fixed sets is no longer a 'thin' curved path. Instead, it starts to appear like a thicker pathway with a wide band of cross-sectional branching on all sides (see Figure 21.2 of section 21.6). These branches give rise to nonempty predictive meta-fixed sets as you traverse along the pathway.

Additionally, from example (v) of the previous section (i.e., with water waves), we have seen that the group itself can vary with time and can contain different set of subcomponents at every instant. The fixed set representation of most abstractions like ball, balloon as well as shapes, sizes and distances are examples of such groups. In fact, all abstract concepts, translated into internal firing patterns, we want to keep track of for understanding consciousness are valid examples of useful groups.

In cellular systems, we can similarly study chemical loops as natural groups. When we have a network of such looped reactions, we can study how a disturbance propagates within this network (see chapter 58). This can give rise to interesting behaviors we observe in cells as we will see later. Other examples of these looped mechanisms are the stable-variant merge loop pattern and the active material-energy loops, which will be discussed later in chapters 53 and 59 respectively. In general, if we choose the groups and their interactions properly, it dramatically reduces the complexity of analyzing a given parallel system and transforms it into a serial system. In the next chapter, I will discuss this from a systems' perspective.

27
Solving Dynamical Equations of Parallel Systems

The differential equations of a complex parallel system (or any sufficiently accurate model) with millions of degrees of freedom are quite difficult to write explicitly. Attempting to solve them is simply infeasible. However, in some cases, it is possible to study the properties of these equations and solutions in a qualitative way. In this section, I will discuss this qualitative aspect in some detail. I will pick a fluidic system like a smoke ring to study this initially. Later, I will extend the analysis to include our brain as a complex parallel system as well.

A smoke ring is an interesting example because the shape of the smoke ring is quite fluidic. It starts circular but distorts to other approximately circular shapes for several seconds before it finally collapses. These distortions are visually interesting to watch. They do not seem to obey any physical laws. A smoke ring highlights an interesting fact that a real physical nonliving dynamical system is capable of exhibiting complex patterns albeit for a short period. I will use this example to highlight that even simple systems (compared to living beings) can be built from components obeying standard laws individually and yet, as a whole, exhibit dynamics that are considerably arbitrary.

I will show how we can derive (qualitatively) the previously mentioned notions of fixed sets and abstract continuous paths from within the dynamical equations of our brain. I will use these results in chapter 28 to show how to develop free will in SPL systems.

27.1 Interactions between system and environment

Consider any grouping in a parallel dynamical system like a smoke ring or water waves on a beach. Each individual molecule within the system does obey standard physical laws. These laws are not arbitrary. They generate unique trajectories for each molecule. If we look at each individual molecule, whether we know these trajectories or not, we know that there is no choice or randomness in their pathways at a visible (or macroscopic) level. However, if we look at the entire grouping itself, like a smoke ring, it does appear to have certain randomness to its time evolution (say, in its shape). When we try to repeat the same experiment multiple times, we hardly ever see the same shape variations. This is the case even with individual molecular trajectories. Yet, somehow, the shape of the smoke ring variations feel more random than the molecular trajectories even though we can ultimately represent these shapes in terms of the individual molecular dynamics.

The reason for this difference in perception is both subtle and important. In the case of individual molecular dynamics of rigid body structures, the interaction between the system (the rigid body) and the environment (everything outside the rigid body) is definite, less time-varying and easy to keep track of. This helps eliminate unknowns in the dynamical equations of motion of a rigid body, helps contain the errors and measure the desired properties. This is the reason why we are able to determine the system dynamics of rigid bodies quite accurately using the fundamental law based approach (i.e., using Newton's laws).

With more fluidic or parallel systems like smoke rings (unlike rigid bodies), even minor variations in the internal dynamics of the system or from the environment produce visible changes in the shape of the smoke ring. In fact, the effects you have decided to ignore like the tiniest disturbances are precisely what gives rise to the important qualitative features in such fluidic and other large-scale complex systems. For complex systems, the ignored effects in your mathematical models idealize the analysis too much that not only is it not useful, but all the interesting features exhibited by such systems lay hidden within what you have ignored.

Furthermore, the ignored effects (like resistance, friction, turbulent forces and various other interactions between molecules in a smoke ring) are partly unobservable and highly time-varying. This imposes severe practical limitations if we want to use the fundamental laws based approach for finite order complex parallel systems. In addition, there are theoretical limitations of using fundamental laws for these systems (see section 1.3). Given these issues, I will study the properties of complex parallel systems qualitatively instead.

27.2 Dynamical equations of a parallel system

Let me now state the objective when studying systems like smoke rings, our brain or chemical cells. We want to understand the physical laws obeyed by the smoke ring as a whole, i.e., its time evolution. For example, we would like to know how and under what conditions smoke rings take a square shape, which then slowly morphs into an oval shape followed by a wiggly circular shape, say. We want to know if it is possible to generate any such reasonable time-evolving pattern of shapes. If these patterns are possible with help from an external system, can we instead make them as part of the main system (i.e., the smoke ring) itself? We would then have the resulting system create its own conditions for its own dynamics, i.e., a desirable feature comparable to free will.

Let us imagine expressing the motion of all individual molecules and their interactions as appropriate large-scale nonlinear vector dynamical equations $\dot{X}(t) = F(X(t), U(t), t)$. This is a mathematical model expressed only in a theoretical sense because the systems of interest have millions of degrees of freedom, which are difficult to write down explicitly anyway. Here, $X(t)$ is a state vector with as many components as the number of degrees of freedom. $U(t)$ are the extra variables from the external environment affecting the dynamics of the system. If we are trying to control the system, we can use $U(t)$ as some of the control variables as well. $F(.)$ is a nonlinear function representing how the dynamics of the system evolves. $\dot{X}(t)$ is the rate of change of $X(t)$. For some systems, the above ordinary differential equations would be replaced by partial differential equations instead (like with Navier-Stokes equations for fluidic systems).

Theoretically speaking, we can write these equations for each molecule separately.

We could then keep track of the forces acting on this molecule from the neighboring molecules and the external environment. We use fundamental laws of physics to account for the physical dynamics to a sufficient degree of precision. These equations are millions in number and hence are impractical to specify in precise terms. Therefore, we imagine this process of writing these equations in a conceptual sense. Within these equations, we have several external inputs that we cannot compute or specify ahead of time like the interactive forces between the molecules and from the environment. As a result, we cannot solve these equations unless each of these millions of time-varying external inputs are specified. As mentioned earlier, even simple approximations can produce entirely incorrect predictions for such large-scale interacting dynamical systems. However, from a conceptual point of view, we can theoretically encapsulate any observed behavior within these equations with an appropriate choice of external inputs.

Therefore, in a theoretical or conceptual sense, we can say that we now have the laws obeyed by a smoke ring or any other complex grouping in a parallel system. However, it looks like we have not gained anything useful with this representation. What we have as equations is too generic and sufficiently incomplete because of the presence of millions of degrees of freedom with many unknowns that we cannot be precise anyway. What we want are answers to questions like if the smoke ring can morph into certain shapes or not. There is still a wide gap between the imprecise equation representation and the types of questions for which we want answers for. Let us try to close this gap. In order to do that, we should identify why there is a gap in the first place. Let me list a few reasons here.

(a) Invariance when specifying the objective: The first important point to understand is how we can represent our objective within these millions of equations of motion. The language used to describe the time evolution of a smoke ring is necessarily imprecise. For example, when I say that I want the shape to be a square morphed later into an oval, I am not extremely keen on expecting the edges to be perfect straight lines or that the angles are perfect right angles (in a mathematical sense). The smoke ring does not have an inherent rigidity to its shape to warrant such precision.

We only need a visible verification of an approximate square or oval shape by a conscious being. This applies not only to spatial dimension, but also along the time axis as well. For example, I cannot detect the change in shape unless it happens within, say, tens of milliseconds or even slower (> 100ms).

Similarly, we only care to specify imprecisely how long we want it as an oval shape. We do not say, for example, that we want the oval shape for exactly 1.5 seconds. If our specification is off by a few hundred milliseconds, we do not particularly care. On the other hand, the above theoretical laws and mathematical equations have very high precision (in both space and time). Our specification, on the other hand, does not require similar precision. In fact, we can say that our imprecision forces certain invariance in the specification of our objectives. The invariance refers to all minor variations in our specification (of, say, the shape) that still gives the same output shape. This invariance is an important property that helps us study the equations qualitatively, as we will see later.

(b) Objectives as abstractions: The above invariance implies that the language used to describe the dynamics versus the one used to describe the objective are different. The former is physical and the latter is abstract. We have identified such a difference for our brain in section 14.4 i.e., a neuron language for dynamics and a higher-level language based on fixed and meta-fixed sets for abstractions and truths. When we defined a fixed set or a truth (see chapter 13 and 14), we mentioned that it is invariant under a set of dynamical transformations. For example, the fixed set for a circle, square or a table is the common set of looped firing patterns within our brain that continue their dynamics even under dynamical transformations like small translations and rotations. This is the correct way to study abstractions, namely, by linking them to the dynamics.

In general, shape is an abstraction unlike the molecules that created this shape. When a conscious being verifies that the smoke ring has an oval shape, the same invariance used in the corresponding fixed set will be applicable now. In the same way, when we describe the time evolution of a smoke ring, we specify it only in the above abstract sense. Every word we use in our description of the dynamics of the system have a large set of the above invariances.

(c) Non-unique solutions due to the imprecision: Even if we try to specify the objective precisely, it is already possible to achieve non-uniqueness in the solution. This is from the inherent symmetries in the solution, i.e., the symmetries of a square or a circular shape of smoke rings. However, the moment we start factoring the above imprecise nature of specifying the objective (as discussed in (a) and (b) above), both in space and time axis, the non-uniqueness in the solution becomes extremely common. This is aided by a lack of rigidity, a large set of external inputs and enormous choices in their variations. This is particularly true when the systems are fluidic and have higher complexity through stable parallel looped dynamics like within our brain and cells. We need a way to exploit these symmetries and state them in qualitative terms at least. In the case of our brain, they do turn out to be the dynamical transformations in the definition of fixed sets.

(d) Easier control of solutions by external systems: The non-uniqueness in the solution makes it appear as if the smoke rings are quite random and unpredictable. In addition, even minor disturbances produce visible changes. Now, if an external conscious being were to control the set of conditions, it becomes much easier to produce a wide range of variations in, say, the time evolution of the smoke rings. For example, we can change the conditions in the middle of the way and morph the shape to produce an entirely different pattern. The control does not have to be extremely precise at a molecular level. Instead, it can be broad and yet local in space and time. Simple mechanisms like blowing wind through several narrow pipes at specific locations or angles gives us enough flexibility already. In the case of brain, we have already seen several such mechanisms like the fundamental pattern and others. With cells, I will similarly discuss several important mechanisms like stable-variant merge loop pattern and active material-energy loops later in chapters 53 and 59.

From the above description (a) – (d), it became clear that an external conscious being could control a complex parallel system to let it produce seemingly arbitrary dynamics and, hence, let it obey arbitrary laws itself. This is a desirable property for free will, as we will see later in chapter 28. This arbitrariness is possible in spite of having a detailed set of coupled dynamical equations representing all degrees of freedom. As a matter of fact, this is partially true even with rigid and low order systems like a ball rolling on an inclined plane. A conscious being can control the angle of tilt in various directions to let the ball move in an arbitrary trajectory. It need not be a random pathway. Instead, it can be well-defined and chosen at-will. The main difference and, in fact, the challenge is to let the system itself control the very conditions for its own dynamics. I will discuss this in chapter 28.

The purpose of the four observations (a) – (d) highlighted here is to study the millions of equations of motion in a qualitatively sense. Let us now apply these observations specifically to our brain's dynamical equations (assumed to have been stated in a theoretical sense) to see if we can infer their non-unique solutions in a qualitative way. The main qualitative results I want to show are: (i) the non-unique solutions of our brain's dynamical equations of motions in a qualitative sense are precisely the fixed sets identified earlier in chapter 14 and (ii) the time evolution of these non-unique solutions for our brain's dynamical equations in a qualitative sense are precisely the abstract continuous paths identified earlier in chapter 21.

Therefore, even though it is difficult to write down the millions of equations of motion for the dynamics that occurs within our brain, we can still qualitatively state the solutions to these equations (if they were indeed written down) as the fixed sets and abstract continuous paths. This is quite surprising because the concepts of fixed sets, abstract continuity and the self-sustaining membrane of meta-fixed sets were introduced primarily to address the necessary and sufficient conditions for consciousness as required by Theorem 23.2. However, we now see that the same concepts emerge as the non-unique solutions to the millions of differential equations representing our brain dynamics itself.

27.3 Equivalence class of solutions = fixed sets

Consider looking at a specific table in your room. We know that a pattern of retinal neurons fire, propagate through the optic nerve and dissipate through the brain via the primary visual cortex. As in the previous section, let us imagine representing the dynamics that occur within our brain as a large set of coupled differential equations (with millions of degrees of freedom) in a theoretical sense. What is an example of one solution of these coupled differential equations?

The key observation and an answer to this question is the following: *one solution to our brain's differential equations can be obtained by directly looking at one realization of the physical dynamics that occur within our brain itself.* In other words, each realization of the real physical dynamics corresponds to a solution to our theoretical model if we claim that the theoretical model is an accurate representation of the physical dynamics of interest. This is the connection between a mathematical model and the true realization of this model in the external world. In fact, this is true with every physical dynamical system.

For example, imagine writing the equations of motion of a ball rolling down an inclined plane along with a set of external forces representing possible obstacles or bumps on the inclined plane. If you let the ball go, the actual physical path taken is precisely one specific solution to the differential equations of motion with an appropriate choice of external inputs. If you let the ball go a second time, you have discovered a second solution to the same differential equations, if it takes a different path. To get a different second solution, the set of external inputs and/or the initial conditions must have been different. Therefore, every time you perform the experiment by rolling the ball down the inclined plane, you are getting a solution to the differential equations of your theoretical model of this phenomena. Let me state this formally as follows.

Claim 27.1: A theoretical model for physical phenomena is said to agree with the experiments if and only if each experimental outcome, which can be seen as a single realization of the physical phenomena, is a solution to the equations describing the theoretical model.

The above claim implies that if you have performed an experiment once, you must have discovered one solution to the differential equations that describe the phenomena. The more experiments you perform, the more solutions you will discover to your theoretical model (under different inputs and/or initial conditions). Some of the solutions are same, but mostly the solutions are different because either the initial conditions or the input parameters/forces are slightly different.

Consider what happens when the experimental outcome does not match the theoretical model for any input or initial conditions. Let us say that we have characterized all the solutions to the differential equations of motion for a given phenomenon of interest. Let us now perform the experiment corresponding to this phenomenon. What happens if this particular experimental realization does not match our previously characterized solutions for the theoretical model? What can we then infer?

The only inference is that our theoretical model is incorrect for this phenomenon. There must be some assumption or approximation in our theoretical model that is either invalid or is important enough but ignored, thereby preventing us predict the specific experimental outcome. This will allow us to reexamine our assumptions and fix the mathematical model. This process is usually iterative until we reach a state where the theoretical solutions and the experimental outcomes are within acceptable error bounds.

This is the typical scientific process we employ when discovering new laws and models. I just formalized it here as a statement so we can use it even in the case of large-scale dynamical systems that have millions of degrees of freedom like our brain, our cells and other SPL systems. The second reason for calling out the Claim 27.1 explicitly is so we can make qualitative statements about the behaviors of large-scale dynamical systems even without solving any dynamical equations. This is particularly useful for systems like smoke rings and our brain because there is little hope of specifying all the internal dynamical relationships much less solving them.

Let us now see how we can apply Claim 27.1 to understand qualitatively the solutions of differential equations of dynamics occurring within our brain. Consider the problem mentioned at the beginning of this section, namely, that of looking at a table in your room. The external inputs are the retinal inputs that trigger cascading neural firing patterns. In addition, several internal inputs that supply energy are also present within our brain. All of this dynamics can be represented as millions of coupled differential equations. What is a solution to these millions of differential equations? If we keep track of all the real physical pathways, we can say that we have discovered one solution to the million-order differential equations according to Claim 27.1. If you tilt your head and look at the same table, you have discovered a second solution to these differential equations (again from Claim 27.1). If you walk towards or away from it, close one eye or jump up and down as you watch the table, you are similarly discovering different solutions to these equations with the corresponding external and internal inputs according to Claim 27.1.

Since this process is identical to keeping track of individual molecular pathways in a smoke ring, what is the equivalent notion of an oval shaped smoke ring in the case of our brain? When we defined the oval shape for a smoke ring, we noted that we could only do it imprecisely (see section 27.1). Furthermore, we defined oval shape by taking a time slice *across* multiple molecular trajectories, not *along* the timeline of the trajectories. We then said that a large number of solutions to the differential equations, all correspond to the same oval shape. Similarly, we now have multiple solutions to the differential equations of our brain corresponding to watching a given table. It is too restrictive to call any one of the precise solution pathway as the table (just as it is not reasonable to call a single occurrence of an oval smoke ring as 'oval'). The correct representation for a given table within our brain is the common fixed set obtained by taking each of these *multiple solutions* into account. This unique fixed set is regarded as an equivalence class (which is like a special grouping) for a given table. The equivalence class allowed us to satisfy similarity and dissimilarity conditions of chapter 7 – two similar variations belong to the same equivalence class while two dissimilar variations belong to different equivalence classes.

This fixed set pattern, though does not have the appearance of a table within our brain, is derived from the non-unique solutions of the differential equations. The non-uniqueness comes from the large set of dynamical transformations (like moving towards or away, jumping up and down or turning our head side to side and others) that make us realize that it is still the same table, just as it makes us feel that it is the same oval smoke ring. This approach is applicable for every fixed set including edges, angles, odors, tastes, concepts and abstract language constructs as well.

Let us briefly look at the meta-fixed sets as well like, say, a table meta-fixed set. In this case, you are looking at different tables with different shapes, sizes, colors and material compositions. Each of them does have their corresponding fixed sets. However, there is a table meta-fixed set, which represents the fact that all of them are tables. In terms of the non-unique solutions, we can view the meta-fixed set as derived from several fixed sets as we view them as 'external' inputs (or indirectly from the retinal inputs of several different tables). The real physical neural excitations beyond the fixed sets, in a space-time sense, are the solution pathways for the meta-fixed set.

In this way, every fixed set and meta-fixed set is an abstraction derived from the non-unique solutions of the corresponding set of differential equations modulo the set of dynamical transformations. We can treat each of them as an equivalence class containing a set of non-unique solutions resulting from reasonable dynamical variations. It is now easy to include the dual space representation as well within this equivalence class. For example, we do not have to look at the external table to trigger the same table fixed set solution. Instead, we can find an alternate solution pathway using differential equations that represent neighboring or related thoughts. We typically trigger this by imagining the context of when and where we saw the table while, say, we close our eyes or we see in our dream.

27.4 Time evolution of equations = abstract continuous paths

Let us now look at the time evolution of these dynamical equations for a time much longer than it takes to identify one fixed set. If we compare this with the smoke ring example, we want to study an equivalent situation to how the shape morphs beyond one shape like, say, from a square to an oval and then to a circle. The time it takes to form one shape is analogous to the time it takes to identify one fixed set. The entire sequence of morphing of shapes is then analogous to long events like those we see with our eyes, sentences we talk in a language, our thoughts, sentences we hear and so on.

The time evolutions we are interested in include every situation that involves linking a large number of fixed sets together to form a meaningful event. Imagine reading or forming a simple grammatically correct sentence. As you read each word one after the other, you are morphing the neural patterns in a specific linear order from one fixed set to the next (just like with the smoke ring). Each solution to the set of differential equations of your brain corresponds, in this case, to one whole sentence (see Claim 27.1). With dynamical events, we have a similar morphing of firing patterns as you try to understand a sequence of tasks or objects' dynamics. With thoughts, this occurs when we imagine fictitious scenarios or analyze some situation and so on.

Are there non-unique solutions to the above differential equations owing to the imprecision in how we specify or understand these events? The claim is that they are precisely the abstract continuous paths. For example, as you read the first word in a sentence, your brain automatically triggers the nonempty predictive meta-fixed sets. We do not pursue most of these parallel excitations. They can vary considerably and can be different each time. We, therefore, treat them as part of the allowed imprecision. Hence, each of these possibilities becomes a valid non-unique solution according to Claim 27.1. When you move to the next word, you do narrow the set of non-unique solutions. Yet, you also add new predictive meta-fixed sets corresponding to the second word. This process continues as you read the entire sentence (see section 21.1). We can now generalize this discussion to all situations involving abstract continuous paths. Recall that to get one solution for the set of differential equations that model a physical system, you simply have to pick one instance of the real physical dynamics itself (see Claim 27.1).

The interesting conclusion from this is that you are able to move along seemingly *arbitrary* abstract continuous paths. We can trigger these paths either externally or through our own internal thoughts. For example, we can think of any grammatically correct sentence or abstract continuous path quite easily. The complexity of the neuron architecture and the ability to form stable parallel loops is directly responsible for our ability to morph from one complex pattern into another.

It is, however, not true that these are 'arbitrary' paths, in an absolute sense, for a given human being. They are, in fact, unique to our memories created through our experiences. For example, people speaking different dialects or different languages have natural accents that are hard to break. In this sense, they are not obeying arbitrary laws. Rather, they are following similar laws based on their past memories. We can view their abstract continuous pathways as the lowest set of 'laws' (especially when we consider the physical neural dynamics) that will be obeyed in most situations. The set of differential equations of his brain will naturally pick them as valid solutions. Eventually, this will lead to the creation of a unique self-sustaining membrane of meta-fixed sets (see chapter 19), which identifies his personality in some sense.

In the next chapter, I will show how the current discussion and the natural groupings in our brain, namely, the fixed sets, abstract continuous paths and the self-sustaining membrane of meta-fixed sets can give rise to free will.

28
Free Will

In chapters 26 - 27, I have described the difference between serial and parallel systems and how time-varying grouping of components in a system help in simplifying the study of parallel systems. In the previous chapter, we looked at how natural groupings like fixed sets and abstract continuous paths are useful for studying our brain. We have also seen the importance of these groupings qualitatively from a systems modeling point of view. Specifically, we saw that the solutions to the millions of coupled differential equations of dynamics that occurs within our brain as we observe an event multiple times with slight variations correspond precisely to fixed sets (section 27.3) and their time evolution to abstract continuous paths (section 27.4). I will show that we can make the laws obeyed by these natural groupings to be quite arbitrary even though their individual components do obey the standard physical laws. In this chapter, I will explore this in further detail and show how we can explain free will.

The key idea is to build a system capable of sensing and controlling the very environment the system operates in. The dynamics of a system in general depends on several external environmental conditions as well. Who sets up these environmental conditions? If the system itself can sense and control the very conditions needed to operate its own dynamics, we would have an interesting loop between the system and the environment itself.

For example, a ball rolling down an inclined plane takes a specific path due to the ball's own physical properties (system) as well as the inclined plane's properties and other external conditions (environment). If the system is capable of controlling part of the environment itself like the angle of inclined plane and its tilt in all directions, we can make the path of the ball go left, right or along any other reasonable curve. While it is nontrivial to take this specific example and make its dynamics self-sustain in the above sense, I will show that we can do this in a complex stable parallel looped system like our brain.

This is, however, not enough for free will. The system needs to *know* the available choices, pick one of them, not randomly but intentionally (in a sense) and know that it is making these decisions. Nevertheless, this self-sustaining sensory-control loop between the system and part of the environment itself is one of the first steps towards the perception of free will in human beings.

28.1 The problem of free will

Let me first list the set of questions that we should answer for the problem of free will.

1. *Obeying arbitrary laws*: The main issue with free will is to address how a system is capable of obeying any law (in a reasonable sense) even though every

subcomponent within the system still obeys fairly well defined physical laws. For example, I am able to move my hand in any direction I want, within a given region. If you claim to have discovered a set of laws that all physical systems should obey and use them to predict how my hand is going to move, I am confident that I can falsify your laws and predictions by moving in a different direction. This decision making process appears well controlled. The issue is not whether our ability to obey arbitrary laws is true in an ideal sense. Instead, the objective is to explain the source of this controlled decision-making and what gives us the ability to obey seemingly arbitrary laws, especially since it happens only for special systems (like, say, humans) and under special conditions (like only when we are conscious).

2. *Choices*: Free will gives the system an ability to choose from a set of possibilities. When we look at simple systems obeying physical laws, the mathematical models for these systems usually exhibit unique solutions. A ball rolling down an inclined plane modeled using Newton's laws does not have non-unique choices, say, to move along two directions. If we view the equations of motion as ordinary differential equations, a slightly strong form of continuity (Lipschitz continuity) guarantees both uniqueness and the existence of a solution. If this were a system with free will that seems to have choices, the mathematical model should first be rich enough to support non-unique solutions. Otherwise, the assumptions and approximations in the model should be considered as too severe. If it did have non-unique solutions at a given instant, how does the system decide which direction to proceed along? Does this decision come simply from randomness, quantum mechanical uncertainties, minor disturbances or some ignored effects? If not, who is evaluating and making this decisive choice?

3. *Knowing choices*: In order to pick one outcome from a set of choices, the system should know what the choices are and what they mean. Otherwise, the process of picking one outcome sounds as a random process originating from minor disturbances, ignored effects and others mentioned earlier. The system should understand that these are valid choices and that it can pick any one of them without violating any physical laws. It should also realize and have an ability to change its choices at any place or at any time at-will. In other words, the system should 'know' its choices, which as we discussed in the previous chapters is related to becoming conscious (or as having a self-sustaining membrane of meta-fixed sets and abstract continuity – see chapter 23).

4. *Making decisions:* From a set of choices, the system should be able to pick one without having to follow any specific logic, rules or laws. The mechanisms to make these decisions should be sufficiently deterministic. How do you even represent the decision making process within the framework of the standard physical laws? It feels like the decision making process is entirely an abstract process with no physical dynamics. During the process of making decisions, what physical laws are obeyed? Is it Newton's laws, Maxwell's laws or some quantum mechanical laws? The language used to describe the decisions does

not seem to be related to dynamics and instead is abstract. Also, who is making the decision and for what purpose? Who defines this purpose and how do we represent it?

With each of these questions, we are not looking at idealized concepts or explanations. They tend to distort the real problem of free will. For example, our consciousness lets us realize events only in a millisecond or higher timescale, corresponding to gamma waves (see section 15.2). If we study our brain's internal dynamics at a nanoscale or at an even lower timescale, it is suggested that there is no free will and that it is just an illusion. However, such an analysis misses the interesting part of the problem. In fact, free will does disappear in several situations and under several idealized assumptions.

For example, we lose free will in any of these cases: (a) when the system under consideration is a simple small order system like most nonliving or man-made systems, (b) when we choose unstructured complexity like turbulence or weather patterns instead of structured complex systems like single-cellular and multi-cellular organisms, (c) when we view components at an atomic scale like when you are tracking individual molecular trajectories or (d) when we study the system at a nanoscale or even finer timescales like when we discretize a parallel system into a serial system.

The problem of free will is interesting only because there are well-defined and common scenarios when systems do exhibit this feature like when a human is awake, even though simple idealized assumptions in our analysis can cause it to collapse. The real problem of free will is to explain why and how it works in these scenarios. Once we have a solution to the very existence of the non-idealized free will, specifying the conditions and limitations will become much easier. Let me start discussing each of the above problems now.

28.2 Obeying seemingly arbitrary laws

How does our brain assist us in obeying seemingly arbitrary laws? What is the root cause of this ability? The answer, as we will see, is in choosing an appropriate grouping to serialize a parallel system. In chapter 26, we discussed several different types of groupings for nonliving systems in great detail (like rigid body grouping, control volume grouping, time-varying grouping in water waves and others). Next, in chapter 27, we showed that the appropriate grouping for our brain is fixed sets and abstract continuous paths. These fixed sets grouping allows us to view the parallel internal dynamics in a serial way as abstractions. Let us look into this in more detail now.

When an image falls on the retina or you hear a sentence through your ears, a large collection of neurons fire in parallel (see Figure 7.4). Each neuron converts an external signal, light in the case of eye and sound vibrations in the case of ear, into an electrical signal of action potentials that fire and propagate along the axon to a collection of neighboring neurons. This process was outlined in detail in chapter 5. The underlying mechanism can be viewed as disturbance propagation within an SPL network (see section 5.3). It is a causal dynamical process that obeys the standard laws of physics. We have a local looped dynamics that disturbs a few immediate neighboring components. This causes them to start exhibiting new local looped dynamics, which then disturbs additional neighboring components and so on. This disturbance propagates quite fast like with sound waves, water waves and even neural firing patterns (see section 5.3). The causal laws of motion of such systems is well studied.

There is no free will in such systems yet. Each disturbance pathway obeys the standard laws of physics. In the case of neurons, it is even more structured because the disturbance propagates only along the axons. The axons are part of a well-defined network pathways. The disturbance does not diffuse throughout the brain in arbitrary ways. The disturbance has to propagate only along these axons unlike sound wave disturbance which diffuses quite arbitrarily across a wide region. As an analogy, in one case, the disturbance propagates only through narrow water pipes, while in another case, the disturbance propagates more broadly through a large pool of water. There are no equivalent 'pipes' for sound waves whereas for our brain these 'pipes' correspond to axons. Similarly, there is no equivalent pool (which exists for sound and water waves) in the case of our brain.

The advantage with having a well-defined network (like with our neuron network) is that we can look at the 'flow' of disturbance *along* this network for each individual neural disturbance. Each such flow pathway obeys the standard laws of nature in an almost deterministic manner. Each pathway is not deciding by itself whether to allow disturbance to propagate in one direction versus another. Causality is preserved. There is no free will when viewed at this individual flow pathway level of detail. If all of our thoughts and actions are eventually built from such deterministic flow pathways, how do we obey arbitrary laws and exhibit free will?

Following the flow pathways i.e., the dynamics *along* the pathways is what I referred to as the *serial view* of this system. However, the number of such flow pathways triggered from our retina when we look at any image like, say, a tree are in the millions. Almost all rods and cones on your retina react in some way when you look at a scenery. This generates millions of flow pathways (i.e., neural firing patterns) that propagate along the optic nerve to the primary visual cortex.

As mentioned before, each individual pathway *along* the flow obeys the standards laws of nature. It is when we look at patterns *across* the flow, there are no obvious laws of nature to obey (i.e., analogous to the images, shapes and patterns in each frame of the movie – across the flow – versus the variations in the scene as time progresses – along the flow). Across the flow pattern requires us to simultaneously keep track of the dynamics of, possibly, millions of individual neurons i.e., millions of 'along' the flow dynamics. Keeping track of this across-the-flow dynamics over thousands of individual pathways is what I refer to as the *parallel view* of the system. In the case of our brain, it is these across-the-flow patterns that we care about the most. Each such across-the-flow pattern is what I refer to as an appropriate *grouping* of interest.

An image of a tree, for example, is one such pattern/grouping across the flow. We do not particularly care about how each neuron within this collection fires (at the retina). Rather, we care about how all these neural pathways together as a pattern (or a grouping) propagates deeper in our brain to give rise to the perception of a single entity called a tree. This is analogous to the athletes example discussed in section 26.3. We do not care about the dynamical patterns of a given athletes' legs, hands and other body parts move individually (as images on our retina). Rather, we care about how the entire athlete (image) moves. We want to use this entire motion to compare how he is moving relative to other athletes in a race.

This is also analogous to the example of motion of balls on an inclined plane discussed in section 26.3. We do not care about the 'cycloidal' motion of individual molecules of a ball as it rolls down an inclined plane. Rather, we only care about the 'linear' motion of the entire ball and use that to compare across multiple balls. These patterns were called the rigid body groupings in section 26.3. There were other types of groupings that were discussed in chapter 26. For example, images on a television screen is a valid grouping. Here too, we do not care about the individual pixels' dynamics. Rather, we only care about groups of pixels' dynamics changing over time as, say, a person walking from left to right.

If we compare the progression of the set of laws obeyed as we start grouping parallel components into serial entities, we can see an interesting pattern. At the smallest scale, the grouping as electrons, atoms and molecules obey the quantum mechanical laws. At the next scale, the molecules grouped together as a ball obeys Newton's laws of motion. The quantum mechanical laws, though present at this next scale, are not critical for the patterns we study. For both of these groupings, the laws obeyed are not arbitrary. Now, when we look at the next grouping at a higher scale, namely, that involves tracking the motion of multiple balls simultaneously, the laws obeyed can look considerably arbitrary.

Similarly, the laws obeyed by the grouping as an individual pixel on a television screen varies according to well-defined laws. However, at the next scale, the grouping as an image of a person, does not obey any preset laws. The grouping at the scale of molecules in a water wave obeys standard Newton's laws of motion. However, the time-varying grouping as a water wave sometimes obeys well-defined laws (like a circular wave front pattern). Yet, it can be made to obey arbitrary laws if we supply energy to this system.

One main reason, the laws start to get arbitrary at a higher scale is when conscious beings start supplying external energy to these systems. For multiple balls rolling down an inclined plane, we supply different energy to different balls to get arbitrary across-the-flow patterns. For television screen patterns too, we control the pixel variations ourselves to get arbitrary patterns. For water waves too, conscious humans supply and control the energy of the waves if we want to produce arbitrary patterns.

In other words, it appears that for most of these groupings, in order for them to obey seemingly arbitrary laws, a conscious human who already has free will, should supply arbitrary amount of energy. Therefore, our free will is 'transmitted' to these systems and make them obey seemingly arbitrary laws as a result. This reasoning sounds circular – in order to explain the origin of free will, we are relying on an already existing free will system, namely, conscious humans to supply arbitrary energy thereby letting a new system also obey arbitrary laws.

However, it is easy to break this circular dependency if we create a system that has an internal source of energy. One of the main purpose of proposing an infinite family of physical and chemical SPL systems described in chapter 2 is to allow the creation of systems that self-sustain without the help of conscious beings. The existence of such an infinite family of SPL systems is one of the main reason why free-will-like systems are possible. Take a smoke ring as an example. Once a smoke ring is created, its shape changes arbitrarily. The shape of the smoke ring, while constantly changing, self-sustains for a few seconds. How can we improve this design or create a new system that self-sustains much longer than a few seconds, while still behaving arbitrarily? The examples of SPL systems suggested in chapter 2 as well as their modifications help us create precisely such systems.

Rather than discuss arbitrary SPL systems here, let us focus our attention on our brain instead. In the case of our brain, fixed sets viewed as across-the-flow patterns are natural groupings to study as a whole instead of the along-the-flow individual neural pathways. For example, a tree image on the retina can have fairly complicated parallel neural firing patterns. However, the unique fixed set for the tree is a well-defined across-the-flow grouping. Using this grouping, we can trigger other across-the-flow groupings like the fixed sets for branches, leaves, a forest, hills, nearby houses and others i.e., those objects that we commonly think of when we think of a tree. These neighboring fixed sets are linked together via abstract continuous pathways as discussed previously. We do not trigger these neighboring groupings (fixed sets) by following the dynamics of along-the-flow individual neural pathways explicitly, even though implicitly this is what happens within our brain. Furthermore, the dual space representation of chapter 25 also suggests us to look at fixed sets and triggering them via other neighboring fixed sets. The same applies with sound patterns and other sensory patterns. The fixed sets for speech are similar groupings that conscious humans track as we listen to people speak. We do not track and control smaller groupings for speech.

The fixed sets is not only a suitable grouping to track, but it also varies quite arbitrarily. A causal pathway from one fixed set to another does not seem to obey any specific laws of nature. For example, a sentence in a language is composed of a linear sequence of fixed sets, which obey abstract continuity as we saw in chapter 21. The rules obeyed by these abstract continuous paths are quite arbitrary and unrelated to the laws of nature though they are fixed. We called these rules as the grammatical rules of the language.

Therefore, the internal physical dynamics within our brain obey the fixed laws of nature, but a suitable grouping, namely, the fixed sets obey arbitrary laws (like the grammatical rules of a language). How do we both sense and control these arbitrary laws? If a system can both sense and control such arbitrary but well-defined variations of these fixed sets (as they are related to our past memories), such a system will appear to exhibit free will.

We have already seen one solution to this problem. The self-sustaining membrane of meta-fixed sets allowed us to both sense and control a large collection of fixed and meta-fixed sets (see chapter 19). The fundamental pattern is an example of one such mechanism (see chapter 16) for creating the self-sustaining membrane. Therefore, the creation of a self-sustaining membrane in an SPL system is sufficient, though not necessary, to give rise to a feature similar to free will i.e., the part related to our ability to obey seemingly arbitrary laws even though each individual subcomponents within our brain and body obeys the standard laws of nature.

Given that we have not identified necessary and sufficient conditions for an SPL system to obey arbitrary laws, let us try to qualify the systems by identifying a few critical properties that these systems should exhibit. In order to obey arbitrary laws, the system should switch from one action to another with minimal energy. For this, the system should sense its own changes and control them using mechanisms like the fundamental pattern. When performing one action, if the system is in a 'ready' state for other possible actions, then it only needs a minimal energy to do the switch. This is possible when the system has abstract continuity within the motor control regions i.e., with partial parallel excitations for each of the other actions as well, not just abstract continuity within sensory regions.

A prerequisite for this is to have an internal source of energy, which we can use to alter the natural flow of dynamical events. We would want the system to setup conditions using its own energy for its own dynamics and self-sustain this for a long time. The self-sustaining property of the membrane is possible primarily with an internal energy source.

As an example, how did we learn to move our hand in any way we want within a given region, making it seem that we can obey any law? A newborn is clearly incapable of doing this as you may have noticed when he struggles to put his hand into his mouth. From the perspective of a simple system, we could say that we only need to (a) sense the position of our hand using one or more of our sensors and (b) control our hand to achieve the desired motion. However, for complex systems, these tasks are quite involved, which we discussed as an application of the fundamental pattern in section 17.7. A newborn is able to control his hand motion without knowing which motor neurons to trigger using the fundamental pattern and other sensory-control mechanisms. He is also able to set an objective of performing this task since the outcome produces an emotional explosion.

Generalizing this, he creates emotional explosions when performing simple primitive tasks (like moving his hand, shaking, lifting, twisting and bending it or turning his head, eyes in all possible directions and so on) using, say, the fundamental pattern. He then performs the same task several times with slight variations as he experiences approximately repetitive scenarios. This helps him achieve a high level of control for each scenario. He does not need any higher form of awareness or consciousness for these mechanisms. He continues to build a large repository of these primitive tasks, linking sensory regions with the motor control regions, during his early childhood.

Once he has mastered these primitive tasks, he can perform other complex tasks by combining them in simple ways. We view this as creating abstract continuous paths within the motor control regions of our brain for the control of different parts of our body (see section 21.11 to see what action-based abstract continuous paths mean). For example, using the above mechanisms, we should be able to move our eyes up, down, left, right or in any pattern quite easily from any current state. The same applies with turning your head, focusing the objects at various distances to create sharp images on the retina and so on. With minimal energy, we can switch our actions because of our action-based abstract continuous paths (see section 21.11).

In this way, we begin to create a large set of both sensory and control type of predictive meta-fixed sets. This turns our brain into a dense network of abstract continuous paths. This leads to the creation of a self-sustaining membrane of meta-fixed sets as mentioned in chapter 19. The self-sustaining aspect of the membrane helps in ensuring that the non-unique solutions (of the model or of the differential equations) morph easily from one fixed set to another along a suitable abstract continuous path (see section 23.1).

Let us now see how we can use the above abilities to make an SPL system obey reasonably arbitrary laws. When you are performing a task like moving your hand from left to right, you have abstract continuity for this task. This implies that, at any given instant, you have an ability to alter your hand motion and move along any other direction. As the fixed sets for the current motion is excited, the neighboring linked neurons corresponding to action-based predictive meta-fixed sets also reach a threshold state. However, these parallel excitations do not produce cascading firing patterns ultimately to generate a motor action. Instead, they stay in a 'ready' state. The important idea we used is to specialize the non-uniqueness of time evolution, discussed in section 21.11, for action-based tasks. These solutions imply that we can switch actions easily. All you need is a small amount of energy to produce a big visible change in your action (cf. with smoke ring example). The abstract continuity and the self-sustaining membrane (based on mechanisms like the fundamental pattern) gave rise to a notion of elasticity to deform the pattern dynamics with a minimal amount of energy. I will discuss various ways to do this shortly.

When you are moving your hand, you continue to predict both the control and the sensory information as you observe external events. Your predictive ability is capable of guessing where your hand is going to reach even before the event has happened. In addition, you also know the other ways to move your hand even though you did not change your action yet.

This predictive nature of the abstract continuous paths helps you maintain dynamical continuity as well. This ensures that the errors do not accumulate significantly. For example, note that in order to detect a given fixed set, it takes a few milliseconds. During this period, you have lost an ability to observe or control. The same situation is true when you switch from one fixed set to the next along an abstract continuous path. However, by having a parallel system with abstract continuity and a self-sustaining membrane, you were able to avoid the accumulation of errors. The physical dynamics, as opposed to the manipulation of abstract symbols, is continuous. Even though you cannot sense or control for a short period, the continuity of the physical dynamics, which gives rise to abstractions or the fixed sets, will ensure that the solution does not deviate excessively.

Once you control the accumulation of errors and stay in a 'ready' state, it becomes easier to alter the pattern dynamics. As mentioned earlier, you only need a minimal amount of energy to produce this change. The non-unique solutions of the abstract continuous paths and the self-sustaining membrane make it easy to switch tasks. Let me now discuss various ways of generating such a minimal amount of energy to alter the pattern dynamics.

1. *Random*: Sometimes there is no apparent pattern to explain the application of internal energy to produce a cascading dynamical effect that we simply attribute it as being random.

2. *Distraction*: At times, we notice that some other external event, either critical or noncritical, distracts us by triggering new fixed sets in a different region of our brain. This distraction is unrelated to the current task, but can result in changing our action while we are in the middle of a task.

3. *Involuntary*: There are situations when we have involuntary motions like muscle spasms or twitches only if we continue our task a certain way.

4. *Memories*: When we perform a task repetitively, we memorize most of the actions as a suitable abstract continuous path. We unconsciously switch our control actions as we sense external inputs. In this case, the energy pushes the pattern along the memorized pathway.

5. *Emotions*: It is quite common that our brain produces an emotional explosion (happiness, sadness, fear, embarrassment and so on) while we are performing a given task. This causes an amplification of the excitations in the neighboring regions of the current abstract continuous path. As a result, the current pathway can deviate considerably and trigger a different set of meta-fixed sets.

6. *External sensory inputs*: When we perform a given task, we typically observe the surrounding objects as well. They can trigger additional predictive meta-fixed sets. This additional information will prompt us take preemptive actions because of the intersection of new fixed sets with the old fixed sets. This intersection narrows the set of choices and even changes them as a result.

7. *Multiple sensory inputs*: When we account for other sensory inputs like, say, a loud sound in the middle of our task, this can trigger a change in our current action. In this case, our brain triggers new meta-fixed sets from the second sensory regions creating an alternate control based pathway. Since we share the action meta-fixed sets, it results in a change in the current action.

8. *External factors*: When we consume chemicals like alcohol that affect the neural firing patterns, they result in different amplification of the parallel pathways than the normal situation. Some choices are eliminated while new ones are created that fundamentally alters the energy distribution and hence our actions.

9. *Internal thoughts*: When we are conscious, we have a self-sustaining membrane of meta-fixed sets. This membrane lets us have internal thoughts most of the time. We have abstract continuity with our own thoughts and hence know them as well. The abstractions in a language obey sufficiently arbitrary set of rules (like the grammatical rules) and are even unique to us. As we switch our

thought to a specific task, we can force a change in our action just from our awareness of the abstract continuity of our thoughts and our actions.

Therefore, we conclude that an SPL system is capable of obeying reasonably arbitrary laws using any of the above simple ways to switch our tasks. This can happen both consciously and unconsciously. In fact, every living being is capable of supplying just such a minimal amount of energy so it can obey reasonably arbitrary laws. This is, however, not the same as exhibiting free will in those living beings. Yet, this is one of the most obvious distinguishing features between living and nonliving systems. While this reasoning is not applicable for living beings that do not have neurons or brain, I will use a similar argument based on disturbance propagation in chemical system in section 58.2 to show that they can also behave as a single entity obeying sufficiently arbitrary laws.

While an SPL system does have non-unique solutions and an ability to pick any of them, I want to show that we can make the process of picking itself nonrandom. In the next section, I will discuss what the notion of choice means and how a system is capable of knowing its choices.

28.3 Knowing choices and the very concept of choice itself

In order for a system to know that it has free will, it needs to be conscious (see Theorem 12.1 – consciousness is equivalent to knowing at least on truth). This implies that we need an SPL system with a self-sustaining threshold membrane of meta-fixed sets (Theorem 19.2). In the previous section, we have already seen how such a system has several choices when performing any given task. The next step is for the system to know these choices and use them to make a decision to pick one. If this system can indeed pick any of them without violating any physical laws, we say that the system has free will.

With choices, there are two topics to discuss: (a) knowing the choices themselves and (b) knowing the concept of choice itself. For the former problem, we can just use Theorem 22.1 from section 22.1 for attaining a feeling of knowing for any given event. The existence of choices themselves come from the presence of nonempty predictive meta-fixed sets (i.e., abstract continuous paths) and their interconnections as discussed in the previous section.

For the latter problem, we can once again use the result from section 22.1. However, we need to create abstract continuous paths for the very concept of choice itself. The explanation I will give here for this problem is quite generic to be applicable for a number of other situations. Some of these are, say, when we need to explain how we know *about* abstractions, how we know about knowing, about space, about time or about other abstract topics. The main feature that makes these problems of knowing about abstractions easier is that we can represent meta-fixed sets, meta-meta-fixed sets and other higher layers of abstraction back into the same framework of fixed sets and looped dynamics (see section 23.1).

Recall that to create abstract continuity for a single choice, you need to have nonempty predictive meta-fixed sets. The typical way we create them is by ensuring the same event (or concept) to occur in multiple different scenarios. For example, to know the concept of the number '3' (i.e., to create a meta-fixed set along with nonempty predictions), the multiple scenarios are those situations that help you recollect and predict this concept. Examples of these are, say, when you interacted with 3 balls, 3 balloons, 3 chairs, 3 crayons and so on. The common region of neural excitations for all of these scenarios is for the number concept '3'. This region can be called as the meta-fixed set for the number '3'. Similarly, to know the concept of choice, the multiple scenarios should involve working with several choices themselves. It turns out that our natural language helps greatly in this case.

Notice that you do not need to know about the concept of choice in order to understand the choices, to use them or to pick one of them. This is clear if you observe a 2-3 year old. The abstraction of choice has little practical use for him. Yet, he works with multiple choices in different scenarios. The same applies with other abstractions like numbers or even language. A 2-3 year old uses them quite effectively, not realizing the abstraction itself (i.e., the concept of choice, numbers and other abstractions). This makes the process of attaining a feeling of knowing for most abstractions a slow and difficult process.

Furthermore, an additional difficulty among the scenarios involving choices is that there may be very few common objects across each choice. When we do have common objects in each choice, the chance of occurrence of both events within a short time gap, necessary for recollection, is low. In addition, for each such scenario with little in common, the corresponding memories (i.e., the fixed sets) are in quite disjoint physical regions of our brain. With sparse interconnections between them, it is highly unlikely for one scenario to trigger the neural firing patterns of the other scenario. Therefore, it is difficult to use such seemingly disjoint examples with the only common factor being the concept of choice as a way to discover the concept of choice itself.

One solution then is to let each such disjoint examples gradually converge to a common region in our brain that multiple scenarios share (see Figure 7.12 where we added new nodes and connections to relate two fixed sets). We can then use the new converged region to create a new meta-fixed set for the concept of choice (or, more generally, to a common abstraction like space, time and others). When several scenarios do overlap, the system can create and identify a new meta-fixed set corresponding to the common concept between the two scenarios. Using the existing abstract continuity from the two scenarios and a common intersection region, we now have an ability to go from one scenario to the other (analogous to Figure 7.12). If we use a language abstraction, we could call this new meta-fixed set by a unique name (like, say, 'choice').

Another way to create this unique meta-fixed set representing the concept of choice is through teaching. An adult constantly identifies the abstraction in seemingly disjoint scenarios and reinforces the concept (assuming he can detect the abstraction). In this way, if one conscious person has discovered an abstraction, it is much easier for other conscious beings to realize the same abstraction even if they would not have been able to discover it by themselves.

Once our brain creates the initial new meta-fixed set, every new future scenario that involves choices and that has common intersection to the previous scenarios will help create more links to the same new meta-fixed set called 'choice'. We begin to relate it to most other tasks as well. The network of everyday fixed sets that are related to the choice-meta-fixed sets starts to grow larger and denser. The choice-meta-fixed set becomes a hub connecting to an increasingly large number of dynamical events.

It is now possible to have abstract continuity around the concept of choice by relating it to several specific scenarios involving choices. These specific scenarios serve as common examples of choices (analogous to recollecting natural numbers, real numbers, complex numbers and matrices as common examples of a mathematical concept called a group). In a new situation involving choices, you will trigger several nonempty predictive meta-fixed sets, which always include this special 'choice' meta-fixed set. In this way, we now have a feeling of knowing for the concept of choice itself (from Theorem 22.1 for abstract continuity).

28.4 Making decisions

We can now naturally represent the problem of making a decision within the above-mentioned framework of our brain. This includes using the abstract continuity of the choices, abstract continuity of the concept of choice, abstract continuity of our actions and the self-sustaining membrane of meta-fixed sets. There is no specific logic or preplanned approach to make a decision. We can even realize the abstract concept of decision making itself using a similar method described in the previous section (for knowing the concept of choice). As mentioned in chapter 17, we have several ways to make the decisions that do not involve consciousness while requiring only a minimal amount of energy. However, with consciousness, you now have an ability to know the choices and pick one using the abstract continuity of action-based fixed sets.

This is how we get a feeling of knowing of free will and choices. With our thoughts, we, therefore, have an ability to talk and think about any topic we want. We feel that we can obey any reasonable law with both thoughts and actions. We now have attained the uniquely important feature called free will.

29
Existence of the External World

The existence of the external world is a fundamental fact that we take for granted. However, in section 24.1, we opened the possibility that we can view the excitations on our sensory surfaces from external objects equivalently as time-varying patterns with some unknown, but hardcoded, mechanisms. For example, if we are looking at a scenery like the houses and trees in our neighborhood, the input to our brain are time-varying patterns of images formed on the retina. These time-varying patterns can be imagined to form in two ways: (a) from variations in the external world, which is how we indeed perceive them and (b) fictitious but preprogrammed variations on the retina without any relation to an external world. In the second case, we assume that some superior conscious being is generating time-varying patterns directly on the retina in a manual way.

This is not as fictitious as it sounds. Dreams are the best examples when we see similar detailed sceneries with houses and trees with no input from the external world. They are formed entirely within our brain, though not necessarily at the retina. These internal time-varying patterns are as real as external variations from the external world as in case (a). Another example is with virtual reality glasses that project movies that are as real as those generated from the external world. In the movie case, the patterns are hardcoded or preprogrammed. Yet, they are as real as case (a).

The reason I proposed these two cases is so we highlight how identical and indistinguishable both of these cases can be. Even though the second case is a thought experiment that can be imagined to be as close to the first case as possible, we should eventually show some inconsistency with the second case at least when the inputs are coming from an external world. Otherwise, we do not need to introduce the notion of an external world at all. Every scenario can be thought of as a dream world or a preprogrammed world. Notice that even though our dreams are not real, we do not find any inconsistencies while we are in the dream state. It is only after we wake up, we find them to be inconsistent. All conscious beings are able to reject the second case and indeed perceive an external world within the first year after birth for precisely those non-dream and non-preprogrammed cases. How are we able to do this?

In this chapter, I will show how this seemingly accurate equivalence breaks down and why we need to create an external world. The answer, as it turns out, is because of our free will. Specifically, I will show that an ability to control the generation of time-varying patterns on the sensory surface using our free will mechanisms plays a critical role towards this. In other words, if we do not have free will as described in chapters 26 - 28, there is no easy way to break this equivalence. We can continue to live in a dream world or a preprogrammed world even with a certain degree of consciousness. In this chapter, I will show that free will forces a certain degree of consistency requiring the creation of an external world.

29.1 Disappearing patterns from the sensory surface

To begin with, what does it mean when a time-varying pattern corresponding to an external object disappears from the sensory surface (say, the retina)? At first glance, it is just like any other pattern, which may or may not reappear back on the sensory surface. For example, when an image of a car falls on the retina and eventually moves out of your field of view, this pattern will disappear. It may not reappear on your retina. Until now, we were only able to refer to a pattern and its fixed set when there are excitations triggered directly from the sensory surface, not when there are none on the sensory surface. In the case of the car example above, there is a reference to a pattern of a car and to its fixed set only if the car image falls on the retina. If there is no car pattern formed on the retina, there is no input trigger to our brain that triggers the fixed set for a car. One way to say this is that we may have seen a number of patterns in the past, but unless the pattern falls on the retina, it is not easy to know about this pattern.

However, if the car pattern reappears on your retina several times, say, in a week, you would form a memory (fixed set) corresponding to this pattern. Now, you can trigger this fixed set through internal mechanisms as well. This is using the dual space representation of a car fixed set (see chapter 25). Several common everyday objects' fixed sets can be triggered in this manner using the dual space interconnections. Yet, the easiest way to trigger a fixed set is through the sensory surface like when an image falls on the retina.

The key point to note in the above discussion is that a pattern that repeats several times will have a unique fixed set within our brain. This pattern and the corresponding fixed set deserves to have an independent existence even when the pattern disappears from our sensory surface. The independent existence has two parts to it: (a) as a fixed set internal to our brain and (b) as a real object in an external world.

Unfortunately, the fact that we have a fixed set for this pattern in our brain is not sufficient to establish its external existence (i.e., (b)). The fixed set is simply a memory of repeating events. It is not possible for our brain to distinguish if it generated this memory through an external object or if it is a preprogrammed behavior. By *passively* observing a pattern, we do not know if we are looking at the same pattern or if a new identical pattern appeared later. Our memory is only internal to us and, therefore, we cannot use this alone to distinguish such cases and conclude that the pattern is from an external world. This is where an *active* control mechanism using free will helps. I will discuss this next.

29.2 Breakdown of imaginary patterns using free will

Our objective is to distinguish if a pattern triggered within our brain as a fixed set is something that is (a) generated entirely internal to ourselves (like as a preprogrammed pattern on a sensory surface or as an internal pattern during dreams) or (b) generated from an external object unrelated to us. By passively observing and analyzing the fixed set, we cannot distinguish this. We need active control mechanisms to separate the two cases.

With an active control mechanism, we can track a moving pattern so that it either remains or disappears from the sensory surface, completely at-will. This is the key point. You can close your eyes temporarily (a control action) and open them to see if the pattern is still visible (i.e., if it generated the same fixed set). You can turn your head away and then turn back to trigger the same pattern. You can perform additional actions like moving ahead or behind the original pattern both to generate new patterns and the original pattern on the sensory surface. Each of these actions cannot be preprogrammed because they are under your control and they rely only on your free will.

Additionally, you can predict how long the pattern remains hidden and when it will reappear back on the sensory surface. You can even compute or predict an independent trajectory for the pattern when it is not falling on your sensory surface. You can then use your active control mechanisms to verify if your predictions are correct or not. You can voluntarily intersect the pattern's trajectory any time you want (even though you many not yet know that you are moving). You do not have to perform each of these free will confirmations intentionally. The existence of these possibilities and the mechanical confirmation/prediction through repeated experiments is sufficient.

For example, if you are looking at a moving car that will eventually disappear from your field of view, you can perform several actions to help you distinguish if this eventually-disappearing car pattern is a preprogrammed one or if it is a real external object. You can turn your head away and then turn back to see if the car pattern is still generated on your retina. The amount of time you turn away and how fast you will turn back to the car is completely under your control. No external system can know ahead of time because of your free will. Therefore, the car motion could not have been preprogrammed. Furthermore, you can predict how fast the car is moving, estimate where it will be in a few seconds from now and indeed confirm this by looking back at that predicted location at that instant to see if the car pattern is indeed triggered again on your retina.

The number of such free will based control actions you can perform are quite large. You can close your eyes, look at the car at an angle, walk towards it, away from it, touch it and so on. For each of these actions, you have a prediction of what the pattern might be when you are not viewing it and later confirm that your prediction is indeed correct. All of these actions can be initiated completely at-will. No amount of preprogramming can capture all cases that you may be willing to try using your free will.

One general mechanism described earlier for performing control actions is to use the fundamental pattern like, say, when tracking a moving object with your eyes (see section 17.3). Now, using your free will mechanism, you can confirm the independent existence of the pattern in any number of ways, whenever, wherever and in whatever detail you want. You can do this without any interference and absolutely no possibility of any errors. If there were errors, your free will lets you repeat the action (or experiment) until you are sure that there are no errors. This clearly shows the need to create an external world.

Since you can experience your free will, you now have two options: (i) that some superior external conscious being knows every detail of your thoughts and uses it to pre-program the patterns on your retina at the exact same time as you are applying your free will actions or (ii) the patterns formed on your retina are independent of your actions i.e., there exists an external world outside of you. Our goal is to eliminate (i) as a valid possibility.

The previous example of a car going out of view, can be generalized to any external object. Our free will mechanisms can confirm the existence of this object whenever, wherever and in whatever detail we want. You may argue that since we are using the external objects and are combining with free will mechanisms to confirm the existence, it is possible that some external superior conscious being can influence the result in some unknown way (case (i)). To eliminate case (i), we should choose other objects or patterns for which the superior conscious being cannot exert as much influence.

What are examples of such patterns you can use to perform at-will experiments? They are the ones generated from one's own body. At first, they may seem no different from patterns generated by any other external object. However, the difference is that body patterns not only provide a wide set of variations like different sizes and shapes, but they are also less influenced by an external superior conscious being. If there is an influence, you should know about it because it is your body. Examples of these are looking at your own legs, hands, fingers and so on. You can make these patterns fall or not fall on your retina at-will while predicting the future states.

The mechanisms described in chapter 18 can now be used to describe how body fixed sets are created, how their motions are tracked and how they can be controlled as well. We also have sensory-control loops (like those based on the fundamental pattern) where we can both trigger and control the pattern produced on the sensory surface. The body fixed sets provide a constant source to confirm your predictions. We can easily revert and repeat the actions (like move your hand to the right and move it backwards) any number of times. As you repeat the events multiple times, you form fixed sets, which help with predictions. You are capable of forming abstract continuous paths through a number of everyday experiences.

Another important source of repeatable phenomena is static objects as mentioned in section 20.2. Even if the object stays at the same place, you, as a system, can move to a different location and back through your actions. With body fixed sets and static objects, the number of such phenomena to confirm with free will are practically infinite. The granularity of repetition with timescales (from seconds to days), distances (from centimeters to kilometers) and speed scales (from very slow to very fast) are also widely generic. As a result, the only abstraction that would appear meaningful is the independent existence of each of these objects. We are, therefore, forced to eliminate case (i) i.e., the influence of a superior conscious being.

Your awareness of the external world, however, is not immediate. You need additional structure to *know* about the existence of a pattern. We have already seen in chapter 22 that the feeling of knowing requires a membrane of meta-fixed sets and abstract continuity for each of these patterns. Your memories, formed as fixed sets and their interconnections, even if they occur in a mechanical way, will eventually create the necessary structure. When you form a threshold network of abstract continuous paths, you will know the existence of the external world as well. We can even distinguish whether our brain triggered these events as internal thoughts or whether it was triggered through external means.

Therefore, the pattern truly has an independent existence. We can now call it as an external object. It has a dynamics of its own whether you are actively observing it or not. You necessarily see an external world. Note that we have not shown, for example, that this external world has a notion of space and time, which the objects are part of. This and several other elementary features will be our focus in the next several chapters.

30
On Sensing Physical and Geometric Properties

In the previous chapter, we have seen how to prove the existence of an external world using our free will. Let us now try to understand the internal structure that exists within our external world. The objects in the external world obey a set of physical laws, which we, conscious beings, are able to discover. All conscious beings are able to see and perceive the world in, more or less, the same way. If we look at just the elementary features, we can broadly classify them into two major types: physical and geometric. These elementary features are indisputable to all conscious beings. The simplest geometric features are like edges, angles, shapes, sizes, space and time. The simplest physical features are like motion, mass, charge, different types of energies and forces.

For example, we all see and agree on the specific shapes and sizes of a number of different objects. The objects have texture or patterns. They have edges that curve around, making well-defined angles. We see the world in three dimensions. We perceive an empty space between objects. The objects obey dynamics and we sense the passage of time. The objects have a well-defined mass. The objects need different amounts of force to move them. All of these facts are either visible in front of our eyes or can be sensed through touch and other senses (if you cannot see). We do not have a theory that provides an explanation for how every conscious being is able to sense and know the existence of each of these elementary features. In the next several chapters, I will cover each of these topics. I will show how the theory presented in this book provides a necessary framework to explain all of them. Specifically, in the next few chapters, I will describe the following mostly geometric and physical features.

(a) *Color, edges, angles and motion*: We consider these features as the most basic ones. Their representation does not have dependencies with other features.

(b) *Shapes, textures and patterns*: These are aggregated features formed from the above basic features.

(c) *Size and distance*: Objects in the external world are not located in one place or are they of the same size. We tend to evaluate these two features quantitatively compared to the previous ones.

(d) *Three-dimensional space, the passage of time and seeing in three dimensions*: Our external world is not tightly packed with objects everywhere. We sense empty space between objects. The feeling of three-dimensional shape of objects is significantly different from a 2-D picture. These objects typically do not stay in one place. We can represent their motion abstractly along a timeline. We perceive the passage of such a timeline even when the objects around us do not seem to move.

(e) *Cohesive perception of the external world*: When we look at objects around us, the entire view is smooth and cohesive. Our brain covers the blind spots in our eyes seamlessly. Our visual view changes in a continuous way as the objects or we move around.

(f) *Force, mass and energy:* We know and sense that our own body has a mass. We know that we need to apply different amounts of force to move external objects or even different parts of our own body. We and the other external objects are not merely objects with geometric properties like shape, size and the above-mentioned ones. We, additionally, have physical properties like mass and energy.

I will refer to these above properties, except color, as the *minimum* (or the threshold) set of geometric and physical properties needed for consciousness. For example, electric charges and magnetic fields are physical properties that I have not included into this minimum set. The above list, from (a) – (e), appears to contain visual features, except for color and texture-less patterns. Yet, all of the above features are equally applicable for people who cannot see with both eyes. This is an important point to emphasize. In the next few chapters, I will provide solutions for both visual and non-visual cases. The similarity with both of these cases is striking when we use the SPL framework described in this book. Whether you can see with both eyes or not, your consciousness and the membrane of meta-fixed sets will still require your brain to encode each of the above features (except color and texture-less patterns) uniformly.

My goal when providing answers to these questions is to be constructive i.e., if you take the solutions presented in the next few chapters along with the rest of the theory and implement it in a synthetic system, it will see and sense the external world the same way as you and I do. Certainly, we have not directly addressed this gap until now. You will see that this is not an easy task. We need to look at each of these features in fundamental terms. There are a number of implied assumptions or loosely stated definitions that cloud the issues considerably (especially, item (e) about how we perceive the world in a cohesive manner).

One way to disentangle these complexities is by analyzing them from the point of view of an infant. We will need to redefine some of these terms so that we do not use abstractions that an infant does not yet know. For example, we invariably use a number (or counting) abstraction whenever we think of size or distance. We talk of ratios of sizes of two objects or count in terms of a standard unit (feet or centimeters) or a nonstandard unit (hand or leg). This is unreasonable even for a two year old. Yet, he senses size and distance of objects similar to you and me. It is important to identify this core essence of size and distance. The same applies with all of the other features as well.

In this chapter, I will discuss the nature of each of these features, their relationships and potential difficulties. Some of the above features like edges, angles and colors are simpler and independent. Others like sensing space, passage of time and creation of a cohesive perception from distributed processing are much more involved and interconnected. The process of learning each of these features in your early childhood helps in creating a highly interconnected network of meta-fixed sets. This network will eventually play a crucial role for forming the self-sustaining membrane of meta-fixed sets, needed for consciousness.

30.1 Common issues with basic and aggregate features

The basic features (item (a)) like edges, angles, colors and motion do not have any other significant dependencies with other systems. We only need to ensure internal consistency. The aggregate features (item (b)) like shape, textures and patterns are also not as complex. There is a direct correlation to the basic features, which we can memorize easily.

The remaining features listed earlier (items (c) – (e)) are not as simple. There are a number of issues when we try to explain how we perceive size and distance (item (c)) using the traditional way. If we open our eyes and look at objects around us, every object appears to be of a very definite size and at a very precise location. The size and location does not vary significantly if we turn our eyes or head around, if we move towards or away from the object or if we remain static. Why does an object not appear smaller but closer (like A in Figure 30.1) or larger but farther (like B)? Clearly, in both these cases, the image on the eye, which is the only source of information into our brain, is itself identical to the original after our eyes auto-focus. Note that if the object is closer but smaller, the focal length of the eye lens needs to be decreased proportionately if we want to ensure that the image to fall precisely on the retina. This relation is $\frac{1}{u} + \frac{1}{v} = \frac{1}{f}$ where u is the distance of the external object to the lens, v the distance of the image from the lens and f the focal length of the lens (see Figure 16.9 as well). Our eyes automatically readjust the focal length using the fundamental pattern as discussed in section 16.1.

Given this, the question we are asking is: why does our brain automatically place a given object at a specific distance and specific size the instant we open our eyes when multiple positions and sizes within the same visual cone (like A and B) do produce the same image I on the retina i.e., when you open your eyes, a table will appear at a position and is of a given size even though other possibilities within the same visual cone are valid, given the information used and processed is the same in both cases,

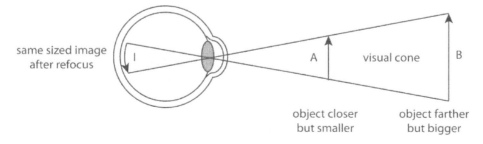

Figure 30.1: Relationship between image size and object's size/distance within the visual cone. If the object is closer but smaller (A) or farther but bigger (B) as shown here in the visual cone, the image I that falls on the retina can still be made identical after readjusting the focal length.

namely, the same image of the table on the retina?

The situation with the size and location of an object being fixed is quite similar even if we close one eye. Furthermore, the size of an image of, say, a chair, on the retina is so small (less than an inch), but when we look at it with our eyes, it *feels* so big (a few feet tall). Who is translating and transforming the inch-sized objects on the retina to their respective correct sizes that are several feet taller? Which part of the brain is 'projecting' the image forward i.e., in front of us to these sizes?

Also, our brain seems to place each of the different objects at their respective correct locations with the correct sizes, instantly. The moment we open one of our eyes, the view in front of us is already accurate. Objects never appear to zoom in or out for a brief period as our eyes and our brain tries to build a visual view by, say, focusing on different objects or by trying to compute the sizes and distances. Our brain has the correct values instantaneously. They also re-adjust correctly, again instantaneously, as we move around.

What is surprising is that even when we dream, objects are of definite sizes and at specific locations. As you perceive your motion in your dream, our brain re-adjusts the object locations automatically as if you are awake and you are physically moved towards the objects. We, therefore, need to explain this size and distance perception in both the presence and the absence of external visual inputs (like, say, when we are dreaming). I will address these specific problems in chapters 34 and 36.

30.2 On detecting emptiness of spacetime

When we showed the existence of the external world, it did not automatically imply that the objects are separated by empty space or that you know the existence of time (item (d) above). What we have seen is that real physical objects generate patterns on the sensory surface. When the system is sufficiently complex, it can store a number of interconnected fixed sets corresponding to each of these patterns. We showed that these patterns have an independent existence by using the systems' self-control or free will mechanisms.

However, space and time are not material objects. There are no patterns generated on any sensory surface directly from space and time. I will use the term 'emptiness' to refer to this fact. There is a lack of sensory systems that directly detects emptiness. If, in fact, a material sensor does generate patterns for nothing, we strongly believe that it is because of some sort of internal noise instead. A sensor can only detect a real physical input or something that has an associated energy. Non-emptiness suggests that an object present in this nonempty region can generate patterns simply by having ordinary dynamical interactions with an appropriate material sensor. In this manner, we can detect nonempty regions as real objects. However, how can we sense empty concepts like space and time? I will discuss this in chapters 35 and 40.

30.3 Existence versus structure of spacetime

Recall that the subject of space and time was a recurring theme in several scientific discussions for hundreds of years. Newton believed in a notion of absolute space when he formulated his laws of motion. Einstein, on the other hand, rejected absolute space when he proposed his field laws. He showed that the metric properties of spacetime geometry precisely represent a gravitational field. According to Einstein's theory, spacetime properties should not be studied independent of the material objects. We use the term background independence to describe such a theory (see Smolin [142]). We have studied the geometry and structure of spacetime in considerable detail since. In this book, what we are interested in is the question of *existence* of space and time, rather than the *structure* of spacetime.

For example, Einstein's general theory shows that, when viewed from a uniformly rotating frame of reference, the geometry of spacetime is non-Euclidean. In particular, the ratio of the circumference of a circle to the diameter of the circle is greater than π in this frame of reference. However, in a Galilean frame of reference, this ratio is equal to π. Note that these computations and questions are well-defined after we assume the existence of space. However, the question of interest in this book is how we can measure the circumference and the diameter of a circle unless we already accept the *existence* of space. How do we even know that the circle drawn exists on a two-dimensional plane? How do we know that we are moving the measuring rod around the circle, in space, while calculating the circumference of the circle? Such existential questions are what we are interested in answering in this book, not whether the structure of spacetime is Euclidean or not. The questions about the structure of spacetime are equally interesting, but are studied in greater detail elsewhere like with Einstein's theory of relativity and others.

30.4 The problem of visual perception

If we combine all the above-mentioned basic, aggregate, size, distance and spacetime features, the net effect is to have a cohesive perception of the world around us (item (e)). However, we do not know how this cohesive perception is created. For example, we do not even realize that we have blind spots in our eyes. To compare, imagine our retina with the blind spot as a television with a damaged dark spot on the screen. The part of the image that falls on this damaged dark region is invisible to us just as the part of the image falling on our blind spot carries no neural signals into our brain. If a movie plays on this screen, we will clearly notice gaps because of the dark spot. Similarly, our retina sends no neural signals from the blind spot regions. Yet, we do not see any gaps in our vision. Our brain seems to compensate for these gaps. How does it do this?

Even if we close one eye, we do not see a hole in our vision corresponding to the blind spot. In other words, the right eye is not compensating the blind spot in the left eye. Even in our dreams, when there are no external visual inputs, we have a clear cohesive perception of the 'imaginary' world. People who cannot see also have a cohesive perception of the external world using the remaining sensory organs. How is this possible?

In chapter 41, I will show how we can combine the solutions of each of the features (a) – (d) to create this cohesive perception. Representing each of these features in terms of SPL architecture (with fixed sets and abstract continuity) turns out to be critical for this cohesive perception. In addition, representing the membrane of meta-fixed sets equivalently as a dual space (chapter 25) and as a threshold network of abstract continuous paths (chapter 23) will help clarify the feeling of continuity and cohesiveness.

30.5 Sensing physical properties like mass

Items (a) – (e) focused primarily on geometric properties of external objects and the external world. However, all external objects also have a number of physical properties like mass and charge. These physical properties play an important role in the dynamical interactions between objects. They are also the source of fundamental forces like gravity or electromagnetic force. It is clear that when we become conscious, we not only know our shape, size and other geometric properties, but we also know our mass or the amount of force needed to move ourselves, at least, qualitatively. There is no doubt that we need these physical and geometric properties to understand the laws of our universe. It is only natural to expect that these properties are needed even to become conscious.

Without knowing the existence of a few physical properties like mass, a synthetic system (like a computer) can never become conscious. Instead, it is just a physical system representing abstractions, not reality. It is, therefore, important to link abstract and physical properties together when representing them in a synthetic system or in our brain. I will show how to do this in chapter 42. Touch is the primary sensor that lets us achieve this goal. Conscious beings are not capable of detecting every physical property using their five sense organs. For example, we do not have a sensor to detect charges, even though our brain functions correctly because of charges like sodium and potassium ions. We detect physical properties from how they interact with properties we do sense or by building artificial sensors (like galvanometer, thermometer and Geiger counter).

Our ability to detect both physical and geometric properties mentioned here leads to other important questions. For example, if we look at an external object, how can we be sure that it is a *real* object? The term reality is entirely relative to our own consciousness. In chapter 44, I will generalize this question to see how we can address other deeper questions like: (a) does 'red' feel the same to you and me or (b) is your sadness similar to mine. In particular, it will lead us to define the term 'equality' of two complex systems (like equality of two different people or equality of two single cellular organisms), as opposed to equality of two simple systems, which is already well known.

Before I discuss these generalizations, let me start by addressing the elementary geometric and physical properties next.

31
Color Vision

Color is a sensation created only from our eyes. No other sensor contributes towards color unlike the other geometric properties mentioned in the previous chapter. Even when all the geometric properties (like size, shape and angles) are identical, a person who can see with his eyes can distinguish two objects purely based on its color. Strictly speaking, in realistic scenarios, it is not common to have all other geometric properties identical except for color. The real advantage of color is when you need to distinguish objects and intricate features quickly or without physical contact. Sometimes it is dangerous to touch an object or the object may be too far away. In these cases, color, additionally, provides an advantage.

I want to point out that color is not critical for consciousness. In fact, vision itself is not necessary. There do exist conscious people who cannot see. Nevertheless, eye is an important sensor, with color as a critical feature. Even with no relationship to consciousness, I will discuss it here because eyes do illustrate a number of useful ideas. I want to show that color uses a common framework that we can generalize to other sensors as well. Most of the discussion in this section will attempt at translating the existing theory on color vision into a form that fits well within the new SPL framework introduced in this book. As highlighted previously, the main departure you will notice from the existing theory comes from studying neural loops, fixed sets and their relationships instead of individual neurons like rods and cones (in the retina) and attributing features like colors directly to them.

What is color? How do we define it? When trying to explain color to a toddler, it appears sufficient to use words like red, blue and green consistently as we point to the respective color objects. Our hope is that he will eventually go beyond recognizing the object (like balloon or ball) and learn to create an abstraction for red color itself. This approach usually works since he is born with an inherent ability to distinguish two different colors. As he grows older, the other explanations we use are: (a) list and define colors using the mixing rules of different colors, (b) mention that white light is composed of different colors, (c) that colors are associated with different frequencies of light and (d) that our eyes have three basic color pigments (i.e., special chemicals sensitive to different frequencies of light) to perceive colors. Each of these explanations seems to capture the mechanical aspects of the color. How do we describe the appearance of color, which is sometimes referred to as *qualia*?

Color is not a property of the external object itself. To see this clearly, consider a circular disk divided into, say, eight sectors. Color each sector alternately with yellow and blue. When we spin this disk, we would see green color instead, which is not present in any of the eight sectors at all. If it is incorrect to attribute color to external objects, it should be incorrect to attribute color to internal objects (internal to our body), like the three pigments within the cone cells of our retina. Analogously, we do not attribute different tastes to chemo-receptors in taste buds or specific odors to chemoreceptors in the olfactory system even though every human being has the same primitive tastes and odors similar to primitive colors. It is clear that the three pigments act as the main source to provide distinct information, but the perception of color itself occurs much deeper in our brain.

Does a newborn start perceiving all colors right at birth, especially since his retinal cells does have these three distinct pigments? He surely can respond to the differences in neural excitations from these three different pigments whether he perceives them as different colors or not.

If we were to build a synthetic system that is capable of experiencing the *qualia* of color similar to us, what should we do? We would definitely include an appropriate distribution of the three different color pigments (like RGB pixels in a liquid crystal display screen). Is this enough for the *qualia* i.e., beyond the detection of colors? Can this system detect, say, green when we mix yellow and blue paint together or when we spin a disk with alternate colors as mentioned above? Can it detect the red color of an apple even in dim light? Are the two states, one for detecting colors mechanically and the other for perceiving the *qualia* of color distinct from each other? I will discuss the representation of color within an SPL system in this chapter using a common framework that generalizes to sounds, tastes, odors and touch as well. Later in chapter 44, I will discuss the problem of *qualia* of color addressing questions related to the appearance of color. Some of these questions are: whether you and I have the same appearance of 'red' and if this appearance of 'red' is similar among people who have only two color pigments (dichromats – considered color blind) or four color pigments (tetrachromats). Given all these questions we want to answer, let us start with the basic physics of light and our eye first.

31.1 Basic physics of light and our eye

When we open our eyes, the input that enters our eye lens and fall on our retina are photons of light. A photon is a discrete packet of energy proportional to a given frequency. The light that falls on your retina can come directly from a source of light like the sun (or a light bulb) or it can reflect from other objects. In both cases, the photons have a range of specific 'component' frequencies (or wavelengths). It turns out that we can take any source of light and decompose into these so-called component frequencies. For example, we can split white light into the respective component frequencies using a prism. We see these as the various colors of the rainbow. We can also combine all of these component frequencies back into white light using, say, another prism. This decomposition of light into individual frequencies is termed as the spectrum, the spectral distribution or the electromagnetic spectrum of light.

When we look at light or photons that have a random mixture of multiple frequencies with a given spectrum, its qualitative nature feels completely different when we compare it with light that has just a subset or individual frequencies. For example, white light appears very different from each of the individual frequencies that appear as different colors of the rainbow.

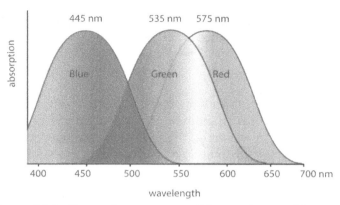

Figure 31.1: Absorption spectrum of three color sensitive cones in the retina of the human eye.

Human eyes can only detect photons from within a small region called the visible spectrum. Our eyes have special nerve cells called photoreceptors to detect light at different wavelengths (or frequencies). There are two types of photoreceptors, namely, rods and cones. These cells contain pigments rhodopsin (in rods) and photopsins (in cones) to absorb light. The chemical structure of rods and cones, how they work and how they interact with other retinal cells like bipolar and ganglion cells are described in thorough detail in standard biology textbooks (see Kandel et. al [82] and Purves [122]).

The spectral distribution of sunlight is said to be a 'continuous' spectrum. For example, it contains photons of all frequencies from the visible region (similar to when you heat a blackbody). When any object is heated, it emits light with a given spectrum of radiation called the emission spectrum. An object that appears green absorbs all other frequencies except from a specific range that creates the appearance of green. Similarly, every chemical element absorbs certain characteristic frequencies from light called the absorption spectrum.

Conceptually, light from a source E like the sun or a halogen lamp with a given emission spectrum travels towards our eyes. This light has a certain qualitative appearance to our eyes like white or yellow color respectively (for sun and halogen lamp). As this light falls on other objects along the way, some of the frequencies are absorbed and the others are reflected. This is according to the absorption spectrum A of the intermediate objects. This gives rise to a resultant light R that falls on our retina. The spectral frequencies in the resultant light have emission frequencies from the original source E minus the absorption spectrum A of intermediate objects. This resulting light R has a different qualitative appearance to our eyes. For example, if the intermediate objects are leaves, then the resulting light appears green under sunlight, but may appear differently under a different source of light (with different spectral characteristics).

The resultant light R with a given spectral distribution falls on the rods and cones of our eyes. The light sensitive pigments rhodopsin and photopsins themselves respond to light with a specific spectral distribution as shown in Figure 31.1. In other words, these complex chemicals themselves have a given absorption spectrum. These chemicals have different absorbance at different frequencies. The relative absorption produces a cascading set of chemical reactions. They eventually create a firing pattern that travel along the optic nerve to the primary visual cortex (see Kandel et. al [82] and Purves [122]).

All rods have only one type of pigment rhodopsin with peak absorption at a wavelength of about 500 nm (nanometers). Cones, on the other hand, have three different pigments. They are referred to as S, M and L cones that have peak response to short (420 – 440 nm: approximately blue), medium (530 – 540 nm: approximately green) and long (560 – 580 nm: approximately orange/red) wavelengths respectively (see Williamson and Cummins [165]). In the human eye, there are about 100 million rod cells compared to about 5 million cones. The relative ratios of S, M and L cones vary significantly among different people. Some common values are: L at about 60%, M at about 35% and S at about 5%. In the fovea, which is the part of the retina responsible for sharp central vision, there are no rod cells or even S cones. It, almost entirely, contains M and L cones.

The distribution of rods and cones are also not uniform within the retina. The mosaic for L and M cones in the fovea, for example, contains large patches in which either set of cones are missing. This distribution is also strikingly different between two humans with normal color vision (see Roorda and Williams [131]) and even between the two eyes of a single human. It was also found that all three types of cones are randomly distributed.

The rod cells are about a thousand times more sensitive than cones and can, sometimes, respond even to a single photon of light. However, they have a slow adaptation process compared to cones and they take as much as 30 minutes to adapt to such low intensity of light. We notice this slow adaptation particularly at night when we enter a dark room. Furthermore, this high sensitivity to low light makes rods predominantly useful at night. Cones, on the other hand, are used primarily during the day.

31.2 Sharp versus blurred image

Let us now see what happens when light, which is either directly coming from a source or is reflected from an object, passes through your eye lens and falls on the retina. Consider light travelling from an external point A towards our eye (see Figure 31.2). The multiple paths from A enter our eye lens at different locations and travel towards the retina. Since our eye lens is convex, all of these light rays converge to a point E inside the eye. The location E where they converge depends on the focal length of our eye lens. By choosing an appropriate value for the focal length, we can ensure that the point of convergence E falls directly on a photoreceptor. The point E is at a precise distance from your eye lens where you have either a photosensitive pigment rhodopsin corresponding to rods or photopsin for cones.

In Figure 31.2, I have shown two other common cases where the point E is either in front or behind the retina. This occurs if, say, the eyeball is too long, the cornea is

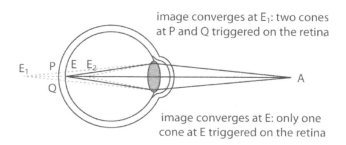

image converges at E_1: two cones
at P and Q triggered on the retina

image converges at E: only one
cone at E triggered on the retina

Figure 31.2: Sharp versus blurred image. If the image of the object at
A converges precisely on the retina at E, then only one rod or cone is
triggered. On the other hand, if the image converges behind the retina
at E_1, then a single external point A triggers two rods or cones at P and
Q respectively. The situation is the same if the image converges in
front of the retina at E_2.

too steep or if the focal length is incorrect. In one case, light continues beyond E and
converges at E_1. In this case, light reaches at multiple points on the retina, shown as P
and Q in Figure 31.2. In the other case, light converges before E at E_2 and extends
beyond E_2 to reach the retina at, say, points R and S. In both these cases, we see that
photons from a *single* external point A reaching *multiple* rods or cones at P, Q, R and
S. This turns out to be the main reason for forming a blurred image. I will discuss this
shortly in further detail. Only in the case when the light converges precisely on the
retina, we see that a single external point A reaches a single rod or cone at E. In all
three cases, the set of cascading chemical reactions that occur after a photon is
absorbed by rhodopsin or photopsin is identical. The number and frequency of
photons may be different, but the response to a given photon is the same.

Let me describe the chemical mechanisms briefly here. Most of the description
presented here is a summary from textbooks like Kandel et. al [82] and Purves [122]
that studied them in greater detail. Consider what happens in the rods and cones
when light falls or does not fall on them. When there is no light, gated sodium ion
channels stay open because cyclic guanosine 3′-5′ monophosphate (cGMP) is bound
to them. This causes the Na^+ ions to enter the photoreceptor, thereby depolarizing it
to about -40 mV. When a photon is absorbed by rhodopsin or photopsin, the retinal
Schiff base cofactor present inside the protein changes configuration from a *cis*-form
to *trans*-form. This structural change causes a series of chemical changes that,
eventually, activates a regulatory protein called transducin (typically, about a 100 of
them). Each transducin activates the enzyme cGMP-phosphodiesterase, which
catalyzes the hydrolysis of about 1000 cGMP molecules into 5′-GMP. This reduces the
concentration of cGMP that had previously kept the gated Na^+ ion channels open.

A reduced concentration of cGMP closes the Na^+ ion channels causing a drop in the
number of Na^+ ions that enter the cell. This hyperpolarizes the photoreceptor, thereby
stopping the release of neurotransmitter glutamate into the bipolar cell (compared to
when there was no light). This decrease in glutamate depolarizes some bipolar cells

461

while hyperpolarizing others. Therefore, the presence of light causes a photoreceptor to release less neurotransmitters compared to when there is no light. There are detailed mechanisms to complete the loop and restore the original concentration of cGMP (see Alberts et. al [1] and Kandel et. al [82]). For example, cGMP is restored when guanylate cyclase is activated when the concentration of calcium ions (along with sodium ions) are decreased.

In the above mechanism, a given wavelength of light on a given rod or cone has a probability of absorption as shown in Figure 31.1. However, after the photon of a given wavelength is absorbed by the photopsin or rhodopsin, the above-mentioned chemical mechanism is not dependent on the wavelength of light anymore. The decrease in the release of glutamate does not change whether the absorbed light has a wavelength of 450 nm, 600 nm or any other value. It is only the probability of absorption that is different for each photoreceptor at each wavelength.

In Figure 31.2, we have seen that light from a given external point A reaches multiple rods or cones at P and Q when the image converges at E_1. Therefore, light from a small neighborhood around A fall on the same rods or cones at P and Q. In other words, a given rod or cone at, say P, receives light from multiple different points around A (see Figure 31.2). These different points around A, typically, have light with different frequency spectrum (say, for example, point A corresponds to green color while point B to blue). Now, even if you do not move your eyes, there is randomness or time-varying absorption of photons depending on whether they arrived at P from A or from B. For example, let us say that the photopsin has changed structurally when A-photon arrived. It is now temporarily inactive when B-photon arrives. When it changes back to an active state, maybe, photon from a new location C arrives. As a result, we have a time-varying release of glutamate simply from the time-varying flipping of photopsin or rhodopsin from an active to an inactive state and vice versa.

On the other hand, when the light converges precisely on the retina at E, light that falls on photoreceptor E has the same frequency. This is because all photons that arrive at E are from a single external point A. Even though, there is a time-varying flipping of, say, photopsin between an inactive and an active state, the photons that fall at E always have the same spectral distribution. The convergence at E has eliminated any time-varying nature in the release of glutamate.

This is precisely the reason why we have a sharp versus blurred image. The blurring is more prominent near the boundaries or edges compared to the interior of an image (since the neighborhood of a boundary point A will have different spectral distributions on either side). The auto-focus mechanism described in section 16.1 that uses the fundamental pattern is, therefore, quite important to produce a steady, as opposed to random time-varying, firing pattern (or, more accurately, release of neurotransmitters) by ensuring that there is convergence, at least, on the fovea. A steady pattern is repeatable enough that we can now memorize it as fixed sets within our brain. We will see how this works next.

31.3 Neural responses from a group of rods and cones

In the previous section, we have seen how we can create a steady pattern of neurotransmitter release from a single photoreceptor as opposed to a random time-varying pattern. We then saw the importance of auto-focus mechanism to ensure convergence of all light rays from a single external point to a single rod or a cone. Strictly speaking, photoreceptors do not 'fire' action potentials like the other types of neurons within the brain. Instead, they release a lower quantity of neurotransmitter glutamate compared to when there is no light. Yet, I will use the term 'firing pattern' with these nerve cells as well, assuming there is no danger of confusion. The fundamental pattern of chapter 16 can be used to ensure the auto-focus mechanism works correctly.

Let us now look at the 'firing' patterns generated across a large set of photoreceptors, both from the fovea and the rest of the retina. Let me assume that we have an accurate auto-focus mechanism throughout the rest of this section (see section 16.1 on how this is possible using the fundamental pattern). Consider looking at an object in front of you in a given direction. Packets of photons, coming from a spatial region R that includes this object and its surroundings, pass through our eye lens. They converge on the retina to form a specific image pattern I. There are three factors to account for when analyzing the firing pattern from neurons within the region I.

1. *Spectral distribution of external light across region R*: Let us take time slices of the beam of light originating from the entire region R as it travels towards our eye. For each time slice $t_1, t_2, ..., t_n$, we have a corresponding spatial arrangement of photons $R_1, R_2, ..., R_n$, of different frequencies across region R. All photons that are part of R_i (for $i = 1, 2, ..., n$) reach the retina to create a firing pattern among a set of rods and cones within the region I. We have thus represented the incoming light with a given spectral distribution as photons R_i with given frequencies.

2. *Spatial arrangement of rods and cones with different absorption spectrum across a given region*: The region I is filled with rods, as well as S, M and L cones in a non-uniform manner. The number of rods and cones are also quite disproportionate, as mentioned in section 31.1 (like, say, 60% L, 35% M and 5% S cones – see Roorda and Williams [131]). In addition, the rods, S, M and L cones within region I have different absorption spectrum for incident light as shown in Figure 31.1. Let $\mathcal{D}$ be the distribution of rods and cones on the retina. The spatial arrangement of rods and cones produce an overall firing pattern across the entire region I.

3. *Time lags between active and inactive states of photopigments across a given region*: The rods and cones have different level of sensitivities. The time lag between an active and an inactive state of rhodopsins and photopsins are not only different between rods and cones, but they are, in general, different for two rods or two cones of the same type as well. This depends on the frequency of incident photon at a previous time instant, the absorbance at that frequency and variations in the rate of chemical reactions for normal operation of the nerve cell.

When we combine all three effects, we have a pattern of glutamate released from the photoreceptors onto the bipolar cells that is unique to the external region R. The photons R_i with a given set of wavelengths fall on the region I of rods and cones, some of which are in an active state and others in an inactive state, to produce a corresponding firing pattern. For a given human being, the mosaic $\mathcal{D}$ of rods and cones is, more or less, fixed though not uniform.

Let me illustrate this with a simple example. If you are looking at green leaves of a tree with a blue-sky background during the day, region R_i has photons with frequencies corresponding to these ranges from the visible spectrum. When you focus your eye lens to create a sharp image, the photoreceptors in region I produce a time varying release of glutamate based on the absorption spectrum of these non-uniformly distributed rods and cones.

Another example is with the spinning disk with sectors of alternate blue and yellow color mentioned in the beginning of chapter 31. Here, a given point in the region R_i keeps flipping from blue to yellow photons along the timeline because of the spinning disk. For example, at a given point in the region R_i, the photon at t_1 corresponds to blue wavelength, then at $t_1+\Delta t$ it is yellow, at $t_1+2\Delta t$ it is blue again, at $t_1+3\Delta t$ it is yellow and so on. The excitations on the cones will individually correspond to blue and yellow effects alternately because the photon frequencies change over time as above (due to the rotating disk). However, the resulting perception of color is unrelated to either of them and instead turns out to be green. This will become clear when I discuss how mixing rules work in section 31.4.

Therefore, the firing pattern captures the spatial and spectral properties of light. These properties are unique to the spatial arrangement of external objects and the source of light used to illuminate the region. The firing pattern is repeatable, modulo minor variations, when both the external region and the light source are the same. This implies that it is possible to memorize some of the most commonly repeating patterns as fixed sets.

If the object, the background and your eyes are static, we have R_i to be nearly the same for all $i = 1, 2, ..., n$. What happens when either you or the objects move? If you have learnt how to track a moving object (see section 17.3) or you move your eyes or head while maintaining focus on the object, the image continues to fall on the fovea. In this case, the tracking object appears static on the fovea while the entire background objects move (like a translation). When you do not track a moving object and instead look at a different object, this object is static on your fovea while the moving object translates in the background. In general, there is a combination of translation and rotation of the images in the background.

Therefore, in both cases, a spatial subregion within R_i varies as i goes from 1 to n i.e., it varies along the timeline. This time-varying region in R_i falls on the retinal region I. Since the distribution $\mathcal{D}$ of rods and cones is non-uniform, the motion of the image excites different patterns of photoreceptors. For example, consider the moving object to be a red colored circular ball. There are a few S cones and no rods when the circular image falls on the fovea. As the image of the ball moves towards the periphery, the number of rods within the circular region increases while the concentration of M and L cones drop. The firing pattern of the neurons is different for three reasons: (a) the non-uniform distribution of photoreceptors, (b) the translation and rotation of the image and (c) the spectral frequencies reflected from the moving object. Therefore, the firing pattern for the motion of a blue ball is considerably different from that of a red ball.

Our brain forms and excites the fixed sets corresponding to the common variations of a pattern from these geometric transformations. For example, you, almost never, experience seeing objects changing their color as they translate or rotate. This fact implies that our brain must create fixed sets analogous to those described in sections 18.5 and 18.8. With these mechanisms, I have shown how fixed sets are formed only for neighboring subsequences or forward and backward sequences. This is reasonable because these are the most common possibilities that occur in nature, not any other random variations.

In the next section, our goal is to define color in terms of a few specific fixed sets instead of the conventional way of defining color based on the S, M and L cones directly. I want to argue that while S, M and L cones help has distinguish spectral distribution of light across spatial regions, they do not give the perception of color directly. The perception of color happens deeper in the brain and they correspond to these special fixed sets that I will discuss in the next section.

In this discussion, I will not focus on what gives rise to the perception of color (i.e., the *qualia* of color). Rather, I only specify what should be used to identify and distinguish colors in a mechanical sense. Later, in chapter 44, I will discuss more on the *qualia* and the appearance of color. We will see in chapter 44 that this question of *qualia*, in general, is not easy to verify directly. There are some fundamental limits to our ability to verify *qualia* across different conscious beings, which I will discuss in detail.

31.4 Colors

In this section, I want to avoid using the term 'color' to identify with specific range of frequencies of light. At the same time, we do not want to associate color with the three types of cones. Both these usages, though common, do not fit within the framework of stable parallel loops. Instead, I will show that the natural way to define color is in terms of a few appropriate fixed sets.

To understand the motivation behind defining colors in terms of fixed sets, let us look back at the first time a human starts to recognize colors. Can a child at birth differentiate and recognize colors? My claim is that while he can use the neural outputs from S, M and L cones to differentiate objects, he has no notion of color at that age. Two objects that are otherwise identical in shape, size, texture, surface and other geometric patterns can still differ in 'color'. Therefore, we cannot use any of these geometric patterns-based fixed sets to distinguish these two objects. We need a new way to distinguish them. If we include the above variations in shape, size and other geometric properties as examples of invariances, they become the set of dynamical transformations that are needed when defining a fixed set. The new fixed set with these above set of dynamical transformations will be called as a 'color' fixed set.

The difference between these two objects with otherwise identical geometric

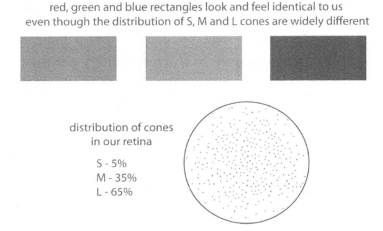

red, green and blue rectangles look and feel identical to us
even though the distribution of S, M and L cones are widely different

distribution of cones
in our retina

S - 5%
M - 35%
L - 65%

Figure 31.3: Distribution of S, M and L cones on our retina. A typical distribution of S, M and L cones are 5%, 35% and 65% respectively. They are widely different from each other. For example, a blue rectangular image falling on the retina only triggers 5% of the cones whereas a similar red rectangle will trigger 65% of the cones. Yet, both red and blue rectangles look and feel identical.

properties implies that both objects have different absorption and emission spectrum of light. As white light falls on both objects, photons of different frequencies emit. Our retina is capable of capturing this difference, thereby producing different neural output patterns. For example, a red rectangle and a blue rectangle produces different patterns on the retina because of different distributions of S, M and L cones (like, say, 60% L, 35% M and 5% S cones – see Roorda and Williams [131]). A red rectangle corresponds to a particular shape of excitation pattern on the retina surface unique to the distribution of S, M and L cones (see Figure 31.3). A blue rectangle has a different shape of neural excitation pattern.

This is analogous to a red rectangle drawn on a television screen with a certain distribution of RGB pixels. Only the red colored pixels will glow producing a unique pattern. This will be different from the original rectangular shape if we look very closely at all pixels. There will be gaps at all places where there are no red pixels. Even the boundary of the rectangle will not be a straight line. Rather, it will be jagged because of these missing red pixels. Imagine overlaying the red rectangle on the circular distribution shown in Figure 31.3. The red edges will not form a straight line. Similarly, for a blue rectangle, the pattern could be considerably different because only the blue colored pixels glow. If the initial distribution of the RGB pixels is quite non-uniform like with our retina (60% L, 35% M and 5% S cones), then these two red and blue rectangular patterns will be considerably different, both at the edges of the rectangle and the interior of the rectangle. Of course, if we look at the television from a distance i.e., with a low resolution, it is typical that these subtle differences are smeared and we see similar shapes but with different colors.

In the case of our retina, we only have 5% S cones. Therefore, blue frequency of light (corresponding to S cones) will only fire a small number of neurons compared to red frequency of light (corresponding to 60% L cones). The edges of a blue rectangle will be considerably jagged and the interior will be quite empty as well compared to a red rectangle. A sparse pattern for blue rectangle versus a dense pattern for red rectangle will pass through the optic nerve into the primary visual cortex.

Given these two patterns i.e., sparse and dense neural firing patterns originating from the retina for blue and red rectangles respectively, our brain needs to distinguish two cases: (i) the two patterns are from the same shape called rectangle but has different colors and (ii) the two patterns are from different shapes corresponding to different patterns of excitations. How does our brain do this? Identifying this difference is the central problem of identifying colors.

Remember that the only input our retina and our brain receives is the 'raw' sparse versus dense neural firing patterns. There is no notion of light and color that is transmitted into the primary visual cortex. Rather, the light and color sensitive chemicals translate light signals to a series of chemical reactions and eventually into neural firing patterns. Once this conversion happens, the information about light being the cause is lost.

A newborn cannot distinguish the above two cases whereas all adult conscious beings reject (ii), namely, that the two colored rectangles are of different shapes. Adults instead identify these rectangles as (i), in which both are of the same shape but different with 'colors'. In other words, adult conscious beings invent a concept of color.

It is easy to see why a newborn cannot reject (ii). A newborn does not know what a rectangle is in the first place. In other words, a newborn does not have a fixed set for a rectangle whereas all adult conscious beings have created a rectangle fixed set. Therefore, a newborn who has almost no fixed sets yet, attempts to create two separate fixed sets for both patterns as two different shapes instead of the same rectangle shape. Now, in order to create a true rectangle fixed set, a newborn should be exposed to hundreds of rectangles, interact with them under a large set of dynamical transformations (see the definition of fixed set in chapter 14). Just two examples are not enough to create a rectangle fixed set.

From this discussion, it follows that color should emerge as a fixed set just as rectangle emerges as a fixed set. There will be a clear transformation for a newborn when each pattern that forms on the retina will start to decouple itself from a concept of true shapes, sizes, angles, edges and other geometric properties to the concept of colors of these objects. Rather than memorizing different patterns formed on the retina as different shapes of objects, our brain groups several patterns into the same fixed set like a rectangle, circle and other shapes while also creating new fixed sets that captures the subtle differences in the patterns as colors like red, green, blue and others. In other words, the original pattern for a red rectangle is decomposed into two fixed sets – a rectangle fixed set and a red color fixed set. A blue rectangle is similarly decomposed as a rectangle fixed set (which is the same as before) and a blue color fixed set.

This is an important *pattern decomposition* that a newborn learns. Two original patterns corresponding to red and blue rectangles are decomposed into three new fixed sets – a rectangle, a red color and a blue color fixed set. Similarly, a red circle, a red rectangle, a red balloon, a blue circle, a blue rectangle, a blue balloon, a green circle, a green rectangle and a green balloon all generate different patterns on the retina. However, these nine patterns are decomposed into just six fixed sets – a circle, a rectangle, a balloon, a red, a green and a blue fixed set. With such a decomposition, there is considerable re-usability of fixed sets as the above example illustrates. Colors as fixed sets is a very important pattern decomposition our brain learns. Note that these color fixed sets are no different from other types of fixed sets discussed previously like edges, angles and shape fixed sets.

If we look at all objects around us, the color-based pattern decomposition is an efficient decomposition that not only re-uses several fixed sets but also maintains accuracy and self-consistency in the external world. We rarely have errors with such a decomposition. When we do have errors, they manifest themselves as, say, optical illusions. I will describe the color fixed sets in more detail later in this section by specifying appropriate dynamical transformations needed for the definition of a fixed set.

Given this motivation for defining colors as fixed sets, let me highlight a set of related questions that we want to answer. I will discuss them briefly here and leave some of the detailed analysis for future work. An important departure to keep in mind from all existing approaches is to not define colors using S, M and L cones directly.

(a) *Similarity and differences in colors*: How can we distinguish two different colors as different? When can we say that two colors are the same?

(b) *Relationship between colors*: How do we address the relationship between colors like with mixing two or more colors to produce a different color?

(c) *Triggering colors purely internally*: How can we see same color when there is a wide variation of external light (color constancy)? What about seeing color when you close your eyes (like during dreams) or when you look at after effects (in some color illusions)?

(d) *Uniqueness of colors*: How do different colors feel the same with respect to sensitivities and appearances even though the relative ratios of S, M and L cones are widely different? How do people with different number and distribution of cones still see the same colors?

(e) *Generalization of color*: How can we generalize color to dichromats (people with only two color pigments), tetrachromats (people with four color pigments), other anomalous cases and for cases with animals that can see in the ultraviolet range?

Let me now give a mechanistic definition of color (i.e., not the *qualia* of color) in terms of fixed sets. This definition of color is motivated from the most basic purpose of color, namely, for finding similarities and differences in the external world using the physics of interaction of light with material objects. For example, we use color to identify fruits and vegetables as well as dangerous and unhealthy situations.

Whenever we have patterns that need to be distinguished, we know that fixed sets can help us achieve this as we have seen in the previous chapters. Each pattern you want to distinguish has a unique fixed set. Can we do the same for colors?

Definition 31.1: Color is defined as the unique meta-fixed set created and excited within our brain with neural inputs from the retina (typically from the fovea, but to a lesser extent from the peripheral neurons on the retina as well) that are invariant under (i) the standard set of dynamical transformations like geometric, spacetime and internal brain structure modifications that are used for defining most object meta-fixed sets as well as (ii) dynamical transformations corresponding to a subset of objects like rectangles, circles and others.

The definition of a fixed set states that the set of neural loops triggered are invariant under a large set of dynamical transformations (see section 14.2). For example, in order to define a rectangle fixed set, we chose a set of dynamical transformations as translations – moving left or right, rotations – turning your head, zooming – moving towards and away from the rectangle as well as changes in your internal brain structure like when you grow older.

What are the similar set of dynamical transformations when defining a color fixed set? In addition to all of the above dynamical transformations, if the set of looped pathways triggered in our brain is invariant even as we change a subset of objects themselves, we call these looped pathways as color fixed sets. For example, we first trigger the same set of looped pathways as we move towards or away from the object, turn our head and others discussed above. Next, we trigger the same set of looped pathways even if we change the object from a rectangle, to a circle, to leaves in a tree, to a car and others. If this happens, we call it as, say, a green colored fixed set.

In other words, we have added additional set of dynamical transformations to the standard set that we use for most fixed sets when defining color fixed sets. These transformations are different objects themselves. This is logical because color is not associated to specific objects. We can have a red chair, a red car, a red balloon and others i.e., red is invariant across a set of objects.

Notice that a newborn forms color fixed sets before he forms fixed sets for geometric patterns. This is because the invariances for colors are far more common than the invariances for a geometric pattern. Color invariances are a superset of geometric pattern invariances. Another invariance for color fixed sets comes from the location of the image patterns on the retina. They can be on the fovea as well as on the periphery where the number of cones are far fewer and sparsely distributed. With all of these invariances, if over a period the same set of loops are triggered for a newborn, we can identify it as a particular color fixed set like a red or a blue colored fixed set.

Let us now consider the photon input pattern R_i for n time instants $i = 1, 2, ..., n$ corresponding to a red colored object. Let n be much larger than 1 i.e., we are looking at inputs for a finite period of time instead of instantaneous input patterns. The distribution $\mathcal{D}$ of rods and cones for a given human being generates a unique firing pattern from the retina for each of these photon patterns R_i. The physics of how this works, described earlier, is the same for a newborn or an adult. What is different, however, is what happens beyond the rods and cones, and deeper in their respective brains. The pattern of release of glutamate when viewing a red versus a blue colored object are considerably different both for a newborn and an adult, modulo the spatial arrangement $\mathcal{D}$ of their rods and cones (see Figure 31.3). A newborn cannot use this distinction generated at the retina to perform any action. These firing inputs do not propagate deeper to produce any cascading effect because of a lack of dense interconnections between neurons in these regions.

The pathways of the firing pattern from the S, M and L cones on the retina were studied anatomically in detail (see Kandel et. al [82] and Purves [122]). The firing patterns from the photoreceptors pass to the retinal ganglion cells and subsequently to the lateral geniculate nucleus (LGN). They then reach the primary visual cortex (V1). The subsequent areas stimulated by differences in color are V4 and some regions located in the fusiform gyrus of the medial temporal lobe. An adult has dense interconnections in these regions compared to a newborn.

Therefore, it is not correct to say that a newborn already has unique fixed sets that let him distinguish a blue colored object from a red one. Surely, the inputs at the photoreceptors are different for a blue colored object from a red one. However, this information is not enough to distinguish the two cases: (i) extracting color as an abstraction and (ii) treating both as different spatial patterns. A newborn has neither a fixed set nor abstract continuity even for a single color. For example, if the same red colored object appears in front of him, then, except at the photoreceptors on the retina, the neural firing paths are considerably different each time. There are no consistent and reliable loops that can be used to uniquely identify a color.

Besides, there is a limited predictive ability with colors for a newborn to create a feeling of knowing. He cannot distinguish if there are color variations or if there are variations within his internal neural network, just from the variations in glutamate release. At his age, the changes in neural network are quite significant because of a large set of visual experiences, most of which are entirely new. Over time, a newborn will create repeatable loops of excitations for colors similar to the other fixed sets. Let us now address questions (a)-(e) mentioned above in greater detail.

Similarity and differences in colors: The unique fixed sets according to definition 31.1 let us answer question (a) above. Two different colored objects will create two different fixed sets that is unique to the spectral distribution R_i (for $i = 1, 2, ..., n$) of light and the photoreceptor distribution $\mathcal{D}$ modulo the dynamical transformations mentioned above. The same colored object will trigger the same fixed set in a repeatable way.

The collection of all spectral patterns we can generate are infinite. So we can view these color fixed sets as a discretization of this infinite set. In other words, for every distinct color we can recognize, there is a unique fixed set within our brain based on the spectral and photoreceptor distribution pair. Furthermore, minor gradations in color like with different shades of red, blue or yellow are just fixed sets that are closely linked to one another. These links produce a feeling of continuity among the colors. The origin of this continuity is clearly the similarity among the firing patterns at the photoreceptors.

There is a minimum size of an image on the retina that lets us identify a color. This is not a single or just a few number of cone cells at the fovea. If we look at a faraway image that has several contrasting colors, we need to scan the image several times to identify as many colors as possible. Besides, the minimum size also depends on how contrasting the adjacent colors (or even border colors) are, i.e., how different the firing patterns are, generated within an image region I. The corresponding fixed sets are naturally distinct and have fewer links between them unlike with gradations of one color.

For example, yellow next to a white background (without a black border) does not let you identify yellow clearly, when you look at it from a distance. Similarly, light blue and light green, or orange and light red have similar issues. The two different colors and their borders, however, do become distinct when you are close enough. This leads us to the next question of how to relate two or more different colors to one another.

Relationship between colors: In the above discussion, I mentioned that light yellow appears closer to white. This is because white color is perceived when all S, M and L cones produce an equal peak response. For light yellow, Figure 31.1 shows that M and L cones both produce a peak response. We have seen that for most human beings, there are very few S cones (< 5%) with almost none in the fovea, i.e., M and L cones dominate fovea. This specific distribution $\mathcal{D}$ produces a neural firing pattern where 'almost all' cones in the fovea produce a peak response. The resulting firing pattern is similar to the one for white color, at least, from a distance or when light yellow and white are next to each other.

This example illustrates one way to relate a firing pattern (i.e., a color fixed set) to one or more other color fixed sets. There are, at least, two common ways to produce a different color by combining two or more colors. I will discuss these mechanisms next though I will be a little vague on the details. Future work is necessary to make this discussion more rigorous. Note that we do not need to use color directly for the problem of consciousness compared to the other visual properties like edges, angles, sizes and shapes. Nevertheless, color is an important topic that fits well with the rest of the concepts within a stable parallel looped system.

The first mechanism is when the input is a mixture of different wavelengths of light with one wavelength of light (or a narrow region of the spectrum) as the dominant wavelength. Such an input light source generates a firing pattern that triggers the fixed set corresponding to the dominant hue. The other spectral components create the equivalent effect of desaturating the input. As an example of this mechanism, consider two spectral frequencies corresponding to blue and red fixed sets. For the red fixed set, S, M and L cones have about 0%, 30% and 50% absorption respectively (see Figure 31.1). For blue fixed set, S, M and L cones have about 95%, 30% and 0% absorption respectively. Here, blue is quite dominant when we mix both of them even though the number of S cones are fewer in the distribution $\mathcal{D}$. Yet, their intense firing pattern makes the resulting mixture, i.e., purple feel closer to blue instead of red. If the absorption percentages are differently distributed, people may perceive purple closer to red instead.

The second mechanism is one where, when you mix two colors, the resulting color falls somewhere in between (on the visible spectrum). This is typically the case when more than one of S, M and L cones has a dominant response unlike the previous case. An example of this is how we trigger a green color fixed set when we mix yellow with blue (see Figure 31.1). Here, S, M and L cones in yellow fixed set have about 0%, 80% and 95% absorption respectively. Blue fixed set has 95%, 30% and 0% absorption for S, M and L cones respectively. The combined effect of these two absorption spectra produces a dominant firing pattern at M cones. The absence of S excitation in yellow and L excitation in blue suppressed both S and L excitation relative to M in the resulting firing pattern. If we compare this with green fixed set, we see that it has about 15%, 100% and 70% absorption for S, M and L cones respectively with M relatively dominant. We can argue that the S cones are more suppressed than L cones because of a smaller percentage of S cones in the distribution $\mathcal{D}$. We can use a similar argument to explain why red (0%, 30% and 50% absorption for S, M and L cones) and yellow color (0%, 80% and 95% absorption for S, M and L cones) firing patterns can produce a resulting orange (0%, 50% and 95% absorption for S, M and L cones) fixed set. The dominant cones here are L cones with M cones slightly suppressed, relatively speaking, in the resulting mixture.

Even though there are relationships between different colors, it is inefficient for our brain to compute the resulting color each time. For example, we can generate most visible colors using three primary colors red, green and blue. Our brain forms fixed sets for each of the different firing patterns that can be distinguished based on the inputs from the photoreceptors. When a given light source falls, it naturally creates a dynamical excitation pattern on the retina, which propagates through the primary visual cortex to excite a fixed set unique to the input light source. Conceptually, our brain may indeed obey the relationships between colors (like yellow + blue = green). However, these are not computed. Instead, the physical dynamics naturally picks the 'truth' without any need to hardcode these relationships using the links of interconnections formed through memory. In this way, we discover these relationships only later. This is another example where the dynamics of our brain architecture is used naturally to form concepts and relationships between them (similar to how abstractions are created – see section 13.1).

Triggering colors purely internally: We can trigger the perception of color even when we close our eyes. Even though there is no external light or there is no excitation at the S, M and L cones on the retina, we can still perceive color. One such case is with color in dreams. It has been shown overwhelmingly that people, in fact, do sense color during dreams even though people seem to remember their dreams in black and white. If we close our eyes and try to recollect an image of leaves or a flower, we can sense the color as well (if we try hard enough). These situations are consistent only with a fixed set representation and not with the traditional definition of color.

In fact, a generalization of triggering the same fixed set purely internally without any external inputs is precisely the dual space representation already discussed in chapter 25. This is applicable with other senses as well as with every fixed set. One way to trigger these fixed sets indirectly is through imagination. For example, you can sense that someone is touching you, or you hear imaginary sounds, or you sense a specific taste or odor, without any external inputs. I will discuss further implications of the dual space representations for color in section 44.2 when discussing 'dual equality' necessary for understanding the *qualia* of color.

Furthermore, synesthesia is a neurological condition where hearing music, perceiving numbers or letters leads to the perception of color. The standard theories of neural basis for synesthesia work equally well with the fixed set representation of colors (see, for example, Sperling et. al [145]).

Let me consider yet another case where we want to determine the color of an object using our peripheral vision instead. Consider looking at an object directly using your central vision. Now, try to determine the colors of all objects outside this central vision i.e., from your peripheral vision. There are two factors that play a role here: (a) your memory of an object and (b) the amount of external input that reaches your eye. A combination of both these factors can help you excite the appropriate color fixed set for the peripheral objects.

For example, the color of a small but distinctly important object that you have previously memorized can be easy to detect (through its fixed set). If, however, it is a small object without distinct memory, you may need to take a quick peek at this object with your central vision to excite the correct color fixed set first (i.e., form a short-term memory). Then, you can look away and rely on peripheral vision to continue to detect the same color well enough. You can similarly analyze other peripheral vision cases with large sized objects.

One prominent scenario where we see the above effect is with color constancy of objects. Objects such as an apple appear red (or green) under a wide range of external light sources because our memory plays an important role to trigger the correct color fixed set. You trigger the abstract continuous path from your memory of an apple along with partial color information to trigger either the 'red' color fixed set or the 'green' color fixed set, as appropriate. We do not see a blue apple. Therefore, there is no abstract continuous pathway to blue fixed set from the shape or feature recognition regions.

It is possible to analyze several color illusions, afterimages (that we view in complementary color when exposed to intense light) and other cases using the above approach. I will not go through these details because, in most cases, it is a standard translation of all the existing research into the current fixed set model. In all of these cases, we see that the fixed set representation for color offers a unified approach.

Uniqueness of colors: If we attribute color to the S, M and L cones directly, as is commonly done, it is difficult to explain why red, green, blue or any other color feels the same in spite of a disproportionate amount of S, M and L cones (S cones are only 5% compared to, say, 35% for M and 60% of L cones) as well as the non-uniform distribution of cones in the retina (like, say, almost no S cones in fovea). If you look at a blue square versus a red square, excitation of 5% S cones versus excitation of 60% of L cones produces identical looking squares. The mosaic distribution does not introduce any holes in our perception for a blue square when 95% of the square is empty (versus 40% for a red square).

Another difficulty with the existing model for color is that the distribution and percentages of S, M and L cones are different in both eyes. Yet, if you at both these squares one eye at a time, they still feel identical in spite of considerable variations. The left eye pattern for red and blue squares are different from the right eye patterns. The only way we can explain this is with a fixed set approach.

In addition, the number of rods and cones, the relative percentages and the distribution of S, M and L cones vary considerably across different people. Yet, everyone seems to perceive the same set of colors in the same way. This can be naturally explained when we use the creation of unique fixed sets as the definition for color. These fixed sets are based on the distinct firing patterns generated by photoreceptor distribution $\mathcal{D}$ from the retina. The specific distribution $\mathcal{D}$ is not as important as long as distinct patterns can be generated among the current set of photoreceptors. For example, dichromats (with only two types of cones, say, S and M cones) do not generate distinct firing patterns for a normal green and red colors. This is because the only difference is in the relative absorption amount of the M cones. This shows up as different shades of the same color.

Generalization of color: I will not discuss more details on cases with color deficiencies like monochromacy (only one color pigment), dichromacy (with only two color pigments) and anomalous trichromacy here. As discussed briefly before, these cases fit quite well within the fixed set representation of color. Each of these cases (monochromacy, dichromacy and trichromacy) map the patterns forming on the retina to different number of fixed sets. For example, for monochromats, the number of distinguishable patterns are far fewer than for dichromats. This translates to fewer distinct fixed sets for monochromats than for dichromats. The same generalizes to trichromats and tetrachromats.

Tetrachromacy (with four color pigments) among human beings is more prominent among women. One common example given to highlight the difference between tetrachromats and trichromats is the following: when a trichromat like you and me group colored clothes in a closet according to similar colors, a tetrachromat reorders them differently claiming that the clothes are not grouped properly. This is because tetrachromats have an ability to distinguish more subpatterns as different color fixed sets. Their four different types of cones result in different fixed sets.

Another example is with other species like birds. They have an ability to see in the ultraviolet range. This clearly gives them an ability to distinguish unique patterns within an object. For example, the additional patterns seen within the petals of a flower appear only when viewed with an ultraviolet filter (see Schmitt [136]). Birds can see them, but humans would only see the flower without these patterns as shown in Schmitt [136].

Some other species have complex internal structures as well. For example, bees have compound eyes with oil droplets that are either clear or filled with carotenoids. They help them see light from the ultraviolet range. Some species of birds have two or more fovea or have fovea as lateral stripes and so on. Some of these cases need further research to translate them into the new framework. I will consider this as future work. Note that I have also skipped the topic of *qualia* for color for now (say, why red feels like red). I will discuss this in chapter 44 instead.

32
Reality of Touch

In this chapter, I want to show that touch is the sensor that makes us feel that an object is real. When we look around us, we see several objects. How are we absolutely sure that these objects are real and not some ghostly images? The simple answer is that we can touch them to feel their existence. Nothing other than touching will make us completely convinced of this reality. For example, most people have had some form of an experience of seeing some illusion or a ghostly pattern in the dark. The first thing we do to clarify if this surprising image is real or not is to reach out and touch it, whatever that is. If we cannot sense it as we try to touch it, we are convinced that the image does not exist. I do want to mention in passing that some people believe that supernatural forces are at work in these cases. But, at this stage, I do not want to pursue this alternate theory because of a lack of repeatability and reliability of such experiments.

Now, what makes touch so special? My objective in this chapter is to provide more details on this. Note that the term 'real' is only relative to one's own consciousness. A conscious being always has physical components, which we normally call as its body. If we were to feel that an external object is 'real', we need to feel that it is indistinguishable, in a certain sense, to our own body, not the shape and the form itself but with respect to some physical and geometric characteristics. The indistinguishability is what I refer to as 'relative' to one's own consciousness (chapter 44 discusses this in more general terms).

In this chapter, I will show that touch is the only sensor that lets us feel this way, as our intuition suggests. No other external based sensors like eyes, ears, taste and odor provide any information about physical properties like mass or forces that we so clearly sense with our own body, using touch. Therefore, they do not help us directly with our perception of reality.

32.1 Direct contact versus noncontact sensors

Let me first make the intuitive notions of contact and noncontact sensors a bit more precise. We know that there is an external world (see chapter 29) and that there are several different objects in it. These objects have geometric properties like shape, size and texture. They also have other physical properties like mass and charge. These objects are placed in the external world at different locations relative to each other in space and time.

Consider the totality of objects in the external world, in an abstract sense, as T. These include objects like chairs, tables, trees, air molecules, photons, protons and electrons. Let us say that we are interested in knowing the geometric and physical properties of just a subset of these objects S. These are typically the macro-sized objects like chairs, tables and trees. In other words, we are excluding microscopic objects like air molecules, photons and, possibly, even individual cells. All objects within T dynamically interact with each other according to the known physical laws of our universe. These objects interact with a conscious being as well. There are several interfaces for these interactions, like, say, each of our sensory organs.

Our objective is to determine the geometric and physical properties of objects in subset S as well as geometric structure (say, relative spacetime locations of objects in S) of the external world relative to the subset S. Since we are only interested in objects from the subset S, we can define the notion of direct contact versus indirect (or noncontact) sensors. For example, an object like a drum generates sound by vibrating air molecules, which reaches our ears. An object like a leaf reflects photons from the sunlight, which reaches our eyes. Drum and leaf are part of our subset S while air molecules and photons are not. In this case, not all dynamical interactions from the source (drum or leaf) to the destination (eyes or ears) stay entirely within the subset S. This makes our eyes and ears as indirect or noncontact sensors. Noncontact refers to the fact that no objects from the subset S are in direct contact with our eyes or ears. Nevertheless, objects from $T - S$ are in contact with our eyes or eyes (i.e., the air molecules or photons are in direct contact but not bigger sized objects). We, therefore, use the terms contact and noncontact relative to the subset S.

Based on these definitions, tongue (for taste) and skin (for touch) are contact-based sensors. Eyes, ears and nose, on the other hand, are noncontact sensors. With eyes, ears and nose, the components that are in contact are photons, air molecules and other chemicals with odor respectively. However, we are not interested in knowing the geometric and physical properties of photons, air molecules or chemicals themselves. Rather, we are interested in knowing the properties of the object that generates or transmits them. These objects, for example, are the leaves (for eyes), drum (for ears) or pizza (for nose), say. In other words, the objects that generate or transmit a signal are different from the objects that are in direct contact with the sensor in the case of eyes, ears and nose. However, for tongue (taste) and skin (touch), these two types of objects are the same.

32.2 Effective sensors for detecting physical properties

Given the above discussion on contact versus noncontact sensors, which of our sensors can be best used to determine the geometric and physical information of the external objects or the external world? Since we restrict understanding geometric and physical information to objects only from subset S (as mentioned in the previous section), let me discuss which of our five sensors helps us the most, in the order of increasing importance.

Taste is the least effective sensor for this purpose. In order to create firing patterns from our tongue, we need to touch any given object with our tongue. As we rub the object with our tongue, we can try to extract geometric properties of the object. However, here we are using touch information (with our tongue) instead of taste itself to determine the geometric information. This dual role of tongue, both for taste and touch, is nevertheless taken advantage of by infants and toddlers. As adults, we almost never use this approach. However, infants and toddlers do place most accessible objects into their mouth. Our tongue has chemoreceptors (taste buds) that help us identify different types of taste like sweet, sour, bitter and salty. Some of these tastes produce emotional explosions, which act as a drive to taste other objects as well. When an infant places objects in the mouth to taste them, our brain will also detect the geometric and physical properties of these objects. Our brain will determine this information even if we are unaware of it. After an infant learns to use his other sensory organs, he will stop using tongue for this purpose.

The chemoreceptor-based approach used by our tongue is, however, used by other species to determine geometric information. Ants, in particular, use pheromones as a way to map complex geometric paths from, say, the food source to their nest. They have chemoreceptors that can detect different types of pheromones to guide them towards these destinations. Using this, for example, they are capable of detecting the shortest paths S (i.e., geometric information) from food source to their nest. They start by dropping pheromones along each path A, B or S (see Figure 32.1). If a given ant or other ants traverse multiple times along a path, the concentration of the pheromones increases. If there are multiple paths from the source to the destination, the shortest path S will have the highest concentration of pheromones during a given time period, as it is the most traversed path during the same period. The pheromones along the other paths like A and B will eventually dissipate. Therefore, the highest concentration of pheromones is the easiest detected signal by the other ants and is also the shortest path from the source to the destination. In other words, ants find the shortest geometric path in an entirely natural way using deposition of pheromones along the path.

With odor, we can use it to determine partial geometric information, but not any kind of physical information like force or mass. For example, you can walk in the direction of a particular source of smell (i.e., geometric information), at least

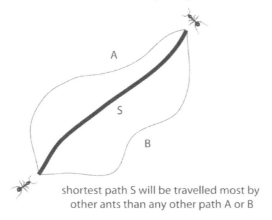

shortest path S will be travelled most by
other ants than any other path A or B

Figure 32.1: Ants discovering shortest paths naturally. In a given time period, the shortest path S among paths A, B and S will be travelled the most number of times by several ants. Each ant deposits pheromones as it travels along a path. The shortest path S will have the highest concentration of pheromones, making it easiest to detect and follow.

approximately, even if you close your eyes. We already know that dogs use their extraordinary sense of smell for just this purpose (and have, for example, trained police dogs).

People who cannot see have a heightened sense of hearing. They use this skill to navigate in known and unknown environments effectively. All humans already have a basic ability to detect directional information using sources of sound. In chapter 17, I have discussed how to use the fundamental pattern to train ourselves with this ability. However, hearing cannot help us with detecting any kind of physical information like force or mass. Bats, for example, use ultrasonic sounds as the primary way to 'see' the external world. Bats can fly effectively in dark environments because they use echoes from ultrasonic sound to create a map of locations of different objects.

This leaves us with vision and touch as the most effective sensors for determining geometric information about objects and the external world. I will, therefore, use these two sensors to discuss all physical and geometric features highlighted in items (a) – (f) at the beginning of chapter 30. The other three sensors (taste, smell and hearing) do contribute, especially for people who can see, but in a less significant way for most other people. For people who cannot see with both eyes, we can analyze the additional role played by these three sensors in an analogous way.

I want to show in the next few sections and chapters that touch is more accurate and real than vision (relative to our own consciousness). Firstly, we cannot use vision to sense any physical information like force or mass just as we cannot do with hearing, smell or taste. Touch is the only sensor that is capable of sensing physical information. Secondly, even with geometric information, vision will appear as real as touch only after we link it to touch. This may appear counterintuitive. For example, when you look at a house, say, 200 ft away, you have a clear sense of the size, distance, edges, shape and other patterns for the house. You know this to be accurate without even touching it. You can indeed verify your visual information by actually going closer and by touching it. Therefore, it may appear that we do not need touch to know geometric properties of distant objects.

However, I do want to emphasize that you need touch even in the above case. The above example is only possible because your prior experiences have created an enormous number of useful links between touch and vision as you interacted with nearby objects during your childhood (see chapter 18). We use these memories to create an accurate representation for distant objects as well, using just vision. Had you not created these links, you would not have been able to estimate the geometric properties of distant objects like the house. In chapter 36, I will show, in detail, how we learn to achieve this.

One interesting point to note is that we do detect forces or masses of distant planets and stars without ever touching them. In these cases, we use the laws of gravity and other physical principles like Doppler effect for light that we have already discovered. However, this requires a much higher form of intelligence and a belief in the concept of generalization applied to abstractions. For example, we do simple experiments at a smaller scale for which we can touch and verify that the principles (like the gravity and Doppler effect) are indeed valid. We can reliably repeat these experiments and be convinced that they are correct principles using our free will (see chapter 29 on how we used free will to show the existence of an external world in a similar way). Now, we are in a position to generalize it to distant stars and planets because we believe in our free will. Note that we acquire this ability to generalize only after we are conscious. Furthermore, this indirect approach is analogous to how our eyes make us believe that all objects we see around us are, in fact, real (with a mass and energy).

32.3 Somatosensory system

Compared to the other four sensors, it is hard to imagine a conscious being born without skin sensors or without information from skin sensors reaching our brain i.e., we exclude cases like complete paralysis 'after' you have become conscious once. In this subsection, I want to describe the basic mechanics of how our somatosensory system works. The intention is to use this description later (in chapters 33 - 36) when explaining how touch helps us with sensing geometric and physical properties. In section 32.4, I want to address the question of why we get a feeling that objects are real when we touch them.

Our somatic sensory system is capable of detecting different mechanical stimuli like light touch, itch, pressure and vibration (using mechanoreceptors) as well as temperature (with thermoreceptors) and painful stimuli (with nociceptors). Each of these stimuli detected using different types of receptors are present under the cutaneous and subcutaneous regions of the skin. In addition, we also have proprioceptors located in muscles and joints to detect positions and movements of different parts of our body. Let me describe the organization and the structure of the somatic sensory system briefly, leaving the details to textbooks like Purves [122] and Kandel et. al [82].

The somatosensory information from all the above types of stimuli is conveyed to our brain via the dorsal root ganglion neurons for every part of our body except the face. From the face, the same information is transmitted by the trigeminal ganglion neurons. The axon of the ganglion is branched into two paths. One end of this path acts as a receptor for the external stimuli, with or without an encapsulated covering (to be discussed shortly). The other end is connected to the neurons in the spinal cord to relay the information into the brain.

The encapsulated mechanoreceptors for detecting mechanical stimuli are Meissner's corpuscles (for touch and dynamic pressure), Pacinian corpuscles (for vibration), Ruffini's corpuscles (for stretching of skin) and Merkel's disks (for touch, texture and static pressure). There are two types of thermoreceptors – one for detecting heat and the other for cold. The nociceptors can be thermal, mechanical or polymodal (i.e., responding to any destructive stimulus). These are always free nerve endings (i.e., no encapsulation of the ganglions). Proprioceptors are muscle spindles (for stretch in muscles), Golgi tendon organs (for muscle tension) and joint receptors.

The basic response from each of these receptors starts by converting the stimulus energy into a neural firing pattern. Recall that for a neuron to fire an action potential, the voltage difference at the apex of the axon should exceed the threshold value for the neuron (see chapter 5). This is possible when the Na^+ and K^+ gated ion-channels on the nerve membrane surface open to cause an influx of Na^+ ions into the cell. In the case of mechanoreceptors, the ion-channels open in response to the force applied on the skin. This force causes the membrane to stretch. The specific ion-channels at these receptors are sensitive to stretch and hence result in the opening of Na^+ ion-channels. In the case of thermoreceptors, the difference in temperature between the inside and the outside of the body causes a similar response (the precise mechanisms are still partially unknown).

Purves [122] and Kandel et. al [82] have described the flow of information from the external stimuli to the receptor neural firing pattern and into the brain, in considerable detail. They have described the processes that occur at every step along this path. For example, the external stimulus energy (associated with, say, the force or heat) needs to be converted into a neural firing pattern. Here, each receptor (say, Meissner's, Ruffini's corpuscles or free nerve endings) is capable of responding to different type and intensity of external stimuli. We look at this as the sensitivity of the neuron to a given stimulus. Once a neuron fires, the voltage spikes propagate along the axon, which is myelinated in some, but not all, of the cases. This results in either fast or slow rate of transmission of information respectively. Let us now look at what happens after the initial receptor neurons fire.

The information from the mechanoreceptors (from dorsal root ganglions i.e., except from the face) for touch and proprioception are transmitted to the thalamus by the dorsal column-medial lemniscal system, ascending ipsilaterally (on the same side) in the spinal cord. The pain and temperature sensation, on the other hand, is transmitted along a different anterolateral pathway to the thalamus, ascending contralaterally (on the opposite side) in the spinal cord. Two pathways convey information about pain: along fast myelinated axons in the Aδ group and along slow unmyelinated axons from the C fiber group. For stimuli other than pain, the pathways are primarily Aα and Aβ fiber groups. In the thalamus, the region of convergence is the ventral posterior lateral (VPL) nucleus for all somatosensory information except from the face.

The touch and proprioceptive information from the face is transmitted from trigeminal ganglions into the brainstem along the trigeminal nerve. This reaches the trigeminal brainstem complex. The trigeminal complex has two components: (a) the principal nucleus (for touch and proprioception), similar to the dorsal column nuclei for relaying mechanosensory information from the rest of the body and (b) the spinal nucleus (for pain and temperature sensation). These neurons reach the ventral posterior medial (VPM) nucleus of the thalamus along the trigeminal lemniscus tract. I have merely summarized the important pathways here. I refer the reader to Purves [122] and Kandel et. al [82] for the skipped details. These textbooks describe them more pictorially as well. The pictures and further details do a thorough job of clarifying the above new terms and pathways introduced here.

The ventral posterior complex (VPL and VPM) of the thalamus contains the complete somatic sensory inputs. Neurons from these regions project to the primary somatic sensory cortex that has four distinct regions known as Brodmann's areas 3a, 3b, 1 and 2. The areas 3b and 1 respond to cutaneous stimuli whereas the areas 3a respond to proprioceptive stimuli (for positions and movements). The area 2 neurons respond to both tactile and proprioceptive stimuli. The neurons from each of these areas project to secondary somatic sensory cortex and to posterior parietal cortex (Brodmann's areas 5 and 7). Area 5 integrates tactile information from skin with proprioceptive inputs from muscles and joints as well as from both hands (for sensory guidance and posture during movement) while area 7 receives visual, tactile and proprioceptive information (for eye-hand coordination).

Each of the four regions of the primary somatic sensory cortex has a complete map of the body surface. These maps are not fixed, but can be altered through new experiences, at least partially. The relative amount of the somatic sensory cortex devoted to each part of the body is proportional to the density of receptors in that part of the body. For example, our hands and face have a larger number of sensory neurons (and hence larger amount is dedicated in the somatosensory cortex) compared to the back of our body.

The nociceptive information (for pain) from the spinal cord has three main ascending pathways: the spinothalamic tract terminating in the thalamus, spinoreticular tract terminating in both the reticular formation and the thalamus, spinomesencephalic tract that eventually reaches amygdala (part of the limbic system responsible for emotions) via spinoparabrachial tract. Thalamic neurons relay this information to the several regions of the cerebral cortex including somatosensory cortex, cingulate gyrus and the insular cortex. The cingulate gyrus is part of the limbic system giving rise to the emotional component of pain while insular cortex is responsible for producing an overall pain experience (see Purves [122] and Kandel et. al [82] for more details).

With this brief summary of the various pathways from the source of different types of stimuli at the surface of your body to the deeper parts of our brain, I will start to explain how touch is capable of sensing geometric and physical properties. The internal structure that results is eventually critical for creating the self-sustaining dynamical membrane of meta-fixed sets needed for consciousness.

32.4 Reality of touch and its relation to consciousness

We believe that the visible objects around us exist even though we do not interact with them directly. For example, are the tables, chairs and other furniture in your room that are visible to you indeed present or are they just a figment of your imagination? You have not interacted with them directly. So how do you know if they exist or if you are just imagining them? If we do want to find out whether these objects truly exist or not, what should we do?

In chapter 29, I have shown that the patterns created by objects on our sensory surfaces (like our eyes) do have an independent existence. We had to rely on our free will mechanism to show this. This, however, does not rule out the possibility that these objects are not a *virtual* object or an illusion. Yet, the moment we touch and feel the object, we seem to be completely sure of its *real* existence. It is not our imagination anymore. In this section, I will discuss why this is the case. Why do we get a feeling that an object is real when we merely touch it?

The answer to this question is closely related to how we become conscious in the first place. This should not be surprising because we know that it is our consciousness that tells us whether something is real (or true). Theorem 22.1 of section 22.1 already highlights the necessity and sufficiency of knowing truths. For example, we have seen that our own emotions are real only if we are conscious (see section 12.3). Reality is itself relative to our consciousness. Therefore, the reason why touch feels real will be related to how touch contributed to our consciousness in the first place.

When we are consciousness, we have a clear sense of our own body at minimum. After birth, as we gain new experiences, we use different parts of our body in different ways. When we become conscious for the first time in our life, our brain's membrane of meta-fixed sets will be built using our past experiences. These will surely involve parts of our own body. Our body, therefore, helps define our very own consciousness. The problem of starting with a non-conscious system and converting it into a conscious system requires the system's body and the external world at minimum. This is especially true if consciousness were to develop naturally (i.e., without help from other conscious beings) and for the first time within a system i.e., we are not just looking at the problem of switching between unconscious and conscious states after the system became conscious once in the past.

What properties about our body do we know after we have become conscious? These are clearly the set of geometric and physical properties mentioned in chapter 30. They are, for example, the shape, size, edges, angles, textures, patterns, mass and forces needed to move different parts of our body. I will describe how we sense each of these properties in chapters 34 - 37. These properties of our body are an integral part of our consciousness. Whenever we want to know if an object is real or not, our consciousness compares the external object with our own body. At minimum, we should sense each of the above geometric and physical properties in the external object that we do sense with our own body. It is only when we feel the external object is indistinguishable to our own body, we can say that the object is real. We should understand the term 'relative to our consciousness' in this sense.

Now, in addition to the above minimum set of geometric and physical properties, there are other properties that we discover using scientific analysis. We are, however, not aware of them when we become conscious. These are, for example, existence of electric charges, magnetic fields, molecules, cells, DNA and even a number of internal organs in our body. They certainly did help us become conscious, except we are not consciously aware of them. Therefore, even if we do not sense these additional properties (like electric charge), it does not seem to matter for our perception of reality of the external object. For example, if we sense that the object has a mass using touch, we are already convinced that the object is real. We do not need to know that the object is behaving like a magnet or that it has a net positive electric charge. We, therefore, only need to sense the above minimum set of geometric and physical properties.

Conversely, if we cannot sense these minimum set of geometric and physical properties, we are not sure of the reality of the external object. For example, with vision we can only sense the geometric properties of an object. We cannot deduce either its mass or whether a force can move the object. Therefore, merely looking at an object does not convince us that the external object is real. For example, what if we try to reach for this object and our hand simply goes 'through' the object like with optical illusions or with 3-D movies? Since our attempt to touch the object has failed, we know that the object is not real even though the visual information makes it seem quite real. With touch, we can eliminate this uncertainty. When we touch and feel the force from the object, we can sense all of these minimum set of geometric and physical properties.

We should note here that besides touch, there are other indirect ways to conclude if an object is real or not. One such way is to use vision, free will and a few other facts or physical laws to make the same deduction, without touching it directly. For example, we can try pushing the object with a stick or a chain of sticks to see if it moves. Here, we have used Newton's third law of motion, namely, that for every action there is an equal and opposite reaction along with the rigidity of the connected objects. Since we can establish that the other connected objects are real using our free will, the object of interest will feel real too. While this approach is still based on direct contact (see the definition in section 32.1), we can also use noncontact approaches. For example, we can deduce masses of planets and stars (and hence their reality) indirectly using, say, their gravitational effects on other systems.

The main reason why touch is critical for reality is because touch results in a direct contact (i.e., with zero distance) with the external object. From the previous section, the somatosensory inputs corresponding to touch are triggered only when the external object is in direct contact (i.e., zero distance) with our skin. The force on the skin is translated into a firing pattern in a reliable way (cf. vision where no such forces are transmitted). If the distance from the external object to the skin is nonzero, there will not be any neural excitations from the skin surface. As a result, we are completely unaware of the existence of an external object (if we close our eyes). In the case of eye, the object can be quite far from us and our body. Yet, a unique corresponding pattern is produced on the retina, which will be memorized as a fixed set. With touch, however, no fixed sets can be created in our brain unless the external object physically touches us and excites a pattern unique to the external object. This is the main reason why touch makes objects real.

If the external object is in direct contact with the sensory surface (i.e., the distance between the external object and the sensory surface is zero at a macroscopic level), the contact surface on the sensor is a direct representation of the surface of the external object as well, provided the sensor has sufficient sensitivity. The molecules from the surface of the external object exert forces on the sensory surface when in direct contact, in accordance to Newton's third law of motion. With vision, on the other hand, there is no force that is transmitted between the external object and our eye.

With direct contact, the size and texture in a small neighborhood around the contact surface, i.e., both on the external object and the sensor, are almost identical.

This is a basic assumption of continuity with macroscopic sizes. This is an important fact that allows us to infer reality using touch sensor.

For example, if we want to estimate the size of a ball, we can use the size of our palm instead, if it fits in our hand. There is a one-to-one relationship between the size of a ball and the size of our hand. This is not the case with vision. The size of the pattern on the retina (i.e., on the sensory surface for vision) is never identical to the size of the ball. We will, therefore, have to infer the true size of the ball through other means. Vision alone cannot help estimate size of objects. I will discuss estimating sizes using vision later in chapter 36.

From the above discussion, given that the information like size and texture of an external object is accurately represented on the direct-contact touch surface, the next step is to transmit this information into the brain without distortion. We want the transmission of the firing pattern to be reliable and effective until it reaches the self-sustaining membrane. In the case of skin sensors, I have already described these pathways in section 32.3.

To help with this problem of undistorted transmission of the sensory signal from an external object, touching our own body instead plays a critical role. Before, we start to infer about an external object, we can first build all the necessary fixed sets related to sizes, shapes, textures, mass and other physical and geometric properties originating from our own body. Then, we will be in a position to compare how touching our own body feels identical to touching an external object. This ability to generalize our inference from our own body to external objects is accurate because touch is a direct contact based sensor.

With dual space representation (see section 25.1), if we stimulate directly in the primary somatic sensory cortex, which has a faithful map of inputs from different parts of our body, it produces the same feeling of touch from the corresponding part even though there is no external stimulus. For example, during our dreams, when we touch the 'imaginary' objects in our dream, they do feel real even though the inputs from our physical body are cut off at the brainstem. Our dual space representation and the experiences when we were awake make the feeling, associated with touch, as vivid as reality.

It is also worth mentioning that there are other ways to make us feel that an object is real using touch. For example, one of the primary ways to experience pain from external inputs is through touch (not vision or hearing). Since pain is entirely personal, it feels real to a conscious being. We can then say that if an external object caused us some degree of pain, that object should be a real object.

In summary, we can say that a direct contact based touch sensor has allowed us to sense and transmit the properties of the external object both accurately and reliably from the sensory surface all the way into the brain. Our own body fixed sets along with our free will has provided ample ways to compare, correlate and verify accuracy of the geometric and physical properties of the external object. Therefore, we can conclude that, when in doubt about the reality of a few geometric or physical properties, we can always use touch to resolve the ambiguity and guide us in the correct direction.

33
Relationship between Sensory Inputs

In the previous chapter, we have seen that touch is a direct contact-based sensor that allows us to capture reality of the external world in more accurate detail compared to other noncontact based sensors like vision, hearing and others. In this chapter, I will discuss how we can relate multiple sensory inputs together so we can extend the reality captured by touch sensor to other noncontact sensors like vision and hearing as well. As our brain builds sufficient number of these relationships, we can infer properties like sizes and shapes of external objects with high precision using noncontact sensors as well. Such accurate inference would have been impossible had we not built relationships between touch and other noncontact sensors.

The specific problem I will consider here is whether touch is necessary for creating a visual representation of the objects we see around us. The trees, the houses and the road you see in front of you seems to have a specific shape, size and is at a specific distance from you. It hardly seems like touch is contributing to this accurate and coherent visual representation. I will, however, show that you cannot create this accurate view with your eyes alone if you had never relied on touch in your life (or before you became conscious for the first time).

33.1 Touch – a necessary input for visual perception

To understand the relationship between touch and visual perception, we should (a) look directly at how touch contributes towards the visual representation and (b) analyze what happens if we do not use touch when sensing geometric properties. For the former problem, I will present detailed mechanisms for each of the geometric properties (like edges, angles, sizes and distance) and show clearly the role of touch. I will discuss them later in chapters 34 - 36.

In this chapter, I will only focus on the latter problem. Let us say that you are just using vision and not touch for determining the geometric properties. If there is a minor discrepancy with your eyes (say, nearsightedness), there will be a corresponding change in your visual representation. Let us take nearsightedness as an example to analyze since it is a bit more common among people. With nearsightedness, you cannot see distant objects clearly. The light that comes in through your eye lens falls in front of the retina instead of falling exactly on the retina. The resulting images are out of focus for distant objects, but are in focus for nearby objects. You wear a corrective lens (a concave lens) to help compensate this error. However, as most people with nearsightedness may have noticed, the eye condition deteriorates, say, after a few years and you may need to change your prescription eye glasses to re-compensate. Even otherwise, nearsighted people tend to change their prescription glasses once every 2-3 years (even for reasons like scratches on eye glasses or when the frame breaks).

One interesting fact people notice when they change their prescription glasses is a sudden and somewhat drastic change in their visual field. The first time you wear them, your visual representation appears considerably different. Objects seem to appear farther and smaller, just by a tiny and but clearly noticeable amount. In section 33.3, I will discuss why this happens. But the interesting question is how your brain fixes this visual representation (not the discrepancy itself). Indeed, everyone notices that their visual representation comes back to the original normal state after a few days. Since it does get fixed, how does your brain know which direction to apply changes? Why do the actions taken by your brain, when trying to fix this view, not make it worse instead of better? I will discuss these questions in detail in section 33.3.

The argument I will use will lead to an important generalization. We can use the reasoning to explain how you perform actions in a correct manner even if you are not conscious, especially when there are multiple conflicting inputs to your brain. For example, if your vision suggests that the object is of a certain size or at a certain distance, your touch suggests differently and your hearing provides yet another estimate, which input would you treat as the correct one?

In the example above where you changed your prescription glasses, your vision says that the objects are smaller and farther whereas your touch inputs, as you extend your arm to reach these objects, provides a different estimate, namely, that they are, in fact, closer. Which input should your brain treat as the correct one – touch or vision? Do you need to make this decision consciously? Can your brain choose this automatically without your conscious awareness?

The answer, as we will see in section 33.3, is that your brain can, indeed, determine this correctly without your conscious awareness, as long as the scenarios repeat multiple times. Also, the most accurate sensor our brain relies on when we have multiple conflicting inputs is always touch in all cases. More precisely, contact based sensors like touch are always the most accurate ones to rely on rather than noncontact based sensors like vision and hearing. The information from the external objects to your body and then finally to your brain is least distorted by contact based sensors. Therefore, touch is necessary to generate an accurate visual perception as well.

33.2 Is vision an illusion?

When we are 'looking' at objects around us, it feels like touch has no role in creating the visual representation in front of our eyes. It seems to be specific to our eyes. In fact, the vivid view of the world makes us feel that we are simply watching what truly and already exists in front of our eyes. Surely, there are a few peculiarities like the blind spot, the inverted images on the retina and several optical illusions. Yet, the number of times we encounter these peculiarities in our everyday life is quite small. Except for these cases, we feel that eye is just acting as a 'transparent window' with a 'conscious you' on one side and the external world on the other side.

There are two distinct questions we can ask when thinking about what we view with our eyes. Firstly, is the external world we see, real or an illusion? We know the answer to this question – the external world is indeed real (see chapter 29). Secondly, is what we *see with our eyes* an illusion, projected by our brain after processing the visual inputs from the real external world? To look for an answer to this question, we need to understand what is happening inside our brain. I have discussed some details in chapter 31 when discussing about color vision. It turns out that as we learn more about the complicated structure of our brain corresponding to vision (like the structure of primary and secondary visual cortex), we begin to realize that our brain is doing much more than a simple translation and projection of the external world. Our brain is indeed working extremely hard to create a view that appears so real.

If you squint your eyes, defocus them so images look blurred, use someone else's corrective lens or look through other optical devices (like, say, a pseudoscope – a device that reverses depth perception: actually, a device that sends left visual field to the right eye and vice versa causing objects to appear inside out), you will see that your visual view of the world is considerably altered. This suggests that these altered views are valid possibilities.

If, in addition, we include a number of eye defects, the effects of drugs on vision, animals with panoramic vision or with compound eyes, we can see that the visual representation of the external world is not unique. Our so-called normal vision is just a special case among many other possibilities that our brain can pick. Some of the above situations appear to be minor variations of our normal vision. Most of these variations do not alter our visual field fundamentally. For example, you do not see a rectangle as a circle. There are several mechanisms that prevent this from happening. In fact, except for optical illusions, there are redundant sources of inputs (like touch and hearing) for the same phenomena that allows us to confirm whether what we see is correct or not.

The other examples mentioned above are a bit more intricate like why we see the world upright even though the images on the retina are upside down. Given these different visual representations, how can we convincingly answer whether our vision is an illusion or not? Unfortunately, to add to the difficulty, our visual field and how the world appears to us is a qualitative experience. We do not have a good way to collect and analyze these qualitative experiences.

For example, our visual representation changes the most at less than 6 months of age. At this age, an infant cannot communicate well enough to let us know how his visual perception is changing. At the same time, we cannot explain or understand his visual perception by looking at what is happening within his neuronal network. As a result, there is no direct way to understand how our visual representation feels before versus after a transition. Examples of such transitions are: inversion of object images on the retina leading to an upright perception, our brain's ability to cover a blind spot, how our visual perception fixes itself when you change your prescription glasses and others.

I will discuss a few peculiar scenarios and leave the rest for the reader to analyze. For example, the reader can easily analyze squinting, blurring of our eyes, random dot stereograms as well as situations when eye defects have been rectified or even when the entire vision has been restored at an older age. In the next section, I will pick a couple of cases in which our visual representation starts from a good state, then becomes incorrect and finally, after a period of time, changes back to a correct state.

33.3 Is touch required for vision?

In this section, I will discuss the generic role of touch for visual perception without focusing on any specific geometric feature. In the later chapters, I will discuss how touch becomes necessary for each individual geometric feature like sensing edges, angles and sizes.

Consider the situation discussed earlier where you are either already wearing a corrective lens or you will need to wear one for the first time for nearsightedness. In both cases, let us say that you are unable to see far objects clearly and that you need a new prescription. Let us analyze how your visual field changes in the first few days after you wear your new prescription glasses.

One distinct feeling is that all objects appear to be located farther than they really are. As you walk in your house, the tables, chairs, the stairs and even the floor feels distinctly farther. They also appear a bit smaller, although the change in size is not as noticeable as the change in distance. The net effect is that you feel taller. In spite of these changes happening with your visual perception, one thing you will not notice is some objects appearing at the correct distance while others appearing farther. Our perception is uniform across all objects, either all objects feel closer or all objects feel farther. It takes several days before our vision feels normal again i.e., objects appearing again at the correct distance and size similar to how we felt before we changed our eyeglasses.

It is not surprising why we feel this way. This is because the images on the retina are sharper with the aid of new eye lenses than without. Therefore, if we include the fuzzy borders, the image without the eye lenses (or with the old prescription glasses) feels a bit larger than with the new lenses. With new glasses, a smaller and sharper image corresponds to an object located slightly farther away. Note that our brain is estimating the size and distance of objects using the corresponding sizes on the retina. Our brain is essentially 'computing' these external sizes and distances using the older memories (i.e., those formed with the older eyeglasses) but taking the smaller and newer retinal image sizes as inputs. As a result, they are temporarily incorrect. This sudden change in eyeglasses (specifically, smaller sizes on the retina) creates an impression of objects being further away.

However, after about a week, our vision restores to the correct state. The objects no longer appear farther and you do not feel taller anymore. How does our brain correct the visual representation? It is clear that we are not consciously aware of how our brain is correcting itself. It is also clear that our consciousness is not helping our brain. Our brain has a natural tendency to fix itself using an unconscious mechanism.

With the new eyeglasses, our brain is exposed to a sudden change in visual inputs. This change acts as a cause to alter and produce new or modified neural connections within your brain. As your brain adapts, why does the change move towards a direction that make the objects appear at the correct distance? What is special about placing objects at the correct distance versus leaving them as is or making them move even further away thereby making it even more incorrect? Vision, by itself, does not have enough information to suggest a direction of change to our brain in this case. All of the above three cases are equally valid if we rely only on vision. We need a second source of input, like, say, touch to provide a conflicting estimate for the objects' distance from you.

Let us now examine what happens when you try to touch an object you are seeing with your new eyeglasses. Since the object appears farther, you extend your hand further too, only to be surprised that the object is closer than you see. We now have two conflicting sources of information about the distance of the object: (a) vision, which says that the object is far away, and (b) touch that says the object is closer. Which of these two inputs is correct? In section 33.1, I have already shown that touch is the real input relative to your consciousness compared to vision. However, why should the brain pick an appropriate direction of change simply because touch is real and not vision? How can our brain know what is real and then treat it preferentially?

First, is there even a reason why the brain should reconcile both these inputs? Can the brain not stay as is, in an inconsistent state with respect to both of these inputs? This is not possible. The reason is that stable parallel looped systems have a natural tendency to synchronize. The existence of a self-sustaining membrane (even a primitive one in the case of animals) connects both vision and touch. The inputs and the actions taken are not disjoint from each other. Your action, say, to catch or hit a ball, to fight for survival and others can be based entirely on visual inputs even though the actions themselves generate a touch input as well. Every time there is an error between vision and touch, the discrepancy is amplified significantly for a task that relies on both these inputs. For example, you may miss catching the ball entirely. Your predictions start to go wrong. Your brain has a natural tendency to create predictive meta-fixed sets based on past experiences from repeatable scenarios. Therefore, the loop based architecture of your brain forces new interconnections between the neurons so that the firing patterns can sustain longer (with correct predictions), when the event repeats several times. This is according to the memory model described in chapter 7. We can view this as a generalization of synchronization in a complex SPL system. Any form of synchronization converges to a specific state. Either the touch input converges to the visual input or vice versa.

As you use your new prescription glasses for several days, you will make a number of mistakes with your estimation of size and distances like when doing simple everyday tasks like reaching for a book, opening a door and so on. With every such mistake, we form new memories, predictions and new explosions that link vision and touch to align with the current situation of new prescription glasses. Strictly speaking, this process is no different now than how it happened the very first time. This, invariably, forces our vision to be changed to align with touch and not vice versa. This is primarily because vision is the input trying to estimate (or predict the distances before you even touch the object) and touch is the one proving it wrong. Therefore, it is the estimation or prediction (i.e., vision) that needs to be fixed and not the one that is validating it (i.e., touch). You are building new abstract continuous pathways to ensure the predictions are accurate.

Our visual representation adapts over a period of a few days to make the objects appear bigger and closer by just the right amount with the new eyeglasses. This is a slow process because our brain needs to re-adjust manually all the links between vision and touch by creating or modifying the abstract continuous paths. Our brain fundamentally creates or modifies our memory and abstract continuous paths as we are exposed to a number of commonly occurring and repeatable visual and touch stimulus during this period. This slow process is partly the reason why we do not notice that our visual representation is changing gradually. The incremental changes, say, over a single day are minor and uniform across all types of shapes, edges, patterns and other geometric properties.

This explains why touch is necessary for vision. Of the two conflicting inputs (touch and vision), we can say that our brain ended up changing our visual representation, not our touch information, over a period of a few days. Furthermore, touch played an active role in helping vision to correct itself. Without touch, our brain does not know how and in what direction to change our visual representation. In this specific example (with new eyeglasses), no other input provided the correct information to help our vision. When I say touch is providing the input to correct vision, I include physical touch as well as *missing* (or a lack of) touch to help you sense the error.

Another example that shows that our visual representation corrects itself using touch was the experiment performed by Stratton [146] and [147]. For a normal human being, the images formed on the retina are upside down. However, Stratton viewed the world with upright images on the retina by wearing a special optical device that inverts the inversion of our eye. The objective of his experiment is different from the question we are asking in this section. He was trying to answer if an inverted image on the retina is necessary for seeing all objects in an upright position. He performed a number of experiments both indoors and outdoors. The errors from vision and touch are much more pronounced in his experiments compared to the example considered earlier (with new eyeglasses). The perceptual and motor coordination was significantly disrupted in his case. However, according to Stratton [146], page 616 "by the third day things had thus been interconnected into a whole by piecing together the parts of the ever-changing visual fields". He also performed additional experiments that disrupted the harmony between vision and touch (Stratton [148]).

His experiments show that the visual representation is the one that is altered to align with the touch information when both these inputs are conflicting. This is similar to the previous example. Therefore, touch is necessary for our visual representation even in this case.

We can now generalize this to other similar cases as well. One special case is when we form our visual representation for the first time, as an infant. This situation is not too different from the above two cases (looking through new eye lenses or wearing special optical devices to invert the inverted image on the retina). At this age, when our vision is not fully established, we rely on touch to help our brain create the correct visual representation. In chapter 34, I will show how touch is necessary for the visual perception of all types of colored edges. In subsequent chapters, I will show the same for each of the other geometric properties. Later, in chapter 34, I will show how we can use touch for estimating size and distance for nearby objects. Then, in chapter 36, I will describe how to generalize it to even farther objects by linking vision with touch. Finally, I will show how our brain creates a coherent visual perception of the external world in chapter 41.

33.4 Linking vision with touch

In the previous section, we have seen that touch is necessary for creating the correct visual representation in our brain. In this section, I will discuss how to link vision with touch. I will discuss one basic link first and leave other links for later sections when discussing each of the geometric properties. These links will help us avoid the need to touch objects to learn geometric properties like edges, angles, shapes, sizes and distances. As a result, we can start using vision directly to estimate these properties. What I mean by linking vision with touch (and vice versa) is that you can use one input to infer information about the other input. For example, if you close your eyes and touch an object, you should be able to visualize a few properties like, say, the shape and size of the object you touched. Similarly, if you see an object, you should be able to guess how some of its geometric properties would feel when touched.

The links between vision and touch are minimal at birth. You do need a set of experiences over a period of several months to create several of them. The easiest and the typical way is when you actively explore the environment. This happens at early childhood. You do not need to crawl or walk in order to gain these experiences. Simply playing with objects around you as well as touching and observing your own body exposes you to a wide range of geometric properties necessary for creating these links. In a sense, we cannot avoid forming these links. Whether we are aware or not, the SPL architecture of our brain will continue to form these links just from the common everyday tasks.

33.5 Seeing what you are touching and vice versa

One of the basic links between touch and vision is to know whether what you are looking at, is indeed what you are touching and vice versa. Other similar problems are (i) you are capable of turning and looking at the object you touch and (ii) you are capable of reaching out and touching the object you look. This basic ability is not dependent on any particular geometric or physical feature of the object in question. After we acquire this ability, we can address the links between vision and touch for specific geometric properties like edges, angles, shapes and sizes. It takes a while for an infant to form this basic relationship between vision and touch. This is because the inputs from your hands, say, and from your eyes take different paths within the brain. They need to converge to a common location to form the necessary links.

Let me explain this in the context of a simple example. Consider a situation where you are holding a ball in your hand. The images of both the hand and the ball fall on the retina when you are looking at them. You want to know two facts here: (i) your eye inputs should deduce that you are looking at both your hand and the ball and (ii) your touch inputs should deduce that you are touching the very same ball. Let me outline the steps taken to create the necessary links.

(a) *Triggering touch and vision*: Mechanoreceptors at the skin surface of your hand are triggered if and only if an object touches your hand. You would not know that a ball touched your hand unless a firing pattern is generated from your hand and you have abstract continuity for this event. Similarly, the photoreceptors on the retina are triggered if and only if you look at some object in the presence of, say, visible light. You may not know the true shape and size of the ball, but you know its existence from the corresponding fixed sets.

(b) *Seeing the ball and the hand with certainty*: You can establish with 100% certainty that you are looking at the pattern of a ball using your free will (similar to how we established the existence of the external world in chapter 29 using free will). What we call as a ball is simply the fixed set corresponding to the same object viewed under a number of dynamical transformations (see fixed set in section 14.2). For example, you can do a number of experiments like turning, bringing the ball closer or farther, walking around and so on. You can perform each of these dynamical transformations at-will to establish that you are indeed looking at an object called 'ball'. Similarly, you can be sure that you are looking at a 'hand' pattern that is different from a ball. The 'hand' pattern is again uniquely identifiable. If you already know that you have a hand, you can also be sure that this hand is yours. You use free will once again to show that you can completely control your hand at-will to produce any reasonable motion. Furthermore, you can sense pain (from nociceptors) and positional information (from proprioceptors) unique to your hand.

(c) *Touching the ball with your hand with certainty*: Based on the touch input, you can know that an object touched your hand whether you see it or not. Even when you close your eyes, you can make sure that your hand was touched with 100% certainty, using your free will in an analogous way. This is the fixed set corresponding to your hand based on touch input. Your brain creates a map of your own body using touch inputs using a number of these fixed sets. When an external object touches your body, you should have an accurate estimate of where it touched you. The ball itself has certain simple features like roundness and texture. While these features, by themselves, do not uniquely identify the ball, you can use them to call it as a ball. For now, the object is a ball until you find a distinguishing feature.

(d) *Touching and seeing both the ball and the hand with certainty*: If we are using information from just the retina, we can infer that the ball and your hand cannot touch each other if the image of the ball and your hand do not intersect. In this case, your skin sensors cannot generate a firing pattern. Next, even if both images do intersect on the retina, it is possible that the two objects are at different distances from each other in the depth dimension. The overlap in the retina is because of occlusion. We cannot guarantee that your hand has touched the ball (see Figure 33.1). Therefore, it seems like retinal input alone using intersection of images alone is not sufficient to deduce that your hand touch the ball. We can clearly see an infant making this mistake as he tries to

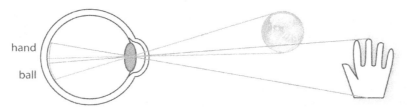

cannot know if your hand touched the ball or not using
just image overlapping information from the retina

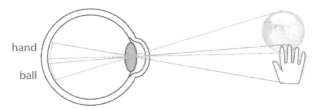

Figure 33.1: Overlapping images on the retina not sufficient to conclude that the objects are touching each other. Here, we see in both cases that a hand image and a ball image overlap on the retina. However, the distances between them in the depth dimension are different in each case. Using information just from the retina, we cannot distinguish these two cases. We need to use other information like our auto-focus ability or touch sensory signal to make this conclusion.

reach out to touch the ball even if the ball is quite far away.

Yet, it is possible to know if the ball touches your hand using just the retinal input. For this, we need to use our refocusing ability to see objects at different distances. In section 16.1, I have described the mechanism for focusing on an object at a given distance to create a sharp image on the retina using the fundamental pattern. When two objects are at the same distance from your eyes, there is no need to re-adjust your focal length (to create a sharp image) unlike objects at different distances. Using this simple observation, we can conclude that *two objects touch each other if and only if the corresponding images on the retina intersect and there is no need to refocus as you switch your central (foveal) vision from one object to the other*. To use this equivalence, we simply need to create fixed sets that can let us detect whether we are changing the focal length of our eyes (say, by triggering the fixed sets for eye muscles when using the auto-focus mechanism of section 16.1). Therefore, if you choose one of the objects as your hand and the other as the ball, the touch input will definitely be triggered if the above condition on the retina is satisfied. In this manner, we are able to deduce information about the touch input with 100% accuracy using just the retinal input.

(e) *Linking both touch and visual inputs*: The above necessary and sufficient condition provides a way to link touch with visual information. With hand as one of the objects, we now have a sure way to know when touch inputs will be triggered using just visual inputs. We can use free will to establish this fact (by changing when and where this happens at-will). This result is true irrespective of the type of objects as well as when and where we performed these tasks.

The fixed sets for your hand and the ball based on visual inputs and touch inputs are in different regions of the brain. For an infant, these regions are not as well linked compared to an adult. We can clearly see the difference in an infant's response compared to an adult, before these links were formed. For example, when you touch an infant on his leg or hand, he does not turn his head to look at where you touched. However, once he learns to track a moving object, say, you will notice that he does look at where you touch. Through our daily experiences, we encounter a number of similar situations where you do touch an object with your hand while looking at the event. The near simultaneity of both visual and touch input generation produces memories that will eventually be linked together through a common region in your brain. You can now establish a one-one correspondence between the firing patterns on your retina with that of movements of your hand as sensed through your proprioceptors. In other words, you can prove that *changes on your retina corresponding to the image of your hand are produced if and only if you move your hand.* This correspondence is built over several months of everyday experiences for an infant.

Once these regions within the brain are linked, the infant can use one of the several mechanisms described in the previous chapters to perform such tasks reliably and at-will. From a number of everyday tasks, we can, therefore, learn to can control the motion of our eyes and our hands either to see what we are touching or to touch what we are seeing.

People who cannot see with their eyes create links between touch and hearing in an analogous way. The links and the representation are limited, however, because of lesser and short-lived firing patterns with sound compared to visual inputs.

When we experience several situations similar to the one described above, we can use the links formed between touch and vision to make useful predictions. These predictions (resulting from small variations of repeatable experiments) give rise to abstract continuity of the event. For example, imagine playing with the ball where your skin sensor inputs are triggered whenever the ball touches your hand. From (d) above, your skin sensor inputs are triggered whenever the ball and the hand images intersect and when there is no need to refocus. As you repeat this event multiple times, you form memories for the entire event. Since the event repeats, the corresponding fixed sets for the 'image size' of the ball and the hand as well as the 'touch size' of the ball and the hand are essentially linked together. These two sizes are clearly different. Therefore, whenever the ball touches your hand, you memorize the corresponding image size of the ball as well.

Now, if the ball is moving towards you, the image size of the ball on your retina keeps growing. When the image size reaches above the memorized value, your past memory links lets you predict the future. For example, you will know that if you now extend your hand, you will be able to reach the ball. This is useful because even in an entirely different scenario, if the size of the ball reaches this value i.e., this fixed set is excited, the stored links from this fixed set are excited to produce an expectation that "if you extend your hand, you will touch the ball". An equivalent way of saying this is that the repeated experiments have created abstract continuous paths.

In the next several chapters, as I discuss each of the geometric properties, I will show how to create additional links between touch and vision. Our visual perception, as it turns out, depends on touch in quite a fundamental way.

34

Sensing Geometric Properties

In this chapter, I will focus on some of the basic and aggregate geometric features like edges, angles, boundaries, shapes, motion, patterns and textures (see items (a) – (b) at the beginning of chapter 30). These properties are quite critical for recognizing objects. Each of these properties are present in every object. Together, they help us identify an object and distinguish it from the background. I will only discuss these topics in brief because there is a considerable amount of work already done by other researchers (see Kandel et al [82], Purves [122] and Ornstein and Thompson [118]). I will translate some of their work into the new SPL framework and focus on the link highlighted in the previous chapter between vision and touch.

To determine each of these geometric properties, we can use either eyes or touch. Of the two sensors, eyes have much higher sensitivities and accuracies than touch. Therefore, properties like shapes, edges, angles, boundaries and patterns are best-determined using eyes. On the other hand, we detect texture through touch. When a person cannot see with both eyes, we know that he can effectively use the remaining sensors better than a person with normal vision. For him, touch sensor is the main source of reality.

It is important to keep in mind that there is an indirect dependency of these features on size and distance. For example, we may detect that a shape is a circle or that an edge makes an angle of 45° on the retina. In order to define a circular shape, we need to know that two different sized circles at different distances have the same 'shape'. The same situation applies with edges of different sized objects. Specifically, we want to know how to interconnect the representations of each of these properties with the corresponding representation for size and distance. I will discuss size and distance only briefly in this chapter, but will discuss them in greater detail in chapter 36. Let me start by comparing how a given abstract geometric property appears on the retina versus how it appears to our conscious self.

34.1 Patterns on the retina versus how they appear to us

If we try to define an edge in simple terms, we say that an edge is a boundary that separates two distinct patterns, with one pattern on either side. The most common way an edge is formed is when you have two different colors that meet at a boundary. There would be no edge within a given region if, instead, you have a uniform pattern or color.

Let us take two simple examples of edges: (a) an edge of a blue rectangle in white background and (b) an edge of a red rectangle in white background (see Figure 31.3). When we look at edges in both these cases, they appear and feel identical to us. However, the firing patterns triggered by the photoreceptors on the retina are considerably different. Recall, from section 31.1, that the S cones, sensitive to the blue region of the visible spectrum, are rarely, if at all, present in the foveal region of the retina. Yet, the blue edge appears identical to the red one as you scan both of them with your eyes. The relative proportion of S, M and L cones differ quite significantly (about 5%, 35% and 60% respectively). These relative ratios also vary considerably among different people. Furthermore, the distribution of M and L cones in the fovea is non-uniform. One would think that this unevenness would cause a number of serious issues. Surprisingly, they do not seem to. As an analogy, we can approximately compare L, M and S cones on the retina to red (R), green (G) and blue (B) pigments on a television screen. How would images on a television screen appear if we have a disproportionate amount of the RGB pigments distributed non-uniformly on the surface (as is done in our retina)? Given these oddities within our retina, the following are the questions we need to answer (which I will do in section 34.2).

(i) How do the edges in both of the above cases (or for any two colors) appear identical and as straight lines? If we draw an imaginary line connecting the photoreceptors that respond to an edge, then we would expect the blue-white edge to be zigzag while a red-white edge to be smoother (see Figure 31.3). However, we never see a zigzag edge in our everyday world unless the line itself is already jagged.

(ii) Even though our eyes see at such high resolution, our consciousness sees at much lower resolution. Our consciousness seems to mask the zigzag lines to make them appear smoother. Why and how does this happen for all sensory organs?

Let me discuss each of these questions next after studying the basic mechanics of edge detection.

34.2 Edge detection

In order to detect an edge, whether we use vision or touch, we need a unique response from neurons along the edge compared to neurons on either side of it. Let me first discuss edge detection with eyes. An important prerequisite to detect edges using eyes is to learn the auto focus mechanism (using the fundamental pattern – see section 16.1) to create a sharp image on the retina. Without this mechanism, we would have a blurred image with no sharp edges. An edge, by definition, has two distinct spectral distributions on either side of it. An edge E is an imaginary boundary that separates the photon frequencies on either side of it (see Figure 16.6). It has a nonzero thickness. If light with such a spatial distribution falls on the photoreceptors of the retina, the neural responses on either side of this boundary are considerably different (see section 16.1).

Consider three neurons P, Q and R in the interior, the boundary and the exterior respectively of the edge E. Take a small neighborhood of neurons at each of these three points as shown in Figure 16.6. All neurons within a neighborhood of P (or R) corresponding to the interior (or exterior) of the object (or either side of the edge) have similar neural responses as glutamate is released on horizontal and bipolar cells. This is because the entire neighborhood P (and R) lies entirely on one side of the edge. On the other hand, neurons within the neighborhood of Q have a mix of two different neural responses, one for each half of the neighborhood. Note that the neighborhood of Q can be split into two parts, one for each side of the edge, each of which generates different neural responses (Figure 16.6).

The neural response from each photoreceptor (P, Q and R) is eventually transmitted to the ganglion cells through the bipolar and horizontal cells. The ganglion cells integrate responses from multiple neighboring photoreceptors. If we integrate the response of all neurons in each of the neighborhoods around P, Q and R, the net response at the boundary Q is clearly distinct from the net response at both P and Q. This implies that a set of photoreceptors along a thin boundary are distinctly different, provided you have a way to integrate neural responses from a small neighborhood around each photoreceptor. Hubel and Wiesel [72] have discovered precisely such neurons. These are the retinal ganglion cells.

In fact, they have discovered something more interesting. They have termed this as the receptive field (see Hubel and Wiesel [72], [73] and [74]). This concept is generalized to a number of other geometric properties including shapes, binocular vision including auditory and somatosensory systems. A receptive field is the region on the retina that alters the firing of a neuron in response to light that falls in this region. The neighborhood region is to be imagined as a center region along with a concentric ring around it. The ganglion cells respond to light in opposite way from each of these two regions, center and surround. There are two types of bipolar cells called on-center cell and off-center cell. If light falls in the center region, the on-center cell responds by firing while the same cell is inhibited when light falls on the surround. The off-center cell reacts in the opposite way. As a result, the on-center cell responds differently when light falls on the center, on the surround or on both the center and the surround. The off-center cell has the opposite effect. Therefore, the ganglion cells transmit information, additionally, about the differences in firing between the center and the surround.

Hubel and Wiesel [72], [73] and [74] observed that this type of organization is present even in the lateral geniculate nucleus, the visual cortex and the extrastriate visual areas. This organization clearly amplifies edges if you consider appropriate center and surround regions. Loops of neurons resulting from these differences in the firing patterns that amplify the edges of an image are precisely the edge fixed sets. Hubel and Wiesel classified receptive fields of cells as simple, complex and hypercomplex cells in the visual cortex. The simple cells are identified with an ability to distinguish a given orientation, with one or a group of cells for each orientation. The complex cells are identified with motion in a particular direction. The hypercomplex cells are identified with definite lengths along each direction. The fixed sets in these regions can be identified with an ability to detect the above-mentioned features, namely, different edges or angles, motion along each direction and different lengths along a given direction respectively.

Let me now switch our attention to touch for detecting an edge, before I start relating both these sensory inputs. When we rub our fingers on the surface of an object (while closing our eyes), if the surface texture is uniform, we do not sense any edge. On the other hand, when we do sense a thin edge, the neural inputs generated from the pressure along the edge are more prominent compared to inputs from either side of it. We can now perform an analysis similar to vision by picking neurons P, Q and R on your hand, on the boundary, the left and the right of an edge respectively, and looking at small neighborhoods around them. As with vision, the neighborhood Q generates distinctly different patterns compared to neighborhoods P and R. This distinct firing pattern from, say, your hand, identified with the edge of a specific texture, is stored as a fixed set, analogous to the case with vision. The notion of receptive fields introduced by Hubel and Wiesel can also be applied to touch in the same way. These give rise to fixed sets in the somatosensory system within the brain triggered from your hand that is unique to the edge. We can now use these fixed sets to detect edges using touch as well.

Given two distinct set of fixed sets to detect an edge, one through vision and the other through touch, we now have an issue to address about how both inputs are consistent with each other. Both these fixed sets are in different regions of the brain. The sensitivity, the amount and the nature of information received about the same external edge using each sensory input is significantly different. Yet, our perception of any edge is unified into a single notion. We do not have two different representations within our brain for an edge, nor do we feel any inconsistency between the two sources.

As mentioned in the previous chapter, the two distinct regions in the brain for vision and touch do get linked together. This happens as early as the first year of your life. This is a slow process. Nevertheless, the links must and do form. The interconnections between the two subsystems let you infer the information about one system using the other. For example, you can look at an edge and sense how it would feel if you touched it. Just by looking at a sharp needle, you know how it is going to feel, sometimes in an exaggerated way, when you were pricked with it. Conversely, you can touch an edge while closing your eyes and imagine how it would appear if you open your eyes. The interconnectivity of the brain is all that is needed to ensure these links can be formed. Let us now turn our attention to the role of touch in the visual representation of an edge and then study the links between touch and vision for edge perception.

34.3 Touch necessary for visual edge representation

In section 34.1, I have raised two questions about how the representations of two colored edges on the retina differ from each other and from a corresponding tactile representation. In this section, I will answer these questions. We have seen above that we can detect an edge both with our eyes and with touch. We also feel that we can link these two inputs to produce a unified representation. However, it is not clear whether touch can affect just the visual representation (note that vision does not affect the tactile representation) even though touch is important for the unified representation. In this section, I will show that touch is necessary for visual edge perception.

For problem (i) of section 34.1, consider two objects, say, two chairs, a red one and a blue one. Let us assume that the chairs are otherwise identical except for the color.

The edges of the red chair appear, say, smooth on the retina because the density of L cones (sensitive to red part of the spectrum) is quite high, namely, about 60% of all cones. The edges for the blue chair on the retina surface, on the other hand, appear jagged because of very low density of S cones (sensitive to blue part of the spectrum), namely, only about 5% of all cones, with almost no S cones in the fovea (responsible for sharp central vision). Therefore, the representation on the retina (not our perception yet) for both the blue and the red chair are significantly different. You would naturally expect your conscious self to perceive the edges of both the red and blue chairs differently. This is, however, not the case. Edges for both the chairs appear identical. The question then is why the jagged representation for a blue chair on the retina change to appear smoother similar to the red chair?

I want to show that touch is absolutely critical for changing the jagged representation. Let us see what happens if we, temporarily, do not use touch. From just the visual inputs, we have two *distinct* firing patterns generated from the retina, one for each colored chair. These inputs propagate deeper into the brain to form distinct fixed sets, at least initially. We can perform a number of actions using our free will like walking closer or farther from the chairs, turning our head sideways, bending while looking at the chairs and so on, except touching the chairs. The fixed sets are invariant to each of these dynamical transformations. Each of these free will tasks help us learn about the "true" nature of the chairs. For example, you can estimate a few geometric properties like the shape, size and certain surface patterns of the chairs. You can even compare these geometric properties of one chair with the other. You can look for answers to a number of questions like, say, is the red chair broken, does the blue chair have a stain, does the blue chair have jagged edges and so on. Your free will actions give you enough variations to be able to answer, at least, a few these questions affirmatively or otherwise.

My claim is that if your visual representation shows that the blue chair has jagged edges, none of the above free will based tasks (excluding touch) can ever disprove this fact. This is true even if the red chair appears to have smooth edges. You do not know apriori whether the two chairs are almost identical (except for the color) or not. The fact that the two chairs have two different appearances for the edges on the retina is not an inconsistency, in itself. The information acquired using your free will tasks about one chair does not help either confirm or contradict the information obtained about the other chair. Therefore, using the information collected so far (without touch), you have no choice but to conclude that both chairs are indeed different from each other beyond just the color as well (i.e., the edges are indeed different for both chairs). As a result, your brain has no reason to change the visual representation of the jagged blue edge to a smoother one.

However, the moment you include touch as a valid free will task, the situation changes significantly. When you touch the blue and red chair at-will (at different times, at different angles and locations and so on), you immediately notice that the edges for both chairs feel identical on your hand. Your touch sends identical signals to your brain. Both feels smooth when touched. This contradicts the visual information. You are now in a state where vision says that the two chairs have different edges (red smooth, but blue jagged) while touch says that they have identical edges (both red and blue are smooth). In section 33.5, I have shown how you can be sure whether what you are seeing is indeed what you are touching and vice versa. This implies that both touch and visual inputs, though contradictory, are referring to the same external chair. Our brain needs to reconcile these errors. How does our brain do that?

Let me first summarize what we have analyzed so far. The facts you discover through your everyday experiences are: (a) blue and red chair edges look different (jagged and smooth respectively), but feel identical when touched (both smooth), (b) using your free will (with touch), you can show that the edges for a number of different types of red and blue colored objects also appear different but feel the same i.e., the experience is not limited to just these specific blue and red chair examples, but even to other objects like blue balloon, blue table, blue ball, blue car and so on and (c) there are objects that appear jagged (with truly jagged edges) and feel rough when touched.

Therefore, from (b), we know that several blue objects have the same contradiction between vision and touch. They appear jagged through vision but feel smooth when touched. From (c), we know that there are several examples of blue and non-blue objects that appear jagged through vision and also feel rough and jagged when touched. In other words, we have both types of examples – those that appear jagged and feel jagged as well as those that appear jagged but feel smooth.

Another way to state this is that when an edge appears jagged, we may feel it as either smooth or rough when touched. However, when an edge appears smooth, it only feels smooth, not rough, when touched. Similarly, when you feel a smooth edge on your hand, it could appear smooth or jagged to your eyes. However, when you feel a rough edge on your hand, it only appears rough, not smooth to your eyes. This argument is only valid for primitive features like edges or angles, not for entire objects.

Given this, our brain tries to reconcile the contradiction between vision and touch by ignoring some of the differences between a red and a blue colored edge with vision. Why should our brain try to reconcile these contradictions? If it does, how does it happen?

The reason why our brain will resolve this contradiction is because of a form of synchronization in complex SPL system discussed earlier. If two sensory inputs trigger at nearly the same time and if these events repeat a large number of times, an SPL system will adapt so that they will synchronize. This is a property of looped systems. In this case, the vision and touch get linked together eventually because of near simultaneity of occurrence of repeatable events (see associative learning or Hebb's postulate summarized as 'cells that fire together, wire together' – Kandel et al. 2000; Gazzaniga et al. 2008; Graham 1990). The net outcome is a single converged state instead of two distinct states, one for vision and the other for touch. This single converged state is the final reconciled representation.

The subtle differences in the shape of a blue versus red edge on our retina is not significant in our everyday experiences. Therefore, changing it to converge to the touch representation results in a consistent and less contradictory view of the external world. In fact, once you have linked your vision, touch and your ability to perform actions based on these sensory inputs (also needed for your free will), you are surprised because of this inconsistency. If you see a jagged blue edge, it creates an expectation (or prediction) of roughness when touched only to be surprised that it is, in fact, smooth. You can confirm the correctness of the touch using your free will several times.

Each time there is a detectable error, our brain creates new memories and explosions that forces our visual interconnections to be re-wired and linked to touch to eliminate the inconsistency. Note that vision is the one that provided the estimate and touch is the input that validates. Therefore, if there is an error, it is your prediction (i.e., vision) that will be altered, not your validating input (i.e., touch). Recall from chapters 32 and 33 that touch is the direct contact sensor that is most accurate for capturing properties of objects of the external world. This is the same reasoning I have used in section 33.3 when explaining how our vision will be altered to correct the estimate in distance when you wear the new prescription glasses. The converse scenario where you estimate with touch and validate through vision is not common, except possibly in illusions, which are not repeatable events in our everyday life. Touch is indeed more real than vision in most everyday scenarios.

As a result, we either lose the ability to access these visual fixed sets (used to distinguish a blue versus a red edge) at a conscious level or break these interconnections, as they are unused. You, therefore, gradually and permanently lose your ability to distinguish these subtle 'visual' differences and instead converge to a common visual representation. You do need a number of repeatable experiences that highlight these errors long enough to create new interconnections (i.e., permanent memories). This is how we see all types of colored edges as identical. This unified representation looks and feels exactly like a straight line to confirm with both visual and touch inputs from the retina and skin sensors, respectively. This answers question (i) of section 34.1. This also explains why people with wide variation of relative proportions and distribution of rods and cones still perceive edges the same way.

To answer question (ii) of section 34.1, first note that the distinct fixed sets for a red and blue colored edge were initially formed because of your high resolution of vision compared to touch. At lower resolution (say, with touch), you feel that they are identical. However, as explained above, our vision moved away from maintaining these two separate fixed sets and instead favored creating just one unified fixed set for an edge. This unified representation appears and feels precisely like a straight line to conform to both visual and touch inputs. Our extremely high resolution of our eyes caused incorrect predictions as the inputs were both linked together. Recall that the rods in our retina have very high sensitivity, i.e., they are capable of detecting even a single photon of light. Yet, we never 'see' photons themselves or see objects at that resolution. You have permanently abandoned the higher resolution and settled for a lower resolution instead, at least, with an edge representation. In the next few sections, you will see that our eyes do the same by choosing a lower resolution for each of the other geometric properties as well. Therefore, we can say that our experiences in early childhood, while helping (and necessary) to become conscious, has moved towards a direction where we use our eyes with a lower resolution and thereby ensure that there is no contradiction between vision and touch.

One great advantage of these links between touch and vision is to generalize our visual edge representation to cases where touch cannot help. For example, drawings or text in a textbook has edges and curves that our eyes can detect. However, we cannot detect them by touch since the paper has a uniform texture and the curves do not stand out. Our visual inputs represent these edges in an exactly identical way. Even though we cannot validate them with touch, you believe that they are just as real. Before you touch the surface of the paper, you do not know if what appears as an edge to your eyes can be detected using touch or not, assuming there is no other auxiliary information to help you, like shadows or neighboring shapes. Therefore, the consistency of your visual representation is guaranteed from the links between vision and touch described earlier. There is no reason to change this representation later because touch is not providing inconsistencies anyway for images drawn in a textbook.

Therefore, we see that initially vision and touch had provided inconsistent views of the external world. Since touch is more accurate and most repeatable events help us evaluate touch more often than vision, our brain had switched its internal visual representation to align with touch thereby eliminating the inconsistency. Next, we are able to generalize the resulting converged visual representation to other objects and images like paintings and drawings on a paper, text on a paper and others for which touch does not provide additional information. Here, the converged visual representation continues to be valid without any inconsistencies from touch. In this manner, we finally have a self-consistent representation for both vision and touch.

Let me now describe a mechanism with edges, which lets us trace along an edge, both with vision and with touch. This mechanism becomes useful later when trying to link different angles together to produce the effect of a continuous variation of angles.

34.4 Tracing along an edge – linking angles

Our sensory surfaces with both vision and touch are two-dimensional. For any two edges, there is a relative angle between them on the sensory surfaces. We are exposed to all angles from 0° to 360° from the objects around us. As mentioned in section 18.12, we do form several fixed sets corresponding to each of these angles. The existence of these interconnected neurons in the primary visual cortex that respond to each individual angle was already discovered (see Kandel et al [82] and Purves [122]). Let us now see how we can link these primitive fixed sets.

Let me begin by discussing how we can learn to control the motion of our eyes and head so we are able to traverse along a given edge (see section 17.4). This mechanism is a combination of the fundamental pattern and our predictive ability, or more generally, abstract continuity. This scenario of traversing along a given edge is quite analogous to our ability to track a moving object. The mechanism is the same for both

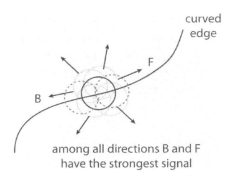

Figure 34.1: Tracing a curved edge. A circular neighborhood is shown on the boundary of a curved edge. Among all possible directions to move away from this position, the only two directions that continue to have the strongest neural firing patterns are the forward *F* direction and the backward *B* direction.

vision and touch.

Consider looking at the boundary of an object. The image of the edge falls on your fovea. When you traverse along the edge, the angle of the edge varies, typically, in a continuous way (until you reach a sharp corner). Recall that section 18.5, we have seen how we form links between fixed sets for neighboring sequences. These links can be viewed as establishing the sequential arrangement of numbers. In an exactly analogous way, we form links between fixed sets for sequential variations of angles. For example, you never encounter the following types of variation of angles in our everyday experiences: 10°, 23°, 15°, 11°, 21°, 53°, 12° and so on i.e., an arbitrary sequence of angular variations. You only encounter either an increasing sequence or a decreasing sequence of angles whenever you traverse along the edges among all naturally occurring objects. Our brain memorizes this fact by linking the fixed sets for angles as explained in section 18.5. Sharp corners do occur in nature. However, on any typical object, the total number of sharp corners are quite few. When you take any two such adjacent sharp corners, they are separated with finite regions of smoother curves in between.

When the image of an edge falls on your fovea, the edge also curves around to fall on the neighboring region of the retina (see Figure 34.1). Even though you are not focusing on the neighboring regions, the parallel excitation paths from these regions provide information about how the angle varies. The links between the angle fixed sets, as explained above, gives you a predictive ability as well. This is the predictive meta-fixed set corresponding to the angle fixed set generated from the fovea. You have such predictive meta-fixed sets in both directions, forward and backward, relative to the edge. This gives rise to abstract continuity for the angle generated at the fovea. You know what angles to expect next. Now, when you turn your eyes the forward and backward directions relative to the edge have explosions of firing pattern compared to any other direction (see Figure 34.1). Therefore, the fundamental pattern of chapter 16 can directly be applied to create control actions that let you turn your eyes along these two directions.

When you turn your eyes along one of these two directions, what was previously the neighboring region moves towards fovea. You now have new neighboring regions that create new predictive meta-fixed sets. These correspond to the new angular orientation outside the fovea. The above process repeats where you pick one of the two directions again using a combination of abstract continuity and the fundamental pattern. This mechanism is quite common, as you may have noticed already. If you look around randomly, you do actually traverse along the edges of objects, almost unintentionally. Your eye motion is not as random as you think.

Even when there is a sharp corner, the nonempty predictive meta-fixed sets generated from the neighboring regions i.e., your abstract continuity, lets you turn 'smoothly' around it as well. This ability requires learning because you need to close the sensory-control loop and form memories, as required by the fundamental pattern.

The above-described mechanism is equally valid for touch inputs where you learn to traverse along an edge that your touch detects. A person who cannot see can use precisely the same mechanism described above to trace along the boundary of an object. An edge fixed set formed for touch inputs has a distinct neural firing pattern compared to neural inputs from either side of the edge (similar to retinal inputs). You feel a sharp sensation of an edge compared to a softer sensation for either side of the edge. These edge fixed sets are linked based on continuous variation of angles, as explained above. This gives rise to abstract continuity. The fundamental pattern lets you pick one of the two directions to traverse based on abstract continuity at any given location of the object surface. Similarly, we can explain sharp corners as well.

34.5 Shapes and other complex patterns

Colors, edges and angles are the most primitive features that enable us recognize a number of complex patterns and shapes for all objects around us. Every pattern we see around us is some complex arrangement and grouping of these primitive features. Let me discuss these complex patterns briefly.

When you have a mechanism to traverse naturally along the boundary of an object, you can use it to go around the object once. You follow your eyes along the edge and take a closed path to reach where you started. This closed path can be grouped as an object. I will sometimes use the term foreground to represent the object. Background then becomes everything except the foreground objects. The ability to distinguish a foreground from a background is a nontrivial task, which we discussed in great detail in section 9.1. In fact, this task is related to our ability to sense three-dimensional space (which I will discuss later in chapter 35).

Shapes are fixed sets corresponding to such commonly and repeatedly occurring closed paths. We do not have fixed sets for every possible shape. For example, we form fixed sets for leaves, flowers, houses, human faces and others in a broad sense based on the similarity of the shapes (i.e., modulo the dynamical transformations in the definition of fixed set).

Even among those complex shapes or patterns that are not as common, the primitive features like edges, colors, angles, closed paths and textures give detailed information (via their corresponding fixed sets) as well. It turns out that after we learn language, we have a language based fixed set for a given shape (complex pattern or primitive features), which is linked to the corresponding image based fixed set.

There is evidence that suggests the existence of such shape specific areas. For example, fusiform face area in the temporal lobe, adjacent to parahippocampal place area, is one such area discovered using PET (Sergent et. al [138]) and fMRI (Kanwisher et. al [83]) and later confirmed at the neuronal level (Tsao et. al [156]).

The shape and complex patterns can similarly be inferred using just touch inputs, especially by people who cannot see. However, this requires higher abilities of the human brain. You need to create a cumulative fixed set using memory from local information as you traverse along the closed paths.

34.6 Motion detection

Motion detection is a primitive ability within our visual system. It has already been shown (see Ornstein and Thompson [118] and Tovée [155]) that a specific set of neurons in the primary visual cortex fire only in response to the motion of an object. I have discussed this mechanism in detail in section 18.12 and also in section 17.3 for explaining how we track a moving object using the fundamental pattern. I have shown that we have an ability to detect the change in relative distance and angles between objects, which let us sense motion. I will, therefore, not discuss this here.

Even with touch, when an object moves when in contact with your body, the firing pattern moves in a continuous way from the initial to the new location. As described in sections 18.3 and 18.4, this lets us create unique fixed sets from the continuous time-varying firing pattern of neighboring neurons. These fixed sets can be used to sense the motion.

We have thus seen that the geometric properties discussed in this chapter so far can be expressed uniformly in terms of appropriate fixed sets. In the next section, I will discuss how size and distance information can also be expressed in a similar way. This will lead us to the important problem of how to sense the three-dimensional space by aggregating color information with geometric information like size, distance, edges, angles, shapes, complex patterns and textures.

34.7 True size and distance of nearby objects

In addition to the geometric properties discussed until now, size and distance are quite important for our perception of the external world. I will separate the discussion into two parts, primarily because the corresponding mechanisms are somewhat different: (i) estimating size and distance of objects that are close enough that you can reach and touch them and (ii) estimating size and distance of objects that are sufficiently far away. I will focus on the former problem in this section and discuss the latter one in chapter 36.

What is the difficulty with estimating size and distance? When we look at objects around us, we see that each object is of a definite size and is at a specific location. Just by looking at an object, in a quick instant, we can estimate both how long it will take to reach it if we walked towards it and what its size is. We can indeed verify our estimate by walking and touching it. Why do objects appear to be of a specific size and at a specific location? Figure 30.1 shows a cone of light reaching your eye from a given object A. The size of the image on the retina is the same for both objects A and B. Hidden in this statement is the assumption that our eye re-adjusts its focal length to make the image fall precisely on the retina. The underlying mechanism of how this auto-focus happens was described in great detail in chapter 16 when discussing the applications of fundamental pattern. The object A is closer and smaller, while object B is farther and bigger. Yet, the image falling on the retina is identical for both objects A and B. In fact, in addition to A and B, every other object that is otherwise identical in shape within this cone produces the same image on the retina.

To be a bit more precise, if u is the distance from the object to the eye lens, v the distance from the eye lens to the retina where the image falls and f the focal length of our eye lens, the optics of a convex lens shows that $\frac{1}{u} + \frac{1}{v} = \frac{1}{f}$. Also, the magnification of an object $m = -\frac{v}{u}$. Since v is the same for both objects A and B while u is smaller for A and larger for B, the magnifications are different. However, if the focal length f does not change, then the first equation implies that if A falls correctly on the retina, then B will fall 'in front' of the retina. Conversely, if B falls correctly on the retina, then A falls 'behind' the retina. However, since we want v to be the same for both objects, we need to refocus our eye lens accordingly. Specifically, if A falls correctly on the retina, then the focal length f must increase so the image stays on the retina for B. If B falls correctly on the retina, then the focal length f must decrease to let image of A fall precisely on the retina.

Given this, we should not be able to predict easily, just by looking at the image on the retina, whether the image corresponds to object A, B or some other identical object somewhere within this cone. This is because the image on the retina is the same in all of these cases. Yet, we perceive the object to be of a specific size and at a precise location within this cone. All other possibilities within this cone are automatically eliminated even though the above analysis suggests that all these cases are indistinguishable. Even if you close one of your eyes, you can still estimate size and distance with some accuracy, especially given that the cone extends all the way to infinity.

In Figure 34.2, I have highlighted the complexity of this situation by showing several cones corresponding to several objects. Given multiple objects A, B, C and D, say, in front of your visual field, what are the true sizes and locations of these objects? Are they arranged in the top, middle or bottom layout of Figure 34.2? Clearly, the number of possible arrangement of objects, as in Figure 34.2, within the cones is infinite, even if you close one eye. Yet, our brain has picked exactly one arrangement among these infinite possibilities to 'project' the world in front of our eyes (more importantly, even with one eye closed). Recall the discussion in section 33.3 about wearing new prescription glasses. I mentioned that all objects would appear to be farther for the first few days when you wear the new prescription glasses. During this period, our brain is projecting an 'incorrect' (though still unique) arrangement in

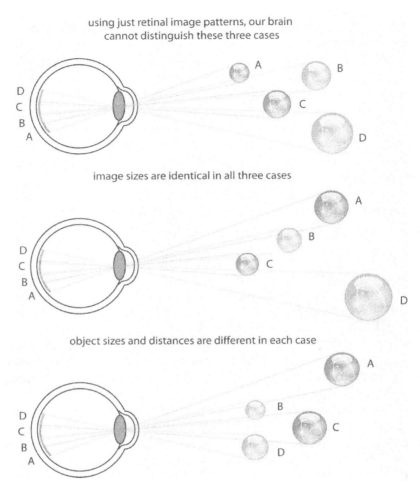

using just retinal image patterns, our brain
cannot distinguish these three cases

image sizes are identical in all three cases

object sizes and distances are different in each case

Figure 34.2: Ambiguity in rendering the layout of objects in our visual field. In this example, we see that the images on the retina of all four objects A, B, C and D are identical in each case. Yet, the sizes and distances of the objects in your visual field are considerably different. Using just the image information, our brain cannot decide which layout to present in front of our eyes.

front of our eyes. This is reverted to the 'correct' arrangement after a few days.

Another difficulty we face when discussing estimation of size and distance is that we tend to think about these properties in a quantitative way (like 6 feet tall or 5 feet away). We use the 'number' abstraction to describe size and distance (using a number of feet and inches, say). However, an infant or a toddler already knows sizes and distances of objects around him almost as accurate as you and me, at least for nearby objects without any idea about number abstraction. He does not know nor does he seem to care that a basketball is about 4 times bigger in diameter than a baseball. Yet, he knows whether a baseball or a basketball is going to fit through a specific sized hole. Furthermore, adult conscious humans a few thousand years ago were able to know sizes and distances without number abstraction like when the concept of numbers were not discovered yet. For nearby objects, if you notice carefully, we instinctively have a 'qualitative' feeling of knowing for size and distance before we consciously try to estimate them 'quantitatively'.

Similarly, most animals have a clear notion of size and distance (as it is even important for their survival) without referring to them quantitatively. Therefore, the underlying mechanism that addresses the issues raised earlier in Figure 34.2 with estimation of size and distance should allow for a qualitative estimation (not quantitative using numbers and counting abstractions). We should also express this mechanism in a way that it can occur naturally because most living beings are capable of discovering it without help from other intelligent or conscious beings. We, humans, do use quantitative measurements after we become an adult. We do this to get better accuracies of sizes and distances when performing complex tasks like when building a house (for otherwise, the house will collapse).

In the rest of this chapter, when focusing on nearby objects, we can rely on touch as the primary sensor, while still linking it with vision. The advantage with touch is that there is a *direct contact* with the external object as discussed in section 32.1. The distance between our sensory surface, i.e., our skin and the external object is zero when we touch the object. This avoids any distortion of information when it is transmitted from the external object to our sensory surface. With eyes, there is no direct contact and is hence error prone, especially when you are an infant. With touch, however, the forces, the size and texture information can be transmitted as accurately as possible through direct contact. These geometric properties like size are nearly identical in the immediate neighborhood of the contact surface i.e., on both sides of it – the external object surface side and the sensory surface side.

The importance of direct contact and the relationship between touch and retinal inputs have already been studied by several other researchers (see, for example, Stratton [148]). My focus in the next few sections is to go beyond explaining the relationship between these inputs. I will give mechanisms for estimating the true size and distances of objects even when we cannot touch them.

34.8 Defining size and distance qualitatively

Consider a person who cannot see with both eyes or imagine yourself closing your eyes and touching objects around you. If you touch the width of a chair with two hands, one hand touching one end and the other hand at the other end, you have a clear sense of the size of the chair. If you are holding a small bottle or a stick between your thumb and the index finger, you can similarly sense their sizes. In both cases, you just have two distinct contact points at two different locations of your body (like one on each hand or two fingers of the same hand as in the previous examples). There is no continuous contact surface between these two points. If, on the other hand, the two contact points for the chair are from one hand and one leg instead of both hands or both legs, we cannot estimate the size easily. The same situation applies with distance or depth as well. How can we explain this? What is the difference between touching with two fingers of the same hand, two hands or two legs versus touching with one hand and one leg?

There are other ways to estimate size and distance using touch, again with your eyes closed. We can have a continuous contact surface instead of just two distinct points of contact. For example, if the object is small enough that you can hold it in your hand, like a ball, then the size of the ball is equal to the size of your palm. This equality is not true with eyes, i.e., image size on the sensory surface (retina) is not equal to the true size of the object. Therefore, with touch, we can use the fixed set corresponding to a given size on the sensory surface to identify the size of the external object (using mechanism described in section 18.6).

Yet another way to estimate size with touch (with eyes closed) is by rubbing along the object, say, when the object is large enough that it cannot fit in your hand. Here, you have to move your hand quickly and, possibly, multiple times along the object. You are relying on speed and your perception of time to estimate how big the object is.

Among these multiple ways to estimate size and distance, the first one described with two widely separated contact points is a particularly interesting case. With this case, we form a feeling of knowing entirely based on experiences. The amount of distance you stretch your hands or legs when you hold or push objects gives rise to abstract continuity for different sizes. You rarely try to hold or push objects with one hand and one leg. Instead, you use two hands, two legs, two fingers, your palm and other combinations of body parts. As a result, we do not have abstract continuity for holding objects with one hand and one leg.

For example, an infant spreads his hands by different distances while holding or playing with different toys every day. Initially, he needs to create fixed sets for the toys themselves along with fixed sets for each different amount of extension of his hands (as he holds the toys). Next, he links both these fixed sets simply from simultaneity of occurrence of events. However, multiple objects produce the same extension of his hands since they have the same size – this is an invariance mentioned as dynamical transformations in the definition of a fixed set. This gives rise to predictions, in a limited sense. For example, when an infant holds an object of one size, the touch inputs can trigger memories of other objects of same size (along with same texture, color and other features). These are the predictive meta-fixed sets for a given amount of extension of his hands. Furthermore, different amounts of extensions correlate well with the corresponding sizes of objects. These links are interconnected so that you have abstract continuity for variation of sizes. Using the result from section 22.1, this gives rise to a feeling of knowing of sizes (once he creates a self-sustaining membrane of meta-fixed sets as well). Therefore, next time he holds a new toy between both of his hands, he can sense the size of the toy by how much he has extended his hands.

This is how two distinct points of contact on two hands, two legs or two fingers, i.e., even when there is no continuous contact surface, can still give rise to a feeling of knowing of size. If one point of contact is on one hand and the other on one leg, you do not have a corresponding feeling of knowing. This is because, in this case, you have never created as many links, as explained above, through your past experiences to generate abstract continuity.

These above-mentioned tasks like holding, pushing, pulling and lifting lets you define size and distance qualitatively. Here qualitative estimation for size and distance refers to the fact that we should attain a feeling of knowing. This above mechanism clearly can occur naturally because it only relies on simple set of experiences. Even animals can sense size and distance qualitatively using the above process, just as infants can. Only later, we estimate size and distance quantitatively as well. In general, this feeling of knowing is not broken down into separate feelings for individual features like color, texture, patterns, size and shapes. It is usually a combined feeling of knowing though some may contribute more than the others. Let us now look at the interplay between vision and touch when sensing size and distance for nearby objects.

34.9 Size and distance using touch and vision

When objects are close enough that you can touch them, you can use direct contact to infer size of external object from the size on the contact surface and from the mechanisms described in the previous section. The situation with vision is different. There are a few problems when using just vision for estimating size and distance. For example, the sizes of images that fall on the retina depend on the distance of the object from our eyes. Yet, for small distances, the perception of the size and shape of an object does not change significantly. When you walk, say, about 10 feet towards an object, the object does not appear to grow in size significantly. This is, however, not the case for large distances. The size of an airplane flying in the sky versus taxiing on the runway appears and feels different in spite of a detailed memory of an airplane at close distances.

There are other additional problems with vision. For example, the sizes on the retina are considerably smaller than the real sizes as detected by touch. The path taken by light, the nerve impulses generated in response, the change in focal length when you look at objects of different distances are fairly complex mechanisms. If we include inputs from both of our eyes, the ability to focus both eyes on the same object, the ability to track a moving object and the fact that the image is inverted on the retina, the underlying firing patterns from the retina become even more involved. This is where touch can help with our estimates.

Therefore, the approach we take involves linking the predictive meta-fixed sets associated with abstract continuity generated from touch with the corresponding visual fixed sets. How do we do this? The first step for using vision is to use the auto-focus mechanism, as described in section 16.1, to create a sharp image on your retina. Next, you can touch the object you are looking at using the mechanism discussed in section 34.3. In this case, your finger and the object are at the same distance from your eyes. As a result, the ratio of the image size of your finger to the object on your retina, is equal to the corresponding ratio of their true sizes. Since you know the size of your finger or hand (see previous section), you have a feeling of knowing of parts of external object that are in immediate contact with your fingers or hand. Your feeling of knowing (i.e., abstract continuity) is now linked to the corresponding image sizes of your finger and the external object. These memory links are formed with several repetitions just from everyday experiences. This implies that the next time you see your finger with your eyes and it has a given size, this automatically triggers a feeling of knowing for both the size of your finger and the nearby objects in contact with your finger.

In other words, vision also creates a feeling of knowing for size and distance by linking the image fixed sets with the predictive meta-fixed sets created for touch, at least for nearby objects. When we try to generalize this for distant objects, direct contact-based touch turns out to be inconvenient. Instead, we use the above created links between touch and vision for nearby objects and generalize it to far objects as well. I will discuss this in chapter 36.

One common approach taken by current researchers to explain how we perceive true size and distance of objects is by using the geometry of similar triangles, subtle differences in inputs from both eyes, the principles of optics and of our convex eye lens. The underlying measurable parameters are, say, the size of image on the retina, focal length of our eye lens and the distance of the image from our eye lens. I claim that such a computational approach is incorrect and incomplete, especially if we look at how infants or even animals are capable of determining them with ease without any knowledge of these computations. Furthermore, if we are trying to build a synthetic conscious system and we use this computational approach, how can we guarantee that the system is sensing true sizes and distances? The computations are purely abstract and there is no easy way to validate the accuracy of the computations (like we have done here with touch).

The computational approach though mathematically accurate is abstract. There is no connection to physical dynamics that occurs within our brain. On the other hand, the approach described in this section, in terms of abstract continuity linking vision to touch, is the correct way to attain a feeling of knowing for sizes and distances of nearby objects. The correct approach should be to first explain how we attain a feeling of knowing for sizes and distances. Only then, we should computational approaches for higher accuracy. The solution should not be expressed the other way around, as is commonly done.

The same situation applies with sensing space, the passage of time and several other topics. It may be that space and time are interconnected or that the structure of spacetime is either Euclidean or non-Euclidean. However, knowing this structure cannot help a synthetic system ever know the existence of space, the passage of time, true sizes or distances. The solution to each of these problems should be such that the simplest possible system that exhibits or wants to exhibit these features (like an infant, a single cellular organism or a synthetic system) can attain them entirely naturally, without the help of any intelligent or conscious being.

35

Sensing Space

In the last few chapters, we have discussed how to create some of the most common geometric properties within the new SPL framework. We began with the problem of sensing several primitive geometric features like edges, angles and shapes using touch sensor. We showed how to attain a feeling of knowing for these features using the membrane and abstract continuous paths. The reality of touch allowed us to show the existence of these features. Next, we used retinal inputs to create a visual representation that is consistent with touch. This helped us show the existence of a visual representation and the feeling of knowing for vision as well. Using the links between vision and touch, we were then able to generalize the representation beyond regions where touch is directly applicable. We were also able to confirm their existence using our free will. If synthetic systems (or even some animals) were to sense these properties or have such representations, they would need to have a membrane of meta-fixed sets and abstract continuous paths for each of these geometric features in a similar way, based on both touch and vision.

These geometric properties help us sense objects in our external world. However, in addition to tangible objects in the external world, we perceive emptiness between objects as well. This emptiness exists not only in the depth direction but also in an XY plane (i.e., between objects next to each other or above one another). The main focus of this chapter is to show how we can sense empty space in our three-dimensional world. Clearly, we see the world around us as containing objects placed in empty space. Our ability to sense space is not limited to vision. People who cannot see can also sense space in a similar way. Also, several animals have the ability to detect space just like conscious humans. Given that our brain does not receive any signals from empty space, how can we become aware of its existence? The problem of sensing space is interesting primarily because we (and other animals) are able to detect space in spite of a lack of neural firing patterns corresponding to emptiness. It is important to distinguish the problem of 'sensing' empty space from the problem of 'seeing' empty space. I will discuss the latter problem in chapter 35.

Let me now present the main intuition. In order to know that there is empty space between two objects, the first important step is to recognize the two objects themselves. This may appear simple to an adult, but this is quite nontrivial for an infant. For this, we need to know that an external world exists and that we can distinguish a foreground (i.e., the object) from the background. Otherwise, we will simply think that the entire picture that forms on the retina is just one giant solid object with moving parts. If we cannot distinguish the foreground from the background, our visual view would appear as one large painting on a 2-D or 3-D canvas with no 'empty space' in it.

Secondly, when one object moves *relative* to other objects, we need to know what will occupy its original place after it moves. This is precisely empty space. We need a way to use our free will to establish this fact, i.e., of empty space as well as its properties, with complete certainty. In section 34.7, I have shown how to use touch to determine the true size of an object. In this chapter, I will show how touch can be once again used to show that there is empty space, namely, by our inability to touch an object. I have emphasized 'relative motion' above because of its critical role for creating the notion of empty space for the first time, say, when we were an infant. When we become an adult, we sense the presence of empty space around us even if all objects around us appear static i.e., without any relative motion. However, this is not the case when we were an infant and we are detecting empty space for the first time. There are a number of other cues for an adult that can help maintain the existence of space even if the external objects do not move. For example, a couple of additional cues we use are: (a) changing our focus to see different static objects at different distances as we turn our head around or (b) moving our own hands or legs relative to the other objects.

Thirdly, if we can attain abstract continuity and hence a feeling of knowing for the above steps, we will sense the three-dimensional space between objects. Now, this intuition seems to rely on our eyes. However, people who cannot see can also sense the three-dimensional space. This is not a particularly difficult case after we have the solution with vision. We can translate most of the same ideas to include the touch sensor. I will discuss this as well in this chapter.

As with the discussions in all the previous chapters until now, the mechanisms described here for creating a feature of sensing three-dimensional space will be generic enough to be applicable even to a synthetic SPL system.

35.1 Distinguishing foreground from background

I will use the term 'foreground' to denote all objects in the external world. The term 'background' then refers to everything else except the foreground. These definitions already agree with our everyday usage.

Light from the external world falls on our retina triggering a firing pattern. This pattern changes over time. The 3-D feeling we have of real physical objects placed at different distances from us is a perception created solely from the 2-D sensory inputs like from the eye, touch, sounds and others. The terms foreground and background do not require vision to define them. For people who cannot see, they are well-defined when they use touch as the sensory system. In chapter 9, I have discussed how a synthetic SPL system can distinguish foreground from background in considerable detail. Let us see how we can use this fact to sense a three dimensional world. Using the terms above for foreground and background, we can make the following claim:

Claim 35.1: If an SPL system senses three dimensional space, it is capable of identifying a foreground and distinguishing it from the background.

To see this, a system that senses the three dimensional space already knows about the existence of an empty space between, say, two objects. This implies that you can group the two objects as foreground. The empty space is not just along the depth direction, but it exists even in the left-right and top-down *XY* plane as well. The empty space acts as a boundary between two surfaces of foreground objects and becomes part of the background.

To understand this claim better, let us also analyze the contrapositive statement, namely, if a complex system cannot distinguish foreground from background, it cannot sense three-dimensional space. Consider the situation with two dimensions first. Let us assume that all objects are at the same distance from your eye, but are placed at different locations on the *XY* plane (i.e., left, right, top and bottom directions). This is not unusual and can be partially simulated if you close one of your eyes and look at, say, your living room. In this case, it is still easy for us to recognize the foreground (say, the tables, and the chairs) from the background. Therefore, you are able to feel that there is space between these objects, even in two dimensions.

To address the above contrapositive statement, let us ask the following question. How would you feel if you could not distinguish the foreground from the background? Almost everything you see around you has recognizable objects. We, therefore, need to be in a state that prevents us from recognizing objects, if we want to attempt answering the above question. It is possible to create a picture that contains random lines and dots in such a way that the image does not appear to contain any recognizable objects in it. As a result, you would not know what the foreground is and what the background is. You would not be able to create a notion of empty space between two lines easily. You, instead, tend to view the entire image as one object. The picture could have been assembled with real paper clippings with a number of holes. Yet, it would not be easy to detect if there is empty space in these holes. The situation is similar to the image of a closed loop shown in Courant and Robbins [25]. The closed loop is specifically designed to be a zig-zap path that it is hard to identify an inside (the loop) from an outside. You do not easily sense a foreground from a background. It even takes a while to know if a given point is inside or outside the closed loop. Other examples are optical illusions in which you cannot quickly distinguish foreground from background. For example, Rubin vase is a painting in which the foreground can be viewed as a vase while the background can be viewed as two human faces facing each other or vice versa as well.

We can extend these examples to three dimensions as well. It is possible to construct a room full of lines and mirrors at different angles with patterns reflecting in several bizarre ways that you lose the perception of depth. You may have seen such a room in science museums already. You feel disoriented when you walk in this room. Sometimes you extend your arm to avoid hitting what you thought is an object only to be surprised to see that your hand passes through the empty space. At other times, you do walk straight through thinking that there is empty space only to bump into a mirror. There would be no recognizable objects to use as reference points to perceive the three dimensions. You cannot easily identify a foreground and distinguish it from the background. Your brain is unable to create the notion of a three-dimensional space between these lines or angles. You are constantly surprised at how many mistakes you make as you walk through this maze. You extend your arm when you do not need to extend and you do not extend your arm when you need to extend. Therefore, as you cannot distinguish a foreground from a background, we truly were not able to sense the three-dimensional space.

In other words, we have seen from the above claim that sensing space is closely related to our ability to distinguish a foreground from the background. Let me now use these notions of foreground and background to address the converse of the claim i.e., of sensing space.

35.2 Fixed sets for foreground patterns

When we look at a paper that contains several images of objects, we quickly recognize most of these objects. Each image has several colors, shapes, sizes, edges, enclosed regions, obstacles and patterns. Yet, we are able to pick the correct set of boundaries for each of the recognized object. We are also able to separate and ignore the rest of the regions in between these images as unimportant background region. In other words, if we identify a familiar pattern in the image, then that pattern becomes the foreground and the rest is the background. We do not make the mistake of interchanging foreground from background and vice versa even accidentally. Also, the images on the paper are in a flat 2D plane. There is no depth with these images to assist us distinguish a foreground from a background. How then are we able to do distinguish them so easily and even unconsciously? When analyzing a static image like the one on a piece of paper, it appears that our past memories are critical for distinguishing foreground from background. Yet, we have several images for which memory is not enough (like 'I spy' images for kids).

In section 35.1, we have seen how a synthetic SPL system can be trained to detect foreground from background. We did this by first creating a number of fixed sets for several common objects around us. In fact, based on the level of detail we observe, we form a large number of fixed sets even for a single object. We link these fixed sets based on the continuity of events. For example, when we look at a car, we do create a meta-fixed set corresponding to the entire car. However, in addition to this, we also create several fixed and meta-fixed sets for the windows, doors, bumpers, wheels, headlights, seats and so on. The abstract continuous paths from our past experiences create dense interconnections between these fixed sets. These predictive meta-fixed sets will be automatically triggered and will let us attain a feeling of knowing for the entire object.

This ability to excite the above collection of predictive meta-fixed sets for a static image also addresses the problem of recognizing a foreground object even when we only see a partial image. Indeed, occlusion of objects is quite common in our everyday experiences. If we cannot form a specific fixed set corresponding to the correct object, we still trigger a generic fixed set called an 'object'. For this 'object' fixed set, the patterns we recognize are just primitive shapes, edges, angles and colors. Similarly, we can extend this notion of foreground even when the image is dynamic i.e., when watching a dynamical event. During the entire event, you can recognize and predict the patterns.

Motion turns out to be very helpful in creating several common fixed sets. The relative motion of a pattern with the background is a natural grouping to create a corresponding fixed set for the moving object (see section 18.12). In addition to static and dynamic images, when we look at real objects in the external world, the patterns formed on the retina also generate the fixed sets in the same way (provided we have already learnt the auto-focus mechanism – see section 16.1). When we auto-focus, we usually do it on the foreground object, not the background. The background is typically left fuzzy or in a blurred state. In fact, as we scan the room or an image, we switch our attention from one object to another. At each time, we pause mostly on different foreground objects and rarely on the background. Our auto-focus mechanism triggers naturally at each of these pauses and creates sharp images for the foreground objects. The discrete switch or jump from one object to the next implicitly involves auto-focusing on a foreground object and triggering a corresponding fixed set. If the object is moving, our eyes can track it as well using mechanism from section 17.3. Using these two mechanisms, auto-focus and tracking a moving object, our brain can detect multiple foreground objects by exciting the corresponding fixed sets. Furthermore, the existence of an external world (chapter 29) is an important prerequisite to identify a given pattern as a valid foreground object. This is because you can confirm the independent existence of the pattern as a given foreground object using your free will.

Let us now consider the firing patterns generated from the 2-D retina surface for two foreground objects separated by a background empty space. In the external world, the gap between the foreground objects is represented as empty space. However, on the 2-D retina surface, we do not have emptiness in a strict sense. Instead, we have blurred image patterns of distant objects. For example, there may be empty space between two chairs placed at the same distance from you. However, between the two chairs and behind them there is a backdrop of, say, a table, a wall, part of the floor and other objects that do send light signals to our eye. These backdrop objects from behind are not focused on to create a sharp image and, therefore, look blurred on the retina.

In other words, on the 2-D retina surface, there is no clear notion of a background or emptiness. Instead, the background is covered by other objects (like a table, a wall and others) but in a blurred state, especially since these foreground objects are actually behind or in front of the foreground object in question (like the chairs). The focal length of our eye lens is only adjusted for now to create sharp image of the chairs. This causes the images of the backdrop table, wall and others to fall in front of the retina instead (resulting in blurred images). When we now change our attention to one of these background objects, some other objects take the place of the background. The process continues as we keep switching our attention from one object to another. What is interesting is that in spite of a lack of emptiness on the 2-D retina surface, which is the only source of input to our brain, we do perceive empty space in both depth and XY directions (even with one eye). I will discuss how we sense this next.

35.3 Sensing space between foreground objects

We have seen that creating fixed sets for foreground patterns on the retina is achieved primarily using motion detection mechanisms described in sections 18.7 and 18.12. When looking at objects around you, we can retrieve and trigger these fixed sets using the pattern produced on the retina combined with past memory. As we are forming these fixed sets, we are also creating a notion of space between objects. Let me now discuss how this happens.

Let me consider, for now, only objects close enough to your eyes that you can touch them with your hands or legs. For such objects, we can use touch to determine the true sizes and distances, as explained in section 34.7. A newborn typically starts with this situation when he is trying to create a notion of space around him in the first few months. With objects so close to him, it is easier to have repetition (using free will) with most of the experiments I will suggest here.

Those who cannot see even at birth do have a clear notion of space around them. They use touch as the sensory mechanism instead of eyes to sense space. Therefore, the mechanism for knowing the existence of space does not depend on which of these sensors we use. Let me summarize the approach first before going into the details of each of these steps.

1. *Identifying what exists at a location when an object moves away*: Let us say you know the location L of a given object A relative to another object. When you move this object A out of this location L, you want to find out what is occupying in its previous location L. You can repeat this event multiple times using your free will by moving object A away in different directions and then back to its original place. Each time you repeat this event, you want to know the properties of the original region L.

2. *Using free will combined with vision and touch to identify properties of the above region*: If you cannot see with your eyes, your event is to touch, move away and touch back again in a repetitive way. If you can see with your eyes, your event is to see an object, move it one way and then move it back in a repetitive way. In both cases, you track the motion of the object sometimes and not track it at other times. When you use free will to repeat the experiment at different places, different times and with different objects, you will be guaranteed to understand the properties of the region and not be tricked by illusions or virtual scenarios created by some unknown or superior conscious being.

3. *Identifying empty space as the above region*: The original region L where the object is no longer present will be identified with empty space. The reason is that no other fixed sets (like shape, size, angles and other geometric patterns) can be used to distinguish this case. The properties of region L is invariant to the objects used in this repetitive experiment as well. These invariances become the necessary set of dynamical transformations needed in the definition of a fixed set. This new fixed set can be called as the empty-space fixed set (cf. color fixed set in section 31.4 that used a different set of dynamical transformations). You establish the 'absence' of an object using touch just as you have established the presence of an object using touch. If your predictions

suggest the existence of an object, but as you reach out to touch and you missed touching the object, you now have a new meta-fixed set that is still triggered. You can identify this meta-fixed set to the empty space.

4. *Using free will to show invariant properties of the above empty space fixed set*: Using free will, you show that the properties of empty space are identical irrespective of which object you use (with different size, shape, color or texture), where you perform the experiment (with different relative distances, angles and directions) and when you do it. Each of these invariances with respect to the above dynamical transformations let us define the properties of empty space. Just to compare, in chapter 31 we have seen that when defining the fixed sets for colors, we used a similar set of dynamical transformations except that the variations with respect to objects are limited to only to those objects that have the 'same' color instead of 'all' objects here (namely, when defining empty space). Every time you miss touching an object or you look at the region after an object is removed, the feeling of knowing is identical. You have now created a new empty space fixed set that is invariant to the above dynamical transformations.

Let me discuss this in more detail. Consider any two objects like, say, a toy and your hand. Since sensing 3-D space does not happen in a day, our experiment of considering pairs of objects will be repeated over multiple pairs. As a result, we will only be looking at the cumulative effect across different pairs of objects. Even the formation of fixed sets as we repeat the experiments with multiple different pairs of objects takes time.

Now, given a pair of objects, we need to be able to recognize both of these objects as foreground objects first. We have discussed how this happens in the previous section. If we are using an eye sensor, this requires us to create and excite the corresponding fixed sets from the retinal inputs. The same applies with touch sensor as well. This problem of recognizing objects is simplified when we have already trained our eyes to auto-focus (section 16.1). The images on the retina are sharper, have good contrasting variations and clear boundaries when you focus. As the image on the retina is rarely static, our ability to track the moving pattern while still maintaining a sharp focus, is also important in identifying the foreground object as the same one (section 35.2).

A few mechanisms that help with forming memories (or fixed sets) are our free will mechanism (chapter 28) and our self-control mechanisms based on the fundamental pattern (chapter 16). When you make the event repeat or when the events repeat themselves, it almost never occurs in the same way as before. These variations help form links between different events. Our brain network becomes highly interconnected. This lets us predict a repeating event at every instant. This, as seen in chapter 22, gives rise to abstract continuity and eventually to a feeling of knowing.

In addition to recognizing the objects, we need to know their true sizes. Using touch, we can determine the true sizes as explained in section 34.7. The true relative distance between the two objects can also be determined using mechanisms of sections 18.6 and 18.11. Again, we are choosing the distance between the objects to be small enough that you can use the length of your hands (or skin sensor) as a good estimate. For this to work, we need to create all size related fixed sets on the skin sensory surface at different regions of our body (see section 18.6). We can similarly get a feeling of knowing for each of these situations when we repeat experiments using free will after we create abstract continuous paths.

Next, we need to examine what happens when an object moves relative to another object. We should consider two cases here, namely, when you can see with your eyes and when you use touch (especially if you cannot see). Consider two repetitive experiments: (a) see an object, move it to a different location and move it back to its original location – where you track the objects motion sometimes and do not track it at other times and (b) touch an object with your hand, move your hand away and touch the object back again.

Let me first consider case (a), with eyes. When you do not track a moving object with your eyes, you are looking at the previous position L, which has no object now (see Figure 35.1). You are aware of the relative motion of the object even though you are not tracking it because you can detect change in relative distances and angles (see section 18.12). When the object O_1 moved away, you change your focus to look at the object O_2 behind it because it is visible now. This new object O_2 was blocked from view when the first object O_1 was in its original location L. As a result, you may feel that that there is no need to introduce empty space in this location L. It may appear that the new object O_2 takes over this original place L quickly when you auto-focused on it (see Figure 35.1).

However, you still need to determine the true size of this new object O_2, which apparently took object O_1's place, at least, on the 2-D surface of the retina. For this, you would use touch in the same way as you did to determine the true size of the original object O_1 (see section 34.7). To determine if the new object O_2 is occupying the same location L as the first object O_1 before it moved away, you would extend your

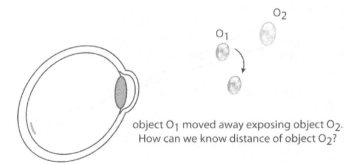

object O_1 moved away exposing object O_2.
How can we know distance of object O_2?

Figure 35.1: Sensing space between two objects. If object O_1 moved away from a location, the object O_2 behind it would form an image on the retina after our eyes auto-focus. If we try to reach for object O_2, we will miss touching it had we made the assumption that O_2 is also at the same location as O_1 from just the retinal image. This discrepancy will eventually manifest itself as empty space between O_1 and O_2.

hand to touch it. You will surely miss it, since the new object O_2 is not the same location L, but behind it. You can repeat this experiment multiple times using your free will only to be surprised that you can never touch the new object O_2 even though you can touch the first object O_1 at this location L prior to moving away. Notice that the image on the retina, modulo your auto-focus mechanism, for both the new object O_2 and the first object O_1, are precisely at the same location.

Using free will and touch, we are now able to confirm with complete certainty that there is no object after the original object O_1 moved, even though your eyes did trigger a new pattern, corresponding to the new object O_2 from the same region on the retina 'after' a refocus. Therefore, the only conclusion our brain can make is that our retinal image information is not correct (or rather it is incomplete). Our direct contact with the first object O_1 did produce a firing pattern of neurons from touch. This cannot be incorrect. With touch, real physical dynamics like a firing pattern occurs within our brain if and only if someone indeed touched us (see chapter 32). With vision, there are no such guarantees with a firing pattern and location of objects. You may have noticed already that an infant does indeed makes this mistake by trying to reach for a toy that is a bit far away thinking that it is just right next to the toy he is currently playing with. We can say that he is just beginning to create the notion of empty space between objects.

The experiment when you cannot see with your eyes (case (b) above) is even simpler. You need to use touch anyway in this case. The difference is that instead of seeing a behind object and missing it when trying to touch, you would need to use your 'imagination' (or specifically, your memory) to move your hand back to the original location. You would still miss touching the object. The tricky problem here is how you know that you moved you hand back to the correct original location. With eyes, you could simply check this by comparing your hand's location 'relative' to nearby objects (see section 18.11). When you cannot see, you do not know where your hand is relative to other objects easily. You need to rely on free will and your own body fixed sets considerably.

Therefore, a much easier experiment instead is to consider both the above objects (O_1 and your hand) as parts of your own body instead of one as an external object and the other as your hand. Through normal day-to-day activities, any person is capable of knowing and controlling the motion of his own hand relative to his own body. For example, even if I close my eyes, I know how far I have extended my hand and in which direction. The internal neural inputs provide this precise information from, say, my elbows and shoulders through my peripheral nervous system. These are represented as the body fixed sets. I can now know whether I moved my hand back to the original location when performing this previous experiment. In fact, once you have created body fixed sets, you can even try the original experiment, namely, of touching an external object. You use the body fixed sets to let you know if you have moved back to the correct original location or not. Now, you can be sure if you have missed touching the object at the original location after it has moved. The situation is not entirely obvious for an infant because he has not yet built the same set of body fixed sets as an adult. You may have already noticed that touching an infant's leg sometimes does not prompt him to look at where you have touched. It takes several months for him to build all these necessary body fixed sets. They are guaranteed to form for all people because the inputs from our body to our brain happens almost every day and is repeatable as well.

Therefore, in both cases, i.e., with or without vision, we have shown the existence of a new 'entity' called empty space using our inability to touch it. What are the properties of this empty space? How does it look? How do we detect the three dimensions? Let us discuss the dimension aspect first. In the above discussion, I have not specified which direction to move the object when trying to detect empty space. Human beings as three-dimensional objects have separate degrees of freedom to move in each of the three directions independently. We can move our hands sideways, up or down and forward or backward. These motions are sensed at the joints, say, at the shoulder. We create fixed sets that let us distinguish each of these three independent motions relative to the rest of our body.

I had initially considered a 1-D sensory surface in chapter 18 to show motion detection along both sides (see section 18.7). This let us detect the existence of single dimension. The same approach is applicable to a 2-D sensory surface like our retina. If a retinal pattern moves up/down or sideways, we can detect this motion and hence the two dimensions using the same mechanisms. This generalizes again to a 3-D sensory surface i.e., for our internal body inputs generated in response to the motion of our hand (see section 18.7). Through repeated motions from our common everyday experiences, we have abstract continuity for the concept of three dimensions. This gives rise to a feeling of knowing of the three dimensions as well. The topic of how the three dimensions appear to us (i.e., the visual perception) will be discussed later in chapter 38.

Now, in addition to knowing that there is a three dimensional external world, we have just seen that there is also an empty space in between all of the foreground objects. The empty space is identical in each of the three dimensions and at different locations or directions. The empty space is also identical when we perform the above experiments with any tangible object, at any time and under any physical circumstance. Furthermore, if we change the sensors from eyes-touch to eyes-ears or to any other combination, the nature of empty space remains the same. We can establish each of these identities (or invariances) again using free will and repetitive observations. Therefore, empty space looks the same at every location, in every direction and at all times.

This is in contrast with physical objects. Empty space fixed set seems to have the largest set of invariances (or dynamical transformations in the definition of fixed sets). We have seen that color-based fixed sets in chapter 31 have similar number of invariances except that the invariance with respect to objects is restricted to a smaller subset. Most other fixed sets for, say, shapes, angles, edges, sizes and distances have even smaller subset of invariances.

35.4 Peculiar scenarios when sensing space

I have already discussed a few cases when we do make mistakes when trying to guess the existence of space (like, say, when we are unable to distinguish a foreground from a background). Let me now discuss a few additional situations when we err in sensing the three-dimensional space. In most of these cases, whenever there is confusion about space and distances, we resort to using touch as the definitive approach.

(a) *Mirror filled room*: Consider walking in a room filled with many mirrors (including plain, convex and concave ones) arranged in every possible angle that you see, literally, hundreds of reflections for any object. When a new person enters such a room, you find it hard to conclude where the person is located because of multiple reflections. You do not know how far the mirrors are. You also do not have the same feeling of empty space as you would in a normal room. In such cases, you naturally extend your hand to touch the person. You cannot trust your eyes anymore. Your only conclusive mechanism is to use touch.

(b) *Dark room*: Another scenario is when you enter a room that is dark. In this case, you cannot use your eyes effectively anyway. As you walk in this room, you extend your hand trying to reach for objects and not bump into them. You form a mental three-dimensional picture of the room based on the objects you have touched. Over time, you will slowly gain familiarity of the room.

(c) *Optical illusions*: Optical illusions are some examples where we get a false impression with sizes and distances (along with the notion of space). A room, for example, can be prepared to make it look as if a smaller person appears bigger. Or a set of lines or curves can be arranged in such a way that parallel lines appear either bulged or converged. This can produce a false sense of sizes, distances and space. Now, if we look closely at the events that occur around us, we observe that there are always redundant sources of inputs for the same information. For example, touch, visual patterns on the retina, relative motion, relative distances between objects, variation in colors, contrast and brightness or sound variations, all constitute different ways to determine the same geometric information. From our early childhood itself, we learn to take all of these multiple inputs into consideration to produce a consistent view of the world around us. However, the optical illusions are specially crafted to produce contradictions with our memory and from one or more of these sources.

(d) *Loss of depth perception*: If you close one eye and look at objects around you, you lose your perception of distance to a degree. You do make mistakes estimating the distance when you try to reach for an object. Your brain has been trained to use inputs from both eyes to provide this information. However, after you touch one of the objects, your brain is quickly capable of estimating the distances of other objects with better accuracy. This is similar to a case when you have a white or a patterned wall with nothing else in front of you. You find it difficult to estimate the distance of the wall from you, as there are no other objects to compute relative distances from you. However, after you touch the wall, it becomes clear that there is empty space between you and the wall. Similarly, stars in the sky never appear in 3-D even though the distance between the stars is enormous. Of course, here you cannot touch them and, therefore, you never can fix your visual perception to change to a 3-D view with empty space in between the stars.

(e) *Breakdown of two-dimensional view*: How can we eliminate the possibility that all objects are floating on a movie screen at a fixed (or zero) distance from your eye? This is reasonable because empty space does not trigger any new firing patterns anyway. We are asking if we can collapse the empty space and make the external world more compact. This can be realistically set up with different sized objects, different time-varying variations and even accounting for both eyes (like in a 3-D movie).

The answer is that we can rely on touch once again as the second input and use it in combination with free will. The same reasoning given in this section applies directly to this case. The introduction of empty space between objects is, therefore, necessary to have a consistent view of the external world with respect to all of our sensory inputs.

In the next chapter, I will discuss how we can extend our analysis of sensing space to objects that are quite far away as well. This will let us perceive the true size and distances of objects in all directions and with a broad range of distances.

36
Determining True Size and Distance

The main objective of this chapter can be stated as follows: if we pick two objects *A* and *B* from the external world, how can we know if *A* is in front or behind *B*, relative to your current location? I will also look at several variations of this question. For example, I want to generalize this even when we have only one object. In this case, *A* and *B* would correspond to two features of this object like two wheels of a given car, two legs of a table, two flowery patterns on a blanket, two leaves on a branch, two dark spots on a banana and so on. Another variation of this problem that I will discuss here is with occlusion – where one object is partially hidden by another object. This also lets us identify which object is in front and which one is behind. Yet another case I will consider is to look at more than two objects or two features. If I have a group of, say, 10 objects (or 10 features), I would like to determine the relative locations for every pair of objects (or features) as well.

One problem that I will not discuss in this chapter is 'estimating' the separation between objects *A* and *B* using numbers (like, say, *A* and *B* are 25 feet apart). I will focus only on whether *A* is in front or behind *B* relative to you, not by how much. Estimating distance quantitatively requires consciousness and other mathematical abstractions. A child estimates qualitatively before he can estimate quantitatively. The latter problem is not considered in this book for now, but it is easy to see that it will surely rely on the results from this chapter.

Now, if two features or objects are close enough that you can touch them, then the above stated problems are easy. All you need to sense is which feature (or object) you are going to touch first, relative to your own body. We have already discussed this in sections 34.7 - 34.9. Therefore, the interesting case in this chapter is when the objects are quite distant from you that you cannot rely on touch. Instead, you need to rely primarily on visual information. Let me discuss this now.

36.1 Occlusion

Let us start by looking at the problem of occlusion and its mechanism first. Consider the following, somewhat obvious, occlusion rule: *if an object A is partially hidden by another object B, then the occluded object A is behind object B*. Our brain is capable of creating this rule automatically. Let me describe these steps. First, we need individual fixed sets for each of the two objects *A* and *B*. These fixed sets (i.e., the memory of these objects) can be created as you interact with these objects or their features several times. You form these memories using any of the previously discussed mechanisms (say, in chapter 18). These fixed sets can be triggered using unique patterns from the retinal inputs. This can happen even if the object is partially hidden. Typically, with most objects, it is quite rare that some other object has the same partial pattern as this partially hidden object. If, however, it does happen, we do find

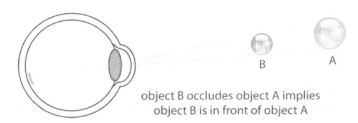

object B occludes object A implies
object B is in front of object A

Figure 36.1: Occlusion. If object *B* is in front of object *A* blocking it from our view, we say that object *B* occludes object *A*. We can use this fact to infer relative distance between objects *A* and *B*.

it difficult to identify the occluded object uniquely.

Every one of us have experienced the following pattern almost every day: if we reach for objects around us, it always turns out that the object *B*, the one in front, is the one we will touch with our hand first (see Figure 36.1). A simple repetition and our free will with a number of different objects show the causal nature of our actions. This causality is memorized as discussed in section 20.6, to create a predictive ability. This will give rise to abstract continuity as well. In this way, our brain memorizes a rule that *B* is in front of *A*, whenever *A* and *B* are objects we have interacted with recently. We may or may not be consciously aware of this fact. Recall that a rule is simply a collection of predictions captured as abstract continuous paths (see chapter 20).

How can our brain generalize this occlusion fact to all other objects as well? This is where meta-fixed sets become critical. One way is to link our predictive ability with the direction of change of focal length of our eye lens. For example, when we switch our attention from a nearby object *B* to farther object *A*, we change the focal length of our eye lens from a smaller to a larger value. This direction of change is always the same irrespective of the type of objects. You create fixed sets to detect this direction of change (similar to the mechanism of section 18.5).

For example, our eyes adjust the focal length by changing the tension of the zonular fibers by either contracting or relaxing the ciliary muscle. This changes our eye lens from a flat to a spherical shape. When we learn to control these muscles, we create two important fixed sets. They correspond to: (a) detecting if the focal length increased when we switched from one object to another and (b) detecting if the focal length decreased during the same switch. These two fixed sets let us detect the direction of change in focal length.

These fixed sets are linked to our predictive meta-fixed sets corresponding to the expectation that we will touch *B* before *A*. We notice this in a child as he starts to extend his hand mechanically to try to touch *B* before *A*. Multiple examples reinforce these interconnections. In this way, our brain is able to generalize the above occlusion fact (Figure 36.1) to all objects. Other factors that can help in establishing this result is the difference in inputs from both eyes and sensing space between *A* and *B*.

When we use our free will to verify this rule (using several examples), we become consciously aware of this fact as well. The mechanism for occlusion just discussed is a trivial translation of what our intuition suggests, into the framework of stable parallel loops of our brain.

36.2 Knowing if an object is in front or behind another

The problem of occlusion discussed above does address whether an object A is in front or behind another object B, at least, in one special case. This case, though common, is restrictive because you require one object to be partially hidden by another. In this subsection, I want to remove this restriction. Specifically, we are looking for a necessary and sufficient condition for the above question of whether A is in front or behind B using vision irrespective of whether A is occluded by B or not. Occlusion turns out to be a sufficient condition, but not necessary. With occlusion, we say that if an object A is occluded (or partially hidden) by another object B, then A is behind B. However, the converse is not true, in general. When A is behind B, A is not necessarily occluded by B.

I will use the change in focal length of our eye lens discussed in the mechanism for occlusion to help find a necessary and sufficient condition. Recall how we determined that an object occludes another object. We first used our ability to focus on an object to create a sharp image using the fundamental pattern as the basic mechanism (see section 16.1). If we use both of our eyes, we also need to converge both eyes onto the same object (again using fundamental pattern – see section 17.2). As we switch our attention from one object to another, typically, we need to refocus to create a sharp image of the new object using the above mechanisms once again.

When we refocus, we either decrease or increase the focal length of our eye lens, which can be detected using appropriate sensory and action fixed sets, as mentioned in section 16.1. If the focal length *increased* when we switched from an object A to object B, then we notice that object B is behind object A, at least for nearby objects (see Figure 36.1). Recall that $\frac{1}{u} + \frac{1}{v} = \frac{1}{f}$ for a convex lens. Therefore, as u is higher for A than for B and with v being the same for both cases (because of auto-focus), the above relationship implies that f should be higher for A than for B. We can, in fact, establish this fact with complete certainty using free will and repetitive experiments for all types of nearby objects. In fact, we can use touch to confirm whether A is in front of B for nearby objects. Similarly, we can use touch and free will to show with complete certainty for nearby objects that object B is in front of A, if the focal length *decreased* when we switched from A to B.

When the objects are far enough that you cannot touch them, you can still be sure that this relationship between (i) the direction of change of focal length and (ii) relative location of objects is maintained. We do have links between these two fixed sets for nearby objects. We can use free will to show that this fact can indeed be generalized with complete accuracy. This reinforces the same links once again. Note that I am not suggesting that we can estimate *how far* a given object is from another object as you switch your focus. Instead, I am only saying that we can be sure whether A is in front or behind B.

A consequence of these links is to create a predictive ability. Whenever we have predictive ability, we will form abstract continuous paths because of minor variations that occur as events repeat. Abstract continuity (along with the membrane) gives rise to a feeling of knowing of the linked event (see chapter 22). For example, the moment our brain detects that our focal length has increased when we switch from A to B, the abstract continuity gives rise to a feeling of knowing that A is in front of B. The predictions and the abstract continuous paths are triggered via these links even for very distant objects that you cannot touch. This is because what triggers the predictions is merely a change in the focal length of your eyes to create a sharp image using the fundamental pattern, which does not depend on how far the objects are.

The feeling of knowing does not mean that we have proven this fact (say, using our free will) as true for very distant objects. It simply means that our brain *generalized* the existing links created when working with both nearby and walk-able distant objects. As these links are triggered automatically in response to how our ciliary muscles change, our brain simply produces the same feeling for very distant objects as well, in an entirely mechanical way. This form of generalization is a natural outcome in any stable parallel looped complex system that is capable of linking fixed sets of different events together. Until there is a contradicting event, we believe that this generalization is true whether we can prove it or not (either logically, mathematically or using our free will). Therefore, we can make the following claim.

Theorem 36.1: The focal length of our eye lens increases (or decreases) when we switch and create a sharp image on our retina from an object A to another object B if and only if A is in front of (or behind) B.

It is important to keep in mind that the above claim fundamentally depended on the concept of generalization. This is no different from accepting, via generalization, that two parallel railway tracks never meet even though they appear otherwise to our eyes.

If you look closely at the previous section, we see that occlusion is just a special case of this result. With occlusion, objects A and B are arranged in such a way that another object blocks part of one object. The general case considered here does not have this restriction.

36.3 Relative locations of multiple objects

Let us see how we can use the above Theorem 36.1 when there are more than two objects. Imagine three objects A, B and C placed in the external world at different locations. Our objective is to determine the relative arrangement of these objects. For example, we want to know if A is in front or behind B and if B is in front or behind C. If we switch our focus from A to B, B to C, C to A, B to A, C to B and A to C, we will know how our focal length changes in each of these cases. If our focal length decreases when we switch from A to B, then from Theorem 36.1, A cannot be in front of B. Similarly, if our focal length increases when we switch from C to B and from C to A, then the resulting arrangement for these three objects should be as shown in Figure 36.2, where C is in front of B which is in front of A.

using occlusion, we can determine that
C is in front of B, which is in front of A

Figure 36.2: Determining relative locations of objects. If our focal length increased when switching from C to B to A, we can use Theorem 36.1 to conclude that C is in front of B, which is in front of A. The relative arrangement of objects in the room can, therefore, be specified sufficiently accurately from just the occlusion Theorem 36.1.

Of course, we do not know, from the above claim, how far the objects are from each other in the depth, width and height directions. The above analysis can now be generalized to any number of objects. We can easily determine the relative spatial arrangement of all of these objects in the above sense (see Figure 36.2).

36.4 Relative locations of multiple features of an object

We can apply the above Theorem 36.1 even for a single object. We can call A and B to be subcomponents or patterns of a single object (see Figure 36.3). Since all objects have a finite size, we can imagine focusing on different regions of the same object. For example, if you are looking at a table, you can focus on different legs or patterns on the top of the table.

This is what we do when we quickly scan an object. Our eyes trace a zigzag path focusing on different regions of the same object. Sometimes we scan along an edge and other times we look for prominent features. In any case, we can continue to use the above analysis whichever path we prefer to trace. It is important to note that in all such cases, the focal length of our eyes increase (or decrease) when a given feature of this object is farther (or closer) relative to the previously focused feature. This is

determine relative locations of features
of a single object using occlusion

Figure 36.3: Relative locations of multiple features of a single object. We can use the occlusion Theorem 36.1 to determine the relative locations of a single object as well similar to Figure 36.2.

just a direct application of Theorem 36.1.

36.5 Determining true size and distance of all objects

In section 34.7, I discussed how to determine true size and distance of nearby objects. There, I was able to rely on touch because there is a one-one correspondence with size on either side of the contact surface, namely, our skin side and the external object side (unlike the case of retinal sensory surface). Subsequently, I discussed how to sense the three-dimensional space, in section 35.3. There again, we saw that touch, or rather, the inability to touch, helped in detecting empty space. In this section, I will discuss how we can generalize our estimation of size and distance beyond nearby objects, to include distant objects as well. Clearly, we cannot rely directly on touch to help with this problem. It is both inconvenient and infeasible to touch every distant object as the objects keep moving around. Vision has an advantage here. Our retinal inputs are triggered just by looking at and focusing on objects. We do not need to be anywhere near the external objects, unlike with touch. Nevertheless, we cannot avoid touch completely. We will see how there is an indirect dependence on touch.

Most of the visual features we have discussed until now like edges, angles, shapes, patterns, size and distance of nearby objects, motion, three-dimensional space and color are, in a sense, primitive features. They did not have too many dependencies with other features. As a result, we were able to explain the core idea for each of them using a small set of mechanisms. However, sensing size and distance of all objects, as well as perceiving the external view in a cohesive way (chapter 41) are two problems that can no longer be called as primitive features. The broad nature of both of these features requires a large number of mechanisms to work in parallel to produce the necessary perception. There is, nevertheless, a core idea, which I will discuss here for the former problem (sensing size and distance). In addition to discussing this core idea, I will also discuss the other important mechanisms that work in parallel, in chapter 37.

How do we determine the size and distance of objects? For people who cannot see, determining these properties for all objects relies primarily on touch. For distant objects specifically, they rely on ears similar to how people with normal vision rely on eyes. I will leave this case for future work.

For people with normal vision, as soon as they open their eyes, they see a large number of objects around them. These objects are placed at different distances both in the XY (i.e., width and height) and the depth directions. The objects are also of different sizes. Since we are specifically interested in distant objects, we can say that the only source of inputs to reach our brain to help us determine the existence of these objects is from our eyes. Our ears play a much lesser role (since we only have a few noise generating objects) except for people without vision. Humans have about an inch-diameter eyeball and a retina with a two-dimensional sensory surface. Yet, we create a perception of even, say, 20-30 feet sized houses or trees and more than 100 feet of three-dimensional distances. The sizes on the retina vary with distance as we refocus with our eyes and yet, our perception of the objects' size does not change significantly. With touch, there is no such shrinking of size and distance on the sensory surface. There are two related problems here.

(a) How is our brain creating 'a' representation of different sizes and distances of objects with such high and time-varying magnifications using our eyes?

(b) How is our brain creating the 'correct' representation with respect to sizes and distances?

Let us look at the second question in some detail. When we see a tree or a nearby house, they appear to be at a fixed location and of a given size. Why should they appear exactly there? Why not closer or farther and bigger or smaller? Strictly speaking, we should ask the same question with height and width (i.e., in the *XY* directions) as well. We seem to overlook this question because the surface of the retina is two-dimensional and it aligns with the external *XY* plane. This seems to make us feel that depth, not the *XY* direction, is questionable.

Yet, it is important to realize that the images on the retina are not as clean as we think. For closer or farther issue, I have already mentioned a peculiar scenario with wearing new prescription glasses in section 33.3. For the initial few days after you wear new prescription glasses, you will notice that the objects do appear farther than they really are. There are also several optical illusions that confuse us about depth perception. For top or bottom issue, for example, the inversion of images on the retina complicates it. If we move a ball up with our hand, the image on the retina moves down. For left or right issue, it is possible to have both of our eyes of different sizes, with healthy distortions, squintedness and others. These distortions can shrink the distances between objects on the left eye, say, more than on the right eye. In spite of all these situations, the visual estimation of sizes and distances are fairly accurate.

To understand the bigger or smaller issue better, recall Figure 36.1 of section 36.1. When we look at the cone shaped region from our eyes to a given object, as shown in Figure 36.1, there is empty space initially. If you imagine extending this cone to infinity, we say that this object *A* occludes all other objects behind it and within this cone. This object *A* forms an image on the retina of a given size assuming that our eye lens has suitably adjusted its focal length to form a sharp image (using the fundamental pattern).

Recall from section 36.2 that if *u* is the distance from the object to the eye lens, *v* the distance from the eye lens to the retina where the image falls and *f* the focal length of our eye lens, the optics of a convex lens shows that $\frac{1}{u} + \frac{1}{v} = \frac{1}{f}$. Also, the magnification of an object $m = -\frac{v}{u}$. If we have two objects *A* and *B* with *A* in front and *B* behind, since *v* is the same for both objects while *u* is smaller for *A* and larger for *B*, the magnifications are different. However, if the focal length *f* does not change, then the first equation implies that if *A* falls correctly on the retina, then *B* will fall 'in front' of the retina. Conversely, if *B* falls correctly on the retina, then *A* falls 'behind' the retina. However, since we want *v* to be the same for both objects, we need to refocus our eye lens accordingly. Specifically, if *A* falls correctly on the retina, then the focal length *f* must increase so the image stays on the retina for *B*. If *B* falls correctly on the retina, then the focal length *f* must decrease to let image of *A* fall precisely on the retina.

Clearly, the image on the retina is not altered if you imagine moving the object closer while shrinking it or farther while expanding it in such a way that the object still fits within the cone. This is particularly true because the focal length is adjusted using the fundamental pattern to create a sharp image on the retina. If you imagine now drawing a number of cones for each object in your visual field, it is no longer obvious which of the three layouts, for example, in Figure 34.2 is the correct representation. In other words, there are an infinite number of seemingly correct representations just by moving objects relative to one other within their respective cones, as shown in Figure 34.2. This situation is particularly true when you use only one eye and close the other eye. How does our brain eliminate all of these infinite possibilities automatically and instantly except for one representation, which so happens to be the correct one we see? Furthermore, our brain seems to update this unique representation continuously while maintaining consistency in a seamless way, as we or the objects move around. Our visual perception does not take time, like, say a few seconds to adjust. You can close your eyes, enter a new room and open your eyes. Instantaneously, all objects appear to be of the correct size and are placed at the correct place.

Note that the issue here is not about estimating sizes and distances quantitatively. It is true that, if necessary, we can make a fairly accurate quantitative estimation. However, even before we start thinking abstractly about quantitative estimation, the objects already appear correctly. You have a unique sense of feeling of size and distance of objects just from a simple casual glance around you, even if the objects are sufficiently far away. You truly are surprised when this feeling turns out to give a wrong estimate. For example, when you wear new prescription glasses, you are surprised at an incorrect view of all objects around you (as discussed in section 33.3). This is also the case with a number of popular optical illusions. Boroditsky [16] shows one such example. In the first picture in Boroditsky [16], we feel that the woman and the girl are about same size. However, in the second picture, it feels as if the girl is significantly smaller than the woman even though you can measure and verify otherwise.

Another issue worth noting has to do with estimating sizes and distances of objects in a small two-dimensional photo (or a video on a television screen) of, say, a landscape. I am excluding pictures that are cartoon-like or purely artistic and, instead, considering photos or paintings of objects from the real external world in this discussion. When we see such a photo, of even an unfamiliar scene, we can still have a clear sense of feeling of true sizes and distances. In this example, we rely on our past memories, on the complexity and on the unique nature of how our visual system works. The information reaching our brain is altered significantly in at least two ways: (i) from the 3-D external world to a small 2-D photo and (ii) from the 2-D photo to our retina and our visual system. Such examples highlight yet another reason why, as mentioned in the beginning of this section, we need a number of parallel mechanisms in addition to the core idea to be discussed here towards estimating size and distance. I will discuss these additional mechanisms in the next chapter 37.

In the rest of this chapter, I will propose mechanisms to explain each of the issues raised above. The mechanisms of section 34.7 that uses touch for nearby objects cannot be generalized to this case. While touch is critical even in this case, we have seen that linking vision and touch offers more advantages (see section 33.4).

36.6 Constraints in relative location from occlusion

Let us focus on questions (a) and (b) of the previous section first. To explain the existence of 'a' representation for size and distance, we can use the mechanisms described in section 34.7 for nearby objects. I will defer further discussion on this to section 36.8. However, the mechanism for creating the 'correct' representation seems nontrivial. As mentioned earlier, it appears as if there are an infinite number of possible locations and sizes of objects simply by transforming the object within its respective cone (see the three layouts in Figure 34.2).

In this section, I will first show that these transformations cannot be arbitrary. They are constrained in a specific way using occlusion rule (discussed in section 36.1) that our brain learns unconsciously. It turns out that this single rule is not sufficient to constrain the possibilities to just 'one' correct representation of sizes and distances. I will discuss that in the next section.

Constraints from occlusion: Consider two neighboring objects *A* and *B* in Figure 36.4. What are the possible locations for object *A*? From the occlusion rule, it follows that *A* can only be in front of *B*. This is because the condition that should continue to be obeyed for correctness is that *B* should to be partially hidden by *A*. This happens only if *A* is in front of *B*. Note that this is just a relative condition. In other words, *A* and *B* can move together as close or as far away from your eyes as long as *A* is in front

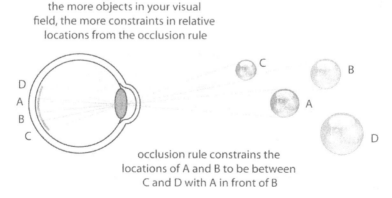

the more objects in your visual field, the more constraints in relative locations from the occlusion rule

occlusion rule constrains the locations of A and B to be between C and D with A in front of B

Figure 36.4: Constraints from occlusion. The occlusion rule in this example constrains the *relative* locations of *A* and *B* to be between *C* and *D*, with an additional constraint that *A* be in front of *B*. The entire set of objects, however, can move closer or farther from your eye. The more the number of objects in your visual field, the more constrains you would have in the relative locations. Eventually, we want to show that the *absolute* positions and sizes are also fixed from these constraints.

of *B*.

If we now consider objects C and D as well, the relative locations of A and B will be further constrained. Figure 36.4 depicts one possible configuration for objects A, B, C and D. In reality, we encounter a number of situations where C is either in front of A or behind A. The same situation applies with D as well. It is also true that we do not analyze this relative arrangement consciously for every object we see in front of us. Nevertheless, our brain keeps track of this information using our occlusion mechanism even unconsciously.

For the configuration shown in Figure 36.4, the possible locations for A and B can only be between C and D, in addition to the requirement that A should be in front of B. Again, as before, while satisfying these relative conditions, the entire group of objects A, B, C and D can move either close or far away from our eyes. Similarly, we can analyze and identify the possible locations for objects A, B, C and D for other arrangements of C and D relative to A and B, say, for example, where C is behind A but in front of B.

As we add more and more objects into our visual field, we tend to add several more constraints for the relative locations of each object just from the occlusion rule. Therefore, even though each object, by itself, can move within its own cone arbitrarily, it turns out that the presence of other objects has constrained its relative location in order to continue obeying the occlusion rule. However, it is clear that these constraints are not sufficient to determine the size and distance of all objects uniquely. The presence of empty space, with no objects in between, seems to present a possibility that the object in front can move within its cone in this empty space. In the next section, I will discuss how we can resolve this issue and constrain the possibilities to a unique location, at least, with the most common cases.

On motion and on the role of both eyes: When estimating sizes and distances, the most common argument we see in textbooks is to use either relative motion of objects or to use both of our eyes. Figure 36.5 shows how both of these approaches produce intersection of cones resulting in a unique size and location for an external object. It is clear that the images on the retina of two eyes are slightly displaced from each other. However, starting from two slightly different neural firing patterns and merging them to create a unique size and location is entirely nontrivial. How does our brain, which simply generates cascading pathways of neural firing patterns, know how to merge both of these neuronal inputs? The millions of neurons that fire will disperse within our brain and do not have one-one correspondence from both eyes. The mechanisms that direct them to merge correctly are never explained in any of these textbooks.

Imagine a machine taking pictures with two cameras at slightly displaced locations. How is the machine going to know if they are indeed displaced and how would it merge the two images by itself (i.e., without a conscious human's help)? Drawing imaginary cones outward into the external world and showing that they intersect in a unique location with a specific size does not mean that our brain or a machine is going to project it the same way. This projection approach is not valid especially because the brain (or the machine) is no longer generating or dealing with light rays after it falls on the retina. Instead, the only inputs that propagate further

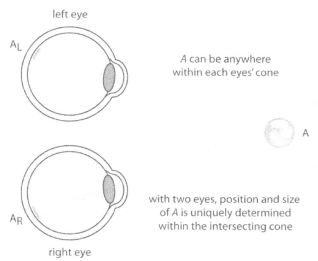

left eye

A_L

A can be anywhere
within each eyes' cone

A

with two eyes, position and size
of A is uniquely determined
within the intersecting cone

A_R

right eye

Figure 36.5: Binocular vision. With a single eye, the position and size of object A is indeterminate because A can lie anywhere within its visual cone. However, with two eyes, the intersection of the two cones from each eye uniquely determines the position and size of object A.

into the brain are electrical neural firing patterns. There are no light rays or straight-line projections anymore within the system. This geometry and projection process is purely an abstraction that human beings have discovered after they have become conscious.

Furthermore, the standard explanation with geometry and projection does not address how people with only one eye also see objects placed at a fixed distance and how they appear of a specific size. They may be incorrect at times. Nevertheless, they are fixed and never change.

In addition, the sizes of the eyes are not identical and hence the number and/or the distribution of rods and cones on each eye are considerably different. Both inputs are considerably different as a result, as is merging both of them. Other examples where we have problems with the standard explanation are for people whose eyes are squinted or have other minor unequal distortions between the two eyes.

The aspect of how our entire visual field just appears in front of our eyes instantaneously as we open our eyes cannot be explained using these standard geometric computational approaches. When we or the objects move around, the entire relative locations and the corresponding cones change. This is not just for the moved objects, but all of them. A global computational approach as required by these approaches is too costly. I will, however, discuss the role of both eyes and motion later in chapter 37 when discussing other useful visual cues.

Let us now return to the occlusion based approach and propose an iterative solution for estimating size and distance. This solution will turn out to work well in most common scenarios without any direct dependencies on motion, usage of both eyes and other peculiar situations mentioned above.

36.7 Iterative approach to determine size and distance

In the previous subsection, we saw that occlusion helped in constraining the relative location of objects around us. In section 36.3, we saw a generalization of occlusion, namely, how to sense the relative location of objects or features. I presented the necessary and sufficient conditions for when an object is in front or behind another object (see Theorem 36.1 in section 36.2). In this section, I want to show how to constrain the relative locations further so we can uniquely determine size and distance of all objects.

The basic idea is to use an iterative procedure along with the notion of continuity. For example, if you are looking at a chair in your living room, there is a continuous pathway from you to the chair that connects other physical objects and features. The pathway is not necessarily arranged in a straight line. The reason for picking curved pathways instead of a straight line is so we avoid 'empty space' everywhere along the pathway. A straight-line pathway between you and the chair surely has empty space and, therefore, we do not want to consider this. In most situations, there are several such pathways for the chair (like by going left, right, along the floor and so on). Now, if you take any two adjacent objects or features along such a pathway, you can know which one is in front and which one is behind in relative terms. You can either use the occlusion rule of section 36.1 or simply determine if your focal length increased or

connected and
continuous pathway

occlusion and change in focal length
used along multiple continuous and
connected pathways determine objects'
locations and sizes uniquely and iteratively

Figure 36.6: Tracing along continuous and connected pathways. Starting from the closest object, if you trace along connected and continuous pathways, we can use occlusion and the change in focal length (Theorem 36.1) to uniquely and iteratively determine the objects' locations and sizes.

decreased as you switch from one object to the other (see Theorem 36.1).

For most part of the pathway, there is continuity with adjacent objects or features. The continuity ensures that there can be nothing in between two adjacent objects, including empty space. Therefore, the two adjacent objects cannot be moved within their cones relative to each other i.e., they are constrained uniquely (see Figure 36.6). For every pair of such neighboring objects along the above connected pathway, the relative locations to each other is uniquely constrained in a similar way (see Figure 36.6).

In this way, the entire pathway is constrained uniquely relative to each other. The only nonuniqueness is if the entire pathway moves as a whole along the depth direction. However, this cannot happen as well because the beginning of the pathway is constrained to be near your hand or leg. For the nearby objects, you already know how to estimate sizes and distances using touch (see section 34.7). Now, for the objects that are hidden or blocked such that there is a break in continuity, you rely on other visual cues, other pathways, memory of its size relative to other objects based on the size on the retina and others. In this way, you can control the error in estimation quite accurately. This implies you can accurately determine sizes and distances of distant objects using just relative comparison and continuity in a simple iterative way.

With this iterative approach, the initial value is the true size and distance of nearby objects, which we already know how to estimate. The continuity will let you extend the solution from this initial value to a neighboring value. We can now iteratively go from this neighboring value to the next one and so on, all the way to quite distant points always ensuring a sense of correctness. This gives one solution for the estimation problem of true size and distance of a given object. In general, there are multiple ways to determine the same solution. Each of these involves traversing along different paths from a nearby object to any distant object. Let me outline all the steps first and discuss them in further detail, as needed, subsequently.

(a) *Initial choice of objects with true size and distance*: First, we need an initial set of objects for which we have already determined true sizes and distances. This set, for example, can be our own parts of our body or any other nearby object that we can touch. I have discussed this in considerable detail in section 34.7. Since we are using vision, not touch, in later discussions, we would rely on the links created between vision and touch for these nearby objects, as explained in section 33.4. The visual inputs are going to trigger, via these links, the same feeling of knowing of size and distance that we gained for these nearby objects with touch. Our memory (i.e., fixed sets) and repetitive scenarios with minor variations are some of the standard features that aid us here.

(b) *Relative size and distance in any given small neighborhood*: Next, using our ability to sense relative locations (section 36.3) combined with the notion of 'continuity' of arrangement of objects in our external world, we will be able to uniquely determine the size and location of objects relative to each other at any given region in space. This implies that if you know the true size and distance of a single object in a given small neighborhood in space, we can determine the

determine relative distances
and sizes within a given
small neighborhood

Figure 36.7: Determine relative distances and sizes within a small neighborhood. For any given small enough neighborhood, we can determine the relative sizes and distances using the occlusion and change in focal length principles (Theorem 36.1).

true size and distance of all objects within this neighborhood (see Figure 36.7).

(c) *Iterative solution to determine true size and distance*: We can now iteratively move from nearby to distant objects by picking overlapping neighborhoods I_1, I_2, ..., I_n as shown in Figure 36.8. As the neighborhoods overlap, any given neighborhood I_k has objects in common with the previous neighborhood I_{k-1}. We can use the true size and distance information from these common objects to determine the same for all other objects as well, using step (b) above. This is certainly true with the initial neighborhood I_1, namely, with the nearby objects using the above step (a). When we move to neighborhood I_2, let us use continuity in the arrangement of objects and a common object A belonging to both I_1 and I_2. Using Theorem 36.1, it is easy to see that object A is behind most objects in I_1. Similarly, using Theorem 36.1, object A is in front of most objects in I_2. These two facts together allow us to uniquely determine true size and distance of all other objects in I_2 using step (b) above. We can now continue this process iteratively to the next overlapping neighborhood I_3, I_4 and so on until I_n.

Clearly, we can choose the overlapping neighborhoods in several different ways. We can pick different paths to reach a given destination. For example, we have shown one path P_1 in Figure 36.8. However, we can pick another path P_2 with a completely different set of neighborhoods to the destination. Yet another path P_3 can be chosen with different sized neighborhoods within the

component A common between
adjacent neighborhoods can be
used to transition sizes and
distances along the pathway

Figure 36.8: Iterative solution to determine true size and distance. We pick several overlapping neighborhoods I_1, I_2, ..., I_n along a connected and continuous pathway P_1 as shown. For each neighborhood, using step (b), we can determine relative sizes and distances uniquely. Component A that is common between adjacent neighborhoods I_k and I_{k+1} can be used to transition sizes and distances across neighborhoods iteratively.

original path P_1 and hence different grouping of objects within each neighborhood. An infant needs to learn to trace along such continuous paths. The basic mechanism to help him initially is by tracing along edges using the fundamental pattern (see section 17.4). However, after sufficient practice, he can quickly find such continuous paths simply from casual scanning and quick glances as he turns his eyes and head around looking at objects constantly. He does not need to trace them systematically or even consciously anymore.

(d) *Feeling of knowing of size and distance as we move across neighborhoods*: I will discuss a simple mechanism that lets us achieve a feeling of knowing of size and distance in all three dimensions. The idea is to use the notion of unequal measuring units. Here, we will choose different sized familiar objects as the measuring rods (unlike the typical approach of using equal sized measuring unit like feet or inches). This collection of familiar objects is not unique. It depends on which continuous path you take or which specific grouping of objects you form along a given path, from the source to the destination object (see Figure 36.8). In essence, what we are doing is just simple counting without the need to use abstract numbers. Counting when expressed in terms of objects connected in a continuous way provides an alternate way to attain a feeling of knowing without the concept of abstraction. Therefore, even an animal with eyes without an ability to abstract can estimate size and distance accurately using this approach. Combining all the above steps will finally produce a feeling of knowing of distances in any of the three dimensions.

Let me now discuss some of these steps in more detail. Step (a) does not need any more description because I have discussed this in sections 34.7 and 34.9.

Iterative solution: The iterative process of step (c) becomes clear whenever we are able to pick overlapping neighborhoods from our eyes to a given destination as I_1, I_2, ..., I_n as shown in Figure 36.8. Here, we need to assume that we are not staring at artificially created scenarios or empty rooms with blank walls. Instead, we need a fairly realistic view, similar to our everyday world. Furthermore, what we call as objects need not be restricted to items made of distinct materials. We can consider patterns on a piece of cloth, different colors, shades, textures, words, drawings and so on as valid 'objects' (within a neighborhood) for the purpose of the current analysis.

With this broad notion of objects, we can choose appropriate neighborhoods between our eyes and a given destination object (see Figure 36.8). In most everyday scenarios, we can pick these neighborhoods such that (i) no neighborhood is made of entirely empty space or (ii) the overlapping region between two neighborhoods is not entirely empty space. In general, there are multiple paths, from our eyes to a given destination object, along which we can choose such appropriate neighborhoods. This is what I refer to as 'continuity' of arrangement of objects, which is common in everyday scenarios.

Relative sizes: For step (b), we can use the results from section 36.3. Recall that we have necessary and sufficient conditions for determining if an object is in front or behind another object using just vision (see Theorem 36.1 of section 36.2). Let us once again assume continuity of arrangement of objects and pick an appropriate neighborhood as explained above. When we pick two objects (generalized as any recognizable pattern within tangible objects, as discussed above), we can know which object is in front and which one is behind by detecting whether our focal length of our eyes increase or decrease as we switch our attention from one object to another (Theorem 36.1 of section 36.2).

Let us pick a sufficiently small neighborhood such that the distance (along the depth) between two distinct recognizable patterns is as small as our visual resolution. Now, the relative locations of all these patterns within this neighborhood are uniquely constrained using Theorem 36.1 of section 36.2, unlike with occlusion. For example, if you have a pen and an eraser on a table adjacent to each other, you can still be sure which is in front or behind even if neither of them occludes the other by knowing whether the focal length increased or decreased as you switch your attention from one to the other. Using the same Theorem 36.1, we know that we should fill the distance between two such patterns using, say, empty space.

The general case of determining relative locations for all objects in a given neighborhood is shown in Figure 36.7. For example, two components A and B are separated both in the width and depth directions. Yet, we are able to determine the relative locations using Theorem 36.1 of section 36.2. We can determine this similarly for any pair of patterns within this neighborhood. As the patterns are closely located within this neighborhood (from the assumption about the continuity of arrangement of objects), we can conclude that we have a unique configuration of all the relative locations of all of these objects. In other words, if we know the true size and distance of one object within this neighborhood, then all other objects' distance relative to this object can be uniquely determined along with their sizes. Note that we can view size as a measure of distance from one end of the object to the other end.

Knowing size and distance: For step (d), we need a way to know size and distance along all three dimensions. When we look through the window to see a house across the road and behind a tree, we do have continuity in the arrangement of these objects. When we walk closer to the house, the 'feeling' we have about the size of the house does not change even though the image on the retina is larger. Furthermore, when we use the above iterative procedure, how does knowing the width of the road help us estimate both the size and distance of the house?

Since the notion of size and distance is perceived qualitatively, not quantitatively, as mentioned in section 34.8, we have the memory of a number of familiar objects to assist us. These memories are not limited to just distinct material objects, but to a generalized notion mentioned earlier of all recognizable patterns within tangible objects as well. For example, these could be familiar edges, colors, angles, textures, simple primitive shapes, shadows, different types of illumination and others.

From the discussion above on relative sizes, we have seen that we can uniquely determine which pattern is in front and which one is behind. Similarly, we can determine which objects are on top or bottom and which ones are left or right. For these two cases, we know that the focal length does not change because these patterns are at about the same distance from our eyes. However, we rely on our ability to move our eyes up, down, left or right. Analogous to the ciliary muscles of our eye lens for changing the focal length, we have six extraocular muscles (superior, inferior, lateral and medial rectus, superior and inferior oblique) that control the movement of our eye. Using the fixed sets from the control of these muscles, we can create links between vision and touch once again (similar to the case with ciliary muscles). For example, we can link the direction of motion of our hand to touch an object with the direction of motion of our eyes to look at the object that we touch (see section 33.5). Therefore, from these links, by turning our eyes to the left (or right), we know that the corresponding object is to the left (or right). The same applies with top and bottom objects. Stratton's ([146] and [147]) experiment discussed in section 33.3 for creating upright image on the retina shows that our brain can alter these links and re-associate them to align with touch.

Therefore, we can say that we have uniquely determined the relative locations for all objects in all three dimensions. The images of nearby objects linked with touch lets us have a feeling of knowing of true size and distance for these objects.

Error correction: As we extend iteratively using overlapping neighborhoods to distant objects, how can we ensure that our estimations do not introduce errors? The more distant the objects are, the more these errors can accumulate, if left uncorrected. To address these errors completely, we need multiple pathways to a given object, and other mechanisms for visual cues beyond what we have discussed until now. I will discuss several of these in chapter 37.

Estimation using unequal measuring rods: Let me discuss one such mechanism for error correction here. The idea is to simply imagine arranging, connecting and counting familiar generalized patterns (those that trigger appropriate fixed sets), mentioned earlier with no gaps in between. These familiar patterns are taken just from objects present in your visual field. You already have a feeling of knowing of the true size and distance of these familiar patterns through your past experiences using, say, touch (like the tables, chairs and other common objects in your room). When you look through your window to a nearby house, your estimate of distance is simply how many familiar objects are present between you and the distant house. It is just a feeling of distance, not necessarily expressed in numbers. For example, we may express this feeling of distance as, say, that there is a road, a walkway, a lawn and finally your room. Some of these intermediate objects can be completely new that you have never seen before. However, they are still composed of primitive patterns (like angles, edges, primitive shapes and others) that are already familiar to you (i.e., their sizes). You can also pick these familiar objects in multiple ways. This gives us different ways to compute the same cumulative size and distances thereby letting us control the accumulation of errors (cf. Figure 36.8 with overlapping neighborhoods).

This simple view of arrangement of familiar patterns can be thought of as a way of estimating distances using unequal measuring rods. Each measuring rod is the familiar pattern. We do not need to actually count using the concept of numbers. Instead, we can think of counting as if we are just looking at connected patterns one after the other in a serial order without any gaps. Our ability to process information serially in spite of a parallel architecture of our brain is already well established.

If you want to know the distance between two objects along the width direction, then we imagine connecting familiar patterns (without gaps) between the source and destination using these unequal sized familiar objects. There are multiple ways to chain these familiar patterns, say, by picking different groups of familiar patterns. The same applies with height and depth direction as well.

If we want to know the size of a single object placed horizontally, vertically or along the depth, then we can imagine composing this object into several other familiar patterns connected linearly (without gaps) from one end to the other. This arrangement gives us a feeling of knowing of the size of the entire object from the size of these smaller patterns. Once again, there are multiple ways to chain familiar patterns.

If you want to estimate the distance or size using numbers, all you need to do is to replace these unequal sized objects with an estimate of their size and adding them together. The error in your estimate of total distance can be minimized if you use multiple familiar patterns.

Quick recomputation due to changes from motion: If we move around as we observe the objects, the arrangement of familiar patterns and the iterative solution is not disturbed significantly. With just a minimal change, we can maintain an accurate perception of size and distances with relative ease. The fact that we only need to update the feeling of knowing of size and distances makes this problem easier than if it were to approach this as a computational problem.

Our brain's stable parallel looped architecture makes it easy and natural to pick a ruler with unequal sized units (like walkway, lawn and road) instead of equal sized units with inches or meters when we want to estimate sizes and distances (like how it is done in a computer). This provides an alternate way to view the three-dimensional external world. Let me briefly discuss this next.

36.8 Dividing space into unequal and dynamical grid

The iterative procedure described in the previous section lends itself to an interesting visual representation of the 3-D world, one that involves unequal and time-varying measuring rods. This representation fits naturally within the brain architecture of a conscious being (as well as for a few other animals). However, we should think of this representation only as a conceptual one.

When we look at objects around us, we quickly recognize most of the objects at a coarse level (i.e., as foreground objects). Using the iterative procedure of the previous section (and visual cues that will be discussed in chapter 37), we have a quick sense of sizes and distances of all objects as well. As we move around, we start to look at some of the objects at a more detailed level. If we try to represent this 3-D world graphically, say, in a computer, we imagine creating a grid with, say, 1 cm grid lines along all three axes X, Y and Z. Any given objects' (x, y, z) location can now be measured from the origin O using these grid lines. Its size can be computed as well using these grid lines. This spatial representation (i.e., the grid) is static that does not seem to care about either the dynamics of the objects or the existence of a conscious being. We refer this as the Newtonian representation of absolute space.

We are not interested whether this representation is right or wrong, although Einstein's theory seems to suggest otherwise i.e., that space and time form a curved surface under the influence of gravity. Instead, we are only interested in presenting a simple conceptual representation that is unique and more natural from the perspective of a conscious being. Imagine drawing grid lines in the 3-D world for each object we look around us. Each grid line we draw corresponds to a recognizable feature or pattern of an object. Initially all of the grid lines correspond to coarse features with unequal spacing between them and quite sparsely located. The unequal spacing is because of the non-uniform nature of different sized objects or patterns that exist in front of us. This is unlike the previous equal spaced grid representation. Empty spaces do not have any grid lines because we recognize nothing in these regions.

As we start focusing on some details, say, the different parts of a car, we can imagine drawing finer 3-D grid lines within these specific regions. We continue drawing them dynamically over a period of time as we keep scanning and focusing on different objects around us. At the same time, some of the finer grid lines disappear as well. This happens as we start to forget the details i.e., when the corresponding fixed sets no longer stays excited as we switch our attention away from an object for extended periods.

The purpose of this representation is to highlight what we remember and recognize at any given time. For example, when we close our eyes briefly, we quickly realize how little we remember about the objects in front of us. We would not be able to redraw the visual representation to the same degree as a computer would. Instead, we only could draw at a coarse level for most objects and finer level for a few objects. The coarse grid lines that correspond to overall features of objects around us are continually excited simply from casual turning of eyes and head as well as from the parallel inputs from the peripheral regions of our retina.

If you did not fully forget the details (say, when you turned your attention away just for a short while), then it is much quicker to re-create the same dense grid lines as before. This is because the excitations of neurons for 'every' path of neurons that are part of the fixed sets for these recognized objects do not saturate. While you may not have a complete path of neurons in an excited state for each fixed set, you will, at least, have several subregions (or loops) along the paths that do stay excited, causing you to quickly lock onto the same past view. In addition, as the objects move around, the grid lines also move with them, leaving with no grid lines in the previous empty region.

This simple conceptual representation brings out an inherent limitation with serializing a large amount of parallel information that reaches our brain. The above representation is natural because of this limitation imposed by the mechanics of consciousness. This representation is particularly useful to understand how an infant views the world around himself as he begins to form more fixed sets. His representation is considerably coarser because he forgets most of the objects as he turns to look at different directions.

36.9 Break in continuity

In section 36.7, we saw that we needed two features to determine size and distance accurately using an iterative procedure – sensing change in focal length of our eyes and continuity of arrangement of objects from your eyes to objects around you. In this section, we will consider what happens when the continuity of arrangement of objects is broken. We will see that this produces noticeable errors in our estimation of size and distance. This suggests the importance of continuity of arrangement of objects for our visual perception. However, we need to keep in mind that since multiple other visual cues do aid in our estimation (to be discussed in chapter 37), the error is considerably compensated though not completely.

One simple example is with very distant objects like, say, an airplane in the sky, a distant island in the middle of an ocean and stars or clouds in the sky. In all of these cases, our perception changes from a 3-D to a 2-D view. We do not perceive the depth as clearly anymore. In general, this is the case whenever we start viewing objects that are farther and farther away. Over large distances, typically, objects are sparsely located in all three directions. The resolution of our eyes is not sufficiently high to capture enough detail. Furthermore, the change in focal length needed to refocus a sharp image on your retina for very large distances becomes smaller and smaller to detect. Therefore, it becomes increasingly difficult to use Theorem 36.1 of section 36.2. Since we lack a continuous path of overlapping neighborhood of objects (i.e., without empty space along the pathway as needed by the iterative procedure), we cannot uniquely determine the relative sizes and distances for these very distant objects.

Another example that is more pronounced in an infant than an adult is watching, say, a bird on a distant telephone wire. At this age, he does not yet detect the direction of change in focal length quickly enough as he scans objects from close to a distant location. Recognizing patterns as well as switching attention between objects, is already a time consuming process for him, unlike an adult. The bird on a wire also has very few paths (maybe just a single one) with overlapping neighborhood of nonempty-space objects. Picking this path and tracing along this path to constrain the sizes and distances uniquely is a nontrivial task for him. This is sometimes true even for closer objects. As a result, we see him attempting to reach for the bird or even a slightly distant toy thinking that it is close enough.

There are several other everyday scenarios where we do see a break in continuity. However, our memory compensates for this error quite easily in most situations. For example, when you look through a window, there is no visible continuity between the inside and the outside objects. However, in these cases, there is only a single discontinuity, i.e., at the window. Since the discontinuity is happening at close enough distance from us, we can estimate the jump in discontinuity using either our memory (through familiarity) or other visual cues (like subtle differences from two eyes, angles and shadows). For example, we have a good estimate of distance between the window and the first closest visible object on the outside, say, the lawn. Once we have this, the rest of the estimates are relatively accurate.

We can use the above example to make the error in estimation artificially more pronounced. Take two papers and roll them cylindrically. Pretend them to be binoculars, but without any lenses. Look through this contraption at objects around you. You can simulate the same scenario by curling your fingers to make the shape of an 'O' and placing your hands in front of your eyes, simulating a binocular. Comparing with the previous example, this setup is analogous to looking through a window, namely the two holes. You have narrowed down your field of view considerably.

What we have achieved here with this setup is that the closest object from your hand (or these cylindrical paper rolls) can be made as far as you want by picking which objects to look at. The continuity of arrangement of objects is clearly broken. In fact, the only nearby objects are just your hand or the paper holes. They do not help in providing an estimate of size and distance of objects visible through the holes. Therefore, your estimates do have errors. The objects will appear closer like with a binoculars even though you do not have any lenses. These errors become more pronounced as you walk towards the object that you are looking at (like a clock or a photo frame on the wall) while still using this contraption. You feel that you are getting closer and closer to the object. However, when you remove your hands and uncover your eyes, you will be surprised that the object is not as close as you thought. Your motion towards the object clearly amplifies the error more noticeably compared to when you are sitting at one place with this setup. The re-computation continues to build the error as you move towards the wall.

Looking for examples where we notice errors along the width and height directions is a bit more subtle (compared to the depth direction illustrated by the previous examples). What we are expecting is to estimate, say, the width of an object, as one value, but it should turn out to be much smaller or larger when measured more carefully. The subtlety comes from the fact that the error in width is considerably minimized the moment we determine the depth accurately. Therefore, in order to have a noticeable error in width, it is better to have an error in depth as well. The above list of examples also produces a slight error in width and height as well.

These examples validate that the iterative solution presented in this section is a central mechanism for estimating size and distance of all objects. In the next chapter, I will discuss the other mechanisms that play an additional role as well.

37

Visual Cues

In the previous chapter, I have described one central mechanism for estimating size and distance while acknowledging that a number of other parallel processes contribute as well. Each of these additional visual processes does rely on touch initially when the objects are closer. Our brain then links the event, experienced through touch, with vision. As a result, when a similar event occurs with distant objects that you observe with your eyes, the above links lets you attain a similar feeling of knowing. We can call this as our brains ability to generalize or create rules from simple everyday examples.

Creating rules or generalizing from everyday examples is quite common and happens even unconsciously. For example, if two similar objects appear different in size, then the smaller object appears to be located further away. We generalize this from everyday examples with nearby objects to distant objects. The stable parallel looped architecture of our brain helps us with such generalizations. We should keep in mind that when it comes to generalizations and rules, there would usually be exceptions and applicability in a limited sense.

In chapter 36, I have already discussed two important visual cues, namely, occlusion and continuity of arrangement of objects. In this chapter, I will discuss additional such examples. Most of the examples discussed below are rather well known (Kandel et. al [82] mentions such a list as well). In fact, artists have been using such visual cues for a long time, to create realistic paintings.

37.1 Familiar sizes

As we interact with objects around us, we develop a feeling of knowing for the sizes of a large collection of objects. For example, most of the objects in your house and even outside do not vary in size significantly, even though there are a number of variations in, say, colors and patterns. As we form fixed sets for each of these objects, they help in quickly estimating distances using the iterative procedure of section 36.7. When you look at a photograph of, say, your room, you have a clear sense of feeling for sizes and distances between objects even though the picture is two-dimensional and small.

One particularly important familiar object is our own body. We form abstract continuity using our own body sizes as reference measurements. For example, we know the distance between our fingers, our hands, the distance when we stretch our hands or legs and how long it takes to crawl, walk or run a few feet (see section 34.7). Here, our own height plays a critical role towards the perception of sizes and distances. If you wear very high heels, walk on your tippy toes or kneel down to the height of, say, a toddler and walk around for a while, you will notice that your size and distance perception already appear considerably altered. Similarly, recall the discussion in section 33.3 about revisiting your childhood town that you have not visited for several years. It will feel as if all objects and distances are much smaller compared to your memory. This is because you had created abstract continuity using your own body as the measuring unit for size and distance, when you were a child. These specific memory links were never updated as you were growing up since you did not revisit the place.

37.2 Detecting empty space between objects

We detect empty space between objects whether we 'see' the empty space or not. For example, if a table partially occludes a chair and you see no other objects in between, you know that there is empty space between the table and the chair because you were able to see the empty space. More generally, it is quite common to see empty space in the width and height directions between objects. On the other hand, if an object is right behind another one in the depth direction, like say, a tree you see through your window, then even though your wall blocks the tree and you cannot 'see' any empty space in between, you still know that there is empty space. I had already discussed how we sense three-dimensional space in chapter 35. It turns out that sensing empty space greatly helps us when we want to estimate size and distance between objects.

37.3 Depth perspective from similar sized objects

If two familiar objects of similar size appear different in size, then the smaller sized object appears farther away. This is commonly used in two-dimensional paintings to create a perception of depth (see Boroditsky [16]). This is formed unconsciously simply from everyday experiences via the links between the corresponding fixed sets and the abstract continuous paths.

37.4 Depth perspective from angles or parallel lines

When we look at two parallel lines, they appear to converge at a distance, like with railway tracks. Similarly, angles between edges of different objects or within the same object provide depth information. When parallel edges of a given object appear to converge, you feel that the object is receding away or is tilting at an angle. If the parallel edges of different objects converge or form various types of angles, this provides information about how the two objects are located relative to each other as well. This is again commonly used in two-dimensional paintings to create a perception of depth.

37.5 Brightness and shadows

If objects have brighter shades of color or are under proper lighting, they appear to be closer than the objects that are dimmer. With paintings, we see the usage of brighter, dimmer and darker colors or shadows more often, to create distances between objects or depth that is more realistic. Creating the necessary fixed sets and the corresponding links between features like brightness and shades is simple using repetitive events from everyday experiences.

37.6 Change in focal length as function of relative distance

We have already seen that the focal length of our eye lens increases as we switch our focus from a closer to a farther object. Consider two pairs of objects both separated by, say, 2 feet from each other. Imagine one pair P_1 to be placed about 5 feet from you and the other pair P_2 about 50 feet from you. When you switch your attention between the objects within each pair, the amount of change in focal length within P_1 is larger than within P_2, even though the distance between the objects is the same. Therefore, the change in focal length as a function of relative distance between objects is nonlinear.

Our brain can memorize this nonlinear relationship, at least, partially for close and for far distances. What I mean by memory here is that if you change your focal length by a given amount at a given distance, you have links that trigger to create a feeling of knowing of the distance of the object. It appears unlikely that you have links for all ranges of distances and all ranges of change in focal length. Yet, it is conceivable that we form such links for distances that you are used to every single day like, say, about 10 feet from you. Similarly, distances that are very far away, say, beyond a few 100 feet, have minor variations in focal length. It seems reasonable to assume that we will create such links for these distances as well. Therefore, these links, where applicable, can aid the iterative procedure described in section 36.7 when estimating the sizes and distances between objects.

37.7 Using both eyes

When we let both eyes fixate on a given object, our retinas generate subtle differences in the two inputs. The additional information, compared to the input from just a single eye, provides us useful cues in estimating sizes and distances. For example, we can identify hidden objects more precisely. This is particularly noticeable at the boundaries of objects. As a result, the problem of occlusion, important for depth estimation, becomes easier. I will discuss more details when I consider the problem of how we see the world in three dimensions in chapter 38.

37.8 Movement while fixating on an object

Our own motion while we fixate with both eyes on a given object provides useful relative depth information. The objects in front of your fixation point moves in the opposite direction relative to the objects behind the fixation point. This is a rule that can be memorized from everyday experiences. Relative to your fixation point, small variations in the inputs from both eyes creates the impression of near or far.

These are some examples of the visual cues that do help us with the problem of estimating size and distance, in addition to the iterative mechanism described in chapter 36. There may be other additional cues not discussed here that plays a similar role. Optical illusions provide a good source of examples to help us discover more such visual cues. In addition to the problem of estimating size and distance, each of these factors shapes our entire visual perception in a fundamental way. I will discuss this in more detail in chapter 41. For now, let me look at how we see the world in three dimensions.

38
Seeing in Three Dimensions

In this chapter, I want to discuss the difference in perception between seeing the external world in three dimensions using one eye versus with both eyes. Of the three dimensions, it is in the depth direction we notice a significant difference in our perception and not as much in width or height (XY) directions. In this chapter, I am assuming that we already have, at least, one visual perception based on the inputs from one or both eyes. In the previous chapters, we have seen how we were able to guarantee such a representation from the primitive geometric features. For example, we established the existence and a feeling of knowing for these features initially using touch. Then, using retinal inputs, we created a visual representation that is consistent with touch for each of these features for nearby objects. We were then able to extend the representation to distant objects through generalization (via the links between vision and touch) and free will. At each stage, we had a feeling of knowing using the membrane and the abstract continuous pathways. This approach works even with a synthetic system and for some animals because we built each of these features incrementally bottom-up using only properties of dynamical systems and not on any abstractions. Using this representation, I want to analyze how the depth perception is quite distinct.

When we look at a nearby object, say, the fingers of our hand with both eyes, we notice subtle variations in the texture of our skin surface. Our skin surface does not appear smooth, by any means. The bumps, dents and the variations in the surface curvature are clearly visible. At the boundaries of our fingers, we even see the skin surface curve around in the depth direction, creating a distinct feeling of a three-dimensional surface.

If we look at the same nearby object like the fingers of our hand with one eye closed, the above feeling of a 3-D surface, both in the interior and at the boundary, is somewhat diminished. If we now look at a distant object, say, a table or a chair, with just one eye, the 3-D feeling is further diminished. With both eyes, on the other hand, our distinct 3-D feeling does not diminish as much and feels about the same. In fact, in general, the farther the objects, the more they appear and feel two-dimensional when seen with one eye versus when seen with both eyes. In this chapter, I want to discuss this distinction, how it arises and the mechanisms behind it.

There are two parts to understanding and attaining a 3-D feeling of our external world. These stem from (a) the physical objects and (b) the empty space. I will discuss how the properties of each of them and the relationship between them effect our perception.

38.1 Consistency with two eyes and feeling of empty space

Before I discuss how and what the consistent view of the external world is when we

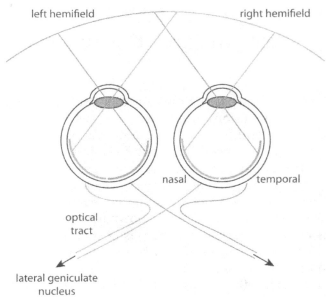

Figure 38.1: Visual field and nasal/temporal pathways. The left hemifield falls on the left nasal and right temporal side of the eyes while the right hemifield falls on the left temporal and right nasal side of the eyes. The hemifield information is transmitted to the lateral geniculate nucleus.

use inputs from both eyes, let me briefly discuss what the individual pathways are from both of our eyes into our brain. I will refer the reader to textbooks like Purves [122] and Kandel et. al [82] for a more complete discussion. The optic nerve carries the entire visual information from the retina. Both the optic nerves, one from each eye, cross at the optic chiasm to terminate in the lateral geniculate nucleus. Specifically, the nerves from the right temporal (near the temple – away from the nose) and left nasal (near the nose) visual field reach the left half of the brain and vice versa (see Figure 38.1). The same set of external objects generate inputs both from right temporal and left nasal fields. They have subtle differences because of the angles at which light rays fall on the retina from the external object. From the lateral geniculate nucleus, the axons carry the firing pattern to the visual cortex along two paths (corresponding to the contralateral superior and inferior fields).

As the inputs from both eyes combine at the visual cortex, it creates alternate bands, one from each eye, at an early childhood. During this age, as our brain is creating these alternate bands, if you close one eye for an extended period, you permanently lose the ability to use the input from the closed eye, even though there is nothing physically wrong with the eye. The perfectly accurate visual information received through this eye is suppressed by deeper parts of your brain making this, otherwise normal eye, blind (see Ornstein and Thompson [118]). Other abnormalities are also possible from defects at each stage along these visual pathways.

Let us assume that there are no such abnormalities for now. Consider curling your fingers as if you were holding a ball. Position your hand such that the thumb is closest to your eyes, partially blocking the remaining four fingers. I want to show how we get a distinct perception of depth when we consider the inputs from both eyes. We have already seen that if we use touch as well, we need to introduce the notion of depth. However, I want to show that depth perception is needed even if we do not use touch, provided we use inputs from both eyes simultaneously.

If you alternate closing the left eye and the right eye while looking at your fingers, you clearly see your thumb displaced relative to the other fingers. For example, your thumb appears to block more of your other fingers when you close your left eye than when you close your right eye. What happens when you open both of your eyes? How should the visual perception look like? Let us call region O_L be the occluded region by your thumb when you open left eye but close the right eye. Similarly, let O_R be the occluded region when you open right eye but close the right eye. Then, $O_L \cap O_R$ is the region occluded when you open both eyes. $O_L - O_R$ is the additional region seen by right eye when left eye is closed (cf. a similar region $O_R - O_L$).

If you did not yet form the concept of empty space in your brain (as discussed in section 35.3), you have a seeming inconsistency when you consider inputs from both eyes simultaneously. Even though the right eye says that your thumb is occluding your other fingers, the left eye is able to provide detailed visual information about 'part' of the hidden fingers (region $O_R - O_L$). The same applies in the opposite situation as well. Here, I am assuming that we are only relying on visual information.

You can resolve this inconsistency provided you define the system as one that includes inputs from both eyes. The resolved solution is one in which you create a new visual perception that places your thumb 'in front' of your other fingers. In other words, you have created a notion of depth and empty space. Since you have the notion of empty space, you can use it to fill in the gap between your thumb (the occluding object) and the other fingers (the occluded objects) along the depth direction. You can verify empty space and depth using touch for nearby objects. We can generalize it to all distances using our free will and the links between vision and touch.

If, on the other hand, you use inputs from just one eye, there is no apparent inconsistency to be resolved (especially if you do not rely on touch). You do not have additional visual information (like region $O_R - O_L$) to contradict occlusion information from your eye. Yet, even with one eye, we do sense depth between our thumb and the other fingers. The effect may not be as pronounced as with two eyes. Yet, we can notice the depth with one eye because we can detect the change in focal length as we switch our attention from our thumb to the other fingers. We discussed this mechanism in detail in sections 16.1 and 36.2. We relied on links between touch inputs and the change in focal length of our eyes to achieve a depth perception with just a single eye.

In addition, the presence of subtle surface texture variations combined with different shadows can aid in the depth perception even with just one eye. I will discuss this in some detail in the next section. Let me now look at another aspect that contributes towards the 3-D feeling, namely, the physical objects themselves.

38.2 Small versus large variations between objects

When we look at nearby physical objects closely, we see detailed surface variations, like the tiny bumps, dents, roughness and surface curvatures. All of these details do add up to give a unique texture to the surface. If we want to detect the depth of the dents and bumps, the mechanism described in section 36.2 that relies on detecting the change in focal length is not as useful anymore. The height of these bumps are so small that there is hardly any noticeable change in the focal length of your eyes as you switch your focus from the top to the bottom of the bump, especially if the object is a few feet away from you.

Let us call the relative distance in the depth direction between A and a neighboring object B as Δx. When Δx is large, the mechanism of section 36.2 can be used. However, when Δx is small, we need an alternate approach. This is where the inputs from both eyes play a critical role. Each of our eyes sends a slightly displaced view of the external world to our brain. Our eyes are separated by about 50-70 mm, for most adults. When we learn to converge both of our eyes onto the same object using the fundamental pattern (see section 17.2), we will be able to match the inputs from both eyes as closely as possible.

The matching of the inputs from both eyes itself uses the fundamental pattern. The explosion needed to satisfy assumption 16.3.4 of the fundamental pattern (see section 17.2) comes from the similarity of inputs from both eyes. A state when the difference in inputs from both eyes are large is not preferred to a state when the difference in inputs is smaller. The regions R_1 and R_2 (of section 16.3) naturally converge the state of the system to create an 'equilibrium' state in which both eyes' inputs are as closely aligned as possible. This error cannot be zero because the inputs from each eye are not identical.

The network of interconnections that creates the explosion is sufficiently static i.e., it does not need to change as you grow older. Once this network is created for the first time at a young age, the changes to it are minor because the distance between your eyes changes very gradually with age. Any object you see with both eyes sends similar inputs from the same regions of the retina as long as the convergence of both eyes onto the same object works correctly. It has been observed that the region in the visual cortex where the inputs from both eyes converge forms alternate bands corresponding to each eye at a young age (see Ornstein and Thompson [118]). The underlying mechanism for the formation of these alternate bands is the fundamental pattern.

When Δx is small and the object is nearby, the subtle surface variations are significantly amplified as you focus both of your eyes on a small region. The smaller the surface variations are (i.e., small Δx), the more prominent are the differences between the inputs from both eyes. If you are looking at the surface of your palm, which has tiny lines or ridges, the input from the left eye has more information from one side of the ridge while the input from the right eye has more information from the opposite side. For smaller Δx and nearby objects, this is the main cause of a distinct 3-D feeling.

On the other hand, when the object is far, these tiny variations are not noticeable.

In such cases, the variation between the two eyes is prominent only at the boundaries rather than the interior of the object. A sharper change in curvature at the boundary of an object produces a noticeable difference in inputs from each eye. This variation becomes even more prominent if the boundary is adjacent to a distant object where you also need to adjust the focal length to see what is right next to it. This is like looking at your hand against distant background objects where the boundary between your hand and the background is so large that you have to refocus your eye lens when you switch your attention from your hand to the background objects. For larger Δx and distant objects, this is the main cause of a distinct 3-D feeling.

In both cases, touch can help verify if the visual representation is consistent or not. Since we can already sense depth with touch, as we rub our fingers on a rough surface to feel the texture, we expect bumps and dents along the depth direction. The visual representation should create appropriate predictive meta-fixed sets as well. Verification with touch is accurate only with sizes and distances comparable to our finger sizes, say. Once we link these fixed sets with those generated with vision, we can generalize, as before, for small and large distances as well. Free will can guarantee that there is no inconsistency between vision and touch.

Yet another physical phenomena that helps with the 3-D feeling of a surface texture when Δx is small has to do with how light is reflected from each of these tiny bumps, dents and rough surfaces creating a variation of bright and dark regions or shadows. These shades of lighting at, say, your skin surface produce subtle variations on the left and right eyes making it possible to highlight the depth perception more clearly. Artists even use this fact, with shades of colors or gray shadows in paintings, to create a three dimensional effect of a 2-D image.

Now, given the repeatability of the 3-D patterns from our everyday objects, it is conceivable that our brain creates fixed sets for a number of common primitive 3-D shapes as well. The 3-D perception of a complex shape can then be composed from these primitive 3-D shapes. These memories are analogous to primitive 2-D edges, angles and shapes that we discussed earlier. In fact, Yamane et. al [176] have found evidence for such an explicit neural representation for complex three-dimensional primitive object shapes.

We have thus seen that the two prominent mechanisms, namely, the iterative approach of section 36.7 and using both eyes creates a distinct 3-D feeling of objects around us. Our 3-D perception, though more prominent with both eyes, is possible even with a single eye. In some cases, the iterative approach, relying on the change in focal length, may not help directly with a 3-D feeling. However, for most everyday scenarios, it does give a partial 3-D feeling, at least, for nearby objects.

There are other class of examples like autostereograms or random-dot stereograms that are specifically designed to produce a 3-D feeling only when we use both eyes. The input that falls on each eye are intentionally random dots. Yet the combined effect results in a 3-D image popping out of this 2-D image. Random-dot stereograms are created to debunk the well-believed hypothesis that each eye needs to have a 'valid' image but slightly displaced in order to create a 3-D perception of an image. In random-dot stereograms, no eye has a 'valid' image of a physical object (they are just random dots) and yet when viewed with both eyes, we have a valid 3-D image.

Let me now switch to the problem of sensing the passage of time. This topic, as it will turn out, requires higher layers of abstraction (like a language) and abstract continuity within these layers. We know that language uniquely distinguishes humans from other animals. Let me study this topic next with a goal of understanding the set of problems that the existence of a language addresses.

39
Language

Until now, we focused primarily on how to attain a feeling of knowing for abstractions and for physical events. This required creating and triggering abstract continuous paths on the membrane for the given event. We saw that we could arrange the information as layers. For example, the external physical dynamics provided inputs to our brain through our sense organs. These propagate within our brain to create fixed sets. This was the first layer of abstraction representing a specific external object or an event (like a specific table or a specific chair). The next step is to create meta-fixed sets using these fixed sets as inputs. This was the second layer of abstraction representing similar objects or events (like different tables and different chairs). We focused on both these layers for the most part until now. However, we can extend our analysis for other layers as well. For example, the third layer of abstraction contains meta-meta-fixed sets and represents an abstraction using the meta-fixed sets itself as inputs (like a notion of dining room that contains specific furniture).

I had discussed one example of meta-meta-fixed sets when discussing free will in chapter 28. This was about how we know the very concept of choices itself. In this example, the fixed sets represent the objects directly involved in the choice-based events. The meta-fixed sets are the actual set of available choices in a given situation. The meta-meta-fixed set is the very notion of choice that is common across several scenarios involving several choices.

We can conceptually continue to higher layers of abstraction. In this chapter, I will discuss how to get a feeling of knowing for these third and higher layers of abstraction. Theorem 22.1 of section 22.1, requiring abstract continuous paths on a membrane would still be applicable for these higher layers as well. I will discuss the difficulties in creating abstract continuity within these higher layers first. Then, I will show how language is necessary to overcome them.

39.1 Too much data at the lower layers

Let me first highlight the difficulties with attaining a feeling of knowing for concepts that belong to the higher layers of abstraction. Let me use the notion of choice as an example of meta-meta-fixed set to illustrate the structure of these collection of fixed sets. You will note that the discussion can be generalized to other examples as well.

Before we create a notion of choice, we first need a set of dynamical events for which multiple choices are possible. Each of these dynamical events involves a number of different objects. Let us call this as the 'raw data' needed to create a notion of choice. Our brain receives raw data for each dynamical event through our sensory organs like eyes, ears and touch. For our actions, the raw data comes from motor neurons. Visual, auditory and touch inputs from the external world constantly feed into our brain as we observe each event in the form of cascading neural firing patterns. The amount of information received by our brain for each event is enormous. The sensitivity of each of these sensory organs and motor control regions is quite high. On one hand, we can view this as an advantage, but on the other hand, the amount of data is too large to process that it appears as a disadvantage. Let us first look at some of the advantages.

(a) The loops can stay excited for a long time because of the similarity of the data across most dynamical events. Simple actions like turning and moving can let you access the same data repeatedly.

(b) The abundant availability of raw data does not require any extra effort from us. For basic recognition of fixed sets, we need a constant source of raw data. This is much easier because the same primitive geometric and physical features like edges, angles, shapes, textures, motion, forces and mass are present almost everywhere. This makes it easier to create fixed and meta-fixed sets.

(c) We also have many variations to create predictions and abstract continuity. The entire context of the event and the scenarios are quite different each time. As a result, we have multiple ways to reach the same fixed or meta-fixed set. This makes the prediction problem much easier.

(d) We can create a dense network of interconnected fixed sets. Since the availability of the raw data is enormous and most of the primitive features are shared across all the subsystems, our brain naturally creates a dense network. This is important for attaining a feeling of knowing.

(e) The above-mentioned dense network makes it easy to create and self-sustain the dynamical membrane of meta-fixed sets as well.

Let us now list a few disadvantages of the information overload problem. This overload is due to large amount of data received from multiple sensory and control regions of our brain.

(a) Processing too much data makes predictions harder, at least, in the beginning. Once we form structural memories within our brain network through physical interconnections between neurons, this becomes easier.

(b) We need a way to serialize the parallel cascading firing of neurons. Loops (or fixed sets) and the interconnections between them as abstract continuous paths make this easier as we have seen before.

(c) We need a way to filter the data and choose important versus unimportant data. The fundamental pattern and other mechanisms are examples that work at an abstract level in spite of detailed neural dynamics.

(d) A large amount of raw data keeps us busy processing this information and prevents us from looking at higher abstract layers. After processing this amount of raw data, we do not have much time left to analyze for or recognize meta-meta-fixed sets directly. This is particularly true with infants and children. They are quite satisfied participating within the dynamics of the event itself (like participating in the activity of playing) that they have little time left to ponder on the entire event or across events in an abstract sense similar to how adults do i.e., they do not have time to analyze 'about' the events. This is also true with other animals.

The presence of enormous raw data solves the problem of attaining a feeling of knowing for most *dynamical* facts i.e., events that involve only fixed sets and meta-fixed sets. The raw data and repetitions at a fixed and meta-fixed level create and self-sustain the dynamical membrane very easily. We can trigger the same physical and geometric features just by turning around in any direction. Every external object is a source of these properties. At the same time, the variations give rise to abstract continuity (or predictions) at every instant quite easily as well. There is an initial learning phase to store these variations within our network linking several of these fixed and meta-fixed sets. However, once we cross a threshold state, predictions are common. We see this difference clearly if we observe and compare an older child to an infant. Therefore, attaining a feeling of knowing for such events is an easy task.

39.2 Too little data at higher layers

If we now look at higher layers of abstraction, the amount of input data starts to drop significantly. For example, even at a meta-fixed layer, beyond the primitive features, we already have limited data. For example, tables and chairs as meta-fixed sets are already not as common as edges and angles. The repeatability of entire events becomes less common as well. If you want to know about choices, you are not subject to situations where you are making choices on a regular basis. The time interval between repeating events also widen considerably.

If the repetitive events drop significantly, your ability to predict drops as well. You are not forming predictive meta-fixed sets at these higher layers of abstraction. The network of fixed sets and their interconnections at these higher layers is not dense as with lower layers. Therefore, creating abstract continuity, and hence a feeling of knowing, at these layers is difficult with limited data and sparse network.

Another source of difficulty with creating abstract continuity at higher layers comes from the high amount of data from the lower layers, as mentioned earlier. The high volume of data from lower layers keeps the system busy with processing both the information and the corresponding actions that there is no 'free time' left to process data at higher layers. As a result, you cannot have predictions at every instant for these higher layers. The real physical event occurs before you had a chance to process data from the lower layers. Infants and young children do not see smaller subtle subpatterns in their daily or weekly routines. They are too busy doing their tasks i.e., interacting with the raw data from the lower layers. Besides, the disproportionate distribution of available inputs from lower layers makes them less likely to think abstractly or *about* events at higher layers. Instead, they simply work with the objects and dynamical systems directly at the lower layers.

The threshold membrane is too small at these higher layers of abstraction to get a feeling of knowing for topics at these layers. We need a way to expand this threshold self-sustaining dynamical membrane to make it dense even at these higher layers. I will show how we can achieve this with a language next.

39.3 Language as a solution – an expanded membrane

Our objective is to have a dense self-sustaining membrane and abstract continuous paths at the higher layers of abstraction just as we already have at lower layers of abstraction. This requirement is necessary and sufficient to attain a feeling of knowing (see Theorem 22.1 of section 22.1) for much more abstract concepts like the notion of time, choices, self or concepts in mathematics. We will see that language is necessary to solve this problem. For simpler abstractions and most dynamical events, the minimal threshold membrane discussed in the previous chapters is sufficient to achieve a feeling of knowing. We see this with children (say, less than three years old) who have not mastered a language and among some animals as well. Once we learn a language, the expanded membrane and a feeling of knowing of meta-meta-fixed sets (and beyond) uniquely distinguishes humans from other species.

From the previous sections, we needed to solve two problems here: (a) reducing the amount of raw data from lower layers and (b) increasing the amount of data at higher layers. The former problem requires some form of compression of the data. Language is precisely one such efficient compression. When we look at a table, the detailed visual representation involves exciting a large network of visual fixed sets. This captures several different features of the table. However, the language representation uses a word like 'table' and just a few additional words for components like legs, base and others. This is a severe compression of information representing the same object. Each individual feature has other words to describe them in a language. When the fixed set for the word 'table' is triggered, we do not even have to look at the table anymore to still think i.e., self-sustain the looped dynamics, about the table. Of course, we cannot directly avoid the raw data itself because of the sensitivity of our sensory organs. However, our brain forms a structural network so that we can quickly trigger the fixed sets for most common objects.

In this way, the sustained dynamics of the visual representation can die down as long as the word in the compressed language representation stays excited as fixed sets (or loops). We can even recall a long complex dynamical event using this compressed language representation by describing it using a short paragraph (like a gist). If you never had a language representation (like with a 1-2 year old or even with some animals), your ability to recall a long dynamical event is severely diminished. When the raw data disappears, there is no easy way to recollect considerable portion of the event without observing or encountering the same event once again. Self-exciting so much data or the corresponding fixed sets is unlikely. This manifests, say, in 2-3 year olds as though their memory and attention span is considerably lower. However, once you have a language representation, even though you may not recollect all visual details accurately, you can use the sentences to reconstruct a visual event mentally. This mental reconstruction is not accurate with respect to the visual details. You use partial images or even different instances of objects like a different table or a different chair than the one you actually saw (i.e., meta-fixed sets instead of specific fixed sets).

For example, a simple sentence like 'you picked up your son from school' is an enormous compression of a complex dynamical event. You can even imagine reconstructing a simple visual representation for such events with low fidelity. Once we create interconnections between the language and visual representations, it becomes easy to trigger one from the other, though not with great accuracy. A toddler creates these links when learning a language. Therefore, language lets you ignore the detailed visual representation and effectively reduce the information overload problem.

The next step is to see how language *increases* the amount of data at higher layers to help you form a dense network at the higher layers. This will let you achieve a feeling of knowing of dynamical events just from sentences instead of requiring detailed visual information. For this, language should be rich enough to express almost every dynamical and abstract event we observe in our everyday world. Most languages, as a result, have nouns, verbs, pronouns, adjectives, adverbs and so on to capture every external event in considerable detail (cf. people reading novels claiming that they enjoyed reading the novel much better than watching the movie, which was based on the novel). The grammatical rules we follow to create sentences are such that they let us have predictive meta-fixed sets at every instant (see section 21.1). This lets us attain a feeling of knowing of abstract sentences as well. I will not discuss many details on how to learn language for the first time, as we can understand this by looking at how a child grows. I will only discuss this as necessary, for the problem of consciousness.

If you notice how a child learns a language, it is clear that he starts with abstract continuity for events using initially the raw data from the detailed visual representation, not by directly memorizing words from a language. He slowly links words to the visual representation for nouns, verbs, adjectives and so on. During this phase, he begins to transition from a feeling of knowing from just visual data to a feeling of knowing with partial visual and partial language data. Eventually, he evens attains a feeling of knowing with purely language data as well once he is capable of having enough predictive meta-fixed sets at these higher layers of abstraction. This process does take a few years. It always works by linking language to the visual data and never by memorizing words without a visual representation. The only exception is with abstractions defined in terms of other abstractions like concepts in mathematics such as Banach spaces, Lebesgue measure and others (i.e., when creating meta-meta-fixed sets and beyond). There is never a detachment from the lower layers of raw data for abstractions representing dynamical events. In this way, our brain can move from one representation to the other in a seamless way.

Another way of saying this is that we have expanded the threshold membrane to include language representation as well. We now have successfully extended our ability to create a dense network at the meta-meta-fixed set layer as well (cf. a dense network already created using the raw data at a fixed set and meta-fixed set layers). This lets us achieve abstract continuity at both lower and higher layers of abstraction. We have much more free time available to predict complex dynamical events using the compressed language representation. We ignore vast amounts of data when we try to make predictions. The self-sustaining nature at a meta-meta-fixed set level lets us have quick and continuous internal thoughts in a language instead of having them in both language and raw data layers like it happens with a child. Even before a child becomes fluent in a language, he clearly knows what he wants whether we understand it or not (as he cannot communicate using a representation that we understand i.e., an abstract language). But he does know what he wants because he does have abstract continuity at the lower layers. We sometimes see his frustration, which translates to crying typically, when we cannot respond to his requests in a timely manner (like not giving him the toy he wants, not taking him to the place he wants to go to and others). The important point to note is that over the years of learning and growth he continues to maintain his original lower-layer abstract continuity as he transitions from a raw data to a language representation.

Another advantage is when you are not even observing any external event. You can still have internal thoughts that re-trigger the same physical events quickly and repeatedly using the language representation. The visual representation will be of low quality, however. You have an advantage when you re-trigger these thoughts repeatedly in a language. You no longer need to spend considerable amount of time parsing the raw data anymore. The compressed data is extremely small that you can have abstract continuity at a meta-meta-fixed set layer without being bombarded with data from the lower layers. You can make quick predictions within these higher layers as a result.

The mechanisms to create abstract continuity in language are same as the ones described in the previous chapters like the fundamental pattern and other mechanisms. In this way, we can improve the structure and network of data content at the higher layers using a language.

It is not always possible to express every event in a compressed language format. The best examples are with feelings, emotions and other topics related to personal experiences and opinions. To a given individual, he does have a feeling of knowing of such specific events as well. However, it is composed of abstract continuity at the lower layers and at the higher layers. Expressing such a complex mixed feeling of knowing (like emotions and others mentioned above) in just language at a higher layer is error prone because of the severe compression.

Consider, for example, asking why a grand master in Chess made a particular move or asking how someone got a 'creative' idea. The creative idea or a chess move is a combination of several complex and subtle ideas built through years of experience. They are not easy to express in a simple language representation. A grand master just 'feels' that it was a good move at that instant. The person with the creative idea (like the one who painted a picture or composed a song) similarly just 'feels' that music or the art. This is true even in more structured topics like with mathematics. A person who solves a mathematical problem using an ingenious idea just 'feels' the problem (in some cases).

When you ask what they felt and how they got the creative idea, the answers given in retrospect are at best a *logical* approximation using a language to a complex *emotional* feeling. These answers may or may not be accurate and can only be seen as a best way to rationalize a complex feeling. Just because they have expressed it in a logical way does not mean that you can follow a similar logic and solve a different problem using such creative ideas.

Therefore, language is not always the best form of communication, especially when we are dealing with complex emotional events like experiencing love or death (of a close relative or friend). It is a good representation for mostly logical events. One reason for this difficulty comes from the fact that we do not use abstract continuity at the lower layers (i.e., in the raw data layers) as an effective means of communication. Speech representing higher-layer data is much easier to communicate than using, say, facial expressions representing the lower-layer data. The interpretation of such facial expressions and feelings are ambiguous. We, therefore, end up constantly correcting our sentences in an effort to be more precise. To an external person, this causes more confusion.

Nevertheless, language solves both problems by compressing data at lower layers and expanding data at higher layers, making it possible to have abstract continuity at the higher layers as well. The new expanded self-sustaining membrane is unique to human beings while animals only have the minimal threshold membrane created from raw data.

39.4 Examples of knowing about concepts

In section 28.3, we considered one example of knowing at higher layers, namely, the notion of choice, which was necessary for free will. In the next chapter, I will discuss another example, namely, knowing the passage of time. Later, in section 43.6, I will discuss yet another example i.e., the notion of self-awareness. The approach for all of these cases is similar. We need abstract continuity for these topics. However, given that these topics are at the higher layers (i.e., at a meta-meta-fixed set layer and beyond), we need a dense network that allows us create predictive ability. Language, as we have seen, makes this problem quite easy to solve.

Let me list a few additional examples to highlight the fact that creating abstract continuity at these higher layers is now a common phenomenon in the presence of a rich language. For example, mathematics, learning a computer language, learning to design electronic circuits, repairing cars, solving engineering problems, reading sheet music for, say, piano, diagnosing medical problems and, in fact, experiences of all sorts are all quite similar. They all require building a dense network at the meta-meta-fixed set layer before you can achieve a feeling of knowing. Each of these examples can be considered to have a unique language of its own (or have a set of abstractions with rules analogous to grammar) different from the natural language like English. Yet, the purpose of these languages, the process to learn them and the way they help expand the dynamical membrane are similar.

For each of them, the initial network at the meta-meta-fixed set level is quite sparse. You build a dense network by solving several examples and from a lot of practice. You create links between the concepts and store them as fixed sets. This will let you have enough predictive meta-fixed sets at every instant even for a new situation and hence a feeling of knowing. High school education is one way to let you build such dense networks in most of the common areas of science, engineering, mathematics and others.

39.5 Differences with animals and synthetic systems

We can now clearly understand the unique difference between animals and humans. Animals are typically in a minimally conscious state with just the threshold membrane built from the lower layers and the raw sensory-control data. They can surely perform complex tasks. However, when it comes to knowing *about* abstractions at third and higher levels like self-awareness, knowing the passage of time, knowing about choices and others, it is increasingly difficult without a language. The raw data is overwhelming and prevents them to make predictions at higher layers without a language.

Just as with a 1-2 year old child, there is less free time available for them to build and trigger abstract continuous paths at the higher layers. As a result, they do not attain a feeling of knowing for these higher concepts. The expanded membrane with a sufficiently rich language makes this possible for human beings. This is the reason why there is a clear distinction between humans and animals with respect to our state of consciousness.

If we were to build a synthetic system with a human level of consciousness, we should solve the same problem of creating dense network at the meta-meta-fixed sets and higher layers using a rich language. This should be created from the raw data generated from dynamical events i.e., from the lower layers. Modern computers skip this step of creating the abstract language using the raw data from the dynamical events. They instead encode abstractions with almost no connection to the raw dynamical events. They work entirely at an abstract language level. This is surely a problem with existing computer designs if they want to solve the problem of consciousness.

The expanded dynamical membrane even for synthetic systems (with SPL designs) should follow the same sequence of steps described in this book to create higher levels of abstractions from the lower levels of raw dynamical data if they were to know concepts like passage of time, self-awareness and space. To be comparable to human consciousness, the first and foremost requirement is to capture reality of the physical world.

The higher concepts like time, self and space may seem purely as abstractions. However, they were only derived from physical reality. We cannot bypass this stage and create the abstractions directly (as we do in the current approach with computers). The purpose of a purely abstract language is merely to know and sense the existence of these abstractions present in the raw data. Yet, the interesting fact is that the existence of these abstractions tends to be lost within a colossal amount of raw physical data.

Abstract language is, therefore, a clever mechanism discovered by human beings to filter the raw data and search for patterns easily. Modern computers seem to be using abstract language instead in its own right, detaching it from the underlying dynamics. As mentioned in section 13.1, abstractions should always emerge from physical dynamics if we want to use them to become conscious. It is only after we become conscious and not before, we can generalize and study abstractions independently like how we do it now with, say, mathematics and other abstract concepts.

40
Sensing the Passage of Time

In this chapter, I will discuss how conscious humans are capable of sensing the passage of time. We will see that the discussion will be generic enough to apply them to synthetic SPL systems as well (to make them sense the passage of time). An important prerequisite is to know and understand the concept of time. Once we know the concept of time, feeling the passage of time becomes quite easy, as we will see. A computer does not *know* the concept of time. Rather, it is just generating a sequence of numbers at regular intervals. Since humans already sense the passage of time, we know that these numbers represent just the measurement of time. The regularity is maintained through periodic physical phenomena (like with a quartz crystal). A 2-3 year old does not know that a computer is measuring time by looking at these numbers, just as animals cannot know either.

The notion of time is an example of a higher layer of abstraction, as discussed in the previous chapter. We cannot detect time directly using our sense organs. We can only feel the passage of time indirectly. In this regard, we will rely on results from the previous chapter. Given this, the main challenge is to explain how a system is capable of knowing the concept of time, after which the problem of passage of time becomes easier.

The approach is as follows. We need to first identify a number of external objects in the world that are part of the first layer (see chapter 14 – this represents fixed sets). Next, we need to detect that the external objects remain the same even when we have variations called dynamics along a specific sequence. In other words, we do not express variations as different individual objects, but as the same object changing in a continuous way. This happens within a second layer of abstraction (again, see chapter 14 – these are meta-fixed sets). Finally, while the variations themselves are different for different dynamical events, the ability to track their continuous variations can be decoupled from the dynamics itself, which we represent it as time. This happens within a third layer of abstraction (i.e., meta-meta-fixed sets). These invariances are the new set of dynamical transformations needed in the definition of a fixed set. No other existing fixed sets capture this difference uniquely.

An alternate way to know time is through repetitions, especially since time is related to frequency (the inverse of time). It is clear that you can know about time if and only if you can know about repetitions. Otherwise, we can treat every object and event as an entirely new one that magically appears whenever and wherever, with no connection to the previously appeared object. Without repetitions, a seemingly valid representation is one in which we create entirely new objects at the new locations at every instant, however absurd it may sound. However, as we know, our brain represents externally sensed dynamics using fixed sets or loops. At first, if the same fixed set is triggered multiple times, we cannot distinguish if it is the same object or if it is an identical but different object. In chapter 29, we have seen that we can use our free will to distinguish these two cases with 100% certainty. Therefore, we can indeed eliminate the above alternate view of recreating a new object at every instant. In other words, the only correct way to describe the external world must use repetitions (i.e., we use the same objects whenever we have continuity).

The real issue then is to explain how we are capable of knowing repetitions and the passage of time. A machine performing the same task or event repeatedly does not sense repetitions and hence the passage of time, unlike a human. Yet, the problem of 'measuring and using' the concept of time seems easy. However, this is unrelated to the question of existence of time. For example, we have created atomic clocks that measures time intervals with high precision. We also have Einstein's theory of relativity that highlights the structure of spacetime and its relation to gravitation in a profound way. Such detailed structure and relationship is meaningful only after a system knows *about* the existence of time, not before. As an example, an unconscious being knows nothing about the passage of time nor does he know for how long he has stayed unconscious like when he is in deep sleep, sleepwalking, under general anesthesia and others.

Therefore, in this chapter, I will only focus on the problem of existence of time, not the structure of spacetime. I will also discuss how we perceive the flow of time in a given direction. To detect the flow of time, we should detect time transitions (from future to past) using predictions for a wide range of frequencies of repetitions, say, from milliseconds to yearly cycles. This will lead to the creation of a feeling of knowing for the concept of waiting, necessary to understand the flow of time.

It should be noted that there are multiple ways to create a feeling of knowing at the higher layers for the problem of passage of time. For example, we can use, say, repetitions, waiting, motion, dynamics and others. The goal with each of these approaches is still to build abstract continuity at the higher language layers. Therefore, I will only focus on one approach here while keeping in mind that several of these processes do occur in parallel.

40.1 Repeating versus non-repeating events

If a system wants to know the concept of time, it should know that between any two events, what has elapsed is time. This definition requires consciousness. An unconscious system or a human being in an unconscious state like deep sleep or under general anesthesia cannot know such time intervals. For this, the system should first recognize and have a feeling of knowing for the two events before it can know that something has elapsed between the two events. Even without counting or measuring, it is possible to sense the passage of time. For example, children do not know how to count or conscious people thousands of years ago did not know counting as well. Yet, both of them do sense or have sensed the passage of time.

Since no sensor can detect the passage of time directly, we have to rely on indirect means using real physical objects or dynamics. Such an indirect approach is quite similar to how we sense other higher layer abstractions like space, choices and our personal interpretations as well. I will often make comparisons with how an infant learns to sense the passage of time in the discussion below. In the case of an infant, he starts with an inability to sense the passage of time at birth and transitions to a state when he starts to sense the passage of time. Such a transition implies that the reasoning used would be applicable to a synthetic system as well to make it sense the passage of time. This is the same even with the previous topics like space and geometric features. In fact, in this book, I have shown that the problem of consciousness itself has such a transition from a state with no consciousness (at birth) to a fully evolved state of consciousness (for an adult). The transition implies that if an infant can learn this process, then so can a synthetic system, provided it has an SPL architecture.

When trying to teach the notion of time to an infant, the biggest problem is that he is too busy observing the real physical event itself and is not interested in thinking *about* the event or across events to look for abstract patterns. The raw data from the lower layers mentioned in the previous chapter is quite overwhelming at this age. This is the biggest disadvantage with non-repeating events. The details of the event are new if they are non-repeating that he is busy processing the raw data.

Therefore, the only useful events for the problem of sensing the passage of time are the repeatable ones i.e., events that start at time instant T_1 and come back to an approximately same state at time instant T_2, as a loop (or a fixed set). The hope is that as the event repeats multiple times, the system memorizes the event so that triggering the corresponding fixed sets is quick and easy for later repetitions. As the system quickly retriggers the fixed set, it would have enough 'free time' available to do other tasks including possibly 'thinking' about the event itself. In particular, the system will have better predictive abilities.

We experience several repeating events every day (see sections 20.1 and 20.2 for a detailed list of repeating events in nature). For example, one of the largest collection of repetitions in our nature comes simply by observing static objects repeatedly i.e., by looking at an object, turning away and looking back at it. This process can be controlled by free will. This would then allow us to achieve almost any time period of repetition, say, from milliseconds to days. The number of naturally occurring repeating events are far too fewer than these static objects observed using your own free will. In other words, all objects around you at almost every instant in your life are repeating events!

Other repeating events are dynamical events like watching the seconds' hand of a clock go in a circle, day-night cycle and other periodic events. Let me pick day-night cycle as an example to illustrate how the fixed sets and their relationships are formed to sense repetitions. An infant starts to create fixed sets for a number of everyday objects or events. These objects appear every day and have several detectable subcomponents. As he plays with them often, he even has abstract continuity for these objects. Let us assume that he is old enough to have a threshold self-sustaining membrane as well. He will now be in a position to attain a feeling of knowing just by looking at these objects or by observing the repeating events (i.e., for the visual representations).

Now, as he plays with these objects, he is able to sense the repetitions as the same, both during the day and during the night in spite of a change in the background. This is a slow process. In particular, he is capable of detecting the difference between day and night just from the difference in brightness levels (as appropriate fixed sets). He even attains a feeling of knowing for this difference between day and night just from the visual representations.

In this way, an almost infinite number of repeating events, as opposed to non-repeating events, exist and start playing a central role for creating abstract continuity. Until now, most of the feelings of knowing are through abstract continuity at a visual level, which I referred to as the lower layers of abstraction. Next, I will discuss how we can trigger the same feeling of knowing through language or higher layers of abstraction.

40.2 Linking language with feeling of knowing

Let us say that we use the words 'day' and 'night' to refer to the visual difference between day and night. As we use these words repeatedly, an infant starts to create and link language and visual representation fixed sets. These are important links connecting the two representations. Using these links, just from hearing the words 'day' and 'night', he will be able to attain the same feeling of knowing of day and night as if he were looking outside to verify if it is visually dark. Language is now capable of triggering abstract continuity for the visual representation stored previously. Language is at higher layers while visual representation is at lower layers (see chapter 39). Note that while a rich language is unique to human beings, the feeling of knowing of, say, day and night is common for a number of other species as well. For example, a cat or a dog can distinguish day versus night even without a language.

A sentence in a language describing a physical event can trigger the same feeling of knowing as if you were visually watching the same event. If you hear a sentence like 'the red colored seconds-hand of the clock in your bedroom is going around in a circle', you attain the same feeling of knowing as if a physical clock is in front of you and its red seconds-hand was indeed moving around. The accuracy of your visual fixed sets triggered (which we refer to as your imagination) via the language links is fairly poor. For example, your imagined clock may not be the same as your bedroom clock. You may not have imagined a red colored seconds hand and so on. Nevertheless, the higher layer language has triggered lower layer fixed sets and abstract continuous paths. In the same way, the words 'day' and 'night' feel as real as the true day and night via these links and the abstract continuous paths. This process is, once again, a slow and time-consuming one that involves creating a large number of links between the visual (or any representation of real physical events) and language layers (for nouns, verbs, adverbs, adjectives and so on).

In addition to the words 'day' and 'night', you use other sentences to describe the events to a child every day. He also creates these repeating words as language fixed sets. These higher layer fixed sets (representing abstractions) are related to the lower layer visual fixed sets (representing dynamical events) in the same way as before. This helps create a structure even in the higher layer language areas. The reused words and sentences start to form a dense network in the language layer. In the previous chapter, we have seen the importance of a dense network at these higher layers using language. Specifically, we have seen how language solves the problem of reducing information overload at lower layers while increasing the scarcity of data at higher layers.

Once the language layer becomes dense, the self-sustaining membrane of meta-fixed sets will be considerably expanded to include them. It now becomes possible to have predictions entirely at the language layer. A rich language and the links to the lower layers like visual, auditory and touch inputs make it easy to have internal thoughts in language itself. These internal thoughts describe physical dynamical events and you do not need to actually watch or sense any physical events with your ears, ears or touch. You can now have abstract continuity entirely at the language layer giving rise to a feeling of knowing just from reading or hearing sentences in a language as if you were observing the real physical event (cf. reading a short manual describing how to assemble a table or a chair – you can visualize this process without actually seeing or touching the parts). This combined with your self-sustaining thoughts now creates an ability to talk *about* the system.

For example, you can use the word 'repetition' to refer to repeating events. As a child uses this word repeatedly, he will initially start to associate the word with the actual event (incorrectly). This is because he does not know the meaning of the word 'repetition'. He may, for example, think that 'repeat' means throwing a ball because you have used the word 'repeat' every time he throws a ball at you. You see such errors made by children. If in a different situation, you use the word 'repeat', you will notice that he will try to throw a ball at you indicating that he has incorrectly memorized the word 'repeat' with the event. However, over time, as you use the word 'repeat' in other scenarios as well, the association to the original, say, ball throwing event, breaks down. You now have a new invariance (or dynamical transformation) for the 'repeat' word fixed set. These transformations are at the higher layers now instead of just spacetime ones like those discussed in chapter 14 when we were defining the concept of fixed set.

With improved predictions and memories, he starts to see patterns entirely at the higher layers across events. In other words, he can form a feeling of knowing across multiple events described in a language itself because of the denser network at these higher layers. He may not be able to recognize the pattern if he is observing real events (i.e., in the lower layers) sequentially. However, he can recognize the pattern if the same events are described in a compressed form in a language.

The compression of data in a language helps in attaining a feeling of knowing of complex and multiple physical events. The system is now capable of knowing about the events using language. The links between the lower and higher layers make the feeling of knowing to be indistinguishable for any given physical event.

40.3 Knowing the concept of repetitions

From physical repetitive events like day-night cycles, clocks and so on, the system is already capable of knowing repeating events. Next, with the usage of words like 'repetition', it becomes possible to associate the lower layer fixed sets with these language fixed sets. Finally, with description of a physical event using a language, you have a compressed representation generating a similar feeling of knowing for the physical event. This lets you move away from the raw physical data and work at abstract language data while still knowing what the abstractions mean. You now have an expanded self-sustaining membrane that includes both the lower and higher layers.

Even though this process is slow and time-consuming, with repeated usage of words and sentences, we help an SPL system build dense network of meta-fixed sets at the higher layers and link them to the lower layers through near-simultaneity of events.

Using an adult's free will, the set of repeating events can be increased enormously for a child to help him learn the concept of repetitions. Using your free will, you can choose simple events and use the same words consistently and repeatedly to teach a child. With sufficient vocabulary and the above time-consuming process, a child can now attain a feeling of knowing for the very concept of repetitions.

Our brain represents the concept of repetitions primarily as an abstract word in a language. The abstract continuity for this word comes from nonempty predictive meta-fixed sets using a sample set of real physical repeatable events. For example, the word 'repeat' triggers predictions like how you repeatedly threw a ball, how you repeatedly brushed your teeth, how you repeatedly went to school, how you repeatedly played video games and so on. The sample set of predictions is also typically represented in the language layer i.e., you do not have to actually perform the repeatable event to trigger these nonempty predictions, but you could if necessary. These predictions give rise to abstract continuity. Abstract continuity in the language layer lets you bypass the actual physical implementation because of the indistinguishability of the feeling of knowing at the lower and higher layers.

40.4 Feeling of waiting and the concept of time

One way to know the concept of time is by knowing the concept of waiting. The advantage with the notion of waiting is that we can easily relate it to physical events. This ability to relate abstractions to physical dynamics is, in fact, the most common and generic way to understand abstractions like time, waiting, repetitions and choices. The reason is that abstract continuity at the physical dynamics is quick and easy initially, at least, until you have a dense network at the language layer (like with using simple and concrete examples initially to learn abstract mathematics, but later bypassing them if necessary). You expand the existing membrane, which is a network of abstract continuous paths, incrementally using new abstract continuous paths for new events (see section 23.1).

Every time the system predicts an external event before the event occurs, it has an extra amount of *free time* available between its prediction and the occurrence of the actual event (see section 20.7). This is capable of leading the system to a concept of waiting. For example, if a ball is rolling towards you, you can already predict that the ball is going to come to your hand from the trajectory of the ball. Therefore, you have a bit of extra time, until the ball reaches closer to his hand, to plan ahead like, say, think about the event or physically extend your hand to catch it, think about throwing the ball after you catch it or simply wait and do nothing. As you wait, you are constantly comparing the actual physical event with your expected outcome until the event completes.

Such situations start to become quite common as you form more fixed sets, more links between them and more predictions that can be triggered well before the event finishes. For example, when talking to some person, your predictions are sometimes quite good that you can finish his sentence. You also can predict where he is going with his argument and start questioning 2-3 steps ahead of his argument already.

The spoken language and the language layer helps with generating more free time in the everyday tasks. From multiple repetitive events, the system will create several abstract continuous paths and hence a feeling of knowing for the actual waiting (at the lower layers), not just waiting for an outcome in a given situation. For example, you not only know that you are waiting for the ball to reach you, but you also know that you are waiting irrespective of whether the ball is about to reach you or not. This is a new invariance (or dynamical transformation) that will be formed for the 'waiting' fixed set similar to how such invariances were created for 'repeat' fixed set in the previous section.

This helps us link this feeling from physical events to the word 'wait' itself. For this, if you use the word 'wait' every time a child is waiting for an event to occur, he will create links between the meta-fixed set for the word 'wait' and the physical events through near-simultaneity. Therefore, the next time he hears the word 'wait', it will trigger (via these links) nonempty predictive meta-fixed sets from the lower layers to attain a feeling of knowing. The word 'wait' will no longer be associated to a specific situation when you were waiting. It is invariant across all instances of waiting (like waiting for a bus, waiting to meet someone, waiting for a ball to reach you, waiting to go the bathroom and so on).

Now, we can use the word 'time' and link it to the concept of waiting for which he already has a feeling of knowing. For example, the notion of time is just another way of saying that you are waiting for an event to occur. Time is the passage of emptiness between now and until the event occurs. The invariances (or dynamical transformations) you sense for creating fixed sets and abstract continuity are that sometimes you wait longer and at other times you wait for lesser time. This can be controlled and made arbitrary (and hence invariant) using your free will. The notion of waiting or time is identical irrespective of which repetitive event you pick, when you pick, where you pick and what you were thinking at that instant. These act as a set of dynamical transformations required in the definition of a fixed set or truth (see section 13.5).

This is just one way to know the concept of time. In general, there are several other ways to create the same feeling of knowing like by using dynamics, motion or just sitting and doing nothing. However, the conceptual approach is the same as described in this chapter. Now, knowing the concept of time once using one approach is sufficient to know or relate it to the other ways of knowing. This is because we can easily relate the concept of time, which we knew using one way like, say, through waiting, with the other concepts (like motion and dynamics) by expanding the self-sustaining membrane suitably. Each of these abstract continuous paths will be linked together. It becomes quite easy to trigger the fixed set for the concept of time by triggering any one of these pathways (of knowing the concept of time).

In fact, the events do not even have to repeat anymore to perceive time even though repeatable events did help in creating the feeling of knowing for the first time. This is yet another invariance and yet another way to generate predictions at the higher layers of abstraction. As you can see, the concept of fixed sets and abstract continuity is quite central for consciousness and for learning other abilities like sensing space and the passage of time.

40.5 Transitions from future to past – passage of time

Once the system is capable of knowing the concept of time, it is possible to sense the passage of time as well. For this, we rely on our inherent ability to predict events, which is related to waiting and time as we have seen before. Recall that our entire consciousness is built from predictions in a fundamental way (cf. the definition of membrane and abstract continuity). Predictions let the system detect 'transitions' from the future to the past in the following sense. If you are able to make continuous predictions as an event repeats, you are waiting for the external event to reach a future state that is closer to your guess at every instant. This implies that *you are waiting for a future event to transition to the past*. When you notice that your guess turned out to be true, you are essentially detecting this transition. You keep noticing these transitions at almost every instant because you are continuously predicting something about the event.

The most important point to remember here is that if the system is not predictive, then it is simply obeying the laws of nature without any useful memory. This means that the system is 'behind' the event even when the event repeats multiple times. In other words, there is no notion of future for such a system. The system is either too simple or too consumed by the present. It simply does not have any free time to guess anything about the future state of the event. For such a system, attaining a feeling of knowing for the concept of future is not possible. The same applies with the notion of past. After the event has passed, the system does not have any free time to analyze what has just happened once again because it is consumed by the present at all times.

Therefore, there are no transitions from future to past for such systems. The machines we have built until now clearly fall in this category. A camcorder recording events from the external world simply obeys the laws of nature. Even though it senses the external world, it does not use this to predict what will happen next. Instead, it simply records the external event as it happens. This is the same with a car. A car simply travels on the road. It does not predict that it is going to take a turn. Even if it does predict (with additional sensors), it does not build an internal structure to keep track of all such predictions. Every man-made system behaves in the same way. None of these systems detect transitions from future to past.

The presence of these transitions is necessary to create a feeling of knowing of the notions of future, present and past. Systems that are not predictive to a set of dynamical events can only have a notion of present (like with all man-made systems). Therefore, they cannot sense the passage of time. Without a mechanism to transition between these three states i.e., future, present and past, these three concepts stay disjoint from one another. These three concepts cannot be combined together into one notion of time.

Predictions tie these three notions together and provide a sense of direction of time as well. We can now have continuity using these transitions at all times. If this continuity is broken, you do not sense the passage of time during that brief period. Continuity of time transitions is one way to attain a feeling of knowing for the passage of time. Let me now discuss examples of situations when this continuity can be broken down even when you are conscious. We will see that in those brief periods, we are incapable of sensing the passage of time.

40.6 Not perceiving passage of time while conscious

There are clearly cases when you do not perceive the passage of time when you have lost your consciousness. We use terms like, have 'lost time' or had a 'black out' period, to refer to these cases. I will not discuss them here because a lack of consciousness trivially explains these cases anyway. The interesting cases are when you are conscious and, yet, you do not perceive the passage of time.

Let me discuss a couple of variations of these now. The examples I have picked will highlight different aspects like repetitions, predictions and memory that were important for the perception of time.

1. *Events occurring faster than you can process*: The propagation of neural firing patterns from the retina into the deeper parts of the brain takes a finite amount of time, corresponding to, say, gamma wave range of about 30 milliseconds. If the external events and the parallel variations change much faster than this, our brain is essentially going to lag behind the actual event, with no continuous predictions. One example of this is with video games involving fast dynamical changes like in extreme racing. Here, you are too busy processing the external information that you do not have any 'free time' available to think about the event. You do not sense the passage of time while you were too involved in the game. It is only when you take a break and you start to consciously think about the event you begin to realize the passage of time to the same degree as before.

2. *Working with intense focus or on completely new problems*: Consider working on a sufficiently complex or an entirely new problem that you need a high level of concentration (like a timed test in college). You do not have past memory to help you make predictions. You are too focused on the details that you do not perceive time temporarily since you are not behaving as a predictive system. We are simply following the direction of time without thinking about the event or about the future. We need to step back and enter into a predictive mode to start sensing the passage of time. We sometimes refer to this as 'time flies' indicating that we did not sense the passage of time as clearly as under normal circumstances. We do not even remember how much time has passed, as a result.

In these examples, the perception of passage of time is reduced to a lower degree compared to the normal situation even though you are conscious. Similarly, there are other examples when time feels to pass slowly like

(a) when we are constantly checking to see if our predictions are true – say, we are frightened and we want time to pass quickly. But since we are constantly checking every second if the situation has improved, time does not seem to pass. We are essentially detecting too many transitions from future to past (at every second).

(b) when we are waiting to meet someone – they do not arrive on time and you are constantly checking every second if they arrived or not. We are once again detecting too many transitions from future to past.

There are other examples when time feels to pass too fast. One common scenario is when you have a low number of predictions. When we are too happy, we do not predict as much and time feels to pass by too quickly.

41
Cohesive Perception

If you close your eyes for a minute and then open them, you instantaneously see a complete and coherent view of all objects in your entire visual field. You may converge both of your eyes only at a specific region and, yet, everything around it feels real and correct too. These neighboring regions are not as clear as your central vision. Nevertheless, a quick glance ends up only enriching your already correct view, due to higher resolutions, without altering your perception in any fundamental way. The original perception already was cohesive. Even as you turn your eyes or walk around, the new visual information that enters and the old visual information that exits your view maintains a certain three-dimensional visual continuity in a seamless way. Your peripheral vision does become less and less detailed as your turn around thereby recognizing lesser number of objects or features. Although this decrease in number happens in a discontinuous way, your perception of the visual view stays continuous and coherent. Similarly, the focal lengths, the angles, the distances between your eyes and the external objects change as you move around while your brain still maintains a visual coherence.

In this chapter, I want to address how a system is capable of having such a cohesive perception. How is this possible starting from a series of electrical firing patterns originating, say, from both of your eyes and transmitting into your brain as chemical disturbances? Such a cohesive perception of the external world exists even for people who cannot see. It also exists during our dreams with no external inputs. Something as big as a blind spot is cleanly patched up or compensated, independently for each eye. Slightly different retinal information from each eye is seamlessly merged together. Inverted images on the retina are fixed to create and project an upright view.

This is a nontrivial task for an SPL system. It is necessary to create all of the features described in the previous chapters like the self-sustaining membrane, abstract continuous paths, free will, geometric properties like edges, angles, shapes, sizes and distances of all objects, space, passage of time as well as mechanisms like the fundamental pattern. For example, these features imply that consciousness is necessary, at minimum. I will only focus on visual perception and leave all other cases like cohesive perception for people without vision or for bats that use echolocation with sound waves as future work.

Let us consider the case in which our brain has a representation for colors (section 31.4) and edges as a straight line or a curve (section 34.2). When you open your eyes now, you will see a number of squiggly lines and colors in every direction. You do not know if these lines mean anything, whether they are all at a fixed distance and what their true sizes are. The first step is to be able to group these lines and colors as objects i.e., you separate the foreground from the background (section 35.1). Then, we need to sense empty space between objects and have an appropriate visual representation of emptiness using both vision and touch, as explained in section 35.3. Now, the presence of empty space suggests that objects can be at different distances from each other. Therefore, we need to know the true sizes and distances of the objects. This requires both visual and touch inputs as discussed in section 33.4. You see the tables, the chairs, the floors, the walls, the doors, the ceiling and everything else to be of the correct sizes, correct orientations and at the correct locations as they should be, using mechanisms from chapter 36. When you turn or walk around, these estimates will be updated continuously as described in section 36.7. In addition, with a number of visual cues (chapter 37), you can now see in three dimensions, as discussed in chapter 38. Finally, you sense the passage of time through the motion and dynamics of objects, which will let you understand and know the external world. If you indeed sense all of the above features, have corresponding visual representations and have a self-sustaining membrane with a feeling of knowing for each of them at every instant, I claim that you already have a cohesive perception of the external world.

In the rest of this chapter, I will discuss three additional topics not covered in the previous chapters but are important for visual perception. Firstly, we need an ability to sustain the parallel excitations as we observe all the objects that are in front of us. As we scan objects around us, the entire representation within our brain should stay intact. Secondly, these sustained parallel excitations should give rise to the recognition of the individual objects or features. Our cohesive perception comes from the objects or features we actually recognize. If we cannot recognize minor features, we just ignore them. Thirdly, we should have a sense of continuity in everything we see around us even when there is an apparent discontinuity. Our brain should be able to maintain this sense of continuity even with several static and dynamical transformations. I will then discuss a few cases for which cohesive perception breaks down when some of the above required features are not maintained.

41.1 Parallel sustained excitations

In chapter 33, I had already mentioned that the true nature of the external world and how our eyes project it are necessarily distinct. A table appearing as a smooth-edged rectangle is not that smooth in reality. Similarly, there are no colors inherent in the external world and so on. Our brain re-adjusts it to maintain consistency between all sensory inputs with touch and other direct contact sensors used as a guide to capture reality.

When we look around us, we seem to be viewing a large region of space. Light from this region falls on our eyes and creates parallel excitations from millions of retinal nerve cells simultaneously. Each of these excitations propagates deeper into the brain. If they do not sustain long enough because of a lack of stable parallel looped structures, then each time we look away and then turn back, we would be processing the same information once again as if it were a new scene (see section 15.1). This is true for an infant because of a lack of dense interconnected network of neurons. However, this is not the case for an adult. When we focus even on a given small region, we do have a good sense of recognition of all the neighboring objects. These parallel-sustained excitations, represented with the self-sustaining membrane, are a basic requirement necessary, not just for a cohesive perception, but for consciousness and most other features of our brain itself.

41.2 Recognition of objects and features

The problem of recognizing objects or of individual features requires, at minimum, the membrane of meta-fixed sets. Let me list some of the critical mechanisms previously discussed that are necessary for recognition and for creating the membrane.

1. *Primitive features*: When we look at objects around us, there are a number of common features shared among all of them. Some of these primitive features shared by all objects are:
   (a) Detecting a wide range of colors (see section 31.4).
   (b) Detecting edges (see section 34.2).
   (c) Detecting different angles (see section 34.4).
   (d) Identifying different common shapes (see section 34.5).
   (e) Detecting different patterns within a shape using colors, edges and angles (see section 34.5).
   (f) Sensing the foreground and distinguishing it from the background (see section 35.1).
   (g) Knowing the existence of an external object or a pattern (see chapter 29).
   (h) Sensing the three-dimensional space between objects (see section 35.3).
   (i) Estimating the size of nearby and distant objects (see section 34.7).
   (j) Estimating the distance of nearby and far objects from you (see chapter 36).
   (k) Sensing the rigidity of an object (see chapter 32).
2. *Correlation between inputs and outputs*: When we receive multiple inputs (like say, touch and vision), we link them together and associate it with an action using some of the mechanisms discussed previously. Here are some examples:
   (a) Focusing on an object to create a sharp image (see section 16.1).
   (b) Detecting the motion of an object (see section 18.7).
   (c) Tracing along an edge or angle (see section 17.4).
   (d) Looking directly at a source of light or a pattern – ability to correlate what you are looking at and the external object itself (see section 17.1).
   (e) Turning towards a source of light or a pattern (see section 17.1).
   (f) Touching the object you are looking at (see section 33.5).
   (g) Tracking a moving object (see section 17.3).
   (h) Moving towards or away from an odor, object or pattern (see section 17.6).
   (i) Correlating edges and angles with touch and vision (see section 33.4).
3. *Knowing and controlling your own body at-will*: Our brain receives a constant source of inputs from our own body. These can act as reference inputs to compare to all other external objects. Free will using our own body lets us repeat the same experiments reliably, thereby letting us form truths (i.e., fixed

sets and hence the membrane). Here are some of these examples:

(a) Detecting your hands, legs and different parts of your body (see section 24.5).
(b) Knowing the lengths of your hands, legs and other parts of your body (see section 34.7).
(c) Knowing the location where you touched your own body (see section 32.1).
(d) Ability to touch at specific locations on your body. This is just an application of the fundamental pattern (see chapter 16).
(e) Moving parts of your body to a desired location. This just requires using the fundamental pattern (see, for example, section 17.7).
(f) Moving parts of your body in a desired direction. This is again an application of the fundamental pattern (see chapter 16).
(g) Moving your whole body to a desired location relative to other objects – an application of the fundamental pattern (see chapter 16).
(h) Touching any pattern or object you want that you see with your eyes (see section 33.5 and use the fundamental pattern of chapter 16).
(i) Determining the size of an object when it touches your skin surface (see section 32.1).
(j) Knowing that you moved an object (see section 18.13).
(k) Determining the difference between you moving versus the object moving by itself (see sections 18.13 and 18.14).

Each of these features and mechanisms creates a large set of interconnected fixed sets. Every object and pattern we see in the external world will itself have an enormously large set of linked fixed sets. For example, we can take any object and break it down into several subcomponents, each with a set of linked fixed sets, in a number of different ways. We can continue this iterative process at least a few times. These fixed sets are themselves shared among several objects thereby giving rise to the notion of abstractions like color, shapes and patterns. They are also dynamically linked with one another. As the number of objects and patterns encountered through our experiences increase, the membrane of meta-fixed sets that is getting formed starts to become an extremely dense network. This dense network helps us recognize as many features or patterns around us and as quickly as possible.

41.3 Continuity even within apparent discontinuity

We perform a certain amount of parallel processing in most everyday tasks. We typically switch between two or three tasks. For example, when we walk, we also think about some other topic. We do not track our leg, hand and other body movements continuously because we do think about one or more topics during the same time. This switching between disjoint tasks makes it appear that there are apparent discontinuities. Yet, we have a clear sense of continuity. If we did not have such a sense of continuity, we would not have a cohesive perception. Before we start switching between tasks, we arrange the millions of neural firing inputs (say, from the retina) into appropriate groups representing objects. We track dynamics of such groups within our brain as moving objects, reading or thinking about words in a language and hearing words in a sentence. The seemingly discontinuous switching occurs, not an the individual firing neurons level, but at the aggregate group level (consisting of fixed sets for objects, words and sounds).

This grouping as entities is not the same across all species. It is dependent on how a given organism processes the sensory inputs and performs an action. For example, birds and honeybees are capable of seeing in the ultraviolet region, similar to tetrachromats in humans. Bats are capable of hearing ultrasound). This lets them sense additional patterns on, say, flowers, that a normal human cannot see. The patterns do appear as distinct edges and geometric shapes, not just as distinct colors, though they are invisible to a normal human (see Schmitt [136]). As a result, these organisms keep track of different aggregate groups.

An infant takes several months to learn this skill of grouping patterns appropriately as fixed sets. Before he learns this skill, how would his perception feel when observing a dynamical event? For example, when he is tracking a moving object, his primitive abilities will create overshoots and undershoots. The objective of tracking is to ensure that the image continues to fall correctly on the fovea even as the object moves away. Since his tracking ability is not precise, he turns his eyes more (creating an overshoot) or less (creating an undershoot) than the desired amount. Which parallel-sustained excitations should he track deep inside his brain as an aggregate group in this case – the new pattern that falls on the fovea because of poor tracking or the old pattern that overshot into the peripheral regions of his retina? If he tracks the new pattern, he breaks the continuity. If he tracks the old pattern, the poor resolution from non-foveal regions, combined with imprecise tracking mechanism produces a poor level of continuity. Over a long period, these discontinuities and errors accumulate considerably. It is no longer obvious which set of patterns to track anymore from your retina. I call this the problem of serializing a parallel system. Therefore, an infant at this stage does not have a cohesive perception. The fundamental pattern is one way to improve serialization when he grows into an adult.

After we develop the ability to serialize the parallel firing patterns as fixed sets, we can work on multiple tasks simultaneously as well. Examples are our ability to talk, think, listen to radio and others when driving a car or when walking. This switching is possible because of our abilities to predict the future dynamics at every instant, which is accurate enough for a short period. Our abstract continuity is critical to estimate the future events locally. For example, if you continue to press the gas pedal and if there is no unusual situation ahead of you, you can predict how the car is going to move even if you do not look at the road, say, 1-2 seconds. You can switch to another task like changing the radio station during this period. If you do have errors in your prediction and they do not magnify quickly, you can correct them by temporarily switching back. Therefore, our quick glances, our accurate estimations of the unfocussed tasks and our abstract continuity allow us to switch between a few tasks without any apparent loss of continuity. We effectively fill the gaps in the continuity for each of these simultaneous tasks using our faster pace of predictions and occasional glances to correct errors.

41.4 Breakdown of cohesive perception

It is possible to breakdown our cohesive perception even when we are conscious. I will discuss a few examples here. The typical approach is to distort basic primitive features like edges, angles, size, depth and shapes.

1. Imagine skewing both of your eyes so you are looking at your nose. Since all objects appear twice, if you focus and analyze harder to recognize them, you will not only lack a cohesive perception, but you will also feel dizzy. This is particularly true if you turn your head quickly around so you do not have the time to analyze what you are observing.

2. Try reading a book immediately after looking at bright sunlight indirectly. In this case, several retinal cells become insensitive and you see bright flashes of light. These flashes of light interfere, making it difficult to read words and sentences. In effect, your eyes have several small patches that are not sending the correct information from the book. Locally within this limited view, you lack cohesive perception.

3. For other similar examples, we use special devices like a pseudoscope or a device that inverts the images so they fall upright on the retina. A pseudoscope is a device, which inverts depth perception by swapping the visual information of the left eye with the right eye. Both these devices produce quite distorted information for complex sceneries. If you do not have time to analyze the information carefully, you experience a similar lack of cohesive perception.

4. Consider a room with a number of distorted and angled mirrors (convex, concave or with other variations) placed in random directions. The result of multiple random reflections of all objects in the room produces a complicated view. You can recognize parts of it but not the whole view. As you try to look closely, the multiple reflect-ions constantly keeps interfering, distorting your view. As you walk in such a room, you lose your cohesive perception as well. We experience a similar effect with some pictures or optical illusions that have unusual and random depth perceptions. Each of these examples suggest that recognition is quite important for perceiving the world in a cohesive way.

Therefore, from the above examples, we do see that the factors discussed in this chapter are not only sufficient, but are necessary for a cohesive perception.

42
Sensing Physical Properties

Until now, we have discussed abstract and geometric properties of objects like colors, shapes, sizes, edges, space and time rather than their dynamical properties like mass, energy and forces. A large set of these abstract properties have helped create a self-sustaining membrane of meta-fixed sets necessary for consciousness. However, a system cannot be conscious without ever sensing dynamical properties like mass and forces. For example, an entirely abstract system like a computer program cannot become conscious because it does not know the existence of mass and force. Interestingly, the reason is that dynamical properties do govern the consistency of rules obeyed by the geometry of external objects. For example, the shape of a spherical bubble (i.e., the geometry) is related to uniform pressure on all sides (i.e., dynamical properties). Our brain uses these dynamical properties whether we know about them, in an abstract sense or not.

In this chapter, I want to show how we can know about the existence of the dynamical properties like forces and masses. The approach will be similar to how we showed the existence of the external world in chapter 29. We will need to create special fixed sets for these dynamical properties as well. The reason our brain automatically creates new fixed sets for mass and force is because we do encounter a number of repetitive scenarios related to these concepts which do not agree with other known geometric and abstract fixed sets like shapes, angles, colors, space and time.

For example, when we push different objects of different shapes and sizes (i.e., with different geometric properties – recall that we do not yet know the concept of mass or force), we notice that they do not move the same way. Even same shape and sized object moves by different distances. None of the existing fixed sets for geometric properties can capture this difference. Our brain naturally introduces a new meta-fixed set that is invariant to these dynamical transformations (cf. new meta-fixed sets introduced by our brain for colors, empty space and time using another set of similar invariances as discussed in chapters 31, 35 and 40 respectively).

Our body also needs a new meta-fixed set which captures the 'reaction' aspect to a push. Recall Newton's third law which states that for every action there is an equal and opposite reaction. When we push an object, our body is pushed back. The amount of push back (registered as a firing pattern from your elbow and shoulder joints) are different even for similar shaped and sized objects. The body-related firing patterns generated from your joints are invariant to geometry of the object being pushed. Therefore, the resulting fixed set cannot be associated to any geometric feature. In this way, we discover a new body meta-fixed set, which corresponds to the concept of force.

We begin to correlate both of these new meta-fixed sets. With a lot of repetitive events, our brain will form abstract continuity by creating predictive meta-fixed sets both for the dynamical properties like forces and mass, but also linked to geometric properties i.e., we begin to predict how heavy an object would be and how much force we need to apply to move it just from its geometric properties. Finally, we can use free will to show the existence of these properties with certainty.

42.1 Body meta-fixed sets for force

As with sizes and distances, we sense dynamical properties qualitatively before we try to quantify them. Among force, mass and energy, we can detect force easily. We can then use it to infer the other two properties. When we apply a force or when a force is applied to any part of our body, our tactile sensors produces a neural firing pattern, which is transmitted to the brain as explained in section 32.3. The more force we apply, the more the number of neurons and the intensity at the contact regions that fire. Newton's third law of motion, namely, for every action there is an equal and opposite reaction, is the fundamental fact that our brain relies on. The system should sense a wide range of forces through the reactions. For example, the same object moves by different distances for different forces. We should keep track of these relationships. Let me outline the steps now. I will only state the mechanisms briefly since we have discussed similar details in the past for other cases like when discussing geometric properties.

Actuation: We have special motor regions within our brain to trigger the movement of a specific part of our body. The mechanisms for exciting these regions are quite well studied (see Kandel et. al [82] and Purves [122]). The advantage is that our brain architecture is the same for actuation, sensing and all other thoughts. As a result, it is quite easy to link these regions using the fundamental pattern and other mechanisms described previously.

Knowing that we moved our body versus someone else moving our body: In addition to producing a motion of a part of our body, say, our hand, we also know that we moved it ourselves as opposed to someone else moving our hand. The motor control regions will not be triggered if someone else moved our hand. We form abstract continuous paths to know about the movement of our hand and link it to the corresponding motor region.

Sensing motion of our body: In addition to making our hand move, we also sense our hand's position as it moves. There are proprioceptors in muscles, tendons and joints that relay information to our brain about our body's position. Irrespective of whether we move our body or someone else does, our brain receives this positional information. The sensory positional information and the actuator control regions are also linked together through our everyday experiences.

Estimating the amount of force needed to push an object: Given the above mechanisms, if an external object hits you, you know the location where it hit you. In addition, different objects produce different degrees of sensation resulting from the impact on your body. This degree of impact cannot be reliably associated to the geometric shape and size of objects, though large objects do typically hurt you more. Therefore, all existing fixed sets (for geometry, space and time) do not capture this unique property. We need a new meta-fixed set to capture this invariance. This is what we call as the force fixed set. The situation is similar when you hit the objects as well. The amount of force needed to push by a certain amount of distance is different for different objects. It is proportional to the mass of the object, not on any geometric properties. We can test this repetitively and with 100% certainty using our free will.

During our childhood, we interact with objects of different masses. We create a number of fixed sets that indicate the strength of forces needed to push different objects. We also link this with some geometric properties like the size of the object and the material texture (say, wood versus metal). We create predictive meta-fixed sets using geometry to estimate the dynamical property like how heavy an object is. We can now predict how heavy an object is just by looking at it, even though it may be sometimes incorrect. This predictive meta-fixed sets give rise to abstract continuity and hence a feeling of knowing of forces (and masses) as well.

42.2 Existence of force and mass

From the external interactions, we have created several fixed sets for different amount of forces. Next, when we move our own hand, say, we do need to exert a force as well i.e., excite the corresponding fixed set. This force on our joints will feel identical to the case when we move an external object with a similar mass i.e., the fixed set triggered are identical in both cases. We can use this similarity property to define mass in a qualitative way for our own body as well as the external objects. Therefore, the advantage is that our very own existence creates ample ways to trigger the same mass and force fixed sets (that were triggered from external objects). Our brain now encodes features like the ability to equate two different objects with same mass irrespective of their geometric properties as well as composition and the ability to equate forces applied on two different objects by the effects they produce, not depending on space and time.

The fixed sets formed subject to these transformations are linked to the corresponding fixed sets for actuation, sensing and our other thoughts. These links and our experiences eventually create abstract continuity necessary to form an abstraction for force and mass and to know them. Our free will ensures repeatability of simple experiments to trigger the same fixed sets reliably to reinforce these notions. Using language, we can even create abstract continuity in the higher layers for these abstract concepts (see chapter 39).

Once we attain a feeling of knowing for forces and masses, it is possible to study these concepts quantitatively. The dual space representation of chapter 25 introduces an alternate way to experience the same feeling without touching or being touched by any object. As long as the same set of fixed sets are excited, you will experience a 'virtual' force even without any physical contact with any external object. One example of this case is during our dreams. We experience the same feeling of walking, running, jumping, kicking, pushing and others even without any physical motion in our dreams.

It is important to realize that this identical feeling (like forces and masses in our dreams) is possible only after the necessary brain structure is formed through real physical experiences. The dual space representation merely provides a different way to excite the same structure, provided the structure already exists. This is similar to how blind people can still have a high level of imagination and feelings through other senses, though they cannot 'see' in the normal sense (Hurovitz et. al [75]). The dual space representation produces feelings in their dreams that are identical to their real physical experiences.

Sensing other dynamical properties like charges and magnetic forces is not as simple as sensing force and mass. The number of repetitive examples in our everyday life for these cases are too few. Our brain does not encounter enough variations that requires us to distinguish these properties using new meta-fixed sets. Once we become conscious, we can create repetitive situations with charges and magnets using our free will. This is when we need a new meta-fixed set to represent these notions. For this we use language as a means instead. In other words, these dynamical properties (like charges) are abstract ones. We do not 'feel' the charges in the same way we 'feel' forces and weights.

43
Self-awareness

In this chapter, I will first summarize all the main results of the previous chapters and show how this leads to consciousness. Then, I will discuss how we can achieve self-awareness. These results are generic enough to be applicable to other SPL systems.

43.1 Becoming conscious and beyond

The central result for the problem of consciousness was Theorem 12.1. It stated that a system becomes minimally conscious if and only if it can know at least one truth. However, to understand this theorem, it became important to understand two topics, namely, what truths are and what knowing is.

1. *Truths:* In chapter 13, we first saw that truths were abstractions. To capture reality, we started linking abstractions to dynamics. This led to creating emergent properties from dynamics. We then defined truth as events that remain fixed under several dynamical transformations. This definition did not emphasize consciousness. We were then able to represent truths within a dynamical system using the notion of fixed sets.

   Once we have a large number of fixed sets stored within our brain network, they form a complex interconnected network of meta-fixed sets. This is what we called as the threshold dynamical membrane of meta-fixed sets. This membrane is interwoven by the natural constraints and relationships imposed by the external events represented by fixed sets. The membrane is capable of self-sustaining for a long time.

2. *Knowing*: In order to know any fact, we needed an ability to make predictions. This gave rise to the notion of abstract continuity i.e., where we have nonempty predictive meta-fixed sets at every instant as you keep observing an event. Abstract continuity combined with the presence of a dynamical membrane of meta-fixed sets gives rise to a feeling of knowing.

Given these two results, we were in a position to use Theorem 12.1 to explain how we attain consciousness. The flexibility we have here is that we can know *any* single truth. We need to combine the structure created from 'knowing' facts with the structure created by the 'truths' or fixed sets. Section 23.1 established this relation where we have seen that we can represent the membrane itself equivalently as a network of abstract continuous paths. Therefore, the conditions needed for attaining a feeling of knowing i.e., the membrane and abstract continuity along with the fact that you can pick any fact to know for consciousness, can now be unified into a single condition. This condition is just the existence of the threshold membrane of meta-fixed sets. This was summarized as Theorem 23.2 in chapter 23 – a system is minimally conscious if and only if it has a self-sustaining threshold membrane of meta-fixed sets for a finite amount of time.

What are the specific meta-fixed sets that are part of the threshold membrane? In the previous chapters, I had discussed several concrete examples. One of the primary set of properties corresponds to knowing your entire body through several internal sensory mechanisms. In addition, there are (a) visual properties (if you can see) like colors, (b) geometric properties like edges, angles, shapes, sizes, patterns, space and distance, (c) dynamical properties like time and motion, (d) physical properties like forces, mass and energy and (e) other properties based on touch, sounds, taste and odor as well as through self-motion, language and internal thoughts. Our brain linked each of these fixed sets in a definite way identified by our unique set of experiences to form a well-defined network.

We noticed in chapter 23 that the minimally conscious state is not unique to human beings. Several animals, especially mammals, are also capable of attaining this state. However, to achieve a state comparable to human consciousness, we needed to expand the threshold membrane considerably using language that is rich enough to describe most dynamical events, as discussed in chapter 39. The expanded membrane makes features like sensing the passage of time and higher-level abstractions possible and unique to human beings. Let me state this as the following theorem.

Theorem 43.1: A system attains consciousness comparable to a human if and only if it has an expanded self-sustaining membrane of meta-fixed sets that includes a natural language capable of describing most dynamical events, for a finite amount of time.

As we grow older, our new memories help us grow our membrane as well. One way to characterize the threshold state of our membrane is by saying that the self-sustaining excitations produce gamma level of brain activity (i.e., brain waves above 30 Hz). Let me discuss more on the characteristics of the threshold membrane.

43.2 Characterizing the threshold membrane

In section 11.2, we have seen a number of altered states of consciousness as well as unconscious states. Using Theorem 43.1, we can now analyze each of these states and classify them broadly in the following way: (a) states with or without the existence of the dynamical membrane, (b) states with or without a membrane self-sustaining for finite time and (c) states with or without membrane reaching a threshold state.

For example, a newborn is not yet conscious because he is in the process of building his dynamical membrane (case (a)). If you are severely drunk or are under general anesthesia, your dynamical membrane is considerably broken using specific chemicals (case (a)) even though structurally the network exists. The loops do not self-sustain long enough (case (b)) because these induced chemicals interfere with how the neurotransmitters normally behave. When you are sleepwalking, or you are in deep sleep, even though structurally you have the required network, the dynamical excitation is low enough that you do not have a threshold level of excitation (case (c)). The excitations are too slow. Specifically, your brain waves are in delta (0.5 – 4 Hz) or theta (4 - 8 Hz) range. When you are knocked unconscious, your blood supply in temporarily cutoff. When you are in state of coma, there is a physical damage in some critical regions of your brain. In both situations, your brain is prevented from forming complex looped patterns. As a result, you are unable to excite your dynamical membrane easily (case (a) and (b)).

Just before you start dreaming, you are in a deep sleep state where you are not conscious. You stay in this state until the excitations reach a threshold state. However, as the excitations reach a gamma level of activity (i.e., brain waves above 30 Hz), you do become conscious (case (c)) and start experiencing dreams (see section 19.6 to see why it reaches gamma level from a delta or theta level). With mental disabilities like dementia, schizophrenia and autism, the formation of the threshold membrane takes a little bit longer because of the structural differences compared to a normal healthy brain.

When you are hypnotized, your brain synchronizes the self-sustaining excitations with an external person. His verbal statements are partially responsible for driving the neural path excitations within your membrane. You are in an intermediate state where you are aware of several facts and, yet, you are not in complete control of your thoughts compared to when you are awake (case (b) and (c)). During meditation, on the other hand, there is no external person and you direct your thoughts in a specific way instead. In both of these cases, the brain wave patterns are in the delta or theta level. Your dynamical membrane is in a 'twilight' state i.e., it is right at the threshold level (case (c)). It is oscillating in a stable way between fully conscious and unconscious states.

In most examples, you are either completely on one side of the threshold state or on the other side. However, by changing the degree of activity, it is possible to enter into an intermediate state. For example, if the general anesthesia is wearing out, if you are partially drunk or if you are just beginning to doze off, you enter into such an intermediate state.

Let me now characterize the threshold membrane of meta-fixed sets in a measurable way (see chapter 19). The external inputs or our internal thoughts produce firing patterns that spread across large regions within our brain. These excitations are capable of self-sustaining for long periods, which we can measure as brain waves. As our brain processes information when performing a certain activity, the neural firing patterns trigger abstract continuity for several facts as well. Each of these facts produces recognition and a feeling of knowing. If this recognition frequency is in the gamma range of brain waves (i.e., more than 30 per second), we can say that there is continuity among recognized facts, in a practical sense. When our brain activity maintains such a level of continuity for extended periods just from common everyday events, we say that our dynamical membrane has grown to be self-sustaining beyond the threshold level.

There are two common ways to achieve such a high level of activity easily: (a) through rich visual information and (b) through a continuous flow of your internal thoughts. Not surprisingly, the regions dedicated for both of these tasks cover large areas within a human brain. Even when a person is blind, the regions for the remaining sense organs should generate a gamma level of activity. If the level of activity drops to delta or theta level, you will lose your consciousness. One exception is during meditation. Here, you train yourself to maintain a self-sustaining membrane even with a delta or theta level of activity. You can achieve the threshold state, say, by localizing the gamma level of activity only to specific regions while still maintaining delta or theta levels within broad regions.

In everyday situations, if we want a gamma level of activity, there should be common features even under widely significant variations in the external world. Fortunately, in a typical environment, objects are rarely uniform. They all share properties like colors, edges, angles and shapes. Until our brain encodes these features as fixed sets or loops, maintaining a gamma level is nontrivial (cf. an infant). There will be no sustained oscillations within our brain network.

Our brain should encode the union of all commonly exposed structures from both external inputs and internal thoughts into the threshold membrane to allow a gamma level of activity. In the previous chapters, I have described how to form several such common patterns within our brain. Let me now discuss some of the implications of the structure of our threshold membrane.

43.3 Correlating with reality – causality

Implicit in the previous discussion is that we form the threshold membrane by correlating what happens in the external world to the internal neural representation. For example, when we walk towards an object, the excitations in our brain correlate with the motion of our hands, legs and the rest of our body. If something hits your hand, you correctly move your hand away, not your leg, in response. This correlation with reality appears similar to the notion of causality of events, i.e., the external cause and your current internal state is directly related to the effect on the system and its response in return. Is there always a correlation to reality and is it important to keep track of it?

Nonliving systems passively follow the causal laws of nature, whether they are predictive or not. In living beings, the active sensory and control mechanisms are linked correctly to yield correlations and causality at both micro- and macro-level, thereby controlling the accumulation of errors. If we want to estimate any physical property like mass, space, time, edges or angles, we need to correlate the external inputs with our internal representation. The notion of repeating events, truths and reasonable variations would then be well-defined. During our dreams, the same external physical laws are not obeyed. This is because the perception is solely based on the dual space representation and not from the direct external inputs. As a result, there are minimal inconsistencies during the dream itself compared to an awoken state.

The stable parallel looped architecture within living beings help in creating these correlations effectively. We use the fundamental pattern and other mechanisms to create the correct set of causal links between sensors and actuators. As an example, consider how an infant learns to control and correlate the motion of his hand with what he sees in the external world. Our body generates and stores excess energy than it consumes. We can use this energy to do a task like moving our hand. As we move our hand, we can observe it as well, as a correlated event. Since our brain has motor regions that can make our hand move, we can identify two such regions: (a) that magnifies the motion and (b) that suppresses the motion. When you are a newborn, you do not have well-structured neural paths to access these two regions easily. However, using the fundamental pattern, say, we can develop an ability to self-control our hand (see chapter 17).

This is how the correct correlations are automatically formed from each such experience. The looped architecture memorizes these structures effectively creating a feeling of continuity within his brain. Two neighboring regions have enough opportunities to be able to link together. With machines, treating these motions purely as abstractions creates discontinuities. We can similarly use suppression of motion to explain how an infant learns to move his hand away if it produces an emotional explosion corresponding to a taste he dislikes. Therefore, using the mechanisms like the fundamental pattern, there is no need to program every task as we do with a computer. The system is capable of self-learning. A complex stable parallel looped system like an infant can learn to correlate the dynamics within his brain with external reality.

43.4 Acting as a whole – maintaining dynamical membrane

One important consequence of the notion of a self-sustaining dynamical membrane is that the entire system acts as a whole, not as selfish or individual parts. In chapter 68, we will see that a stable parallel looped system is sufficient for a system to behave as a whole instead of requiring consciousness, as with single cellular organisms.

With a self-sustaining membrane, the stability of the membrane as a whole manifests itself as a self-healing system. The definition of stability here implies that the system will resist its own destruction when subject to various disturbances even if it means that it has to sacrifice some parts of the system at the expense of the overall system stability i.e., it acts as a whole. This is possible because the membrane internally senses and controls every aspect of itself. If a region of your body is under stress, other regions will work to negate the effect. For example, language and logic-based structures can yield new abilities to stabilize the system more effectively.

A couple of analogies sometimes help us understand some aspects of the abstract dynamical membrane. One is that of a soap film that is capable of self-sustaining subject to a diverse set of disturbances. The other is a random dot autostereogram that creates a visual illusion of a three-dimensional scene from a two-dimensional image. When you turn your head away briefly and look back at the autostereogram, the 3-D scene breaks down. You need to refocus once again. As an analogy, imagine that our brain has locked itself into a state where everything you see around you is like an autostereogram that does not breakdown under ordinary operations like blinking your eyes, turning your eyes or head, walking, running and others. For example, this is indeed true with the autofocus mechanism of your eyes. Your brain automatically adjusts focal length to lock onto objects at different distances for a clear image. Similarly, both of your eyes automatically converge and lock onto the same object. One eye does not see one object and the other eye, another object. The only way the locking mechanism seems to breakdown is when you lose your consciousness like with sleepwalking, general anesthesia, deep sleep and others. Note that these are only incomplete analogies to aid your intuition and are not related to the theory itself.

Our self-sustaining membrane can 'tear down' in a number of different ways, corresponding to how we do become unconscious. For example, we can cutoff external inputs from our body like with knocked unconscious, coma, deep sleep or sleepwalking. We can chemically suppress our ability to form fixed sets (or loops) in our brain like with general anesthesia and severely drunk cases. We can alter normal levels of excitations through controlled mechanisms like with hypnosis or meditation. We can also have sparse network in several critical regions that the membrane is not self-sustaining as with infants.

Now, after your membrane is broken, you still have an ability to rebuild it provided the fixed sets corresponding to, at least, your internal body parts as well as critical sensory and control pathways are not damaged severely. Some examples of this are how you transition from a deep sleep to a dreaming state, from a sleeping to an awaken state or when the general anesthesia wears down.

Next, there are cases when it is not even possible to create a self-sustaining dynamical membrane. One obvious case is if the system does not have a stable parallel looped architecture. A less obvious case is if you have sufficient symmetry within your brain structures like the left-right hemispheric symmetries. Complex interconnected loop structures are possible only with a sufficient level of asymmetry. Symmetries like bicameral and other local bilateral ones are too stable to form rich network structures (i.e., enormously complex and interconnected meta-fixed sets), necessary to abstract a number of external and internal phenomena. Another less obvious case for not forming a dynamical membrane is when the recognition rate is not fast enough i.e., not in the gamma range (> 30 Hz).

43.5 Knowing single truth = knowing membrane of truths

For a given statement, if we ask ourselves how we know if it is true or false, we typically expect that we only need a few additional facts to answer it. We rarely find statements that we cannot explain in terms of a few simple statements. Most everyday statements can be broken down into simpler statements, which are already too obvious to seek additional explanation. However, this is true only if you take your own consciousness for granted. See chapter 13 for a detailed example where the explanation is not obvious if you do not assume consciousness.

Our ability to understand complicated statements and theories do require creative analysis. However, this analysis is built on the foundation of truths resulting from our very own consciousness. Theorems 12.1 and 23.2 clearly shows that knowing even a single truth requires knowing the entire dynamical membrane of truths, which gives rise to our consciousness. There is no quick shortcut to knowing even a simple statement by avoiding knowing consciousness itself. Therefore, to know just one truth, you truly need an immense network of interconnected truths.

43.6 Self-awareness

Since knowing one truth automatically lets you know an entire membrane of truths, you become self-aware as well. The threshold membrane of meta-fixed sets already contains knowing your own body, space and time. We can now imagine that every action you perform and every physical body movement excites a special meta-fixed set that we can call as the 'self'. This meta-fixed set is interconnected with all of your internal body fixed sets as well as with the external objects using the notion of relative distances and directions through the three-dimensional space. Therefore, this special meta-fixed set is not concentrated in a given region of your brain. This meta-fixed set is spread across wide regions within your brain. The interconnections also allow you to create the necessary abstract continuity to know yourself.

You know yourself relative to the external world using external inputs and through your internal inputs. All of this information is integrated qualitatively to create the feeling of knowing of your own self. An infant also has such a feeling of knowing once he has a self-sustaining dynamical membrane. This is true even if he has not yet acquired language abilities. However, he may not be able to communicate about his self-awareness clearly. In fact, he may even not realize self-awareness as an important and interesting abstraction. He cannot see how to use this abstraction and relate it with other external events to know about the external world in the same way an adult does. The threshold membrane is not big enough to include these higher layer abstractions.

Now, once he acquires language abilities, he will begin to create formal abstractions for each of the fixed sets, thereby expanding his threshold membrane. He can then formulate several abstract relationships including self-awareness much more easily using the abstract continuity at the higher layers (see section 39.3). Therefore, knowing your consciousness and becoming conscious do not happen simultaneously. Knowing your consciousness happens only if your threshold membrane is expanded to include the truth and abstract continuity about your own consciousness in a higher language layer.

There is strong evidence (see Jaynes [79]) that language existed in the past without the concept of 'I' for a long period. If this were the case, it is possible that you are conscious but never asked the difference between the external world and yourself in an abstract sense. With infants and even older kids, we do see a similar situation, especially when they do not have 'I' in their limited vocabulary.

44

Comparing Conscious Systems

Until now, we have been discussing various aspects of the problem of consciousness. My objective now is to discuss how to compare these features across two human beings. Is my appearance of red color same as yours? Is my feeling of sour taste same as yours? Is my consciousness or a given emotion same as yours? Even simple questions like how a circle appears to me are nontrivial to compare between two different people. The necessary and sufficient conditions for a minimally conscious state and the feeling of knowing described earlier are two examples where we can make this comparison directly. However, this approach does not generalize to all cases. We would need a generic comparison method instead.

The comparison problem becomes important if we start building synthetic systems that exhibit features similar to consciousness. How would you compare if the synthetic consciousness were similar to human consciousness? To address such questions, I will discuss the generic problem of how we would compare any two complex dynamical systems itself. It turns out that for complex systems, the conscious person who is doing the comparison is equally important. The conscious person always compares everything relative to his own consciousness. If he is not conscious, the comparison problem is ill-defined. We also need a way to avoid the subjectivity that exists when two conscious people are comparing two similar events.

The new comparison approach I will introduce in this chapter is based on a notion called 'dual equality', which has similarities with dual space representation of chapter 25.

44.1 Difficulties defining equality for complex systems

When you and I look at the letter 'A' with a wide range of handwritings, we both recognize them the same way even though the internal structure of your brain is not similar to mine. How are we able to do this? I will discuss the issues with such comparisons starting from simple systems to complex systems like human beings.

Equality in abstract systems: If we look at an abstract system like mathematics, equality (or isomorphism) of two structures like fractions, functions, groups, rings and fields are defined in a precise and verifiable way. In addition, it is quite common to define equality in multiple ways, some stricter than the others. There is no emphasis on the conscious person, who is necessarily evaluating the conditions of equality. We choose the definition of equality such that the person's subjectivity either is eliminated or is concentrated to just the axioms. The subsequent evaluation itself is just a mechanical task, which can be done repeatedly and consistently. This definition is too strict.

Equality in simple physical systems: Let us see how we define equality for simple physical systems like electrons, protons, atoms and molecules. Firstly, we pick properties that are quantifiable like mass, charge and other quantum numbers. The property called mass is similar for all these systems and their values let us distinguish them. Secondly, we look at an abstract internal structure of these systems, which may or may not be precisely quantifiable. For example, we use structural arrangement (like the angles between chemical bonds or the shape of molecular orbitals) and the same set of subcomponents (like two hydrogen atoms and an oxygen atom in a water molecule or similarly for two hemoglobin molecules) for comparison. Now, to compare physical and chemical properties, we attribute these properties back to the above atomic and molecular structures. A conscious being can verify the equality of these properties using his free will with 100% certainty because of the inherent measurability and precision of the definition. This definition is once again too strict.

Equality in aggregate physical systems: When we start aggregating thousands of simple or complex molecules into solids, liquids or gases, we quickly realize that the above strict definition of equality is not useful. We know that no two tables or chairs will ever be identical using the previous definition. Yet, we want to be able to say whether two tables are the same or not. Our common usage of equality in these cases allows for sufficiently large margins of error. In fact, it is necessary to include imprecision into our definition of equality, for otherwise, any precise definition will guarantee that no two macroscopic real physical systems are equal.

While the precision with the previous definition removed subjectivity of the evaluating conscious being, the imprecision with the current definition brings the subjectivity back. The imprecision is present with quantifiable features (like the number of elements and the thickness of objects) as well as with qualitative features (like the appearance of a circle or an oval). When evaluating nonliving systems like tables and chairs, the subjectivity is manageable. This is because all conscious beings already process the information in a similar way. We can readily identify any distinguishing feature and the errors are within reasonable bounds.

Equality in complex physical systems: Let us now compare features of complex systems like cells or the human brain. The initial strict definition with 'point-wise' comparison is irrelevant because two human beings are not the same with this definition though their features appear similar enough. At the other extreme, definition comparing 'aggregate' structures like hippocampus, thalamus and amygdala is too imprecise because it would claim that even several animals are similar to us, contrary to our intuition.

Our inability to quantify features like tastes, odors and feelings makes the errors unmanageable. At the same time, diverse brain networks do allow the similarity of these features as well. For example, we have seen in chapter 31 that the distribution of S, M and L cones on the retina are significantly different. In addition, the subsequent visual pathways are entirely different for any two given people. If you and I see a red colored object, the set of neurons on the retina and the subsequent paths taken within our brains are extremely different from each other. How can we then say that my appearance of red color is identical to yours? The same applies with every feature discussed until now like space, passage of time, edges, angles, sounds, tastes, odors, consciousness, emotions or any other feelings. The subjectivity as well as the internal structures is far too different between two systems. I will now discuss how to avoid this problem using a notion called dual equality.

44.2 Dual equality – relative to our consciousness

The definitions given previously for comparing two complex systems are what I call as the *primal* definitions. They rely on comparing the components of complex systems in a direct way. In the case of our brain, we directly use the specific neurons that fire and their interconnections responsible for a given feature to compare if two people have the same feature. For example, the primal representation of 'red' color tries to identify and compare the exact neurons triggered and their pathways when your brain experiences a red sensation by watching a red ball, red balloon or a red rose. For cellular systems, we would compare the specific chemicals and their pathways for a given feature. We saw that this approach is not useful because no two complex systems have the same substructures.

Dual representation, on the other hand, tries to define a feature entirely relative to other features (see chapter 25). While it is still possible to look at the neurons ultimately responsible for the dual representation, we do not track at such detailed level. The specific neurons would change over time, depend on your mood, depend on the event you are currently experiencing, on your past experiences, where you are, who you are with and so on. For example, the dual representation of 'red' color tries to define it in terms of the other red objects you can remember like a red ball, a red balloon, a red shirt and so on. When you are just imagining a red color, you are using the dual representation. When you see red color in your dreams, you are using the dual representation. On the other hand, when you are directly looking at a red colored object, you do have both a primal and dual representation for 'red'. For feelings or *qualia,* in general, the dual representation is well defined only when we are conscious.

Therefore, the primal representation can be viewed as a pathway reaching a given fixed set directly from the external inputs or from the inputs that are directly responsible for creating the fixed set in the first place. The dual representation, on the other hand, can be viewed as a pathway reaching the same fixed set indirectly through other unrelated fixed sets (i.e., events or thoughts). You are essentially reaching the same fixed set using the network of fixed sets built through everyday experiences (see chapter 25). For example, the primal pathway to trigger a fixed set for a specific chair is by triggering fixed sets directly for the features of the chair like its color, its shape and other special properties. The dual pathway instead is to trigger the same fixed set by recollecting other unrelated fixed sets like when you saw it, where you saw it or if something interesting happened to remind you of this special chair.

The true advantage of dual representation is when we apply it to feelings and *qualia* itself. Surely, a primal way to experience a feeling like happiness, sadness or fear is to trigger it directly through a real physical event (like a real spider actually crawling on your hand). What is interesting is that you can trigger the exact same feeling like happiness, sadness or fear, using the dual way, purely by imagining some fictitious scenarios (like scenes from a scary movie).

The fact that the dual representation is applicable to all situations – all real physical events as well as all feelings or *qualia* – makes it ideal to use it for comparison. You no longer need to care about the structural and dynamical differences between two brains in order to be able to talk about equality of colors, angles, feelings, emotions and even consciousness. Fixed set is the correct abstraction that lets us work with the dual representation easily. For both you and me, a red colored object will trigger a unique fixed set that we identify with 'red'. This fixed set has no structural or dynamical similarity in both of our brains. Nevertheless, the way we consistently trigger it and link it to other fixed sets is similar for both of us. The relativity of the dual representation to the other fixed sets is fundamental when studying equality within complex systems.

In fact, in this book, I have also shown that all *qualia* can be derived from within physical systems using the corresponding dual representations. For the case of feeling (or *qualia*) of knowing, I have shown this using the necessary and sufficient conditions of chapter 22. For the case of the qualia of emotions, I will derive the corresponding necessary and sufficient conditions in chapter 49. For the case of qualia of realizations, I will derive the corresponding necessary and sufficient conditions in chapter 48. This implies that even though the primal representations are disjoint for two different people, their dual representations are similar.

Furthermore, the appearance of color or, in general, all *qualia* exist only when you are conscious. These properties are, therefore, relative to your membrane of meta-fixed sets. This inherent relativity suggests that we can only use dual representations to compare two people. The primal representations of the membranes for both people are disjoint and are not comparable. For example, as long as you and I have had similar experiences with colors, your dual representation of red is the same as mine relative to our respective consciousness. If you were never exposed to, say, green color during your first year of life, we can say that we do not have the same set of experiences with colors. Even if you have the same color pigments as me, your appearance of color will no longer be the same as mine.

As we grow older, the entire set of meta-fixed sets created from both of our childhood experiences will be interlinked to each other. For example, the appearance of color will be interlinked to the shapes, sizes, sounds, textures, tastes and odors of specific colored objects we encounter (like red apples and red roses). We never have a fixed set disconnected from all other fixed sets in our brain. There is always a 'dual' pathway to reach a given fixed set just as there is a primal pathway (see chapter 25). These fixed sets are linked to your self-sustaining membrane. They become the absolute truths like your emotions and all of your *qualia*. No one can lie to you about them. Therefore, the dual representation is central for defining equality for complex systems. Let me now state this formally.

Definition 44.1 (dual equality): We call a feature that exists in two complex systems as equal in a dual sense whenever the dual representation for this feature is similar in both systems.

I will use the term *dual equality* when I want to explicitly distinguish it with the previous strict definitions of primal equalities. For conscious beings, we can start with simple features first and define equality in the above sense. We can then extend this definition to even more complex features by using the interlinked network of fixed sets. Eventually, the membrane of meta-fixed sets will become the basic network that will be 'similar' for all conscious beings. The similarity of the network is in terms of the relationships between fixed sets, not in terms of the neuronal interconnections. For example, the relationship between the legs of a dining table, the food on the table, the chairs around the table and others are the same for all conscious beings even though the exact set of neuronal loops for each conscious person may be located in different regions of their brain. This network acts as an underlying framework for all conscious beings to ensure similar dual representations. As a result, the above definition of dual equality becomes well defined across all conscious beings unlike any notion of primal equality.

We should note that the above definition of dual equality is quite general. If we look at abstractions, the reason why numbers and other mathematical abstractions are considered same for all conscious people is because the dual space representation of each of these concepts are the same for everyone. For example, everyone learns to count numbers the same way. Everyone uses the same concept of numbers consistently whenever we refer to any set of objects. Hence the relative relationships between these specific fixed sets in our brains, giving rise to dual space representation and abstract continuity, is the same for everyone, even though the primal representations are all different for different people. This is true with logic and every abstraction in mathematics. This dual equality and the similarity of the membrane of meta-fixed sets is the precise reason why such abstract concepts do not depend on conscious beings as well.

Yet, I want to discuss if there are any limitations with the definition of dual equality next.

44.3 Limitations of dual equality

Dual equality defined above appears convincing when we use it to compare measurable features like edges, angles, shapes, sizes, tables, chairs and other objects across two conscious beings. Yet, there is an issue with dual equality when we apply it to feelings and all *qualia*, especially in a philosophical sense. On one hand, viewing consciousness and all feelings as properties of special stable parallel looped dynamical systems, as outlined in this book, convinces that dual equality is well-defined and realistic. On the other hand, viewing consciousness and all feelings as unique to humans and to a given individual, makes us feel that dual equality is not realistic. I want to discuss if there is a way to make the above subtle difference, when applied to *qualia*, more concrete.

Let me use the term *true equality* to refer to the latter view. True equality, then, refers to our desire to know if your feelings are precisely the same as mine in a more convincing way than what dual equality achieves. In other words, is there a more convincing reason for why your appearance of color is same as mine beyond saying that the root cause of this similarity comes from our similar experiences with colored objects? We have a desire to know if your appearance of red is exactly same as mine or if instead your red appears the same way as my blue. We can ask the same question with all *qualia* as well.

These questions become more important when we start building a synthetic conscious system. Does the synthetic system sense and feel red color the same way as we humans do even if we believe that it has acquired a human level of consciousness? One situation when true and dual equalities are indeed the same is when the internal structures of both systems are identical. Even here, researchers have created a notion of philosophical zombie that they argue is possible. A philosophical zombie is another being that is otherwise identical to normal conscious being except that it does not have *qualia* or the feelings. According to the theory presented here, construction of a philosophical zombie is impossible. It is only a theoretical possibility because until now we did not have necessary and sufficient conditions for attaining a feeling of knowing. According to the theory presented here, if you attempt at constructing a philosophical zombie and if that system acquired abstract continuity and a dynamical membrane of meta-fixed sets, it will necessarily have a feeling of knowing (see Theorem 22.1). The existence of a philosophical zombie is just a logical statement that may have appeared convincing. It could not have been proven primarily because until now we did not have a consistent theory of consciousness. The fallacy with philosophical zombie is with the assumption that you can construct such a system when, in fact, there is no known way of constructing a stable system that has millions of moving parts until now. The SPL constructions given in chapter 2 are the first and the only way to do this (for now).

Even otherwise, this unique case where two systems have identical internal structures is uninteresting because it is impossible to create two large-scale complex systems with this property. It is even *theoretically* impossible to create two large-scale complex SPL systems that have identical static and dynamical structure. Recall that free will is a property of large-scale parallel dynamics (see chapter 28). Therefore, even theoretically, if you construct two identical static large-scale SPL structures and let them go, their dynamics will necessarily become different. It is only an imagination and hope in your mind that you are indeed controlling every molecule's motion precisely and identically. Your very presence (i.e., the system trying to control) is different enough to make the resulting dynamics to be different in both systems. One system may have looked at your right eye as you were controlling it and the other system looked slightly to the right of your right eye. This difference is sufficient to produce a completely different cascading neural dynamics.

Now, if there are indeed differences between dual and true equalities for feelings, perception of color, emotions and others, how can we determine them? By definition, dual equality requires that two conscious people have a similar set of experiences if they were to have the same *qualia*. Therefore, if we ever want to know the difference between true quality and dual equality, we should first ensure that the two people being compared do 'not' have the same set of experiences. In other words, I want to consider two conscious people who are not dual equal. Then, I want to see if they both have the same *qualia* or not i.e., if they are truly equal.

Specifically, we have to assume that the membrane of meta-fixed sets formed for both conscious beings are very different from each other if we were to claim that they are not dual equal. Now, a newborn forms his membrane of meta-fixed sets during his first few years. For two newborns to have significantly different membranes, we should place them in considerably different environments and let them experience different and possibly disjoint set of events during the first few years of their lives.

For example, if we want to check if both people have the same perception of color, one person should be exposed to one set of colors or colored objects while the other person is exposed to a completely different set of colors during the first few years of their lives, i.e., never let both of them experience the same color. If person A is brought up in a room without green color for the first few years after birth, I claim that he will not perceive green the same way as a normal human being. This is true even if person A is not colorblind and has all three types of cones (S, M and L) distributed in a similar way to a normal human being. Similarly, if person A, who is otherwise normal, is exposed to only sounds of certain frequencies, tones and intensities and not others, I claim that he will not feel and hear all sounds similar to a normal human being. We have already seen examples of such cases, in a partial sense, with nonnative speakers of, say, English or Mandarin. These speakers are not exposed to the same set of sound frequencies (pitches and tones) as a native speaker. As a result, they find it difficult to distinguish words and, hence, understand them clearly. They also pronounce the words differently (which we term as having an accent). However, once they interact with native speakers i.e., have a similar set of experiences, they will be able to distinguish these different sounds clearly. A better example is with a particular person hearing music or songs in another person's native language. The subtle sound nuances themselves, not the emotions they trigger, cannot be felt by the non-native language speaker while the native language speaker does detect them clearly.

Let us now look at emotions like happiness, sadness or fear. Are there any differences between true and dual equalities here? Creating situations where two people do not have the same set of emotional experiences during the first few years of their lives is a bit difficult. There are ethical implications when trying such experiments with human beings. Nevertheless, there have been some purely accidental situations, which were well documented. They are referred to as feral children. These children have grown up with minimal or no human contact at all. Of course, there has only been a limited study of the psychology of such children. Besides, not everybody agrees with the analysis and the documented details. Yet, several people have reported that these children lack basic social skills. Their emotional development is significantly different from normal human beings (see http://www.feralchildren.com). As an example, except for fear, Amala and Kamala (one case of feral children) did not seem to show any other human emotions. Other feral children also seem to lack some of the human emotions. For example, feral children almost never seem to exhibit joy or sadness, initially. There are some disagreements among researchers with these studies. Yet, from a broad agreement, they seem to suggest that dual equality is indeed identical to true equality i.e., experiences are same (dual equality) $\Rightarrow$ feelings are same (true equality) and experiences are not the same $\Rightarrow$ feelings are not the same.

There is one difficulty to keep in mind when we try to study specific feelings or *qualia*, especially, say the perception of a color. This is that our brain adapts quickly within a few days to the new environment making the experiment tainted. For example, Stratton [146] has documented from his inverted image experiment that he was able to adapt to the new inverted view of the world in just a matter of three days. Even the change in our perception of sizes and distances when we change to a new prescription glasses happens in just a matter of few days (see chapter 33).

There is yet another way to evaluate if true and dual equalities are the same, at least, for some features. This is with taste and odor. The set of experiences for vision, sound and touch are difficult to control for different people. A person needs to be isolated from the external world to have widely different set of experiences. However, with taste or odor, it is not uncommon for different people to have different sets of experiences. There are challenges here as well to identify the root cause of differences in tastes and odors. How can we claim, in a reliable and repeatable way, that the different set of experiences is the cause of different perception of tastes and odors? Are eating a few chocolates at a young age and/or feeding on breast milk a little longer important? Furthermore, genetic factors clearly matter more with tastes and odors compared to vision and sounds. Each of these factors seem to play a critical role to give rise to the wide set of differences among human beings with respect to tastes. We should design experiments carefully to separate each of these factors.

Therefore, as a first approximation, we can say that for now there is no difference between true and dual equalities. However, in the future, it may be possible to understand this difference by setting up any of the above experiments in a controlled and humane way. The generic approach we should follow is quite simple. We pick a feature (including *qualia*) and eliminate all set of experiences linked to this feature for one person. We now check if this feature still exists for this person.

Since dual equality generalizes very well to a wide range of features for complex systems including some *qualia*, we can use it to compare or claim equality of two systems or their features. For now, as a first approximation, dual equality is the closest to measuring true equality of complex systems. For example, a statement like 'a synthetic system is conscious to the same degree as human consciousness' should now be viewed in the dual equality sense. In a philosophical sense, it appears that true equality is just our deepest desire to understand 'why' (as opposed to 'how') we have feelings. In this sense, true equality may, in fact, be an unreachable concept.

44.4 Uniqueness of consciousness and feelings

One consequence of the definition of dual equality, worth highlighting as a special case once again, is that consciousness and feelings are unique across all human beings. For example, the concept of fear for you and me is the same even though what causes the fear may be different for both of us. Similarly, my feeling of consciousness including space, passage of time, free will, self-awareness and others are the same as yours. The comparison should be understood in the dual equality sense now.

What is surprising is that some of the common feelings are the same even if you compare them with most of the animals. The fear or pain of a dog or a cat appears the same as ours. This is true even long before humans ever existed on earth. We can now understand them in terms of the dual equality even with entirely different brain networks.

Briefly, the main idea is that since every human thinks, does, observes and learns a similar set of things, our feeling of knowing is identical as well (i.e., from the dual space representation). This is applicable with some specific features in animals as well. Therefore, the detailed network structure of our brains is not relevant to the feeling of knowing. Instead, the two structures, namely, membrane of meta-fixed sets and the abstract continuity for any given event is sufficient, according to Theorem 22.1.

These two structures rely on the formation of meta-fixed sets and the relationships between them. The links at a meta-fixed sets level is all that is needed, not the detailed interconnections between individual neurons. In other words, the stable parallel loops are important here, not the specific nonlooped interconnections between neurons.

Every species can form stable parallel loops within their individual brain architectures. These give rise to the notion of fixed sets for all living beings, as described in chapter 14, just as they did for human beings. A given fixed set corresponds to a given abstraction of the external world (like colors, edges, shapes, motion and sound wave patterns), its own internal parts (like movement of its body parts) and the interactions between them. The exact set of neurons that make up a given fixed set, the number of neurons, their locations within the brain, the neighboring interconnections and their dynamical properties are no longer relevant. These precise set of neurons are widely different both across different species and within a given species as well. Yet, whatever the neuron representation is for a given event, all living beings observing this same event will trigger its own version of fixed set neurons in a very reliable way. This is the dual representation discussed in chapter 25.

Next, the relationships between the fixed sets are also formed entirely from the same set of events from the external world for every living being. The laws of nature are the same for all living beings. Since consciousness and the feeling of knowing requires only a particular network of fixed sets from the same set of dynamical events, each feeling, if mapped in a given brain, will be identical across living beings. The claim is that this generalization with other species using the necessary and sufficient conditions is valid even with synthetic systems we will build in the future.

44.5 On reading your mind

With the theory of consciousness proposed here, is it possible to use it to read your mind, your thoughts and predict your future actions? This question sounds reasonable given that we do use other theories like Newton's laws, Maxwell's laws, Einstein's laws and quantum mechanics to predict the outcome of future events. For a given configuration of masses, charges and forces, I can use the above fundamental laws to predict how the dynamical system will evolve over time with sufficient accuracy. Can I now use a combination of all known theories including the one proposed here for consciousness to predict your future, treating you as the dynamical system? One difficulty unique with predicting a human's future is from his free will. If you are aware that I am attempting to predict your future, you can simply change your actions to falsify my predictions. We have already seen that free will is a property of parallel dynamics in large-scale SPL systems (see chapter 28). Yet, we would like to know if it is possible to predict your future in a theoretical sense.

Now, among your everyday tasks, we have several situations when your actions are indeed predictable, especially if I know your objective. The set of choices are minimal in these cases. You typically perform your actions based on your past experiences and memory. The current theory can help keep track of your past and form the relationships in a broad sense. Using this, we can predict your actions with good accuracy than without the theory. However, this only works as long as you are not aware that someone is predicting your actions and you are not altering them using your free will.

It still appears that the above difficulty with predictions stem from using an inferior 'simulated' system than your own brain. The question remains if it is possible to create a highly advanced simulated system in the future and use it to predict your future actions including your free will. Clearly, we can improve our predictions using such an advanced simulated system beyond the above obvious cases. However, are there any theoretical limitations that prevent us from predicting your future even after we have the theory of consciousness? In this section, I want to show that there are indeed fundamental theoretical limitations for this problem.

The root cause of this limitation is that an advanced simulated system capable of predicting a conscious human being's free will would itself need to be as complex as the original conscious being. For example, the simulated system should simulate, at minimum, the self-sustaining dynamical membrane and the free will of the original conscious being. For a high enough accuracy in real-time, the simulated system would need to be dynamical. Therefore, the simulated system would start to have a self-sustaining membrane and will exhibit free will. When this happens, the simulated system will deviate with its predictions by using its free will for its own purposes.

As a conscious human performs his everyday tasks, we want the external simulated system to observe his internal structure and his past actions to predict his future actions. The simulated system needs to solve two problems in order to achieve a high degree of accuracy: (a) recreate the static structure and (b) ensure that it can maintain the dynamical state of the original conscious being in an almost identical way. For the former problem, we can create the static structure theoretically to a great degree of accuracy. This includes creating the fixed sets and the associated abstract continuous paths. Maintaining the static structure in the simulated system as the conscious human alters it with new experiences is a tedious process. However, we do not need to update it at every instant because the changes would need to accumulate enough before it would start showing visible external effects.

For the latter problem, we start with an accurate static structure and simulate the dynamics using the same external inputs as the conscious human. The set of choices for the simulated and the real system will be similar because of the similar static structures. How can we now ensure that the actual choice is the same in both systems? For this, we need to simulate his free will with sufficient accuracy. The self-sustaining membrane of the conscious human is sensed and controlled by him at-will. The simulated system would need to align with the dynamics of the conscious human. While the simulated system can track the external inputs accurately, the errors with the internal inputs of a conscious human will accumulate. Even small errors produce large output variations. The resulting internal 'thoughts' are different for both the conscious human and the simulated system. Since the human is capable of altering what, where, when and how much of the inputs are generated and received, minimizing the internal errors is important.

The choices and free will implies that we cannot compute the outputs *apriori*. Therefore, the simulated system needs to monitor the conscious human's external actions as well as his internal dynamical neural firing patterns constantly. The only way to achieve high accuracy and yet compute the outcomes in real-time is to run the parallel physical dynamics even in the simulated system. This forces the simulated system to behave like a dynamical system as well. In other words, it will have a self-sustaining membrane itself, which will give rise to free will. Now, its own free will would cause a deviation from what it is trying to predict. Therefore, even theoretically, there is a limit to what we can predict about a conscious being. Nevertheless, we can use correlations instead of causality to predict some of your future actions with some degree of accuracy.

45

Origin of Consciousness

Let me now briefly discuss the problem of how and when humanity, as a species, acquired consciousness for the first time. We have seen in chapter 23 that the 'minimal' consciousness, requiring just the threshold membrane, is possible within animals and other species as well. The expanded membrane with a rich language, however, is unique to human beings (see chapter 43). In this sense, human consciousness is considerably different from minimal consciousness.

In this chapter, we want to understand how nature created human consciousness by itself without the help of any intelligent being. There is one difficulty that precludes us from discussing the nature's approach in detail. This is the fact that there are limited documented accounts of events that occurred thousands of years ago, which are themselves subject to interpretation. Yet, I claim that the theory presented here is compatible with Jaynes' [79] work on the origin of consciousness. Let me discuss this briefly.

45.1 On the origin of consciousness in humanity

After nature discovered nerve cells as an effective way to communicate, it became possible to link sensory and control inputs easily. When the number of interconnected neurons became extremely large, maintaining dynamical stability becomes a nontrivial problem. One effective approach taken by nature to deal with the problem of stability is to have structural or dynamical symmetries (cf. crystals, polymers and other complex structures).

These symmetries existed even among early life forms. For example, Beklemishev [9] identified four types of symmetries among all living organisms: spherical, radial, bilateral and triaxial. These are common among animals, say, for example, jelly fish's radial symmetry, crustacean's bilateral symmetry and human's triaxial symmetry. Along with Geodakyan [50], they proposed that not only animals but also plants evolve from symmetry to asymmetry. This occurs even at the organ level like in flowers, leaves, fruits and seeds. Zygomorphic flowers with bilateral symmetry like orchids are more evolved than actinomorphic (radial symmetric) flowers like Primula and less progressive than triaxial symmetric ones like Cannaceae. The same trend exists in leaves: spherical symmetry of chlorella, radial symmetry of pine needles, bilateral symmetry of Magnolia leafs and triaxial asymmetry of Begonia or Elms leafs. Or, in embryogenesis: spherical zygote, radial gastrula, bilateral embryo and triaxial asymmetric child (see Geodakyan [50]) .

Complex brain structures also needed stability for basic survival. Bilateral symmetries were once quite common in our human brain, according to Jaynes [79]. Bicameral mind, as discussed in detail in Jaynes [79], is one such symmetry unique to our brain. Unfortunately, these symmetries hinder the growth of complex structures and the creation of highly interlinked fixed sets. As a result, a large and intricate self-sustaining threshold membrane of meta-fixed sets is not as easy as it is now. Our current brain's asymmetric architecture helps greatly with creating a dense and intricate network structure as described in this book. Therefore, we can say that the structural requirements for human consciousness as discussed in this book and the hypothesis proposed by Jaynes [79] that the breakdown of the bicameral mind is critical for developing consciousness seem to align.

Furthermore, language that is expressive enough to include the concept of 'I' is quite critical for people to know about their own consciousness, as suggested by Jaynes [79]. In section 43.1, I have also shown how such a rich language helps expand our membrane beyond the threshold state. I now refer the reader to Jaynes [79] for more historical details and thorough analysis of these along with other possible interpretations.

46
Necessity of Stable Parallel Loops

With the discussion so far, we have seen that stable parallel looped framework is sufficient for studying the problem of consciousness. We were able to express features like free will, self-awareness, sensing space, passage of time and others within this framework. However, an interesting question is whether this framework is necessary for understanding each of these features of consciousness. The framework used by neurons within our brain (and within the brains of other animals) is already a stable parallel looped one. We do not have any other concrete example to suggest whether the framework is necessary or not. If we want to build a synthetic conscious system, it is unclear if we should use this framework or if some other framework might work, equally well.

One typical framework we use with synthetic systems is the architecture of serial and parallel computers, Boolean algebra, electronic memories using hard disks, RAM and others. With tremendous technological advancement we have had in the last 30 years, it appears logical to ask if we can continue using the same architecture to build a synthetic conscious system as well. Surely, there have been attempts to build human-like robots, artificial intelligence systems and others using the current architecture. I want to mention here that the architecture we currently use for computers is not a stable parallel looped dynamical framework. However, if we advance the technology further, miniaturize the components, build more parallel CPUs, more RAM, hard disk space and others, will we one day build a synthetic conscious system, assuming we *do not* change the architecture to a stable parallel looped one? If that is not possible, in what direction would the architecture evolve? Would it converge to a stable parallel looped architecture?

In this chapter, I will address each of these questions. I will specifically show that any architecture that attempts to build a synthetic conscious system would necessarily converge to a stable parallel looped one described in this book. The existing computer designs simply cannot create a synthetic conscious system.

46.1 Principle of existence

I will derive the necessity of stable parallel looped framework for consciousness from a seemingly obvious principle. The principle states conditions under which it is impossible to know the existence of a given physical object. A person can know the existence of a given object only if an input from the object ultimately reaches one of his sensory organs through a well-defined causal physical mechanism. If no inputs or effects exhibited by the object reach his sensory organs, there is no loss of generality in assuming that the object does not exist.

For example, if light from distant stars have not reached the earth yet and if they do not produce any measurable effects on earth (gravitational or otherwise), then these stars do not exist for you. This does not mean that these stars do not exist at all. Rather, it means that 'you' have no way of knowing about its existence, at least, for now. In the future, it may be possible to invent a device that can detect some signal like gravitational waves from the stars. When we transform this signal into a form that our senses can detect (like a ticking sound or an image), we would then know about its existence. Similarly, if you close your eyes, the existence of objects in front of you can no longer be guaranteed by you (as they could have been moved or even destroyed) unless you can, say, touch or hear them.

Definition 46.1 (existence principle): The principle of existence states that a system infers the existence of an object if and only if the object produces a causal dynamical effect that can be sensed by the system.

In section 24.6, I had discussed this principle when defining a sensor. When the sensory surface (like the retina) receives inputs, it is nontrivial to determine whether the source is from the outside, inside or at the boundary of the surface. We needed considerable additional structure, a form of free will and others to make this determination. If, on the other hand, the sensory surface does not even receive any input, it makes little sense to assume that the real physical object exists. We can state the principle of existence loosely as "you cannot create something (physical) from nothing".

It is likely that the principle of existence stated here may not represent the external world in an accurate way. For example, we cannot see or sense air in a calm room nor can we see or sense germs. Yet, we cannot say that they do not exist. Here, we additionally believe in some form of continuity and generalization. If you have sensed air in the room once, then as long as continuity and generalization is not broken, we assume its existence even in the future. Similarly, the principle of existence excludes ghosts and supernatural beings unless they can be reliably and repeatedly detected. Therefore, the principle of existence states that if a system does not receive any type of signal from an object, then by assuming that the object exists, we only run into logical inconsistencies. Instead, it is safer to assume that the object does not exist until the system senses a measurable effect from the object.

In fact, we cannot create even a thought with no causally related changes in our brain. You can think of any random thought and switch between thoughts as quickly as you want. However, there should be dynamical changes in the form of neural firing patterns that are directly related to each of these thoughts and the switching of thoughts. If there are no such dynamical changes, you can never have such thoughts or, more precisely, it is even ill-defined to say that you have thoughts under those circumstances. The reason is that if there are no dynamical changes in your brain, then it is impossible to distinguish between these two cases: (a) there is nothing happening in your brain or (b) any one of the imaginary thoughts is occurring in your 'mind' (i.e., not in the physical world) like, say, you are thinking about some place or some event, walking somewhere, touching some objects and others. Since there are no causal dynamics to confirm or deny both of these possibilities, it is accurate to assume that there is 'nothing' happening in a causal sense in your brain instead of saying that 'everything' (or 'any one' of the possibilities) is happening in your brain.

The reason for picking choice (a) above is because our consciousness arose causally from physical dynamics. A conscious human is created causally from a non-conscious embryo. This process is repeatable. It takes a few years from birth to your early childhood. During this period, the process is entirely causal. This process has been occurring and still occurs reliably for every human being as he grows from a non-conscious embryo. Now, even going a step back, the non-conscious embryo or single cells, in general, are also created causally using stable parallel looped architecture from inanimate objects (see chapters 52 and beyond, on the origin of life). Therefore, since we have causal mechanisms from inanimate objects all the way to human consciousness, any thought that ever occurs within our brain must necessarily be causally related to the internal dynamics of our brain.

The principle of existence is a simple principle that forces a certain causal structure in systems, especially if we want to build systems that are guaranteed to be conscious. Let me now discuss what these structures imposed by this principle are.

46.2 Necessity of SPL structure for synthetic systems

Consider memory as an example. Let us compare how a human and a computer stores and accesses memory to highlight the importance of the principle of existence. Let us say that both you and your computer have stored the memory of your son's birthday party. The computer stores a video of the birthday party in a hard disk whereas you have stored your personal experiences as memory (see chapter 7) within your brain network. What makes it possible for the computer to detect the existence of this video at this instant? The fact that it is stored in the hard disk is not enough. Similarly, in the case of a human brain too, the fact that you have had a memory of your son's birthday party within your brain network is not enough to say that you know about this memory at this instant.

Let us apply the principle of existence to both of these cases, the computer and your brain. The term 'system' in the principle of existence is your consciousness (or the computer) and the term 'object' is the memory of your son's birthday party. The principle of existence says that you do not know the existence of your son's birthday party as a memory unless there is a causal dynamical effect from this memory location within your brain to your consciousness. If we think in terms of a sensor as an analogy, the principle of existence says that a sensor does not know about the existence of an object unless some signal from the object (like light or sound) reaches the sensory surface (like eyes or ears).

In the case of our brain, the sensory surface is the self-sustaining membrane of meta-fixed sets, which represents our consciousness. The object is the region (or a collection of linked meta-fixed sets) corresponding to the memory of your son's birthday party. The principle of existence implies that unless (a) the meta-fixed sets corresponding to your son's birthday party are excited and (b) the resulting neural firing patterns reach the self-sustaining membrane, you do not know the existence of your son's birthday party. You may have memorized your son's birthday party somewhere within your brain. However, unless neurons from those regions fire and these signals reach your consciousness layer, you do not know about your son's birthday party.

This is true with any memory stored within your brain. Every human has an enormous number of memories stored in his brain simply through his everyday experiences. Yet, the only memories we can remember at this instant are the ones that have reached (or intersected) the self-sustaining membrane. The principle of existence applies to all other memories. If our consciousness were to become aware of a given memory, it is not enough to trigger those regions in our brain (i.e., item (a) above). They should extend and intersect our self-sustaining membrane as well (item (b) as well).

Therefore, the regions containing the memory should self-sustain for a sufficiently long time so our self-sustaining membrane can expand to this region as well. Stable loops are the best representation of memories as a result. They have the property of self-sustaining their dynamics using minimal energy (and they also have several other useful properties as described in chapter 3). In this sense, stable loops are necessary for memory.

Let us see the necessity of parallel loops as well. When we recollect our memories, we seem to think that all of our related experiences are simultaneously present in our thoughts. There is an illusion that all of them are instantaneously recollected and in no specific order. In fact, the word memory has a sense of pervasiveness or simultaneous existence instead of a linear process of individually recollecting parts of it. For example, if you think of recollecting a person like your son (or an object like your home), there is a sense of knowing the entire human instantaneously. It does not seem to occur as a slow process in which you recollect his eyes, his ears, his nose and other features in some order, until you eventually create the memory of your son.

The very definition of memory of any object already has the notion of simultaneous existence even though each object is composed of a large number of subcomponents. Similarly, circle as an abstraction is memorized and recollected as a single entity even though millions of neurons (and several abstract subsets) would need to be triggered. Cohesive visual perceptions have similarities to this form of existence as well. This applies to any abstraction including thoughts, words in a language, sounds, concepts and others. For example, we know a word as a single entity in one instant and not as individual letters. This is especially true if we had already formed a memory for the word as opposed to a never-before-heard word.

Therefore, the principle of existence implies items (a) and (b) above, which then implies that stable parallel loops is the correct model for our memories.

How do we apply the same principle to the computer memory case? Just as our brain stores several of our memories within the SPL network, a computer stores memories within the hard disk. The term 'object' in the principle of existence refers to these memories in the hard disk. What is the equivalent notion of 'system'? It is not the entire computer just as it is not the entire human. In the case of human, it was just the self-sustaining membrane of meta-fixed sets i.e., our consciousness.

In the case of a computer, the CPU is the subsystem that access data from RAM, registers and the hard disk. In the simplest case, a CPU has an instruction pointer that reads data in a serial order. Until the CPU reads a given memory from the hard disk, we can say that the computer is unaware of the memory. The principle of existence applies in this case too. The computer cannot know about any memory unless the CPU accesses it. Since our computers only have a few CPU's, the amount of information that they can access simultaneously is too little. Even if the memory is loaded from the hard disk to the RAM, unless the CPU starts accessing this data, the existence of this data is unknown from the principle of existence.

At any given time, a computer has access to just a few bits of data, namely, those bytes that are currently accessed by the instruction pointer. The rest of the data, even if present in the RAM, is not accessible at this instant. When the CPU accesses new memory, it has to remove the existing memory. The previous memory is lost. The principle of existence implies that the CPU is unaware of past memory as well as future memory. At any given time, the CPU is only aware of the current memory of, say, about 32 bits of data. The sensor being the CPU's instruction pointer receives data from the future memory locations and removes data from the older memory locations while keeping about 32 bits of data at any given instant. When the existing 32 bits of data is deleted in order to load new data, the CPU has lost all information about this deleted data. This happens at every few clock cycles. Therefore, the computer never has any past information at a given instant except the current 32 bits of data.

Even if a computer is playing a video of your son's birthday party, at any given time, it only has access to 32 bits of data. The continuous video played is only understood by conscious humans because we can simultaneously access an entire self-sustaining membrane of data. This is millions of times larger than 32 bits. The computer, on the other hand, does not. As mentioned previously, you can recognize your son instantaneously in parallel. You do not recognize him in a serial order like first his eyes, followed by his ears and so on. A computer, on the other hand, does process 32 bits of data simultaneously and everything else in a serial order.

A human processes millions of bits of data simultaneously in parallel while it processes some parts of the data in a serial order as well. When we process data in a serial order (like a thought), we do not delete all the existing memory at once (like how a computer deletes all 32 bits of data at once to load a new 32 bits of data). We keep significant portion of the millions of bits still in an active state through the stable parallel looped excitations while expanding our dynamical membrane to other regions. In this way, we maintain continuity whereas existing computers do not have continuity. Some of the excited loops corresponding to the memory triggered half-an-hour ago, say, are still part of the dynamical membrane unlike with computers.

Therefore, to truly make a computer aware of large number of memories, it should have a large number of CPU's which can simultaneously load large amount of data. The computer design needs to be modified to have the equivalent of a self-sustaining membrane that holds a large amount of data simultaneously in a ready-to-access state. The existing computer designs have a few limitations as a result: (i) too little data is in an active state like, say, only 32 bits of data, (ii) all of this data is deleted when new data is loaded and (iii) there is no continuity of the data that transitions from an active to an inactive state. The principle of existence implies that all of the data from the past and the future are completely inaccessible with such a design leaving only 32 bits of useful memory.

Such a computer design can never become conscious. Any computer design that attempts to become synthetically conscious must have a large amount of memory in a simultaneously excited state. The principle of existence should be applied to a system that has such a large amount of simultaneously excited state unlike just 32 bits of data. The only known way to achieve this is by using SPL architecture as described in this book. The loops should be interconnected to form the equivalent of a self-sustaining membrane of meta-fixed sets. This membrane then becomes the system with millions of bits in a simultaneously excited state.

Every other memory stored in hard disks and other regions can be loaded incrementally into this region while partially deleting some other memories. These newly loaded memories must self-sustain long enough that the equivalent membrane has a chance to interconnect with it. A stable looped design is, therefore, necessary to achieve this. The principle of existence can now be applied as the newly loaded memories intersect the membrane causing the membrane to become aware of the new memories. Abstract continuity can be maintained between all of these past and future memories because they are all simultaneously excited and are part of the self-sustaining membrane.

The principle of existence, though intuitive and simple to state, forces the existence of a stable parallel looped design and a self-sustaining membrane that expands to new regions of memory and contracts from older regions of memory.

47
Synthetic Consciousness

In the previous chapters, we have seen what the necessary structures are to understand the problem of consciousness. In Theorem 23.2, I had stated the necessary and sufficient conditions for a system to become conscious. In this chapter, I want to briefly discuss how we can use these results to create a synthetic conscious system. If you recall the discussion in the previous chapters, you would notice that most of the mechanisms and constructions described are not unique to our brain. The memory model, the fixed sets, abstract continuity, free will and others are properties of SPL systems instead of just the brain. Even though, we have been associating consciousness only to a human being, all the discussions until now suggest that they are applicable to more broader SPL systems. In this chapter, I will address consciousness for systems other than human beings. This is the problem of synthetic consciousness. In this chapter, I will only discuss synthetic consciousness from a theoretical standpoint, not the practical issues of actually building it.

47.1 Consciousness – a property of SPL systems

An important conclusion from the solution to the problem of consciousness is that consciousness is a property of a specific SPL system (recall the discussion in chapter 23). All necessary and sufficient conditions of Theorem 43.1 only require a specific structure within these SPL dynamical systems. Two main structures are the self-sustaining membrane (either the threshold or the expanded one) and abstract continuity, both of which depend on the notion of fixed sets. Using these structures and additional dynamical processes like the fundamental pattern, I have derived all other features like attaining a feeling of knowing, free will, sensing colors, physical and geometric properties, sensing space, passage of time and self-awareness. Most of the discussion in the previous chapters only relied on how to use these stable parallel looped substructures and how to combine them in unique ways. The necessary and sufficient conditions imply that if these structures do not form, the system will not exhibit the feature.

These structures are not arbitrary. Not every species have these specific structures within their brain. Only human beings have the necessary asymmetry that allows the creation of each of these unique structures. Now, if we want to build a synthetic system that exhibits the same set of features, the necessary and sufficient conditions derived here will come to our rescue.

The abstractions I have identified here within the stable parallel looped systems are the correct set of substructures that allow us to study the problem of consciousness and other human abilities in a generic way. These abstractions essentially decoupled the dependency of the problem of consciousness on the unique and detailed structure of neurons and their interconnections.

For example, in chapter 19, we have seen how to represent all the hierarchical layers of abstractions into just two levels. One of this is the low-level neuron dynamics and the other is with fixed sets, represented as stable parallel loops. This result and the subsequent analysis implied that consciousness depended heavily on the higher level (i.e., fixed sets) within the stable parallel looped dynamics and not so much on the low-level neuron dynamics. In other words, we could replace the detailed chemical neuron dynamics and even the neurons themselves at the lower level with any other synthetic materials or subsystems as long as we can guarantee a similar stable parallel looped dynamics using the new materials at higher levels. If the synthetic materials guarantee fixed sets and the ability to create and interconnect them using the new low-level dynamics, the rest of the solution for the problem of consciousness would continue to work.

This should not come as a surprise. It is not true that life on earth and consciousness among humans is critically dependent on every detailed chemical within our cells or neurons. Rather, the functional aspects i.e., the abstractions, derived from the dynamics are the most important attributes. Abstractions without the underlying dynamics cannot help with the problem of consciousness (see section 13.1). Other variations at the lowest level are possible as long as you guarantee features like fixed sets and abstract continuity using a stable parallel looped architecture.

A simple analogy is to say that the concept of magnetism was initially discovered using naturally occurring magnetic materials and iron. However, once we have discovered magnetism, there are other ways to create magnetism using the basic principles of electricity and magnetism. Similarly, neurons may be the first way to create conscious beings. However, once we have discovered the basic principles for creating consciousness, which are the SPL designs, fixed sets, abstract continuity, dynamical membrane of meta-fixed sets and others described in this book, the limitation on neurons can be overcome. The lower layers to create fixed sets and others can now be changed to use newer materials (like solenoids and others to create magnetism). Unfortunately, with consciousness, we have not discovered anything other than human nerve cells for now, among both naturally occurring and synthetic ones.

The main advantage with nature's solution to human consciousness relying on neurons is that a neuron can be discovered entirely naturally from a single cell without the help of any intelligent being. I will describe several of these later when discussing the origin of life from inanimate objects. The underlying mechanisms are similar, though they rely on chemical SPL designs.

One other reason to convince ourselves that neurons are not necessary for consciousness is to realize that neurons in all species do not use the same set of chemicals and DNA. Yet, their functional aspects are identical. These functional aspects are what we have abstracted as the key principles (like fixed sets and others) in this book for creating consciousness. This already suggests that multiple variations within a neuron, while preserving the abstract functionality i.e., fixed sets and stable parallel loops, can still produce similar features. For example, fear in an animal is comparable to fear in humans as discussed in chapter 44. Similarly, a dog avoiding obstacles just like a human should have a cohesive visual perception comparable to humans even though their neurons are sufficiently different (i.e., they are not interchangeable between the two species).

Therefore, synthetic consciousness, though a theoretical possibility for now, is feasible using the theory proposed in this book. The unique structure of SPL systems and the necessary and sufficient conditions has turned consciousness and all other human abilities into true properties of dynamical systems. We do need to overcome the practical difficulties of building a system with millions of dynamical components. A stable parallel looped architecture truly helps here as discussed in chapter 2. As another example to create large-scale dynamical systems, I will show in chapters 52-68 how to create life for the first time from inanimate objects using the same stable parallel looped architecture.

47.2 Following sequence of steps of a human infant

Using the results derived until now, let me summarize some of the critical processes necessary for building a synthetic conscious system. I will be brief here because most of the steps are similar to what an infant follows in the first year of his life. These steps are well documented in most books (see Hathaway et. al [68]).

We start with a stable parallel looped system with several sensors and actuators. We use them to measure and interact with the external world. It is best to arrange them structurally similar to how it exists in a human brain, as our initial design. This is because these specific arrangements did favor and proved the existence of one solution to the problem of consciousness. The fact that there were only minor changes in the relative structural arrangements of brains across widely different species for millions of years may just be an artifact of chemical evolution. It was not with a distant future objective of creating consciousness in humans.

The system's architecture should include the memory model discussed in chapter 7. The structural and dynamical combination should now allow the formation of several fixed sets using the sensory and actuator inputs. Some of the mechanisms for forming the links between them are, say, the fundamental pattern and others discussed in chapters 16-18. Included within this design is an ability to grow, change, delete and create new fixed sets and alter relationships between existing fixed sets. Predictions and abstract continuous paths should also be possible using the set of mechanisms described in the previous chapters. This makes the internal structure of the stable parallel looped system sufficiently more complex as we expose it to new external inputs.

The next critical step is to allow interactions with the external world using its existing sensory-control mechanisms to create a set of 'experiences' memorized within the stable parallel looped design. In effect, the set of fixed sets begin to grow enormously to include all everyday sensory inputs like visual, sound and tactile.

During this stage of development, the synthetic system essentially follows the same sequence of steps as a human infant. A human infant is born without consciousness. He becomes conscious after about a year. At this stage, variations in the precise sequence of steps do not matter significantly as can be seen with infants reaching similar developmental milestones with entirely different experiences.

While it is possible to list the sequence of steps followed by an human infant during the first year of his life, I will instead refer the reader to more popular textbooks on this subject (like Hathaway et. al [68]). The dual space representation implies that following the same sequence of steps are sufficient to produce the same relationships between the concepts of the external world. They are both necessary and sufficient to build a synthetic conscious system. Let me briefly summarize a minimal set of developmental milestones below. This list is grossly incomplete. In fact, all the details mentioned in the previous chapters are more important than the ones listed here. This list can, therefore, be viewed as those items that I did not get a chance to go over in greater detail in the previous chapters.

- Start with an initial system with several innate abilities analogous to mechanical reflexes like sucking reflex, startle reflex, crying for hunger and an ability to move hands, legs, eyes and head randomly. These reflexes combined with emotional explosions produce a drive to do additional tasks initially.
- Next, create a number of fixed sets and the relations between them as memory through near-simultaneity of events, control through emotions, the fundamental pattern and other mechanisms. Examples of these are visual fixed sets like simple edges, angles, colors, shapes and other geometric features linked to a tangible object.
- The initial random movements of, say, hands and legs become more structured using the fundamental pattern. Examples are tracking a moving object with eyes, turning to a sound and others. Self-learning is now possible (happens at about 3-6 months of age).
- The ability to decouple the external language from the internal neural dynamics is now possible (about 6 months of age). The drive to perform actions slowly transitions from emotions to logic.
- Linking emotions and actions to words begin to form at about 9 months of age. For example, 'yes' and 'no' are associated to positive (say, happiness) and negative feelings (say, fear or sadness) respectively. However, language quickly decouples from emotions. Society pressures start to cause conflict between emotions and language-based actions. These are initially unique to him. Only after several years, this logic starts to agree with other people. A choice between emotions and logic starts to emerge from these conflicts.
- As described in chapters 34-40, it is possible to sense space, passage of time and all other geometric features as well. A self-sustaining membrane of meta-fixed sets begins to stabilize for several hours a day. Using internal language, it is possible to expand the membrane beyond the threshold state. The feeling of knowing becomes possible for a number of fixed sets like space and passage of time using the large repository of meta-fixed sets and abstract continuous paths. Consciousness begins to emerge as a property of the complex system.
- The realization of consciousness is a much slower process. Learning with help from an adult, language, speech, creating and formulating grammatically

correct sentences is now possible. The power of self-reference of the language and the ability to define new concepts or definitions in addition to performing tasks creates the possibility of self-awareness. The feeling of knowing for a concept of 'I', developed initially through cause and effect, say, by pointing to itself, becomes uniquely distinguishable. Knowing oneself and self-awareness through abstract continuity is now possible.

Each of the above steps and several others described in the previous chapters can theoretically occur within a synthetic system, especially if the dynamical system has an SPL architecture.

48

Realizations

In the previous chapters, we have primarily focused on the problem of consciousness. Here, among the different types of feelings that we experience, I have presented necessary and sufficient conditions for the feeling of knowing (see chapter 22). However, there are other types of feelings that are quite different from the feeling of knowing, the prominent one being emotions. In this chapter, I will discuss another closely related feeling, namely, realization before I discuss emotions in the next chapter.

Realization is one of the most basic human ability that distinguishes us from other living beings. When we see circular ripples in a pond, we 'realize' that it is actually circular, the same circle that we encountered a number of times before, under different circumstances. It is more than just knowing that the ripples are circular. Typically, we say that we have 'realized' some event or fact when it happens for the first time. For all other subsequent times, we say that we 'know' the fact instead. The one aspect that appears important to capture whenever we think of realization is a notion of simultaneous occurrence of events leading to an explosive effect. For example, when we see the circular ripples, a number of other recently stored memories of a circular shape are triggered simultaneously to form an explosion giving rise to a feeling of awe. When this happens, several unconnected neural networks link to form bigger and cascading loops of firing patterns.

It is meaningful to talk about realization only after you become conscious. When you are in deep sleep, under general anesthesia or other unconscious states (and even hypnotized state), you do not have a feeling of realization. Intuitively, an explosion produces a feeling only if it produces a 'shock' to the self-sustaining dynamical membrane of meta-fixed sets. If there is no membrane i.e., no consciousness, there is no 'shock'. Hence, there will be no feeling of realization. In this chapter, I will also discuss why abstract continuity is critical for this 'shock' to be effective, just as it was necessary to attain a feeling of knowing (see chapter 19).

48.1 Realization scenarios

Experiencing realizations is actually quite common in our everyday life. For example, when watching a movie, if you uncover the plot yourself, you experience the feeling of realization. It typically accompanies a feeling of happiness or content that you discovered the inner meaning. However, sometimes it does result in sadness or even fear like with tragic or horror movies. When a person realizes the inner meaning of a joke, or when a doctor identifies the correct cause of a critical condition after having tried several possibilities, they experience a feeling of realization. You also experience realization when you solve a difficult mathematical, engineering or scientific problem. This happens when all the pieces of the puzzle finally fall in place, after several failed attempts. Even with games like chess or a jigsaw puzzle, when several small moves all add up to produce a complete solution, you experience the feeling of realization. This feeling has a quality that is unrelated to the details that caused it.

When we think of realization, we are looking beyond the details of the event. We are thinking 'about' the events and we are looking for a broad pattern common across several events. There are two essential aspects for realization. Firstly, we need non-uniqueness i.e., the same event or object should have been already encountered previously. Secondly, we need to attain a big picture from small details with a large cumulative effect. After we experience realization, we store both the big picture as well as the details into our neural structure as memory. It is important to keep in mind the distinction between realization and simple recognition. In each of the above examples, you attain the feeling of realization only when you successfully relate several seemingly disjoint facts that seem to have occurred under entirely different circumstances.

There is a considerable difference in the types of events that produce realizations and the degree to which they produce in humans versus other animals. A dog walking along two different paths every day, one shorter than the other rarely realizes this difference unlike a human. These types of realizations require abstract thinking which animals lack to a similar degree to humans. Of course, animals do have realizations of their own. However, they are relatively less frequent. A diverse set of abstractions, in general, provide humans with an enormous number of possibilities for realizations unlike any other species.

Realization clearly has an evolutionary advantage as well. It is conceivable that the discovery of a sharp spear as a tool to hunt for food probably came from a realization by looking at a number of scenarios where animals were killed when they were hit by a sharp object. Furthermore, some of the highest forms of creativity in humans like, say, discovering Newton's laws of motion and Maxwell's laws of electromagnetic theory has produced powerful technological advancements that gave us significant advantages over animals. Without realization, we could be repeating the same task multiple times without identifying unique differences.

This suggests that realization is actually a nontrivial skill. It is quite common that humans do not realize several facts even though we encounter and perform similar tasks repeatedly. For example, you would have read so many novels and books. Yet, you probably did not realize that the even numbered pages are always to the left and odd numbered pages are to the right. In fields like mathematics, science, engineering and business, it is even more apparent that we do not realize a large number of facts. In fact, every discovery and invention can be considered as such an example in which some individual realized the connections between seemingly disjoint topics.

Now, even if others have already realized a fact, it is sometimes not easy for us to realize the same fact without a great deal of training, practice or experience. Furthermore, the degree of realization of the same event is different for different people and is different between kids and adults as well. For example, abstractions in mathematics are simple for some, but requires formal mathematical training for others to realize this. There are similar situations with modern art, classical music, politics, business and others. Realization is, therefore, quite personal with both emotions and consciousness playing a role in specifying the intensity of it.

48.2 Feeling of realization

In section 9.11, I had introduced an important pattern that can form with parallel loops i.e., explosions – a small cause producing a very large effect. We have seen that it can help store memory with just a single experience instead of the traditional approach where we need to practice several times. Explosion also results when two or more disjoint regions of our brain corresponding to different topics are linked together.

Let me now describe the process for all scenarios of realization. When we perform a given task in multiple different ways, in the beginning, we are not necessarily aware of the similarities among the multiple approaches. Yet, we store memories of all these multiple solutions in a mechanical way. The network of parallel looped paths (or fixed sets) start to converge at multiple locations. The converged fixed sets correspond to the common topics or themes for each of the multiple solutions. For example, some of these common topics among multiple events could be at similar words, similar logic, similar images, similar objects involved, similar sounds produced and others. These common topics act as hubs for multiple memories.

You begin to form a complex network of multiple solutions with junctions of fixed sets at these common topics, possibly, at distant regions of the brain. The interconnections between multiple such junctions keep getting closer through everyday experiences. Suddenly, one day, a specific event bridges all the remaining gaps by interconnecting them across a large set of these junctions. When this happens, you have an explosive effect of firing pattern for this specific event. It is not easy to predict when or which event will trigger an explosion for a given person. The strength of explosions are also different for different topics.

Explosion is a natural mechanism for a stable parallel looped system. We have already seen it used in the fundamental pattern of complex systems (see chapter 16). The structure of the fixed sets and the brain network that is forming, the sequence of multiple solutions that you tried and the frequency with which they were practiced, all contribute to create the explosion. It is this explosion that makes you 'realize' that there is a connection between all the facts that this explosion is composed of. Realization does not occur if you did not experience and memorize each of these facts already.

When an explosion occurs and you have a membrane of meta-fixed sets, it causes parallel excitations across several regions of the membrane. Since the dynamical membrane is self-sustaining, a large number of new simultaneous excitations across several regions is produced, resulting in a strong cascading disturbance of neural firing patterns. This is what I termed loosely as a 'shock' for complex stable parallel looped systems, in the beginning of this chapter. The parallel nature of the above disturbance makes it hard for us to focus on one or more subtopics at that instant. It creates a feeling as if you know all of them simultaneously.

When you do realize something, you actually have a feeling of knowing of each of the facts that are part of the explosion. This implies that you have abstract continuity for each of them. This is necessary for the feeling of realization as well. If you are glossing over a book without attempting to understand each of the sentences (i.e., no abstract continuity or a feeling of knowing), then you never have a feeling of realization with respect to what you are reading.

In fact, abstract continuity for each of these topics itself helps in creating an explosion. This is because abstract continuity for each topic excites a wide band of neighboring fixed sets 'orthogonal' to each abstract continuous path itself (see Figure 21.2). When you do not have abstract continuity, these neighboring fixed sets are not excited and only the fixed sets along the main path are excited. This is too weak to produce an explosive effect even if you have multiple such converging paths.

The dense network of excitations from multiple abstract continuous paths resulting in explosions, therefore, leads to the feeling of realization. On the other hand, if you did not have a self-sustaining dynamical membrane of meta-fixed sets (i.e., you are not conscious), there will be no such shock to result in an explosion. In this case, you will not be able to have a feeling of realization. Let me state this result as the following theorem.

Theorem 48.1: A system experiences a feeling of realization if and only if it has (a) a self-sustaining threshold dynamical membrane of meta-fixed sets, (b) abstract continuity within several critical junctions and (c) explosive dynamical excitations at different sub-regions of these critical junctions.

In the next chapter, we will see how the feeling of emotions is closely related to the above theorem. Specifically, we would specialize the above critical junctions to sensory-control regions of our brain (see Theorem 49.1). In fact, when discussing realizations, it is quite common to exclude these sensory-control junctions. We, instead, focus on regions corresponding to higher-level thoughts and abstractions.

What happens after a realization has occurred in our brain? The explosion itself has sufficient energy to create new interconnections between neighboring neurons that were previously not connected, in addition to reinforcing existing connections. This produces a strong memory of each of the individual sub-event. One outcome of this is that you now have the ability to move from one topic to an entirely different, possibly disjoint, topic through these interconnections. This produces possibilities of new relationships between events that you simply could not guess before the explosion (or realization). In a sense, you feel that your knowledge and experience has taken a big leap.

Furthermore, you will be able to remember the event easily even if you experience it only once. As a rule of thumb, if you want to memorize something quickly, you need to bring your brain's state such that it simulates a scenario capable of creating an explosion and, hence, attain a feeling of realization. You can do this, say, by thinking about a number of closely related topics, alternate approaches, several analogies or be emotionally involved so you can create the necessary non-uniqueness. This will trigger multiple parallel loops in your brain which can lead to an explosion, realization and memory. The above suggested approaches like emotional involvement and analogies are already well known and practiced by several people.

Realization mechanics is, therefore, simply explosions. There is no specific location in the brain responsible for realization because explosions can happen anywhere in the brain. It can also occur across distant regions in the brain. By the very nature of the explosions, realizations are intense and effective for a short period of time. Events and tasks performed days or weeks before, have no direct relationship to realization even though they are responsible for memory and, hence, have an indirect relationship. The actual realization feeling peaks fairly quickly and dissipates just as quickly. Subsequent recollections of the same event is not as explosive as the first realization. This is because the connections and the 'structural' parallel loops are already formed. There is no effective magnification of small excitations that lead to large sets of parallel loops in the second and subsequent times.

49
Emotions

The most common set of emotions we experience regularly during our life are happiness, sadness, fear, anger, surprise and disgust. However, the set of emotions and their classification into appropriate types is far from clear. In an attempt to understand emotions, researchers and philosophers alike have studied emotions from several different points of view. Some have tried to classify them as primary versus secondary (or blended) emotions. The above list correspond to primary emotions. Some others have studied emotions from an evolutionary point of view to see how they give a survival advantage. There were additional interpretations based on perception and cognition. From a neurobiological framework, we have identified a group of structures called the limbic system as the primary region for emotions. Later, regions like amygdala, prefrontal cortex, anterior cingulate, ventral striatum and insula have also been identified as important. In addition to these studies, there were other qualities of emotions that were well documented like the healing power of emotions, emotional instinct and the effectiveness or persistent nature of the drive to do a specific task when emotionally engaged.

In spite of significant research on emotions, there are several open questions that have been left unanswered. Our objective in this chapter is to answer them. I will not be discussing the problem of classifying emotions or several of the above problems addressed by others. I will instead look at how emotions originate within a specific class of stable parallel looped systems.

49.1 Open questions

Let me now list a set of questions that have not been satisfactorily answered until now. My main focus in this chapter is to address them.

1. *Emotions for dynamical systems*: Emotions seem to occur only in complex systems and not simple systems. However, what are the properties of these structured complex systems that make them suitable to exhibit emotions? Are there other complex systems like animals that also experience similar emotions? Can we artificially create a structured complex system so it will have emotions similar to what human's experience?

2. *Origin of emotions*: Let us pick one class of complex systems like human beings that already experiences emotions. First, we want to understand the origin of emotions for a specific human being. Is it before birth i.e., during the pregnancy itself, or is it a few months after birth, or is it after we become conscious? How did emotions emerge or transition from a state where there were none, say at an embryonic stage, to a state when we do have them as an adult? Next, we will explore the origin of emotions for the entire human species itself. How did emotions emerge in human beings, by itself, when there were other species

that did not have any emotions (multi-cellular organisms like worms)? How did this transition from no-emotions to emotions across species occur?

3. *Nature of emotions*: Next, we want to look at specific emotions. When we look at a newborn, it is quite clear that he cannot experience anger, surprise or disgust. The only emotions he can 'potentially' experience are happiness, fear and to a lesser degree sadness. Even these are hard to judge. Yet, as we grow older gaining new experiences in our lives, we do *acquire* the remaining emotions. How does this happen?

    The degree of perception of a given emotion also changes over time. For example, fear of going under a crawl space, fear of heights, fear of darkness, fear of snakes and spiders, fear of pain like needles or injections, fear of your own death and your loved ones and others are all essentially classified as fear. However, based on experience, we know and sense them to be subtly different. In fact, to distinguish them verbally, we have created several types of *phobia* even though the emotion itself is fear. It is also interesting to note that what causes a specific emotion like fear can change over time, but the emotion itself is relatively static. How do emotions feel different from each other? For example, happiness is different from sadness or fear. Are there people who experience more emotions than others and earlier or later than others?

4. *Mechanics of emotions*: We would like to understand the mechanisms for emotions by analyzing what happens within our brain just before, during and just after an emotional experience. What conditions are needed to create and trigger a given emotion? How can we keep the emotions sustained for a long time? How can we make the emotion decrease or stop?

5. *Feeling for emotions*: The interesting aspect of emotions is clearly the feeling. As mentioned in clinical studies like Crile [29], emotions have a distinct characteristic where several parts of the body are in an excited state simultaneously. For example, for fear, as mentioned in Crile [29]: "they are palpitation of the heart, acceleration of the rate and alteration of the rhythm of the respiration, cold sweat, rise in body temperature, tremor, pallor, erection of the hair, suspension of the principal functions of digestion, muscular relaxation, and staring of the eyes". Other emotions like happiness, sadness and surprise similarly produce a large number of such simultaneous excitations throughout the body. Whether it is a snake crawling by your legs or something else, why would you *feel* the fear? Why does your body not respond mechanically as mentioned above with palpitation of the heart and others, for fear? What is the cause of the feeling itself?

    Even if you are consciously aware of the situation, why should you still feel the fear? Another example is when someone hits your hand with a stick very hard. Why do you not simply retreat your hand and perform everything else related to the mechanics of this event while skipping the feeling of pain entirely? You can be consciously aware that someone hit you. You can even watch yourself screaming and jumping in response. Even from a survival and evolutionary point of view, it is quite clear that screaming, jumping, faster heart rate and other mechanical effects appear quite sufficient as an advantage. Is your experience of feeling offering you an additional advantage beyond this? If so, what is the source of this feeling? It is this seemingly inexplicable nature of emotions that we want to answer in this chapter.

From this last question, we have, therefore, a very interesting question that relates consciousness and emotions. Can we be conscious and yet not have any emotions? If so, under what conditions can this happen? Conversely, can we have emotions without being conscious? This converse was already answered in the negative in chapter 12 (see Theorem 12.1). It is for this reason that we need a precise understanding of consciousness before we can address the above similar questions on emotions.

The answers to questions in (1) about relating emotions to complex systems, (2) about the origin of emotions and (3) about the nature of emotions will be quite similar to the answers of the corresponding questions for consciousness discussed in the previous chapters. I will discuss them briefly after addressing issue (4), i.e., the mechanics of emotions. In section 49.5, I will discuss the necessary and sufficient conditions for the feeling for emotions (item (5)). Yet, we do need future work to clarify the differences between the dual and true equalities for emotions as discussed in chapter 44.

49.2 Mechanics of emotion – sensory-control explosions

From clinical studies like Crile [29], it is quite apparent that every emotion involves simultaneous excitations throughout the body. With fear, I have already listed them (see item (5) above), from Crile's work. While it is true that fear produces simultaneous excitations like palpitation of the heart, acceleration of the rate and alteration of the rhythm of the respiration, cold sweat and rise in body temperature, is the converse true? Namely, will we experience fear when all of them are simultaneously excited? There is clearly a practical difficulty in ensuring that all of these simultaneous excitations are indeed produced, excluding via the direct experience of fear. There are other ways to produce this as well which I will discuss now. The key to understanding these additional mechanisms is to realize that what we need is an *explosion* within certain critical regions in our brain. An explosion, by definition, is a neural firing pattern in which small firing signals at a few locations results in a large and cascading firing patterns across, possibly, distant regions of our brain. The explosion combined with the self-sustaining membrane of meta-fixed sets will produce a large set of simultaneous excitations like the ones listed above for fear.

This is precisely how we characterize the mechanics of all emotions i.e., as explosions that produce a 'shock' to the self-sustaining membrane of meta-fixed sets. The actual expressions that result from each explosion might vary based on the specific event you are experiencing. We can characterize the mechanics of emotions for all experiences involving happiness, sadness, fear, anger, surprise, disgust, love and others as explosions. I will examine these explosions in more detail now.

Explosions in the sensory-control regions: In the previous chapter, I mentioned that realization mechanics are explosions as well. While realization mechanics are explosions in regions of the brain responsible for higher mental abilities like thoughts, intelligence, planning and others, emotion mechanics are explosions in a different set of regions of our brain. These are the sensory-control regions. Encyclopaedia Britannica, Inc [182] shows part of our brain having several special substructures with different responsibilities. Let us explore two important special areas in our brain: those that are responsible for control and those that are responsible for sensing (see, for example, the primary motor area and the primary somatosensory area in Encyclopaedia Britannica, Inc [182]). Control regions take actions and are active. Sensing regions observe and gather information from the environment and are passive. For example, we have areas that control the motion of externally visible body parts like hands, legs, fingers, joints, head and neck, eye movements, mouth, tongue and facial movements. For the internal control regions, however, there are two types: (a) that control internal body parts like, say, your heart, muscles, lungs, hair, body temperature, digestion and others, and (b) that control your internal brain regions responsible for abstractions, thoughts, planning, speech, memory and others. Sensory regions are already well known like eyes, ears and others.

The advantage with an identical architecture for both the control and sensing regions (i.e., both represented as networks of neurons) creates interesting possibilities of loops within these networks. The axons from one region grow to the other during your lifetime to form highly interconnected regions. We train both the control and sensing regions independently during our early childhood to gain sharper, fine-grained control for fine movements as well as high sensory sensitivities. Let me now discuss what happens when there is an explosion in the sensory-control regions of our brain after they were interconnected to form dense network of loops.

What triggers the explosions? There are several ways to trigger these explosions. Explosions for internal body parts are triggered when there is either damage to the parts or when there are unusually large deviations from normal conditions like hunger, bowel movements and others. Explosions for externally visible parts like eyes, head, hands and legs are usually driven by intense sensory inputs like shining bright light, loud and unexpected sounds that startle, sharp pain on the body causing you to retreat and others. For human beings, language and higher-level thoughts trigger explosions too, say, like happiness, jokes, imagining a sad situation, strategies that are either deceiving or perfectly executed and others. In all of these situations, our brain produces parallel loops of excitations that propagate to result in cascading firing patterns. This causes an explosion at the control regions of our brain.

How many control regions are involved for each explosion? Some simple explosions occur in only one control region. Some of the examples are like hunger in an infant causing him to cry, sharp object causing a sudden retreat of your hand and others. Recall that for it to be even called an explosion, a small cause should produce a large effect. The above examples do obey this property. There are other situations where multiple control regions are involved in each explosion. Our brain has several such junctions. For example, the motor cortex (for control) and somatic sensory cortex are located very close to each other in the brain. As a result, several coordinated loops between these organs are possible. This can lead to explosions involving multiple control regions. Hypothalamus is another junction that regulates autonomic nervous system and controls body temperature, hunger, thirst, endocrine system via the pituitary gland and circadian cycles. The reticular formation that extends the brain stem is another central area that receives a number of sensory inputs. A specific region here called the reticular activation system (RAS) is known to be responsible for arousal, motivation and maintenance of an alert state in the cerebral cortex. This is an important junction where inputs from sensory regions converge and outputs are relayed to several parts of the brain and the body. These and several other structures like the limbic system form important closely linked regions that control a number of functions in the brain and the body. We can have explosions in these regions of the brain as well. The more the number of control regions involved in an explosion, the more closer the resulting outcome feels like a specific emotion. This is especially true when you already have a self-sustaining membrane to aid the propagation of the explosive effect to very wide and distant regions within our brain.

What is the qualitative nature of these explosions? Explosions in these central regions (like RAS) of the brain triggered by, say, external sensory inputs produce qualitatively the emotions that we experience. Fear, surprise and disgust are examples of such emotions triggered in a relatively short duration. Sadness, joy and sometimes anger are a bit slow to trigger. Here, several events, memory and thoughts are combined in such a way as to produce an explosion that affects multiple control regions of the body.

Since these regions are centrally located with extensions that reach even as deep as the cerebral cortex it makes their explosions yield quite powerful externally visible effects. The energy of such emotional explosions can cause a drive that produces very strong results like fight-flight response, even intelligence and other higher abilities. For example, it is already well known that emotional drive and emotional engagement in any work produces more effective results in educational fields as well as sports, music and others.

With the development of a higher-level language in humans, there is an interesting loop established in our brain that lets us talk to ourselves. This creates a new mechanism that connects thoughts to control region and vice versa. As a result, multiple thoughts can all lead up to the central regions of the brain mediated through Broca's area (responsible for language processing, speech production and comprehension). Explosions resulting from this thought-mediated convergence are like fear resulting from thinking about your death, fear of losing someone you love and others. These are examples of emotional explosions triggered by a conscious self.

What are the after-effects of these explosions? The after-effects of realization explosions are quite different from that of the control related explosions. Energy from a realization explosion is used to create memories by forming new interconnections between regions of simultaneous excitations occurring at that instant, as described earlier in chapter 48. In other words, we channel the explosive energy back into the brain and effectively capture it within the neural structure of our brain as new memories. With control explosions, however, a significant portion of the energy of an explosion is consumed to generate large amounts of force, motion, shouting, jumping and other physical effects. Some other ways this energy is used is the equivalent of emotional drive to continue pursuing the task beyond one's abilities. We use the remaining available energy to store the experience in the brain as memory. This is one of the reason why we do not remember too many details of emotional explosions other than a few facts like whether you enjoyed it, disliked it, what caused it and others. The positive aspect of not storing them as memory is that you can experience the same emotional explosion several times with a similar intensity. This is unlike realization explosions, which are not as effective the second time you experience. Of course, prolonged exposure to these emotional explosions does degrade the intensity. For example, we do get used to hunger, sadness, too much happiness and others if they sustain for, say, an hour or more.

Let me discuss two specific emotional explosions briefly here, namely, happiness and fear. Happiness emotion usually produces an explosion with an after-effect drive to continue doing more and more of the same. We can initiate it internally by thinking about specific events or drive it externally through encouragement. Emotion of fear usually produces an explosion with an after-effect drive of strongly retracting away from doing the task. A mere recollection of the memory of an event or the start of the occurrence of the event is sufficient to trigger both these explosions and the corresponding drives. These two emotional explosions are present even at birth unlike other emotions. In fact, the structure of the primitive brain naturally creates this possibility within the complex system.

49.3 Properties of emotional explosions

Let us now tie in the mechanics of explosions with the real emotions. Consider two scenarios for emotions: (a) fear that you experience while watching a horror movie versus fear of spider crawling on our leg in a dark room, or (b) sadness that you experience when watching a tragic movie versus sadness that results from losing a beloved one. In each scenario, the first example is a 'simulated' situation while the second example is a 'real' situation. Yet, the emotions you feel in both cases are real. Let me identify a few differences between these two scenarios by looking at the properties of emotional explosions like how fast we can trigger an emotional explosion, how long will it stay active and how strong is it going to be.

Rate of triggering an emotional explosion: The emotion of fear or sadness in the simulated case builds up slowly. The situation depicted in the movie appears real to you and it triggers a number of thoughts and parallel excitations to produce a cascading effect that can eventually lead to an explosion. However, you can distract yourself and shutdown this explosive buildup by simply turning away. For horror movie, you close your eyes or turn away. For tragic movie, you think of an entirely different topic or other happy situations. This causes the neural loops to dissipate away. When you switch back and watch the movie again, the emotion starts to build up once again fairly quickly. This is because even though the explosion dissipated, the continued loop excitations in several parts of your brain still exist and/or the neurons are near the threshold state of excitation.

In fact, distraction as a tool, is an effective way to control your emotions. We use it with toddlers when they are sad, crying, in pain, angry, afraid or even too happy and excited. They, however, do not have as many sustained loops in their brain compared to an adult because of limited memories. As a result, their emotional explosion dissipates quite fast when distracted.

However, with the above non-simulated (real) examples, the process of buildup of emotion is very quick. A mere act of sensing the spider on your leg or the news of the death of your beloved one produces a sudden explosion of fear or sadness respectively.

Duration of emotional explosion: With the simulated scenarios of fear and sadness, the duration of the explosion is quite short lived. You get back to your day-to-day life afterwards. As mentioned above, you can even reduce this duration by self-distraction with some other topic. However, for real scenarios like when facing a near death experience, the fear is long lasting. In these cases, you trigger several firing patterns by thinking about a large number of events and scenarios important to you, which results in a big explosion. It also feels as if the passage of time is slower (like with drowning – see chapter 40).

Similarly, when you hear about the death of a beloved one, the explosion is long lasting. You continue to remember a number of past experiences and also those that you can never experience anymore, on a regular basis, for a sustained period of time. These sustained explosions for long periods can cause severe depression as well. You cannot distract yourself easily with any other topic. Even simple day-to-day activities that you do afterwards remind you of past experiences that you had enjoyed with him or her. They continue to produce explosions. The duration of emotional explosion is an important property that changes the qualitative experience of emotion.

Strength of an emotional explosion: The strength of an emotional explosion for simulated scenario is not as much as for the real scenario with fear and sadness. Fear of death or sadness from the loss of loved one produces an explosion that causes a much large number of simultaneous excitations in your body. The strength of the initial explosion is quite high as well. There are several subsequent explosions triggered constantly for extended period just from simple day-to-day activities. With the simulated cases, since you are aware of the non-seriousness of the emotion (like sadness and fear i.e., you know that the situation is not life-threatening, say), the strength of the emotion is suppressed by simple distraction.

49.4 Examples of emotional explosions

Let me now discuss several other examples of emotions and describe them in terms of the above mechanics. Happiness is a strong emotion and is usually short-lived. A sustained state of happiness is not easy. One way to sustain happiness is to continue recollecting the happy event. Jokes, as an example, can be a simulated scenario that produces a feeling of happiness. Simulated scenarios are short-lived and produce lighter explosions. Similarly, simulated scenarios for disgust are not as strong and are short-lived. Real experiences with disgust, on the other hand, are unusually strong. You usually self-distract and move away very quickly.

Anger is an example, which is usually strong in a simulated scenario than in a real scenario unlike the other examples considered so far. When there are no real events happening simultaneously, you are not distracted. Your thoughts and logical analysis are clear which fuel your anger significantly more in a simulated situation. In a real experience, fear and a lack of clear thinking takes over and lessens the anger, especially if it is premeditated. If it is not premeditated, anger is usually strong because of the impulsive nature. It usually becomes short-lived because fear takes over instantaneously after the first act of explosion, especially when you are dominating. When you are not dominating, you become a victim, which leads to different set of feelings altogether.

Now, an explosion can be triggered in your brain with the aid of other people as well. In the case of happiness, these are from positive reinforcements like encouragement or showing love and affection. In the case of anger, others can further fuel your anger by pointing out additional facts that you do not like. Alternately, others can help suppress your anger by reminding you of happy situations.

There are also ways to switch from one emotion to another easily like with switching from anger to calmness using other happy thoughts. You can now perform this analysis systematically and build the relationships between the emotions within the current framework. Most of the facts mentioned here are already well known. The unique aspect is the connection with the mechanics within the neural loop architecture.

Now, from above discussion using emotional mechanics, it clearly follows that we have more than the standard set of, say, ten emotions listed earlier. Much more fine-grained classification is possible. The variations in the emotional after-effects are subtle that we do not bother to distinguish them in fine detail. We also do not have different names to distinguish each of them. In the case of fear, we do have different types of phobias to distinguish the different types. The serial nature of consciousness blurs such parallel explosions of emotions. Even though our brain is inherently discrete, we can talk about an approximate continuity of transitions because of millions of neurons involved. This semi-continuous nature of emotions, as a result, has smoother transitions from one emotion to another. This is the source of fine-grained nature of emotions and their transitions. For example, too much happiness can lead to boredom. Sadness can transform to happiness, multiple experiences of fear or sadness can transform to acceptance and so on. The smaller set of about ten emotions is the discretely distinctive ones while there is actually continuity between these emotions to yield different *feelings*.

As mentioned for realizations, events and tasks performed days or weeks before are not directly related to emotion itself. Instead, emotion is a quick sudden explosion. On the other hand, we have seen that consciousness takes almost a year of experiences. We needed to build a self-sustaining membrane of meta-fixed sets from several different sources of inputs. This difference has been one reason why consciousness turned out to be quite an elusive problem compared to emotions.

49.5 Emotional feelings

In chapter 22, when discussing the feeling of knowing, I mentioned that Theorem 22.1 does not address why we have feelings or what they are in a physical sense, even though we did derive the necessary and sufficient conditions. The same will apply even with the current discussion about emotions. Furthermore, we can only compare these feelings in the sense of 'dual equality', not 'true equality', as discussed in section 44.2. Given these caveats, let me state the necessary and sufficient conditions for experiencing the feeling of emotion as follows.

Theorem 49.1: A system experiences a feeling of emotion if and only if it has (a) a self-sustaining threshold dynamical membrane of meta-fixed sets, (b) abstract continuity within several critical sensory-control regions and (c) an explosive dynamical excitations at different sub-regions of these critical sensory-control regions.

It is interesting to compare Theorems 22.1, 48.1 and 49.1 that gives necessary and sufficient conditions for the feeling of knowing, realizations and emotions respectively. The feeling of knowing is the mildest form of feeling, which requires just the membrane and abstract continuity. Realizations are a bit stronger with an additional requirement of explosions in regions corresponding to thoughts and abstractions. Emotions are the strongest form of feelings, which requires explosions in the sensory-control regions instead. The conditions (a) and (b) are essentially common for all three types of feelings.

If we look at explosions needed for emotions, different possibilities affect either different set of regions and/or affect at different intensities. These possibilities result in different types of emotions like, say, happiness, fear, sadness and others. The condition (a) requires that the system be conscious (see Theorem 23.2) before it can experience an emotion. This is an important result that I state it as the following corollary.

Corollary 49.1: A system experiences a feeling of emotion only if it is already conscious.

The conditions (b) and (c) emphasize critical sensory-control regions. The arguments used for this theorem will be similar to the ones discussed for realizations in chapter 48. Firstly, the self-sustaining membrane i.e., consciousness is necessary for experiencing emotions. Even if there are explosions in the critical sensory-control regions, the dynamical effect does not propagate far enough within the brain. The lack of a self-sustaining membrane implies that there is nothing to bind the neural excitations across distant regions of our brain. The existence of a membrane ensures that most of the neural excitations work in unison and as a single entity. The membrane acts as a connected set. It is not even reasonable to call these excitations as explosions in the absence of a membrane. There are no reverberating effects. When you are under general anesthesia, severely drunk, deep sleep, coma and in other unconscious states (see Figure 11.1 in sections 11.2 and 11.3), the explosions simply fizzle out. You may only have minimal startling or other reflex reactions, not strong ones. This is the reason why you do not feel an emotion when you are not conscious (condition (a)).

Secondly, there is another way the explosive effect diminishes even if you are conscious. This is when you do not have abstract continuous paths within the critical regions. Recall that abstract continuity requires nonempty predictive meta-fixed sets at every instant as an event occurs. For example, if you do not have abstract continuity with vision or with your ability to turn your eyes or head, this means that you are only observing what is directly in front of you. You are not processing neighboring visual information. In effect, you keep staring for long periods, one object at a time. If it is a moving object instead, you are simply tracking it without processing neighboring static objects. You may have noticed this behavior with infants. There is a period at, say, 6-9 months of age, when an infant either stares at familiar objects or tracks moving objects ignoring everything else around them for several seconds. At this time, he is still learning to form abstract continuity with common sensory and control mechanisms. In these situations, he does not have explosions of fear like when someone tries to scare him because he is not even observing events around himself.

The nonempty predictive meta-fixed sets (for abstract continuity) give rise to a dense network, in general. However, his lack of abstract continuity within these critical regions of his brain resulted in a sparse network instead. This is another reason why you do not have explosive firing patterns even if you have self-sustaining membrane. To experience emotions, the intensity of stimulus should be either high or more directed.

There is a similar situation with adults, say, when you are working on a specific task with intense focus, much like an infant staring at only objects in front of him. When someone calls you or merely touches you from behind, you are suddenly startled. Unlike an infant, you have already created the necessary abstract continuity within these critical regions. Your dense network creates a startle response unlike for an infant.

Finally, each type of explosion produces a different emotion. You never experience an emotion without an explosion. As mentioned earlier, there are several variations and degrees of emotions to do an exhaustive classification of the corresponding explosions. I will, therefore, mention briefly some of the types of explosions and then relate it to an emotion next.

49.6 Types of explosions and emotions

To understand the type of explosion for happiness, let us first look at some general properties of a self-sustaining dynamical membrane. A self-sustaining dynamical membrane for a complex system is never uniform. There will regions of local 'peaks' and 'valleys' that keep changing over time constantly. Peaks correspond to regions that have excess energy than the average value. They enhance the positive firing signal onto the neighboring neurons. They can also be viewed as neurons generating strong excitatory responses. As a result, these peaks are capable of producing further cascading neural excitations. The valleys correspond to regions with lower energy than the average. These valleys consume energy to produce excitations. They correspond to either high threshold neurons regions (i.e., they consume a lot of firing activity before they can fire) or neurons that produce strong inhibitory responses.

As the membrane self-sustains, our brain keeps transferring energy from peaks to valleys and consume new energy, as needed, in order to maintain stability. If the brain is unable to maintain stability, the membrane ruptures and you lose your consciousness. A normal healthy way to rupture the membrane is when you are tired. You fall asleep as a result.

This is an abstract view of the membrane – as a continuously oscillating membrane surface with a continuously altering peaks and valleys. Now, the external sensory inputs as well as our own internal thoughts play a critical role towards changing the peaks and valleys of our membrane surface. Even the very stability of the membrane surface is controlled using these external and internal input sources. One such stabilizing mechanism is the fundamental pattern of chapter 16.

Given this abstract view, several interesting patterns are possible. For example, there will be periods when there are more valleys than peaks. Furthermore, these valleys are deep enough that they are not as easy to convert them into peaks (using simple internal thoughts or external stimuli). However, occasionally, after a sustained period of time and effort, our brain does succeed in generating an explosion that converts these deep valleys into peaks. The external and internal stimulus that causes this to happen varies depending on the specific situation.

In the current discussion, we should view peaks as causing a 'good' feeling whereas valleys as causing a 'bad' feeling. With this interpretation:

(a) Happiness is a feeling you will experience whenever an explosion converts valleys to peaks.

(b) Sadness, on the other hand, is the feeling you experience when you have an explosion that generates a significantly more number of deeper valleys.

(c) Fear is the feeling that corresponds to an explosion, which is quite deep, but more localized. This differs from sadness where the valleys are deep and spread out across broad regions of your brain. As a result, sadness takes much longer to dissipate compared to fear. Happiness explosion is also, typically, present for a short duration. Once the valleys convert to peaks, your happiness feeling dissipates. Furthermore, in this situation, the self-sustaining nature of the membrane automatically creates smaller valleys at other regions naturally (i.e., it behaves like a partial conservation of energy between peaks and valleys). This is one reason why we do not stay happy for too long. You start to feel bored. You want to try something new once again.

(d) Surprise is a feeling that corresponds to an explosion resulting in new peaks and/or valleys in an entirely unexpected region of your brain and, hence, makes us feel good or bad respectively.

(e) Disgust is a feeling corresponding to an explosion that takes a peak and turns it over into a deep valley. This creates deep valleys not just in this region but also with other interconnected fixed sets. Sometimes, they are not necessarily peaks converted into valleys, rather they can be shallow valleys converted into deeper valleys.

(f) Anger is a feeling that corresponds to an explosion where the predictive fixed sets, which you expect to peak, is sometimes severely overturned into a valley. The corresponding situation is one in which you expect to do a task a certain way, but in reality, the opposite happens. You get irritated or angry, as a result.

If you look closely at each of these explosions, you will notice that the new peaks and valleys created are not always related to the previous peaks and valleys. This is, for example, true with surprise and with some types of fear. Besides, when we consider predictive meta-fixed sets as well, we get new and interesting patterns as discussed with anger above.

As mentioned earlier, other types of explosions also give rise to the same set of emotions. The above description only highlights a simplified and a small set of patterns out of an enormous number of other possible patterns that occur within our brain. Other additional explosions that give rise to similar emotions can be seen with the dual representation. What I have described here is just a glimpse of the complexity that can occur within our brain.

49.7 Origin and nature of emotions

Let me now briefly return to issues (1) – (3) raised earlier in section 49.1 after having discussed issues (4) and (5) in considerable detail in the previous sections.

The questions about how emotions can be created in any other dynamical system (issue (1)) can now be answered easily using Theorem 49.1. Any complex system with a stable parallel looped architecture can be structured such that it can obey the necessary and sufficient conditions outlined in Theorem 49.1 needed to create emotions. From a practical standpoint, this is not an easy task, just as it is not easy to create a synthetic conscious system (see chapter 47). However, it will be possible soon enough to create a synthetic system that is capable of exhibiting emotions, at least, in a dual sense (see section 44.2).

The questions about the origin of emotions in humans (issue (2)) are again clear from Theorem 49.1. The conditions stated in Theorem 49.1 are represented in a form where a system can naturally evolve to reach this state without the help of any intelligent being. This is true with consciousness as well. An infant truly starts experiencing emotions only after he becomes conscious. Until his brain reaches this threshold state, his emotional experiences are, at best, intermittent just as it is with consciousness (i.e., for about a year of age).

His membrane is not stable enough to self-sustain for more than 12 hours at a stretch, as is the case with all adults. This is one of many ways to understand the importance of sleep. An infant's ability to self-sustain his membrane keeps growing with age as his network becomes denser with new experiences. During these conscious states, he will experience emotions as well. However, he may not remember most of these experiences nor will he have a feeling of continuity of events until after a year of age.

Notice that since Theorem 49.1 only requires the threshold membrane, self-awareness is not a prerequisite to experience emotions. Therefore, humans may not know about their own consciousness in an abstract sense until they developed language. Yet, during this period, they were able to experience emotions. In this sense, origin of emotions in humanity is different from the origin of consciousness in humanity discussed in chapter 45. There is, however, very little documented evidence on the origin of emotions in humanity.

The nature of emotions (issue (3)), when seen as different types of explosions, provides a way to explain how we acquire new emotions as we grow older. Our new experiences increase the complexity of the network and the structure of the dynamical membrane. As a result, new types of explosions are possible that were not possible before. Disgust, guilt, anger and other emotions are examples, which require several new experiences, say, from social interactions. For example, guilt requires knowing that you did something wrong. To know right from wrong, you need to have experiences related to living in a society. A toddler does not have guilt as a result, until his network expands and he has explosions within this expanded network. The same applies with anger as well. In general, not all humans acquire each of these emotions at about the same age because the rate at which their membrane expands is different. For example, falling in love is one example of a complex emotion where the age difference across humans is quite apparent.

One interesting question is whether some people experience new emotions that others do not. If it does happen, which according to Theorem 49.1 is quite easy and likely, there is an inherent problem with communicating this experience. If I cannot experience an emotion that you do, then I simply cannot understand what you truly feel. You may be able to describe it in terms of other common emotions, but you and I both know that it is only a vague approximation. Language is not an effective way to describe an emotion because of the inherent parallel nature of an explosion compared to the serial nature of language. For example, a male human being simply cannot understand the feelings and emotional experiences of a women's pregnancy, the birth of a child or her menstrual cycle. He can surely use analogies to compare with his own similar experiences, but this only means that he understands it in a dual sense.

49.8 Uniqueness of emotions across systems

We can only compare emotions across systems in a dual sense even when two systems experience the same emotions. I have discussed the topic of dual equality (and true equality) for all other types of feelings in chapter 44. Let me briefly discuss if a given emotion is the same across all systems that do experience it.

Since comparison of feelings can only be done in a dual sense for complex systems, we can use an alternate approach to address this question directly as we have done for the feeling of knowing. This approach is to look at the necessary and sufficient conditions for the feeling of emotions directly (i.e., Theorem 49.1). Theorem 49.1 expresses these conditions purely in relative terms or in a 'dual' sense. You need a membrane, abstract continuity and explosions for experiencing emotions, all of which are defined relative to an individual's stable parallel looped architecture of their brain. Since these features are unique across all systems in the dual sense, as described in chapter 44, we can say that emotions are also unique across systems in a dual sense.

However, it is important to note that there are different ways to reach a given emotional state. These are different for different people. For example, a child is happy for some silly expressions or events unlike an adult. Once you have the network created that connects the external or internal fixed sets to these critical sensory-control regions, it becomes possible to have different types of explosions. Each explosion within these critical regions are (a) triggered in different ways for different people and (b) propagated across different regions for different people. In other words, the causes and the effects are different for each type of explosion for different people. The fact that the nature of explosions is similar for each person implies that the emotion itself is unique across different people or systems, in the dual sense.

In this way, the new mechanics of emotions as explosions in the sensory-control regions and Theorem 49.1 can be used to address several of the open questions mentioned in section 49.1. The *qualia* of emotions is only addressed in the dual equality sense as with other feelings (see chapter 44).

50
Specific Abilities of the Human Brain

Let me now discuss a few applications of consciousness as well as the other higher abilities of the human brain. The discussion will be brief because the topics are common enough that we already understand and convince ourselves in several other ways. Therefore, my only objective is to explain and translate these familiar topics into the new framework of stable parallel loops introduced in this book.

50.1 Approximating parallel system with a serial system

Human brain as well as the rest of our body is a massively parallel system. The neurons that keep firing simultaneously are always in large numbers and at all times. Whenever we have such parallel systems, it is quite common to approximate them as serial systems to make the analysis easier. Most people believe that all parallel systems can be approximated as serial systems, at least conceptually, without any loss of generality by discretizing the time scale into sufficiently small intervals. The belief is that there are no new properties of parallel systems that cannot be captured by the above serialization and vice versa. However, in chapter 28, we have seen that free will is a unique feature that only exists with parallel systems and not with the corresponding serialized representation. In fact, each of the topics discussed in chapter 10 on consciousness relies on parallel nature of dynamics that simply cannot be explained using serial representations. In this sense, parallel complex systems are fundamentally different from any serial system.

Serial analysis of a parallel system breaks down, even with sufficiently small discretization of time intervals, because it is possible to form a large number of different types of groupings (see section 26.4) of subsystems. For example, if you are driving a car, at any given instant, there will be neurons that will be simultaneously active for several tasks like, say, pressing the gas pedal, watching the road, listening to songs in your car stereo, talking over a cell phone and others. Even when the time interval for discretization of the parallel system is extremely small, how would you know how to group the 'correct' set of neurons as a valid pattern and then study its time-evolution? An example of a correct set of neurons to track is the collection of neurons on your retina corresponding to an image of a tree. These neurons do change to a different 'correct' set at the next instant as the firing pattern propagates from the retina to the primary visual cortex. At the next instant, the correct set of neurons once again change as the firing neurons propagate to different regions from the primary visual cortex and so on. It is simply impractical to keep track of all neurons instead of just the correct set of neurons. At the same time, finding out the correct set of neurons at each instant is an equally challenging question.

Should we study just the firing neurons that are close together and ignore the sparsely distributed firing neurons? Even among the closely related firing neurons, should we group all of them as just one object or are there different groups representing different objects that we should keep track of? In any case, we surely need to supply energy to study each of these parallel tasks. Otherwise, these neurons will not stay excited beyond their natural dynamics.

This grouping problem is uniquely different between serial and parallel systems. There are a theoretically infinite number of 'correct' set of groups to analyze serially, even though the number of neurons are finite. This is because of a hierarchy of patterns. For example, you have (a) patterns of firing neurons corresponding to a given task, (b) groups of patterns and (c) groups of groups of patterns and so on. This is analogous to fixed sets, meta-fixed sets, meta-meta-fixed sets and so on. As discussed in chapter 47, it is not unrealistic to have more than 4-5 levels of hierarchy in common everyday scenarios.

Let N be the total number of neurons in our brain (or we could only look at the subset of neurons that are currently active, but this set keeps changing over time). The number of possible ways to groups these neurons into each of the above patterns or groups are roughly 2^N for case (a), 2^{2^N} for case (b), $2^{2^{2^N}}$ for case (c) and so on. The collection of all layers becomes theoretically unbounded even if N is finite. This is because the number of layers are infinite even though the number of neurons are finite.

For example, if we pick at the first layer i.e., case (a), then some of the 2^N patterns could be the subsets on the retina that can be grouped as, say, a tree, a table, a human, a car and so on. The neurons that are part of each such shape on the retina is a valid subset among 2^N of these subsets. Next, for each such subset (of shapes), we have 2^N of subsets once again. For example, if we picked tree, then a subset within this subset could be a branch, a leaf, a flower, a fruit and so on. If we picked a table, then a subset within this subset could be the top surface, the legs, the shape of the top surface and so on. If it is a car, the subset within this subset could be a wheel, a door, the bumper, the window and so on. These are 2^{2^N} possibilities that are part of case (b), which is the second layer. We can look at the third layer with $2^{2^{2^N}}$ possibilities and then the fourth layer and so on.

We do not know *apriori* which ones to track and study. Of course, in reality, we do not care about infinite levels of hierarchy. Furthermore, we do not have to analyze every group. Only an abstract system like a computer would analyze so many groups. We could look at the actively firing neurons to guide us which the important subsets to study. However, even these are too many in number as we go to the second, third and subsequent layers. In this book, we have seen that loops and fixed sets and their relationships represented by abstract continuous paths are the correct subsets to study. Also, the number of groups drops down significantly when you include dynamical and physical laws. However, the difficulty is precisely in how to know which groups to ignore and which groups to keep track of (if we do not use the theory presented in this book).

From a practical point of view, when N is of the order of millions of active neurons at a given instant, even case (a) would correspond to $2^{1,000,000}$ possibilities. This is already impractical to deal with. You only have a finite amount of time and energy to quickly determine the 'correct' group of neurons for case (a). In the next instant, the important sets of neurons to keep track of have already changed (possibly, a different set of million neurons). If you consider the other levels of hierarchy, the amount of processing needed becomes extremely large in an infinitesimally small time interval. The time intervals become shorter and shorter while the computation needed within this interval becomes larger and larger. Therefore, you have no choice but to pick some of them and ignore the others, more or less at random if you were doing this abstractly in a computer. This would mean that the serialized representation that you are currently studying is no longer comparable to the parallel system that you intended to analyze.

This additional task of determining the correct set of groups of subsystems in each layer of hierarchy (like case (a), (b) and (c)) is the root cause of unique difference between parallel and serialized system. This difficult task can be completely eliminated if you study parallel systems directly without discretizing it, at least, in some special cases. One such special case considered in this book is with complex systems that have a stable parallel looped architecture. For example, we have seen how to study the parallel system directly with the problem of free will in great detail in chapter 28. In addition, in the previous chapters, I have shown how the class of SPL dynamical systems do indeed exhibit complex features like consciousness, emotions, free will and others by looking at the parallel system directly instead of arbitrarily discretizing it. The concepts of fixed sets, abstract continuous paths, dual space representations and the mechanisms behind each of these patterns, as described in the previous chapters are all quite critical to study parallel systems in its own right.

50.2 Phantom limbs – dual space representation

When a person has an amputated limb, he senses as if it is still attached and is moving appropriately relative to the other body parts. This is termed as a phantom limb sensation. Even though the limbs themselves were amputated, the corresponding mapped and internetworked regions within the brain are not disturbed, i.e., the membrane of meta-fixed sets is not appropriately shrunk. Now, with the absence of limbs, there is no direct way to excite these fixed sets in your brain from the body (when you are awake). Instead, you have memories of these fixed sets created in your brain prior to the accident. These memories are interwoven with the rest of the fixed sets and with the dynamical membrane.

Whenever you re-experience an event that involves using these limbs, it now becomes possible to excite the corresponding fixed sets via these interconnections. The oscillations within the membrane produce a phantom sensation similar to the other feelings (like knowing, realizations and emotions). This phantom sensation also occurs with the removal of other body parts as well, not just with limbs.

This is a good example to show the reality of dual space representation of chapter 25. Dreaming is another situation when you feel the presence of your limbs from the dual space representation. Recall that the real inputs from your body are explicitly cut off at the brain stem during sleep. Sensing your body during dreams is different from phantom sensation because there is considerable pain associated with phantom sensation. It was hypothesized that the pain is due to excitations and irritation in the severed nerve endings. When the membrane oscillations dictate your body to move your limb, there is no feedback from the missing limb. You may sometimes remember this awkwardness from a lack of feedback even in your dreams, say, when you moved your hands, legs or tried to pedal a bike, although there is no pain associated in these cases. Ramachandran et. al [124] have developed mirror box technique by creating an artificial visual feedback that 'shows' the limb in a mirror. With proper training using the mirror box technique, they have shown that subjects with phantom limbs were able to 'move' the virtual limb and, in some exceptional cases, did not feel the pain at all.

50.3 Synchronization with external inputs

External inputs through our sensory organs trigger neural loops within our brain as we have seen earlier. This can sometimes produce synchronization between the external system and our internal inputs that can sustain for a while. This synchronization is not very stable in adults compared to infants. Infants have fewer loops and interactions that can destabilize the coherence.

Consider the case of rocking an infant to put him to sleep. The rocking motion is usually in 1-1.5 Hz range (about 1-2 rocking motions in one second). This corresponds to delta wave frequencies, associated with deep sleep, in our brain. The neural firing patterns within our brain can synchronize with this external rocking frequency. If the infants' eyes are open, the visual inputs produce retinal firing patterns that oscillate in the same external rocking frequency range of 1-1.5 Hz. These firing patterns produce cascading excitations that also have an approximately similar frequency range of oscillations due to the loop structure of our brain network. Additionally, external sounds that he hears, say, from a fan or if we hum a lullaby in soft tone, combined with the rocking motion, similarly produce firing patterns that oscillate in the same range. If we have a fan that produces a gentle breeze, the air touching his entire body (via skin sensors) as we rock him also triggers firing patterns that oscillate in this same 1-1.5 Hz range. Each of these sensory inputs produces cascading and large-scale neuron oscillations, due to the stable neural loop structure in his brain. These 1-1.5 Hz oscillations occurring in multiple and broad regions of his brain produces delta wave patterns and hence results in deep sleep for an infant.

This approach does not work well for an older kid or an adult. We constantly preoccupy ourselves with a number of internal thoughts and other distractions. These thoughts are also distributed across broad neural regions of our brain. They destabilize the oscillations induced through external synchronization like rocking motion. We do need to stop thinking about these events if we want the rocking motion to induce delta wave oscillations (and, hence, deep sleep).

Another similar example is with infants falling asleep sucking milk, provided he takes a while to drink milk. The sucking motion is also in the 1-1.5 Hz range. The external oscillations in this case are induced through the neurons in the taste buds of his tongue (cf. 1-1.5 Hz oscillations from eyes, ears and touch in the previous discussion).

There are other complex examples involving synchronization between two or more people. For example, you and your infant son can sometimes synchronize laughing for as long as 2-3 minutes, say, by making a silly face and laughing to which he laughs back in return. If you continue laughing back in response, he also laughs back. This can continue sometimes for as long as 2-3 minutes. If you are feeding your son with a spoon, the opening of his mouth can trigger you to open your mouth almost unintentionally. This is sometimes explained using 'mirror' neurons in the existing literature. Similarly, we can consider meditation and hypnotism as a form of synchronization as well.

50.4 Meditation and out-of-body experiences

Let me now discuss two somewhat related cases in an ideal sense i.e., meditation and out-of-body experiences. In both cases, the fixed sets on the dynamical membrane will be directly controlled and self-sustained with minimal, if at all, external inputs. With meditation, you only generate delta and theta brain waves, which typically correspond to deep sleep. Yet, you need to control your brain to stay conscious. With out-of-body experiences, you have alpha or beta waves instead, corresponding to a more relaxed state. As discussed in chapter 25, both these cases rely on dual space representation of fixed sets.

Some people associate out-of-body experiences with near-death experience while others claim to have induced this state through meditation. Consider the situation when all motoneurons are almost perfectly inhibited like, say, during sleep. In this case, no truths related to parts of your body are excited if you are still conscious. You have no way to know about your own body. If, now, your body fixed sets are directly excited in the brain (i.e., not from your actual body via the brainstem), your consciousness will give rise to a feeling of awareness of your body even though you do not sense your body directly. This is the dual representation of body fixed sets. Therefore, under ideal situations, you do not need your real body to 'sense' and know your body. This is true only after the dual space representation of your critical body fixed sets are established, say, during your first year of life. If you can train yourself to directly trigger these body fixed sets using meditation even as your body inputs are cut off at the brainstem, you will have an out-of-body experience.

This is different from watching an image of yourself but as a different person like in a dream. You are not merely watching an 'image' of yourself performing tasks during an out-of-body experience. You truly sense the body of this 'image' as well. For example, if someone hits that body, while you are in this state, you actually feel the pain. This is not the case with images of other people in your dream, even though in both cases the images that you see are entirely generated within your brain without the help of any external inputs. You can similarly predict what that image of yourself is going to do next, unlike images of the other people in your dream. This is, therefore, a nontrivial state, especially if induced through meditation.

During meditation, the recognition frequency (see section 15.2) take as high as 500 milliseconds on average because of the delta and theta waves of your brain. With such slow recognition, it is quite difficult to stay conscious. The dynamical membrane cannot form at such a low number of fixed sets per second. In general, as mentioned earlier, a loop sustains for about 50-100 milliseconds when it does not receive any external inputs. If, however, the recognitions are in the 500 millisecond range, the only way you can still remain conscious is if you can control your loops to stay them excited without dying down for much longer than 50-100 milliseconds.

One way to achieve this is by echoing the same thoughts over and over to yourself. This is a common technique when you practice meditation. This requires significant practice and focus. People who practice this form of meditation sometimes refer to it as a higher form of consciousness than the normal awake state. When you decrease your brain waves from alpha or beta waves to delta or theta waves, the natural mechanism tears down the dynamical membrane i.e., you lose your consciousness. One example of such a natural mechanism when the dynamical membrane tears down is when you fall asleep. From such an unconscious state, you recover to a conscious state of dreaming later through self-excitations. The process during dreaming goes from conscious awake state to an unconscious deep sleep state and then to a conscious dreaming state.

During meditation, you skip the equivalent of unconscious deep sleep state even though your brain waves are similar (delta or theta waves). You never lose consciousness during the entire period. Instead, you lower your brain wave activity very gradually without ever tearing down the dynamical membrane i.e., without becoming unconscious. This is one reason why people practicing meditation regard it as a higher form of consciousness than the normal case.

50.5 Distinguishing true from false statements

A computer cannot distinguish a true statement from a false one unless a conscious being explicitly hardcodes this difference. We, on the other hand, can easily spot this distinction. A statement and its negation are not different to a computer. How do we know which statement is true and which one is false? In this section, I will look at this question briefly.

The notion of a false statement is not as concrete as the notion of a true statement. This is because different conscious beings have different set of truths (and hence false statements), in general, within their brain architecture. Besides, the same statement can be true under certain conditions while it is false under other conditions, even for the same conscious being. How then are we able to pick a specific statement as true while a computer cannot? For example, if I have two statements: 'a ball falls down when dropped' and 'a ball falls up when dropped', we can easily pick the correct statement unlike a computer.

Consciousness is critical for picking a true statement. When we have a threshold membrane of meta-fixed sets, we already know a large set of truths. When we encounter or observe any new event that is not yet integrated into the threshold membrane, we interlink the new fixed sets with the existing membrane of fixed sets. No new event is stored in a region disjoint from the existing membrane. As the event repeats in an approximately same way multiple times, we form the fixed sets and link them with the truths of the threshold membrane. Therefore, these new fixed sets also become truths via the links with existing truths (see chapter 21 on abstract continuity). In this way, the threshold membrane of truths keeps growing bigger and bigger with new experiences. The set of truths grow even into areas that are entirely abstract like with mathematics. As a result, any given statement can be broken down until we can ultimately express it in terms of primitive truths that were part of the original threshold membrane, the one responsible for giving rise to your consciousness.

Contradictions occur when your predictive meta-fixed sets and the actual event are significantly different from each other. Let us say that you are observing an event like a ball falling down when dropped. If for some reason the ball instead floats, you will notice this difference immediately. This is because your abstract continuity produces a set of predictive meta-fixed sets for this event while the actual event is opposite. This disturbs your future predictions and sometimes makes them even empty. It now produces a feeling of contradiction. This is true with any event that deviates considerably from your memorized abstract continuous paths. Examples are gibberish words in the middle of a well-formed sentence, any sort of erratic behavior, bizarre events and so on.

An infant or a toddler does not notice as many contradictions as an adult. He does not yet have as many predictive meta-fixed sets (i.e., abstract continuity) for each of these events yet. You can easily trick him and he would not even know that you are tricking him. As he grows older and gains more experiences, he forms more predictive meta-fixed sets. It now becomes difficult to fool him.

A computer does not have this structure and dynamics within its architecture. When we detect contradictions, we can use our free will to ensure repeatability, if necessary. We can either perform an actual experiment or we can imagine a thought experiment using our expanded set of truths (beyond the threshold membrane).

Even for abstract topics like mathematics, we start with physical dynamical events as the starting point for truths. For example, geometry and numbers provide a vast framework, by relating them to physical dynamical events like counting real objects, drawing lines, angles and curves using a pen and so on. Since these abstractions and operations are based on dynamical events, we can convincingly test if they are true or false. We have, therefore, expanded the threshold membrane, which was based on dynamical events for now, to include abstract concepts and operations like numbers and geometry. We can further expand the membrane now, by using numbers and geometry to include like groups, rings and vector spaces. In this way, we continue to grow the membrane iteratively ensuring truths at each instant. Our consciousness guarantees our ability to distinguish true from false statements because the entire expanded membrane rests on the physical dynamics as the foundation.

The link between abstractions and dynamics is, therefore, not only critical for consciousness (see section 13.1), but it also helps us expand our knowledge, distinguish true versus false statements and ultimately expand our set of truths. Using the above approach, conscious beings can create new definitions and valid rules or operations all by themselves unlike a computer. Conscious beings are, therefore, capable of discovering the laws of nature by themselves.

50.6 Gist

When the event you are trying to remember is a simple or short event, you can remember several details of it. However, our experiences are seldom simple. We may be reading an entire book, watching a long movie or taking a long vacation. In each of such events, we can only hope to remember a small subset of the details, which we call as gist. The rest of the details either are absent or can be partially recollected as you re-experience the event.

Our memory of a given event is typically branched at several turning points (or landmarks) as discussed in section 9.9. If an event corresponds to an abstract continuous path from region A to region B of our brain, then we have, in general, several turning points along this path. We remember each of these turning points quite clearly, while we do not recollect the path in between the turning points instantly. These set of turning points correspond to our gist for this event. When we re-experience the event, the details of the external event excite the path between the turning points as needed and we start to recollect significantly more details than our gist. We can think of the turning points as the conscious memories while the paths between the turning points as hidden unconscious memories.

For each experience, our brain naturally forms this structure within our stable parallel looped architecture. The turning points, typically, correspond to events that trigger an emotional explosion or a realization. As a result, these are unique to an individual. Every time you repeat the event, you create more turning points at different locations along the abstract continuous path. Initially, when you recollect an event, you do so in the same sequence as it occurred. However, the natural variations and multiple turning points as the event repeats will eventually alter the linearity of abstract continuous paths and your memory-recollection process.

50.7 Knowing the concept of knowing and not-knowing

Among the set of events we experience, we know some of them very well while we do not know some others at all. Using just these two types of events, we do not need to know the concept of knowing (or not-knowing). On the other hand, we do have events, which we knew once in the past, but do not know them now. You have forgotten parts of the event. For such forgotten events, we do not have a continuous pathways, either abstract or dynamic. In other words, our forgotten memories created discontinuous pathways in our brain.

If we look at all events we experience, the events themselves always have continuity in the dynamics. The discontinuity mentioned above is only stemming from our internal brain's architecture because of our inherent ability to forget parts of an event. The interesting question now is: how can observing only continuous events give rise to a notion of discontinuity, especially since discontinuity is an artifact of our brain, not of the external events? What aspects within our brain can give rise to a feeling of knowing (i.e., abstract continuity) of the concept of discontinuity while only observing and knowing continuity?

An infant or a toddler knows several facts even though he may not know that he knows them. To know the concept of knowing, what we need is to form abstract continuous paths for the concept of abstract continuity itself. You need nonempty predictive meta-fixed sets for the concept of prediction itself. I have already discussed some details for answering this in chapter 21 on how we can know 'about' an event. Let me discuss a few additional details here.

As mentioned in the previous section, our memory of an event as a linear abstract continuous path has several turning points, which we remember as gist. As the events become long and complex, we will have several turning points from start to finish of the event. The recollection process takes a finite amount of time as well. Consider an abstract continuous path time-ordered from region A to region B of our brain with several turning points as well as excitations orthogonal to the main path from A to B. To attain a feeling of knowing of the entire event and at every instant, we need abstract continuity at most of the intermediate regions along the path from A to B as well as along the entire path.

From the way loops are interconnected along the abstract continuous path of an event, it is possible that we will excite some regions along the path faster than other regions. They are not necessarily excited in a linear order anymore because of the parallel cascading excitations orthogonal to the main abstract continuous path. For example, when we are recollecting what we did on a given day, we sometimes recollect a later event before an earlier event. The time sequence of our recollection is sufficiently random, but can be correctly ordered, if necessary.

Now, our neural loops will be overwritten with new interconnections from new experiences. This will make us completely forget significant portions along the path from A to B even though we may remember several turning points of the entire event. In this situation, we have a feeling of knowing for only portions of the event, but not the entire event. In an attempt to fill in these gaps, we start to recollect the sequence of events by retracing the steps.

This is the situation when you feel you know that you knew the event once, but you are unable to recollect all the details at this instant. The discontinuities with your recollection produce a unique feeling in itself that is not the same as knowing the entire event. When your inability to recollect several details occurs several times, under different situations and from different topics, you start to form predictions (or abstract continuity) of your inability itself. These different situations and topics are the invariances needed in the definition of fixed sets (i.e., for creating the not-knowing fixed set). We will have several non-empty predictions whenever such discontinuities occur. You start to know the difference between knowing and not knowing.

50.8 Language and mathematics

The main point of this section is to argue against the claim that there exists an universal grammar common to all languages and that is innate to humans. The notion of abstract continuity and the discussion of several mechanisms in the previous chapters clearly suggests the opposite. There appears to be two groups of linguists, namely, nativists and empiricists. Nativists argue that there is a universal grammar common to all languages, which is innate to humans. They have proposed a poverty of the stimulus (POTS) argument (see Chomsky [21] and Laurence and Margolis [92]) to claim that the grammatical rules of a natural language cannot be learned by a child because of a limited amount of data and is, therefore, partially innate. Poverty of the stimulus means the following: a child learning a new language is exposed to too little external stimulus i.e., as too few examples of correct sentences and within a short period of time (like, say, a year). The logical possibilities of correct and incorrect sentences are too high that a child has too little time to be exposed to all such variations in grammar and, yet, he is able to generalize and formulate correct sentence structures with ease. This implies, according to nativists, that there should be an innate structure called the universal grammar. The empiricists oppose this argument and claim the opposite.

The research on this topic by both groups of linguists is quite extensive. Note that it is difficult to prove if one or the other group is correct using just logical analysis. The only convincing answer would come from any attempts at building a synthetic system. Unfortunately, there were no such attempts. In that sense, I will also be brief in my reasoning here. I will only point a few potential sources of inconsistencies between the nativists hypothesis and the theory proposed in this book. A few of these are:

- One assumption made by both nativists and empiricists which is at odds with the theory proposed in this book is that the analysis is linear and serial (as is typically the case with analysis in most branches of science, engineering, mathematics, language and even with our very own thought process). Our current computers are also modeled in the same way. However, the dynamics within our brain, as we have seen in this book, is inherently parallel and has loops. Even though the sentences in a language are linear, our consciousness only creates this illusion of linearity and serialness as we align our reasoning along a single timeline.

  The outcome of this is that the set of easy problems within the SPL architecture of our brain are quite different from the ones that are easy in a linear and serial system like a computer. As an example, logic and generalization (similar to the ones used in mathematics) is hard for humans in the sense that you need either a formal training or a significant amount of experience. Instead, analogies and intuition are easy for humans. These are purely parallel and looped properties.

- One important result of this book is that we should fundamentally relate abstract language to real physical dynamics (see section 13.1). Grammatically correct sentences are a consequence of abstract continuity in an SPL system. A child never creates grammatical rules by generalizing 5-10 similarly structured sentences. It is possible that some adults do, but not all adults can. Creating a rule from examples not only requires considerable amount of training, but it also is not natural within an SPL framework of our brain. Abstract continuity, predictions and associating physical and chemical dynamics to abstractions as described in this book is the only known way to ensure we create a conscious system.

- Learning a language by creating fixed sets and linking them to everyday dynamical events happens even before the child becomes conscious, as shown in this book. This process happens much before a child starts to speak the language. For example, a child has a well-defined 'internal' language even before 1 year of age when he can hardly say a word. The internal language corresponds to his abstract continuity of everyday dynamical events.

  The central result of chapter 22 is that a conscious child always has a feeling of knowing of all events at least to himself whether he can communicate to us in a natural language or not. Initially, his abstract continuity is based on images, sounds and other feelings. Over time, his spoken language aligns with this previously created abstract continuity. As he gradually transitions from this internal language to utterance of a few words and eventually to sentences in a natural language like English (say, by 2 years of age), the continuity of the feeling of knowing is never broken. This is when the rules he learnt and created, based purely on abstract continuity of dynamical events, align with our grammatical rules of an abstract natural language like English as well.

  Without a relation to physical dynamics, the ability to generate grammatically correct sentences becomes a mathematically hard problem and the possibilities become exponential. With the fixed sets structure and the links between them based on experiences, we do not have such theoretical difficulties.

This is the same situation even with mathematics viewed as an abstract language. Mathematics, similar to natural language, is a very formal language where there are a number of seemingly simple rules. Just as with natural language, we rarely discover the rules of mathematics or mathematical theorems from examples and axioms directly. Rather, we do the opposite. We discover the rules first using analogies and intuition instead. How did we create this intuition? We did this by building imprecise gist, mostly, from several examples and theorems that you learn over a period and creating abstract continuity as a result. We then use this intuition to discover and sense an approximate form of a theorem. We use formal mathematics to refine it and prove it within the axiomatic framework. Subsequently, we use additional examples to validate and apply our theorems. This is an easier approach within our brain architecture, not the process of discovering through generalization (though some people are capable of doing this easily as well). We can see this clearly with a 5-6 year old, even with simple rules like commutativity, associativity, identity and closure rules of addition and multiplication. He is surprised to know that these rules are indeed true when explicitly told about these, even though he has been using them all along.

This is the same with discovering grammatical rules of a natural language by generalization. The rules in language mentioned in Laurence and Margolis [92], for example, 'take the first auxiliary verb and put in the beginning of the sentence or place the verb from the main clause to the beginning', is a hard rule even for an adult to discover just by looking at several examples. If we have to rely on applying such rules every time we need to create grammatically correct sentences, it would be too difficult to formulate a grammatically correct sentence quickly.

Of course, there are people who do have an extraordinary level of intuition to discover such rules. In fact, they even discover a completely different set of rules and patterns as well, usually imprecise, than the ones we learn in school. This is true with chess grandmasters, mathematicians, musicians, artists and people with extraordinary computational abilities. For example, in Berndt [13], the author notes the following:

> But certainly Ramanujan's prodigious output of theorems would have dwindled had he, with sounder mathematical training, felt the need to provide rigorous proofs by contemporary standards. As an example, we cite Entry 10 of Chapter 3 for which Ramanujan laconically indicated a proof. His "proof," however, is not even valid for any of the examples which he gives to illustrate his theorem. Entry 10 is an extremely beautiful, useful, and deep asymptotic formula for a general class of power series. It would have been a sad loss for mathematics if someone had told Ramanujan to not record Entry 10 because his proof was invalid.

This example shows how our unique stable parallel looped brain architecture exhibits abilities that a linear and serial system cannot possible address.

One point to keep in mind is that the statement 'universal grammar is innate' simply pushes the problem around. It does not address how the universal grammar was created within a human brain (but not in other animal brains) for the very first time. In this book, I have shown that it is possible to create a synthetic conscious system starting from a non-conscious system. Learning a natural language is surely a simpler problem than creating a synthetic conscious system. Yet, I would say that my discussion in this section is not complete or satisfactory. It may be that the only clear way to address the innateness of universal grammar is by attempting to construct a synthetic system. I will treat this as a future work for now.

51
Chemical Systems

Until now, we have focused primarily on how the higher abilities of human brain work. We studied the structure of our brain containing an enormous number of 'physical' dynamical interactions. Here, physical refers to the fact that the complex features of our brain result largely from the physical dynamical network, not from the variety of chemical reactions that occur within different sets of neurons. We saw how stable parallel loops provided a framework to explain features like memory, perception of space-time, consciousness and emotions. I will now focus on the structure of 'chemical' systems and discuss how they form stable parallel loops. I will use it to explain how a cell and all of its behaviors work. The eventual goal in the next few chapters is to explain how life itself originated for the first time from inanimate objects without any help from conscious beings using the SPL architecture at a chemical level. In this chapter, I will briefly outline some similarities and differences between physical and chemical systems.

51.1 Cellular life

Most complex systems we want to study have both physical and chemical components. For example, with brain, the underlying dynamics is chemical in nature while the higher-level features come from the physical interaction of these chemicals within the neuronal network. The firing patterns, the different paths taken and the network of fixed sets formed are what I refer to as physical processes.

The dynamics within cells, on the other hand, are mostly chemical in nature. Each of the behaviors exhibited by all living organisms comes from a unique set of chemical reactions occurring within the cells. The physical dynamics correspond to transport of these chemicals. This is true despite the fact that some of the behaviors appear quite abstract or qualitative. A reproducing bacteria, a moth struggling to survive, a set of antibodies released to fight an infection and a plant or an animal evolving to adapt under harsh environmental conditions are some simple examples that involve, primarily, specific chemical reactions. I will discuss several such behaviors and their corresponding chemical reactions in more detail later.

If we look at a high-resolution picture of cells, we will notice an intricate internal structure within the cell membrane. There is a cytoskeleton structure made of long chains of molecules arranged in a seemingly arbitrary and complex network. This is not just a rigid static structure to support the shape of the cell. It aids in the transport of molecules from one location to another. There are also other substructures called organelles like mitochondria, ribosomes and nucleus that are involved in specific tasks. In addition to these, there are hundreds of types of molecules present within the cell participating in a series of chemical reactions.

These reactions appear to be perfectly coordinated and they seem to occur as if there is an overall purpose. The cell behaves as a *single entity* performing reactions that appear to be for the greater good of the cell, sacrificing some or changing the order of others, to ensure that the entire cell is ultimately benefitted in some sense. This is, of course, a human interpretation. The cell itself is simply obeying the physical and chemical laws of nature. Yet, these unique set of behaviors makes us feel that the cell is alive. This is unlike any nonliving system. In nonliving systems, the physical or chemical dynamics seem to occur in a rather predictable and independent manner without many variations. They simply react to the internal or external dynamical changes.

For example, if you push a rock or even a dead body off a steep cliff, it falls down in a predictable way. However, a living being sliding off a cliff struggles in every possible way to avoid death. Therefore, its motion is quite unpredictable. In general, any living being, including a single cellular organism, subject to a stress attempts within its means to resist the stress and avoid death. This is much more than saying that the system is trying to be stable. Stability suggests that the system is merely 'reacting' to maintain this state when some external stress tries to destabilize it. However, living beings play an 'active' role for survival and for other behaviors, as if they know what they want. My objective is to show how the stable parallel looped architecture of cells (or chemical systems) would give rise to such active behaviors.

Among all living beings, it is conceivable that those having a brain (i.e., with a complex physical dynamics) and with a certain level of awareness do perform tasks with a purpose. This purpose is, in general, unique to a given living being. However, it is not obvious to see how we can extend this intuition to chemical systems like, say, a single cellular organism. Without a brain, we do not have examples among purely chemical systems that exhibit some level of awareness. Yet, it is possible to re-use some of the ideas from the previous chapters when studying chemical systems. I will describe them where applicable in the next few chapters.

51.2 Systems approach

Before we can propose how life originated on earth from nonlife, we need to understand the structure and complexity of existing life in an abstract sense. We need to identify the aspects of life that are truly difficult to develop by nature itself, based on our current knowledge. We can then look for alternate ways to explain the complexity of single and multi-cellular organisms. For example, one way is to see if we can start with simple systems and iteratively add complex features.

We resort to two approaches when trying to explain the complex features of life: (a) that a given feature is created all-at-once or (b) that the feature is created incrementally and iteratively. Given the complexity of life, some people assume the all-at-once approach for, at least, the initial set of features, even though they believe fundamentally in the second incremental approach after the first cell is created. For example, in extreme cases, people have suggested that an entire (primitive) living being, with a large set of features, is created all-at-once. Some of these suggestions for creating the first 'complete' life-form all-at-once are: spontaneous creation of a primitive cell from simple amino acids and auto-catalytic RNA, creation of a protocell, assembly of a primitive cell from a primordial soup, creation with help from an intelligent being or a creator like God, creation of a primitive cell in an alien world and transferring it through meteors and others. In each of these cases, there is little clarification on either the complexity of life or the complexity of transition from nonlife to life. Each of these approaches assume a certain unknown or random mechanisms that start from a disordered universe and create a complete, well-structured and perfectly coordinated primitive life-form with a clear structure.

Another example is the creation of intricate organs like human eyes or ears. Not only do they seem too perfectly designed, but also the idea of creating them incrementally, giving a survival advantage to the organism at every intermediate stage, seems to be a far-fetched idea for some people. The number of prerequisites for such an incremental chemical design are: discovering the correct set of enzymes as well as several other organic molecules, incorporating the creation of most of them into the DNA, linking them through a series of chemical reactions using the correct set of transcription factors and other allosteric enzymes, creating a detailed structural design (of a particular shape and size – like a birds wings should be symmetrical and should eventually make them fly), encoding and decoding this structural design within the organism so the feature can be reproduced later and tweaking the designs as if the organism understands the function. I will discuss more about this in chapter 70.

Irrespective of which approach we choose, all-at-once or incremental, the amount of complexity in these systems demands us to use a system's approach towards understanding them. System's approach is quite common in most areas of research. The idea is that we focus on aspects that are common across a group of systems instead of a single object. Geometric, physical and dynamical properties of systems are clearly examples common across several systems. With chemical systems, this approach is not as widespread. This is because the specificity of enzymes, DNA, amino acids and other organic chemicals make it seem that the set of common properties are either small or less useful. It appears that taking a system's approach might not yield useful results. Yet, we did identify several useful common properties like transcription factors, allosteric enzymes, DNA binding domains, operons and others. Starting from chapter 52, I will show how to identify more properties based on stable parallel loops and use them to answer the problem of the very origin of life.

In the case of brain, in chapters 10-50 I have used systems approach to explain complex features like consciousness, memory, emotions and free will. Dual space representation (see chapter 25) becomes the natural language to describe systems when taking a system's approach. This is because you are now studying the relationships between systems and their properties, not the specific chemicals within each system individually. The primal representation can be seen analogously as looking directly at each specific enzyme, the specific chemical reaction and the specific pathway when explaining a given task. The dual space representation, on the other hand, looks at the relative relationships between the set of chemicals, their reactions, the repeatable looped reactions that occur, the direction of propagation of a chemical disturbance and others when explaining the same task. This is somewhat analogous to the brain features as well.

The dual space representation also naturally leads to a notion of dual equality (see section 44.2). This dual equality was used to compare when two brain structures are similar even though the exact number of neurons, network connections and other structures are considerably different. A similar dual equality is useful and necessary even for chemical systems. In chapters 52-70, we will see that each of these concepts and, more generally, the system's approach, starts to play a fundamental role when understanding the problem of origin of life, transition from unicellular to a multicellular world and others.

51.3 Comparing physical and chemical systems

Before I discuss the problems like the origin of life at a chemical level, I want to list a few key similarities and differences between the physical and chemical systems.

1. *Input triggers for chemical dynamics*: Both physical and chemical systems require a trigger to start a cascading chain of processes. With physical systems, these are typically forces and motion of subcomponents transmitted within its internal substructures. With chemical systems, they are either chemicals that enter the system like hormones or chemicals that are generated internally like an enzyme.

2. *Role of geometric and physical properties for chemical dynamics*: The properties like shape, size and strength of the physical structures that are part of a physical system do assist the internal and external dynamics in a positive way. This can be, say, in the form of specific directional forces or pressures. On the other hand, in a chemical system there are no such obvious substructures to help a chemical reaction proceed in a given direction.

3. *Physical structures and memory mechanisms in chemical systems*: In a physical system, the physical structures contribute towards memory of a past event in a reliable way. For example, in the brain, as an event occurs several times, a new set of looped connections within the network of neurons capture the 'static' part of the memory quite effectively. Subsequent excitations allow the firing pattern to take the previously established pathways. In chemical systems, there are no such previously established pathways in the same sense. This makes the problem of memorizing a past event considerably different. In chapter 58, we will see that there are multiple ways to store memory in cells as well. For example, some of these are within the DNA, within a network of looped reactions and within the threshold concentrations of chemicals.

4. *Control of chemical systems*: The notion of control in physical and chemical systems has subtle differences. In physical systems, the control of any event is possible by moving parts in space and time as well as through simple attachment or detachment of components. You push objects around or you break and rearrange them. In chemical systems, in addition to controlling events using the above mechanisms, there is another important type of control present. This is with the creation of new chemicals or with controlling

the production of existing chemicals. For example, by generating an appropriate set of transcription factors, by altering the rates of chemical concentrations or by discovering new antibodies, it is possible to produce entirely different behaviors. These processes cannot be viewed as simple attachment and detachment of components or as spatial and temporal movement of chemicals.

5. *Using conscious human notions when describing chemical systems*: When working with physical systems like brain, we use abstract notions like choice, decisions, desires, purpose and intentions. However, when we start looking at chemical systems like single cellular organisms, we cannot use the same notions to describe them, primarily because these chemical systems are not conscious. Unfortunately, using human notions is quite common. Most research descriptions utilizing Darwin's theory suggest, for example, a survival advantage, i.e., an organism detects a chemical change and 'wants' to survive, that a prey 'discovers' a chemical feature (within its DNA) to outsmart a predator and so on. These are imprecise and incorrect statements since the only mechanism proposed for these chemical systems are random mutations, while the above statements imply a sense of awareness for the organism and an ability to use this awareness to discover new and correct chemicals. There is no known way to convert an intent to survive into useful mutations and correct chemicals in an offspring even if we wait long enough. As a result, such descriptions start to get problematic when these subtle assumptions start to add up. For complex problems like the transition from unicellular to multi-cellular organisms, the boundaries between such assumptions, our analysis and the real chemical mechanisms starts to blur.

In the next few chapters, I will explain how the problem of origin of life i.e., the creation of the first life forms from inanimate objects without help from conscious beings, the transition from unicellular to multi-cellular organisms, the transition from prokaryotes to eukaryotes, the diversity of life, the direction in evolution and other related problems can be answered using the stable parallel looped architecture at a chemical level.

52
Origin of Life

In this chapter, I will discuss the problem of how life originated from inanimate objects for the first time without help from conscious beings (or a Creator). I will show that a stable parallel looped structure is once again critical in solving this problem. In addition, I will discuss how evolution proceeds in a directed manner i.e., how nature creates multiple features so they become cumulatively better and better. In this chapter, I will only formulate the problems and leave the detailed discussion to subsequent chapters.

Let us start by asking what a possible definition of life is. If a system were to be classified as a living being, it should exhibit several uniquely important features that researchers have already identified. These are, for example, structural and dynamical organization of the system, reproduction, metabolism, growth, interaction with the environment, adaptation and self-regulation. In addition to these, we do seem to make several other *hidden* assumptions when thinking about living beings. For example, it would appear that we would not accept an inorganic system (made of, say, electronic components) as a living being nor would we accept if the system does not have substructures like proteins, DNA and other complex molecules, even if it exhibits the above features.

In addition, we expect an unspecified amount of complexity if the system were to be a living being. If it is too simple and/or it has a limited number of subcomponents or substructures, we would think of it as a mechanical system instead. If we human beings can create the system easily, we would hesitate to call it as a living being. Furthermore, we attribute abstract qualities to life that we cannot quantify easily. For example, a living being wants to live and it opposes death. This feature appears nothing more than a notion of stability i.e., a stable system naturally opposes an input that tries to destabilize it. However, dying is qualitatively unique, beyond becoming unstable or collapsing.

In spite of these issues with the definition of life, I will just use both the accepted definition as well as our intuition when addressing all features exhibited by living beings. I will focus on how this generalized meaning of life emerges starting from a state where there are only inanimate objects on earth (or any other planet). The goal is to understand if life can be created both naturally and in a repeatedly reliable manner. Where possible, we want to avoid random chance as a possible explanation.

52.1 Primitive systems and conditions on early earth

It is estimated that life evolved on earth about 3.8 billion years ago. Earth, on the other hand, was believed to have formed about 4.6 billion years ago. During the first 800 million years after the formation of earth, the physical conditions must have been quite violent and toxic for living beings to survive.

Given these harsh conditions, researchers have been looking at (a) how molecules necessary for the formation of life, like amino acids and nucleic acids, can be created under such hostile conditions and (b) where the possible sources of favorable conditions could be with some suggestions being hydrothermal vents, under sea bed or near the poles. For the former question, Miller-Urey experiment (Miller [106] and Miller and Urey [108]) is one of the first encouraging result. They showed that amino acids (and other organic compounds) can be created from water (H_2O), methane (CH_4), ammonia (NH_3), hydrogen (H_2) and a simulated lightening in the form of electric sparks. For the latter question, researchers have found microorganisms called extremophiles that survive under extreme conditions like in a boiling sulphur pool and under intense cold temperatures at the north and south poles.

While we will never know what happened 3.8 billion years ago, the puzzling question is how complex molecules like proteins and DNA did get created and remained stable for millions of years in those harsh environments? The locally stable environments suggested by researchers (like hydrothermal vents and others) must remain favorable for millions of years for life to develop and sustain.

In this book, I will discuss several primitive SPL systems that iteratively improve and become stable over time. I will show how such an iterative class of chemical SPL systems will eventually lead to the creation of life from inanimate objects. I will, however, not discuss any of the conditions of early earth in more detail.

52.2 Processes to create life from inanimate objects

What sequence of steps occurred on earth that gave rise to the first set of primitive cells from inanimate objects? As mentioned earlier, even after explaining how molecules like amino acids and nucleic acids are formed, it is nontrivial to explain how they coordinate to form a complete living cell. There is little explanation, if at all, on how this happens. Answering this question will be one of the main focus of the next several chapters.

For example, if we take a cell and either crush it or heat it up (like boiling an egg) to breakdown its structure, essentially killing it, then recreating a living cell from these 'favorable' molecules is quite impossible for now. Since mass is conserved, all the original atoms are present, though as different molecules. Conceptually, what is involved for recreating a living cell is to break all the incorrect chemical bonds, recreate the correct bonds and, hence, all the important complex molecules. We also need to create the necessary physical structures like a cell wall and nucleus and arrange all the correct molecules at the correct physical locations. During this process, we do need to supply energy to run most of the chemical reactions in the cell. These steps are not only difficult, but most of the reactions are not even thermodynamically favorable to occur naturally. A seemingly great level of 'intelligent' coordination is necessary to make this happen.

The problem of origin of life requires a similar level of re-arrangement and coordination. Of course, we do want to avoid help from an intelligent designer for this problem and for explaining the origin of life. While this is already a nontrivial problem, how can we explain how to create life from inanimate objects when we do

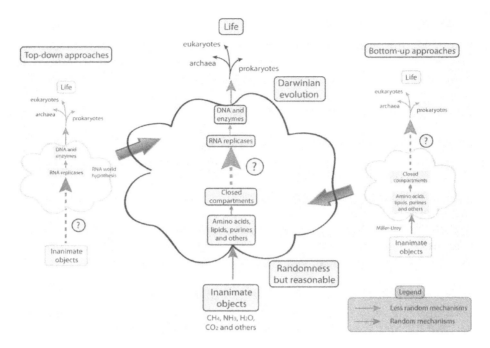

Figure 52.1: Top-down and bottom-up approaches to the problem of origin of life. Top-down approaches create structures closer to existing life forms while bottom-up approaches use prebiotic conditions to propose simpler structures that eventually lead to primitive life forms.

not have any of the above favorable molecules (cf. a boiled egg that still has some useful proteins and DNA).

Two basic approaches are commonly used here. One is to rely on an intelligent designer of some sort (including references to God). The other is to explain the complex designs of life without a designer. The latter approach, however, requires some form of randomness or luck. For otherwise, the occurrence of a sequence of thermodynamically unfavorable chemical reactions violate the laws of nature unless a designer assists with setting up the favorable conditions.

All scientific theories take the second approach including Darwin's theory and the theory presented in this book. Within this approach, researchers have looked at the problem in a top-down and a bottom-up fashion (see Figure 52.1). Top-down approaches try to identify common chemicals across all existing domains of life and attempt to understand how they could have assembled to create the first life-forms using random chance mechanisms (see Figure 52.1). An example is RNA world hypothesis, which aims to create RNA replicase. Bottom-up approaches try to identify chemicals that are easier to form under prebiotic conditions and studies how they form compounds present in all current life forms. Creation of amino acids under prebiotic conditions by Miller and Urey [108] is a typical starting point for bottom-up approaches. The hope is that both approaches converge to form a complete theory.

The theory proposed in this book is one such theory.

Let us briefly look at a couple of proposals – Darwin's theory and RNA world hypothesis. Darwin's theory ultimately relies on random mutations to explain every simple and complex feature. Unfortunately, randomness is a less convincing explanation for all types of life-like features, especially for the creation of life from inanimate objects when we do not yet have DNA, RNA, replicase or other complex molecules. Since we have never seen life arising from dead naturally, why should we accept the same for the first time (i.e., for the problem of origin of life)? Using randomness for the entire transition process i.e., from nonlife to life is simply unreasonable. Darwin's theory falls short during this period.

Another example is with RNA world hypothesis. Here, the assumption is about the existence of self-replicating and self-catalyzing RNA molecules. Even if we succeed in creating a large number of such molecules in a lab and if they self-assemble to create an entire living cell, we still have a question of understanding the elegant structure of the chemical world. The current experimental approaches for creating life from inanimate objects seem to 'skip' the transition problem (from nonlife to life) by waiting for an entire cell to form by itself through random self-assembly. The theory presented in this book is a significant departure from such theories. It will truly clarify the detailed structure of a chemical world necessary for creating life from inanimate objects, not just on earth but on any other planet.

52.3 Repeatable randomness – a new framework

The only known mechanisms suggested for addressing the problem of evolution of life for now is randomness and Intelligent Design. Darwin's theory, for example, uses the former mechanism, namely, random mutations within a hereditary chemical DNA (followed by natural selection). What is the core idea of the new theory? I call it repeatable randomness as opposed to arbitrary randomness used by Darwin's theory (see section 71.3 for a detailed account of this). It is unrelated to Intelligent Design, however. Let me briefly discuss what this means.

With Intelligent Design, the assumption that someone else created a given feature does not give any insight into the underlying mechanism. Similarly, with arbitrary randomness like with Darwin's theory, the underlying structure of how to create complex interacting dynamical systems is not clarified. The arbitrary nature of randomness masks too many details. We cannot use both of these theories to create life in a lab from elementary molecules. We also cannot explain how life is created entirely naturally with no help from conscious beings. Contrary to what most researchers suggest, natural selection is not a mechanism or a driving force for the 'creation' of any feature. According to Darwin's theory, the feature is created by random mutations, which is the only suggested mechanism. Natural selection plays a role only subsequently i.e., after a random mutation produced an useful feature in an offspring, it gives these organisms an advantage over others making them survive 'longer'. In chapter 71, I will discuss the problems with Darwin's theory quite thoroughly.

Some examples in which arbitrary randomness is unreasonable are: (a) for cases where the random mutations will accumulate over hundreds of generations to create an overall improvement, (b) for creating a complex feature or (c) for explaining a favorable direction in evolution. Something more fundamental is necessary to explain how the random mutations can create a long sequential set of gradual improvements spread across multiple generations. A new mechanism should be discovered and it should clarify why the direction goes against the second law of thermodynamics (i.e., the law that suggests that systems evolve in a direction of increasing disorder – resulting in a natural random distribution or a collapse).

The new theory presented in this book also relies on randomness, but in a very different way. I only rely on randomness that is repeatable in a reasonable sense – not on arbitrary randomness. The events do not have to be reliable and predictable. Instead, they just need to be repeatable in a reasonable sense. For example, if I say that I want wind to blow in the northern direction to create a given feature, then even though the event is random and unpredictable, the wind surely does blow in the correct direction for a sufficient large number of times. Similarly, if I want bubbles with a few holes form in sticky mud or clay for some feature, this can surely occur in a repeatable way on a rainy day even though we cannot predict when it might happen. Another example is the creation of a closed membrane using phospholipids. This indeed happens quite naturally because of the properties of phospholipids. Other examples are with looped physical or chemical systems.

What is interesting is that even though the repeatable randomness is not reliable initially, I will show how to improve reliability as the system accumulates more and more stable interacting looped systems using repeatably random events. In other words, I will show how to improve reliability of creation in an iterative manner by linking more and more stable parallel looped subsystems together.

Let me list a few examples of random events that simply cannot be accepted as repeatable. The creation of enzymes for the first time on earth or any enzyme without the help of, say, DNA, RNA polymerases and ribosomes, creation of a specific DNA or RNA sequence, creation of a self-replicating and self-catalyzing RNA molecule (as suggested by RNA world hypothesis) are definitely not reliable or repeatable random events. All existing theories on the origin of life assume these events as reliable, which is incorrect. These examples are only acceptable if we discover a way to break them up into much simpler and repeatably random events of the type described above.

52.4 Minimal structures

Consider a sealed room filled with several types of simple molecules like CH_4, NH_3, H_2O, H_2O_2, H_2S, CO_2 and others. Let us assume that it is supplied with sufficient amount of randomly distributed energy for as long as we are performing this experiment. Initially, the molecules in the room are in a state of complete disorder. The question we ask now is: what would we expect to see in this room after, say, 100 million years? If the same set of molecules stays in the same state of disorder, we clearly do not require any further explanation. On the other hand, if we see considerable order like L-form amino acids, D-form sugars, DNA, enzymes and even primitive life forms, we have a difficult question to answer. How did such high degree of order get created from the initial disorder? What natural laws and mechanisms make this happen, assuming there is no help from conscious beings? The only answer for now is random chance-based mechanisms, which is clearly unsatisfactory for the high degree of order. Our goal is to present a rigorous alternative by identifying new structures, new dynamics and less chance-based mechanisms to create order from disorder.

Take, for example, the familiar notion of stability of molecules. Those molecules that are stable under a given set of environmental conditions (like temperatures and pressures) remain after a long time while the unstable molecules disintegrate and disappear (or become trace elements like H_2O_2 compared to H_2O). Therefore, in the above sealed room, we see that stability alone is capable of producing order from disorder. After 100 million years, only those molecules from the initial set that are stable will remain. We want to know if there are additional patterns and properties beyond stability that can bring further order within this sealed room.

If we place highly ordered systems like enzymes and bacteria into the sealed room as the initial set, it is not surprising that more order will be created subsequently (see Darwin's theory). Therefore, the above problem of evolution of order is interesting only if the initial structures in the sealed room are the most elementary ones. I refer to them as **minimal structures**, using which we want to explain how nature creates highly ordered systems.

Consider another example of order, namely, the long ordered sequence of chemical reactions that occur within our body at every instant for every task (like when walking or when digesting food). If we place the same collection of molecules in a beaker, as in a dead cell, the same cascading sequence of reactions cease to occur. What static and dynamical arrangement of molecules allows such coordination to occur in living cells but not in dead cells? The basic survival of an organism requires such coordination. While we have abundant proof for the physical existence of such high degree of order (i.e., every living organism), what we do not have are the theoretical structures and mechanisms that explain the creation of large-scale stable complex systems like living beings.

From these examples, we have two distinct questions to answer. One is an *analysis* problem in which we want to understand the laws and mechanisms that allow the existence and evolution of large-scale highly-ordered complex systems. The second is a *design* problem in which we want to create such large-scale complex systems both naturally and synthetically. Every existing theory that attempted to provide a framework for such problems started analyzing and designing a system with a small degree of order. However, they failed to scale to systems involving thousands of reactions or dynamical subcomponents. As a result, there is a huge gap in the frameworks used by theorists and experimentalists in biology and complex systems. Experimentalists argue that the models chosen by theorists are not realistic and that they do not capture the scale of complexity observed within living cells.

In the previous chapters, we have already introduced a new theoretical framework that allows the creation of large-scale (millions or, in fact, an infinite number) and highly interconnected dynamical systems with 'any' specified complexity, which as a whole does not collapse for a sufficiently long time. We called them as stable parallel looped (SPL) systems. They address the above discussed issues and acts as a common framework for both theorists and experimentalists. They not only capture the complexity of observed reality of living systems, but they also let us create complex dynamical systems that cannot form naturally.

One objective that is important to answer is as follows: *describe the structures, patterns and natural mechanisms within existing physical laws using which the probability of creating large-scale stable dynamical systems with high-order from less ordered components keeps increasing over time.* A few points to emphasize are: (a) in addition to synthetic mechanisms, with help from conscious beings, we should provide natural mechanisms for the evolution of order, (b) the theory should allow analysis of existing large-scale systems like cells as well as design of new large-scale synthetic dynamical systems and (c) the probability of creation of order should increase over time – this eliminates pure chance-based mechanisms. For example, there is no increase in probability of bringing the molecules in a small enough region within a short interval, a necessary requirement before we can create order, in the second million years after, say, 'randomly' trying for the first million years. It is apparent that the problem of origin of life is a subset of the above problem of evolution of order. The less ordered components are inanimate objects and the high-order stable dynamical systems are the primitive life forms.

For the above problem of evolution of order, one fundamental question typically ignored or postponed is the following. When a highly-ordered system is created naturally, we focus on identifying the minimal structures and the sequence of steps that caused the creation of order. However, what laws and mechanisms guide all necessary subcomponents to indeed follow these identified sequences of steps? For example, according to RNA world hypothesis the minimal structure for the origin of life is the creation of an RNA replicase enclosed in a closed membrane (Szostak et al. 2001). However, to create replicase molecules, several prerequisite molecules and steps are needed. These are, say, lipids, nucleic acids, sugars, arrangement of monomer chains to create unique polymer sequences, inorganic catalysts like clay montmorillonite and others (Hanczyc et al. 2003; Hanczyc and Szostak 2004).

The above question translates to (see Figure 52.2): (i) how can all of these prerequisite molecules naturally accumulate in a small spatial region (cf. air molecules reaching one corner of the room) so they get a chance to react chemically? This seems to oppose the second law of thermodynamics. (ii) How can they accumulate within a short period? Otherwise, they may disintegrate or move away before they can be part of a long cascading chain of chemical reactions? For a few reactions, abundance of molecules or chance mechanisms as an answer may seem satisfactory (like the spontaneous self-assembly of phospholipids to form a closed membrane). However, this is unsatisfactory for processes of life as they involve long cascading chain of reactions like with creating a unique RNA replicase sequence. (iii) If structures like RNA replicase are created by chance, how can they be created reliably and repeatedly so the order can be recreated in the event they are starved (like during ice ages or meteor strikes) or when they disintegrate? The mere presence of all these prerequisite molecules is not enough to create high degree of order (cf.

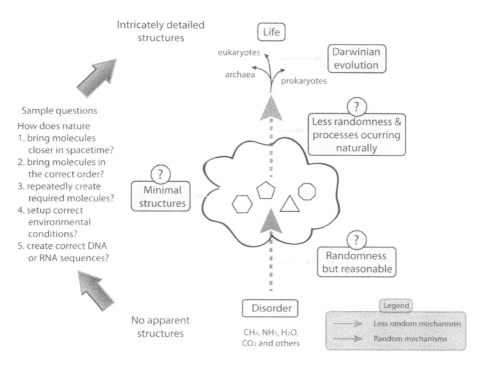

Figure 52.2: The problem of creating life from inanimate objects. Our goal is to explain how we start from an initially disordered state of simple molecules like CH_4, NH_3, H_2O and CO_2 to create intricately detailed structures like living beings. The transition from initial random mechanisms to less random mechanisms happens through the creation of *minimal* structures. These minimal structures should be simple enough that they can occur easily and naturally (i.e., simpler than specific RNA sequences that can both self-replicate and self-catalyze). Our task is to propose mechanisms for each of the steps indicated by '?' by answering several questions listed here.

dead cells or all required chemicals dumped in a beaker). There seems to be a natural sense of regulation or coordination at a large-scale that avoids collapse, though it appears highly likely, in living beings. The above questions are applicable even for processes that occur in our body at this instant.

52.5 Laws of nature to create a design

If we want to create a car or a computer, what set of laws do you need? Are they Newton's laws, Maxwell's laws or some other laws that we have already discovered? I am not asking about the laws a car or a computer obeys after it is created. Rather, I am asking about the laws necessary to create the design of a car itself. Therefore, we are looking for laws (like Newton or Maxwell's laws), which when applied gives the geometric curve in the shape of a car as the solution.

For now, the laws obeyed to design a car or a computer is hidden within a human brain as 'creativity'. We do not have laws for creativity. Yet, we do try to explain how a person is able to come up with a creative solution. However, each such attempt at describing creative solutions seem to have no connection to Newton's and Maxwell's laws that every other system seems to obey. Our brain, which is the ultimate source of creativity, does obey Newton's and Maxwell's laws (at the level of individual neurons). Yet, we do not know how to convert these low-level dynamics to high-level features like creativity. It is obvious that we did not discover any laws that will give the shape and structure of a car, a table, a computer, a tree or a cell as the solution.

It is surprising to see that the laws we truly need, can be grouped together as 'intelligence', not as Newton's laws or any other laws we have discovered until now. Yet, not all humans are capable of designing cars or computers. We do need laws of creativity. Surely, we are uncomfortable at suggesting that the laws of intelligence and creativity are merely randomness. Unfortunately, for now, randomness is the only scientific solution for the *creation* of any feature. A cell self-assembles for the first time using random chance. A bird discovers a feature like wings using random chance mutations and so on. A human brain is an example of a system that does exhibit the laws of intelligence and creativity (whatever they are, in an abstract sense). It is not a random structure, nor are the features it exhibits like consciousness, intelligence and creativity. Yet, strangely, our brain is made of components that ultimately obey Newton's laws, Maxwell's laws and other quantum mechanical laws.

This is a serious problem with the knowledge we have accumulated in science so far. The domain of laws for the creation of any system or feature is completely disjoint from the domain of laws obeyed by the system after it is created. The former laws are the outcome of encoding within the structure of a complex system like a human brain. The latter laws are the standard laws of nature (like Newton's, Maxwell's and other quantum mechanical laws) we have discovered until now.

The theory of complex systems introduced in this book using stable parallel looped dynamical systems are the correct set of laws that unify both these domains together. For example, I will show how life itself can be 'created' entirely naturally from inanimate objects using the same stable parallel looped architecture. In the previous chapters, I have already used the very same architecture to explain how our brain and all of its features like consciousness, intelligence and others work as well. In this sense, the SPL architecture unifies features like intelligence and creativity, which are eventually needed to explain the creation of man-made designs like cars and computers. For chemical systems and for the origin of life, a similar SPL structure can create new designs, as we will see.

52.6 First time creation versus subsequent creation/usage

Living beings have a large set of features, each of which are quite complex to create naturally. Nature needs to discover a long sequence of steps spread over, possibly, hundreds of generations just to create one feature like an eye or a heart. Yet, after it is created for the first time, the subsequent creation or usage of this feature happens in every single generation quite easily. The difficulty is not with why it takes such a long time to discover for the first time versus every subsequent time. Instead, the difficulty I want to highlight here is in the significant difference in the underlying mechanisms between the two instances.

For example, the mechanisms we use to explain the creation of a multi-cellular organism for the first time when the world only had single cellular organisms are completely unrelated to the embryogenesis mechanisms used for all subsequent times. Similarly, the mechanisms used to explain the creation of the first life forms from inanimate objects, namely, chance occurrence of self-replicating and self-catalyzing RNA molecules, say, is completely unrelated to the subsequent creation of life forms from existing life forms. Another example is the creation of sexual reproduction for the first time when there were only asexual reproduction mechanisms versus creating two sexes subsequently. Other examples are with the creation of brain features like human consciousness for the first time, excluding the theory presented in this book, versus how it is created now.

The theory presented in this book uses the exact same mechanisms for both the first time and all subsequent times. This is unlike any of the existing theories. The existing biological framework does seem to provide similar mechanisms between the first and subsequent times in some situations. These are, for example, with the creation of a functional eye, ear, heart, lungs or any other organ in our body. Both mechanisms, in this case, require the discovery of a correct set of enzymes and subsequently carrying out the correct sequence of chemical reactions. However, what is not clearly explained is (a) the discovery of the correct enzymes and (b) how the location and structure of chemicals store the sequential order of execution of chemical reactions as memory.

52.7 Direction in evolution without a plan

If we look at the timeline starting from the origin of life, we see that there is a specific direction in the evolution of life. They follow a general pattern of increasing complexity until some unforeseen event occurs and disrupts this direction. Even after a dramatic event like a mass extinction, natural mechanisms take over once again to produce the same pattern of increasing complexity. The term increasing complexity is hard to define in precise terms and I am only using it in an informal sense. For example, informally, we feel that the diversity of life seems to produce life forms that are more stable than before. One of the important questions in understanding the origin of life is how and why we have such a directed evolution. Does the directed evolution happen only within living beings or does it extend before the origin of life in nonliving systems as well?

Examples of increasing complexity: Let me discuss a few examples to emphasize the pervasiveness of the directed evolution. First, at a molecular level, we see that a complex sequence of events should occur for any cycle of chemical reactions to work correctly. Some of the important examples are Kreb's cycle for cellular respiration, Calvin cycle for photosynthesis and cell division cycle. Take Kreb's cycle, for example, that occurs in the mitochondria for generating energy that the cell needs. This cycle requires eight enzymes to work correctly and a few other enzymes to create appropriate inputs needed to start the cycle. Even if one enzyme is not yet discovered, the cycle would not occur in nature.

In addition to these enzymes, let me briefly describe what other cell machinery is needed, in generic terms. Each of the required enzymes will need to be created through a corresponding DNA sequence. We need several enzymes for the transcription and translation mechanisms needed to create a protein from DNA (or gene). For example, for transcription, we need an enzyme to open the DNA strands, uncoil the DNA and create mRNA. We need enzymes to transport mRNA from the nucleus into the cytoplasm. The translation of mRNA into protein sequence requires several enzymes as well. The freshly created enzymes need to be transported to mitochondria using another set of enzymes along microtubules, where they will be used. Subsequently, the mRNA's (and even some enzymes) should be broken down by other enzymes so that the components can be reused.

Next, the production of enzymes from the DNA needs to be controlled, for otherwise, you would have starvation of chemicals and loss of cellular energy. This control can occur anywhere within the process – from transcription to translation. For transcription, there are specific DNA sequences called promoters, activators, repressors and corepressors, referred collectively as operons. Special proteins called transcription factors are bound to these DNA regions that either helps or inhibits the transcription mechanism. The transcription factors themselves respond to, say, environmental stimuli, hormones and other chemicals, the existence of which are equally important. In addition, there are other ways to control the activity of a gene like methylases and chromatin remodelers, besides transcription factors. Now, each of the enzymes mentioned above as well as the transcription factors would require the discovery of specific DNA sequences.

How did nature discover every chemical described above in a directed manner for running chemical loops like Kreb's cycle? Almost all looped chemical reactions within the cell need most of these intricate mechanisms, though several of the enzymes are shared across these reactions. Besides, it is not enough to just create each of these chemicals. It is also equally important to discover the correct sequence of mechanisms and processes, again in a directed manner, to make the entire set of events to work correctly. The description above illustrates such a sequence of steps, some of which are long and others sequentially linked together. Cell division, as another example, has a specific sequence of steps too like dividing the centrosome first, followed by copying the DNA, splitting the nucleus and so on.

It is unreasonable to assume that all of these chemicals and mechanisms were discovered by nature simultaneously. Instead, they must have been discovered in stages, not necessarily in the order discussed above. For example, there would be a period when only two of the eight enzymes needed for the Kreb's cycle would have been discovered. In that case, the desired sequence of reactions cannot occur, making the entire Kreb's cycle nonfunctional. In general, if chemicals were discovered in stages, then at any given instant, the desired sequence of reactions cannot occur. If this is the case, how is nature able to make incremental progress for literally hundreds of enzymes in a directed manner? Or is it just pure random luck in every case, for millions of years?

Another example where the complexity of discovering the entire directed sequence of steps is apparent is with the development of an adult multi-cellular organism from a single fertilized egg. It has been shown in detail how, for example, the concentration gradient of morphogens released at one stage helps prepare the production of proteins at the next stage. The number of such perfect sequence of stages is far too many for complex multi-cellular organisms. If at any stage, there is a discrepancy in the location, amount or types of chemicals released, you will see an abrupt change in the body plan, sometimes viable but mostly leading to death. Rather than give more details here, I refer the reader to Nüsslein-Volhard [115] in which the author discusses different body plans for *Drosophila* by changing the concentration and timing of release of chemicals like morphogens. How is nature able to discover such a perfect and long sequence of steps? Specifically, how was the entire body plan including detailed designs of each organ encoded into a single cell? How is the relationship between a multi-cellular feature (like heart or lungs) to the single cell that started its creation are encoded?

Other such examples that exhibit a similar direction of increasing complexity is the evolution of eukaryotes from prokaryotes and multi-cellular organisms from single cellular ones. We even see this specific direction of evolution at an individual feature as well. For example, there must be organisms without any immunity against diseases before the immune system was developed. Similarly, vascular systems, heart, lungs, neurons, sensors like eyes, ears and touch, jaws, hands, legs, ability to stand upright, intelligence and so on are examples that follow the same general direction of increasing complexity. Even among plants, we see a similar direction like with leaves, trees, roots, ferns, gymnosperms (with seeds) and angiosperms (with flowers).

If we look at each of these examples, we realize that there are literally infinite possible ways nature could have evolved using random mechanisms. Why did nature pick such a narrow set of directions instead? Besides, the choices that nature picked appear much easier to make than the ones that were excluded. The main point with these examples is that there is a clear direction in evolution at a molecular level, cellular level, organ level, organism level, species level and even at the ecosystem level.

Degree of randomness and the existence of a designer: Whenever we consider any mechanism for explaining how evolution works, we should analyze to see if the mechanism can be discovered by nature automatically or if nature needs help from a conscious or intelligent being for this mechanism to work. If we believe that nature discovered life and evolution by itself without help from an intelligent designer, the mechanisms it used must be simple enough that they can occur naturally. We should keep searching for only those mechanisms that fall in this category. The examples described above on evolutionary direction shows that the sequences of steps are long and intricate for almost every feature. This makes it nontrivial to propose mechanisms that nature can discover them by itself.

Looking directly at the proposed mechanism, is there an easy way to say whether it can be discovered by nature or not? In this regard, the degree of randomness (or an average behavior across a population) can be used as a rough measure. At one extreme, if there is no randomness within the proposed mechanisms for evolution, it clearly suggests that there is a plan and a designer. Without randomness, a system following a long specific sequence of steps implies either (a) that the steps obey a physical law or (b) that the system is following a preplanned sequence of steps mechanically. If we did not discover any new laws that support (a) or if there are many choices along the way during the execution of the above sequence of steps, this excludes (a). For each choice a system encounters, some form of decision is needed. The decisions are not unique. Since we already excluded randomness, (a) is not a possibility in these cases. For (b), someone has to setup the sequence of steps correctly so that the system can simply trace them mechanically. This is the role of an intelligent designer. Notice that I am excluding the case where the living being has a certain level of awareness and knows what it wants or what direction to take. I should exclude this until we present a mechanism that explains how even a single cellular organism attains such a level of awareness.

At the other extreme, if there is sufficient randomness in the mechanisms, it is entirely nontrivial to explain how a specific direction arises in evolution. It may be possible to get lucky occasionally. However, to be lucky for 4 billion years and to recover from a catastrophe as well and subsequently following the same specific direction of evolution appears nontrivial if only random mechanisms are allowed. The amount of luck seems exponential.

In summary, if there is little randomness, it seems to imply that there is an intelligent designer. If there is too much randomness, we need to explain how evolution is inevitable and not merely lucky. In other words, if we want to avoid the role of an intelligent designer for directed evolution, we need a certain level of randomness combined with a structured set of physical laws. In this book, I will use the latter approach by proposing a set of minimal structures using randomness and then combining them with a set of structured physical laws, namely, stable parallel looped structures.

Nature of difficulty with directed evolution: The difficulty with a specific direction or sequence of events is that at every stage of the sequence, you need to ensure two things simultaneously. Firstly, the system should not lose the structure it has accumulated from the past sequence of steps when it is applying new random mechanisms at the current step. Secondly, the system should only pick those possibilities from the current step that necessarily improves the overall state of the system. If both of these steps do not happen simultaneously, you will not have a specific direction to evolution. This is an exponentially hard problem if the length of the sequence increases.

A simple way to understand the difficulty with this problem is by looking at the following analogous problem discussed by other researchers. Let us say that you want to create the entire work of Shakespeare using only random mechanisms. Your mechanism may have you randomly pick letters from the alphabet and concatenate them one at a time sequentially. If, at any stage, you happen to get lucky and get the correct sequence of, say, the first 100 letters (which you do not know, for otherwise, the problem is much easier and has less relationship to the problem of direction in evolution – you should have some other measure of correctness), your next sequential step should ensure that (a) you keep these correct 100 letters and (b) try to determine the correct 101st letter. However, since your mechanism has randomness, it is difficult to ensure that the first 100 letters stay correct even if you succeed at getting the correct 101st letter (especially since you do not know that you did get the first 100 letters correct). The probability that the first 100 letters become disordered is quite high.

This example is not a good approximation for all types of directed evolutionary problems listed earlier, for at least a couple of reasons. Firstly, you do not need to ensure a specific sequence of events. Rather, you only need a specific direction for evolution. Any other path is just fine as long as the complexity seems to be increasing. This might mean, for example, that you only need to create a book with grammatically correct sentences arranged as a meaningful story instead of a specific work of Shakespeare. Secondly, your mechanism is not completely random. There is some level of past memory to guide your future direction. However, you have to be careful to ensure that the memory is not too selective. For otherwise, you would need an intelligent designer to explain why you are selective or why you have made these specific choices. Nevertheless, one class of problems that do appear closely related to this fictitious problem is the birth of an adult multi-cellular organism from a single fertilized egg discussed earlier. In this case, the specific sequences of steps are indeed critical. For otherwise, you could have grown an additional leg somewhere else on the body of the organism (see Nüsslein-Volhard [115]).

52.8 Role of specific chemicals and reactions in evolution

Let us now look at the set of chemicals and chemical reactions needed for a single enzyme to perform its function correctly. Our objective here is to understand the complexity of discovering each of these chemicals and the corresponding processes by nature itself. I will use *lac* operon for lactose metabolism in *Escherichia coli* as an example since this was studied in considerable detail (see Jacob and Monod [78]). These researchers have outlined the underlying genetic regulatory mechanism for the breakdown of lactose into glucose and galactose. Let me list a few of these steps here for clarity.

The *lac* operon consists of three structural genes: *lacZ* (which encodes β–galactosidase responsible for cleaving lactose into glucose and galactose), *lacY* (which encodes lactose permease that is bound to the membrane and is responsible for transporting lactose into the cell) and *lacA* (which encodes galactoside O-acetyltransferase responsible for transferring acetyl group from acetyl-CoA to β–galactosides). In addition to these three structural genes, the *lac* operon consists of a promoter, a terminator, a regulator and an operator. The production of each of these enzymes are well regulated using two mechanisms. Firstly, if there is no lactose present, the production of the above enzymes (particularly, β–galactosidase) are suppressed through the *lac* repressor. Secondly, if there is a better source of energy like glucose, an unphosphorylated form of EIIA^{Glc} binds to lactose permease preventing it from transporting lactose into the cell. In this case, the cell consumes the existing glucose instead of relying on lactose.

The difficulty in explaining the above process has two parts: (a) on how the process works and (b) on how nature discovers all of the above mechanisms for the first time. The former question is already studied in great detail in Jacob and Monod [78]. However, for the latter question, not even a single approach has been suggested. One reason might be that there are too many steps that must be coordinated too perfectly across several thousands of years while the necessary molecules are being discovered, in order for this process to work correctly. Let me now briefly outline some of the perfect steps that are needed here.

1. The three-dimensional shape of the enzyme β–galactosidase needed for cleaving lactose into glucose and galactose is quite specific. If this shape is either different or does not form correctly, it cannot catalyze this reaction.

2. To create enzyme β–galactosidase, we need a very specific DNA sequence *lacZ*.

3. Next, we need a way to control the production of the enzyme. For this, we need other proteins like repressors and activators as mentioned above, which have specific three-dimensional shapes as well. The repressor prevents these enzymes to be created in the absence of lactose. The activator (like catabolite activator protein) enhances the expression of *lac* operon when lactose is present while glucose is absent.

4. The repressor and activator proteins need specific DNA sequences as well. For example, the repressor protein is encoded in *lacI* gene.

5. In addition, there are specific promoter sequences that help RNA polymerase, necessary for transcription of *lacZ*, to bind effectively.

6. The specific DNA sequences for promoter, activator and repressor proteins should be placed in front of the DNA sequence of the enzyme for them to be effective. Even if these sequences are discovered randomly, they should be attached in front of the β–galactosidase enzyme sequence (*lacZ*), not at any other place.

7. The shape of the repressor protein should also be special enough to allow it to bind to (a) the above specific repressor DNA sequence, and (b) to lactose which β–galactosidase catalyzes. The repressor protein should satisfy these two distinct conditions simultaneously.

8. The repressor protein should bind to lactose itself, not to any other random chemical. The purpose of repressor protein is to prevent lactose from breaking down (using β–galactosidase). It is important that the repressor protein bind to the very same molecule that it is about to prevent from breaking it down (i.e., lactose).

There are additional proteins and processes involved in the metabolism of lactose, which I have omitted here (see Jacob and Monod [78] for more details). Nevertheless, it is clear that discovering each of the above steps by nature itself through random mutations (followed by natural selection) is less convincing. We need additional mechanisms to ensure that the sequential steps across thousands of generations will cumulatively produce useful chemicals.

Besides, the relationships described above are quite important. Some other chemical cannot easily take over the place of a given chemical. There are cascading repercussions if you change even, say, one chemical. We would, then, need to change several dependent chemicals would need to be altered to achieve the same final functionality. If the chemical is not lactose, then the repressor transcription factor, the breakdown and the ability to pump lactose into the cell will not work. The only way to make it work is by explaining how the past dependencies can be maintained while creating future changes. In this way, you are guaranteeing that the future systems are not worse off than the past systems. Therefore, the accumulation of changes will eventually lead to an overall positive improvement. In the next several chapters, I will present a few such mechanisms (see chapter 59).

52.9 DNA and its role in replication

Creation of DNA allowed the possibility of self-sustaining life easier. There are a few distinct issues that we should answer when understanding the role of DNA in evolution. Some of these are: (a) how nature created specific DNA sequences, (b) how DNA was used to make individual features of the cell (i.e., from genotype to phenotype), (c) how DNA is critical for making an entire copy of the cell and (d) how various properties and features are encoded back into the DNA or within some hereditary system (i.e., from phenotype to genotype, if at all).

According to the central dogma of molecular biology, the last problem (d) never occurs in a strict sense. The flow of information typically happens from DNA to RNA and then to proteins. It never propagates in the reverse direction i.e., from a protein back into the DNA (although RNA to DNA is possible with retroviruses). What I refer to problem (d) instead is how a cell conceptually stores 'meta' information (explained shortly) within the DNA. For example, to create complex organs like heart, lungs or eyes, the organism should first create a large number of correct set of proteins. Each of these proteins and the actual assembly of individual cells should occur in a specific sequence and for specific durations. Otherwise, the organ will simply fail to form. Even minor changes in the sequence or durations either can result in the collapse of the organ or can generate serious abnormalities. The term 'meta' information for, say, the creation of lungs, refers to this specific sequence of steps and the durations of these chemical gradients. It is absolutely critical to encode this meta-information either in the DNA or somewhere within the cell, if nature were to create such organs repeatedly. This is what I refer to as the essence of problem (d).

For problem (b), the existing research has already outlined this process quite well, say within the central dogma of molecular biology. For example, transcription and translation is used to create necessary proteins to carry out the chemical reactions needed for a specific structure or a given external feature. However, what is not explained is how the complex sequence of events and specific durations are carried out in a well-coordinated fashion to perform a number of complex tasks. When you are exposed to a bacterial or viral infection, complex sequences of events occur within your body to fight off the infection. When you have an open wound like a cut on your finger, a series of chemical events occur like blood clotting, inflammation and healing of the wound. This is further linked to transfer of information to the brain through neural pathways as well. Why are these events linked together? The very origin of life as well as its existence and survival critically depends on an enormous number of such linked events at the molecular level. It is important to explain how DNA and these linked events are interrelated.

For problem (a), the specific sequence is critical because the three-dimensional shape of the protein created from the DNA is critical for its enzymatic properties. Sometimes even minor changes like point mutations in the DNA sequence can be detrimental to the cell. Not only does nature create such specific sequences, but it also maintains high fidelity of these sequences across multiple generations.

For the cell reproduction problem (c), replication of the DNA is not sufficient. There should be additional processes and sequence of steps that should be coordinated as well to create other structural and chemical components. It is important to outline these steps and the role of DNA in executing these steps.

52.10 Enzymes for chemical reactions

The rates of chemical reactions, in general, are too slow for a large number of them to occur in living cell. Enzymes play a critical role in speeding them up (catalysis), sometimes by a factor of a million. However, it is not obvious how you would discover an enzyme for a given chemical reaction. An enzyme is a complex three-dimensional protein with a very specific shape to catalyze a given chemical reaction. Catalysts made from RNA are termed as ribozymes. Even a small structural or chemical change renders an enzyme ineffective most of the time. The specific amino acid sequence of an enzyme is encoded within a specific DNA sequence. A sequence of physical and chemical reactions take the specific DNA sequence and creates an enzyme. A number of other specific enzymes like RNA polymerases and ribosomes catalyze each of these steps.

There is no apparent correlation between the amino acid sequence and the specificity of the enzyme for a given chemical reaction. The linear amino acid sequence appears to be present merely to generate the enzyme reliably and repeatedly. It is, however, possible to infer a correlation once the protein folds into a three-dimensional shape. In spite of this disconnect between the functionality and the linear sequence, most of the chemical reactions that occur inside a living cell are catalyzed using appropriate proteins or enzymes. Life would have been impossible if not for these enormous set of proteins and the corresponding chemical reactions.

There are, at least, a couple of difficulties here. Firstly, how did nature discover the correct DNA sequences corresponding to a given enzyme that catalyzes a given chemical reaction, when we know that the sequence information is uncorrelated to the functionality? Secondly, how did nature create each of the enzymes necessary for the very enzyme creation process itself (i.e., the chicken-and-egg problem for enzyme creation itself)? Briefly, to create a single enzyme, you start with a specific DNA sequence, transcribe it into an mRNA sequence and then translate it into an amino acid sequence using a specific genetic code. However, each of these steps requires unique enzymes themselves like RNA polymerase for transcription, ribosome for translation and, at least, 20 aminoacyl-tRNA synthetases for the genetic code (i.e., a code that specifies how triplets of nucleic acids correspond to a unique amino acid like, for example, UUU or UUC for phenylalanine, where U stands for uracil and C for cytosine).

If each of these enzymes is itself created using the same process, we have a serious problem. How can you create the set of enzymes necessary for the creation of themselves? It is unreasonable to say that they were created randomly or by luck through some form of self-assembly. Even if nature was lucky, what happens when any one of them degrades? This is like a chicken-and-egg problem: we need DNA to create even a single enzyme, but to create (or replicate) DNA, we need several enzymes themselves – creation of DNA needs enzymes and creation of enzymes need DNA. RNA world hypothesis was suggested as a way to avoid this problem. Here, RNA was thought of as the single type of molecule that can both replicate itself (like DNA) and catalyze its own creation (like an enzyme). However, the problem is not solved because we still need to claim that several self-replicating and self-catalyzing RNA molecules were created randomly. The theory presented in this book highlights a simpler, but special chemical structure necessary to explain the transition from a process that does not require DNA, RNA or enzymes to the above enzyme-DNA coupled process using just natural mechanisms.

52.11 Complexity of life

After answering each of the above questions, one important objective is to make sure we understand how an enormously complex set of chemical reactions are capable of occurring within a cell in a seemingly perfect and coordinated way without the entire cell collapsing. You can understand the difficulty if you ever try to juggle, say, five balls. It is quite difficult to make all of them stay coordinated without collapsing. How can our cell 'juggle' thousands of chemical reactions without ever collapsing? Furthermore, how did nature create such a high level of complexity by itself?

The second law of thermodynamics states that in a closed system, the entropy (or disorder) always increases. A living cell has almost perfect order and, therefore, the entropy is extremely low. How did nature manage to 'defeat' second law of thermodynamics, at least, locally and create such a perfectly ordered life without anyone's help? The total entropy of our solar system, considered as an approximate closed system, does increase. Yet, a living being, though not a closed system, has discovered a way to avoid collapse and, hence, defeat the second law locally in some sense. I will discuss this in chapter 53.

With each of these problems, I will show how stable parallel looped architecture helps in creating complex coordinated systems. We have already seen the advantage of this architecture for the problem of consciousness and other higher human abilities of our brain. It will be interesting to see that the same architecture is precisely the reason why life even exists and originated on earth by itself.

I want to mention a few additional problems that are equally important even after the creation of the initial life forms. These are the problems associated with the diversity of life during the last 4 billion years. The knowledge acquired through the current research in biology has already examined this problem of diversity of life in detail. Yet, there are several gaps in our understanding. I will only discuss some of them in the subsequent chapters.

For example, how did the first multi-cellular organisms evolve from single cellular organisms? I will discuss this briefly in chapter 70. A related question is on how the complex process of embryogenesis of a multi-cellular organism is encoded in the DNA and subsequently executed when creating the organism (section 70.8). This reproduction process is much more involved compared to a single cellular organism. For a single cellular organism, I will discuss the problem of reproduction in chapter 67. Another example is on how we can explain the origin of two sexes from asexual reproduction. I will not discuss this problem in detail and instead refer to Geodakyan's [51] theory of sex in chapter 69. Yet another example is the transition from prokaryotes to eukaryotes.

In fact, in general, the problem of diversity of life is too broad. Every transition from one form to another (or from one feature to another) in organisms and species is a nontrivial problem. The transition from water to land animals and subsequently to birds that fly is an example of such a nontrivial transformation that took thousands of years and created several new species. Darwin's theory is regarded as the central theory that explains this complexity of life. In chapter 71, I will discuss a few short-comings of Darwin's theory and show how the new framework of SPL systems help in overcoming these issues. In this chapter, I have primarily focused on describing several open questions within the problem of origin of life. I will now start by discussing each of these topics in greater detail.

53
Creating Order from Disorder

Even though there is no clear measurable definition of order, living beings are still considered the most complex and highly ordered naturally occurring systems we have ever seen. If we include man-made systems as well, a car, a computer and other systems do have sufficient order. However, it is clear that man-made order is only possible because of consciousness, intelligence, creativity and other features of human beings. For such systems to be created naturally, there is no known way without first discovering conscious human beings. Therefore, for the purposes of our discussion on creating ordered systems, we can exclude man-made systems and instead focus on how nature creates living beings naturally i.e., without any help from other conscious beings.

The main source of difficulty with creating ordered systems comes from a well-known physical law called the second law of thermodynamics. This law states that in a closed system disorder always increases over time. Opponents of Darwin's theory of evolution or any other existing theory on origin of life use the second law of thermodynamics as one main argument to support their reasoning. Their basic question is: if disorder increases over time (in a closed system), how can nature create highly-ordered living systems for the first time and how can the order keep increasing subsequently as well to create a diverse set of species? It is helpful to think of the analogy mentioned in section 1.4.2 for building a public transportation looped network and using it to create order as opposed to expecting order to form naturally by random chance.

A typical answer given to counter the opponents' argument is that earth is not a closed system, since Sun supplies energy. Unfortunately, this is not an useful answer because if we treat earth and Sun together as a closed system, the basic question still remains unanswered. It is not satisfactory and enough to say that the extra order created on earth is countered by a large amount of disorder created in the Sun. Such an answer is simply a way to avoid addressing the fundamental question. Specifically, the proponents' simple answer does not offer any insight into how highly-ordered living systems can ever be created. For one, we have never seen living beings created from dead bodies even on earth (i.e., a non-closed system). Nevertheless, it is unfortunate that the opponents use the second law to conclude that the only solution to this problem is the need for a Creator or some form of Intelligent Design. Even the opponents are not looking for other physical laws or easy-to-create structures and processes and, hence, a new theory to provide an alternate answer. While the opponents' question is valid, their answer is equally (or more) unsatisfactory. In essence, both teams of researchers have essentially ignored the core problem of creation of order.

In this chapter (and in this book, in general), I will provide a new alternative explanation to address the difficulties arising from the second law of thermodynamics. Specifically, in this chapter, I will show how to 'use' the very second law of thermodynamics to create order from disorder. But this can only happen locally since we cannot violate the second law in a closed system.

In a sense, this is precisely what we need – an explanation for how we can create an 'enormous' amount of order naturally and *in a local sense* while still obeying the second law of thermodynamics in a global sense. I have emphasized the word 'enormous' here because I want to exclude the creation of trivial order like sugar crystals and snowflakes that the proponents typically offer as examples. While disorder must increase in accordance to the second law in the entire closed system, we should show how to increase order by an enormous amount within some pockets of this closed system to account for the creation of living beings. The mechanisms I will use to create order are themselves simple enough that they can occur naturally i.e., we do not need to rely on conscious beings, Intelligent Design or a Creator.

Specifically, I will show that the SPL designs discussed in this book allow us to answer this issue in a clear and constructive way. The notion of repeatable randomness (see section 52.3) and a set of minimal structures (see section 52.4) will be used to create increasingly ordered systems in an iterative manner starting from simple easy-to-create natural SPL systems (cf. the analogy of building a public transportation looped network). In this chapter, I will start with a few such mechanisms first, but over the course of next several chapters, I will show how to create life for the first time from inanimate objects without opposing the second law globally.

53.1 Minimal structures

Recall the discussion in section 52.4 where we considered a sealed room that contains simple molecules like CH_4, NH_3, H_2O, H_2O_2, H_2S, CO_2 and others and is supplied with random energy inputs. If we see complex molecules and, possibly, even primitive life forms in this room after waiting, say, a million years, we have a difficult question to answer. How did such ordered structures get created from an initial disordered and random set of molecules? This seems to go against the notion of the second law of thermodynamics, though in this case, the room is a not a closed system. Yet, the interesting question is how order is created from initial disorder in this room. The only answer for now is random chance-based mechanisms, which is clearly unsatisfactory for the high degree of order. Our goal is to present a rigorous alternative by identifying new structures, new dynamics and less chance-based mechanisms to create order from disorder.

Take, for example, the familiar notion of stability of molecules. Those molecules that are stable under a given set of environmental conditions (like temperatures and pressures) remain after a long time while the unstable molecules disintegrate and disappear (or become trace elements like H_2O_2 compared to H_2O). Therefore, in the above sealed room, we see that stability alone is capable of producing order from disorder. After, say, a million years, only those molecules from the initial set that are stable will remain. We want to know if there are additional patterns and properties beyond stability that can bring further order within this sealed room.

If we place highly ordered systems like enzymes and bacteria into the sealed room as the initial set, it is not surprising that more order will be created subsequently (see Darwin's theory). Therefore, the above problem of evolution of order is interesting only if the initial structures in the sealed room are the most elementary ones. I have referred to them as *minimal structures* in section 52.4, using which we want to explain how nature creates highly ordered systems. In this section, I will identify four such minimal structures, all of which we have already been discussing in great detail in the previous chapters for physical systems (though not as much for chemical systems). For example, in chapter 2, we have already seen how to create an infinite family of dynamical systems with any specified complexity using these four structures. I will continue to discuss more such minimal structures and mechanisms in the subsequent chapters with focus on chemical systems (as opposed to physical systems in the previous chapters) to explain the problem of origin of life.

Stability: The first already well-known minimal property is the stability of molecules (with long half-life). A system steadily accumulates non-random stable collection of molecules even if we do start with a random collection of stable and unstable molecules. Some of these stable molecules are CH_4, NH_3, H_2O, H_2, HCN and others used in prebiotic experiments like Miller-Urey (Miller 1953; Miller and Urey 1959; Miller and Cleaves 2006), Oró (1961) and others (Hargreaves et al. 1977; McCollom et al. 1999; Miller and Cleaves 2006).

Closed compartment: The second well-known structure is a closed compartment. Reactions occurring within a closed compartment have better conditions like temperatures and pressures than the ones occurring in the open environment. A closed compartment will, therefore, help create order from disorder as well. How can we create a closed compartment naturally? It was shown that the synthesis of lipids and fatty acids, important for forming closed compartments, can occur under prebiotic conditions (Hargreaves et al. 1977; McCollom et al. 1999; Miller and Cleaves 2006). Using these molecules, it was shown that vesicles (i.e., closed compartments) do form under prebiotic conditions using fatty acids, organized as micelles (Hanczyc et al. 2003; Hanczyc and Szostak 2004). Clay montmorillonite catalyzes the vesicle formation and the vesicles are stable for days to months. In addition, it was shown that when fatty acid micelles are added, the vesicles even grow and split using a process called flipping (Hanczyc et al. 2003; Hanczyc and Szostak 2004).

In spite of an improvement in the problem of creation of order when using a closed compartment over an open environment, there is one major problem. The products of chemical reactions are generated only if all reactants are able to enter this small spatial closed compartment within a short period in the correct cascading linear sequence. The necessary molecules floating around the closed compartment is not enough. All the required molecules do need to enter this small compartment at the correct time, in the correct order and at the correct spatial location. This is less likely. It is analogous to expecting air molecules to reach one corner of a room (at a specific time and after a specific event). If we have to rely only on random chance, there is almost no possibility of this happening. Therefore, we need additional structure to offer better than random chance-based passive and diffusion mechanisms.

Active dynamics: The third minimal structure is an active power source with two distinct purposes: (a) to power the chemical reactions within the closed compartment like the equivalent of ATP in a cell and (b) to suck inputs into the closed compartment from the external environment. The latter requirement is a rather important one. Most researchers do not mention this. They only specify energy needed to run chemical reactions. However, what is the value in having energy to run chemical reactions if the necessary chemicals are not at the correct place, at the correct time and do not appear in the correct sequence?

The third structure that has an active power source should first have an ability to bring all necessary molecules closer to the compartment. A simple diffusion into the compartment is too random and unlikely (like air molecules reaching a corner of a room by themselves). It is unlikely because of the second law of thermodynamics i.e., random collisions along cannot bring air molecules to one corner of a room. Instead, if you have an 'active' mechanism to 'push' molecules to one corner of a room like a fan blowing air to one corner, a strong wind through a window or a nozzle, then air molecules do have a high probability of reaching one corner of the room. In the same way, the closed compartment should 'actively' suck inputs into the system (instead of passively waiting for molecules to diffuse into the compartment) using an active power source, in addition to using the power source to power the chemical reactions.

All systems relying on a collection of specific inputs require this feature, at minimum. Indeed, we see that all living beings did retain this feature – they actively fetch the necessary inputs or food. Figure 59.1 described later is a minimal 'pump' design that achieves this goal. With several active suction-based pumps, chemical reactions no longer need to wait for all reactants to passively accumulate in a small spatial region (the closed compartment) within a short interval by chance or wait for better environmental conditions to appear randomly. The pump (Figure 59.1) is the first new structure identified in this book for the purpose of creating order, causing the new theory to deviate from all existing theories on the origin of life.

Next, even if favorable chemicals (like O_2, ATP and others) are created naturally using the three structures identified so far, how can we address the steady accumulation and repeatable creation of these molecules? The current models account for this by trying to identify unique conditions when the half-life is high. For example, at 100 °C, half-life of decarboxylation of alanine is 19,000 years (White 1984). However, molecules like nucleosides, peptide bonds and phosphate esters only have a half-life ranging from a few seconds to a few hundred minutes (White 1984). Therefore, expecting the same environmental conditions to work for most molecules is unlikely. Yet, creation of high order requires not only the stable molecules, but also less stable and low half-life ones as well (like nucleosides and phosphate esters).

Looped dynamics: The fourth structure needed for creating order from disorder is a way to continuously regenerate molecules, namely, a looped set of reactions, analogous to metabolic networks (see Jeong et al. [80]; Nicholson [114]). Each reaction within a chemical loop can be assisted by enzymes or by using a collection of

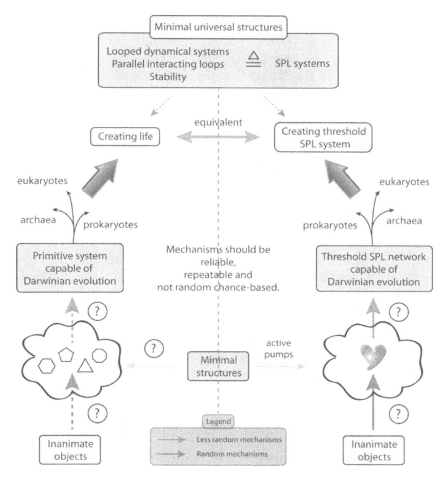

Figure 53.1: Overview of the new approach to the problem of origin of life. The problem of creating life can be equivalently viewed as creating a threshold SPL network that is capable of Darwinian evolution. We use active pumps and other minimal SPL structures to help achieve this goal. The details on how to do this will be discussed in subsequent chapters.

active pumps mentioned above (Figure 59.1). When the loop is completed, all chemicals involved self-sustain and have steady concentrations i.e., they are self-regulated. The looped system as a whole requires abundant energy and chemicals as inputs. As the required set of chemicals are present, the pumps suck them into the closed compartments and constantly regenerates them each time a chemical loop completes with the primary additional input being energy.

In addition to chemical loops, physical loops like circulation, water cycle, seasonal cycles and vortices cause steady mixing to generate diversity, but in a repetitive manner. These looped dynamical structures are the second new set of structures identified in this book unlike all existing theories on the evolution of order. Indeed, we see that living beings have retained the looped dynamical property through the creation and growth of complex metabolic networks (Jeong et al. [80]; Nicholson [114]).

We will discuss these minimal structures in greater detail in the next several chapters. It is important to emphasize that the approach we take using the minimal SPL structures for the problem of origin of life is considerably different than the traditional approaches. Figure 53.1 shows how the new approach can be simplified using active pumps as the minimal structures of Figure 52.2. With the new approach, the problem of creating life for the first time can be viewed equivalently as creating a threshold SPL network capable of Darwinian evolution using just active pumps and other minimal SPL structures with mechanisms that are both less random and more reliable/repeatable.

In chapter 2, I have already shown how to create an infinite family of physical and chemical dynamical systems with any specified degree of complexity using loops, namely, the SPL dynamical systems. Those were not necessarily powered by pumps, but in chapters 59 - 60, I will show how to create SPL systems using pumps. In fact, the problem of evolution of order now translates to the problem of creating increasingly complex looped networks.

Note that concepts similar to 'chemical' looped systems like hypercycles were studied previously and applied to the problem of origin of life, though they were specialized to autocatalytic systems with self-replicative catalysts (Eigen and Schuster 1979; Hofbauer and Sigmund 1998). However, the creation of an infinite family of physical and chemical dynamical systems, the unification of both physical and chemical SPL systems and the design of SPL systems using minimal structures like pumps instead of complex self-replicative units like enzymes and replicases, all of which are discussed in this book (some in previous chapters and some in the next few chapters) is entirely new.

To see that loops are necessary for the creation of order, consider Figure 53.2 which shows the energy change for a few sample reactions. Some reactions proceed naturally from an uphill energy state to a downhill state. They can be either entropy driven or enthalpy driven reactions – see how Gibb's free energy can be used to determine whether a process or a chemical reaction proceeds spontaneously in either the forward or the reverse direction (see Pauling [120]). If after reaching the downhill state, the molecules with lower energy state do not react and turn into higher energy state by consuming energy, the system, as a whole, would eventually settle down at the lowest energy states i.e., near absolute zero temperature. The only way to avoid this situation is by supplying external energy, thereby completing loops (Figure 53.2). There are multiple ways to bring the molecules uphill: (a) using passive random-chance mechanisms, (b) using enzymes, replicases or other catalysts, which are non-minimal structures used by all existing theories and (c) using active looped mechanisms with pumps i.e., minimal structures used by the new theory in this book. Using the four structures identified here, the primitive world would start to exhibit detailed chemical networked structure over time creating several complex molecules, as we will see. However, as the looped network begins to create polymers, a new set

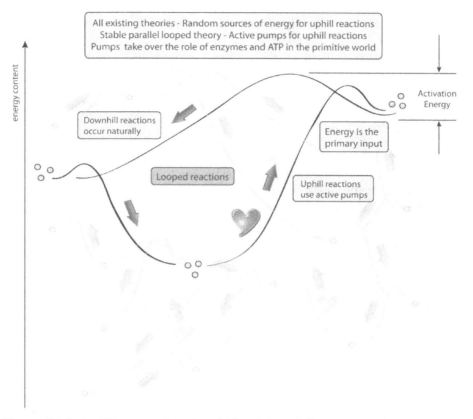

All existing theories - Random sources of energy for uphill reactions
Stable parallel looped theory - Active pumps for uphill reactions
Pumps take over the role of enzymes and ATP in the primitive world

energy content

Activation
Energy

Downhill reactions
occur naturally

Energy is the
primary input

Looped reactions

Uphill reactions
use active pumps

Figure 53.2: An SPL network with uphill and downhill reactions. If we represent different molecules at different heights based on their energy content, we can view the downhill reactions as ones that occur naturally. However, reactions proceeding uphill require either passive random energy or an active source of energy. All existing theories on origin of life assume a random energy source for uphill reactions while the new theory suggests using special SPL pumps instead. With active pumps, self-sustainability, reliability, repeatability and continuity can be guaranteed (unlike with random energy sources). In the primitive world when we do not have enzymes and ATP, we can view these active SPL pumps, which we will see are easy to create and self-replicate, as a good replacement.

of problems appear. For example, creating random proteins or random RNA sequences from a pool of amino acids and nucleic acids have little purpose unless the polymer sequences are precise. What laws or structures direct the creation of specific polymer sequences for the first time and repeatedly? With RNA world hypothesis, these are merely random chance mechanisms to create RNA replicases (Gilbert [55]; Rasmussen et. al [125]).

If the random collisions occur in parallel, the process of creating either a randomly permuted RNA sequence or a correct RNA sequence are indistinguishable from each other even though the functional differences between the two RNA sequences are significant. The 'usefulness' of a polymer sequence does not make it any easier to 'create' it reliably and repeatedly. Therefore, we need a serial mechanism using a template sequence. One of the challenging problem is to explain how nature can do this without the equivalent of an RNA replicase, enzymes and other non-minimal structures. In chapters 62 - 65, I will show how additional structures like a helical geometry and complementary nucleotide bases can be combined with active dynamical pumps mentioned above to provide a theoretical design/mechanism to solve this problem.

Therefore, the central result is that using only the minimal structures identified in this section, collectively termed as active SPL systems, I will show (a) how to create order of any specified complexity and (b) how to understand existing ordered systems among both living and nonliving systems. For (a), we have already seen in chapter 2 how to create an infinite family of both physical and chemical SPL systems of any specified complexity. However, we have seen that this infinite family of SPL systems typically needed help from conscious beings to setup favorable conditions. The goal in the next several chapters is to remove this restriction of requiring conscious beings and instead let nature itself create any specified complexity using SPL systems. In this regard, one of the minimal structures identified as active dynamics with pump-like designs discussed briefly above will play an important role to address this issue.

53.2 On using second law to create order locally

Second law of thermodynamics, in loose terms, states that the entropy always increases in any closed system. Entropy, informally, is a measure of the amount of disorder in a system. In other words, the second law states that disorder always increases in a closed system. Therefore, nature, left alone obeying the standard physical laws without any intervention from conscious and intelligent beings, apparently creates disorder in the entire closed system.

There are three points to keep in mind when trying to understand the second law: (a) the law is about macroscopic systems, (b) it states that the change in entropy is for the system as a whole and (c) it deals with closed systems. These imply, for example, that you can have a decrease in entropy or disorder locally within the system. The law then states that the local order you have achieved will be compensated by a higher disorder elsewhere within the closed system. Similarly, the system has to be closed or isolated for the second law to be applicable. If energy can flow into the system from an external source like the Sun, we cannot apply the second law directly. We would have to include this external source as part of the closed system. In addition, we do not apply the second law to microscopic systems that have just a few molecules. We need enough atoms to be able to define concepts like heat and temperature.

All living beings, as we know, are highly ordered systems with billions of dynamical subcomponents arranged and interacting in quite specific ways. This order appears to contradict the second law. In the nonliving world, a highly ordered situation like, say, for example, one in which all molecules move towards one corner of a room is nearly impossible using only natural mechanisms. The birth of a living being, on the other hand, is several orders of magnitude more ordered (in some imprecise sense). For example, not only should several molecules be at the right place and at the right time, they should react chemically at the right time, with the right concentrations, in a right sequence of steps and so on. Discovering such perfect order and coordination through natural means seems entirely nontrivial.

Nevertheless, the second law seems to allow for small pockets of highly ordered structures locally while the system as a whole is more disordered. Our planet contains many living beings i.e., *locally highly ordered* subsystems. It appears, at first glance, that there is too much order on earth than disorder, which only seems to be growing over time compared to the disorder. How can we then reconcile this growing order with the second law of thermodynamics?

In this chapter, I will not only explain how such highly ordered systems are compatible with the second law but I will also show how to 'use' the second law itself to create order locally using entirely natural means.

53.3 Difficulties with the second law of thermodynamics

Let us distinguish two types of problems that arise from the existence of living beings in relation to the second law. The first involves creation of life for the first time when there was no life on earth, say, about 4 billion years ago. Creating a highly ordered primitive cell from conditions similar to that on moon or Mars using entirely natural mechanisms seems nontrivial.

The second involves sustaining life for 4 billion years. It appears that life and highly ordered systems are only growing in number. Their structural and dynamical complexity is also increasing over time – see, for example, creation of eukaryotes from prokaryotes, multi-cellular organisms from single cellular ones and so on. With living beings, creating more order from existing ordered systems is also quite common. This has happened several times in the past like with creating different types of organs and other features that makes an organism survive better in new environmental situations.

Since the second law suggests that disorder should increase in a closed system, we would expect it to cause life to slow down and bring disorder eventually. However, both of the above problems appear counterintuitive from observations in the nonliving world.

In an attempt to answer these issues, researchers have used two types of explanations. Firstly, a few examples of highly ordered systems like snowflakes and sugar crystals are mentioned. These examples do highlight that locally within a closed system you can create order. Yet, the claim is that there is considerably more disorder in other regions, say, through loss of latent heat, to compensate the order created, thereby satisfying the second law. However, using these examples and generalizing the reasoning to the creation of complex multi-cellular organisms, using only natural mechanisms, is definitely a giant leap in logic. If we cannot create life from dead cells or dead organisms, how can we be convinced that a high level of order present in living beings can be created by generalizing examples like sugar crystals and snowflakes?

Recall that simple wear and tear is one of the biggest causes of collapse of engineering structures, if left untouched. We know that our cars or airplanes will break down if not regularly serviced. We use rigid structures where possible and have a minimum number of moving parts in most of our designs. The more moving parts we have, the harder it is to stabilize the system from unknown external environmental conditions. In other words, we are constantly fighting the effects of second law (like wear and tear) within our engineering designs. On the other hand, all living beings behave like systems that can self-service themselves automatically, even without their knowledge. If designing a system with hundreds of moving parts is difficult, with considerable care by a conscious being, how can the snowflake-like examples capture the intricacies of creating a billion subcomponent well-coordinated and highly interacting dynamical system, entirely through natural means, without collapsing for billions of years?

Secondly, it is mentioned that earth is not a closed system. Sun is constantly supplying energy to earth for billions of years. The high order created on earth through living beings is less than the disorder created within the Sun and possibly the atmosphere. Therefore, when we treat earth and Sun as a closed system, there is a net higher disorder in accordance with the second law. However, it is not clear how the order created on earth causes the disorder in the Sun or vice versa. With less connection between the two, the growth of disorder in the Sun should be much faster than the growth of order on earth in the last 4 billion years. This may very well be true, but it does not explain how and why the order is growing on earth – a smaller pocket within a closed system.

Our objective is to understand how to create such enormously complex order of living beings using natural means. The standard explanation that asks us to wait for millions of years lacks enough details on what happened during this period. Why is the waiting so important? Why can I not redesign a similar level of complexity in a laboratory through natural mechanisms without having to wait for millions of years, just as I can recreate snowflakes and sugar crystals in the lab?

I will now discuss how to not only avoid the difficulties with second law, but also use it to create highly ordered systems locally despite a global disorder. This will be possible only under special conditions and for a special class of systems, namely, SPL systems.

53.4 Stable-variant merge loop

In this section, I will propose an iterative process called stable-variant merge loop (SVML) for creating increasingly higher locally-ordered systems using entirely natural means. This process will work under a few assumptions, which I will discuss in detail in section 53.5. Briefly, they are (a) that the system has a stable parallel looped architecture and (b) that when subjected to unexpected variations, the system brings much slower disorder than the time it takes for the system to heal itself. If these assumptions are satisfied, the system will give rise to an increasingly ordered local structure.

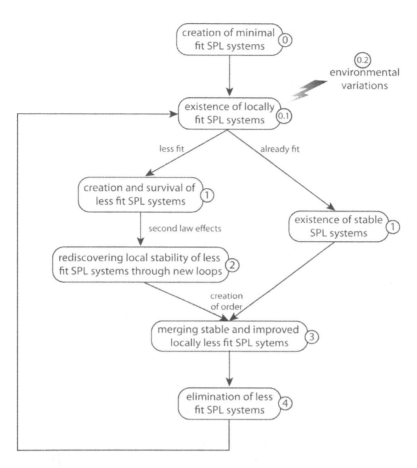

Figure 53.3: Stable variant merge loop (SVML). Here, we outline an iterative process to create more order from disorder within SPL systems. When a collection of locally fit SPL systems are subject to environmental variations, the systems partition into less fit and already fit systems. A subset of the less fit systems will become stable through the second law effects by discovering new loops. The next step is to merge the stable and the improved less fit systems to create new better fit systems in the new environment. The iterative process continues after the elimination of less fit SPL systems.

In this section, I will focus primarily on physical systems. However, the more interesting case is the corresponding explanation for chemical systems, which I will discuss later in chapters 54 and 55. The reason the application of SVML to chemical systems is interesting is because it will answer a long-standing question of how evolution has a definite direction in evolution in spite of the fact that the core mechanism for evolution comes just from random mutations. Furthermore, the SVML pattern for chemical systems generalize well to evolution within a given species, across different species with interactions like predator-prey, parasitic and symbiotic

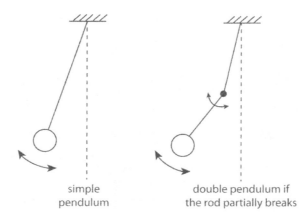

simple
pendulum

double pendulum if
the rod partially breaks

Figure 53.4: A simple SPL system. A simple pendulum oscillates for a long time with just random energy inputs like wind gusts. If the pendulum partially breaks because of second law effects like external disturbances, then the resulting system is still an SPL system. It has more degrees of freedom than before. We can say that the second law of thermodynamics has effectively increased the complexity in SPL systems while maintaining stability.

relationships as well as for more complex interactions across systems within the entire ecosystem itself.

The application of SVML with physical systems discussed in this section is not as elaborate as the one with chemical systems to be discussed in chapters 54 - 55. The assumptions needed to be satisfied in order to apply SVML pattern, discussed in greater detail in section 53.5, are quite critical. While it is possible to have systems that do not satisfy one or both of the above conditions (a) and (b), I will show that the systems that do satisfy them are quite common as well. The overall iterative mechanism represented as a flowchart is shown in Figure 53.3.

Step 0 (Initial setup – creation of simple minimal and fit SPL systems): We start with a simple looped system (a pendulum) like the one shown in Figure 53.4. This looped system takes input in the form of external energy to keep it working for a long time. We assume that the substructures within this system are made from simple materials that are available in the environment. Let us say that these substructures are neither too rigid nor too flimsy. The ability to form such a single looped system naturally is not too difficult (cf. simple oscillating systems). One particularly new and important example of a single looped system is the active material-energy looped design (referred to as active pumps in section 52.4). These material-energy looped designs will be discussed in greater detail in chapter 59. As long as the external energy is available, such a system keeps running. As a result, we can call these active material-energy looped designs as stable or 'fit' systems in the current environment.

The interesting question is what happens as these 'fit' systems move about in the external world. They may no longer be fit in the new environment. In most cases, systems do not stay at the same physical location for all time. Instead, they move around due to random variations or from disturbances in the external world of various types. The other possibility is if the external world around a system changes considerably (say, when a large set of external subsystems move around). This is when the overall disorder starts to increase. Even existing order present locally in the form of existing fit systems start to decrease i.e., these fit systems start to become less fit. This can be seen as the effect of second law trying to increase disorder.

The goal is to ensure that the existing order is not destroyed. Systems should be able to adapt themselves to realign to the new environment and become fit once again. The steps outlined below try to achieve this naturally within a special subclass of dynamical systems (i.e., not across all systems).

Step 1 (Effect of the second law – creation and survival of less fit systems through environmental variations): As the system moves around, there are two ways it can become less fit or less stable (see Figure 53.3). Firstly, the system becomes less fit from changes in the environment. The external environment can change independent of the systems' involvement like when it relocates to a new place or due to some random events (like volcanic eruption, ice age, tornadoes, earthquakes and others). These variations are typically unforeseen. Secondly, the system is subject to normal wear and tear. This causes its internal mechanisms to operate inefficiently and hence can lead to less stability. We can view both of these as second law effects that try to create disorder within the originally stable and fit system.

As an example, these two effects might produce a 'dent' in the substructure. The dented system is now less stable than the original system. Its dynamics is a bit wobbly as a result, say. If the system is already a complex interconnected looped system, we can say that these second law effects have produced a local collapse of the structure even though the system as a whole is still functioning. As an analogy, you may have hurt your leg, but you can still walk. A car may have a flat tire, but you can still drive it.

From assumption (b) above, we can say that such second law effects are small and infrequent. It does not result in the collapse of the entire looped system. If it did collapse the system, we will have to just pick another looped system. We will continue picking different looped systems until we find one that does not collapse soon enough. The resulting dented looped system can be called as a less fit system that has survived in the new environment.

Since several variations of such looped systems exist, not all of them will be less fit. Some of them will be quite stable even in the new environment possibly because their internal substructures are stronger or they were exposed to less harsher environmental changes. In other words, among the collection of all looped systems subject to a new environment, most of them will be less fit, but a few of them will be reasonably fit.

Step 2 (Slow effects of second law – rediscovering local stability within less fit systems through new loops): The next step is to assume that the effects of second law, bringing more disorder into a system, are slow enough that the less fit systems have a chance to repair themselves. Take the less fit dented system mentioned above that continues to operate its wobbly dynamics. Over a period, the gravitational forces and other structural variations start to expand this defect more and more. If the materials used for constructing the system are available in the environment, the system will accumulate some of these materials just from random effects like wind and other natural changes. This can potentially make the single looped system change into a two-looped system as the defect grows, especially since it is an active system that consumes energy to continue its looped dynamics (see Figure 53.4 where a 'simple' pendulum breaks to become a 'complex' double pendulum).

We see such natural examples in materials that have crystallographic defects. For example, there are edge or screw dislocations that are quite stable and that grow gradually over time. Many of the properties of materials like malleability, plasticity, strain hardening, annealing and others are well understood through the study of edge and screw dislocations. These can be viewed as defects created by the second law, but are stable. In these materials, if the defects are introduced slowly, they help in creating more detailed structure within the system. Generalizing this, a single degree-of-freedom system becomes a two degree-of-freedom system with the defect acting as a joint (see Figure 53.4).

In stable looped dynamical systems, such defects are not undesirable after all. Each such defect creates a new degree of freedom provided the defect grows slower, giving the system a chance to heal than let it collapse. It is necessary to ensure that the system does not completely collapse. In this way, a single wobbly looped dynamical system becomes a two-looped system, a two-looped system into a three-looped system and so on. In general, when a looped system buckles, it evolves into a more ordered stable looped system naturally, as it continues to accumulate materials from the environment (see Figure 53.4). Therefore, the system rediscovers local stability through new dynamical loops. The fact that the system is dynamical and looped is critical for this to happen as opposed to a system that is more rigid and static. For example, it is difficult for a defective rigid structure built with steel bars to fix itself through natural mechanisms because of the less fluidic nature of the materials involved and a lack of dynamics.

As an analogy, imagine a juggler who, while juggling correctly in repeated patterns, makes a few mistakes occasionally. If the mistakes are not too severe to cause a collapse of the loops, the juggler creates new and interesting patterns. These patterns are unique, primarily, because of the subtle 'mistakes'. The situation is similar with a hula hoop performer. Several of the interesting hoop patterns come from subtle mistakes (or whenever they exhibit variations that have more degrees of freedom).

If a system were to continue its looped dynamics in spite of minor defects, we need the disorder created from the second law to be much slower. The system should have sufficient time to self-sustain its looped dynamics thereby having an opportunity to heal itself entirely through natural mechanisms.

As emphasized in the previous step, you will not always get a more complex, ordered and stable system from the less fit system. It is quite likely that the less fit system eventually collapses if the random events could not help fix it in time. Since there is a large population of similar systems, the chance that a few of them can become stable, not necessarily with higher order, becomes likely.

Note that in the current discussion, it is only sufficient that the system finds additional stable loops, which integrates within itself. There is no requirement on how useful these new loops are, whether the system is able to fight predators in the new environment and so on. The system merely rediscovers stability and becomes fit once again in the new environment. Yet, it is possible to interpret and identify how these new loops have given the system an ability to 'survive' better. These interpretations, however, are typically subjective human reasoning. They are, in most cases, not related to the physical laws obeyed by the system.

***Step 3 (Merging of altered locally less fit systems with past stable systems – creation of more order)*:** When the original less fit system alters itself to become a new locally more stable system, we can classify the set of all systems into two groups. The first group contains systems that are more stable, at least, near the region of the initial defect. Each less fit system has evolved into a stable state with potentially different set of enhanced features through new sets of interconnected stable loops. It is possible that this stability is achieved at the expense of lower stability elsewhere within the system. I call this group of systems as 'altered locally less fit' systems or 'variants'.

The second group contains systems that did not have any defects to begin with. These systems were already reasonably stable in the new environment. There is less reason for these systems to change. They will typically not introduce any new degrees of freedom or more complex interconnected loops unlike the previous set of less fit systems. I call this group of systems as 'already stable' systems.

If we now find a way to merge two systems, one from each group, then the combined system can potentially have features of both systems. This process of merging can be viewed as a way to 'reproduce' or generate a new system from existing systems. With physical systems, there is no equivalent of sexual or asexual reproduction. A notion of merging of two systems and splitting of two systems can be seen as a simple way to 'reproduce', although it is strictly not a reproduction because the split systems are not identical (in any reasonable sense) to the original systems. At this stage of evolution, a simple split of closed compartments is a good approximation to reproduction. See the discussion in section 53.1 where we saw that when additional fatty acid micelles are added to vesicles created under prebiotic conditions, the vesicles grow and split using a process called flipping (Hanczyc et al. 2003; Hanczyc and Szostak 2004). In chapter 67, I will discuss additional mechanisms that can favor more accurate reproduction than a simple splitting in both physical (based on active material-energy looped systems) and chemical systems.

The combined system, if possible to do so naturally, is more stable than both groups of systems, in the old and in the new environments. With physical systems, we can think of the merging or joining as a simple attachment (or benign entanglement) of the two systems when they come next to each other. This can happen through natural means like wind disturbances and others. The chance that two different types of systems coming next to each other is quite high when we have a pool of both types of systems. As the two systems stay together for a long time, the looped dynamics from one system transfers energy to the other system and vice versa. As both the systems are fluidic in nature, they can eventually adhere to each other. They will gradually transmit forces and energy evenly between the two systems causing both the systems to unify into one.

With chemical systems, the merging of both groups of systems can be much deeper. The chemicals from both systems can fuse together and start to work as a whole. Some examples of this merging in living cells are sexual reproduction and horizontal gene transfer mechanisms (see chapters 54 and 55). Even here, the merging process can occur through natural means.

In this way, using just the effects of the second law (i.e., environmental disturbances or wear and tear), it is possible to create new stable parallel looped systems which were more ordered than the original systems. We did this by merging of two systems – the altered less fit system (or a variant) and the already stable system.

Step 4 (Elimination of the less fit systems): The looped dynamics of any system continues for a while and eventually dies down. This can happen for any number of reasons. For example, the system may no longer have access to useful energy. The system has become old and has worn out considerably. The environmental conditions may have changed significantly. The system has had one or more critical accidents and so on. The less fit system is subject to same type of negative conditions as a more fit system. Less fit systems will die down faster, excluding the unforeseen scenarios like accidents. This is because they have a more number of defects or their looped dynamics are not coordinated as perfectly. In fact, I will define less fit systems as those looped dynamical systems whose dynamics run for relatively low amount of time, excluding unforeseen situations, compared to other similar looped systems. Therefore, over a period, several less fit systems stop working. This step will happen naturally as well.

The iterative process of repeating the above four steps can now continue with the remaining stable systems. Strictly speaking, the iterative process could have continued even without the elimination of the less fit as well (see Figure 53.3). This last step does not necessarily occur in every iteration. However, the assumption that the effects of the second law are slow will naturally eliminate several less fit systems before we restart step 1 (i.e., change of conditions) of the iteration.

Consider what happens if this step 4 occurs before step 2. In that case, most of the less fit systems will be eliminated before they have a chance to find a more locally-stable state at the location of the defect. This is not a good situation. This is because the iteration goes from step 1 to step 4 back to step 1, skipping step 2 and 3 (see Figure 53.3). If the skipping of these steps does happen a few times, the effects of second law bringing disorder into the system will accumulate considerably. It may be too late to fix all of these accumulated defects. This is one way to see how the entire iterative process can come to an abrupt stop. In the worst case, even the stable systems will start to die down, with a real threat for the entire group to become extinct.

Under normal circumstances, after the completion of these four steps, the iterative process can now continue with the available set of combined systems as the starting point for step 1. In summary, these steps start once again with a change in the environmental conditions or internal wear and tear. The combined system becomes less fit in these new conditions with a few defects. The defects will heal slowly and naturally. The healed system (a variant) merges with the previously stable system to create a more ordered system. The less fit systems will be eliminated and the process repeats. I call this process as the stable-variant merge loop (SVML) pattern – a stable looped system and a variant looped system merges to create a more ordered system.

53.5 Assumptions with stable-variant merge loop

Let me outline the assumptions made when explaining the stable-variant merge loop (SVML) pattern. These assumptions are quite critical for the iterative process to work correctly. Yet, we will see that these assumptions are not unusual or too difficult to occur naturally.

Assumption 1 (Stable parallel looped dynamical systems): This is the single most important assumption needed for the SVML pattern to work correctly. You cannot pick any type of physical or chemical dynamical system and expect to create highly ordered substructures within them. This is true even if you have a chemical system and use only organic chemicals. What is necessary instead is for dynamical systems to be able to form stable parallel loops. This is the only known structure that can occur naturally and result in a directed evolution with increasing order and complexity, namely, by using the SVML pattern.

We have already seen in great detail in chapters 14 - 19 how order is created in the case of brain using stable parallel looped physical dynamics. I will describe examples among chemical systems starting from chapter 57. This is also the only naturally occurring case that takes randomness and still creates local structure. Without this feature, there is a severe limit to what randomness can do. The most you see are systems with simple structures like snowflakes or crystals that do not evolve in complexity beyond a certain point. In effect, you can take randomness and create structure entirely naturally within a system if you let them form SPL systems, in spite of the second law. All complex systems like our brain and living beings are only possible since they belong to this important class of dynamical systems. Identifying and studying this important subclass is critical for any theory on complex systems.

Assumption 2 (Slow effects of second law): Wear and tear, presence of disturbances, say, from varying weather conditions and other sources of disorder are not only quite common, but are inevitable in nature. When we see ordered structures around us, we can broadly classify them from relatively static ones to highly dynamical ones. The static ones are typically solids with less number of degrees of freedom like crystals, snowflakes and other rigid structures. The dynamical ones are typically fluidic systems within flexible substructures like cells. For these structured dynamical systems, we need to constantly supply energy to make sure they stay intact in spite of the effects of the second law that try to break it down. This is when we needed to assume that the second law effects are not too severe and quick that the system collapses completely. Instead, they should only produce small dents within the stable parallel looped systems. This, fortunately, will result in adding new degrees of freedom instead.

Assumption 3 (Availability of raw materials): The stable parallel looped dynamics need energy as well as different types of raw materials to operate continuously. These raw materials should allow the system to form more and more chemical or physical loops. In general, there will be several structured systems within a given region, some more complex than the others. It is important to realize that there is nothing sacred about picking the most complex system within this set as the system to study or focus on. This might very well break down at a later time, while some other initially less complex system might pick up its raw materials to become more evolved (like, say, when the dinosaurs became extinct, primates were at an advantage). The evolution of structure among these systems is itself quite fluidic. The best-evolved structures are constantly changing from one system to the next within a group or from one group to another, thereby recycling raw materials within themselves.

53.6 Additional features of stable-variant merge loop

As mentioned earlier, the iterative process can come to an abrupt stop whenever the systems are unable to discover a stable set of loops to fix the defects. This can happen if the environmental conditions change too fast or if the looped systems are too slow to reorganize their internal dynamics. When this happens, the system collapses even after it has evolved positively for a while. What happens when the iterative process breaks down? Nature may have to start all over again. However, most likely, the broken down system has a memory in the form of smaller broken physical substructures, existing chemicals or, better yet, as fragments or even entire DNA.

The iterative process also makes the system quite robust under a number of environmental conditions. At each stage of iteration, the evolved system is robust under all the previous environmental conditions and from a few modes of internal failures through wear and tear. The evolved systems keep a history of past successes as memory within its internal structure. The stable-variant merge loop (SVML) pattern allows such memory to be possible because you are always merging a previously 'stable' system with a 'variant' system.

In our previous discussion, we have given less emphasis to the merging of two different 'variant' systems or two different 'stable' systems, though they do sometimes result in a more stable system as well. One reason is that stable systems are fewer and variant systems are larger in number. As a result, two stable systems merging is less likely to happen. Also, two variant systems merging, though common, are not necessarily more robust in both old and new environment (since they are both less fit to begin with). Therefore, merging a stable and a variant system is both more likely (in number of possibilities) and results in a more stable system in the old and new environments compared to the other two cases.

Another interesting conclusion is that the natural evolutionary process of living beings should be a slow process if the iterative process were to succeed. The environmental conditions or internal wear should be slow enough to cause mild less-fit systems instead of severe less-fit ones (that breakdown in just one generation). The less fit systems should be able to sustain without collapsing for a long time as well. Furthermore, the less fit system should have sufficient time to find a new solution to each of these defects.

In this way, we see that the second law has actually helped in the evolution of life itself because of its slow impact on stable parallel looped systems. This is a rather special situation with second law. This can only happen if the looped system is quite stable and the rate of disorder introduced by second law is much slower than the amount of time these loops can sustain. The looped system is capable of incorporating the disorder introduced by second law effectively to become more complex and a higher degree of freedom system that continues to self-sustain without breaking down for a long time.

54

Directed Evolution with Sexual Reproduction

In this chapter, I will apply the stable-variant merge loop (SVML) pattern to the case of living beings for creating locally higher order systems. In particular, the systems of interest for now are those that have DNA and that can reproduce sexually. I want to show how nature can use the SVML mechanism to drive evolution in a specific direction within this subclass of systems.

If a living being is already viable, how does its offspring's continue to stay viable and discover new features when subject to new stresses and random mutational changes? How can we maintain viability over several generations? If only one organism generates another organism as an offspring (say, by splitting into two – asexual reproduction), the random mutations cause more damage than help it stay viable. To see this, consider the changes that occur to the metabolic maps (see Jeong et al. [80]; Nicholson [114]) when a mutation occurs. A mutation in the DNA sequence does one of three things – see Figure 54.1: (a) it renders an existing enzyme non-functional, (b) it results in a new enzyme that catalyses a different chemical reaction or (c) there is no observable effect of the mutation. Case (a) has higher probability than case (b) because discovering a new and useful enzyme through random mutations is experimentally observed to be rare phenomena. Therefore, over a period of, say, thousands of generations, the number of mutations making dysfunctional enzymes will be much higher than the ones that catalyze new reactions. The accumulation of these deleterious mutations in each generation will threaten the basic survival of the organism, unless there is a hidden mechanism that guarantees the opposite. Therefore, if we accept only Darwin's theory, the metabolic (i.e., stable parallel looped) network will disintegrate and eventually cause the organism and species to become unviable even with no competition or other external stresses i.e., just from random mutations during normal reproduction. In this chapter, I will show how SVML pattern can overcome this seemingly inevitable extinction of all organisms, species and life itself.

The idea behind the new SVML mechanism is that it works with two systems instead – a stable system with almost no random mutations and a variant system with several mutations. When they merge to create an offspring, the chemical network from the stable copy will guarantee that the offspring is at least as viable as the stable parent (Figure 54.3). The variant copy will create new interconnecting 'bridges' of chemical pathways within the stable parallel looped chemical network. These bridges introduce new, previously non-existent, features with some probability. In this way, because each organism has two copies of, say, DNA, with one copy ensuring that the original viability is not degraded and the other copy ensuring that new features are added on top of existing features (i.e., by not replacing existing features), the SVML pattern overcomes the issues with randomness destroying viability. In this chapter, I will discuss one way to perform the merge operation – using sexual reproduction. However, in the next chapter, I will show another way to perform merge, namely, using the horizontal gene transfer (HGT).

Even though it is well understood that sexual reproduction brings high level of robustness for survival than asexual reproduction, it was never claimed before that sexual reproduction and HGT is the root cause for guaranteeing a direction in the evolution of life. In fact, I will show in the next chapter that the underlying mechanism, namely, the stable-variant merge loop guarantees a direction in evolution with (i) all types of interactions like cooperative (symbiosis) and competitive (predator-prey) and (ii) at all levels of interactions between organisms i.e., within a given species, across species and at the entire ecosystem level itself.

54.1 Difficulties with directed evolution

Consider a chemical SPL system like a living being that undergoes sexual reproduction. As discussed in section 2.6, the various chemical reactions that occur within such a system can be represented as metabolic networks as shown in Nicholson [114]. Figure 54.1 is a directed graphical representation in which the nodes are either the reactants or the products of a chemical reaction whereas the edges represent the direction in which a chemical reaction proceeds. Each directed edge from node A to node B is a chemical reaction that takes reactants A and convert them into products B. Most chemical reactions are catalyzed by specific enzymes. These enzymes are specified on top of the edge using an EC (enzyme commission) number. For example, 2.7.7.43 is the enzyme CMP-sialic acid synthetase that catalyzes the conversion of N-Ac-Neuraminate into CMP-N-Acetyl neuraminate (see the top left part of Nicholson [114]). Sometimes more than one enzyme catalyzes a given chemical reaction in which case we would see two EC numbers on an edge in Nicholson [114]. For example, 3.1.3.29 is the enzyme N-acylneuraminate-9-phosphatase that also catalyzes the above reaction.

In Nicholson [114], we can clearly see several looped directed pathways i.e., in addition to Kreb's cycle and Calvin-Benson cycle. In order to run a series of chemical reactions corresponding to a specific metabolic pathway, the cell needs to generate all the necessary enzymes from DNA (using the transcription/translation process) as well. A metabolic network can, therefore, be viewed as a static representation of the complexity of chemical reactions that occurs within a living being. As another example, Jeong et. al [80] presents a similar metabolic network for a yeast.

The DNA in an organism contains the information needed to create all of the enzymes represented by EC numbers in such metabolic networks. The survival of an organism depends heavily on its metabolic network. Therefore, when an organism replicates (or creates offspring's), it is critical that the metabolic network is preserved if the species were to survive well. One way to do this is to ensure that the DNA replication happens with high fidelity. This then implies that each of the enzymes in the metabolic network can be created using the replicated DNA reliably. Hence, the metabolic network as a whole would be preserved for the offspring provided an initial concentration of the remaining non-DNA based chemicals within the metabolic

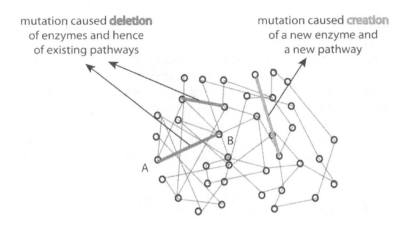

mutation caused **deletion**
of enzymes and hence
of existing pathways

mutation caused **creation**
of a new enzyme and
a new pathway

Figure 54.1: A sample chemical SPL network shown schematically, subject to random mutations. A random mutation sometimes deletes existing enzymes causing the existing pathways to break. Similarly, in some rare cases, a random mutation might give rise to a new enzyme causing it to create a new pathway in the SPL network.

network (i.e., chemicals at the nodes of the graph) are transferred from the parents as well.

Recall from section 3.7 that in any chemical SPL network, there is a well-defined notion of *self-replication of the entire collection of molecules* i.e., not just a smaller subset like DNA and proteins that is commonly attributed to the term 'replication'. We have seen in section 3.7 that a few chemicals within a loop are sufficient to generate all chemicals within the loop (see Claim 3.1). This is an useful property when thinking in terms of replication. We do not need to replicate these remaining non-DNA chemicals with great fidelity. They just can be transferred partially from one parent during a split and the remaining molecules will be generated on-demand via the looped metabolic pathways. Therefore, the metabolic network is a good model for studying the problems of replication, mutations and the direction in evolution within and across species.

What do random mutations over several generations do to a metabolic network? This is an important question to answer because the phenotypic and other properties exhibited by an organism indirectly comes from its metabolic network. When there is a random mutation to the DNA, one of the following possibilities happen: (a) none of the existing genes, and hence the production of existing proteins, are altered, (b) some of the existing genes are altered considerably to prevent the production of the corresponding proteins – this results in the breakdown of some of the existing pathways in the metabolic network, (c) a new gene is formed creating a new useful protein i.e., one that catalyzes a previously uncatalyzed reaction or a new way to catalyze an existing catalyzed reaction, (d) an existing gene is altered so that it is no longer possible to create an existing protein, but it is simultaneously replaced by an ability to create a new protein that catalyzes a new chemical reaction and (e) other variations of these.

In summary, in terms of the graphical model, a random mutation (see Figure 54.1): (i) deletes a directed edge in the graph, (ii) creates a new edge because a new enzyme is created, (iii) reinforces an existing edge with a new protein, or (iv) creates various combinations of the above ones. What is the probability of occurrence of each of these possibilities if the mutations are sufficiently random? Clearly, deletion of a directed edge has the highest probability because of the second law of thermodynamics. All other cases have low probabilities. The reason is that creating a new edge requires creating a new enzyme reliably and repeatedly, which needs a large number of steps (i.e., *all* of them) to work correctly whereas deleting an edge just requires breaking down just *one* of these steps. Let me outline a few steps here to highlight this difference.

To create an edge in the metabolic network, we need a specific enzyme E to catalyze the corresponding reaction R. The enzyme's three-dimensional shape is quite critical to catalyze reaction R. If the shape is not just right, enzyme E can no longer catalyze reaction R. Yet, it is possible that enzyme E could catalyze some other reaction. To create enzyme E, we need a specific DNA sequence. For now, the only known way to create new DNA sequences is through random mutations of existing DNA sequences. To transmit these random mutations reliably across generations, it is necessary to look at only those mutations that occurs during the reproduction of an organism. Using this DNA sequence, a detailed process of transcription and translation (see Purves et. al [123]) needs to occur to create the sequence of amino acids that are part of the enzyme. These transcription and translation processes require help from several other enzymes like ribosomes, which we will discuss later as well. The final amino acid sequence folds into a specific three-dimensional shape giving the enzyme its catalytic properties. Misfolding of proteins can be fatal. Misfolded proteins are called prions when they become infectious agents (see, for example, bovine spongiform encephalopathy, also known as 'mad cow disease').

Therefore, *all* of these steps should occur in a stable and reliable manner in order to create a new edge in the metabolic network. On the other hand, just *one* of the above steps need to go wrong to delete an existing edge. In other words, a high level of coordination and the creation of a precise sequence of DNA makes possibilities (ii)-(iv) very unlikely, whereas (i) highly likely to happen.

Given this, if we wait for a long time, what will happen to the metabolic network of an organism as several random mutations accumulate across generations (see Figure 54.2)? From the above discussion, since deletion of edges is highly probable than creating new edges, the metabolic network starts to become sparser and sparser in each generation. In Figure 54.2, I depict this iterative looped process schematically. When a parent creates an offspring, the random mutations cause a few enzymes and the corresponding pathways to be deleted. In addition, with low probability, a new enzyme and, hence, a new pathway is also created. The resulting offsprings' metabolic network is sparser than the parents (see Figure 54.2). The iterative looped process now continues for several generations. With high probability, the metabolic SPL network becomes sparser pushing the species towards inviability. Also, there would

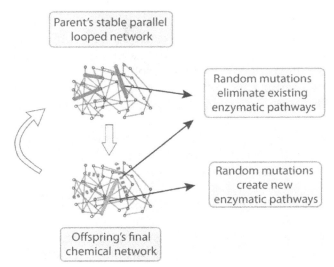

Parent's stable parallel looped network

Random mutations eliminate existing enzymatic pathways

Random mutations create new enzymatic pathways

Offspring's final chemical network

Figure 54.2: SPL chemical network of a species with random mutations. With random mutations more existing enzymatic pathways are lost than gained. As the offsprings reproduce, the chemical network keeps degrading over several generations. This eventually makes the organism inviable.

be no direction in evolution as a result. If more number of edges are deleted in each generation than are created, eventually the organism and the species itself will become extinct unless there exists some other mechanism that 'preserves' the metabolic network. The stable-variant merge loop pattern is precisely one such mechanism.

Darwin's theory of evolution, by itself, does not seem to explain the problem of direction in evolution and the problem of extinction of species (discussed above) well enough. Let me briefly discuss why this is the case by studying the effect of Darwin's theory on the evolution of metabolic network over time. Let us assume that an organism started with a given metabolic network M_1 at time T_1. At time T_2, the metabolic network for most organisms is sparser that the original network M_1 because of the above analysis. Consider the fittest organism at time T_2 and its metabolic network M_2. It is likely that M_2 is sparser than M_1 in spite of survival of the fittest.

Note that the process of 'survival of the fittest' does not change the metabolic network. The only cause to change the metabolic network is random mutations during the creation of an offspring. Survival of the fittest is an 'effect', not a cause.

The sparsity of the metabolic network continues over several generations even if we take survival of the fittest into account. If $M_1, M_2, M_3, ..., M_n$ are the fittest metabolic networks in generation $1, 2, 3, ..., n$, then the size and density of the metabolic networks, in general, keeps decreasing over time. They get sparser and sparser. However, occasionally, in some generation $k < n$, it is possible that a set of random mutations produce fitter species than the previous generation i.e., M_k is potentially better than M_{k-1}. But the number of times this happens i.e., the network becoming denser than the previous generation, are far fewer than the number of times the opposite happens i.e., the network becoming sparser than the previous generation.

Therefore, if we take a long sequence of metabolic networks M_1, M_2, ..., M_n over n generations, the typical pattern we observe if we use Darwin's theory of evolution in its current form is (see Figure 55.7): (a) the sparsity of the metabolic network monotonically increases across several generations until an outlier happens by chance and (b) a few outliers happens occasionally at which time the density of the network jumps higher than in the previous generation. As Figure 54.2 suggests, the sparsity increases monotonically for most species except at a few outliers. Such a pattern contradicts the direction in evolution we observe in the ecosystem.

Figure 54.2 also seems to suggest that most species converge towards extinction unless some other hidden mechanism exists to prevent this from happening. The goal of this chapter is to show that the stable-variant merge loop (SVML) pattern is capable of *reversing* the pattern of Figure 54.2 itself i.e., the sparsity will not increase over generations. Instead, the density of the metabolic network increases across generations – the exact opposite of Figure 54.2. I will show that SVML is the hidden mechanism that clarifies most of the unanswered questions within Darwin's theory.

54.2 Intuition for stable-variant merge loop pattern

In the previous section, we have seen that it is easy to delete edges in a metabolic network than to create new edges (see Figure 54.2) if a cell is isolated and allowed to split and mutate randomly. Over time, the metabolic network becomes sparser and sparser. This would not give rise to a direction in evolution and, in fact, makes it easy for a species to become extinct. However, in reality, we do notice that this is not the case. There is a clear direction in evolution in which increasingly new and interesting features are discovered over time. How does this happen? What is the mechanism that ensures that the metabolic network does not become sparser over generations? In this chapter, I will show how SVML addresses this issue in one case i.e., among species that use sexual reproduction. In the next chapter, I will apply SVML to other cases as well. Let me first present the intuition in simple terms.

Since random mutations delete edges more easily than create new edges of the metabolic network, which in turn causes the network to become sparser, one approach to prevent this would be to use 'two' copies of the metabolic network i.e., two copies of the DNA (see Figure 54.3). Consider the situation in which one copy of the DNA mutates very little (a *stable* one) while the second copy of the DNA mutates much higher than usual (a *variant* one). In this case, the first copy will preserve the original metabolic network. The second copy, since it mutates quite heavily, will not only destroy several of the existing edges of its metabolic network, but it also increases the chances of creating new edges as well (see Figure 54.3). However, since an organism has both the copies of DNA, the resulting metabolic network will be a 'union' of the two metabolic networks corresponding to each copy of the DNA. This is because in order to catalyze a reaction R that converts reactants A to products B using an enzyme E, even though the second DNA has mutated that prevents it from creating

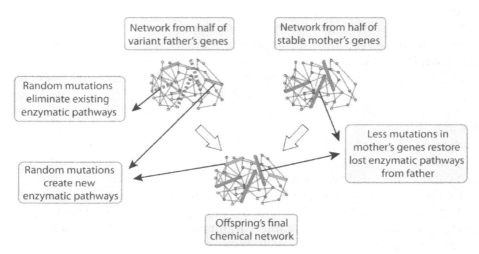

Figure 54.3: SPL chemical network of an organism that has two parents. With two copies of genes, an offspring's chemical network contains original enzymatic pathways from the stable parent as well as new pathways from the mutated parent. Note that we are keeping track of the entire chemical network, not just the genes (cf. molecular mechanisms for Mendel's laws).

enzyme E, the first DNA has not mutated and, therefore, can generate enzyme E for the offspring (see Figure 54.3).

The generalization of this explanation is the concept of dominant and recessive genes caused by having two copies of genes, one from each parent (see Purves et. al [123]). Each copy is termed as an allele. An example of the above case where only one good enzyme E from one parent is sufficient is with the protein MC1R. Just one good allele allowing us to create protein MC1R in small quantities is sufficient to prevent us from having red hair. Only when both alleles are mutated will we have red hair. The situation is similar with Other such examples are brown versus blue eyes, attached versus hanging earlobes and so on.

Strictly speaking, there are several other reasons why a given allele is dominant or recessive. These depend on the function of the corresponding enzyme E. For example, it is possible that the absence rather than the presence of an enzyme E produces a given trait. Also, the amount of concentration of enzyme E produced should be high enough to produce a given trait (i.e., we need both alleles to generate sufficient concentration instead of just one allele). This is termed as haplo-insufficiency. An example is a neurological disorder called William's syndrome caused by haplo-insufficiency of genes at 7q11.23. Muller [113] has studied several other cases of dominance and recessive genes and has coined terms like amorph and hypomorph that produces a 'loss-of-function' and hypermorph, antimorph (also termed as dominant negative mutations) and neomorph that produces a 'gain-of-function' (see Muller [113] and Wilkie [164] for more discussion on these). Other situations are when an allele is dominant for one trait and recessive for another, codominance, incomplete or semidominance and so on.

Therefore, going back to the metabolic network and seeing how the deletion of an edge from random mutations in one copy effects the overall fitness of the organism, we see that for every edge in the original metabolic network (i.e., every original enzyme), the first copy of the DNA will most likely be successful at creating the corresponding enzyme (because it is a stable one) even though the second DNA cannot (because it is a variant one). At the same time, the second DNA can create a new edge with higher probability because it is no longer required that the random mutations be less for this variant copy. In other words, by keeping two copies of the DNA i.e., a stable one and a variant one along with a way to merge the two copies into one organism, we can solve both problems: (a) preserving the original metabolic network and (b) creating new edges in the network. This implies that the density of the metabolic network can increase over time. Therefore, the species can have a clear direction in evolution. This is the core idea of the stable-variant merge loop (SVML) pattern.

In the case of species that have sexual reproduction, it is observed that the female is the stable one i.e., whose DNA does not mutate as much while the male is the variant one i.e., whose DNA mutates considerably. The process of merge is the process of sexual reproduction. The term 'loop' in SVML refers to the iterative process using the above steps across several generations that eventually gives rise to a direction in evolution. In the next chapter, I will discuss how SVML generalizes to all other species, across species (like with predator-prey, parasitic and symbiotic relationships) as well as across the entire ecosystem.

It is interesting to note that we do use the basic idea of SVML (i.e., by maintaining two copies – a stable and a variant one – to make directional progress) with several other examples in our everyday life as well. For example, when we solve a mathematical problem systematically (as opposed to coming up with a creative or an intuitive solution), we do tend to pick a generic 'stable' idea and try many different 'less fit' variations around this, each time merging with the stable one to see if there is any improvement. We eliminate the poor combinations and continue to make progress in a directed manner. If we do not make much progress, we abandon the idea (i.e., the path becomes extinct). We backtrack to the most stable one, try new variations and iterate this process. With some luck, we arrive at the correct solution using this systematic approach. Otherwise, we start the iterative process once again with another stable idea.

This approach is quite typical, not just with mathematical problems, but with several systematic scientific problems as well. This approach relies on consciousness. It will not work with all types of problems, especially those that require an intuition. Nevertheless, we see that SVML is a systematic way to approach such problems.

Let me discuss another fictitious example to help with the intuition of SVML. Consider mashing three-dimensional structures randomly to create interesting shapes. If we do not follow any particular pattern, then most of the shapes created through such random mashing of structures will look quite random and less interesting. However, if we use the SVML approach in which we choose a stable approximately interesting structure and then merge it with several 'less fit' variants, it is likely to see directional improvements over time. For example, if the initial stable structure looks like a 10-story stable building and the variant one is a randomly 'mutated' structure, then it is likely that the merged structure can introduce new features that were not present before. An example of such a new feature could be a primitive form of staircase between, say, the 3rd and the 7th floor (assuming the original building never had stairs). Sure enough, random mutations (or mashing) will destroy a system much more than fixing it. However, a combination of an already fit system with a mutated one can introduce, at least a few, entirely new features like the stairs.

One last problem I want to mention briefly is the one discussed in section 52.7 on how a computer (or a monkey) creates, say, Shakespeare's work by randomly typing letters. Disbelievers of Darwin's theory use this example as an analogy to suggest that it does not explain the observed direction in evolution or that it does not explain how to create such high degree of order using just random mutations. The analogy is not perfect because living beings do not evolve towards an *apriori* known future state as suggested here (like converging to a specific Shakespeare's work). Nevertheless, the basic issues raised above (like the direction in evolution) are valid enough that they deserve a more convincing explanation.

I want to claim that we can give a vague answer to this analogy as well using the SVML pattern. The idea is that instead of working with a *single* computer that types letters randomly, we work with *pairs* of computers, one of which is responsible for preserving the correct sentences from Shakespeare's work while the other is responsible for exploring the alphabet space randomly. The resulting output sequence of letters comes from a notion of merging the output from both computers. Of course, I have not specified how to discover a correct sentence, how to preserve it and how to merge the two copies to produce a final result. We need a way to translate the useful sequence of letters from Shakespeare's work into a similar notion of edges i.e., enzymes of a suitable metabolic network. Rather than elaborating on this analogy (which I consider as future work), I will discuss the details for the 'real problem' instead, namely, with the evolution of living beings.

54.3 Stable-variant merge loop with sexual reproduction

Figure 54.4 shows the stable-variant merge loop pattern of section 53.4 specialized to the case of living beings capable of sexual reproduction. These classes of living beings already satisfy the initial assumptions mentioned in section 53.5 because they are self-sustaining and stable parallel looped systems. Let me apply the steps specified for physical SPL systems in section 53.4 to the special case of chemical SPL systems with sexual reproduction.

Step 1 (Creating two sets of systems from an environmental stress): An environmental stress is any variation that forces the system to operate its dynamics in a fundamentally different way than its current stable form. For example, a significant set of chemical reactions that normally occur within the system are either incapable of occurring or are taking a different set of chemical pathways. This may or may not be severe enough to threaten the survival (i.e., collapse of the internal stable

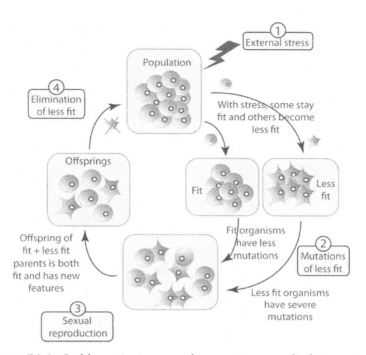

Figure 54.4: Stable-variant merge loop pattern applied to systems capable of sexual reproduction. The iterative process of SVML with sexual reproduction allows us to have a direction in evolution. An external stress triggers the iterative process, which makes some organisms fit while the others as less fit. More mutations occur within less fit organisms. When there is a sexual reproduction between less fit and fit organisms, the offsprings have both parents' features.

dynamics) of the system. In addition, whether the system is conscious or aware of these chemical stresses is also irrelevant. This is because the subsystem that provides awareness or consciousness and the subsystem that is under chemical stress are not always synchronized to the same extent.

When the environmental conditions are altered considerably, some of the systems within the collection become less fit while others remain fit even in the new environment. This splits the collection of systems broadly into two types. This is the most typical cause for a change in any organism. For example, when the availability of food gradually changed from softer seeds to harder ones, some finches, say, with long beaks in the Galapagos Islands became less fit. Even if there were no finches with short beaks to crack these harder seeds, at least some of the long-beaked finches are quite fit. This is because not all softer seeds have disappeared entirely.

The probability for the external environment to change is much higher than the system to change internally. This is because the environment is much less stable with far fewer loops compared to the collection of stable parallel loops within a living being. Yet, not all types of species experience the environmental stress in the same way or at the same time. The advantage in one species is typically at the expense of a disadvantage in another species, at least initially. If two stable parallel looped systems with such differences begin to interact, there is a natural tendency for both systems to synchronize. Such natural synchronization through interaction is possible even if only one of the systems is a stable parallel looped system with the other one as the environment. The stable-variant merge loop pattern of section 53.4 for physical systems shows how this synchronization is possible. However, we do not yet have a mechanism for synchronization if neither system is stable parallel looped one. We have to wait for at least one of them to become a stable parallel looped system.

Step 2 (Mutations in the less fit systems): It is important to realize that for evolution to proceed in the correct direction, the less fit systems should survive. This seems to contradict natural selection, which claims that the less fit systems will be eliminated. For example, the long-beaked finches should *not* be eliminated through natural selection as they are less fit in the new environment. Instead, they should survive long enough to be able to mutate. Rather than discussing the comparison with natural selection or Darwin's theory here, I will postpone the discussion to chapter 71 (see also section 55.8).

The requirement discussed in section 53.5, namely, that the changes in the environment attempting to destabilize the system should be slower than the life span of the stable parallel looped system is quite critical. This is because the less fit systems have a natural tendency to mutate much more than the more fit systems. Hall [61] has suggested a notion of hypermutable state that organisms enter into when they are in an environmental stress. This leads to an overall increased rate of mutations within the entire genome, not just in specific subsets, compared to when they are not under stress. Had the less fit systems be eliminated quickly as suggested by natural selection, we would not have hypermutable or variant organisms with a large number of random mutations.

In fact, we can convince ourselves that the notion of hypermutability should be true using simple logic as follows. When an organism is under stress, the set of chemical reactions that can and do occur are significantly altered. This is true both during normal growth of the organism as well as when the organism reproduces. The amount of energy available, say, as ATP, is considerably lower now. As there is a competition among the set of chemical reactions within the cell for the consumption of ATP, those that are triggered through long cascading sequence of reactions are at a disadvantage. There are several examples of these like, say, a hormone binding to the cell surface protein triggering a cascaded set of reactions involving amplification of the signal, transporting chemicals into the nucleus, binding to transcription factors, triggering transcription of a number of proteins, transporting mRNA's out of the nucleus, translating and producing enzymes, transporting enzymes to the correct place and so on. You can imagine that this long sequence of reactions will have less chance of occurring when the cell is starving. Somewhere along this chain, the reactions will die down because of competition or a general lack of ATP to power all of these reactions.

When such a stressed cell is undergoing mitotic or meiotic cell division, its chemical processes are also under considerable risk. Successful reproduction of a cell under normal conditions involves running a long sequence of reactions similar to those described earlier and a considerable amount of energy. After the energy intensive processes of making a copy of the DNA, most cells execute elaborate DNA repair and error correcting mechanisms as well. Under starved ATP conditions, it is reasonable to suggest that these subsequent error correction reactions are less likely to occur. As a result, errors during DNA replication, which are otherwise corrected under normal stress-free conditions, now accumulate considerably more because of the presence of stress. This results in a higher mutation rate than normal. This can be called as the hypermutable state resulting from an environmental stress.

It may even be possible to identify which set of reactions are not occurring due to the stress if we have a complete map of all chemical reactions that occur within the cell. For example, we can determine the amount of correlation between the specific environmental stress and the set of genes that do mutate.

Organisms that are better fit in the new environment, on the other hand, are not significantly stressed compared to normal conditions. As a result, the shortage of ATP and other adverse conditions that the less fit organisms are experiencing are not quite applicable. Therefore, the rate of mutations in these organisms is not too different from normal conditions.

It is not necessarily true that the mutations in the less fit organisms will immediately benefit the offspring or that they will make it more fit in the new environment. Instead, what I mean by random mutations being useful is in the following sense: if the new chemicals created from random mutations either form new chemical loops or help existing chemical reactions within the system to form loops, then the mutations are useful. We want to achieve new chemical reactions that integrate with existing cellular reactions and form new stable parallel loops. In a single less fit organism, this may be unlikely to happen.

In fact, in the early stages of life and in single cellular organisms, we should allow usefulness of mutations even when they do not form chemical looped reactions right away. It is possible that only after a few generations the older mutations i.e., mutations that occurred in the past generations start to integrate with the newer mutations i.e., mutations that just happened in the current generation to form a looped reaction. For example, Kreb's cycle (for cellular respiration) requires at least eight enzymes to complete the entire looped reaction. It is likely that not all of them i.e., the corresponding DNA sequences for each enzyme were discovered in the correct sequence, all at once. Such examples are quite typical with most chemical processes that occur within a cell. In the early stages of life's origins, it appears as if there is value in preserving almost every mutation long enough instead of eliminating the less fit ones immediately. This is because a large variety of enzymes are necessary to form a large collection of chemical loops.

In other words, until a chemical loop is formed there is no obvious phenotypic advantage even though other important internal chemical structures are being formed. After a period of seemingly no advantageous traits, one 'last' useful mutation generates the required enzyme (like the eighth or the last enzyme necessary to

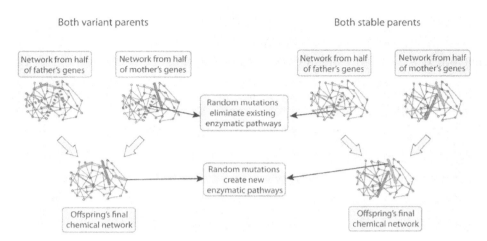

Figure 54.5: Stable parallel looped network of an offspring when both parents are either stable or variant. When both parents are variant with several mutations, the offspring will lose a few enzymes even though it may gain new enzymes. There is no clear direction in evolution in this case. The resulting network is unpredictable and potentially inviable. When both parents are stable with few mutations, the offspring does not develop new features and possibly does not lose existing features either. However, in a stressful situation, such an offspring is not the fittest and cannot overcome the stress in a directed manner when no new features are introduced. In spite of a lack of direction in evolution in both these cases, it is clear that sexual reproduction gives rise to robustness in survivability (since the same deleterious mutation should occur in both parents) compared to asexual reproduction.

complete the Kreb's cycle) that can now complete the chemical loop. It will feel as if the cell has created seven enzymes with no visible advantages. However, when the eighth enzyme is discovered, the cell now forms a complete chemical loop (like, say, the Kreb's cycle). This suddenly creates an explosive jump in evolutionary features because the entire loop becomes a highly advantageous phenotypic feature. Therefore, we can say that phenotypic features evolve in *discrete* jumps rather than in a continuous fashion. This is because each phenotypic feature requires several *chemical loops* to be formed, in general.

Whether these new loops give rise to new phenotypic advantageous behaviors, like shorter beaks in finches, is not a requirement that should be imposed for the usefulness of most mutations. The approach suggested here is uniquely different because most researchers call a mutation as useful only if there is a useful phenotypic feature. In fact, deriving the connection between behaviors and stable parallel loops is itself an important aspect that I will focus on from chapter 55 onwards (though some of this connection was partially discussed in chapter 53). In contrast with current research, mapping these new stable parallel loops to new behaviors can and should be done separately. I will discuss a few of them in this and the next several chapters.

When a system has many complex chemical loops, it naturally exhibits more behaviors (see chapter 59 and later in chapter 63). Several of these behaviors are defined relative to the structure of loops and not on the specific chemicals used. I characterize them as the chained sequence of chemical reactions whose resulting outcome is the given behavior. The precise set of chemicals within this chain can be altered, if necessary without altering the behavior itself. For example, if a series of reactions are triggered to produce an effect like retracting your hand away, then the exact set of chemicals involved in this series can be changed while still achieving the desired behavior. With stable parallel looped structure of our brain, we have already seen that the dual space representation of chapter 25 captures the same notion.

In reality, when random mutations do occur, they may actually end up breaking some of the existing useful features (see the discussion in section 54.2 on dominant and recessive genes). Even as we have a large number of less fit organisms mutating simultaneously and over long periods, expecting that some of the less fit ones will become better fit just through these random mutations is less likely. It is more an exception than a rule. The SVML approach suggested here does not make this assumption unlike Darwin's theory. The steps to be discussed below will highlight one way to guarantee that you get better-fit organisms in the future using these less fit ones. All of the steps of stable-variant merge loop are indeed necessary. Therefore, this iterative process fundamentally deviates from Darwin's theory. In fact, contrary to what Darwin's theory suggests, it is the less fit organisms that tries to find novel solutions (see chapter 71 for more detailed comparison with Darwin's theory).

Step 3 (Sexual joining of mutated less fit and previously fit systems): Let me now restrict our attention to organisms that undergo sexual reproduction. In this case, the offspring's are created from two parents, males and females. However, under stressful environmental conditions, not all combinations of sexual reproduction between males and females will be beneficial. In Step 1 above, I mentioned that there are two sets of systems, the less fit and the already fit ones. The potential beneficial offspring's come from sexual reproduction among males and females with one within each group.

If the male is from the less fit group, the female should preferably be from the already fit group and vice versa. The other two possibilities (i.e., both male and female from the same group) will surely happen. However, excluding exceptions like, say, both already fit or both less fit can give rise to more-fit offspring in unusual cases, such combinations would not be more beneficial, in general (see Figure 54.5). There are at least two reasons for this. Firstly, from Step 2, the highest chances of mutations are among the less fit group than within the already fit ones (Figure 54.5). Secondly, by ensuring that one of the organisms is quite stable with almost no mutations, we can guarantee that evolution proceeds in a definite direction instead of a random or disorderly manner. The offspring's will have features from both parents. As a result, it will get all the stable unmutated features in addition to some mutated features (see Figure 54.3 and section 54.2 on how the merged metabolic network becomes better). If the original environmental stress that triggered this iterative process is not too severe, then surely offspring's with both parents from already fit group will be useful for better survival.

For the offspring's, we can even be confident that there will be new features developed with high probability (see the merged metabolic network in Figure 54.3). For example, the number of mutations in the less fit parent is far more than under normal conditions. Therefore, there is a high chance of discovering new sets of proteins that never existed before. They can act as enzymes making it possible to carry out a few new chemical reactions. These can be viewed as creating new edges in the metabolic network of Figure 54.3.

Secondly, when you combine these new proteins (from mutated less fit parents) with existing chemicals generated from the stable parent, you have the possibility of new chemical pathways from existing reactions. The importance of stable parent is apparent here. It preserves the original metabolic network (see Figure 54.3). You want the new chemicals from the less fit parent to act as a 'bridge' (or an edge in the metabolic network) connecting existing chemical pathways (coming from the stable parent) to some other existing or new chemical reactions. This bridge would not have been possible without the mutations in the less fit parent. At the same time, the offspring as a whole is not a viable organism without the stable parent.

When the mutated less fit and the already fit organism sexually reproduce, the offspring is at least as fit as the stable parent. To see why this is the case, recall that the offspring gets two copies of a chromosome, one from each parent. Now, the original fit set of chemical reactions that occur in the stable parent can continue to occur correctly in the offspring as well i.e., the stable parents' metabolic network is still possible to generate. Even though the genes from the mutated chromosome from the less fit parent might be defective, the offspring can simply use the chromosome of the stable parent instead. Therefore, sexual reproduction between two such specific organisms offers a unique advantage. One parent can take heavy risks with several random mutations while the other parent offers stability and viability. You now have a high likelihood of discovering new chemical pathways that lets evolution proceed iteratively in a directed manner yielding progressively better organisms.

It is interesting to note a few special cases here. Some species like slime molds, algae, fungi and many plants reproduce both sexually and asexually. Apparently, these species reproduce asexually under favorable conditions and change to sexual reproduction under adverse conditions. This observation fits very well with the stable-variant merge loop pattern described here. These adverse conditions precisely trigger the above Step 1 and the entire iterative loop of SVML. This iterative loop depends fundamentally on sexual reproduction to ensure directed evolution, which agrees with the switch to sexual reproduction by these species.

Another special case is with hermaphrodites i.e., those organisms that have both male and female reproductive organs. Some examples are pulmonate land snails, earthworms and even some types of fish. These species have an ability to mate with *any* other organism within the same species. As a result, the number of possible sexual combinations required for the current step of the iteration is considerably more.

Step 4 (Elimination of the less fit systems): With various sexual combinations occurring in the wild among all types of organisms (not just between mutated less fit

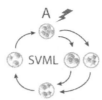

Figure 54.6: A simple schematic representation of the SVML pattern. We will use this shorthand representation as a quick way to talk about SVML as we discuss other applications.

and already fit organisms), there are clearly some offspring's that are severely less fit than the rest. Sometimes they are eliminated even before they are born, like with miscarriages. From our assumption that the change in environmental conditions are slower than the average life span of organisms (i.e., the second law effects are slower), we can be sure that not all less fit systems will be eliminated. This step is much slower compared to the other steps of the stable-variant merge loop pattern.

The iterative loop now continues, going back to Step 1. The set of organisms now at Step 1 are the ones that remain after Step 4 of the previous iteration (see Figure 54.4). The iterative process continues as long as the organisms remain stressed in the new environment. The iteration proceeds for several generations before you start seeing noticeable changes in the species. In this manner, the stable-variant merge loop pattern among sexually reproducing organisms can give rise to a direction in evolution among these species.

Figure 54.6 represents the above iterative process schematically. The species that is undergoing directed evolution is shown as *A*. The environmental stress that triggers the process is shown on the top while the iterative loop is simply shown below. I only highlight one aspect of this loop, namely, that it is based on sexual reproduction.

Figure 54.6 hides most of the details of the iterative process. The reason for such a representation will become clear as I start discussing other generalizations of stable-variant merge loop pattern later (see chapter 55). I will combine several such schematic pictures of different interacting species to show how more complex forms of directed evolution can occur. The simplified schematic will aid us visually in understanding the directed evolutionary processes of complex inter-species processes more clearly. Some examples of these are predator-prey, parasitic, symbiotic relationships as well as more complex relationships observed in our entire ecosystem (see chapter 55).

54.4 Properties of SVML

One new outcome of the stable-variant merge loop pattern is that it shows how central sexual reproduction is in evolution and how well the basic idea behind the advantages of sexual reproduction, namely, the stable-variant merge loop pattern, generalizes beyond sexual reproduction to include asexually reproducing species, between species and, in fact, in the entire ecosystem itself (see chapter 55).

While the advantages of sexual reproduction have been well studied and understood by researchers, one uniquely new outcome of this discussion is that sexual reproduction and, more generally, SVML is the cause of a direction in evolution itself. A somewhat related notion to SVML in the existing literature, though not generalized to other cases like with SVML, is the concept of heterosis. Heterosis is an ability to produce improved or enhanced traits in an organism by mixing genetic contributions of its parents (see Shull [141]). In addition to generalizing this notion of heterosis, the new SVML pattern also explains how evolution is capable of overcoming the effects of the second law of thermodynamics, which typically manifests through environmental stresses and causes systems to collapse or bring disorder.

Let me briefly summarize the properties of SVML pattern below. Firstly, the less fit organisms do not eliminate quickly and are, in fact, the primary ones to help create new features, rather than the fittest ones. Secondly, quarantining the less fit as abnormal and avoiding them to reproduce sexually with fit ones does not help create a direction in evolution. Thirdly, sexual reproduction and horizontal gene transfer (to be discussed in section 55.1) are necessary for a direction in evolution. Fourthly, the fittest organism shuffles within the group because the new features come from a combination of one stable parent and one less fit one. The offspring's of the both fittest parents do not necessarily remain as the fittest offspring's in the future, as is seen here. The SVML algorithm is not a 'greedy' solution in that fit organisms' offspring's are not necessarily fit after several generations (unlike Darwin's theory). Fifthly, the complexity of the SPL chemical network keeps increasing steadily and cumulatively while creating new features within one or more species i.e., the maximal SPL network monotonically increases.

54.5 Testing SVML experimentally

How can we test if this SVML process works in nature? The existing experimental data to validate Darwin's theory is directly applicable to validate the above mechanisms as well. However, in addition, the structure identified here clarifies the direction in evolution. One interesting example is with species like slime moulds, algae, fungi and many plants that reproduce both sexually and asexually (heterogamy). It is known that these species reproduce asexually under favorable conditions and change to sexual reproduction under adverse conditions (see Cole and Sheath [23], Reekie and Bazzaz [128] and Mehrotra and Aneja [104]). The behavior of such species fits well with the stable-variant merge loop pattern described here.

Another simple way to test SVML is to consider three artificial environments and introduce a specific stress to trigger the stable-variant merge loop. In one, we isolate the less fit organisms from the already fit ones in each generation. We do not allow sexual reproduction between the two groups. In another, we intentionally magnify the sexual reproduction between the less fit and the already fit organisms in each generation. The third one is closer to a natural environment where random sexual reproduction between the organisms is allowed. If we compare the rate of discovery of new solutions in some measurable way, the stable-variant merge loop predicts that the second environment will produce the most diverse set and will have several new features. The first environment will have the least set while the third will be somewhere in-between the two, in a measurable sense. Recall that in a natural environment, there are other ways to avoid the non-persistent stress, the prominent ones being physical migration and isolation. The above mechanisms are for guaranteeing a direction in 'chemical' evolution when attempting to discover new evolutionary features like eyes, heart and others.

If the correct conditions specified in section 53.5 are met, the iterative process will, more or less, end up in two states after a long while for a given environmental stress. One state is where the iterations do converge with the discovery of new enzymes that ultimately help the species overcome the adverse conditions. The other state is where the new enzymes do not link with existing chemical reactions well enough. This does not yield beneficial states for the organisms. Ultimately, the second law effects will cause the system to collapse. In this case, the species proceeds towards extinction.

54.6 On the evolution of sexual reproduction

There is some similarity with the iterative procedure outlined here to the evolutionary theory of sex outlined by Geodakyan [51]. I will not discuss this in great detail but will refer the reader to Geodakyan [51]. He suggests that two sexes evolve eventually from two types of systems. The first type attempts at *conservation* and tries to maintain stable features. This will lead to the evolution of female sex. The second type attempts at *change* that involves taking more risks. This leads to the evolution of male sex. He says that useful features are passed on to the females from males after several generations. For more details on his theory, I refer the reader to Geodakyan [51]. The similarity with the iterative procedure of SVML can be seen if you compare (a) the less fit organisms that mutate considerably to the systems that attempt at 'change' i.e., males according to Geodakyan and (b) the already fit systems to the ones that attempt at 'conservation' i.e., females according to Geodakyan. However, the difference is that I only use the *existence* of two sexes for the iterative procedure of SVML instead of discussing the *evolution* of sex itself.

54.7 Chemical evolution versus physical alternatives

The ability for some species to physically move around in the environment already gives them better ways to survive. For example, they can migrate to new locations with better environmental conditions without having to discover new enzymes through the above SVML iterative process. Another possibility from migration is for the less fit species to encounter other organisms that are already better fit in the new location. These interactions could be with organisms from the same species or even with a different species. With different species, the above sexual reproduction Step 3 should be changed to either (a) bacterial or viral infections to transfer useful enzymes (see section 55.1) or (b) to a symbiotic relationship (like with *lactobacilli* bacteria in humans, gut fauna in herbivores, rhizobium, nitrogen fixing bacteria in leguminous roots and several others – see section 55.6).

With migration ability, organisms have extra options to alleviate the effect of stress. They can (a) stay in the same location and discover a novel solution using the above stable-variant merge loop pattern, (b) walk away to a new environment and avoid the stress or (c) exhibit some kind of intermediate behavior in which they have partial chemical modifications using the above iterative procedure while they also migrate to new locations resulting in some overall benefits. What is critical to note here is that there are fundamental limits on how well you can survive using purely physical means like migrating to a new environment.

When the environmental conditions in a sufficiently large region are adverse, the systems have no choice but to look for chemical means instead, for better survival. This is the chemical evolution. With chemical evolution, the above iterative procedure (stable-variant merge loop) becomes the preferred approach. This is yet another way to convince ourselves that directed evolution is a slow process. The less costly approach of physical migration will naturally be attempted first. After exhausting this, the systems typically will attempt the above slow chemical SVML iterative process.

54.8 Types of environmental stress

Various types of stresses trigger the above SVML iterative process. These stresses directly threaten the survival of the organism. The most common ones are: scarcity of food and the presence of predators.

Stress in the context used here refers to a prolonged state of deprivation of normal conditions, not just temporary changes to normal conditions for short periods. In particular, the stress should be present for, at least, a few reproduction cycles. It should force a fundamental change in the set of chemical reactions that can occur during reproduction and when passing genetic information to the offspring's. Furthermore, it is necessary that this change occur repeatedly and for extended periods of time. The simplest way a given species can have a persistent stress is when some of the dependent species migrates to a different location.

You may have already questioned why a systems' survival is important, if threatened. For example, an organism, being just a chemical dynamical system, could simply collapse and die. Why should a living being struggle to survive under stress while a nonliving system does not? The answer to this question is quite nontrivial. It relies fundamentally on the fact that living systems use a stable parallel looped architecture. Non-looped systems will not struggle to stay stable or alive. I have already discussed this in great detail in the context of physical systems throughout the book, but I will discuss this in the context of chemical systems in more detail in chapter 57 and beyond.

A stable parallel looped system has no control over when, where, how or what types of environmental stresses occur. They could be either purely unexpected or introduced by minor variations from a number of living beings, say, owing to their ability to move randomly. In the latter case, we expect the minor variations to collectively produce a considerable impact on the system of interest.

55

Applications of Stable-Variant Merge Loop

In this chapter, I will generalize stable-variant merge loop pattern of the previous chapter to cover all other cases of evolution within our ecosystem. In chapter 54, we had focused primarily on how evolution proceeds in a directed manner for species that reproduce sexually. However, we do observe that such a direction in evolution exists more broadly across all species. In this chapter, I want to show that the underlying mechanism for all cases is still the stable-variant merge loop pattern.

55.1 Directed evolution through horizontal gene transfer

If we look at the stable-variant merge loop closely, the step that involves merging of the altered less fit with an already fit system does not require both systems to be of the same type. In section 54.3, I replaced this step with sexual joining of two organisms from the same species. We can now generalize and say that (i) the two organisms need not be from the same species and (ii) that the joining need not be sexual reproduction. This may seem unreasonable or less common at first glance. However, it is observed that some form of joining of two organisms is quite common with unicellular organisms like bacteria. In fact, such mechanisms are now considered prevalent even beyond unicellular organisms. This is termed as horizontal gene transfer (see Syvanen [150]). An example is with *Bdelloid Rotifers* species that stopped undergoing sexual reproduction for at least 35 million years. Yet, they were able to survive and acquire genetic diversity using HGT (see Gladyshev et. al [56]).

Even though bacteria undergoes binary fission i.e., asexual reproduction where a single bacteria splits into two, there are several other ways a bacteria acquires foreign DNA (see Baron [7]). For example, there is *transduction* where DNA is transferred between two bacteria by a virus. A bacteriophage (i.e., the virus that infects the bacteria) helps in transferring a portion of bacterial or its own DNA into other bacteria. Another process is called *transformation* where a bacterium acquires DNA from the environment through its cell wall, provided it is not part of any other cell. Yet another process is called *conjugation* where two different bacteria transfer genetic material between each other through some form of contact (either direct or by forming a bridge). Plasmids or transposons act as conjugative elements in this case.

Each of these three mechanisms, namely, transduction, transformation and conjugation can be seen as a form of infection, since there is a transfer of foreign DNA into a given bacteria. These processes (at least, conjugation) are not viewed as sexual reproduction. What is, however, interesting is that these mechanisms are also applicable (a) between two different species within bacteria or prokaryotes, in general, (b) between bacteria and viruses and (c) between eukaryotes including multi-cellular organisms and bacteria/viruses (see, for example, Baron [7] for more details).

For (a), antibiotic or antimicrobial resistance is considered as a common example. When bacteria mutates and develops antibiotic resistance, it transmits this drug resistance to other bacteria within the same species as well as to other bacterial species through plasmid exchange. For (b), bacteriophages are examples mentioned earlier. They are viruses that transmit their DNA into bacteria. They are one of the most abundant organisms in our biosphere. Similarly, there are virophages (or satellite viruses) that transfer genetic information between viruses. For example, the Sputnik virophage multiplies only in viral infected amoeba and not in normal amoeba. This Sputnik virophage infects a virus called mimivirus. For (c), an example of a plant pathogen called *Agrobacterium* that uses bacterial conjugation as the mechanism is well known. This pathogen has an ability to transfer its DNA to plants causing tumors. There are other examples within eukaryotes like, say, fungi acquiring genes from infecting bacteria and others. Other more common examples are retroviruses or more generally, proviruses. Retroviruses are RNA viruses that first convert their RNA into DNA using an enzyme called reverse transcriptase and then incorporate this DNA strand into a host DNA using another enzyme called integrase. Endogenous retroviruses are examples of such retroviruses that have remained in the genomes of humans and other vertebrates, after infecting germline cells sometime in the past. As a result, they are now inherited. It is estimated that about 5-8% of the human genome is comprised of human endogenous retroviruses (Belshaw et. al [11]).

Using the above cases for horizontal gene transfer, we can now modify the stable-variant merge loop as follows (see Figure 55.1 and compare it with Figure 54.4 of section 54.3).

Step 1 (Creation of two sets of systems from an environmental stress): In this step, we start with two species. One is the species that infects, which is typically bacteria or viruses. The second species corresponds to the organisms that are infected. I am using the term infection in the sense mentioned above, even though there are other forms of infection that do not involve any transfer of DNA. The second species can include multi-cellular organisms like humans as well, as described previously with endogenous retroviruses. When there is an environmental stress, both of these species will be under stress. Recall the discussion in Step 1 of section 54.3 on what constitutes an environmental stress. As a result, under the new conditions, for each of the species, some organisms will be less fit and others, already quite fit. The two sets of interest in this step are the less fit set and the already fit set. Therefore, within each of these sets, we will have some organisms from the first species and some from the second species.

Step 2 (Mutations in the less fit systems): Both these species within the less fit set will have different types of mutations. They respond to their individual stresses. One interesting possibility that can occur here since there are two species involved is that the mutations in one species can benefit the other species. The new chemicals of one species can chemically react with the existing chemicals of the other species. This gives rise to the possibility of several stable parallel looped reactions in each species. Furthermore, unlike the case in section 54.3 (with sexual reproduction), the

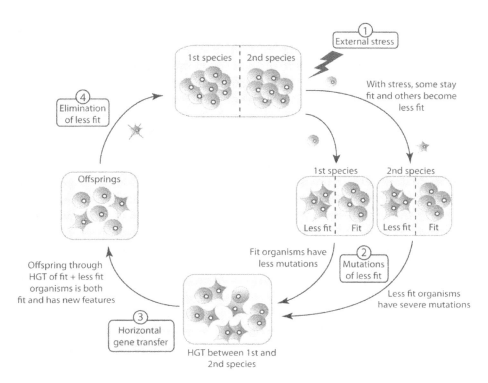

Figure 55.1: Stable-variant merge loop pattern applied to systems capable of horizontal gene transfer (HGT). The iterative process of SVML with HGT allows us to have a direction in evolution. The process is basically identical to the one in Figure 54.4. An external stress triggers the iterative process, which makes some organisms fit while the others as less fit. More mutations occur within less fit organisms. When there is HGT between less fit and fit organisms, the offsprings have both parents' features.

mutations in the less fit systems may not benefit each species individually as much. In this case, the combined effect, to be discussed next, becomes critical to create several looped reactions in a given species.

Step 3 (Horizontal gene transfer between modified less fit and already fit systems): The modified less fit and the already fit systems contains organisms from both species. I will refer to the infecting species (those that perform horizontal gene transfer) like bacteria and viruses as the first species and the infected species as the second species. We now need to analyze four cases.

(a) *First species modified less fit + second species already fit*: With this combination, we can have transfer of genetic information using any of the horizontal gene transfer mechanisms like transduction, transformation and conjugation discussed earlier. Since the second species is stable, we will have a directed evolution within the organisms of this species using this iterative process.

(b) *First species modified less fit + first species already fit*: Since the first species is typically bacteria and viruses, this combination occurs through horizontal gene transfer mechanisms as well. In this case, the first species can have directed evolution (like with antibiotic resistance).

(c) *Second species modified less fit + first species already fit:* With this combination, there is little benefit if at all. The second species is already less fit with several mutations. When it acquires DNA from the first stable species, even if there are new looped reactions, the chances of this second species surviving for a long time is not high because of other deleterious mutations. As a result, there will rarely be directed evolution in the second species here.

(d) *Second species modified less fit + second species already fit:* This combination is possible within the same second species in two different ways. One way is in which the species allows for horizontal gene transfers within themselves. This is typically possible in prokaryotes. This combination is quite unlikely within the same species in eukaryotes or multi-cellular organisms if the underlying mechanism is horizontal gene transfer. The second way is with organisms like eukaryotes and multi-cellular ones, capable of sexual reproduction. If one of the organisms is male and the other is female, sexual reproduction results in directed evolution in the second species as well.

In section 54.3 when we considered sexual reproduction, we required that one of the set be mutated less fit and the other be already fit. However, since there are two species involved in this case, there are chemical benefits from horizontal gene transfer even when both the sets are already fit. The possibility of new looped chemical reactions that can occur when you add chemicals from the infecting species into the second species is already higher irrespective of whether one of them is less fit or not. Of course, the infections that are dangerous enough to kill organisms of the second species will be automatically eliminated as less fit. In addition, when both sets are mutated less fit, the chances of having chemical benefits are low. They will also be eliminated as less fit.

The current discussion on various possibilities of merging will result in high probabilities for directed evolution in the second species. This is primarily because we have an enormous number of infecting organisms like bacteria and viruses (that actually incorporate their DNA into the second species' genome) on our planet compared to multicellular organisms. In fact, scientists estimate the number of bacteria and viruses to be of the order of 10^{30} and 10^{31} respectively (cf. humans who are only about 7×10^9).

Step 4 (Elimination of the less fit systems): This step works the same way as in section 54.3. Less fit systems in both species will be eliminated over time even though there are differences in the relative rates. For example, the first species are typically bacteria and viruses. Their genomes are considerably small and hence only a small set of reactions occur within them. This small number makes it easy for them to continue their looped reactions even under widely varying environmental conditions. They are also ubiquitous for this reason. Furthermore, when the favorable conditions reappear, they can spring into existence and multiply once again (especially viruses). Therefore, these organisms are not eliminated as quickly as the second species. The second species, if sufficiently complex, will have less fit systems that may or may not continue their looped dynamics long enough. These less fit systems will eventually be eliminated.

It is important to realize that the above iterative process can occur easily only until the second species develops a feature called immune system. With an active immune system, there is a natural tendency for the second species to fight horizontal gene transfer from the first species. This is when the second species treats this process as an infection. Yet, the iterative process occurs quite often because of the sheer number and diversity of organisms belonging to both the first and second species.

Besides, if one infecting system (say, bacteria) learns to avoid the immune system of a species through some mutations, another infecting system can acquire the same trait using the very same iterative process described above (relying on horizontal gene transfer). This is analogous to how bacteria acquire antibiotic resistance.

The iterative process discussed here that uses horizontal gene transfer is the most prevalent mechanism in the early stages of life. In fact, for the first two billion years after life evolved on earth, organisms were neither multi-cellular nor had sexual reproduction. There were only micro-organisms on earth like the single cellular prokaryotes (bacteria and archaea). Therefore, during this period, iterative process of section 54.3 based on sexual reproduction will not be applicable. However, with horizontal gene transfer as the critical step for directed evolution (actually, another one based on symbiosis is prevalent as well – see section 55.6), the notion of tree of life is extremely fuzzy and even incorrect during this period. Once we have sexual reproduction and an immune system, horizontal gene transfer based stable-variant merge loop pattern becomes less common. Sexual reproduction based stable-variant merge loop pattern of section 54.3 takes over as the important mechanism for directed evolution, which would naturally lead to a tree of life. It is interesting to see that the stable parallel looped architecture of life (and complex systems, in general), gives a deeper insight into the underlying structure as well as the observed diversity.

Figure 54.6 of section 54.3 can be viewed as a similar schematic representation of the above iterative process based on horizontal gene transfer. The main sequence of steps are as follows. The environment puts a stress on both species A and B. Species A is typically bacteria or viruses. They evolve in a directed way using horizontal gene transfer as shown. Species B, on the other hand, can be complex multi-cellular organism as well. They evolve using either sexual reproduction or horizontal gene transfer. As both species A and B evolve, there is a horizontal gene transfer from A to B.

There is yet another way to represent the directed evolution based on horizontal gene transfer. Here, the environmental stress in only on species A. This evolves in a directed way using horizontal gene transfer. During this evolutionary phase, A performs horizontal gene transfer with species B as well. In this case, organisms in species B are not under any severe stress and they are not undergoing any severe mutations. Instead, they are quite stable organisms and are yet genetically 'infected' by species A.

55.2 The role of consciousness on directed evolution

Human consciousness and intelligence seems to provide additional ways to avoid considerably large-scale environmental stresses. However, with advances in genetic engineering, would we modify ourselves to introduce new features like gills to breathe under water, genes that allow us digest cellulose or grow extra limbs or eyes? Nature would have introduced such features slowly, if necessary, depending on whether the stresses persist for thousands of years. With genetic engineering, the changes would be abrupt and too soon. Also, there are ethical questions if we were to do this on a set of embryos in a laboratory.

The less abrupt and less risky approach is to use the natural stable-variant merge loop mechanism even within humans. Using such an approach for common environmental stresses would be the same as with other species. Let us briefly discuss other special situations unique to conscious humans instead. Firstly, consciousness has elevated even each individual's survival as important from an evolutionary point of view unlike chemical evolution within other species. For example, our empathy, morality and other feelings introduce issues that other species did not have to consider when trying to survive under severe stress. Secondly, classification of humans into new groups based on strong individual preferences is less likely with other species. These preferences like, say, based on dietary restrictions, religious and other cultural beliefs sometimes introduce new sources of stresses that persist across multiple generations within these groups even though they are not hereditary. As a result, these nonhereditary stresses can naturally create the two groups – less fit and already fit people – necessary for the SVML pattern to work. The iterative process can now continue making subtle genetic changes and through sexual reproduction between the two groups. Strangely, this process continues even from socially or culturally driven stresses.

Let me briefly compare the differences between Darwin's theory and SVML when applied to evolution within conscious humans. One point to note is that Darwin's theory would not suggest interbreeding between the less fit (or severely stressed) and already fit humans, contrary to what stable-variant merge loop pattern suggests. Darwin's theory would suggest we wait for the correct mutation to be discovered for a given stress. If we were to apply the survival of the fittest argument from Darwin's theory, we would need to isolate (or quarantine) the least fit individuals, as they will be eliminated anyway. However, the claim here is that it is less likely to have both stability and new features in a species unless we allow sexual reproduction between the two groups. The healthy group typically does not mutate enough to overcome the stress by themselves. This is certainly true with other species. One reason is that Darwin's theory does not emphasize the role of sexual reproduction in guaranteeing a direction of evolution sufficiently. In chapter 71, I will discuss other ways the proposed theory deviates from Darwin's theory.

55.3 Directed evolution through inter-species interactions

Until now, we have treated environmental stress as a generic stress that is affecting one or more species simultaneously. We have considered the variations created by the environment to be random. We did not look into the possibility that there is a structure within the environment. Nor have we looked at what happens to the environment while the systems are evolving. In this section, I will treat the environment itself as a set of evolving systems. For example, you have two or more systems interacting in a specific manner with one system introducing stresses on the others and vice versa. Therefore, in this case, not only do the systems that are subject to the stress evolve, but the systems that introduce the stresses will evolve as well.

The living world around us is a lot more interconnected than what we have considered so far. No species exists in isolation. Each species depends on and is influenced by a number of other species in quite fundamental ways. Any stresses or relative imbalances even in one species will have far-reaching repercussions across a wide range of species. In the next few sections, I will look at these interactions in detail. I want to show how we can start understanding the question of directed evolution within these species via these interconnections.

There is a complex 'invisible' network that ties together all living and nonliving systems. In spite of the second law of thermodynamics, this invisible network exhibits robustness and stability when viewed as a complex dynamical system. In the next few sections, my objective is to show how the stable-variant merge loop pattern not only maintains this stability and robustness, but also drives the entire ecosystem to evolve in a directed way. I will start the discussion with simple cases and build upon them to achieve arbitrary level of complexity.

The SVML pattern is triggered through a stress in some species. We will need to keep track of the various types and sources of stresses when understanding how the interdependent networks of species evolve in a directed manner. I will start with three of the most common types of relationships and study how the stresses originate and propagate between the species. These are: (a) predator-prey relationships, (b) parasitic relationships with a host in which there is little benefit or even harm done to the host species and (c) symbiotic relationship with mutual benefit for both species. Using these three relationships, we can build more complex relationships between different groups of species. In general, any given species will have more than one of the above relationships with other species. Analyzing such complex relationships will be a direct application of the same underlying ideas that will be presented here.

The current discussion is not on how these complex relationships work or how the relationships are formed in the first place (i.e., not on the problem of evolution of these features). Rather, I will be using the existing relationships to show how and why evolution will proceed in a directed manner. I will also show how these relationships help create new features over very long periods. The discussion on the creation of these interrelationships itself will be deferred to section 55.7 instead. What may come as a surprise is that the ideas required for explaining the creation for the first time and the usage for all subsequent times for each of these relationships are identical (see section 69.4).

55.4 Directed evolution via predator-prey relationship

Consider lions and deer's as an example of predator-prey system. I have represented this relationship in Figure 55.2 using an arrow from the prey (deer's) to the predator (lions). The direction of the arrow implies that the predators (A) depend on the prey (B) for survival since it is a main source of food for the predator. There are several examples of predator-prey relationships in the animal, plant as well as single cellular organisms.

We already know that a predator and prey system naturally converges to an equilibrium state, in terms of the total population (see, for example, Hofbauer and Sigmund [70]). Let us now see how both the predator and the prey species develop new features over a long period in a directed manner. For example, the predator or the prey species develops features like discovering poisonous chemicals, growing strong jaws, teeth or claws, an ability to vary skin color for camouflage, improved vision, highly sensitive ears and so on. Each of these features may take thousands of years and hundreds of generations. Nevertheless, there is a sequential and directed evolution of each of these features within both species. The equilibrium state mentioned above reached by predator-prey populations occur at a much faster pace than the evolution of each of these features in a directed manner. In this section, I want to show how we can use the stable-variant merge loop pattern iteratively to explain the evolution of each of these features in both species.

Since the predator kills the prey for food, the prey is constantly under stress. In Figure 55.2, I have represented this stress from predator to prey as the first step in directed evolution (Step 1 of SVML). This stress on the prey species B triggers the stable-variant merge loop iterative process. Figure 55.2 represents this as the second step (using a loop) similar to how we schematically represented this in section 54.3. This directed evolution in B occurs either through sexual reproduction or through horizontal gene transfer, as the case may be. This iterative process occurs over several generations resulting in better adapted species B. This means that the resulting species B is better capable of avoiding the stress from predators A to a certain extent.

Of course, there are physical ways to avoid the stress as well like hiding or not making any sounds. These can be learnt, say, by imitation and can be memorized in the brain instead. However, these mechanisms do not yield a chemical evolution, which we encode in the DNA. Nevertheless, these physical mechanisms are quite sophisticated as we observe with, say, lions, deer's and among other animal species. In fact, the responses from these living beings to a given situation are quite specific. We can explain this form of physical evolution using ideas and concepts discussed in chapter 18 on how specific fixed sets are formed within a brain network. I am excluding this kind of evolution (i.e., evolution through nurture) in the current discussion even though it is equally important.

Since predators A depend on B for survival, the improved species B capable of avoiding A places a stress on the predators as shown in the third step of Figure 55.2. The next step is for the predators A to evolve in a directed manner using the stable-variant merge loop. As with prey, the predators can use either sexual reproduction or

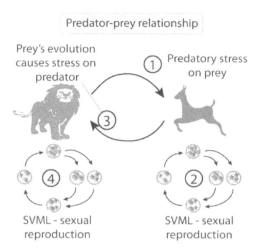

Figure 55.2: Directed evolution via predator-prey relationship. A predatory stress triggers SVML on the prey causing it to discover new evolutionary features through sexual reproduction. This in-turn causes a stress on the predator, which results in discovering new evolutionary features via SVML in the predator as well. The process continues resulting in a direction in evolution in both species.

horizontal gene transfer in this step. The result of this iterative process is to have modified predators *A* that outperform the modified prey species *B* and are better able to survive once again. This causes a stress on the modified prey species *B* and the process repeats. In this manner, the predator-prey dependency relationship has caused both of the species to evolve in a sequential and directed manner using stable-variant merge loop as shown in Figure 55.2.

It is important to note that there are other ways to introduce stress in this predator-prey system, namely, from unpredictable environmental changes. Even though there is a natural stress introduced through the predator-prey relationship, we can have external unpredictable environmental stresses that can trigger stress in either the predator or the prey species. Some examples are volcanic eruptions, ice age and other less accounted relationships with other species.

55.5 Directed evolution via parasitic relationship

A parasite is an organism that survives within or on another organism called the host. It benefits from the host and sometimes even harms the host. Some common examples are parasitic tapeworms or flukes living within the gastrointestinal tracts of vertebrate hosts. Protozoan of the genus *Plasmodium* causing malaria is another example of a parasite that lives within *Anopheles* mosquito. A parasite depends on the host for survival and it rarely kills the host even though it may harm the host (see Baron [7]).

Figure 55.3 shows an example of parasitic relationship. It is represented as a

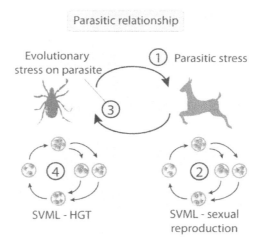

Figure 55.3: Directed evolution via parasitic relationship. A parasitic stress triggers SVML on the host species causing it to discover new evolutionary features through sexual reproduction. This in-turn causes a stress on the parasite, which results in discovering new evolutionary features via SVML in the parasite as well. The process continues resulting in a direction in evolution in both species.

dotted arrow from the host (human) to the parasite (tapeworms) showing the direction of the dependency relation i.e., parasite requires the host. The dotted arrow signifies a weaker relation in that the parasite does not kill the host. This is unlike the predator-prey relationship of section 55.4 where the predator actually kills the prey (and is, hence, represented by a straight arrow). Nevertheless, the parasite creates a persistent stress on the host species. As in section 55.4 for predator-prey systems, our objective here is to show how new chemical features are created in both the parasite and the host in a directed manner over a long period of time.

A parasite consumes the resources of the host, for example, by living in the gastrointestinal tracts. This creates a stress on the host species. If this stress sustains for long periods of time, the host species responds by undergoing a directed evolution using the stable-variant merge loop pattern based on sexual reproduction (see Figure 55.3). Host defense mechanisms are slower to evolve but include examples like discovering specific toxins or complex immune system responses. This introduces stresses on the parasites as well.

Parasites have complex lifecycles relying on one or most hosts for survival and reproduction. In addition to evolving using physical mechanisms like proliferation, dispersion, switching between different hosts and different ways to infect the hosts, parasites also undergo chemical evolution (see Roberts and Janovy Jr. [129]). This includes discovering chemicals that give out specific odors to attract hosts, ability to sense specific chemicals and others. This directed chemical evolution uses the stable-variant merge loop pattern based on sexual reproduction for most parasites. Now, the newly evolved parasites trigger new stresses on the hosts. In this way, the

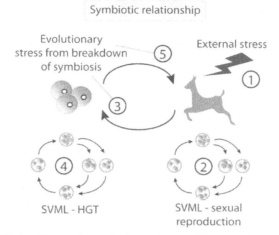

Figure 55.4: Directed evolution via symbiotic relationship. An external stress triggers SVML on one of the species, causing it to discover new evolutionary features through sexual reproduction or HGT. This in-turn causes a stress on the symbiotic relationship, which results in discovering new evolutionary features via SVML in the second species as well. The process continues resulting in a direction in evolution in both species.

evolutionary cycle repeats and proceeds iteratively in a directed manner.

55.6 Directed evolution via symbiotic relationship

In a symbiotic relationship, two species fundamentally depend on each other and mutually benefit from each other's interaction. There are several examples of symbiotic relationship like, for example, nitrogen-fixing bacteria *rhizobia* in leguminous plants, pollination of flowers by bees, mutualism between clownfish and sea anemones, gut flora in herbivores to digest plant food and others.

Figure 55.4 shows this mutual cooperative relationship between two such species A and B using straight arrows from one to another. While I defer the discussion on how such a relationship is formed for the first time to chapter 70, I want to show how both species evolve in a directed way. Unlike predator-prey or parasitic relationships discussed earlier, there is no natural stress introduced by one species onto the other, since both species cooperate. As a result, it appears as if there is less reason to evolve. However, these species are subject to external stresses either through other species or through random environmental stresses. Let us assume that species A (say, the deer's in Figure 55.4) experiences this external stress.

The external stress will create an internal stress between the two species as well. This is because the external stress strains the strong cooperative relationship between them. In this case, species A strains species B. Furthermore, species A evolves using the stable-variant merge loop pattern due to the external stress based on sexual reproduction (and in some cases, using horizontal gene transfer if A is a symbiotic bacterial species). Species B now becomes stressed from the imperfect symbiotic relationship from the modified species A compared to before it experienced stress. This causes species B to evolve using stable-variant merge loop pattern based on sexual reproduction or horizontal gene transfer, as the case may be.

The net result is that either both species will better align once again or one species will form a new symbiotic relationship with an entirely different species. Once the external stresses are relieved, there will be less reason to have significant directed evolution. The above iterative process will start once again as soon as one or both of the species encounter a new external stress.

55.7 Evolution via complex interactions in our ecosystem

In general, every species is subject to many stresses both from the external environment and from several other species. Furthermore, organisms within a species create stresses on other species. We can classify each of these individual relationships as predator-prey, parasitic or symbiotic ones. For example, a deer has a prey relationship with lions and tigers. Ticks on deer's have a parasitic relationship with deer's. Furthermore, several bacteria that have symbiotic relationship with deer similar to those that help digest cellulose. Therefore, deer's have all three relationships with different species. A significant stress from any of these relationships can trigger a stress on deer's (see Figure 55.5).

If we imagine mapping these species and their relationships as a graph, we get a complex and dynamical network that represents various dependencies between species (cf. dependencies on deer's from lions, ticks and bacteria as discussed above). This is what we view as our entire ecosystem. One such example is shown schematically in Figure 55.6. Here, I have also shown one scenario of how stresses

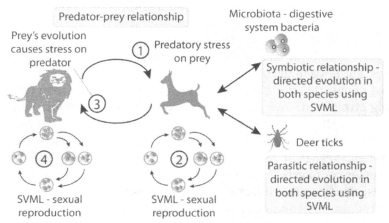

Figure 55.5: Interspecies interactions and directed evolution using SVML. Directed evolution occurs in all species using SVML based on either sexual reproduction or horizontal gene transfer (HGT) for stresses from predator-prey, symbiotic, parasitic and other interspecies relationships.

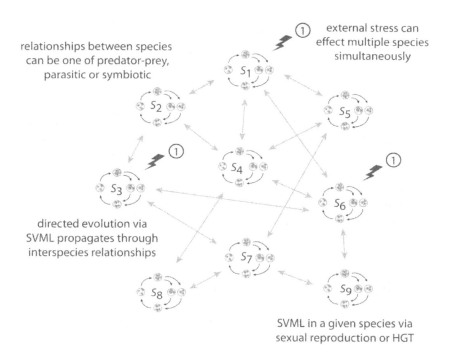

relationships between species
can be one of predator-prey,
parasitic or symbiotic

① external stress can
effect multiple species
simultaneously

directed evolution via
SVML propagates through
interspecies relationships

SVML in a given species via
sexual reproduction or HGT

Figure 55.6: Directed evolution using SVML via complex interactions in our ecosystem. When we include multiple species (S_1-S_9), there already exist several relationships between them like predator-prey, parasitic or symbiotic. These hidden relationships can be depicted schematically in a graph like shown here. If an external stress is introduced in multiple species simultaneously, the initial response is for those specific species to evolve via SVML to better fit in the new environment. However, this disturbs the existing relationship between multiple species causing them to evolve via SVML as well. This evolution via SVML produces ripples within the ecosystem causing all connected species to evolve new features over time.

might occur. If these stresses persist for long periods, we can keep track of how they propagate between the species. The stress on species propagates to a distant species via these ecosystem level interdependencies as the intermediate species start to evolve in response to the stresses. Each of the stresses will trigger stable-variant merge loop pattern within each respective species based on either sexual reproduction or horizontal gene transfer, as the case may be. Such evolutionary improvements will cascade through the entire ecosystem. I will not discuss this in detail. Instead, I will refer the reader to the description given in Figure 55.6. You will notice that we just needed to combine multiple mechanisms of directed evolution discussed in the previous sections in the correct order of propagation of various stresses to understand the directed evolution in the entire ecosystem itself.

Each scenario for external stresses can similarly be analyzed. In addition, we can analyze other more involved network of relationships using the same approach shown in Figure 55.6. Representing complex relationships within our ecosystem graphically, similar to Figure 55.6, aids us in understanding how evolution proceeds in a directed manner.

55.8 Darwin's theory as a special case of SVML

The discussion on various generalizations of stable-variant merge loop pattern is incomplete without comparing it with Darwin's theory. In this section, I will show that Darwin's theory can be derived as a special degenerate case of the stable-variant merge loop pattern. The similarities should already be apparent by now.

Let me first represent Darwin's theory in a form that makes it easy to compare it with stable-variant merge loop. Darwin's theory has three steps acting together in an iterative manner (see Figure 55.7). The first step is the introduction of a stress that results in random mutations in the DNA. In the second step, random mutations give rise to useful features in, at least, some offspring's, which are termed as the fittest. In the last step, these features are inherited resulting in a population in which the fittest survive and the less fit are eliminated. The iterative process now continues with the remaining organisms, creating new random mutations and, hence, potentially new features.

The comparison with Darwin's theory is now immediate if we ignore the specific structure within merge step of stable-variant merge loop. The main difference is that Darwin's theory does not identify the role of sexual reproduction and horizontal gene transfer (HGT) as the hidden structure responsible for creating a direction in evolution. It only lists the role of sexual reproduction or HGT as important for creating a diversity of organisms giving a better chance for overall survival. For developing inter-species features like from symbiotic, predator-prey and parasitic relationships, Darwin's theory states that these features develop because each species wants to survive better. In fact, in all cases, the notion of survival or struggle to survive in Darwin's theory hides the underlying structure within organisms and species.

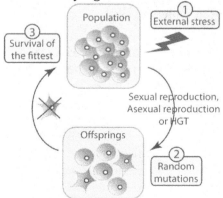

Figure 55.7: Darwin's theory of evolution expressed as an iterative process similar to SVML. This iterative process does not explicitly highlight the direction in evolution unlike SVML.

Darwin's theory does not explain why or how an organism's desire to survive translates to a positive direction in chemical evolution (not physical evolution, which itself fits within Darwin's theory). On the other hand, stable-variant merge loop clarifies this hidden structure to give rise to chemical evolution within organisms as well as within the ecosystem.

The stable-variant merge loop degenerates to the above special case under the following two conditions.

(a) When a species is subject to a new environmental stress, let us assume that there are a negligible number of already fit organisms within this species. In other words, almost every organism is less fit and is under severe stress, mutating considerably as a result.

(b) We also assume that the variations in the environment are changing quite fast both to create and maintain the stress in almost all organisms of a given species. Without this assumption, there is a natural tendency for some organisms to find a fit state just by using physical mechanisms like migration as discussed earlier.

Given these two assumptions, I will show that the stable-variant merge loop degenerates to the one shown in Figure 55.7. From assumptions (a) and (b), Step 1 (of section 54.3) would need to be altered to have only one set of systems, all of which are less fit (unlike the regular case where you have two sets – less fit and already fit). Now, this and assumption (a) implies that Step 2 should be altered to say that mutations exist in all systems. This degeneracy occurs since most of the organisms are less fit. Merging of two systems, namely, between the mutated less fit and already fit systems, i.e., Step 3 degenerates simply to merging of all combinations of organisms. We do not have special subgroups here because the two subgroups are degenerated to just one complete group. The last step, namely, the elimination of the less fit stays unaltered in this case.

For brevity, I have represented this degenerate case as Figure 55.7. When represented in this form, it becomes easier to see the relationship with Darwin's theory. The iterative process starts when the environmental variations creating stress in *all* organisms. This causes mutations in *all* organisms. They use *all* heritable mechanisms like sexual reproduction, asexual reproduction or horizontal gene transfer to transmit these mutations to the offspring's. The net result after several generations of this iteration is that the less fit ones are naturally eliminated, namely, the fittest will survive. This is, in short, what Darwin's theory of natural selection states. With Darwin's theory, new phenotypic features are introduced through random genetic mutations. These new features make some organisms more fit. They will transmit this fitness to their offspring's (using DNA) making all of them survive better. The less fit ones will be eliminated naturally over a period of time. The process is iterative until the stress disappears, until the species discovers a solution or until the species becomes extinct. Therefore, using the above degenerate case of stable-variant merge loop pattern, we can express Darwin's theory as a special case. In chapter 71, I will perform an in-depth comparison of the new theory with Darwin's theory.

It is important to note that the degenerate case has some severe limitations. While it is reasonable to assume that this approach gives rise to a new chemical feature through random mutations, it is unreasonable to suggest that we can create a long series of features all related to one another, creating one after the other, sequentially improving the system and producing a complex feature like heart, eyes or ears. For such complex sequential features, we need to rely on one or more of the directed evolutionary mechanisms described in sections 55.1 – 55.7 in addition to a stable parallel looped architecture. The above degenerate case is not a directed evolutionary mechanism.

Nevertheless, we do have a few examples where new useful features are introduced just from simple mutations (like, say, point mutations), rather than a long sequence of them. For example, the gene responsible for causing sickle-cell anemia involves just a single point mutation. In Africa, people who have only one copy of this mutated sickle gene i.e., from just one parent, does have resistance to malaria (see Behe [8]). In these cases, the above degenerate case offers a sufficient explanation (as do the more general mechanisms of stable-variant merge loop).

Other examples for this degenerate case are with viruses and single cellular organisms like bacteria. These organisms are prolific and reproduce several times in a day. In fact, bacteria can reproduce several times an hour as well. As a result, with a large population of these organisms that mutate in parallel, we have an enormous diversity of mutations in a brief period. These random mutations eventually will introduce features that can help relieve the environmental stress, at least, after thousands of generations. In this case, just the degenerate case is sufficient to understand the discovery of several enzymes (see Blount et. al [14]).

Even though it appears that the degenerate mechanism is convincing for simple single cellular organisms, a more correct explanation for *directed* evolution among them is to use horizontal gene transfer-based SVML iterative process of section 55.1. For multi-cellular organisms, on the other hand, this degenerate case cannot handle the complexity and the diversity of life we observe around us. It is only possible to explain this diversity using the more general version of stable-variant merge loop pattern as we have already seen in sections 55.1 - 55.7.

55.9 Self-sustaining for the longest time

Let me now list advantages and generalizations of stable-variant merge loop applied to stable parallel looped systems beyond what Darwin's theory offers.

Firstly, note that no physical laws are violated if living beings, viewed as dynamical systems, do not survive or evolve towards diversity. Survival as a reason for creating a new 'chemical' feature (like eyes or shorter beaks in finches) not justified. Instead, the hidden structures responsible for survival are: (a) self-sustaining property of stable parallel looped dynamical systems, which automatically enforces survival and (b) stable-variant merge loop with sexual reproduction or horizontal gene transfer, which enforces stability, survival, diversity and a direction.

Secondly, 'survival of the fittest' notion hides details about both survival and fitness. It is unclear what the statement means when we view living beings as dynamical systems. Instead, the statement applicable to all dynamical systems is the notion of 'dynamics self-sustaining for the longest time'. This new notion of *self-sustaining for the longest time* is a clearer and more widely applicable notion that generalizes and replaces the 'survival of the fittest' notion. Stable parallel looped systems described in this book, which contain living beings as a subset, are the correct class of systems for which this statement applies naturally. We can now state loosely that any dynamical system, which continues its dynamics for the longest time to be the fittest system that survived. This notion of 'self-sustaining for the longest time' is a far-reaching generalization of the 'survival of the fittest' in at least a few ways: (a) it extends beyond living beings to include all dynamical systems, (b) it makes the notion of survival well-defined and applicable for all dynamical systems and (c) the concept applies to features of our brain including consciousness, free will, self-awareness, sensing space, passage of time and others described in great detail in this book.

Thirdly, by definition, every looped reaction must necessarily have at least one reaction that requires energy (i.e., the uphill reaction; the downhill reaction occurs naturally – see Figure 53.2). Hence, the only way a feature can be decomposed as a linear sequence of smaller and fittest chemical discoveries spread across hundreds of generations is if the uphill reaction is at the very last step. Otherwise, the energy consuming reaction when the loop is not yet closed is not an advantage to the system. In other words, the probability of guaranteeing a greedy solution (i.e., offspring's of fittest organisms are fittest) is almost zero, except trivially in one or two-step cases, when the underlying architecture is based on stable parallel loops. The fitness must shuffle within the population. Therefore, the less fit organisms *necessarily* contribute towards fitness, as described here, for the discovery of any feature unlike what Darwin's theory suggests. As an example, Darwin's theory would suggest that we quarantine the less fit from the fittest as they will be eliminated anyway, contrary to the mechanism proposed here, which necessarily requires sexual reproduction between fit and less fit.

Fourthly, unlike Darwin's theory, the stable-variant merge loop pattern is applicable for any stable parallel looped dynamical systems whether it has DNA or not. Therefore, this is applicable before the origin of life as much as afterwards. As an example, consider a 'physical' stable parallel looped dynamical system. When subject to a stress, if there is a local damage (from impurities, dislocations, structural weaknesses and stress concentrations), but not a complete collapse, it causes an increase in the degrees of freedom by creating new 'joints' at the cracks (see section 53.4). If the second law effects are slower, some stable parallel looped systems heal by the accumulation of new materials and energy. In addition, the physical repeatable dynamics of the loops adds to the healing process. The merge step is just a simple joining (like entanglement followed by benign integration of the dynamics) of the fit and modified less fit systems. After the elimination of less fit stable parallel looped systems, the iterative process continues (see chapter 53).

Fifthly, genes as memory are not sufficient for a direction in evolution without the larger stable parallel looped chemical network itself (see Claim 3.1 and section 57.1). In addition to the stable parallel looped network as chemical memory, we require two additional sources revealed by the new dynamical network structure of life proposed here, unlike Darwin's theory. The first source is to memorize a long sequence of steps, not just the molecules (like enzymes) within the DNA. For example, genes do not memorize, say, the entire sequence of steps for embryogenesis, geometric memory (like shape of a heart or a memory that hand should be attached at the shoulders not at the stomach) or cellular activity in everyday scenarios. These sequences of steps should be repeatable and is necessary even for basic survival. The cascading chain of steps stops easily or changes to a different pathway if the localization and timing of accumulation of chemicals is altered. The second source of chemical memory is for the creation and adaptation of complex features like multi-cellular organs, for the first and subsequent times. Here, the above source of memory of the sequence of steps is not enough. The rates and timings of the chemical reactions within the sequence should also be memorized for repeatability of creation (so, for example, offspring's of short-beaked finches will continue to have short beaks and the same with long-beaked finches). I will describe the specific mechanisms for both of them in chapter 58.

56

On Creating Life for the First Time

In the previous chapters, we have been discussing different variations of SVML patterns to explain directed evolution in living beings. All of these mechanisms assumed the existence of DNA and its role to create the same proteins repeatedly. More importantly, they already assumed the existence of life. Furthermore, we relied on random mutations in the DNA as the only source to generate new proteins. Our objective now is to start looking at these issues and to start finding better solutions than random chance to explain the origin of life on earth.

Since all features of living beings are primarily manifested through specific chemical reactions executing in a specific order within specific regions and at a specific time, let us pick the most basic question to answer first. This is on how nature runs a single chemical reaction. Specifically, how can nature discover better ways to control the execution of a set of chemical reactions (like glucose combining with O_2 to give CO_2 and H_2O) when there is no life yet? We humans control chemical reactions in a laboratory by setting up the correct conditions like temperatures, pressures and other external conditions as well as by adding suitable catalysts using our consciousness. These conscious choices ensure that a series of chemical reactions occur in a desirable manner to produce desirable outcomes. However, nature does not have this option and yet only a specific reaction needs to occur at a specific time in a specific region to exhibit life-like features. How can nature control and choose which reactions to run without consciousness or awareness? How can nature do this repeatedly and reliably? This is a necessary prerequisite step before nature can create life.

To solve this problem, there are at least two critical topics to address: (a) how can nature control the fetching of the correct molecules to the correct location at the correct time and (b) how can nature control setting up of additional conditions individually for each reaction like temperatures, pressures, volume, pH, energy as well as additional chemicals like catalysts. Only if nature has better control over both these steps can we expect life to form without random chance.

In chapter 59, I will focus on problem (a) and propose minimal structures called active material-energy looped systems to address the issue. In chapter 60, I will discuss problem (b) by providing designs for primitive enzymes that do not require DNA or proteins. Later, I will provide additional designs that help us iteratively create detailed chemical SPL network structure until we create DNA, enzymes, features like reproduction and eventually primitive life forms.

Once we reach this state, we can use any of the above SVML mechanisms to explain directed evolution and the diversity of life. Next, I will discuss the gaps within the problem of diversity of life like, say, with embryogenesis, origin of multi-cellular organisms from unicellular ones, origin of sexual reproduction and the creation of complex relationships between species.

We will see that for both sets of problems, namely, the origin of life and the important transitionary problems within the diversity of life (those that are not entirely addressed by stable-variant merge loop), the SPL architecture will play a fundamental role in providing solutions. In this chapter, I will first discuss several subtle issues that should be addressed with chemical evolution when there is no life on earth. I will present just a few mechanisms in this chapter and defer the discussion on the remaining topics to subsequent chapters.

56.1 On creating primitive cell using intelligent beings

One important requirement when providing solutions to the above problems is to create a system that functions like a living cell *entirely using natural means*. The typical approach we have been taking until now is to use our intelligence and consciousness to create systems as complex as life (see Szostak et. al [151] and Gibson et. al [53]). This approach is limiting because nature needs to create life without 'knowing' the very laws of nature itself. I will show how stable parallel chemical loops can help avoid these limitations. In the current section, I will briefly explore how one might use human intelligence and consciousness to create life-like features though that is not the primary objective.

In general, there are several ways to propose designs using our consciousness. I will pick designs based on stable parallel loops for the current discussion. One reason is that later when we enforce the requirement to avoid intelligence and consciousness, a stable parallel looped approach becomes necessary. Let me begin with the designs discussed in section 2.7 on how to create a few life-like features of a primitive cell. The approach I took was to start by building the simplest, stable and self-sustaining system using looped dynamics. A conceptually simplest one is a closed system that takes external inputs and uses them to generate energy (see Figure 2.12). One important requirement we imposed on this system is that it generates more energy than it consumes. This system consumes energy in several ways. A few of these ways are when pulling the external inputs into the system, when generating and delivering energy (typically, in a circulation loop) to different parts of the subsystem and when pushing all unused byproducts out of the system.

In this example, I did not specify the external inputs, the mechanics for pulling external inputs into the system, the process for generating energy, the mechanics of circulation and exhaust in any detail. I only described the system in an abstract sense. Yet, it is clear that at least one real physical system obeying the abstract description should exist. In fact, there are several ways to physically design such a system, some more efficient than the others. Some could be mechanical systems, others can have electronic components or they could even be biological systems with organic or inorganic components. Each such system can self-sustain as long as the external inputs are present, provided there are no severe disturbances.

An internal combustion engine in your car, at least the one that is based on carburetor instead of those based on the modern fuel injection systems, is one such example that can self-sustain after it starts running, for as long as there is fuel and oxygen rich air. Fuel is sucked into the air stream simply from pressure differences by using appropriate designs for venturi and throttle. A living being though more complex, is quite obviously, another example that obeys the above design principles. In fact, we can design several looped systems that satisfy the conditions and requirements outlined above.

I call such a design an active material-energy looped system i.e., the system actively takes material inputs and generates energy in a looped manner. A few key points about this system are: (a) it actively *fetches* inputs it needs though just from local regions and (b) it generates more energy than it consumes. It is important to distinguish this system with other similar looking systems (say, a motor connected to a wall socket, refrigerators, HVAC systems, which only run their looped dynamics when energy is supplied to them passively). We say that examples like motors and refrigerators are not active material-energy looped systems. The above features (a) and/or (b) are not satisfied for these systems. Some examples like, say, wind turbines do generate energy, but not by actively sucking inputs i.e., (a) is not satisfied even though (b) is satisfied. Human beings, on the other hand, actively pull air into their lungs through breathing, instead of wind or air pushing itself into our body to generate energy.

With the requirements (a) and (b) above for the simple material-energy looped system, we already see that it exhibits primitive and partial life-like behaviors. For example, actively fetching its own energy (or food) from the environment is considered unique to living beings. In addition, since the system has excess energy available (than it consumes), we can easily alter the design (using stable parallel loops) to make it move from locations with a scarce amount of food to locations where there is an excess amount of food (see chapter 63).

Another life-like feature this system exhibits is resisting its own destruction (or 'death') when choked, at least, at a very specific location. The system wants to 'survive' in a mechanical sense. We already saw in section 2.7 that when you choke the inlet or the exhaust, a back pressure is built up within the system that tries to either kick the obstruction away or move itself away from the obstruction (similar to kickback you experience with power tools).

We then discussed briefly, in section 2.7, how an intelligent human could improve this design in subsequent versions to add more life-like features using stable parallel loops. We simply needed to copy as many features from existing living beings as we can. For example, we can add the equivalent of hands and legs, more sensors and actuators, more ways to detect damage, more ways to generate energy by trying to consume other types of food and so on. We have already taken such an approach, at least partially, when designing robots and other mechanical systems. Yet, this approach has the problem that we need an intelligent designer to coordinate all of the complex chemicals correctly. Our objective is to avoid this assumption.

56.2 Transport of chemicals before existence of life

Let us consider the period when there was no life on earth, say, about four billion years ago. The types of organic chemicals and their respective quantities are negligible compared to the present day earth. Furthermore, any organic chemicals present must have been quite dispersed. We want to examine this state of earth to see how continuous occurrence of chemical reactions is important and how having a partial set of enzymes is actually detrimental to the creation of life for the first time.

To start, how can molecules, both favorable and otherwise, physically move from one location to another under these conditions? If these molecules were part of a living being, we already have several elaborate mechanisms to move them from one location to another. For example, animals walk, birds fly, fishes swim, some components of plants are transported through pollination by bees and unicellular organisms have flagella to help them move around. By being part of these systems, the molecules move in a well-structured way over long distances. For short distances too, living beings have several mechanisms like, say, blood circulation systems, neural networks or xylem and phloem (in plants). At microscopic distances, the cells have microtubules and other motor proteins for transport of individual chemicals from one region to another within the cell.

Unfortunately, none of these transport mechanisms are useful 4 billion years ago since there was no life on earth, even though there might be several organic chemicals scattered around. The only transport mechanisms possible at that time are based on random molecular motions, say, through wind, water waves, earthquakes, meteor impacts, volcanic eruptions and so on. Even if we had important chemicals necessary for one of the above transport mechanism, but are dispersed, how can they come together close enough to run and complete a single 'useful' chemical reaction (like, say, a reaction that is part of Kreb's cycle)?

Physical transport of chemicals is critical for both creation and survival of life. At almost every instant in every living being, the correct set of chemicals should be present in the correct concentrations at the correct locations in order to run the correct sequence of chemical reactions. More generally, life is a highly dynamical process. It is simply not enough to discover the correct set of chemicals, place them in a small enough region in the correct concentrations and hope that the system will carry out all the tasks that a living being performs. To avoid such precision, one might think that we could replace a number of chemicals within living beings as long as we do not alter the original network relationships of looped chemical dynamics. For example, consider three chemicals A, B and C in a metabolic network and a corresponding chemical pathway $A \rightarrow B \rightarrow C$. If B is not interacting with any other chemicals in the network, we can alter this pathway to another one like $A \rightarrow B_1 \rightarrow C$. In this case, all we need to do is to discover two new enzymes E_1 and E_2 that catalyzes $A \rightarrow B_1$ and $B_1 \rightarrow C$ respectively.

If transport of chemicals is so critical for life, a lack of mechanisms during this period (before life exists) introduces a nontrivial problem. The only available random transport mechanisms are inherently slow if they have to work correctly. Even the simplest cell requires thousands of chemicals and chemical reactions to occur within a small region of space.

56.3 Discontinuity from few enzymes detrimental to life

The second problem to address before the existence of life is when we have a partial set of enzymes or other types of catalysts. The presence of a subset of catalysts for chemical reactions makes the problem of having the correct chemicals at the correct location much worse before the existence of life. To understand why this is the case, notice that the conditions we face during this period are (i) a shortage of organic chemicals (both in quantity and type) and (ii) a lack of efficient transport mechanisms or the availability of only slow natural random transport mechanisms. If under these conditions we have a fast catalytic reaction, it leads to damaging consequences. All reactants are almost instantaneously converted into products in regions where the catalysts are present. This is similar to an explosion (without the destructive effects from high energies). All the reactants have disappeared completely. For example, CO_2 and H_2O combining to form H_2CO_3 in the presence of the enzyme carbonic anhydrase occurs at about a million times faster than without the enzyme. This is a serious problem because it would have taken a very long while for sufficient concentration of each of the reactants to accumulate using just random transport mechanisms. Yet, the mere presence of the correct enzyme completely annihilates all of the reactants in that region, converting them into products H_2CO_3. We now have to wait for another long period to let all the reactants re-accumulate in this region.

In effect, we have sudden bursts of fast catalytic reactions that occur within seconds followed by no chemical activity for days while all the necessary reactants are accumulating through random mechanisms. This makes it impossible for any interesting set of chemical reactions to occur, let alone anything as complex as life. Even to run 3 or 4 chained set of reactions would be impossible because all the correct reactants should be present at the correct space-time location.

Therefore, we need a notion of 'continuity' among the chained set of chemical reactions occurring within a system in order to exhibit life-like features. The discrete bursts of chemical activity followed by complete inactivity, i.e., a discontinuity in the flow of reactions, are not useful for life. Continuity is not only preferable, but it is necessary under these conditions. It is better for the chemical reactions to occur slowly so the system can have continuity instead of having fast enzymatic reactions that occur in discrete and abrupt jumps. Until we solve (a) the problem with active transport of chemicals and (b) the simultaneous presence of almost all useful enzymes or catalysts, the fast catalytic reactions cannot help with the creation of life.

For example, if DNA helicase unwinds the DNA within seconds while it takes days for the complementary pairs of nucleotides to accumulate and bond to create copies of both helices (say, due to a lack of remaining enzymes), the replication process is bound to fail. The unwound DNA would not stay in this state for so long. They may disintegrate. Or if some motor proteins help chemicals move from one location to another within seconds, but there are no chemicals present at the destination to continue subsequent reactions, these quickly arrived chemicals will get dispersed by the time other important chemicals accumulate slowly (in days) at the destination.

Correctly timing the arrival of chemicals within a given region is quite critical for life-like reactions to occur. Unfortunately, a shortage of chemicals (both in quantities and types) about 4 billion years ago makes it difficult to ensure correct timing if some of the reactions occur extremely fast compared to the others. The differences in the rate of reactions makes it appear as if the system is static most of the time and dynamic only when these fast enzymatic reactions occur. The presence of catalysts when there is a broad shortage of organic chemicals splits the spacetime dynamics into two parts: one containing relatively static events (i.e., slow reactions relying on random transport mechanisms) and the other containing relatively fast dynamical events (i.e., fast catalytic reactions). On the other hand, if either all reactions occur slowly or all reactions occur fast, it seems logical to suggest that the dynamics of chemical reactions are 'smooth' and 'continuous' relative to the local timescales.

To say that a subsystem is dynamical as opposed to static, requires continuity in the rate of reactions among a wide range of reactions occurring over a long enough time period. Since the presence of a partial set of enzymes or catalysts break the continuity due to the shortage of chemicals, life cannot form easily in these conditions. We can, therefore, state *a necessary condition for creating life from inanimate objects as requiring that continuity of chemical dynamics be maintained locally at every instant during the transition to life.*

This statement may sound contrary to our common belief because enzymes are indeed critical for current life forms. What we are claiming is that the transition from the absence of enzymes before the existence of life to the presence of enzymes after the existence of life should be continuous and smooth.

57
Minimal Structures and Mechanisms for Creating Life

The necessary condition stated at the end of the previous chapter implies that continuity is ensured if evolution proceeds by discovering stable chemical loops rather than chemical non-looped pathways. Examples of these loops can be simple reversible reactions at equilibrium or more complex cascading chain of reactions. The former types of loops are common initially when we do not have a diverse set of organic chemicals, but the latter loops become possible as nature discovers several functional groups of organic chemicals.

In this chapter, I will discuss several minimal structures and natural mechanisms that are necessary for creating life from when there was no life on earth. Our goal is to describe processes that occur naturally i.e., we do not want to rely on help from conscious and intelligent beings. We will see that chemical loops have several useful properties and will, therefore, play a critical role towards achieving this goal.

57.1 Chemical loops as basic units of discovery

The smallest unit of discovery by nature is a chemical loop. The rate of reactions in the chemical loop need not be fast because we do not require the existence of enzymes yet. For example, a chemical loop like $A \to B \to C \to A$ is more favorable than a non-looped chemical pathway like $A \to B \to C$ if we are observing for a sufficiently long time. This is because in the former case, the chemicals A, B and C and their concentrations will continue to exist in nature at a given location for a long time even if the rate of reactions are slow. Even if only one of the chemicals, say, C exists while the others i.e., A and B have disappeared at a given location, it is still possible to regenerate A and B eventually using the above looped reactions $C \to A \to B \to C$. In other words, the existence of a 'single' chemical is all that is needed to regenerate 'all' chemicals within the loop.

In the latter case (non-looped chemical pathway), chemical A at a given location will eventually disappear because it will be converted to B and finally to C even if it happens slowly. In other words, the only chemical that will ever remain after a long time is just chemical C.

More generally, if we have a set of chemicals that form a chemical looped network as shown in Jeong et al. [80] and Nicholson [114], it is possible to regenerate all chemicals within this network using just a few types of chemicals. This is true even if none of these reactions within this network occur fast with help from enzymes.

Let me state this result in a more abstract sense. Consider a strongly connected directed graph representing a chemical network (see Figure 57.1). The nodes in the graph represent chemicals while the directed edges represent the direction of

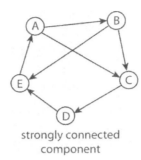

strongly connected
component

Figure 57.1: Sample strongly connected component in a graph. It is possible to reach every node from every other node in this graph. This graph clearly has at least one loop *ABCDEA*.

chemical reaction that is most favorable under a given set of environmental conditions. A graph is said to be strongly connected if there exists a path from every node to every other node in the graph. Such a graph clearly has several loops.

Theorem 57.1 (Self-replication of a collection of molecules): A chemical looped network that is strongly connected can be completely regenerated using just a single chemical within the network.

A single chemical in the above claim does not mean a single molecule. Rather, it means a single chemical type (like, say, glucose) representing a single node within the network. In other words, we can say that a strongly connected chemical looped network becomes extinct if and only if all chemicals within the network have become extinct. We use the term extinct loosely to refer to the disappearance of the chemicals either globally or locally within a region. The presence of even one chemical (in the above sense) would completely regenerate the entire network albeit very slowly (in the absence of enzymes or other catalysts).

If there is a catastrophic event in nature like a meteor strike, volcanic eruption and others that threatens primitive life, Theorem 57.1 suggests that we only need a single chemical for each strongly connected component of a looped network. We can say that life has a natural resiliency because its internal structure is an SPL network.

57.2 Self-replication of a chemical looped network

Theorem 57.1 is novel and useful especially before the existence of life on earth and before the existence of proteins and DNA. Theorem 57.1 can be viewed as *the self-replicating property of a collection of molecules*. The collection of molecules that are self-replicated are the ones that are part of a strongly connected network (see Claim 3.1 of section 3.7). We know that DNA can self-replicate using an elaborate process within living beings (see Purves et. al [123]). However, the problem of discovering the DNA replication process for the first time is a chicken-and-egg problem. This is because DNA replication requires several enzymes to be present while each of these enzymes require DNA to create themselves, i.e., a chicken-and-egg problem. To avoid this chicken-and-egg problem, researchers have been looking for other types of molecules that can self-replicate themselves. RNA world hypothesis is one such model. Researchers have found chemicals termed as RNA replicases that can not only replicate themselves but can self-catalyze their own replication. In other words, RNA replicases act both as catalysts and carry hereditary information. Creating such unique RNA replicases naturally is still an unsolved problem.

Theorem 57.1 suggests yet another type of self-replication that does not involve and require DNA, RNA replicases and other such complex chemicals. Why is this important? There are several reasons why researchers have been requiring special molecules with special properties to address the creation of life for the first time. Some of them are: (a) to address the problem of creating a wide variety of organic chemicals, (b) to create molecules reliably and repeatedly so they can exist for thousands of years or generations, (c) to run a large set of chemical reactions quickly using catalysts instead of slowly otherwise, (d) to generate a wide variety of catalysts in addition to a wide variety of reactants and products of chemical reactions and so on. DNA and enzymes together address the above problems (a)-(d) naturally except that we have a chicken-and-egg problem. RNA replicases and other suggestions also try to address problems (a)-(d) by avoiding the chicken-and-egg problem. Yet, the difficulty is that these special RNA replicases are still too complex to form naturally. In spite of these difficulties, the only suggested mechanism for creating these special molecules is just random chance under suitably identified favorable conditions. The goal of this book is to show an alternate non-random-chance mechanism to create life for the first time. Theorem 57.1 is an important first result in this regard.

Theorem 57.1 addresses item (b) above for now and, therefore, allows us to avoid the chicken-and-egg problem. In the subsequent chapters, I will propose an iterative process that helps us address items (a), (c) and (d) without requiring complex and special molecules like DNA and proteins. The novel aspect of Theorem 57.1 is that we can define and generalize the property of self-replication to other types of chemicals as well, not just restrict this concept to DNA, RNA and other special molecules. The difference, however, is that no single chemical within this strongly connected network is replicated by itself (unlike DNA), but the collection as a whole can be replicated.

If an existing cell needs to replicate, we would need to replicate its DNA. However, if a primitive cell that does not yet have any DNA, needs to replicate, what can we do? If such a primitive cell contains a strongly connected network of chemicals, then Theorem 57.1 shows one way to replicate it. All we need to ensure is that this primitive cell splits to create two primitive cells such that both copies has at least one chemical copied over from the parent. Theorem 57.1 would then imply that all other chemicals within the strongly connected network can be recreated within each copy. In other words, we have successfully replicated the original primitive cell into two copies without the presence of any DNA.

Therefore, chemical loops form an important subset to study during the period

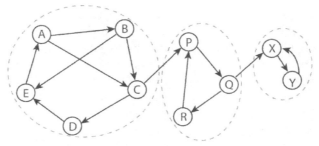

partitioning a graph into strongly connected components

Figure 57.2: A sample graph partitioned into strongly connected components. In this example, we have three strongly connected components as shown. Such a graph partitioning can be found in linear time as well using, say, Tarjan's algorithm.

when there is no life on earth. A single loop of chemicals is the simplest example of a strongly connected network. Therefore, the basic unit of discovery of chemicals by nature are chemical loops, not non-looped chemical pathways. In non-looped chemical pathways, all reactants will become extinct and only the products will remain. A simple analogy in the modern world is that all components that humans use should be recyclable i.e., be part of a strongly connected graph. Otherwise, we will generate a lot of waste, namely, the products of a non-looped pathway.

Let us revisit the requirement of chemical continuity/discontinuity of section 56.3. What will happen if nature has only discovered a partial set of catalysts in this strongly connected network? The part of the strongly connected chemical network that has catalysts will proceed at high rates, converting all of those reactants to products. The remaining chemicals in the network will proceed at a slow rate. In other words, the chemicals that remain are the ones that react slowly. The fast and slow rates of reactions produces discrete jumps in the chemical dynamics. While this is not desirable for exhibiting life-like features, the problem of extinction of chemicals and the production of wastes identified in section 56.3 does not happen anymore. Therefore, chemical loops and looped networks does not eliminate the useful chemicals discovered until then. We can call this as a form of chemical memory.

Until now, we have primarily considered the case of a strongly connected network. How can we generalize the discussion to the case when we do not have a strongly connected network? It is quite common not to have a strongly connected network as nature evolves chemically before the existence of life on earth. It is likely that the chemical network during the last 4 billion years at different spatial regions was non-uniform, sparse and time-varying (see Figure 57.2 for an illustrative example). Any such chemical network can be partitioned into strongly connected regions as shown in Figure 57.2 since strongly connected components form an equivalence class. Let N be the number of strongly connected components in a directed graph. In graph theory, Tarjan's algorithm can be used to find the strongly connected components of a graph (see Harary [65]). For each strongly connected component, Theorem 57.1 suggests that we only need a single chemical to regenerate it. Therefore, the entire network can be regenerated using as many chemicals (i.e., N chemicals) as there are strongly connected components. If a primitive cell contains an arbitrary chemical network, it is possible to replicate it by a simple split as long as each split contains one chemical from each strongly connected component. Even after a natural calamity like meteor strike, volcanic eruption and others that threatens existence of life on earth, primitive life can regenerate itself because we only need to keep N chemicals intact (one from each strongly connected component), not the entire primitive cell.

The evolutionary process of discovery of new organic chemicals proceeds incrementally by expanding the chemical looped network gradually and iteratively. In a theoretical sense, if there is a way to grow the chemical looped network over time, nature can begin to have a directed evolution even in the absence of complex chemicals like DNA and enzymes. In the next several chapters, I will discuss how to create minimal structures (see chapter 59), which subsequently allows us to create more elaborate set of chemicals and looped networks.

57.3 Nonuniform stability – primitive direction in evolution

In section 53.1, we have identified several minimal structures like stability, closed compartments and chemical loops to create order from disorder. Minimal structures are those that have a non-negligible probability of forming naturally. The conditions needed to create them may not be present at all times. They may be present only intermittently. The important requirement is that they should not be unlikely. These conditions may be random, but they should be repeatable, though not necessarily reliably. If the probability of natural occurrence of these favorable environmental conditions is close to zero, then we will exclude any properties that depend on them. In this and the next few sections, I will elaborate some of these minimal structures to show how, using these structures, we can achieve a direction in evolution even when there is no life on earth.

Let us consider a time period when there was no life on earth. We assume that only simple molecules like oxygen, water, carbon dioxide and others are present. From our knowledge of chemistry, we know how different atoms combine to form molecules with, say, ionic or covalent bonds. The molecules have a three-dimensional geometric structure. They have unique physical and chemical properties different from the atomic counterparts. Any standard textbook on chemistry (like Atkins [4], Pauling [120] and McMurry [103]) covers these topics in detail.

One interesting property of a molecule is its stability. The existence of a molecule under a range of environmental conditions (like different temperatures and pressures) implies that it is locally stable. The shape, structure and composition of the stable molecule correspond to a minimum energy configuration under those conditions. The same applies even with the structure of an atom. For example, using quantum mechanics, the electronic configuration within an atom can be described by wave functions specifying the most likely locations of electrons around the nucleus. This spatial and energy distribution is represented by atomic orbitals like s, p, d and f. Molecular orbitals similarly describe electronic configuration of a molecule. Here, different atomic orbitals are mixed to yield stable energy configurations (which includes hybridized orbitals in some cases).

In either case, what is important to realize is that the entire atom or the entire molecule is not uniformly stable. There exist some spatial regions within an atom or a molecule, which are more stable than the other regions. The reason why we even consider non-uniform stability is that each of these regions within an atom or a molecule is actually dynamical, not just a static arrangement of electrons. Systems with moving parts have some that will be more stable than the others. For example, electrons closer to the nucleus of the atom are tightly bound and, hence, more stable than the valence (i.e., outermost) electrons. Similarly, we have σ-bonds that are much more stable than π-bonds in compounds with covalent bonds.

This non-uniform nature of stability is much more pronounced as we look at larger and more complex organic molecules (with higher molecular weight). We, typically, can organize or order the regions within a molecule as increasingly stable regions, in relative terms. For example, electrons closest to the atom are more stable. As you move away, relatively speaking, the stability decreases. When we supply energy in increasing amounts to any molecule, the less stable regions are the first to react to this change. The more energy we supply (or release), the deeper you go within this ordered sequence of stable regions. Atoms and molecules, as a result, find ways to stabilize these less stable regions by reacting with other chemicals in the environment. Chemical reactivities of various molecules can be attributed to this non-uniform stability feature.

If the atoms or molecules are subject to random collisions (say, from large energy supplied through high temperatures and pressures), chemical reactions occur. New products are produced when the atoms and molecules rearrange and lock themselves into these locally stable states. It appears that the set of transitions to locally stable states occur in discrete jumps. The intermediate states between any two such discrete stable states are not stable.

A chemical reaction can now be viewed as a process that takes one set of locally stable chemicals and transition them over to another set of locally stable chemicals under suitable conditions. These conditions are created through random mechanisms, i.e., they are not directed or guided by an intelligent being, especially when there is no life on a planet. Some typical examples of these conditions are: availability of sufficient amount of energy, suitable temperatures, pressures, catalysts and appropriate concentrations. Similarly, the underlying process of the chemical reaction is also through random collisions, except when enzymes are present to assist the reaction (which happens only after life exists).

Even though the conditions for a chemical reaction are randomly created and the mechanism of the chemical reaction itself is random, the molecules themselves that are created are not random at all. We can only have stable molecules. The unstable molecules are ephemeral and are not common. In essence, randomness has given rise to a primitive direction, at least. This can be viewed as the first source of directed evolution (or coordination) in our nature when there was no life on earth. Using the terminology of section 52.3, we can say that non-uniform stability gives rise to a repeatable random event. We would not know which precise set of chemicals would be formed using just these random mechanisms. Nevertheless, we definitely know that all of these molecules are stable under those conditions.

57.4 Formation of chemical loops – direction in evolution

The second important source of primitive direction in evolution, even when there is no life on a planet, is the formation of chemical looped reactions as discussed in sections 57.1 and 57.2. The chemical loops will give rise to repeatability of reactants and products even though they are random events. The actual occurrence of chemical looped reactions, as mentioned earlier, is only intermittent and slow. Nevertheless, the set of all chemicals in the world after millions of years will be in a state where most chemicals will be part of some looped reaction.

Consider the collection of all chemicals present on earth before the existence of life. Each of these chemicals must have been formed from another subset of chemicals within the same collection, by definition. There exists a chemical reaction, i.e., a well-defined mechanism, responsible for generating each of the chemical within this collection. For example, if a chemical A can react with at least one other chemical, the possibility of the reverse reaction ensures that the original chemical A can be generated. Except for the initial set of atoms, all subsequent chemicals have such a generating mechanism obeying the laws of nature. The set of chemical reactions provide a sense of continuity (or interrelationships) between the chemicals within this collection.

The collection of chemicals and the associated chemical reactions, if represented graphically, will form chemical loops similar to the one shown in Nicholson [114]. The rates of these reactions are, in general, slow. There are no enzymes to make these reactions go faster. Yet, what we can be sure of is the existence of chemical looped reactions relating several chemicals.

Among such a chemical looped network, we have three distinct groups of chemicals: (a) chemical inputs i.e., almost no chemical reaction generates them except through reverse reactions, (b) chemical outputs i.e., they do not react with any other chemical except through reverse reactions, and (c) chemical intermediates i.e., products that are linked together via a few chemical loops. The simplest loop is a degenerate one, namely, a reversible chemical reaction under equilibrium conditions. The fact that we only have a finite number of chemicals guarantees the existence of chemical loops. Chemical inputs of type (a) will eventually disappear, say, after a million years, since they will be converted into products, provided the reverse chemical reactions do not persist. Similarly, chemical outputs of type (b) will remain as large amounts of nonreactive waste (assuming the reverse reaction does not happen).

In the beginning, there will be more chemicals of type (a) and (b) i.e., those that do not react with other chemicals as much. They tend to behave like inputs and outputs mentioned above, but not as intermediate products i.e., chemicals of type (c). Over a period, random physical mechanisms like wind, waves, volcanoes and meteor impacts will move chemicals from one location to the next. This will give rise to the possibility of forming more types of chemicals and hence more number of loops, though only using random mechanisms. While we cannot guarantee if a specific chemical loop will form or not, we do know that there will always be a nonempty set of chemical loops.

The existence of looped chemical reactions implies that there is a specific direction of evolution in spite of only random mechanisms. When a chemical loop is formed, each of the chemicals within this loop will be constantly regenerated as long as the external inputs and external energy are present. This is unlike any non-looped chemical reaction. As a result, the chemicals part of a looped reaction end up surviving for a long time (see Theorem 57.1 and sections 3.7 and 57.1 where we mentioned that nature discovers new chemicals in increments of chemical loops).

In other words, even though each chemical is stable enough to exist individually, the existence of looped reactions will force one set of chemicals to be eliminated, namely, those that are part of non-looped reactions. Therefore, earth preferentially has only chemicals that are part of looped reactions. Other chemicals within the collection are eliminated. This gives rise to a specific direction in evolution, although this is a slow process in the beginning. This direction is derived just from randomness as seen in the previous section with non-uniform stability.

57.5 Interactions within looped reactions

Let me now briefly discuss how a set of interacting looped reactions can help create new chemicals that are not part of these loops. I will discuss this abstractly for now. Let chemicals P and Q react to produce chemicals A and B. The chemicals A and B would not sustain for a long time through this chemical reaction alone. If, however, chemicals P and Q can be created through two chemical loops L_1 and L_2 as shown in Figure 57.3, we now have a way to generate A and B as long as the two looped reactions continue to occur.

Another way to look at this is by saying that when multiple chemical loops (say, L_1 and L_2) interact as shown in Figure 57.3, we will create new chemicals like A and B that are not part of the original loops. The creation of chemicals A and B will sustain as long as the looped reactions occur. In other words, the interactions between the looped reactions have enabled the creation of a number of other chemicals that branch off from these loops. These branched chemicals (like A and B) accumulate in sufficient quantities for sufficient time creating the possibility of new chemical looped relationships (see Figure 57.3).

If the loops L_1 and L_2 did not exist, chemicals like A and B would not have had the opportunity to sustain long enough to become part of some new chemical loops. Random transport mechanisms and random conditions are not enough for improving the chance of bringing chemicals A and B to the correct place and at the correct time to allow the possibility of reacting with other existing chemicals. However, since A and B were generated through branches from other chemical loops L_1 and L_2 as shown in Figure 57.3, they are now able to sustain long enough that random mechanisms can allow them to participate in new chemical loops.

As mentioned in the previous sections, we cannot guarantee a 'precise' set of chemicals or the chemical loops with this argument. Rather, we can only guarantee

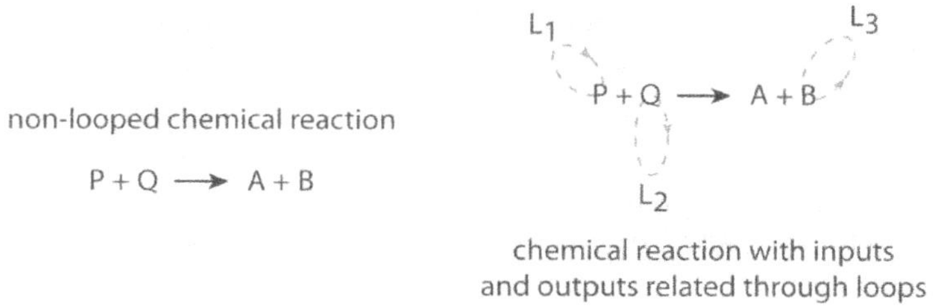

non-looped chemical reaction

$$P + Q \longrightarrow A + B$$

chemical reaction with inputs
and outputs related through loops

Figure 57.3: A chemical reaction related with or without other chemical loops. In the first example, we have a chemical reaction that does not have any interactions with other chemical loops. In the second reaction, chemical P is generated using a loop L_1 and chemical Q using a loop L_2. The outputs A and B can now be generated continuously in the latter reaction because of loops L_1 and L_2. This gives an opportunity for A and B to be part of additional reactions like loop L_3.

the ability to form 'some' new chemicals and new loops that increase in quantities and types over time. The state of chemical evolution moves in a direction to contain stable parallel loops, though slowly in the beginning. It only increases the chance of guaranteeing repeatability as nature incorporates more stable parallel loops.

The presence of carbon and water greatly helps in creating a variety of chemical (organic) compounds owing to the unique properties of these compounds. In section 57.10, I will discuss how stable parallel loops can directly help in creating new chemicals. We will see that structural complexity of organic compounds facilitate in creating new chemicals through stable parallel loops. The complexity created when several stable parallel loops interact has a certain unique and robust structure that is not possible if not for the loops. It turns out that only an SPL architecture allows us to create stable systems that have millions of interacting dynamical subsystems (see chapter 2). This ability is a necessary feature for creating life on earth naturally without any help from an intelligent being. In the next few sections, I will discuss a few additional properties of chemical loops, several of which were already discussed in the previous chapters.

57.6 Regenerating chemical looped network

In section 57.1, we have already addressed the question of regenerating a chemical loop in some detail. The basic question is as follows. If a chemical loop, which is important for a given feature, stops working because of a lack of chemicals, how easy or difficult is it to restart it? Theorem 57.1 showed that we just needed one chemical to be present if the chemical network formed a strongly connected component. The chemicals for all the remaining reactions can be recreated from any of reaction through the cascading chain of the chemical loop (assuming energy is available in abundance). Generalizing to an arbitrary looped network, we just need a few chemicals to be present to regenerate the entire functionality. Without looped reactions, this would not be possible. This convenient property helps us during adverse conditions.

For example, Nicholson [114] shows a highly interconnected network of metabolic pathways in living beings. If we imagine picking just a handful of chemicals from Nicholson [114] and start proceeding along the different chemical pathways, we will see that we can create a large number of chemicals. Even without the enzymes, the reactions do occur, but very slowly. The system will not form a living being without enzymes, but will behave as a large network of slowly occurring reactions. This is one way to see how earth could regenerate most chemicals necessary for life (i.e., an elaborate network of stable parallel looped reactions) even when it faces a catastrophic event that causes the extinction of as much as 99% of all of life.

57.7 Coordinating a sequence of chemical reactions

When we do not have looped reactions, random transport mechanisms as well as the random environmental conditions have a very short spacetime window for the chemical reactions to occur. The presence of looped reactions changes this fundamentally since the chemicals within the loops will now be present for a very long time as the reactions keep repeating slowly and continuously. It is even possible to coordinate multiple chemical reactions using loops within a much larger spacetime window. For example, we have already seen how looped chemical reactions help in coordinating the firing pattern in neurons in section 5.3 via disturbance propagation in a looped network (see Figure 5.5). I will discuss additional examples of disturbance propagation in looped networks in chapter 58.

Consider the problem in which we require two chemicals A and B to be

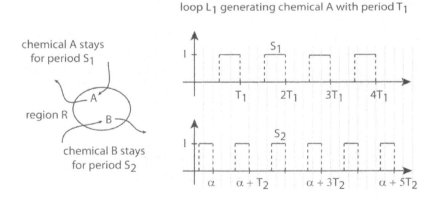

Figure 57.4: Evaluating when two chemicals A and B can be present simultaneously in a given region R. If both chemicals A and B are not generated in a looped manner, the chances that they will be in a given region R is quite low. However, if they are generated repeatedly using two chemical loops L_1 and L_2 respectively with periods T_1 and T_2, then they can be present simultaneously a large number of times.

simultaneously present in a small region R in order to create a specific feature (see Figure 57.4). Most practical examples require considerably more than two chemicals to be simultaneously present. Furthermore, these chemicals need to be present in region R in a specific order. For example, if you cut your finger, you need both the blood clotting mechanism and the antibody generation to occur simultaneously to heal the wound. When creating a specific protein, we need a number of chemicals to be simultaneously present (like ribosomes, specific enzymes, amino acids, DNA and others). In fact, most processes that occur during embryogenesis (when creating an adult from an embryo) and during normal everyday chemical activity requires a great deal of coordination of multiple chemicals spread across different regions of our body.

If chemicals A and B are generated through non-looped mechanisms, then ensuring that both reach the same region R at the same time is highly improbable. This issue gets even worse if we were to require several chemicals A_1, A_2, A_3, ..., A_n to be simultaneously present (like, say, during the creation of a protein). The probability of all of these chemicals reaching the same region R at the same time in the correct order is almost zero if each chemical is produced using non-looped mechanisms. Nature has to be too lucky for this to happen. The other option would be if nature gets help from an intelligent designer. A mere abundance of chemicals is not sufficient because the order of coordination is equally important. How can we solve this problem without relying on luck and random chance? The answer is by using looped processes.

Let us assume that chemicals A and B are generated through a looped process with a time period of repetition as T_1 and T_2 respectively. Let us assume that chemical A stays in region R for a period of S_1 seconds while chemical B stays for S_2 seconds (see Figure 57.4). Let us plot the timeline for each chemical in Figure 57.4 as the processes repeat. In Figure 57.4, I have depicted the y-value as 1 when a chemical reaches region R and 0 otherwise. Therefore, chemical A stays as 1 for a period of S_1 seconds and chemical B for a period of S_2 seconds. Our goal is to understand how often both A and B are at the region R simultaneously i.e., how often the 1-value intervals overlap.

The interval for chemical A when it reaches region R is $[nT_1 - S_1, nT_1]$ where n is the nth repetition. Similarly, for chemical B, it is $[\alpha + mT_2 - S_2, \alpha + mT_2]$ where m is the mth repetition. We want to find n and m such that these two intervals overlap. This happens if and only if $\alpha + mT_2 - S_2 \leq nT_1$ and $nT_1 - S_1 \leq \alpha + mT_2$. These conditions can be stated as: $\alpha - S_2 \leq nT_1 - mT_2 \leq \alpha + S_1$. This condition is clearly satisfied by an infinite number of n's and m's (see Figure 57.4 which shows the non-empty overlap regions using the grid lines).

Therefore, by having active looped processes for bringing chemicals A and B to region R, there are several opportunities for both of them to be simultaneously present and hence create a specific feature, unlike non-looped processes. In reality, the time periods T_1, T_2, S_1 and S_2 vary with time. We also have require multiple chemicals to be present at the same time in region R. Nevertheless, our intuition (and using a similar computation), we can see that by having active looped processes, the chance that these chemicals meet in a given region R can be quite high.

More generally, if the problem is to coordinate a sequence of operations, then having each operation be part of a loop along with having a set of overlapping loops allows us to solve this problem using just natural mechanisms even in the presence of sufficient randomness. As the chemicals in one loop repeat, the randomness in the environment might disperse several of these chemicals in unpredictable ways. The same happens with the chemicals in the neighboring loop. However, since both loops repeat several times, we still have a sufficiently high probability for chemicals in the overlapping parts of the loop to be present simultaneously at least a few times (see Figure 57.4). This causes the transition to occur from one loop to the next with a sufficiently high probability. In this way, the entire sequence of operations can occur a bit more reliably.

We have, therefore, increased the probability of occurrence of the combined system by choosing interconnected chemical loops than by running the loops separately and disjointly. The low probabilities are eliminated because these chemicals will get a chance to meet again when the loops repeat even if they have missed once. The original repeatable random event starts to become increasingly reliable. Had there been no loops for individual sequence of operations, the chance of chemicals occurring exactly once (since there are no looped repetitions) at the correct location and at the correct time to ensure the entire sequence proceeds in a coordinated way using random mechanisms is almost zero. We usually dismiss the natural occurrence of a long complex sequence of tasks as a chance or a lucky event. However, when we have a looped network, we have seen that we no longer have just one lucky opportunity at performing a complex sequence of tasks.

57.8 Life-like features derived from stable parallel loops

The most important difference between the approaches suggested here to create life (both synthetically and naturally from inanimate objects) versus the approaches we have been taking when imitating features of living systems through man-made machines is that we are now requiring stable parallel loops as the fundamental architecture. Any abstract feature that the system exhibits should be the 'outcome' of our design rather than an attempt to introduce the feature into the design intentionally. For example, features like seeking your own inputs (or food) or resisting your own destruction (or death) should not be actively designed for. Instead, they should be the natural outcome of a design like, say, the self-sustaining active material-energy loop (see chapter 59).

If we do not take such an approach and instead 'design' each feature, then there is a natural question of who the 'designer' is. In this book, the goal is to never design a feature explicitly. Every life-like feature like survival, death, seeking your own food, evading danger, breathing under water, flying and so on will be created as natural outcomes of a suitable SPL architecture. I will discuss several such features later in chapter 63. With stable parallel looped architecture, we try to consider all dynamics by recycling inputs as much as possible. In this sense, we will see that an appropriate set of interconnected stable parallel loops span all complex features of living beings.

57.9 Invariance with looped systems

The fact that we are deriving abstract features of life using interconnected looped systems implies that life as it occurred on earth for the first time is not special. The designs can be based on a completely different set of chemicals or, more drastically, even on electronic or mechanical components. There is no doubt that organic chemistry is the best known and the easiest way of generating complexity entirely naturally through stable parallel loops. Yet, there are other natural and synthetic ways to create complex stable parallel looped structures.

The natural ways occur through a different combination and spatiotemporal distribution of favorable and harsh conditions over a period of billions of years than how they occurred on earth. On an alien planet, this can alter the choice of complex molecules and hence result in a different set of life-like features through a different stable parallel looped network. The synthetic variations in evolution are introduced when an intelligent being interferes with natural processes and consciously modifies them in a specific direction. The extreme case involves designing complex looped systems using new materials altogether. The designs suggested here (see chapters 59 and later) are at a systems level. They highlight an invariance on environmental conditions more clearly.

In fact, when searching for extra-terrestrial life, we can start looking at the complexity in the looped dynamics on the planet instead of specifically looking for organic chemicals or a specific cellular structure. This will give an estimate of how evolved or how far along the planet is from creating features similar to what we identify as life.

The properties of looped chemical reactions discussed in sections 57.1 - 57.8 are useful provided we have enough chemicals to begin with. If the mechanisms for creating new chemicals are based on random environmental conditions alone, then the processes with be too slow to accumulate enough chemicals in a reasonable amount of time. What we need is a coordinated, non-random and active mechanism to create new chemicals. This will be the focus in chapter 59. I will now discuss how to use chemical loops to *create* new chemicals.

57.10 Iterative way to create new chemicals using loops

In the previous sections, we have seen how chemical loops can give rise to a direction in evolution even in the primitive world. However, there is one problem. The set of chemicals before life began on earth are quite few in number. All living beings, on the other hand, have a large variety of organic chemicals. Until we have a way to explain how nature can go from such a small number of complex chemicals to such a wide collection of chemicals, we will not be generating a large chemical looped network necessary for life. Therefore, creating life is nontrivial until we have a threshold set of chemicals. We want to understand how this transition happens quick enough. Organic compounds have very useful properties that make them suitable for creating both a large variety as well as complex sized molecules. However, if all we have to rely on are random transport mechanisms and random environmental conditions, the possibility of creating such a diverse set of chemicals that sustain long enough to react constructively to form a primitive living cell is extremely unlikely.

We need a coordinated mechanism for creating new chemicals instead of relying only on randomness. In this section, I will discuss an iterative mechanism to create new chemicals using the very chemical dynamics of stable parallel loops. Therefore, we use chemical loops to create new chemicals, which allow them to form more loops, which in turn creates further new chemicals and so on, in a self-sustaining recursive way. Using this mechanism, nature can now build a large repository of organic chemicals necessary for the creation of life over, say, millions of years. This process is very slow because of random transport mechanisms during the period before life existed. However, as the number of new chemicals accumulates, the process accelerates. The repeatability of looped reactions makes the probability of creating new chemicals higher and higher.

One mechanism we humans use to create new chemicals is by adjusting the environmental conditions like temperature, pressure and concentration of various chemicals causing them to react and create new products. Here, we are proposing another way to create new chemicals, namely, using dynamical looped chemical reactions. With the looped reactions based approach, the new chemicals created are neither the products nor the reactants of any chemical reaction within the loop. Instead, they are analogous to traces of other chemicals present when a chemical reaction occurs. For example, when H_2 and O_2 combine to form H_2O, we also see traces of H_2O_2 formed as well.

In general, these trace elements like H_2O_2 are in negligible quantities. Nevertheless, the formation of trace elements is actually quite common from almost every chemical reaction. Sometimes, they do not stay in those states for a long time. They either breakdown or react with other chemicals to form more stable compounds. The important result is that if we generate the trace elements through looped chemical reactions, they will no longer be in negligible quantities. As the looped reactions repeat multiple times, these chemicals will slowly start to accumulate into significant amounts, especially, if the loops occur for hundreds or thousands of years. Besides, they will be constantly regenerated as long as the looped reactions continue to occur, i.e., without ever being eliminated completely.

This is an important process for creating new chemicals. Let me now briefly discuss the underlying mechanisms for creating these trace elements naturally.

Imperfections with random collisions: One obvious way these trace elements are created is simply through random processes. Random collisions among the reactants to generate products are not always perfect. Sometimes the orientations, the number of molecules and the energies of the collision may not be within the desired range. This results in imperfect collisions giving rise to new stable chemicals. These chemicals are not created in bulk unlike the normal products of a chemical reaction. Instead, these are minority products that you hardly notice. Nevertheless, they always exist.

Amplification of non-uniformly stable parts of molecules through dynamics and loops: A second more important mechanism for creating these trace elements comes from the existence of non-uniform internally stable parts of all molecules. Recall the discussion in section 57.3 on non-uniform stability of atoms and molecules. Any atom or molecule has regions that are relatively more stable than the others. For example, orbitals of electrons closer to the nucleus are relatively more stable than the ones that are farther from the nucleus. Similarly, σ–bonds are more stable than π–bonds in covalently bonded molecules. For complex organic molecules, this non-uniform stability is extremely common.

Therefore, as molecules repeatedly keep reacting to generate products as part of looped reactions, the atoms and molecules involved are in constant and vigorous motion. As the molecules wiggle and jiggle continuously, you can expect the relatively less stable regions to breakdown or react with neighboring molecules. This is similar to how a physical system with a lot of moving parts, some more stable than the others, start to interlock with each other at the relatively less stable locations. Had these systems not been reacting chemically or physically, i.e., had they been just relatively static in one region, you would rarely see additional breakdown of molecules or unusual reactivity.

For example, if an organic compound has a double bond (one σ–bond and one π–bond), the π–bond has a chance of reacting with, say, a hydrogen molecule. This can happen even if the current chemical reaction does not involve reacting with hydrogen molecule. The new compound is just a trace element in this case. The dynamics of the chemical reaction has exposed the non-uniform stability of the double bond in the original compound. The amplification of non-uniformly stable parts of molecules is the second nonrandom mechanism for creating new chemicals.

To be clear, there are two parts that makes this mechanism work. The first is the dynamical nature of chemical reactions as opposed to nonreactive static nature of chemicals like when they are sitting in a flask. The dynamics causes the relatively less stable parts of the molecules to react and yield minute traces of new chemicals. Without dynamics, you would need to resort to random mechanism discussed above. The second is the presence of chemical loops that creates repeatability of these chemical reactions. This repeatability ensures accumulation of the trace elements. The looped reactions will eventually generate a significant amount of new chemicals. Without loops, the traces of new chemicals will be insignificant to participate in any interesting reactions.

Since the formation of chemical loops is quite natural even before the origin of life, as seen in the previous section, the second mechanism (amplification of non-uniformly stable parts) will be more prominent than the random mechanism (from imperfect collisions). Yet, together we now have a natural way to create new chemicals. I want to show that using the amplification mechanism above, there is both a directed and an iterative way to create a large number of new chemicals.

57.11 Non-uniform stability to create new chemicals

Let me now list a set of useful features of the non-uniform stability based mechanism discussed above for creating new chemicals.

Loops and dynamics versus statics: If we take a test tube filled with chemicals, one of the first things that will happen is some of the chemicals within the test tube will react to produce products. If the chemicals are not as reactive under normal environmental conditions, we would rarely see new chemicals produced spontaneously from the existing ones. For example, if you place a number of amino acids, glucose and other organic chemicals, they would rarely react with each other, especially since we did not include any enzymes into the test tube. We would rarely see amino acids combine spontaneously to form peptide bonds. Even if we wait for several days, it is still highly unlikely that we would see any new chemical within the test tube. Only the set of chemicals introduced initially will be present in the same state for all of these days. Even if we supply energy in reasonable quantities, we do not see significant chemical activity to produce new chemicals. The supplied energy is not essentially consumed.

Instead, let us say that we have the test tube filled with chemicals that do react to form dynamical looped chemical reactions. In this case, we would need to supply energy and/or inputs constantly and incrementally so the looped reactions can continue for a while. The existence of chemical dynamics as well as loops in this case will give rise to new chemicals in the test tube, at least say, after a few days. This is simply because of the amplification of the non-uniformly stable parts of molecules i.e., the mechanism discussed earlier in section 57.10.

When studying the problem of origin of life, we need to perform realistic experiments by including chemical dynamics as well as looped reactions. This will increase the chance of creating new chemicals compared to when we just adjust environmental conditions in a beaker filled with static and less reactive chemicals. When simulating conditions of the primitive earth, say, corresponding to 4 billion years ago, we must account for this important mechanism.

Feasibility even in the absence of enzymes: Before the existence of life, the looped reactions are slow because of a lack of enzymes. Yet, even in this case, the combination of two parts, namely, the dynamics to create traces of new chemicals followed by amplification of these trace elements through loops, gives rise to new chemicals. The slow rate of looped reactions makes the creation of new chemicals using this mechanism correspondingly slower. Nevertheless, this mechanism is more dominant than the random collision mechanism described earlier for creating new chemicals.

Analogous physical mechanism in our brain: A physical system like our brain exhibits features like intelligence and consciousness only if it creates complex interconnected loops of neurons as discussed in the previous chapters. Similarly, a chemical system like a cell exhibits life-like features only if there are complex interconnected chemical loops, which in turn, necessarily requires creating new chemicals. What is interesting to see is that both mechanisms (in the brain and in a chemical system) for creating interconnected loops are similar.

In the brain, I have already discussed this mechanism in chapter 6. We need an ability for growth cones to extend the axons from one neuron to a neighboring one. These interconnections enable the formation of physical loops in our brain. In a chemical system, we need the mechanism discussed here for creating new chemicals in order to form new chemical loops. The growth cones are initiated from non-uniform surface defects on the axons amplified by a highly dynamical environmental conditions (see section 6.2 for more details). This is analogous to the amplification in the non-uniformly stable regions (or defects) in chemicals through the highly dynamical looped reactions.

Set of new chemicals created from a looped reaction: Let us pick a looped reaction and the corresponding set of chemicals *I*. We want to look at the set of new chemicals created using the mechanism described in this section. Initially, the dynamics and the looped reaction creates a set *A* of new chemicals. As the looped reaction repeats, we are not limited to creating just set *A* of chemicals. The set *A* and *I* of chemicals can give rise to a second set *B* of chemicals using the same mechanism since all we needed was dynamics and loops to create the first set *A* of chemicals. The dynamical collisions between chemicals *A* and *I* can give rise to set *B* of new chemicals and so on. In this way, as long as the looped reactions continue for a long time, the mechanism automatically explores to create a wide range of new chemicals. After a while, either no more new chemicals are created or the looped reaction stops due to lack of energy or inputs.

Interaction with the second law of thermodynamics: The effect of the second law of thermodynamics is to bring disorder into a system. If the system is a large collection of complex molecules, the second law effects will slowly breakdown these molecules into simpler constituents. This is analogous to taking stable molecules like alkanes and breaking down into less stable molecules like alkenes or alkynes, i.e., creating more disorder by some sort of wear and tear.

The mechanism of this section, using dynamics and loops to create new molecules, is capable of performing the opposite of what second law does. While we also create less stable molecules, what is interesting is that we can take the less stable parts of a molecule and react with neighboring molecules to create new and more stable molecules. This mechanism is analogous to shaking up a system (using dynamics) to kick parts locked in a locally stable state and move them into a better and, possibly, globally stable states. This method is similar to annealing. The second law and the mechanism described here, together forms an iterative process to generate new chemicals continuously.

Iterative process for growing new chemicals: With the above non-uniform stability-based mechanism for creating new chemicals, we now have a cascading effect in the following sense. The looped dynamical reactions create new chemicals and the new chemicals give rise to more interconnected looped reactions, which once again create new chemicals and so on. This iterative process now repeats almost indefinitely (see Figure 57.5). We can think of this iterative process as a way to grow new chemicals. To initiate the iterative process, we do need a threshold set of chemicals and, hence, a sufficient number of looped reactions. The set of chemicals created and the complexity of looped network can potentially grow exponentially, especially for complex organic molecules.

There is a limiting factor though, beyond the obvious requirement of the availability of energy and inputs for running the necessary chemical loops. When there are no enzymes, this limit comes from the slow rate of reactions. When there

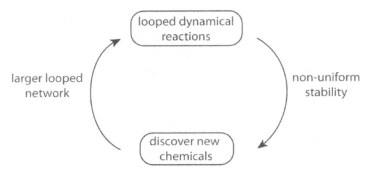

Figure 57.5: An iterative process for growing new chemicals using non-uniform stability mechanism. As the chemical reactions occur repeatedly, the nonuniform stability will create additional trace molecules whose concentrations improve over time. They begin to react with other chemicals to increase the size of the looped network, which in turn allows more looped reactions to occur. The process now repeats.

are enzymes, the limit is from the discovery and the timely creation of the required enzymes to keep up with the fast rates of enzymatic reactions.

Creating chemicals in a directed way: There are several types of trace elements that will be generated using the mechanism described here. Which specific trace element is picked and amplified seems quite random for now. However, I want to suggest that this process is, in fact, directed in a specific way. To understand this, let us first recall the analogy with the brain discussed above. When neighboring axons fire, they influence the surface defects and hence the corresponding growth cones of a given neuron to move preferentially towards that direction. The growth cone itself may initially be created at random from surface defects. However, as it grows, it moves not in random directions connecting to some random neuron. Instead, it moves towards neurons that fire simultaneously with this neuron. This is the reason why simultaneous events are linked together in our brain (see chapter 6).

With creation of trace chemicals, the non-uniformly stable defects will be exposed when the loops continue running for a long time. Consider a situation when a given chemical looped reaction occurs at the same time as another set of neighboring looped reactions, at least, a few times repeatedly (cf. simultaneous firing of neighboring fixed sets in our brain). The simultaneity causes the neighboring looped chemicals to preferentially influence the creation of only those specific trace chemicals that can react or align with these existing chemicals. Conversely, the given looped reaction will generate trace chemicals that react with neighboring looped reactions.

In essence, there is a natural tendency to link these two sets of looped reactions by preferentially creating new appropriate trace chemicals. Of course, this approach is not guaranteed to work for all loops and all chemicals involved. However, with a wide range of organic chemicals, the chances of this happening are quite high (see Nicholson [114] of metabolic pathways). This is a natural direction in evolution to create increasingly complex stable parallel loops.

Given all of these features, it is clear that dynamics and loops are critical for life-like complexity to form. A dead body has lost most of its dynamics and looped reactions. Therefore, restarting those using natural mechanisms is almost impossible, though they may be possible with help from intelligent conscious beings. If we truly want to slow the dynamics, we do it in a proper way using, say, hibernation, resting or sleeping instead of death.

In addition to guaranteeing a clear direction and developing an iterative process to create new chemicals, I will discuss yet another important advantage of looped systems next, namely, on how to chain a long sequence of events using disturbance propagation within a chemical looped network. This next advantage will help create complex features that involves chaining a long sequence of causal events in living beings, which we seem to dismiss for now as coincidences.

58
Memorizing Long Sequence of Steps

We have seen in section 53.1 that formation of stable chemical compounds is one of the first primitive directions in evolution using just random collisions. The next important direction in evolution came from the formation of chemical loops that occurs naturally when you have enough variety of chemical compounds and a sufficient amount of energy (section 57.1). In section 57.10, we have also seen that the dynamics of chemical loops itself can help create new chemicals in a directed and iterative way. Each of these advantages of loops discussed in chapter 3 and 57 suggest that creation and evolution of life is not as random as is commonly believed. The goal of this chapter is to identify more such structure.

The problem we consider here is how living beings are capable of following a specific ordered sequence of steps when performing most tasks. For example, the firing of a single neuron requires a long cascading sequence of steps to be followed until the release of neurotransmitters to the neighboring cell (see Figure 5.6). Similarly, to create a single enzyme, a cell follows a long sequence of steps involving transcription and translation. Other examples are cell division, operation of any organ in our body and birth of a human being, with some more involved than the others.

Whenever we view a task as a sequence of steps, we are tempted to ask how nature created or designed such a sequence for the first time. There seems to be a big difference between creating this sequence for the first time versus simply following it for all subsequent times. Creation of a sequence of steps seems to require a level of awareness or intelligence. Random coincidences to create a long sequence of steps sound unconvincing. Most of the times, a precise sequence of steps appear necessary for the corresponding task. Alternate ways of doing the same task rarely exist, as can be seen with each of the examples given above. Therefore, every time we ask how a specific sequence of steps was designed for a given task, we start to feel that life is too complex to form by itself using natural means. It appears that too many steps need to be coordinated too perfectly even for a single feature in a living being.

In this chapter, I want to show how to overcome this difficulty by expressing the same task (and hence the same sequence of steps) as a disturbance that propagates through an appropriately chosen network of loops. Stable parallel loops will once again provide an alternate solution to random coincidences. The choice of network of loops when propagating disturbance will be obvious when we identify a given sequence of steps. I will show how to do this translation from one representation (sequence of steps) to the other (disturbance propagation in a network of loops). This translation is not difficult or new. What is new and powerful is the representation as *disturbance propagation in a network of loops* for *all* features of cellular systems. Though the translation may seem like a trivial exercise in the beginning, you will soon see that it is powerful enough to explain the natural mechanisms of both the origin and the creation of several complex features of life. Unlike the case of our brain's neural network where we have a clear SPL network, for chemical systems like cells, we only have a 'virtual' SPL network (i.e., metabolic pathways of Jeong et al. [80] and Nicholson [114]). The disturbance propagation referred here is within this virtual SPL network.

One new outcome of translating a sequence of steps into disturbance propagation in a network of loops is that chemical SPL systems have a new way of *memorizing* a sequence of steps within its SPL network without depending on DNA. Chemical memory is usually attributed to DNA in living beings. To reproduce a cell, to fight off a bacterial infection or to digest food, we think that DNA memorizes these steps and reliably transmits them to an offspring. However, we will see that this is an incomplete view. The chemical SPL network and disturbance propagation within an SPL network (which includes memorizing DNA as well) is the true and complete source of chemical memory for any sequence of steps. This addresses one of the main puzzle with the origin of life: how does nature memorize useful but long sequence of steps even if discovered accidentally when there is no DNA (which acts as memory during replication)? Memory of a long sequence of steps is *embedded* within the same SPL network.

For nonliving systems, representing a linear sequence of steps within a suitable network of loops is either not always possible or is unwieldy. For example, taking the TV remote, turning the TV on, opening the channel guide, picking a specific channel and so on are a linear sequence of steps to watch a TV show. We do not express them as disturbance propagation within a network of loops. Such sequential steps are quite common in our everyday life. Rather than viewing them as a network of repeatable loops, we leave them as a linear list of items or represent them graphically as flowcharts. More generally, for systems like humans that use free will, it seems rather odd to represent a sequence of steps within a network of looped systems as the events are rarely repeatable. On the other hand, features exhibited by single and multi-cellular organisms at a chemical cellular level are naturally repeatable and, hence, can be represented in the language of loops. Let us start with a few examples of how to do this.

58.1 Examples of dynamics within a network of loops

I will now discuss a few examples briefly to illustrate how the dynamics of a sequence of steps can be translated into the propagation of a disturbance in an appropriately chosen network of loops. As mentioned earlier, identifying the network of loops should be obvious for a given sequence of steps. One of the example is the same one discussed in section 5.1, i.e., the sequence of steps that occur when a neuron fires and releases neurotransmitters onto a neighboring neuron (see Figure 58.1). These steps are as follows:

1. A stimulus, say, a chemical, pressure, stretch or a membrane potential, triggers several ion channels on the nerve cell membrane to open. This causes a flow of ions into the neuron.
2. Ions accumulate in the cell when several neuron sources open multiple ion

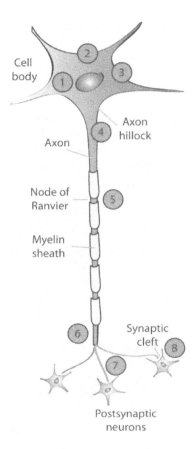

Figure 58.1: Looped processes within a neuron. The details of its operation is described both in Figure 5.3 and here as steps 1 through 8.

channels at about the same time. These ions reach the axon hillock.

3. The density of sodium ion channels is greatest at the axon hillock. If more Na^+ ions enter than K^+ ions leaving, the membrane potential increase and can cross a threshold value. When this happens, several more Na^+ channels open at the axon hillock that generates an action potential.

4. The jump in the Na^+ ions causes the voltage spike to travel along the axon. At the first node of Ranvier, the membrane potential once again exceeds the threshold value causing a large number of Na^+ channels to open. This produces a voltage spike of the same magnitude.

5. In this manner, a voltage spike appears as a disturbance that propagates along the axon. At each subsequent node of Ranvier, a jump in voltage potential causes Na^+ ion channels to open locally creating a voltage spike of the same magnitude. This propagation continues until the voltage spike disturbance reaches the axon terminal.

6. At the axon terminal, the voltage spike causes several voltage gated Ca^{2+} channels to be opened. This causes a sudden influx of Ca^{2+} ions at the axon terminal.

7. The Ca^{2+} ions bind to the surface proteins of vesicles. They trigger reactions that make the vesicles to open up and release neurotransmitters into the synaptic cleft.

8. The neurotransmitters diffuse through the synaptic gap and bind to the surface proteins of the post-synaptic neuron. This forms a stimulus on the post-synaptic neuron to open a number of ion channels. The above steps now repeat on the post-synaptic neuron.

Let me now outline the same steps above as disturbance propagation in an appropriately chosen network of loops (see Figure 58.2). When choosing the loops, we should ensure that each loop is a stable one. By this, I mean that if we perturb the loop by a small amount, there should be enough mechanisms in place to help the system restore its original state.

1. When the ion channels open up and let Na^+ ions flow into the neuron, this causes a disturbance to the existing stable state of the neuron. The original balance will eventually be restored using Na-K pumps in the neuron. This is one of the looped systems. The stimulus mentioned earlier creates an initial disturbance to the stable condition of this looped system (see Figure 58.2).

2. However, before the local equilibrium condition is restored, the disturbance is propagated within the neuron to the axon hillock. A local stable loop here uses the same mechanism of Na-K pumps to maintain a steady-state condition. In this state, the accumulation of the ions at the axon hillock does not result in increasing the membrane potential beyond the threshold value. In this case, the stable state will be eventually restored.

3. Once again, before the locally stable state is restored, the membrane potential sometimes exceeds a threshold value. The rate of accumulation of ions in this case turns out to be much faster. The disturbance that has reached the axon hillock now gets to propagate beyond, by opening a much larger set of Na^+ ion channels at the axon hillock. This disturbance shows up as an action potential. At the axon hillock, there is a natural stable loop as well. Under normal conditions, the Na^+ ions entering and the K^+ ions leaving the neuron are balanced to maintain a resting potential.

4. At each node of Ranvier, a local loop maintains a stable state of ion flow in and out of the axon. However, before the effect of the disturbance at the axon hillock is nullified, the influx of Na^+ ions propagates along the axon. They reach the first node of Ranvier. This causes a disturbance in the stable loop at the node. This disturbance opens a number of Na^+ ion channels locally and creates a sudden increase in membrane potential. This results as an action potential of the same magnitude at this node.

5. The local disturbance at each node of Ranvier propagates to the next node. The Na^+ and K^+ ions are then restored back to their normal state after the disturbance propagates further along the axon. In this way, there is a local

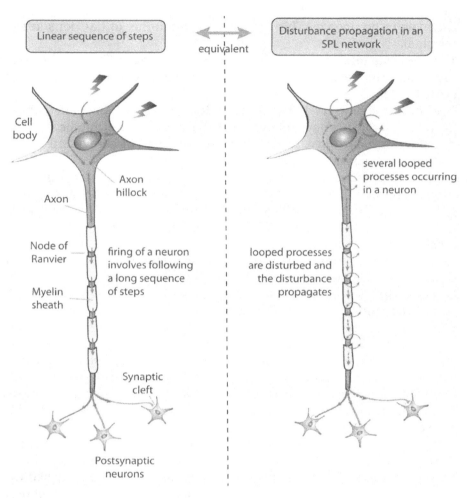

Figure 58.2: A linear sequence of steps for the operation of a neuron represented equivalently as disturbance propagation in an SPL network. On the left, the operation of a neuron is represented in the traditional way as a linear sequence of steps. On the right, we represent them as disturbance propagation within the SPL network of a neuron.

stable loop at each node that helps the disturbance reaching a node to be propagated to the next node. The propagation happens before a given node has a chance to restore to its original stable state. This is analogous to how water wave disturbance propagates. The voltage spikes represent the disturbances that appear to propagate as a wave along the axon until it reaches the axon terminal.

6. At the axon terminal, there exists a local stable loop that maintains a certain concentration of Ca^{2+} ions. However, the disturbance that travelled along the axon creates a significant potential difference with the exterior of the cell. This

membrane potential triggers new voltage gated Ca^{2+} ion channels to open up. In other words, the Na^+ ion disturbance was converted into a Ca^{2+} ion disturbance at the axon terminal (see Figure 58.2).

7. Before the effect of the Ca^{2+} ion disturbance could be nullified, the Ca^{2+} ions bind to the membrane proteins of the vesicles that contain neurotransmitters. Now, the creation and the maintenance of the vesicles at the axon terminal is itself a stable loop. The neurotransmitters themselves are created within the nerve cell and are transmitted along the axon via microtubules. This process is controlled through a series of stable loops. Once these chemicals reach the axon terminal, they are packaged as vesicles. There is a stable state maintained for this process. However, the disturbance in Ca^{2+} ions causes the vesicles to open up and release the neurotransmitters into the synaptic cleft. In other words, the Ca^{2+} ion disturbance was converted into a disturbance in the vesicle looped system.

8. Several parallel processes occur now from multiple looped systems. One looped system is with the post-synaptic cell where the disturbance from the neurotransmitters opens ion channels as they bind to the surface proteins. This is same as Step (1) above but occurring on the post-synaptic cell. Some of the neurotransmitters unbind from the post-synaptic cell just from thermal effects. The second looped system is from the disturbance in the re-intake of neurotransmitters by the presynaptic cell. A related looped system involves metabolic breakdown of the neurotransmitters caused from the same disturbance. In this way, we see that one initial disturbance can trigger multiple parallel disturbances as well (see Figure 58.2). The disturbance now propagates through the post-synaptic neurons.

As we can see from this example, we can translate a linear sequence of steps as disturbance propagation within a suitable network of physical and chemical loops. We then showed how an initial disturbance is converted into suitable forms as it propagates within this network in a specific way.

We have seen two instances of parallel propagation in this example. In one scenario, the axon can have multiple branches leading to several different neighboring neurons. This natural parallel cascading effect is at a physical level because of the branching of the dendrites. Another scenario we have seen is at a chemical level. The production of vesicles, creation of neurotransmitters, uptake or breakdown of neurotransmitters from the synaptic cleft and the ion channel imbalance in the postsynaptic cell are triggered in parallel from a single disturbance, namely, from the release of neurotransmitters into the synaptic cleft.

I have skipped several detailed sets of chemical reactions that occur within the neuron for the sake of clarity. This includes creation of several enzymes and other necessary intermediate chemicals. However, each of those steps can similarly be mapped as disturbance propagation in an appropriately chosen network of loops. As an example, it is an interesting exercise to try to identify a network of loops, along with the appropriate disturbances, corresponding to the sequence of steps for the creation of a specific enzyme. I will briefly list a few of these loops to get the basic idea.

The sequence of steps followed to create an enzyme include transcription and translation. These steps also have a set of interconnected and regulated stable loops. For example, these are the opening and closing looped process for separating and joining two DNA strands, the transcription factors binding and unbinding looped process at the promoter region, RNA polymerase binding and unbinding loop, mRNA lifecycle loop, mRNA transport loop using transport factors called karyopherins (i.e., importin/exportin loop to mediate movement of chemicals in and out of the nucleus), ribosomal loop to create an enzyme, tRNA synthesis loop and so on. It is possible to outline this entire sequence of steps in detail as we have done for the case of neuron. I will not discuss them here because the steps are almost identical to the above neuron example. Using these sequences of steps for the enzyme creation process, we could use the above looped systems to create an appropriate network of regulated loops. Once we have such a network of regulated loops, we can study how a disturbance propagates within this network in response to an initial disturbance that forces the creation of a specific enzyme.

As mentioned earlier, the identification of a network of loops corresponding to a specific sequence of steps is not entirely new. Scientists have been detailing these series of chemical reactions for a long time (see standard textbooks like Garrett and Grisham [48] and Alberts et. al [2]). What is, however, new is to see how this disturbance representation in a network of loops is central to most of the features of life itself. This will become clear later in this chapter as well.

Using these two examples (neurons and enzyme creation), we should now find it a bit easier to translate more generic sequence of steps into a disturbance propagation within a suitable network of loops. Most chemical features in a single or a multi-cellular organism can now be imagined to be represented within a suitable network of loops. A given dynamical feature, like walking, immune response, digestion and pumping mechanism in a heart, can then be seen as the outcome of an appropriately chosen disturbance, which propagates within this network of loops.

58.2 Disturbance propagation within a network of loops

The new result discussed through a couple of examples in the previous section, namely, that memory of a long sequence of steps is embedded within a chemical looped network, is central to the problem of origin of life, especially when DNA is not yet discovered for memory. Before DNA is discovered, evolution proceeds iteratively by discovering increasingly complex chemical looped network (as discussed in the previous chapter and will be explored in greater detail in subsequent chapters). The new result shows that the same chemical SPL network also helps memorizing useful and long sequence of steps, even if discovered accidentally, across generations or thousands of years when there is no DNA.

Let me summarize the basic approach we need to follow in order to translate a generic sequence of steps into disturbance propagation within a network of loops. The advantage of this translation is that we have changed the problem of random coincidences when creating a long sequence of steps into a repeatably random and

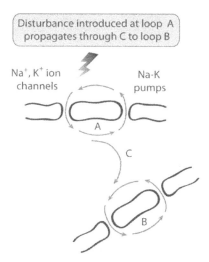

Figure 58.3: Disturbance propagation in a simple 2-looped system. Here loops *A* and *B* are in a stable state until a disturbance is introduced. The disturbance at *A* causes more Na+ ions to flow into the system, which leaks via *C* into the loop *B*. In this way, the disturbance propagates to loop *B*. Energy is consumed by both loops during this process. Loops *A* and *B* will eventually return to the steady-state conditions.

reliable mechanism within an appropriate network of chemical loops. The formation of such a network of loops uses only natural mechanisms as we have seen in the previous chapter. The problem of origin of life from inanimate objects is no longer a random event as we will see.

Let us first consider a simple two-looped system interconnected as shown in Figure 58.3. Imagine each loop to be a physical loop with molecules moving back-and-forth through a pipe, a conduit or even through two openings – one for intake and one for exit (see Figure 58.3). The looped dynamics itself can be simply viewed as a steady-state condition analogous to a state of dynamic equilibrium with reversible chemical reactions. In the case of a neuron, we can think of this loop to be the flow of sodium ions in-and-out of the neuron, one through the opening of Na+ and K+ ion channels and the other through Na-K pump. This produces a steady-state condition referred to as the resting potential. Therefore, under normal conditions, the concentrations of chemicals flowing in-and-out as a loop remains stable and steady for both loop *A* and loop *B*. Let us assume that the diffusion of chemicals from *A* to *B* via the conduit shown in Figure 58.3 is negligible under normal conditions.

Let us now examine what happens if this system is subject to a disturbance, namely, a sudden increase in the flow of chemicals, at *A* as shown in the Figure 58.3. This disturbs the resting state for the loop *A*. As a result, *A* has an increased flow of chemicals within the loop. The disturbance now propagates from loop *A* to loop *B* via the conduit connecting the two. This causes loop *B* to operate above its resting state.

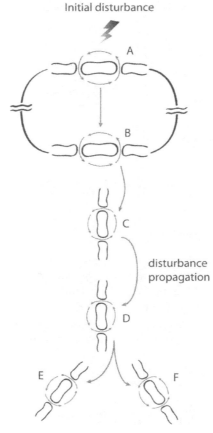

Figure 58.4: Disturbance propagation across a chain of loops. Here, each of the loops are in a steady-state condition initially before a disturbance is applied at loop *A*. This disturbance will propagate along the chain from *A* to *D* and eventually to *E* and *F*. After the disturbance propagates, all loops return back to their original resting states. During this process, energy is consumed by all loops.

In other words, the disturbance at loop *A* has propagated to loop *B*. After a while, the excess molecules leak through the system and each loop returns back to its respective resting states. In this simple example, the disturbance at *A* has propagated through the loops and the system returns back to the original steady-state condition after a while.

A generalization of this example is one in which we have a large number of loops interconnected linearly as shown in Figure 58.4. A disturbance at the first loop propagates all the way to the last loop linearly. Each of these loops return to their resting states once the disturbance propagates. This generalization can be seen as the mechanism behind the propagation of voltage spikes across the axon of a neuron. The initiation of the disturbance is the excess Na^+ ion accumulation at the axon hillock (i.e., beyond its threshold value). This causes an action potential to be generated at the axon hillock, which is beyond the resting potential. At each node of Ranvier, there is a resting state. These are disturbed serially, one after the other, as the disturbance propagates along the axon. The process is identical to the examples discussed in Figure 58.2 and Figure 58.4.

Nicholson [114] shows another example of an interconnected looped system (i.e., the metabolic network). Here, several looped processes are regulated locally within a cell. The concentration of chemicals in each loop can be considered to be in a state of dynamic equilibrium instantaneously. Energy is certainly consumed to maintain this steady-state. In most cases, small amounts of input chemicals are also required just to maintain steady-state conditions because of diffusion and other leaks, though these effects are minimal for a well-regulated system. Even though all loops are drawn in Nicholson [114], the dynamics of these loops are not observable as much. Some reasons for this are: (a) the steady-state conditions are not visibly dynamical (for example, water and water vapor in a closed bottle are in a state of dynamic equilibrium with no visible conversion from one form to another) and (b) the steady-state concentrations of chemicals for some of the loops can be quite low. These effects seem to hide the existence of these loops. Nevertheless, it is best to keep track of them at all times. We never know when, where and what types of disturbances are introduced in the system.

If we now introduce a disturbance in this system at a specific location and among a specific set of chemicals P, it is easy to understand how it propagates through the system before the loops return back to steady-state conditions. This is similar to how the disturbance propagates in a physical system as shown in Figure 58.4, even though in this case, we have a collection of chemical reactions that occur in a cascaded manner (like hormonal signaling in humans. We can analyze similarly if the disturbance is initiated at another location like Q and/or at multiple locations simultaneously. From these examples, it should become clear that once we identify and draw the SPL network, it is a simple matter to understand how the disturbance, introduced somewhere within this network, propagates.

As an example, if you have a cut on your finger, you have introduced a disturbance to the looped network represented within your cells and your body. The blood oozing through your finger, the blood clotting mechanism triggered and so on are just the propagation of disturbances. I refer the reader to Alberts et. al [2] for details of which chemical reactions occur during the healing of the cut.

The speed of disturbance propagation within SPL networks depends on several factors like (a) how long it takes to transfer the disturbance from one loop to the next, (b) the time period of each looped process, (c) the length of the interconnected looped pathway, (d) if the pathways occur in a linear way or if there are several parallel pathways as well and (e) the timely availability of extra energy as each looped process responds to the disturbance.

In the previous examples, we considered physical loops with molecules moving within a loop. The disturbance was also a physical one in that it altered the concentration of chemicals. We could have considered either a chemical loop or a disturbance that is based on a chemical reaction. For example, a disturbance could be triggered by a series of chemical reactions, which open ion channels at the dendrites causing a sudden influx of ions. An example of a regulated chemical loop is Kreb's cycle. A disturbance in the form of excess glucose intake causes more ATP to be generated.

Let me now use the two examples from previous section as a guide to summarize the translation of a sequence of steps into disturbance propagation in generic terms. First, a disturbance causes the local stable conditions to be altered. In order to eliminate the disturbance and return to normal conditions, this disturbance needs to be dissipated. However, before the disturbance dissipates, it causes a new disturbance in the neighboring interconnected looped systems. In general, the network of loops has several pathways from a given loop. In this case, multiple disturbance signals are generated for each of these pathways. They all propagate simultaneously along different directions within this network of loops. This disturbance propagation continues in a cascading fashion until all the looped systems involved, return to their normal stable states.

The stable parallel looped architecture and the dynamics within them in the form of disturbance propagation is, therefore, capable of representing a complex sequence of tasks. If we want to understand the nature of a sequence of steps, we can now look at the corresponding network of loops instead.

More generally, given a network of loops, we can even artificially introduce the disturbance anywhere within this network and study how it propagates. You can imagine the looped network to be a locally stable dynamical system. The disturbance is then a minor perturbation in some region that alters its local steady-state condition (see Figure 58.4). As the changes (i.e., above or below the average local steady-state conditions) propagate, we observe the outcomes as different types of physical responses. For example, if the disturbance is a cut on your hand, the outcome could be clotting of your blood and/or an immune system response from a potential infection-based disturbance around the open wound.

58.3 Energy distribution within a network of loops

Sometimes, it is easier to think of water waves as an analogy to understand how a disturbance propagates from one region to another. With water waves, the energy needed to make the waves travel large distances is essentially supplied at the source of the initial disturbance. The difficulty, therefore, is that the energy from the disturbance dissipates without reaching very far. You cannot get interesting behaviors if the sole supplier of all the energy for the entire sequence of events is just the initial disturbance. The event will not occur both quite far and for a long period.

On the other hand, with disturbances propagated within a network of chemical loops, the energy is constantly supplied locally. Remember that each of the looped dynamics locally has to continue to occur (and stay in a regulated state) even before the system was subject to any disturbance. For this to happen, each local looped subsystem must have an internal source of energy. For example, every cell has mitochondria to generate ATP. This ATP is distributed within the cell quite abundantly. Therefore, even though the initial energy of the disturbance is negligible, living beings that use stable parallel loops can make the disturbance travel very far and for a very long time by consuming energy locally. You sometimes notice this as a small cause producing a large effect.

For example, a small needle pinch on your foot can produce a large effect like jumping up in the air. Examples of this sort can be best understood when you view the sequence of steps as disturbance propagation within a network of loops where local energy is naturally necessary to maintain each of these looped reactions. Treating them as looped systems with constant dynamics through internal energy sources is fundamental to explaining this instead of treating them as statically connected sequence of steps whose dynamics is only triggered after the disturbance occurs.

Furthermore, without this new view, we need to give a reason why energy is supplied at each step. Several questions are not satisfactorily answered. For example, questions like who is supplying the energy, how is it supplied at the correct place and at the correct time, is it random luck that all the required chemicals are present at the correct place, at the correct time and in the correct order and so on, are unclear. However, with this looped network view, these questions naturally disappear. In fact, these questions become relevant only when explaining how to create of the network of loops and how to maintain these stable (or regulated) dynamics locally, but not afterwards. The subsequent processes explained above with the disturbance propagation within this network, follows quite naturally.

In essence, we move all the hard questions into the *creation* of the network of loops instead of asking them repeatedly for every instance of seemingly complex, unrelated and disjoint sequence of steps. This is central for solving the problem of origin and diversity of life without help from any conscious or intelligent beings. All difficult questions are essentially pushed into the creation of suitable chemical SPL networks. However, the creation of a specific network of loops for a given feature is not as easy. Each case requires a special way of identifying and modeling the feature accurately. In the later chapters, I will elaborate more on this. As a concrete example to illustrate how to create a specific network of loops, I will discuss how to *create* the first catalyst (like an enzyme) when we do not even have any DNA to begin with. I will do this in chapter 60. There are other examples that I will discuss later as well like, say, the creation of the first multi-cellular organism (see chapter 70).

58.4 Chemical memory and variations in sequence of steps

When we have a network of loops, we have seen that the disturbance propagation within them can produce a set of behaviors by executing a specific sequence of steps. This implicitly assumes that the concentrations of chemicals that are part of these loops are at a given steady-state levels. As these steady-state levels are perturbed by a sudden influx of one or more types of chemicals, it creates disturbance that propagates within the network.

For example, in the case of the brain, if you see a bright red colored object with your eyes, there is a sudden influx of ions into the retinal cells. This triggers a

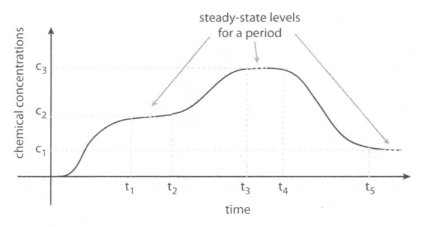

Figure 58.5: Steady-state concentrations of a given chemical over time. The steady-state levels of chemicals change over time due to various external or internal factors. Some typical examples are when you consume alcoholic drinks (aspartate, glycine and glutamate levels change), during normal growth, when consuming high-calorie foods (chocolates and ice cream) and others.

disturbance that propagates within the brain network in a certain way to redistribute the disturbance energy. The concentrations of the ions prior to the influx of ions have a certain steady-state levels. These steady-state levels are not fundamentally altered due to the influx of ions. Rather, the excess ions beyond the steady-state values can be considered as a measure of the size of the disturbance that was triggered at the retinal cells (see Figure 58.4). The bigger the disturbance size (or the associated energy), the farther these effects are transmitted. However, at the end of the propagation of the disturbance within the brain network, all the ion concentrations will return to their original steady-state levels. As an analogy, after water waves finish propagating due to a disturbance, the water levels return to their original steady-state levels i.e., the water level will not be higher than before the disturbance.

However, in some situations, the very steady-state levels of these chemical concentrations are itself altered to newer threshold levels, analogous to high tides and low tides in oceans (see Figure 58.5). This can happen in a number of different ways. For example, when you consume large amounts of alcoholic drinks, you alter the steady-state levels of several neurotransmitters like aspartate, glycine and glutamate, at least for very long periods of time. This is not a permanent change in the threshold steady-state levels though. In other cases, like, when you eat a lot of high-calorie food (like chocolates and ice cream), your body accumulates large amounts of fat altering the steady-state levels fundamentally. It is harder to return to the original levels from the newer steady-state values (i.e., to reduce your weight, as we all know – we restrict our diet, exercise and so on to reduce the steady-state levels back to the original state).

A growing child offers a lot more generic examples of change in steady-state values. For example, a growing child increases the size of various organs and body parts to become bigger. In order to maintain chemical stability with the new adult sizes of organs, the threshold steady-state concentrations of various chemicals involved in each of these organs have to be fundamentally altered. Otherwise, the body does not operate optimally with higher energy needs. In the brain, the set of interconnections between neurons constantly keep changing causing the threshold steady-state conditions to be altered in fundamental ways.

The disturbance propagation within the exact same network of loops before versus after the newer threshold steady-state values produce entirely different outcomes. More importantly, this is an entirely new way of *memorizing* the outcomes without changing any part of the network of loops itself. Therefore, we can say that we now have three unique memory mechanisms within our cells at a chemical level:

(a) memory in the DNA sequences,
(b) memory within the network of loops accessed through disturbance propagation and
(c) memory through altered threshold steady-state concentrations of chemicals within the network of loops.

The second and third types of memory can be grouped into epigenetic memory, i.e., not involving genes itself (see section 58.2). These two memory mechanisms are entirely new. They are a direct consequence of the theory of stable parallel loops introduced in this book.

I have used the word 'threshold' to highlight the fact that this memory occurs in discrete jumps in chemical concentrations. Minor and continuous changes in chemical concentrations itself do not contribute to the memory. The jumps have to be in discrete quantities that are beyond the normal (or average) variations that are allowed in these chemical concentrations from everyday usage. These threshold quantities are different for different types of chemicals, in general. For example, our body accumulates chemicals gradually, which slowly alters the steady-state concentration levels. However, these minor changes do not fundamentally produce altered behaviors. In this sense, we can treat the changes to be within the allowed variations. Over time, the accumulation will be considerably high that the system starts to exhibit different behaviors. This is when we should treat the steady-state concentrations to have jumped to the next discrete or threshold state.

It is this third type of memory (i.e., altered threshold steady-state values) that is important for the problem of transitioning from single-cellular to multi-cellular organisms (see section 70.8) or transitioning from prokaryotes to eukaryotes (see section 70.9). This is important because even though you have not changed any chemicals, you have altered the 'phenotypic' outcomes in significant ways just by altering the threshold steady-state conditions. For example, one steady-state condition in, say, calmodium (CaM) and bone morphogenetic protein 4 (BMP4), will produce thin long beaks while a different steady-state condition will produce thick short beaks in finches. I will discuss the advantages of this new type of memory in chapter 70 in more detail.

58.5 Creating versus executing a sequence of steps

As mentioned in the previous section, one interesting feature of living beings is that they perform fairly complex and well-defined sequence of steps. Furthermore, one sequence of steps (for one feature) does not seem to have any relation to another sequence of steps. When we think we have understood and explained how a given sequence of steps works, we typically encounter another unrelated sequence of steps that life exhibits. This new sequence of steps usually does not fit within the framework of the original explanation. This is a problem if we are attempting to discover a common explanation for all cases. Each feature and the corresponding sequence of steps are so incredibly complex that there seems to be nothing in common between them. When asked how such specific sequence of steps are *created* naturally for the first time, any detailed explanation of how the system *executes* them subsequently, i.e., after it is created for the first time, is not satisfactory.

The result discussed in this chapter presents a clear solution to this problem of both creation and execution. Each complex and unrelated sequences of steps can now be represented as appropriate disturbances propagating within specific networks of loops. This appears to be true, at least, whenever free will is not involved.

We, no longer need to claim that one sequence of steps in one species is much different from another sequence of steps in another species. All diverse sequences of steps are now uniformly represented within a single framework. Different network of loops may have different level of complexity. Nevertheless, our task is to find a specific network of loops to capture a given sequence of steps.

If we look at examples of man-made systems, there is no connection between creating the design versus how the design works after it is created. Consider how we create the design of a car. The laws that explain the creation of the design of a car are hidden within the notion of intelligence, consciousness and creativity of a human being. The shape and design of a car are not the outcome of any known physical laws. Rather, they are the outcome of human creativity. However, after we create the car, it obeys the standard physical laws that we have already discovered. Surely, our brain obeys the same laws as well. However, we have a huge gap in our understanding of how to create designs. We know how to use Newton's laws and others to explain how a car works. But, we do not know how to use Newton's laws and other quantum mechanical laws to explain how our brain created the design of the car. Until we have clear explanation that closes this gap for the creation problem, complex events like life appears just as random coincidences.

What happens within our brain to create a specific design of a car and what happens as the design runs its dynamics (i.e., operates this car) by following a subset of the laws of nature, appear too disjoint and far from one another. The same problem applies and is, in fact, exponentially magnified with living beings. Creating even a single feature in a living being involves following a long sequence of steps (like steps followed by a neuron when it fires action potentials, steps followed to create an enzyme, steps followed to heal a cut on your fingers and so on). The execution of these steps is now easily understood with our current research in biology and chemistry. However, the same laws and knowledge we have acquired does not explain the *creation* of each such sequence of steps for every feature. For example, the transcription and translation mechanisms do explain how to create an enzyme (see

Alberts et. al [2]), but the same laws or mechanisms do not explain how nature discovered the very transcription and translation mechanisms in the first place.

In the case of man-made designs, it was easier to tuck this difference in laws within a conscious human brain as creativity. However, with the creation of life for the first time and the creation of other complex features of life, there is no such equivalent intelligent or creative system to help. Therefore, if we were to claim that nature created living beings without any help from an external intelligent brain, it is all the more important to ensure that the laws for creation and the laws for execution (after creation) be closely related.

For example, we should explain using similar laws the creation of life for the first time versus creating it subsequent times, creation of two sexes for the first time versus subsequent times and creation of a multi-cellular organism for the first time versus subsequent times. Other examples are with the creation of the first eye, ear, wings, heart or any other organ, creation of the first enzyme, the first DNA and so on. I claim that these set of creation problems can only be answered with the result discussed in this chapter.

The SPL framework is the only correct architecture that uniformly answers both the creation and the subsequent execution problems. I will discuss each of the above problems in detail in subsequent chapters. The problem of creating a given task for the first time and the problem of performing that task after it is created are no longer unrelated to each other. The former problem involves creating a specific network of loops and the latter problem involves letting the dynamics run within this network of loops i.e., both problems involve working with a special chemical SPL network. Without this result, the problem of creation appeared to be outside the realm of the problem of executing the created dynamics as discussed above.

The representation of a sequence of steps as disturbance propagation within an appropriate network of loops is a nontrivial result with unique advantages. Even though, at first, it appeared as though we just moved the problem around with this representation, this new result, nevertheless, revealed the true structure of the problem of creation. It showed that the problem of creation is identical to the problem of execution even though both of these problems have stayed disjoint until now.

It should, however, be noted that the problem of expressing each new feature as a network of loops is not an easy one either. Each feature requires a unique set of loops, though nature could reuse several of them. How to create such loops will be the focus of the next several chapters. One interesting example is how to create the first primitive enzyme when we do not even have a DNA to start with. In chapter 60, I will present the appropriate set of loops to solve this problem. Another example is to create multi-cellular organisms for the first time, not after the shape and structure of a multi-cellular organism is already encoded into the DNA of a single cell (that subsequent creation of a multi-cellular organism follows the process of embryogenesis).

59
Active Material-Energy Loops

Until now, we have focused on the importance of looped systems for the problem of origin of life. We have discussed several interesting properties of such systems. These included showing how to create new chemicals from looped dynamics, how natural it is to create looped systems, how to represent a linear sequence of steps as disturbance propagation within an appropriate network of loops and how creating for the first time and creating/executing the feature for all subsequent times can be unified within the same SPL framework. These appeared to be abstract properties of looped systems.

We will now turn our attention to concrete designs using looped framework. I will start with a simple looped designs (i.e., a minimal structure – see section 53.1) that are easy to form even when there was no life on earth, say, under conditions of primitive earth 4 billion years ago. We do not need the existence of enzymes or other complex organic molecules for these initial designs, although the presence of, at least, some these complex molecules (say, phospholipids) makes it easier for these designs to form quickly. We will not assume the existence of enzymes in any of these designs. In fact, creating the first set of 'primitive' enzymes when there is no DNA is itself one of the main objectives of the next few chapters.

For the above creation problem, one fundamental question typically ignored or postponed is the following. When a highly-ordered system is created naturally, we typically focus only on identifying the necessary structures and the sequence of steps that caused this creation of order. However, the true creation question is: what laws and mechanisms *guide* all necessary subcomponents to indeed follow these identified sequences of steps? For example, according to RNA world hypothesis the minimal structure for the origin of life is the creation of an RNA replicase enclosed in a closed membrane (Szostak et al. 2001). However, to create replicase molecules, several prerequisite molecules and steps are needed. These are, say, lipids, nucleic acids, sugars, arrangement of monomer chains to create unique polymer sequences, inorganic catalysts like clay montmorillonite and others (Hanczyc et al. 2003; Hanczyc and Szostak 2004).

The above question translates to: (i) how can all of these prerequisite molecules naturally accumulate in a small spatial region (cf. air molecules reaching one corner of the room) so they get a chance to react chemically? This seems to oppose the second law of thermodynamics. (ii) How can they accumulate within a short period? Otherwise, they may disintegrate or move away before they can be part of a long cascading chain of chemical reactions? For a few reactions, abundance of molecules or chance mechanisms as an answer may seem satisfactory (like the spontaneous self-assembly of phospholipids to form a closed membrane). However, this is unsatisfactory for processes of life as they involve long cascading chain of reactions like with creating a unique RNA replicase sequence. (iii) If structures like RNA replicase are created by chance, how can they be created reliably and repeatedly so the order can be recreated in the event they are starved (like during ice ages or meteor strikes) or when they disintegrate? The mere presence of all these prerequisite molecules is not enough to create high degree of order (cf. dead cells or all required chemicals dumped in a beaker). There seems to be a natural sense of regulation or coordination at a large-scale that avoids collapse, though it appears highly likely, in living beings. The above questions are applicable even for processes that occur in our body at this instant.

The goal of this and the next several chapters is to address the above creation-related questions (like, say, what laws bring the correct molecules to the correct location at the correct time and in the correct order). Towards this end, we need to propose concrete designs, each of which should be easy to form under primitive conditions. Each such design is termed as a 'minimal' structure (see section 53.1). I will begin by describing one of the simplest minimal structures. This is what I call as the active material-energy looped system. It is a simple three looped system: one loop at the inputs, one loop at the outputs and the third loop for converting the inputs to outputs and vice versa, while generating a net positive energy. It is termed as material-energy loop for the following reason. Material inputs are taken into the system which react to generate energy. The energy generated is then used to get more material into the system, thereby continuing the process as a loop.

This is an extremely important and yet simple looped system that can form naturally. It not only exhibits several life-like features but also solves several of the creation problems mentioned earlier (as we will see). I have already discussed one way of creating a material-energy looped system in section 2.7. However, that design required the help of a conscious being. In this chapter, I will remove this restriction. I will propose other natural ways of creating material-energy looped systems. We can think of this simple material-energy looped system as an example of a system that can be repeatedly created using entirely random mechanisms.

59.1 Motivation – passively versus actively seeking inputs

Let me restate the main motivation for proposing the active material-energy looped system. This was already discussed briefly in section 2.7. If we look at one fundamental difference between all nonliving and living beings, it is the fact that living beings actively seek their own inputs and energy (i.e., their own food) in order to continue their internal dynamics and their very existence (i.e., to survive). Nonliving systems, on the other hand, passively consume the inputs and energy when they are available. If the inputs are unavailable, their dynamics simply cease to occur. When the inputs re-appear, the dynamics that was occurring before, most likely, will not occur again, but sometimes, it does. In any case, the nonliving system will never seek the inputs it needs from the environment to run its dynamics. In a sense, the inputs themselves have to find their way into the system from the environment through diffusion or other random effects.

Every living being, from a single celled to a multi-cellular organism, has some form of mechanism to actively seek inputs into the system. This is the uniquely distinguishing feature between nonliving and living beings. The main objective of the active material-energy loop design of section 2.7 was to capture the essence of this feature within an extremely simple system. The human respiratory system considered in section 2.7 seemed like a good mechanism to study, although, in hindsight, any other example would have sufficed just as well. Specifically, notice that when we breathe, we actively suck air (filled with oxygen) through our nose and release carbon dioxide. Air does not simply diffuse through our nose and enter our bloodstream. If we try to hold our breath for a while, we experience a huge buildup of some sort of pressure that forces us to breathe again. The same applies with eating food with our mouth. However, modeling the food intake process is a bit messy because it requires us to capture lot more detailed mechanisms like an ability to walk, an ability to control the motion of our hands and so on. The resulting design would not have been simple. Therefore, we chose the simpler respiratory system as the example to model in section 2.7.

In section 2.7, we stripped down the example of our respiratory system down to its essentials by just having lungs, a heart and a simple circulatory system (see Figure 2.12). When the lungs expand, air is sucked in and when it contracts, air is pushed out. The air with sufficient oxygen enters the blood circulatory system that is pumped through the heart. The heart and lungs gets energy from ATP generated through Kreb's cycle. This consumes the oxygen and gives out carbon dioxide. It is possible to simplify this design considerably. For example, we do not need Kreb's cycle to generate the energy. Any simple exothermic reaction would have sufficed.

This simple design showed how a system can actively seek inputs for its continued operation (or survival) versus passively getting inputs. This is the central difference between living and nonliving systems. If the material and energy inputs the system actively acquires is higher than what it consumes, the system has an ability to exhibit and perform additional functionality. These additional features will make it appear as though the system is 'alive' as we will see later. Let me begin by simplifying the design of section 2.7 so nature can create such active material-energy looped systems without help from conscious beings.

59.2 Pumps – active material-energy looped systems

In chapter 2, we have seen the 'existence' of a family of chemical SPL systems via a suitable mapping from physical SPL systems. Now, we will start looking at how we 'create' such families of systems using natural mechanisms. I will now propose one approach using special chemical pumps, which we mentioned as one of the minimal structures in section 53.1. These minimal structures avoid the use of enzymes, DNA, replicases and other highly ordered molecules, as we have not yet shown how to create them from a disordered collection.

Figure 59.1 shows the design of a simple material-energy looped system (Design A). It has an elastic outer membrane, similar to the one created naturally under prebiotic conditions (Hanczyc et al. 2003; Hanczyc and Szostak 2004), with no internal structure. It has an opening I for inputs (as a dent). We also need a way to get

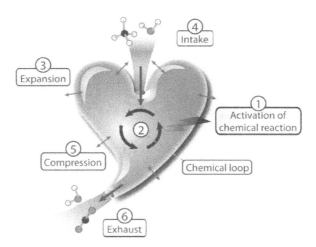

Figure 59.1: Design A – a simple material-energy looped system. This system behaves like an active pump that has an intake and an exhaust. Once the system is activated, it sucks external inputs and operates continuously in a looped manner for a sufficiently long time.

rid of unnecessary waste. This happens through another opening for exhaust E (as a bump). The geometric shape looks approximately like a heart to facilitate suction and exhaust (discussed shortly).

This system has an internal source to generate energy. The easiest way to create this internal energy is to use material inputs rather than taking raw energy directly from an external source. Inside this system, an exothermic reaction (say, a single reversible reaction) occurs. The energy released causes the membrane to expand. When it expands, more inputs are sucked in through the intake opening (Figure 59.1). The shape of the intake I is like a nozzle to promote the sucking of the inputs. The elastic property of the membrane causes it to contract assuming that the system does not burst from expansion. During the contraction cycle, the outputs are pushed out through the exhaust. The new inputs that were sucked in causes the internal exothermic reaction to proceed once again generating new outputs and energy. Some of the energy generated is consumed for the necessary expansion and contraction of the membrane.

One important condition required for this version of the material-energy looped design (even with the design of section 2.7) is that the energy generated from the inputs is more than the energy consumed by the membrane expansion plus any other wastage of energy. Furthermore, for any concrete design, we want the mechanism to repeat for a long time. Therefore, we should ensure that all the components and processes described here use looped subsystems. Indeed, there are two loops: the physical compression-expansion cycle (at the intake and at the exhaust) and a suction-chemical reaction-exhaust outer loop. The name 'material-energy loop' or, simply, 'chemical pumps' refers to the fact that materials drive the energy creation and the energy drives the material intake, as a loop. The term 'active' refers to the

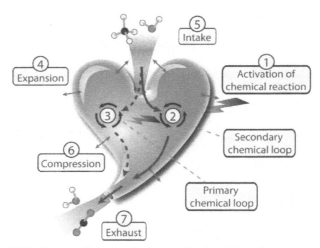

Figure 59.2: Design B – a variation to Design A in which a secondary chemical loop assists a primary chemical reaction. With this active material-energy looped system, it is possible to run otherwise less favorable chemical reactions as well.

ability to generate and consume energy actively as opposed to passively acquiring energy. This design clearly satisfies definition 2.2 and is, therefore, an SPL system.

Combining all of these features let me now describe how the entire process works. After the material-energy looped system is created through random mechanisms (discussed in section 59.3), it stays idle until there is a way to jump-start the dynamics for the first time (like a spark). After this initial step, energy will be generated within the membrane from the chemical reaction. This energy will be used to expand the membrane at *I*. This sucks more inputs into the membrane. The elastic property of the surface membrane causes the entire system to contract after a while. This pushes the outputs through the exhaust *E*. The inputs sucked in earlier causes the chemical reaction to continue to occur, thereby constantly generating energy. The process now repeats as long as there are inputs available in the environment.

The pump is a simple generalization of a closed compartment with additional compression-expansion dynamics and input-output channels. The system exhibits a tendency to continue its dynamics for a long time. When we choke the intake or the exhaust, sufficient backpressure will be generated within the system from the expansion. This will try to dislodge the obstruction (like kickback in power tools). This self-sustaining ability does not apply for other damages like a tear in the membrane, as there are no loops to detect and oppose them.

Let me suggest a few natural variations to design A. Design B (Figure 59.2) actively fetches additional reactive chemicals like acids, bases or even ATP, if present in the environment. These secondary spontaneous chemical reactions occur in conjunction, releasing energy for the primary, possibly slower, chemical reaction. The excess energy will also make the compression-expansion cycle continue for a while. In Design C, the intake and the exhaust openings are approximately molded to allow only

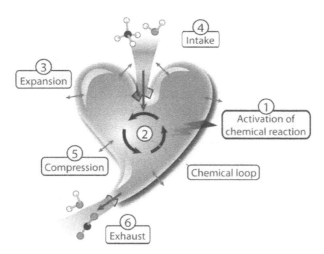

Figure 59.3: Design C – another variation of Design A. In this variation, the intake and the exhaust openings are molded to allow only specific molecules to enter and leave respectively. This material-energy looped system begins to specialize to specific subset of molecules.

a specific set of molecules (Figure 59.3), similar to the groves formed from erosion through repetitive operation of active water flow. Design D has a twist-untwist motion in addition to inward-outward motion in each compression-expansion cycle. This is common when we have non-uniform mass distribution and irregularities. As the system expands, the left-half turns counterclockwise while the right-half turns clockwise and vice versa analogous to holding a balloon with both hands, twisting it as we compress and untwisting it as we expand. Additional variations are combinations of the above designs and other active mechanisms that use just the above abstract features.

In the above designs, even though we have assumed the reactions to be exothermic that generate more energy than what the chemical and physical parts of the system consumes, there are two ways to relax these assumptions. The first way is to use the excess energy available in the environment like within hydrothermal vents or in a hot primitive earth. The second way is to use design B in which a secondary exothermic reactive reaction is coupled with an endothermic primary reaction.

What are the chances of creating the structure and running the dynamics naturally? If we include help from conscious humans, we already have efficient designs like internal combustion engines, pumps with similar intake and exhaust openings and nanotube nanomotors (Fennimore et al. 2003; Laocharoensuk et al. 2008). Even without such help, it has been shown experimentally that vesicles (i.e., closed compartments) do form under primitive earth conditions using fatty acids, organized as micelles (Hanczyc et al. 2003; Hanczyc and Szostak 2004), instead of relying on phospholipids. It was also shown in Szostak Lab that clay montmorillonite catalyzes the vesicle formation and that the vesicles are stable for days to months. In addition, they have shown that when fatty acid micelles are added, the vesicles even grow and split using a process called flipping (a primitive reproduction mechanism for just the membrane).

The designs A-D proposed here are simple and do not require anything else other than the outer membrane with two irregularities (openings for intake and exhaust). Therefore, the current experimental results from Szostak Lab (Hanczyc et al. 2003; Hanczyc and Szostak 2004) already show the 'structural' feasibility of these designs under primitive conditions. The reproduction mechanism of flipping is sufficient to recreate these designs repeatedly for a long time.

However, the above designs are useful only when they execute their dynamics. Recently, catalytic nanomotors (Laocharoensuk et al. 2008) that have self-propulsion and functionally similar properties to pumps were built by inserting carbon nanotubes into gold and platinum nanowires synthetically to achieve high speeds (50-60 μm/s). They even combined hydrogen peroxide fuel with hydrazine to achieve higher speeds (94-200 μm/s) comparable to natural biomolecular motors. However, creating functional pumps under natural primitive conditions is not yet done.

If the material-energy looped system were to continue the dynamics for a while, the following conditions must be satisfied: (a) the looped process should be initiated by jump-starting the system, (b) the energy generated should be more than the energy consumed for the compression-expansion cycle and other losses, (c) the inputs should not run out, (d) the intake and exhaust should not be severely blocked and (e) there should be no damages to the system. These conditions lower the probability of occurrence of dynamics, though they continue to occur repeatedly.

Most existing molecular-sized pumps in living beings (not organ-sized pumps like heart) like sodium-potassium pumps, calcium pumps, proton pumps, large class of ABC pumps and others, use specialized proteins. They are efficient, as are synthetic nanomotors (Laocharoensuk et al. 2008). In comparison, the pumps suggested here only need to be feasible with a small enough efficiency of operation. This is reasonable owing to its simplicity, but needs experimental verification. The likely locations for these designs to occur naturally would align with geological evidence like within hydrothermal vents and hot springs.

Given the simplicity of the design in Figure 59.1 at a conceptual level, coming up with a concrete design is definitely an easy task. I will discuss one such design in further detail shortly. However, what is entirely nontrivial is to understand the implications of this seemingly trivial design. For example, in section 60.7, I will show that it is possible to create a primitive catalyst even without the presence of a DNA using just such a trivial material-energy looped design. Before I discuss these nontrivial results, let me first explain how nature could create such simple material-energy looped systems using just repeatable random mechanisms.

59.3 Creating material-energy looped systems naturally

The active material-energy looped system of Figure 59.1 requires two types of looped systems. The first one is a physical loop caused by the expansion and contraction of an elastic membrane. The second one is a chemical loop, the simplest being a reversible chemical reaction. Now, both types of loops can be formed naturally even before the existence of life using only repeatable random mechanisms. Let me explore various choices for each type of loops as well as for the entire material-energy looped system in a somewhat theoretical sense (see also the experimental results from Hanczyc et al. 2003; Hanczyc and Szostak 2004 discussed in the previous section).

Energy generation loop: The system has a chemical reaction R, represented as a loop that generates energy. There are several possibilities for this chemical reaction, the simplest being a reversible exothermic reaction that is in a state of 'dynamic' equilibrium, namely, the rate of forward and reverse reactions are nearly equal. Hence, the reactants and the products are constantly switching back and forth (in a loop) even though there is no visible change in concentration of either set of chemicals. Local dynamic quasi-equilibrium conditions should be relatively common even on primitive earth.

Another possibility is to have multiple chemical reactions chained together to form a loop. Some or all of the outputs of one chemical reaction act as the inputs to the next linked chemical reaction until you have a loop. A special case with just one chemical reaction within this chained loop is the degenerate case mentioned earlier, namely, a reversible chemical reaction in equilibrium. Of course, the possibility of creating two or more chained looped chemical subsystems by nature itself is not as common as a single reversible chemical reaction, at least, until we have several enzymes and/or complex organic molecules.

Yet another generalization is to chain subsystems to form a loop instead of chaining individual chemical reactions (like at an ecosystem level). Each of these subsystems can have several chemical reactions, chained, looped or otherwise. One subsystem takes inputs and produces outputs to be used as inputs by the next chained subsystem until you form a loop of subsystems. Again, this is increasingly harder to occur before the existence of life.

Choice of input and output chemicals: For a given chemical reaction, there are several choices for the reactants and products, among both inorganic and organic chemicals. For now, we do not require catalysts for these chemical reactions. The set of chemical reactions that we are excluding for now with this design are those that proceed to completion i.e., where most of the reactants convert into products. Even though I am excluding them as candidates for reactions R running within active material-energy looped system of Figure 59.1, they will turn out to be useful as the material for membranes. I will discuss this shortly.

Another consideration when picking reactants and products is to choose the appropriate states of matter. The best choices are fluids (liquids and gases) instead of solids. This is primarily because they are easier to transport from one location to another. An abundant availability of input chemicals localized within a given region, a useful requirement for our material-energy looped design, becomes much easier with fluids than with solids.

System membrane: After picking appropriate reactants and products for a reversible chemical reaction, we want to ensure that the reaction can reach, at least, quasi-equilibrium conditions. For this, it is best to have a closed system. It has been show experimentally that vesicles (i.e., closed compartments) do form under primitive earth conditions using fatty acids, organized as micelles (Hanczyc et al. 2003; Hanczyc and Szostak 2004) without relying on phospholipids. It was also shown in Szostak Lab that clay montmorillonite catalyzes the vesicle formation and that the vesicles are stable for days to months.

It is likely that other closed systems that do not use fatty acids created through natural means will have some holes. For example, if solids like clay, sand and pebbles enclose a liquid, you now have a natural partially closed container. However, it is desirable to have a flexible surface with a few pores as the surface membrane. Clay does have some sticky and elastic properties when it is wet. A generalization of clay is just other types of 'waste' products. When a chemical reaction proceeds to completion, the products generated can be viewed as waste products, in the sense that they are not recycled. These wastes can sometimes accumulate to form the required enclosures for our material-energy looped design. Clay can simply be viewed as wastes made of fine-grained mineral deposits. In fact, in general, we can form a closed surface by taking extremely fine-grained particles and letting them adhere using any viscous liquid. The surface tension and viscosity can give rise to some elastic properties as well.

Yet another way to form a closed surface is by taking highly viscous liquid and trapping less viscous immiscible liquids within them. The high viscosity introduces elastic properties to the membrane surface. When a different liquid or gas gets trapped within this viscous membrane, it can expand or contract. This can stretch the membrane surface much like an air bubble or a soap film. If we are lucky and lipid molecules exist, then they can spontaneously coalesce to form a lipid bilayer membrane much like the cell membrane. This being less likely before the existence of life, I will not emphasize the need for a lipid bilayer membrane for now.

Shape of the system: There are several possible shapes for this membrane, depending on the imperfections and the non-uniform compositions of the surface. They are not just ideal shapes like spheres as with water bubbles. Instead, the most likely shapes are irregular. Yet, the features of the heart shaped membrane shown in Figure 59.1 can be easily imitated by them. The heart shape is merely an idealization to explain the feature in simple terms. Most irregular shapes will already exhibit the very same properties needed for the material-energy looped designs. For example, for each of these irregular shapes, the expansion of the surface membrane can cause the inputs to be sucked in at some locations I (that look like dents) and the outputs to be pushed out through other locations E (that look like bumps). In general, there may be multiple pores for both inputs and outputs. The irregularity naturally can give rise to nozzle-like shapes (bumps and dents). This will help with an ability to suck the inputs into the system.

In fact, as a precursor to this design, we can even assume that the inputs simply diffuse through the membrane just from the concentration gradient between the inside and the outside. As a result, we may have some systems with favorably shaped surface membranes that actively suck inputs and others that use diffusion. After a while, through just random means, the systems that use diffusion change shape to provide active suction mechanism. In fact, by definition, whenever I use the term material-energy looped system, I always require that it should have an active suction mechanism, not simply a diffusion mechanism instead.

With the necessary physical structure formed and the required chemicals present in the environment, we can say that it is quite easy to create material-energy looped systems, entirely using natural means.

I have focused on the processes themselves instead of the precise set of chemicals needed for the required functionality of the material-energy looped system. This makes it easy for us to see how such a system can form again and again, by nature itself. When better and more efficient chemicals are discovered later, the material-energy looped system can be rebuilt even more easily. Furthermore, these newer and better designs will make the system sustain much longer than before. For example, when the conditions are right for amino acids and lipids to form, these designs become far more efficient and easier to occur.

59.4 Life-like features of material-energy looped systems

As mentioned in section 59.2, the dynamics of the material-energy looped system sustains for a long time when (a) the external inputs are freely available, (b) the chemical reaction R proceeds in the exothermic direction and (c) the energy produced through R is more than the energy consumed for the intake of inputs, for the exhaust of the outputs and other losses through heat. There will be physical oscillations of the entire system because of the flexible or elastic behavior of the surface membrane. These sustained oscillations continue as long as external inputs exist or until the second law of thermodynamics causes the system to wear out even when there are no unexpected external destructive influences.

From the previous discussion (see section 59.2), it is quite clear that each of the above three conditions (a) – (c) are not as restrictive. When the conditions are met and the physical system is created once, the system dynamics can self-sustain for a very long time. Let us now look at a few life-like features the material-energy looped system exhibits.

Actively seeking the inputs: One of the most important life-like features of this system is that it actively sucks the inputs it needs instead of letting the inputs diffuse through the system passively. Of course, this only happens as long as the external inputs are available. Unlike living beings, this system does not 'walk' around to find its inputs at a different place if they are depleted at the current location. Nevertheless, this is a big improvement in the way most chemical reactions have been occurring before this design existed, i.e., only randomly. Now, a set of chemical reactions can occur actively, not just passively using random collisions.

Survival and an attempt to prevent the systems 'death': If you choke the system at the inputs I or at the outputs E, the system generates a sufficient backpressure because of the energy and expansion produced within itself. This creates a large enough force that attempts to dislodge any obstruction at the inputs I or the outputs E. This is very similar to how we are forced to breathe because of a backpressure if we attempt to hold our breath for a long time. This can be viewed as if the system is trying to prevent its own 'death', at least, in a limited sense. The limiting aspect is that the system only exhibits this behavior when choked at these two special locations. It does not fend off other destructive forces elsewhere on the system similar to how living beings do (like a tear in the membrane). The system, therefore, exhibits a limited survival instinct (especially, if you take the active suction of inputs into account as well).

Alive and resting states: Let us now look at what happens when the external inputs do run out. The chemical reaction R will eventually cease to exist because of a lack of new inputs. Most of the inputs will get converted to outputs and will diffuse through the two openings I and E. The energy generated by the reaction R will eventually be consumed through the expansion and contraction of the membrane. Since there is no more new energy generated within the system, even the membrane oscillation will come to a standstill.

When the external inputs reappear at a later time, they would need to diffuse into the system initially. They would also need to be jump-started through some random mechanism so the chemical reaction R can initiate. For this to happen, external energy greater than the activation energy of the reaction should be supplied through random means. When this happens, the system will come back to its original state. It will self-sustain its dynamics as before until the new inputs get depleted once again. In this way, the material-energy looped system will switch between active (alive) and inactive (resting or hibernating) states depending on the presence or absence respectively of the external inputs. The physical structure can also move around quite randomly to new locations and cause the dynamics to start again, as long as it is not destroyed.

Growth: Over a period of time, it is possible for the material-energy looped system to grow bigger in size. If the material needed to create the membrane structure exists, it will randomly attach itself to the existing system over time. For example, as the system expands, it becomes bigger in size. At this time, more membrane material can get randomly attached. The inputs entering the system can be higher in quantity relative to the outputs leaving the system, resulting in a net increase in the amount of material within the system. This will make the system bulge over time with a corresponding expansion of the membrane as well. This can be viewed as a primitive version of growth of the system over time. It is, therefore, likely that different material-energy looped systems within a given region will be of different sizes and shapes.

Reproduction: Notice that since there is no DNA in this simple design, it is not possible to talk about reproduction in the traditional sense. Yet, we can see how nature can make approximate copies of this design. As we see that the material-energy looped design is nothing more than running a single chemical reaction within an enclosed container, the notion of reproduction is quite trivial. Reproduction, in this case, means that (a) we should make a copy of the membrane with an approximately similar shape that has input and output cavities and (b) enclose initial concentrations of the input and output chemicals for the chemical reaction to occur. Both of these steps are quite easy for this simple material-energy looped system.

As the system grows, as explained above, it starts to bulge over a period of time. At some point, the size becomes too big that it can no longer stay intact. The membrane is not strong enough to withstand the increase in weight. Gravity and its own self-weight will cause the system to break down into two or more pieces. The elasticity and viscosity of the membrane will help in letting the split happen in a gradual and nondestructive way. This is analogous to a large soap bubble splitting into two or more smaller soap bubbles. The shape of the membrane is, however, quite irregular with several bumps (for exhaust) and dents (for intake). As a result, the split pieces will have a few cavities as well. Since there is only one chemical reaction that is occurring within the original system, the inputs and outputs will definitely be present within each of the split pieces. Therefore, we can conclude that each individual split piece is a valid active material-energy looped system. We can now say that the original material-energy looped system has successfully reproduced. In general, we will have more than one copy when it splits.

When there are enough inputs and membrane materials in the external environment, we see that this sort of primitive reproduction occurs quite frequently. The process of growth and splitting into multiple pieces occur in a loop. This type of splitting looks like a primitive version of asexual reproduction (say, binary fission in bacteria or archaea).

In addition to systems splitting into two or more parts, we can have two or more systems join to form a bigger system as well, analogous to how two or more soap bubbles coalesce. This joining process followed by a split looks analogous to sexual reproduction (joining of egg and sperm, followed by cell division).

If both joining systems have the same single chemical reaction, the resulting 'offspring' of this primitive 'sexual' reproduction is not interesting. An interesting possibility occurs when two systems that have different chemical reactions join and then split later. In this case, the split pieces run two different chemical reactions. The possibility of this happening is quite random and, in general, you do not get a great variety of systems using just this mechanism. Typically, one chemical reaction will become more dominant than the other even if the system has two chemical reactions.

In addition to splitting and joining as ways to create several new material-energy looped systems, let me revisit the original random mechanism to see how to create this system from scratch. Several instances of material-energy looped systems can, in fact, be created entirely independently and naturally from scratch. This is realistic because the design is too simple and easy to form naturally. While each of these independent designs do not look similar in shape, it is still reasonable to call them as copies of one another. In fact, this is one typical way of creating several different material-energy looped systems that run completely different chemical reactions within them.

We can now imagine a primordial earth bubbling with activity creating several material-energy looped systems with different internal chemical reactions, several of them growing bigger, some joining together, others splitting into multiple pieces and so on. In other words, it is likely that material-energy looped systems are common, abundantly present and form naturally under minimal conditions. Of course, they do not appear to be useful enough to take special notice. We, typically, dismiss them as simple bubbling activity.

For example, we do not notice them forming around us under present conditions because they do not sustain longer than other typical activity we observe around us with the diversity of life. However, in the primitive world, say, 4 billion years ago when there is no other life, this bubbling activity, though a very slow process, is observable in some regions relative to inactivity elsewhere. In the subsequent chapters, I will show how to use these active material-energy looped systems in an iterative manner over millions of years to eventually create primitive life itself.

It appears that the active-inactive behavior (discussed above as alive and resting states) of material-energy looped systems can continue forever without the need to add or change new behaviors. If this is the case, what drives the evolution of new features within this system? As mentioned in chapter 57, new behaviors emerge when multiple looped systems interact and integrate to form complex stable parallel looped systems. I will discuss a few more such natural designs in the later sections (like chapter 63).

In the above discussion about life-like behaviors, we should keep in mind that nature never attempts to design such behaviors with a specific intention. There is no inherent consciousness within these systems to evaluate what is desirable and what is not. This implies that nature would not develop a feature just to make the system survive better. Rather, the stable parallel looped architecture would naturally create a feature with an outcome of better survival i.e., better survival is an effect, not a cause or a requirement. We have seen that the stable-variant merge loop (SVML) pattern is one way to create a directed evolution using purely random mechanisms. We have seen several variations of the SVML mechanism, some requiring DNA (see chapter 54) and others without DNA (see section 53.4). Let me now focus on the problem of creating primitive enzymes (or catalysts) using material-energy looped designs.

59.5 SPL structure of primitive earth using active pumps

Let me present a brief overview of how the new approach of SPL systems lets us answer how life can be created naturally from inanimate objects. Figure 59.4 shows this overview and compares it with RNA world hypothesis. The details of this new approach will be discussed in the next several chapters. The main difference between the new approach and other existing approaches is that we use minimal structures like the active pumps discussed in this chapter and several other SPL systems that are both natural and easy to create.

If the primitive conditions on earth favor the creation of a large number of active material-energy looped systems (in the primordial soup), one highly probable scenario is that earth would accumulate a large variety of simple organic molecules over time. The collection of molecules accumulated will be sufficiently random, simple ones and are also not as directed. Nevertheless, we would expect to see a wide variety of organic molecules simply because of the active mechanisms of material-

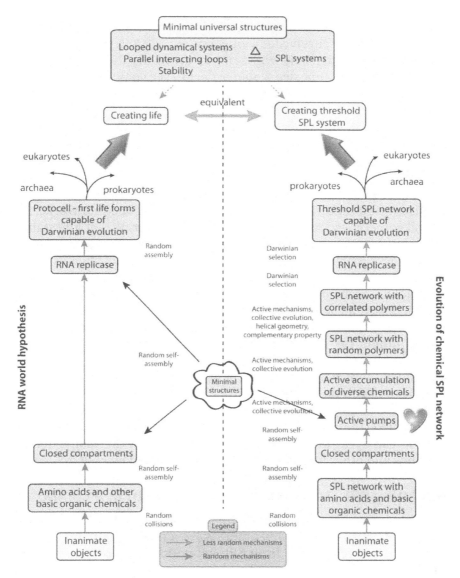

Figure 59.4: An overview of important steps comparing RNA world hypothesis and the new SPL based approach. The main difference is the creation of a collection of minimal structures like active pumps and others that enable creation of life for the first time entirely naturally.

energy looped systems as opposed to passive mechanisms, which most researchers have typically alluded to (see Figure 59.4).

The active mechanisms and the material-energy looped systems are not themselves complex. They do not require DNA, RNA, replicases and other complex molecules unlike proposals from other researchers (like with the RNA world hypothesis). We are only proposing simple primitive pumps, but a large number of them, each of which are easy to create naturally under primitive conditions and each of them do not necessarily survive for a long time as well. Yet, the collection of organic molecules generated as a whole will start to grow simply because of the active mechanisms. At any given time, we would expect to see some bubbling activity in the primordial soup because of the pumps or the active material-energy looped systems (see Figure 59.4).

A variety of organic molecules accumulated opens up the possibility of forming stable looped chemical networks. These stable looped networks will not be within a single system, as is the case with living beings now, but will be spread across a sufficiently large region. It would be difficult to predict exactly which chemicals will be present, but we can be sure that the abstract structures formed with whatever chemicals discovered until then will be stable parallel looped networks. These SPL networks will sometimes grow in size and sometimes shrink. Yet, the stable units of discovery of chemicals are looped networks. An important turning point will be when these SPL networks reach a threshold state of size, variety and stability that complex organic molecules can begin to form more easily. In the subsequent chapters, I will discuss additional mechanisms that show how nature can iteratively reach and go beyond such a threshold state (see Figure 59.4).

A 'room' initially filled with a random collection of basic molecules like CH_4, H_2O, CO_2, NH_3 and others (as discussed in section 53.1) will begin to contain a wide collection of simple organic molecules that form SPL networks provided we have a large collection of these active material-energy looped systems over a long period. In other words, order will be created from the initial disorder in this room using the newly introduced minimal structures, namely, the simple and the naturally-easy-to-create systems like the active material-energy looped systems, albeit slowly. The same situation can occur in special regions like the hydrothermal vents, under sea bed, hot springs and other locally stable regions suggested by researchers.

59.6 Simulating SPL structure of primitive earth in a computer

We mentioned above that a 'room' initially filled with a random collection of basic molecules like CH_4, H_2O, CO_2, NH_3 and others will eventually contain a wide collection of simple organic molecules that form SPL networks. It is possible to simulate the evolution of this 'room' in a computer. This is an important area of future work, which I will be working on next. For now, I will briefly describe the process.

An important first step is to represent the operation of a single pump (active material-energy looped system) using a light-weight thread in a computer. This light-weight thread (or the pump) actively sucks inputs from a locally-shared collection of 'molecules' from the 'environment'. Each molecule here can be represented by a suitable data structure in the computer, while the environment is represented as a suitable global collection that is shared by all light-weight threads. The light-weight thread executes one or more chemical reactions from a dictionary of possible chemical reactions to generate appropriate output molecules. The light-weight thread (i.e., the pump) pushes the generated output molecules back into the

environment collection. The process repeats for each such light-weight thread.

In the simulation, we can include thousands of such pumps as light-weight threads, possibly distributed across multiple computers in the cloud as well. We also maintain a large collection of chemical reactions as the dictionary of possible reactions that can occur along with their energy requirements. Each light-weight thread looks at the inputs present in the pump (after it actively sucks them from the environment), compares them with the allowable chemical reactions from the dictionary and picks a suitable set of reactions to execute based on the available energy as well. If a large number of chemical reactions are possible, we can randomly pick a few of them to execute and the light-weight thread generates the new output chemicals accordingly.

We can run thousands of these light-weight threads for, say, a few days and check the final output chemicals generated. The expectation is that the initial molecules like CH_4, H_2O, CO_2, NH_3 and others and the final molecules generated after a few days will be considerably different. The expectation is also that the resulting output molecules will form an intricate SPL network of chemical reactions to self-sustain the chemical dynamics for a long time. This is an interesting area of future work, which I am actively pursuing now. I am personally curious to see what the final output SPL network would be after running the computer simulation for a few days. In addition, the simulation can be expanded to include additional features discussed in the subsequent chapters as well.

In the next few chapters, our goal is to further reduce the randomness and bring in more structure/order from disorder in an iterative manner using the accumulated collection of molecules, the looped networks and additional naturally-easy-to-create minimal structures. Even though I will not discuss in great detail later, it is important to keep in mind that these additional iterative processes can be suitably incorporated into the above computer simulation as well. I am actively working on simulating the SPL structure of the primitive earth.

60

Primitive Enzymes

In the previous chapter, we have seen how to construct one of the first useful minimal structure, namely, active material-energy looped systems using entirely natural mechanisms. These systems help us solve one of the difficult problems necessary for creating life, namely, that of actively bringing a collection of molecules into a small region. This is also one of the simplest designs that exhibit several life-like features even without the presence of complex organic molecules (see section 59.4). We have also seen how these systems help create an SPL structure under primitive earth conditions using simple organic molecules and slow non-catalyzed chemical reactions for now. Our goal in the next several chapters is to show how to take such simple active material-energy looped designs and iteratively create complex organic molecules, enzymes, DNA, larger chemical SPL networks and eventually primitive life.

An important step towards achieving this goal requires nature to discover better ways to control the set of chemical reactions it can run. Humans control the set of chemical reactions to run in a laboratory by setting up the correct temperatures, pressures and other external conditions as well as by adding suitable catalysts. These conscious choices ensure that a series of chemical reactions occur in a desirable manner to produce desirable outcomes. However, nature does not have this option and yet only a specific reaction needs to occur at a specific time in a specific region to exhibit life-like features. How can nature control and choose which reactions to run without consciousness or awareness?

To better understand the issue, let us analyze a more basic question: what makes a single chemical reaction R generate products from its reactants (like H_2 and O_2 gases combining to generate H_2O). How can nature control how much water (H_2O) to produce and when to produce? The first step is for nature to bring H_2 and O_2, which is spread across randomly, to a specific region. The active material-energy looped systems helped solve this problem by actively sucking inputs into the system rather than wait for passive diffusion to accumulate sufficient amount of H_2 and O_2 at a given location. Nature now has better control of this step (thereby avoiding extremely low probability scenarios like bringing H_2 and O_2 into one corner of a room). Of course, the prerequisite for this step is to create active material-energy looped systems easily and naturally, which we have already discussed in chapter 59.

The next step, after nature brings molecules like H_2 and O_2 to the desired location, is to setup additional conditions so they react to produce H_2O. Examples of these conditions are suitable temperatures, pressures, volumes, pH, energy as well as additional chemicals like catalysts. Different chemical reactions require different conditions (or environment) to be setup. How does nature 'know' what these are for each chemical reaction, let alone set them up at the correct location and the correct time without help from conscious beings? The goal of this chapter is to show how nature can acquire, not perfect, but enough control to solve this problem when there was no life on earth, using once again active material-energy looped systems. As mentioned in section 59.6, computer simulation of these active material-energy looped systems is an active area of my future work.

60.1 Nature's ability to control chemical reactions

In this section, I will provide the intuition for the problem of how nature better controls the execution of a specific chemical reaction. The basic problem is the following. In the absence of conscious beings, how can a collection of molecules, moving about quite randomly, setup their own conditions like temperatures and pressures and find their own catalysts to run their chemical reaction? The molecules do not have an ability to communicate with each other in the same way conscious humans do. They cannot plan and decide to accumulate in a small region to generate a specific pressure or plan and decide to move faster to generate a specific temperature and so on, with each of the environmental conditions needed to run their chemical reaction.

In addition, there is a chicken-and-egg problem here. The reactant molecules that want to run a chemical reaction require a specific set of environmental conditions like temperature and pressure, while the environmental conditions require the molecules to be of certain energies as well. Therefore, changing one (the system) changes the other (the environment) and vice versa. The system and the environment are not decoupled from one another. In the case of environmental conditions setup in a laboratory, conscious humans perform the task of decoupling the system from the environment (see Figure 60.1).

If nature has to do this by herself, it needs a different solution to break the coupling between the system and the environment. The key insight here is the following. The 'language' used to describe the system (i.e., the chemical reaction in this case) primarily consisted of chemicals like the reactants and the products whereas the 'language' used to describe the environment was composed of (a) chemicals like catalysts and (b) conditions like temperature, pressure and volume that introduced strong coupling. Therefore, to break the coupling, we need the environment to be changed into the language of chemicals once again (i.e., not express them in terms of temperatures, pressures and others). In other words, the environment itself should be expressed in terms of a set of chemical reactions instead of temperatures and pressures. This is possible through catalysts or enzymes.

Therefore, if nature needs to solve this problem cleanly without help from conscious beings, it needs an enormous variety of catalysts (or enzymes) and, hence, remove the dependency on hard-to-control concepts like temperatures and pressures (see Figure 60.1). This allows us to view the environment itself as a system. Both the system and the environment are now expressed in the same 'language' of chemicals and chemical reactions.

One set of chemical reactions C create the catalysts, which helps a second set of chemical reactions R to run (see Figure 60.1). But the converse does not typically happen, i.e., occurrence of the second set of chemical reactions R does not usually

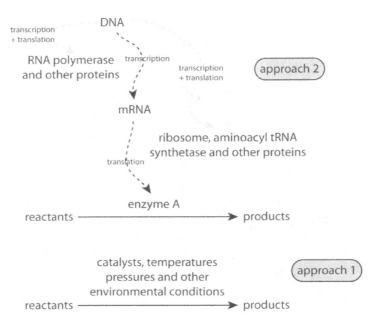

Figure 60.1: Two ways to run a given chemical reaction R. In approach 1, suitable *environmental conditions* like temperatures, pressures, catalysts and others need to be setup for a given chemical reaction R to occur. The reactants and products themselves disturb the very conditions that are being setup, making it nontrivial to control and maintain these conditions. In a laboratory, a conscious human carefully manages them in addition to bringing the necessary molecules at the correct place and at the correct time. How can nature set these up automatically? Nature uses approach 2, which eliminates the need to setup the most difficult-to-control environmental conditions. Nature replaces them with *chemical reactions* instead.

have a direct effect on the first set of chemical reactions C (for creating catalysts). For example, the enzyme creation and breakdown process occurs through a looped set of reactions. The reactions R also occur through a looped process. Together, they form part of a larger SPL network. Both sets of chemical reactions can occur independent from each other without a strong coupling. The coupling does exist, but is manifested as disturbance propagation in an SPL network (as we will see). The other environmental conditions like temperatures and pressures are largely irrelevant as long as they do not vary too much. In this manner, nature has eliminated the circular dependency on conditions (b) above.

This is the key idea that allows nature to self-discover and self-control tasks without help from conscious beings. Of course, this idea works only if nature can discover an enormous variety of catalysts and a way to generate them through chemical reactions. Nature needs active mechanisms to create catalysts, not just rely on finding existing simple catalysts passively. Examples of such catalysts are proteins and other organic enzymes. Their creation is through the process of transcription and translation from DNA. In the next several chapters, we will see how to create primitive enzymes and eventually transition to the processes that rely on DNA.

Given the importance of enzymes (or catalysts) with the new solution, let us begin by looking at enzymes, their properties and how to create primitive versions of them when there is no life on earth.

60.2 On enzymes

Before I show how to take a material-energy looped system and turn it into a primitive enzyme, I want to discuss what an enzyme is in detail. This is quite critical for understanding the complexity of life and for proposing alternate ways to achieve the same function of an enzyme. This is ultimately necessary to explain how to create life from inanimate objects. In this section, I will discuss what enzymes are, their role in chemical reactions and how they help in speeding the rate of a chemical reaction. I will only briefly touch upon how enzymes are created from DNA in a living cell. These processes (the transcription of DNA to mRNA and translation of mRNA to a protein) are thoroughly studied and covered in greater detail in standard biology textbooks. I will, therefore, refer the reader to textbooks like Alberts et al [1] and T. McKee and J. McKee [102].

If we take a quick look at physical loops versus chemical loops (that we have discussed until now), we note that a physical loop gets considerable help from the environment. The particles that make up the dynamics of physical loops move along surfaces arranged in a rather special way like, say, pipes, roller-coaster-like peak and valley surfaces or brain-like complex structures (see chapter 2). Without these external structures, the particles would move in quite random directions and the possibility of complex loops is significantly diminished.

With chemical loops, the only help from the environment that we have discussed until now is one where raw energy is supplied. With external energy and high reactivity of chemicals, you are relying on an approach where the chemical bonds of reactants are forcefully broken, through random collisions, and recombined to form products. Unlike physical loops, chemical reactions cannot have special physical surfaces to enhance the rate of reactions. One common external mechanism to help a chemical reaction proceed faster is with chemicals called catalysts. The catalysts are not arranged on special surfaces like with physical loops, but their presence is critical to improve the rate of reaction by several orders of magnitude. For biochemical reactions, the catalysts are typically proteins and they are referred to as enzymes. We will see that complex chemical loops become possible with help from the environment through special enzymes.

In living beings, almost all chemical reactions are run with the help of enzymes. Besides, each enzyme is quite specific to a given set of reactants and products. Any changes to, say, the structure of the enzyme or to the set of reactants or the products will make the enzyme unsuitable for this chemical reaction. It is, therefore, quite important to understand how these specific enzymes are discovered for each of these reactions. I will discuss this in section 60.7.

A chemical reaction is inherently a parallel mechanism. When molecules of the reactants combine to form products, they do so in parallel and, at least, partially independent from each other. The chemical reactions that we perform in a laboratory typically deal with an Avogadro number (6.023×10^{23}) of molecules. On the other hand, the reactions that occur inside a cell works with far fewer number of molecules, sometimes as few as a thousand molecules (or just a single polymer in the case of DNA).

There are several factors responsible for making a chemical reaction proceed in a given direction. Some of these are:

    (a) an ability to form collisions between reactants or products with sufficient energy, namely, above the activation energy for a given reaction,

    (b) correct orientation of these molecules,

    (c) suitable pressure, temperature, volume, relative concentration of chemicals, pH and others, and

    (d) catalysts.

Let me discuss two scenarios to highlight how each of these factors contributes for a given chemical reaction in widely different ways. One is with inorganic reactions that occur in a laboratory and the other is with organic reactions that occur within a living cell.

For the first scenario, it is quite easy to control the first three factors (a) – (c), primarily because we work with a large number of molecules (an Avogadro number). The possibility of collisions as well as the chances of having the correct orientations during collisions is quite high. In addition, we can supply large amount of energy (and, hence, energetic collisions), setup high pressures, high temperatures and high relative concentrations to make the reaction proceed faster. As for the fourth factor (d), discovering the correct catalysts is quite nontrivial for inorganic reactions. We do not know catalysts for many different inorganic reactions. This is not a serious drawback because we can compensate this deficiency by providing favorable conditions, as necessary, with the first three factors (a) – (c) in a controlled manner using our consciousness.

For the second scenario, consider organic reactions that occur within a cell. In this case, as mentioned earlier, we do not have a large number of molecules for each chemical reaction. Furthermore, we cannot use high pressures, high temperatures or high energies. They would simply destroy the cell. The set of chemical reactions that occur within a cell are quite large and all of them operate under similar external factors like temperatures and pressures, collectively termed as homeostatic conditions. Now, the low energies as well as fewer molecules also imply that the number of collisions is few and the collision energy is not typically above the activation energy. Therefore, the chances of having the correct orientation of the molecules during these few collisions are quite low. In other words, the first three factors (a) – (c) does not help the chemical reaction go faster in any significant way. Besides, with so few molecules present, a cell trying to control these three factors, by itself, for each of the chemical reaction is almost impossible. The fourth factor, catalysts (or enzymes), is the only way you can make each of these reactions occur at high rates. This is significantly different from the previous scenario.

Discovering catalysts, though difficult for inorganic reactions, is crucial for the existence of life and the chemistry of organic reactions. Let me now discuss what an enzyme does and how we can discover it for a given organic chemical reaction.

60.3 Different ways to increase the rate of reactions

Recall that a chemical reaction, ultimately, needs to breakdown the reactant molecules into suitable parts and rearrange them to form the correct products by either consuming or releasing net energy. Let me use the analogy with physical objects to see various ways we can break, rearrange and join them into desired objects just like how chemical reactions do as with addition, elimination, substitution and rearrangement reactions. I will discuss a few ways of performing these tasks next.

1. *Precision mechanisms*: With physical or mechanical systems, we can take the equivalent of small pliers to cut the structure at several interior locations. This will eventually make the structure weak and can then be broken into two pieces easily. To reassemble the broken pieces, you re-orient them in the correct way and glue them together. This approach has a fundamental problem when working with chemical systems. It requires very precise and detailed mechanisms for both splitting and reassembling the two pieces. Strictly speaking, you need consciousness and awareness to make this work correctly. This level of perfect coordination is difficult to occur naturally. Another difficulty is that the small pieces i.e., the chemical bonds that were cut can join quickly because of electrostatic affinities.

2. *Brute-force approach*: We can use brute-force or a lot of energy to hit two structures together or with another hard surface. This will eventually break them into pieces, especially when they hit the hard surface with correct orientation that expose the weak points of the structure. This is the collision approach of chemical reactions discussed earlier. When the broken pieces come together, they naturally form locally stable energy configurations (cf. a crushed building forming some random stable configurations). We choose these locally stable products as the desired products. With chemical compounds, there is a certain level of repeatability and reliability of these locally stable configurations unlike with physical structures. For the breakdown to be effective: (i) the temperatures and pressures should be high and (ii) the collisions should occur with proper orientation and with energies high enough to cross the barrier of activation energy. As mentioned earlier, this approach is not suitable in living cells as they would damage the cell.

3. *Using complex structures with gigantic sizes*: Consider building a large and stronger machine that clamps and encloses in a snug way the structure that you want to breakdown and reassemble. The structure can then be broken into two pieces by applying a torque to twist the clamped machine. If the machine has a unique shape like a mold, it is possible to create desired output structures by twisting and turning the machine a few times in a specific sequence.

   With chemical reactions, the equivalent mechanism is the so-called induced-fit model of enzymes. Enzymes are complex and gigantic molecules compared to the molecules they assist with the chemical reaction. The reactants enter the active site of the flexible enzyme and mold it until both the enzymes and the reactants attain a stable transition state. The large structure of the enzyme is advantageous because of, at least, a couple of reasons. Firstly, the energy needed to twist or turn the enzyme is not too high because of the multiplicative leveraging effect owing to the enzymes' size. Secondly, you do not need to be extremely precise on where and how the energy is applied on the enzyme to produce the desired catalytic effect. In terms of the previous analogy with large mechanical machine, this means that you can hold the machine in any way and anywhere you want when twisting, breaking and reassembling the interior enclosed structure. Both these advantages cause the enzyme to work very effectively without any specific guidance and precision. Incidentally, the induced-fit model also explains why a given enzymatic chemical reaction does not proceed in the reverse direction. This is because the 'products' do not form an induced-fit with the same enzyme, unlike the reactants.

Hence, the chemical reaction can proceed extremely fast even with a mere presence of *random* energy. Of course, for this approach to be effective, the shape of the enzyme is critical to hold the substrates at the precise positions as both of them mold together appropriately. You can see that this approach effectively lowers the activation energy needed to make the chemical reaction proceed in a given direction.

It is possible to have other internal mechanisms to help lower the activation energy of a chemical reaction when using catalysts. However, the approach discussed in (3) above is how most enzymes operate within a cell. Of the three approaches discussed above, the most efficient approach, in terms of energy requirements, is with enzymes. This was a major milestone in evolution. In section 60.7, I will show how the material-energy looped system can be made to behave like a catalyst or a primitive enzyme and perform the same function. This primitive design would not be as effective as an enzyme. Nevertheless, it is a simple design that can occur naturally and yet provide a means to generate a large variety of organic compounds in the absence of DNA and proteins. I will also show how to transition from this primitive enzyme (i.e., a material-energy looped system) to the amino-acid based enzymes that are currently used by all living beings.

Enzymes are also ideal for *controlling* a chemical reaction, as discussed in section 60.1. Without enzymes, the only ways to control a chemical reaction is by adjusting the external conditions like pressure, temperature, volume, energy and concentration of reactants and products. These external factors would achieve a maximum number of correctly oriented collisions that exceed the activation energy. However, nature cannot rearrange the system and the environment effectively to create these favorable external conditions if it were to rely only on random mechanisms for each reaction. One of the reasons is that a chemical reaction both effects and is effected by external conditions. Therefore, this coupling makes it quite difficult to control the external conditions. Fortunately, enzymes eliminate the need for such precise control of external conditions. A chemical reaction controlled by another 'chemical' (i.e., an enzyme) provides several ways to have loops of reactions that self-regulate effectively, which I will discuss later.

Let me briefly discuss one example of an enzyme to show both their complexity and effectiveness in living beings. A single molecule of enzyme carbonic anhydrase can process about a million molecules of CO_2 every second to convert CO_2 and H_2O into H_2CO_3. Without this enzyme, the forward reaction runs at less than once per second. This reaction is required for red blood cells to transport CO_2 from the tissues to the lungs. Without enzymes like carbonic anhydrase, these chemical reactions occur at speeds that cannot help life exist in its current form.

60.4 Which came first – DNA, proteins, RNA or others?

In all present-day living beings, enzymes (or proteins) are created only from DNA. They are structurally far too complex to be created by any other means (see Alberts et. al [2] for several examples). The creation process, namely, the transcription and translation steps, are too complex to occur by themselves naturally for the first time. In fact, a large number of other enzymes assist these processes like: RNA polymerases to unwind and translate a DNA into mRNA, enzymes to splice non-coding segments in mRNA, proteins to transport mRNA out of the nucleus, proteins to add polyA tail to protect mRNA from getting degraded by ribonucleases, ribosomes for translation, creation of twenty tRNA's and so on. Even though some of these steps are absent in prokaryotes, we still require enough number of proteins to assist the creation of an enzyme.

There are no known mechanisms to create a protein without the assistance of these special enzymes. However, each of these special enzymes is itself created from unique DNA sequences using once again the very same processes. This is a chicken-and-egg problem. How are they created? Each of these special enzymes would need to be created from DNA sequences using the entire collection of the very same special enzymes.

Given this self-dependency (or chicken-and-egg problem), the question is which molecules came first when life is just beginning to form for the first time, about 4 billion years ago – DNA or the enzymes? We need the DNA to create any enzyme. At the same time, we also need the special enzymes to create any enzyme, including the special enzymes, from the DNA. This is the chicken-and-egg problem. In other words, you need all of the special enzymes and the corresponding DNA sequences *simultaneously* to jumpstart the enzyme creation process itself.

Researchers have suggested RNA-world hypothesis as a way to resolve this dilemma for the origin of life. It turns out that some RNA molecules can both store genetic information like DNA and have catalytic properties like enzymes. Such RNA molecules are called as ribozymes. The hypothesis then is that primitive life forms based on RNA, instead of DNA and enzymes, predates the current life forms. Only later, they were replaced by much more efficient DNA-enzyme based life forms. We use the catalytic-RNA's in place of the necessary special enzymes as the initial primitive set to jumpstart amino acid based enzymes.

Since then, other similar hypotheses have also been suggested for the primitive life forms that predates the RNA world. These are based on molecules like peptide nucleic acids (PNA), threose nucleic acids (TNA), glycol nucleic acids (GNA) and polycyclic aromatic hydrocarbons (PAH).

In spite of all these hypotheses, one main problem is not addressed, namely, how each of these molecules came about for the first time. The only suggestion so far is through random mechanisms or self-assembly. The approach has always been to pick simpler and simpler molecules at each stage and use randomness as the only way to create them. However, DNA and enzymes with the above-mentioned specific properties are too complex to form together randomly (or through self-assembly). Therefore, the suggestion is that RNA, with the same specific properties, is a simpler alternative. Whenever these molecules are difficult to form randomly under primordial conditions, other molecules like PNA and PAH have been suggested. There is no attempt to outline any underlying structure within these molecules or within the processes that makes them occur in a directed, nonrandom and in a natural progressive sequence of steps.

I will now discuss the problem of creating primitive enzymes by generalizing material-energy looped systems in different ways. For this, we first need to relax the restrictions discussed in section 59.3 for material-energy looped systems.

60.5 Relaxing the restrictions in material-energy loops

One interesting property of material-energy looped systems is the following. It actively sucks input into itself, lets a single chemical reaction run, pushes out its outputs and continues this process in a loop. The same chemical reaction, if it were occurring outside the material-energy looped system, would run entirely randomly instead. The specific structure of the material-energy looped system has enabled a chemical reaction to run not only efficiently, but also actively in a controlled manner for a much longer time, than if it were happening outside through random collisions.

The chemical reaction within a material-energy looped system should be thermodynamically favorable to occur. This is an inherent assumption in all of the discussions. Just having active suction mechanisms and a closed system is not sufficient. The advantage with the material-energy looped system is that the same thermodynamically feasible chemical reaction occurring outside can now occur even more easily and actively inside the material-energy looped system.

We also saw that this system lets the chemical reaction run as long as there are enough external inputs available. When the inputs are exhausted, the physical structure will still remain intact even though the reaction stops. When the inputs re-appear, the chemical reaction and the entire looped process gets a chance to restart again just as before. What is interesting is that since the physical structure is still intact, an entirely different set of inputs can appear the next time. They can take the place of the original set of inputs. As a result, a different chemical reaction could run within this structure until once again these new inputs also run out.

There is one subtle difference though. In the original material-energy looped system, I have made an assumption that the chemical reaction is an exothermic one and that it generates more energy than what the entire system consumes. The new chemical reaction that occurs may not satisfy this condition. If this were the case, the chemical looped system may not continue to occur as before even if abundant inputs are available. The same physical structure but with lesser internal energy will be unable to perform the original suction and exhaust mechanisms. For example, the new chemical reaction may be endothermic instead, requiring extra energy instead of releasing energy.

With cases where there is less internal energy, one way to compensate for this is by consuming energy from the external world. It is generally assumed that there is considerable amount of freely available energy under primitive earth-like conditions. This external energy can be used to compress the membrane, thereby pushing the outputs out of the system (and possibly suck inputs into the system). When the chemical reaction inside the structure does occur, the elasticity of the membrane can cause it to expand and suck inputs into the system.

Therefore, even for an endothermic chemical reaction and, in fact, for all types of chemical reactions, the abundant freely available external energy acts as a replacement for making the material-energy looped system continue to occur for a long time. In this sense, we can still call this system as a generalized material-energy looped system with the understanding that the external energy acts as a substitute when the internal energy is not sufficient. In other words, in all cases of chemical reactions, the generalized material-energy looped system can continue to occur as long as there are inputs available for intake, with the energy supplied either internally or externally.

Also, there is no reason for the generalized material-energy looped system to stop and restart. It can keep on running even when the current inputs run out simply because some other inputs and, hence, a new chemical reaction will take its place. The primordial conditions with excess freely available energy would make the system continue its dynamics actively for a much longer time than when we insisted that the chemical reaction be exothermic. Therefore, the generalized material-energy looped system has become more repeatable and reliable than the previous design.

Multiple loops within material-energy looped systems: Another possibility that can occur is the material-energy looped system runs two looped chemical reactions. One is the main chemical reaction of interest and the other is a generic chemical loop whose main purpose is to generate energy to power the compression and expansion as well as the main chemical reaction. This was a variation to the material-energy looped system, which we called as Design *B* in section 59.2. This generic chemical loop is common across a number of material-energy looped systems while the main chemical reaction is unique to each system.

As the system actively sucks inputs, it takes inputs needed for the generic reaction as well as the main reaction. When you allow the possibility of a reaction whose sole purpose is to generate energy for everything else, there are other interesting possibilities as well. Such a reaction is analogous to the ATP reaction within all modern cells. The inputs and outputs for this energy reaction can be shared between two systems. For example, one system can generate outputs that are used as inputs for the other system and vice versa, while both systems use the same shared mechanism to generate sufficient internal energy. This is analogous to how plants and animals exchange material inputs for Kreb's cycle and Calvin-Benson cycles. This is yet another generalization to make the material-energy looped system become increasingly more reliable and repeatable.

60.6 Transition from nonliving to life-like systems

With abundant amounts of freely available energy, one would think that the same set of chemical reactions could occur even without the help of the generalized material-energy looped system, namely, just from random collisions. However, the most important and significant difference is that one is a passive and random mechanism while the other is an active and directed mechanism. This latter feature is the single most critical difference between living and nonliving systems. The generalized material-energy looped system is the simplest and a naturally easy-to-create system that offers this feature unique to living beings before the origin of life.

This generalized material-energy looped design marks the beginning of the separation of the world into two groups: one with nonliving systems and the other with life-like systems. This is especially aided with features like primitive versions of survival, growth and reproduction discussed in section 59.4. One group operates entirely randomly, while the other group actively begins to seek its inputs, grow, multiply and survive in a completely nonrandom way.

A simple form of randomness has given rise to a nonrandom generalized material-energy looped system as the minimal structure (see section 53.1). The reason why nature was able to reduce the randomness is because of the *memory* of stable dynamical and physical structure of a generalized material-energy looped system. The same situations apply even with the formation of simple stable chemical molecules or with the formation of looped chemical networks (see section 53.1). The definite structure of an active material-energy looped system is simple and quite natural to form. It is powerful enough to start dividing the world into two groups: a nonliving and a life-like world.

Let me start looking at how this division will differentiate more and more, creating detailed substructures within the group that contains life-like systems.

60.7 Primitive enzymes using material-energy loops

While the generalized material-energy looped systems of the previous section has life-like features like an ability to survive, grow and reproduce, they are still too primitive compared to existing life forms. In order to evolve to the level of complexity of current life forms, a wide repository of enzymes (or catalysts) and organic chemicals become necessary. In this section, I will address how to create catalysts for most chemical reactions using the generalized material-energy looped systems.

I will call them as *primitive enzymes* even though they do not have any amino acids. They will, however, behave like catalysts, just like enzymes. These primitive enzymes are a good substitute for catalyzing chemical reactions in the primitive world when there are no complex proteins or DNA yet. They are not constructed in the same way as the modern enzymes i.e., from DNA. Yet, they can be constructed reliably and repeatedly, just like the modern enzymes. They are also not as efficient as modern enzymes. Our goal in chapter 61 and beyond is to show how we can replace the primitive enzymes with the modern amino-acid based enzymes using a definite and directed process.

To begin, first note that the pump designs A-D (see section 59.2) behave functionally as catalysts, similar to enzymes in a cell. This would allow us to replace the entire family of chemical SPL designs that relied on enzymes (like Nicholson [114] and Figure 2.11) with ones that use active pumps.

To see the catalytic property, note first that any chemical reaction occurring passively in the environment through random chance, self-assembly or even within a sealed flask in a lab can now occur within the enclosed pumps proposed here. Secondly, the rate of reaction is much faster using these designs (due to active suction mechanism, confinement within the membrane and accumulation of inputs and energy through repeated cycles, better temperatures, pressures and other conditions) compared to the alternative of occurring openly in the environment. Thirdly, the designs themselves do not undergo any change while running the internal chemical reaction.

These properties together define a catalyst. These pumps, however, are *universal catalysts* as they are not specialized to a given chemical reaction (unlike enzymes). The same pump can run one chemical reaction now and a different reaction a few minutes later. If we use design C instead (see Figure 59.3), in which the shape of the intake and exhaust openings are morphed to allow and even assist a specific set of molecules to enter or leave during the compression-expansion cycle, the design becomes analogous to the induced-fit model of an enzyme (Garrett and Grisham [48]). The pumps continuously and autonomously draw inputs and push outputs for a long time as an active looped process. This operation, though simple and is theoretically feasible, would need to be verified experimentally. Let me now discuss more details on design C since behaves as a primitive enzyme.

Material-energy looped system executing a single chemical reaction: In order to understand the catalytic function of a material-energy looped system, we should first specialize it to run just a single chemical reaction, namely, the one being catalyzed. In the previous section, we have seen how a material-energy looped system can be generalized to cover all types of chemical reactions i.e., both endothermic and exothermic reactions. Let us now look at a special case in which a material-energy looped system runs just a single chemical reaction.

Consider a material-energy looped system shown in Figure 59.3 discussed in section 59.2 as a variation called design C. This system does not stay in one place over a long period of time. As it is subject to random motions from external effects like wind and other disturbances, it accumulates a number of small particles on its surface, much like how dirt settles on most systems. As the compression and expansion continues to occur, there is a slight increase in stiffness. This will not be as significant on most of the surface because the same random wind and rain effects will periodically wash the dirt away. However, the places where this does have a significant effect are near the openings for intake and exhaust.

It is quite natural that holes tend to be covered from random effects. However, the system is powered, either internally or externally, to let a flow of molecules in and out of the openings. Part of this energy is consumed to clear these openings. Therefore, even if we started from a perfectly circular opening, the shape of the opening will become quite random over a long period of time (if the system survives). Nevertheless, the opening will be shaped in such a way that molecules can flow in and out of this opening.

Therefore, the second law effects cause a restriction in the flow of molecules by altering the shape of the openings. If the same types of molecules flow in and out of a given opening for a long time, then the shape will become molded to eventually allow just those molecules (see Figure 59.3). The specific shaped input molecules will form an impression on the surface membrane (see design C). The elasticity (or stiffness) of the membrane can potentially harden the shape of the opening to conform to the shape of the molecule modulo different orientations (as they will be moving in and out of the openings). While this explanation is theoretically reasonable, we still need to confirm design C experimentally.

To have additional confidence that design C is reasonable, consider natural examples like with the shape of openings in rocks, stones, rivers, streams and so on. Any flow-based system typically conforms to the shape and other properties of the fluid particles that flow through it. For example, if there is a landslide, mudslide or any other flow that has definite shaped particles within the fluid, the shape of the openings created is a clear indication of the shape of the particles that were present in the stream.

Therefore, we can now say that the material-energy looped system has become specialized to allow only a single chemical reaction to occur within itself. If these inputs are exhausted, the system stops its dynamics as before. However, it may not be able to restart its dynamics if a different set of input chemicals become available. While designs A and B (of section 59.2) would have allowed this possibility to occur, design C would not. In the worst case, only the same input molecules must re-appear if the system were to spring back and start its dynamics once again.

In reality, this system may allow other chemicals to flow in as well, especially when the molecules are not too big. For simple molecules, the shape of the opening is not too constraining. Small molecules typically have high root mean square (rms) speeds. As a result, the openings tend to get larger and less specialized. The high energy of the motion of molecules tear opens the hole much larger than the size of the molecules as they flow through. However, when the molecules are large, complex and much slower (like larger organic molecules), the opening has a tendency to conform to the shape of the molecules flowing through them. Therefore, design C is somewhat leaky and inefficient. It specializes to more than one chemical reaction, though not to all of them.

Material-energy looped system as a primitive enzyme: The specialization design C mentioned above is an important one. It can occur entirely naturally and was, in fact, aided by random second law effects. This system can also be created repeatedly and with increasing reliability. Let me reiterate the properties of the specialized material-energy looped system as it pertains to catalytic ability.

1. *Increasing the rate of chemical reaction without itself undergoing any change*: Design C actively pulls inputs and pushes the products out after running a specific chemical reaction. Without this system, even though the same chemical reaction can occur naturally, it only happens through random collisions and entirely passively. Therefore, we can say that the specialized material-energy looped system is assisting and speeding the rate of a chemical reaction through active mechanisms compared to the random collision based mechanisms. This is analogous to how an enzyme or a catalyst does. Furthermore, the system itself does not undergo any chemical change. Only the chemical reaction occurs at a faster rate. This ability can be considered as the very definition of a catalyst.

2. *Induced-fit*: The shape of the input and output openings do not need to be rigid. They can be elastic as well. For example, as the input molecules begin to enter the opening, the membrane can stretch and conform more precisely to the shape of a group of similar molecules. Once the shape is altered, the molecules would be able to enter the system much more freely. The flexibility of the membrane can even assist in actively pushing the molecules into the system. The induced-fit model (see section 60.3) suggests a similar mechanism for how an enzyme works.

3. *Specialization to a given set of inputs and outputs*: Designs A and B and their generalizations discussed in section 59.2 were capable of running any chemical reaction within the system. These systems were merely acting as a shell to carry out any chemical reaction in an active and much more controlled manner than random collisions. Design C, however, is specific to a given set of inputs and outputs. This is analogous to how modern enzymes have a similar level of specificity towards input and output chemicals.

4. *Reliable way of creating the specialized material-energy looped system repeatedly*: We have already seen that the material-energy looped system can be reproduced reliably in section 59.3. However, the specialization discussed here (design C) required additional structure at both the intake and exhaust openings. We have seen that this can form naturally as well using just random effects. Therefore, every time the system reproduces (by simple split for now – see section 59.3), the 'offspring' needs to operate long enough to specialize itself. The catalytic properties evolve to become efficient during this time. This is not how modern enzymes are created from DNA. Yet, you can reliably recreate the same specialized system, although quite slowly and inefficiently compared to modern enzymes. The resulting 'offspring' will function similar to the 'parent' system.

   It should be noted that the material-energy looped system acts as a catalyst even without this specialization. Therefore, the additional structure is strictly not needed and the reproductive abilities discussed in section 59.3 is quite sufficient for repeated creation of this primitive enzyme. Either way, unlike the modern enzymes, the repeated creation happens without any template like the

DNA, though in a reliable way.

Given these properties of the specialized material-energy looped system, we can say that this specific structure (which is analogous to a protein structure) is itself a primitive enzyme for the chemical reaction that runs within itself.

Each chemical SPL system from the infinite family identified in section 2.6 were composed of several uphill and downhill reactions. The pumps, now acting as primitive catalysts, are capable of running each of these reactions, albeit slowly and inefficiently. More generally, the same infinite family of chemical SPL systems in section 2.6 can now be modified to use pumps instead of using enzymes (for the inclined plane in the mapping), at least for reactions involving simple non-polymer molecules.

If we have millions of these pumps in a sealed room, the outputs from the pumps will be taken as inputs to generate new chemicals iteratively. This is another stable looped process that continues to occur within the sealed room. As a result, we begin to see new collection of chemicals beyond the initial random collection we started with. The new chemicals and the pumps acting as catalysts will start to form chemical SPL networks. This system exhibits a self-replicating property (see sections 3.7 and 57.2). This is the beginning of generation of order from disorder in which random chemical collections begin to lock themselves to form looped networks. As mentioned in section 59.6, computer simulation of these active material-energy looped systems is an active area of my future work.

60.8 Features of the primitive enzyme

Let me briefly outline a few features of this primitive catalytic system. Some of these features are similar to existing enzymes, but with a reduced efficiency.

1. *Complexity of primitive enzyme*: The material-energy looped system is more complex than the input reactants or the output products of the chemical reaction running inside the system. As mentioned in section 60.3, certain types of complexity makes the system efficient and easier to achieve catalytic features.

2. *Multiple input molecules processed simultaneously*: This material-energy looped system processes several input molecules together at any given time. The closed membrane provides better conditions for each reaction to occur. On the other hand, the modern enzymes typically work with one set of input molecules (or just a few, at most) at a time.

3. *Efficiency as a catalyst*: The material-energy looped system is not as efficient as modern enzymes for several reasons. Firstly, the random mechanism for creation and reproduction creates uncertainty and defects from the ideal design. Secondly, the shapes of the openings are not too perfect to allow only one set of input chemicals. Thirdly, the chemical reaction and the distribution of energy occurs less efficiently. In fact, at every stage, starting from sucking inputs to pushing outputs out of the system, including the design flaws, there are inefficiencies compared to modern enzymes.

4. *Multiple specialized openings and multiple chemical reactions within the system*:

It is quite possible for this system to have multiple specialized openings to suck different types of inputs. This could have happened in two ways, serially or parallelly. The system could have been exposed to different external inputs serially, at different times (with one input at a given time), or in parallel at the same time. Each of these specializations will alter the openings for one or more inputs. The chemical reactions can also occur within the system simultaneously or one at a time. Some modern enzymes already exhibit this feature.

5. *Looping the system and the environment*: If nature wants to make a given chemical reaction occur favorably, it needs to adjust the temperatures, pressures and other environmental conditions. This is not easy. A material-energy looped system, on the other hand, can evolve and specialize itself suitably to favor a given set of inputs. Different copies of such systems favor different sets of chemical reactions. We now begin to see the evolution of systems that create conditions to control the very environment itself. Without these systems, the environment stays a certain way. However, with these systems, the environment begins to alter considerably. This is just the beginning of a more elaborate control. In other words, the system and the environment together become a looped process.

With all of these additional features, the specialized material-energy looped system i.e., a primitive enzyme fundamentally begins to alter the landscape on earth. Several new chemicals begin to be discovered using these primitive enzymes. This is because you no longer need to rely on random collisions to make a series of chemical reactions to occur. Each reaction can now occur reliably and favorably using these primitive enzymes. Multiple reactions now have a chance to be linked together with the products of one reaction acting as reactants to another reaction. Previously, we would need all of these chemicals to be generated either as loops or be present at the correct place at the correct time in the correct order.

The primitive enzymes now act as a memory to the chemical reaction it specializes. This is similar to how enzymes are considered as memory for chemical reactions in a living being. Of course, DNA turned out to be a better source of memory now. However, this is just the beginning of the creation of more complex structures. The SPL architecture was critical to initiate as well as continue the process of creation of more complex chemical systems and life itself.

60.9 Disturbance propagation in material-energy loops

From section 58.2, we have seen that any linear sequence of steps can be represented as disturbance propagation in an appropriately chosen network of loops. The material-energy looped system is one such network with just three loops. I will represent the definition of a catalyst as disturbance propagation within this simple three-looped network. This connection is important because, as we have already seen in chapter 58, it will pave way for the creation of complex sequence of steps like the transcription of the DNA, creation of an organ and others.

By definition, a catalyst is a system (or a chemical) that speeds the rate of a chemical reaction without itself undergoing any chemical changes. In simple terms, there are three parts to this: (a) a subsystem that actively takes inputs, (b) favorable conditions to carry out the chemical reaction and (c) a subsystem that actively release the outputs. The rate of reaction will be considerably increased compared to a random collision based approach because of the active subsystems and other favorable conditions. To discover a design for this feature, we want to represent each of the above three parts as a loop and then interconnect them appropriately.

We can now express the above three steps (a) – (c) of a catalyst to an equivalent three loops of a material-energy looped system as follows: (i) compression and expansion at the intake that actively suck the inputs, (ii) a chemical loop like a reversible reaction within a favorably enclosed elastic membrane and (iii) compression and expansion at the exhaust that actively push the products out of the system. Each of these three loops can operate independently. However, these loops are linked in a specific way. For example, the intake loop is linked to the chemical loop and the chemical loop is linked to both the intake and the exhaust loops. The similarity between (a) – (c) of catalyst features above and (i) – (iii) of the material-energy looped system is now clear.

To complete the analysis, notice that when there is a disturbance to the intake loop, say, like a faster (or slower) than average compression and expansion, this results in more (or less) intake of inputs. This disturbs the chemical loop to proceed in a direction that opposes the change in inputs, according to Le Chatelier's principle. For example, more inputs make the forward reaction to occur, producing more products and vice versa. This disturbs the exhaust loop. More outputs produced make more of them to leave through the exhaust and vice versa. Therefore, an initial disturbance that slows the oscillation at the intake loop uniformly slows down the three loops because of the linear arrangement of the network of loops. Similarly, faster oscillations at the intake uniformly make it faster.

In this manner, an important feature like a catalyst (or a primitive enzyme) is discovered and expressed as a network of loops in a specialized material-energy looped system. Given this looped network representation, it is possible to control the operation of a primitive enzyme by propagating a disturbance through the looped network.

With these specialized material-energy looped systems, we now have primitive catalysts for most chemical reactions. Since they are active processes that convert reactants to products, a large number of chemical reactions can now occur easily compared to the case when we were to rely only on random collisions based approach. The chemical landscape on the planet begins to change drastically. The slow chemical reactions based on random collisions prior to material-energy looped systems are gradually replaced by these primitive catalyzed chemical reactions. This transition happens uniformly across all chemical reactions. We now have relatively faster rates of reactions uniformly across all reactions rather than non-uniformity that would have resulted in discontinuity in the collection of molecules generated. We have seen that such a discontinuity is detrimental to the existence of life in section 56.3. The primitive enzymes using special material-energy looped systems avoids this issue. Computer simulation of primitive enzymes is an active area of my future work.

61
Monomers to Polymers

In the previous chapter, we have seen how to create primitive catalysts for a set of reactions that occur within a local region. The collection of reactions begins to form a network of loops. The presence of loops, as we have seen in section 57.10, can help create new molecules. If we have organic chemicals, nature would be able to create a large repository of molecules, chemical reactions and primitive catalysts to work with. With organic chemicals, we have a number of functional groups like carbonyl, carboxyl, aldehydes, ethers, esters, alcohols, ketones, amides, amines, sulfhydryl and the ability to chain them to form a large set of complex compounds (see Garrett and Grisham [48], Koolman and Roehm [89] and T. McKee and J. McKee [102]). These functional groups have special properties making them suitable to create quite complex polymers like DNA, RNA, proteins, polysaccharides and lipids.

In this chapter, I will discuss how to take a collection of monomers and create polymers. All living beings require, not random polymers, but specific chains of polymers. In this chapter, I will discuss dynamical and geometric properties of polymers that give them unique abilities towards creating and sustaining life. In the next chapter, I will show how material-energy looped systems can once again help in creating specific chained polymers like DNA reliably and repeatedly, though not with the same precision as with current enzymes and DNA. Nevertheless, these material-energy looped systems can perform a primitive version of replication and transcription.

At this stage in evolution i.e., before a primitive life-form exists, the errors when copying the DNA are, in fact, quite helpful. They generate a variety of enzymes, DNA and RNA since the errors manifest as random mutations. Evolution needs this variety when life does not yet exist. Recall that with existing life-forms, random mutations are less common. Living beings have elaborate DNA repair mechanisms to ensure high fidelity of DNA replication. This aligns well with the objective of preservation of life. However, before the existence of life, it is more critical to have a variety of DNA, RNA and proteins rather than trying to preserve specific molecules. It is interesting that the low fidelity of material-energy looped systems do help with this goal quite naturally.

The low fidelity copies of DNA, RNA and proteins also solve another problem before the existence of life. Currently, the only known natural ways to create and maintain a large variety of these polymers is through the processes of life. The material-energy looped systems, which cannot be called as primitive life yet even though they exhibit a few life-life features (see section 59.4), are entirely new mechanisms proposed in this book as an alternative to creating a variety of polymers. Current research, on the other hand, offers random chance mechanisms as the primary approach. They will cause the breakdown of complex molecules into simpler constituents (because of the second law of thermodynamics) with no obvious way to rebuild them reliably and repeatedly. Therefore, that life resulted in a diversity of organic chemicals is just as true as the diversity of organic chemicals gave rise to life, without the newly proposed active material-energy looped systems.

Let us start by discussing the dynamical and geometric properties of polymers that help chemical reactions occur favorably. I will refer the reader to standard textbooks on organic chemistry to see examples and properties of diverse organic chemicals (see Solomons and Fryhle [144] and McMurry [103]). As discussed in these textbooks, different types of bonds (ionic, covalent or hydrogen bonds), the bond strengths and the bond lengths give rise to a variety of stable molecules. The shape of molecules that correspond to the most stable configuration has interesting geometric properties. This has been studied in some detail for simple molecules like CH_4, C_2H_4, C_2H_2 and C_6H_6 (i.e., alkanes, alkenes and alkynes). Even for complex molecules like enzymes, we use geometric shapes of molecules as an equivalent way to describe dynamically stable configurations. While it is hard to characterize these shapes in precise terms, it is generally believed that they obey a minimum energy principle. This principle guarantees that the same shaped molecules are created repeatedly most of the time. Let us explore these concepts in more detail now leaving the computer simulation of these ideas (see also section 59.6) as future work.

61.1 Dynamical properties – locality and directionality

In general, large complex molecules have several degrees of freedom. Whenever forces or energies are supplied to the molecule, they are not typically spread uniformly across the entire molecule due to its large size. Instead, these forces are only localized to a part of the complex molecule. This causes the molecule to be locally twisted, bent, compressed or expanded relative to the rest of itself. We cannot treat it as an idealized point particle when thinking about how the forces and energies act on it. Rather, the molecule behaves like an elastic object. It seems reasonable to make such idealizations for simple molecules like O_2, CO_2 and H_2O but not for complex molecules like proteins, DNA and RNA.

This is actually an advantage because such molecules are capable of absorbing considerable amounts of energy and force without breaking down. They restore to their original shapes and structures afterwards, owing to their elasticity. While they absorb, deform and restore their original shape, the molecules are capable of performing useful work as well. For example, this localization property is critical for the enzymatic properties of complex three-dimensional proteins.

In addition to this localization property, complex molecules exhibit directionality property as well. Directionality here means that a molecule behaves efficiently along a few specific directions but not along others. For example, a force that locally twists a molecule does so easily in one direction than in the opposite direction. This is because motion along certain directions are easier than other directions. For example, the DNA helicase molecule uncoils the DNA as it moves along the length, during replication. When replicating the DNA, it is necessary to uncoil the double helix DNA. This specific functionality is derived from the directionality property of helicase, as ATP supplies the energy. Coiling helical molecules are easier in one direction than the other direction. Similarly, compression in different directions, stretching, bending and a combination of different stresses produces different strains along different directions.

In order to explain features like transport of molecules, transcription and translation of DNA using natural mechanisms, we use the locality and directionality properties of molecules in combination with different types of chemical reactions. When different large molecules temporarily bond together, say, during transcription, the energies or forces are small that they do not produce rigid body translation or rotation of the entire complex. Instead, you get a directional and local motion. The local degrees of freedom are constrained by these large molecules with different bond lengths and bond strengths. We can then express these mechanisms as a sequence of steps, which can later be translated into disturbance propagation in an appropriate network of loops.

61.2 Geometric properties of complex molecules

In addition to dynamical properties, complex molecules have useful geometric properties as well. This is more apparent in organic polymers. Polymers are made up of several smaller and repeating chemical subunits. These smaller units (like, for example, amino acids and nucleic acids) are typically connected together using covalent bonds. Some of the most common examples of naturally occurring polymers are polynucleotides (DNA and RNA), polypeptides (proteins) and polysaccharides (carbohydrates). These polymers have unique physical and chemical properties. Even though creating most polymers require help from specific proteins, there are some cases where they are capable of growing spontaneously under minimal conditions. Phospholipids that are part of a cellular membrane is one such example. Irrespective of how they are formed, the geometry naturally introduces several important life-like features, as we will see.

When we think of how chemical polymers grow, it is sometimes useful to think in terms of the equivalent growth in physical systems. Crystals and snowflakes are some simple examples of such solid physical structures that can grow to large sizes. Here, a number of smaller subcomponents are linked together to form complex repeating patterns. Crystals are quite common with most materials and the subunits can be connected with either ionic bonds (salts) or covalent bonds (diamond and graphite). These physical structures also have complex geometric shapes. However, the fact that they are solids, as opposed to fluidic polymers, makes them more difficult to manipulate and work with. For example, it is quite common that biopolymers like proteins are constantly broken down into constituent units (individual amino acids) and reassembled to form other polymers with different physical and chemical properties. This is not easy to do with crystals even though there may be some geometric similarities between the two. Let me now list a few geometric shapes for chemical polymers commonly found in nature:

1. *Linear*: Saturated hydrocarbons can form long chains of linear structures while some of the unsaturated hydrocarbons form branched-chain structures.

2. *Circular*: Phospholipids form circular shapes by forming lipid bilayer. The polar and nonpolar molecules arrange in two layers with the hydrophilic regions on the outer side and hydrophobic inner regions. These are very critical for all cellular membranes in living organisms. Phospholipids spontaneously form liposomes, which are spherical vesicles of bilayer membrane, when introduced in an aqueous solution.

3. *Helical*: Chains of nucleic acids like DNA and RNA molecules form helical shapes. Other examples of helically shaped polymers are microtubules, actin and intermediate filaments that form the cytoskeleton of cells. Microtubules also play a critical role in the transport of chemicals within the cell. I will discuss how microtubules are suitable for actuation later in section 63.1.

4. *Complex three-dimensional shapes*: Amino acids chains, namely, proteins, are the best examples of polymers that fold into complex three-dimensional shapes spontaneously. Considerable help from other proteins is needed to assemble these amino acid chains. However, once the polymer chain is assembled, the complex 3-D shape of proteins seems to form with little or no help. For some proteins, chaperones were discovered recently to assist in the assembly of these structures (see Ellis and van der Vies[37] and Ellis [38]). This complex 3-D shape is critical for them to act as catalysts enhancing the speed of chemical reactions, sometimes a million fold.

In the next few sections, I will use these geometric and dynamical properties to explain how we can alter the material-energy looped systems to create new additional features like transcription and replication of DNA molecules (see sections 62.1 and 62.2).

61.3 Parallel versus serial mechanisms to create polymers

The material-energy looped systems considered simple molecules as both inputs and outputs until now. However, with organic molecules, it becomes possible to create large and complex compounds like, say, polymers. Our objective is to see how we need to modify the material-energy looped system when creating such complex polymers. I will do this from the next section. In this section, I want to discuss the possible ways to create polymers and highlight the differences between serial and parallel mechanisms.

For simple molecules, the problem of creating the same molecule repeatedly is not too difficult. All we need to ensure is that the simple molecule is the product of one or more chemical reactions. As long as the inputs for any of these reactions are available, we can generate the molecule of interest by running any one of these reactions. We can even have an active mechanism for a given reaction using the corresponding material-energy looped system. The geometry of the molecule would be simple due to its small molecular weight. As a result, trying to explain this geometry using minimum energy principle is quite manageable.

The repeated generation of one simple molecule can be expressed in terms of the repeated creation of one or more other simple reactant molecules. We can now continue to express each one of these input reactant molecules once again in terms of another set of simpler molecules until together they all form a series of networked (or looped) reactions. The loops are necessary for repeatability. A mere abundance of chemicals is not sufficient if you want to continue to create the same molecules for billions of years. Active material-energy looped systems discussed so far are, therefore, sufficient to create *simple* chemicals repeatedly.

The situation with polymers is, in general, very different. If the polymer consists of repeated long chains of a *single* simple monomer molecule *A*, then the difficulty of repeated generation of this polymer is not much more complicated than the repeated generation of the simple monomer molecule, which was already discussed above (using a material-energy looped system like Design *A* of Figure 59.1). The chain reaction that generates the polymer once can be repeated several times to produce the same polymer (like *A-A-A-A-A*...). The length of the polymer sequence may be different each time. Yet, we consider the polymer to be the same primarily because the physical, chemical and geometric properties are not significantly dependent on this length as long as it is large enough. Examples of such polymers are silicones or plastics like polyethylene, polypropylene and polyvinyl chloride (PVC).

The difficulty comes when two or more monomers make up the polymer (like *A*, *G*, *T* and *C*). For example, DNA and RNA sequences have four monomer (nucleic acid) subunits and protein sequences have 20 monomer (amino acid) subunits. Even in these cases, the simple monomer molecules can be repeatedly created the same way as before. However, the polymer molecule created from a corresponding chain reaction is no longer the same every single time or even most of the times. The precise sequence of monomer subunits is usually different each time (like *A-G-G-T-A-C-*... versus *T-A-G-C-C-A-*...). Furthermore, the physical, chemical and geometric properties are widely different each time. Therefore, repeated generation of the same polymer molecule is no longer easy.

Even though we considered two polymers with different sequences to be similar when it was made of just a single monomer subunit, we can no longer do that for polymers made of two or more monomer subunits. Any change in the monomer sequence, its length and its geometric shape affects the polymers' properties. This is especially true with proteins, mRNA and DNA. Their unique behaviors and properties require us to create these polymers with high precision if we want repeatability. Continued survival of living systems depends on the repeated creation of useful polymers. For simple molecules, we have already seen how material-energy looped systems and a network of looped systems can guarantee this. To guarantee the same for polymer molecules, nature needs to modify the material-energy looped system.

For simple molecules, the creation of the entire molecule happens in just one or a few stages. This is true with addition, substitution, elimination or rearrangement reactions. There are only a few intermediates, if at all, when creating the final products from the reactants. However, for polymers, there are a large number of intermediate stages (typically, of the order of the number of monomers) before you create the final chemical like a protein, mRNA or DNA. At each stage, a chemical reaction occurs using the existing monomers or partially created polymers.

The random collision based mechanism for a chemical reaction is inherently a parallel process. If these parallel collisions start with a pool of monomer molecules, then within a short period, you will have a mixture of monomers, 2-chained, 3-chained or a few chained monomers. As the parallel collisions continue, the small length polymers can grow to larger sized polymers and, at the same time, the larger sized polymers do breakdown into smaller sized ones. This limits the maximum length of polymers. The smaller chains quickly attain minimum energy configurations preventing longer chained polymers to be formed, unless you supply large energies to breakdown or open up these smaller stable configurations. The accessibility of the critical regions and correct orientations are not easy with random collisions.

The bigger challenge with random collisions is to ensure that the order of joining the subsequences be the same with each attempt in order to create the same polymer. Furthermore, the resulting geometric shape or the minimum energy configuration will differ considerably even if the monomer sequence is identical in two situations. For example, the folding of $A_1A_2A_3$ joined with $A_4A_5A_6$ is different from A_1 joined with $A_2A_3A_4$ followed by A_5A_6 even though the entire sequence is identical in both cases (cf. protein folding problem).

One way to avoid these difficulties is to eliminate the parallel mechanisms and, instead, discover a purely serial sequence of steps. With a serial sequence of steps, the idea is that you always join a monomer to the growing chain of the polymer sequence, never two polymer subsequences together. Since the sequence of steps will be the same every single time a given polymer is assembled, the geometric shapes will be identical each time. We only need to ensure that the monomer sequences are identical. This is the solution nature chose with, say, proteins, DNA, mRNA or tRNA. Each of the difficulties mentioned above due to parallel mechanisms can be eliminated by using a serial mechanism. For example, the energies are lower because we are only adding one monomer at a time. We also add it before the growing polymer takes a fully stable configuration. Identical sequences will guarantee identical geometries unlike the example with $A_1A_2A_3A_4A_5A_6$ discussed above.

This implies that the serial process should use an active mechanism, much like the material-energy looped system instead of random mechanisms. This will be the focus of discussion in the next section. The only difficulty we still cannot avoid easily with serial processing is to guarantee the precise sequence each time we create the same polymer. The solution employed by nature here is to use an assistive template to guide the creation of the polymer. For example, DNA strand acts as a template for mRNA, tRNA and for DNA itself. Similarly, mRNA acts as a template for the protein sequence. On the other hand, microtubules (made of α- and β-tubulin proteins) and phospholipids are examples where a precise polymer sequence is not required for the desired functionality. For these, we do not need an assistive template. Let me now discuss how we can modify the material-energy looped system to assist the creation of a polymer in a reliable and repeatable way.

61.4 Creating random polymers from monomers

We can use the material-energy looped system to create polymers in the same way we used it to run a simple chemical reaction. It actively sucks the monomers into the system. With favorable conditions and sufficient energy available either externally or internally through another chemical loop, the monomers combine to form a polymer incrementally. In every expansion-compression cycle, it is possible to chain one or more monomers (or previously created partial polymers) to a growing polymer. The material-energy looped system acts as a primitive enzyme even in this case (see Figure 59.1).

There are two ways the polymer sequence created can exit the system. In one scenario, the growing polymer sequence fits through the exhaust opening. In this case, the entire polymer can be pushed out in one or just a few compression-expansion cycles. This is more common with linear or helical polymer sequences like DNA, mRNA or even microtubular structures. The other scenario is when the polymer is unable to fit through the exhaust opening. In this case, multiple polymers are generated within the system while it grows bigger in size. It then explodes thereby releasing all of the polymer sequences. This scenario is more typical with polymers that have complex three-dimensional shapes like proteins, tRNA or even extremely long DNA or mRNA sequences.

I will refer to this mechanism simply as *monomers-in-polymers-out*. For example, amino acids enter the material-energy looped system and the outputs are protein polymers. The protein can fold into three-dimensional shape either inside the system or outside. With nucleotides as the inputs to the system, the outputs are mRNA, DNA, tRNA and other types of RNA molecules. Various different polymers can be created using just this simple mechanism.

The resulting proteins and other polymers may or may not be useful in any sense. Also, the processes for creating long chained polymers are much slower than the current enzymatic mechanisms. Yet, at this stage of evolution, usefulness of these polymers is less relevant than creating a large repository of polymers. Another drawback is that it is not possible to create the same polymer repeatedly using this mechanism. The mechanism does help create a large repository of organic chemicals actively, except without any guarantees of regenerating them repeatedly. If we get lucky and created DNA polymerase, RNA polymerase or a ribosome, this advantage will be lost once these molecules disintegrate. There would be no guarantees on when these molecules will be recreated once again.

Yet, it is important to realize that the creation and the subsequent breakdown of polymers, viewed jointly, is a stable looped process. This active creation-breakdown cycle of molecules can continue for a very long time without any specific direction in evolution. We do not even need a living being for this looped dynamical process to self-sustain actively for a long time. We can have a pool of these monomers and polymers and keep converting from one type (monomer) to the other (polymer) continuously. The mechanism for the breakdown of polymers is simply from instability and second law effects, while the mechanism for creating polymers is from a collection of active material-energy looped systems.

In spite of a lack of guarantees to regenerate the same sequenced polymer using the monomers-in-polymers-out mechanism, we should not regard this as a worthless mechanism. This is because there are several polymers where the precise sequence is not critical for the useful functionality they exhibit. Some examples of these are phospholipids and polysaccharides. The random sequenced phospholipid structures can spontaneously concatenate to form a lipid bilayer. This can act as a better surface membrane for material-energy looped systems than clay or other primitive materials. Similarly, the random polysaccharides can act as cellulose (required for plant cell walls). It is with DNA, RNA and proteins this lack of both repeatable regeneration and nonrandom and, in fact, precise polymer sequence pose a serious problem. A nonrandom DNA sequence is necessary for all critical functionalities like creation of a specific protein and mRNA sequences. For polymers that help creating cell membranes and microtubules (for transporting chemicals), a random polymer sequence is quite sufficient.

62
Creating Polymers Repeatedly

One of the main drawbacks of creating polymers from monomers using the monomer-in-polymer-out mechanism discussed in the previous chapter is the inability to regenerate the same polymer sequence repeatedly. Yet, not all polymers need precise polymer sequence (like phospholipids, saturated hydrocarbons, saturated/unsaturated fats and microtubules), but some of the important ones like DNA, RNA and proteins do. While nature uses this mechanism to generate a wide variety of polymers, there is no clear direction in evolution. Creating useful polymers like ribosomes and others using this mechanism may be useful temporarily, but they will soon be lost as the polymers eventually breakdown into monomers from second law effects. The only useful polymers that may remain for a long time are replicases that are both self-replicating and self-catalytic. However, discovering replicases through this mechanism, though appears more likely because of active mechanisms of material-energy looped systems and from an explosion of random polymer generation, is still not as convincing.

If we look at the theoretical solution proposed until now for creating life for the first time, it was quite systematic and iterative. It revealed new and useful chemical SPL network structure. It allowed nature to generate a variety of organic molecules iteratively and with minimal randomness. It was able to overcome difficulties from the second law of thermodynamics and several others using a small set of minimal structures rather than relying on complex minimal structures like replicases. The transitionary processes though slow and takes millions of years were less random, increasingly reliable and repeatable compared to existing solutions that relied purely on random chance. Of course, we do need experimental justification, but the mechanisms described until now appeared simple enough and are synthetically feasible. However, if we now claim that the transition from monomers to polymers using the mechanism of the previous chapter can generate replicases, the proposed solutions' continuity and iterative approach breaks down. Rather, we would need to rely on random chance to take this next leap in evolution. This is unsatisfactory.

Therefore, the goal of this chapter is to propose a more gradual, iterative and systematic solution to the problem of creating polymers reliably and repeatedly. We do not want our solution to make discontinuous jumps in the argument by relying on random chance mechanisms that are nontrivial like with generating replicases. In order to do this, I will propose a new mechanism that creates polymers using other polymer templates in the context of material-energy looped systems. The idea is to create a correlated polymer to a given polymer template, not a perfect replica as is possible now with DNA transcription and replication.

As mentioned earlier, the errors in replication are quite helpful at this stage of evolution. However, the goal is to ensure that the newly created polymer is correlated to the original polymer template. This improves the chance that if a useful DNA chain is created, there is a good chance of recreating an approximately similar polymer repeatedly because of the correlated replication mechanism. The monomer-in-polymer-out mechanism of the previous chapter does not create correlated polymer sequences. It generates entirely random polymers. Therefore, the evolution after polymers are created is not directed. On the other hand, the new mechanism proposed in this chapter will guarantee a direction in evolution because of the correlated polymers it generates.

The new mechanism is based on a variation of material-energy looped system. It relies on the dynamical and geometric properties of long chained polymers discussed in chapter 61. It is also a minimal structure that requires randomness, but not to the same degree needed to create a complex structure like a replicase. This minimal structure is once again a repeatably random structure (see section 52.3) that can be created naturally with finite probability similar to the other designs proposed earlier in this book. These ideas can once again be simulated in a computer as mentioned in section 59.6 and is an active area of my future work.

62.1 Creating polymers using polymer templates

In order to have well-defined dependencies or correlations between molecules, nature needs to discover monomer molecules with special chemical affinities first. Purines and pyrimidines are examples of chemicals with such properties. For example, adenine pairs with thymine using two hydrogen bonds while guanine pairs with cytosine using three hydrogen bonds. The affinity of such pairs of monomer molecules naturally creates a structure in the primitive world.

If we introduce a large collection of such pairs of molecules in a sealed room (see section 52.4), we will begin to see order created from disorder naturally (not life itself, but order). In this sense, such pairs of molecules can be thought of as minimal structures as long as their discovery using natural means is easy. Since these molecules are only monomers, their complexity is not too high. Therefore, they can be created using active material-energy looped systems just as other simple organic molecules can be actively created.

In fact, there has been considerable research done to show that purines and pyrimidines can be created under prebiotic conditions. I refer the reader to Miller and Cleaves [107] which summarizes a large number of experimental results for creating almost all critical monomers necessary for life under prebiotic conditions. For example, Oró [119] showed the abiotic formation of adenine, which can be considered as a pentamer of HCN ($C_5H_5N_5$), using aqueous solutions of HCN and NH_3 with up to 0.5% yield (for details on mechanism of synthesis, see Oró [119]). Since then, several other experiments have shown the production of adenine under different primitive earth conditions, including at low temperatures like between –10°C and –30°C (see Sanchez et. al [135]). For example, it was also shown that the presence of formaldehydes and other aldehydes, accelerates HCN polymerization to produce adenine (see Voet and Schwartz [160]). For the production of guanine, it was shown that polymerization of concentrated NH_4CN can occur both at –80°C and –20°C (see

Levy et. al [95] and Levy et. al [94]).

Prebiotic synthesis of pyrimidines was also studied extensively (see Miller and Cleaves [107]). It was shown, for example, that abiotic synthesis of (a) uracil is possible from malic acid and urea – see Fox and Harada [44] and (b) cytosine from cynanoacetylene (HCCCN) and cyanate (NCO⁻) – see Sanchez et. al [134]. I refer the reader to Miller and Cleaves [107], which summarizes several other experimental results and the corresponding mechanisms for the synthesis of pyrimidines (see Ferris et. al [40] and Robertson and Miller [130]). Therefore, we can say that the same set of chemical reactions that the above experiments refer to can occur inside an active material-energy looped system. This can result in the creation of a purines and pyrimidines naturally and repeatedly with a high enough yield under prebiotic conditions.

The challenge here is to show how to create a specific polymer sequence of purines and pyrimidines under prebiotic conditions. There is little experimental work on this aspect. In the discussion below, I will relax the requirement of generating a specific polymer sequence and instead only require the creation of an approximately correlated polymer sequence, but repeatedly. The errors in this primitive form of replication are desirable as they generate a wide variety of molecules necessary for creating life for the first time.

When a material-energy looped system actively sucks inputs, the presence of long chained polymers in the environment, say, from the monomer-in-polymer-out mechanism causes them to be caught or entangled at the intake openings. Besides, the surface irregularities make it easier for the polymers to be entangled. This is an example of a situation that is random though quite easily repeatable. These long chained polymers rarely enter the system through the intake opening. Even as the polymer sequence is entangled to a material-energy looped system, it continues to compress and expand as usual running its internal chemical reaction. If the polymer caught is helical in shape (i.e., like the threads of a screw), the system and the polymer together exhibit a unique behavior, at least sometimes. In each compression-expansion cycle of the material-energy looped system, the polymer can twist and advance along its length, similar to a ratchet like mechanism on a linear (not circular) rack – see Design D of section 59.2.

If the material-energy looped system is perfectly symmetrical, it is not as easy to achieve a linear motion of the polymer. However, since the system's surface has natural imperfections, there will be a slight torque generated as the system compresses and expands. It is not a perfect inward-outward motion during each compression-expansion cycle. Instead, the material-energy looped system will naturally twist-untwist slightly along with an inward-outward motion during each compression-expansion cycle. This twist-untwist torque generates a force that pushes the helical polymer (cf. the threads of a screw) along its length (or axis) while the inward-outward motion pushes the polymer sideways (imagine Figure 59.1 twists-untwists in addition to expand-contract as well). The net effect is that the polymer advances along its length and is caught in the next helical grove in each compression-expansion cycle. This is analogous to a ratchet advancing one thread pitch at a time while getting caught in each compression-expansion cycle. The active suction mechanism and natural imperfections are quite critical for this type of behavior with helical polymers. This movement of the helical polymer is not necessarily efficient in that it would not happen in every compression-expansion cycle.

Yet, this would give rise to a directional property (see section 61.1) to the material-energy looped system. If the polymer is not helical, a considerable amount of force is needed to slide and drag the polymer along its length. This is not as easy. Besides, the amount of translation is not uniform during each compression-expansion cycle. Therefore, the helical shape is quite critical for ease of operation with an active material-energy looped system. With a helical shape, the incremental amount by which it advances during each compression-expansion cycle is not only uniform, but is proportional to the pitch of the helix (i.e., width of one complete helical turn, measured along its axis). Besides, there is a natural advantage where a twist and a push cause it to advance along its length with minimum force.

Clearly, the above description does not work with polymers that fold into complex three-dimensional shapes like tRNA, proteins and other three-dimensional RNA molecules. DNA and mRNA molecules are typically the only favorable helical shapes for this mechanism to be useful with this class of material-energy looped systems. We can view the relative motion between the material-energy looped system and the helical polymer in one of two ways: (a) the system is static and the helical polymer moves or (b) the helical polymer is static and the system moves. The latter view is popular when we look at the operation of DNA polymerase or RNA polymerase along the helical DNA polymer (primarily because we tend to think of the DNA sequence as the reference point during replication).

With the new material-energy looped system that attaches and pushes a helical polymer in each compression-expansion cycle, let us examine the types of inputs that it will actively suck in. When the helical polymer is not attached, any of the constituent monomer molecules that fit through the opening will be actively sucked in. The set of monomers in the environment near the opening are quite random. As a result, the system will pull them in, in a random order. This was the reason why the polymer generated internally using the monomers-in-polymer-out mechanism, is a random sequence as seen in section 61.4. As an example, when no mRNA polymer is attached to the material-energy looped system, the mRNA generated internally within the system, through active suction of nucleotides, is a random sequence.

However, we now have a helical polymer attached to the material-energy looped system, which advances in each compression-expansion cycle (with some errors). Therefore, the set of monomers in the environment near the opening are no longer random, provided some monomers have natural affinities towards each other. The most common example is with the nucleotide bases that make up DNA or RNA. Adenine (A) has a natural affinity to thymine (T) and forms two hydrogen bonds. Similarly, guanine (G) has a natural affinity with cytosine (C), forming three hydrogen bonds. This is referred to as complementary base pairing. If the helical molecule caught is an mRNA or a DNA strand, the free bases (monomers) in the neighborhood of the intake opening, close to the attached mRNA, will be complementary to them.

The free complementary bases (as monomers) will naturally attempt to hydrogen bond with the attached mRNA. This tendency automatically creates a correlation between the monomers actively sucked in and the attached mRNA sequence. As the attached mRNA sequence advances in each compression-expansion cycle, the corresponding set of monomers near the opening change as well. They will match to be complementary once again to the bases of the new advanced sequence in the neighborhood of the opening. There will be sufficient shuffling of inputs at the intake opening due to the suction action. If the frequency or rate of compression-expansion cycle is not too high, it will give enough time for the environment to reach steady-state conditions with the correct complementary bases reaching the intake opening once again.

Since the order of monomers drawn in aligns with the attached or entangled mRNA polymer sequence, the intake monomers combine internally to generate a polymer, which is closely related to the attached mRNA sequence as well. The attached mRNA is now the template for generating an almost identical polymer within the material-energy looped system.

There are several sources of errors between the template attached and the actual polymer generated within the system. Some of these are: (a) the inputs pulled in are not always complementary to the attached mRNA for several structural, dynamical and environmental reasons, (b) the advancement of the attached mRNA sequence can be non-uniform because of the entanglement and other imperfections, (c) additional monomers beyond the complementary bases can be pulled in during each compression-expansion cycle and (d) there can be errors within the system when the monomers combine to form the product polymer. All of these sources of errors suggest that the output polymer sequence may only be partially accurate. Yet, the output polymer sequence can be considered as sufficiently correlated to the attached mRNA polymer template unlike the previous monomers-in-polymers-out mechanism. This is sufficient to guarantee a direction in evolution at this stage of evolution. We can even regard these errors as analogous to mutational errors during DNA replication.

62.2 Primitive replication and transcription

The mechanism above is a way to create a polymer using another polymer as a template, though not with high fidelity compared to the modern mechanisms. It is now apparent how this mechanism corresponds to primitive replication (for making a copy of a DNA strand) and transcription (for making mRNA sequence from a DNA strand).

If the template is a single DNA strand and the input monomers are adenine, thymine, guanine and cytosine nucleobases attached to deoxyribose sugar, the resulting polymer created is an approximately accurate complementary DNA strand. This is the primitive replication process.

If the template is a DNA strand and the input monomers are adenine, uracil, guanine and cytosine nucleobases attached to a ribose sugar, the resulting polymer created is an approximately accurate complementary RNA strand. This RNA strand can be an mRNA, tRNA or any other type of RNA molecule. This is the primitive transcription process.

These primitive replication and transcription mechanisms do not require any modern enzymes. They, however, behave like primitive DNA polymerases (for replication) and primitive RNA polymerases (for transcription). Each time a given DNA strand is attached to a material-energy looped system, the output DNA or RNA strand generated is approximately complementary i.e., it is not a random one.

We have avoided the chicken-and-egg problem of DNA replication using active material-energy looped systems, helical geometry of DNA strands and complementary base pairing properties of purines (adenine and guanine) and pyrimidines (cytosine, thymine and uracil). This is an alternate theory to RNA world hypothesis. The process described here is natural to occur, reliable and repeatable, though with considerably more number of errors than the alternatives suggested with replicases. The difference, however, is that replicases are too complex to form naturally. Also, the primary mechanism for creating a replicase is random chance.

Another outcome is that the evolution of polymer molecules is now directed compared to when we relied only on monomers-in-polymers-out mechanism of section 61.4. This is a significant improvement. The direction proceeds as follows. Initially, random DNA and RNA molecules will be created using monomers-in-polymers-out mechanism. Next, using the template-assisted mechanism described here, the randomly created DNA molecules will be approximately replicated in large numbers. Similarly, using the same random DNA molecules as template, a large number of RNA molecules will be generated. The distribution of DNA and RNA molecules will now be far from random. If there are unreplicated DNA or RNA strands, they will disappear eventually. We can partition all the remaining DNA and RNA polymers into equivalence classes of approximately similar replicas.

Another stable looped process is the switching between creation and breakdown of polymers. These processes can self-sustain for a long time even if without living beings. Nature is capable of creating a structured distribution of DNA and RNA polymers even in the absence of life.

It is interesting to note that creating an ordered collection of similar DNA and RNA polymers utilizes the available energy more effectively than creating a random collection of DNA and RNA polymers using monomers-in-polymers-out mechanism. To see this, consider the total available energy in a given region. It can be decomposed as unused external energy + effectively-used internal energy of the material-energy looped systems. With the monomers-in-polymer-out mechanism, the useful work comes from just the internal energy of the system. It effectively ignores additional available external energy. On the other hand, the template-assisted mechanism relies on chemical affinities of the complementary bases as well. This requires consuming additional available external energy to perform useful work, which is then consumed by the material-energy looped system. Therefore, the total useful work by the new mechanism includes additional external energy + internal energy of the material-energy looped system.

This is another way to say that the new mechanisms have created order from disorder – part of the disordered external energy is exploited and used effectively to create additional order, namely, complementary bases and, hence, approximate copies of existing polymers. Creating order from disorder, in general, is not easy. In this specific case, subtle and simple properties like helical geometry and complementary bases are exploited by active material-energy looped systems to steal a little extra disordered, but freely available, external energy to create just a bit of extra order. As a result, we will naturally find more similar polymers now. The distribution between random and similar DNA (and RNA) molecules will be considerably uneven, with random ones being far fewer than the similar ones.

It should be noted that the above mechanisms are compatible with any of the hypothesis like an RNA world, a DNA-enzyme world, PNA or other possibilities. The current analysis does not shed light on which of these hypothesis actually did occur. Further research into the actual biochemistry might clarify this better than just these generic mechanisms.

It is worth reiterating the point that the new template-assisted mechanism is not just a theoretical design. We can create these systems easily using synthetic materials and larger sized physical systems. In fact, several man-made systems functionally identical to the mechanisms described here already exist (see section 62.1). The feasibility and the physical existence of such systems are not in question here. What we do require is experimental proof of cellular-sized versions of these synthetic designs and proof that they can be created entirely naturally (i.e., without help from conscious beings). Towards this end, I have emphasized the simplicity of these designs (to give us confidence on the feasibility) and presented a few options in chapter 59 on how nature can indeed create these designs.

62.3 Improved primitive enzymes for polymer inputs

The material-energy looped systems proposed as primitive enzymes in chapter 60 were quite suitable for catalyzing chemical reactions for which the reactants are simple molecules. However, at this stage in evolution, we have seen that we also have a large collection of polymers. The same material-energy looped systems begin to become less efficient because these large polymers are difficult to be sucked into the system. In this section, I want to suggest that there is a small but finite probability that these material-energy looped systems naturally adapt to become more specialized to a given polymer (or large molecules), analogous to modern enzymes. If this happens, the resulting systems also tend to have increased efficiencies and exhibit enzymatic properties more closely i.e., the system is closer to an induced-fit model of an enzyme.

In the previous section, we have seen that long chained polymers with shapes like linear, circular, helical and complex three-dimensional ones start to become quite common, irrespective of whether they are generated randomly or if they are template driven. Each of these different shapes pose different sets of challenges during the intake and the exhaust flows of material-energy looped systems, making them less efficient at catalyzing these chemical reactions. When the molecules were simpler and smaller, these problems did not exist.

Therefore, a natural way to cope with these problems is by not sucking the polymer inputs into the material-energy looped system. Instead, the surface membrane should be adapted to conform to the shape of the complex polymers. This may be likely, especially since the available choices of molecules like phospholipids and others for the surface membrane have increased. This can happen naturally as well, similar to how the openings for inputs and exhaust can adapt for smaller molecules (design C of section 59.2). The intake and exhaust openings do continue to exist. However, they will primarily be used for smaller molecules. The smaller molecules will be part of the energy generation chemical loop. Sometimes, it is more common to have the energy generation for powering the compression-expansion cycle to be decoupled from the main chemical reaction that runs within the system (see design B of section 59.2). In that case, we have two chemical loops: one for the secondary energy generation loop (analogous to the ATP-ADP loop) and the other for running the primary chemical reaction of interest.

When working with polymers, we can now say that the secondary energy generation chemical loop can still run within the system through an active suction-exhaust mechanism. However, the primary chemical reaction of interest no longer runs within the system because of the large sizes and shapes of these polymers. Instead, they are actively attached to the surface membrane as suggested by the induced-fit model for enzymes. The shape of a part of the surface membrane will conform to the shape of the complex molecules that will be attached. The compression-expansion cycle deforms the shape of the surface membrane to allow the creation of the products on the surface itself. The rearrangement, substitution, addition or elimination reactions happen on the surface membrane. Sometimes, the deformation of the surface membrane can cause these complex molecules to be enveloped, at least, considerably during the compression-expansion cycle.

This model is analogous to the induced-fit model of the enzyme – (a) the system actively sucks inputs to the surface membrane, (b) the surface membrane deforms just enough during the compression stage to fit the molecules more accurately, (c) the secondary internal energy generation reaction powers the primary chemical reaction that occurs on the surface and (d) the products of the chemical reaction are released during the expansion stage. While the energy generating chemical inputs are common across several systems, the surface molecules are unique to a given system. We can identify the new specialized material-energy looped system by the set of molecules that bind to the surface.

While the explanation given here is theoretically feasible, it is far from clear if such specialized systems are widespread. Experimental verification is more critical in this case. Furthermore, reproducing this system with the above unique structure on the surface is also a slow process, similar to standard material-energy looped system (see section 59.2). After the system breaks down into two or more subsystems, the system would need to be exposed to similar inputs in order to form a surface impression closely related to the original representation. This is not perfect, in general.

One advantage with this system over the previous version is that it is more energy-efficient. This is at the expense of a richer structure and slower process of regeneration. The template-assisted mechanism of creating a polymer from another helical polymer using this new design is more reliable and accurate because the monomers no longer need to enter the system. The sequential process of combining them according to a given template has flaws, as mentioned in section 62.2, if the monomers enter the system and randomly move about. If, on the other hand, growing the polymer chain occurs close to the surface membrane, the sequential ordering can be maintained more accurately.

In the next chapter, I will discuss how to use this new material-energy looped system for the problem of creating a primitive genetic code.

63

Creating Sample Life-like Features

In the previous chapters, we have seen that the collection of organic molecules discovered by nature form an SPL network. We have seen how nature starts with simple molecules and creates more complex organic molecules using different variations of active material-energy looped systems. The discovery of molecules proceed iteratively in a directed manner even in the absence of life. Indeed, we have seen the creation of most polymers, some of which can be replicated with good enough accuracy. The chemical SPL network in an entire region grows larger over time in an iterative manner.

In this chapter, I want to take a few concrete examples of life-like features and show how nature can discover them by herself. I will present designs using the active material-energy looped systems. The goal is to show that variations of active material-energy systems are quite important and do result in generating several sensory-control features. These features are typically attributed to living beings. Yet, we will see that in the absence of life, nature can create them quite easily. As mentioned earlier, there will be several other ways to create the same feature reliably and repeatedly. My emphasis is on the fact that an SPL architecture offers a wide diversity of designs and features using natural means.

63.1 On sensors and actuators

Two of the most important features of living beings that make their dynamics self-sustain for a long time are sensors and actuators. The sensors detect the environment and the actuators make components or even the entire system move in a specific manner. If the actuation is in response to a sensory input, the resulting motion becomes nonrandom as well. In this way, you form a looped dynamical system that can be either stable or unstable. Yet, if we can explain how to create such looped systems naturally, the world evolves in a direction to have more structure and order.

Sensors, in general, can detect the internal states within the system in addition to sensing the external environment. Similarly, actuators can move components and chemicals internally within the system as well as make the system move about in the external world.

Let me first list a few sensors that already exist, say, in our body. In section 24.6, I have defined what a sensor and what an actuator is and discussed a few of their properties. I have also discussed what it means to create a sensor that 'knows' what it senses in section 24.9. In the context of that discussion, we can say that there are several sensors in our body that detect chemical gradients, concentration gradients, specific chemicals, height, position, pressure and temperature gradients, light and sound gradients and others. These sensors are used both internally and externally in a number of situations.

Let me list a few examples here. Our body uses hormones to regulate certain chemicals and reactions, thereby preventing the system reach toxic states. Specific neurotransmitters like acetylcholine can be considered as chemicals that neurons sense causing them to open Na^+-K^+ ion channels. During a mitotic cell division, a series of chemicals are detected causing the cell division to proceed in a controlled manner. Synthesis of chemicals like amino acids and others are controlled through positive and negative feedback, which is a typical sensory-control mechanism. Transcription factors for controlling the gene expression are triggered in response to specific molecules. The end result is either an activation or repression of transcription of certain genes. Human sensory systems like eyes, ears and tongue generate nerve impulses in response to specific external stimuli. Our immune system is capable of detecting several foreign bacteria, viruses and other pathogens. In general, there are several more such examples in living beings.

Similarly, there are several examples of actuators used in our body. For example, microtubules and motor proteins move chemicals within the cell from one location to another. Heart is a rather powerful pump that circulates blood and other important nutrients within the body. Muscular contractions and other mechanisms exist for the movement of hands, legs and other body parts. Even though plants cannot move as a whole, they have capillary mechanisms that let water and other chemicals move internally, say, from roots to leaves.

Now, a conscious and intelligent human being can create a sensor or an actuator easily using our understanding of the laws of nature. A machine or even a single cellular organism, on the other hand, does not create sensors or know what they mean. Rather, they simply use the sensory or actuator subsystem in a mechanical way.

Clearly, systems with sensors and actuators coupled in a certain way can sustain their dynamics for a much longer time than without. For example, if a cell can detect a harmful chemical, defined as one that destabilizes the existing looped reactions, and if this detection is combined with an actuator like a motor protein that rejects this chemical (say, by spitting it out), the system will continue to run for a much longer time than if it did not reject the chemical. Similarly, if a cell can detect a low level of stored energy or food, then by combining this detection with an actuation that makes the cell move around and search for food, the system has an increased chance of continuing its dynamics. Such a system would have a higher probability of discovering new sources of food when the resources deplete in a given region compared to a system without these sensory-actuator mechanisms. In fact, every example of sensory-actuator subsystems in living beings, if linked appropriately, has clear advantages to make the entire system sustain its dynamics for a longer time.

The looped coupling between sensors and actuators can be made stable using the fundamental pattern as explained in section 16.3, when the system has, say, a brain. For chemical systems, on the other hand, the specific advantageous sensory-control loop would need to be memorized within a network of loops using one or more of the mechanisms mentioned in chapter 57. The cascading effect of disturbance propagation within this memorized network of looped reactions then produces the desired behavior at a macroscopic level. The disturbance propagation will appear as a sequence of steps obeying causality. When we view this sequence of steps at a systems level, we will recognize it as the sensory-control loop (see chapter 58).

63.2 Actively searching for food

Let me consider an example of how a system can actively search for food. This is an example of a system that both senses the environment and has a minimal actuation. Here, I want to illustrate how nature can design the system by herself using stable parallel loops. Let me begin by describing how a system creates a primitive sensor, which detects 'food' needed to continue its dynamics. For simplicity and generality, I will use a simpler material-energy looped system (see Figure 59.1) as the starting point. In this example, we will show how an active material-energy looped system can be modified to create a feature that detects the shortage of food and causes it to move away and start looking for food. The design is created in a reliable and repeatable way though some of the necessary mechanisms are random.

Figure 63.1 shows such a system. It contains a polymer that twists around its length by consuming energy and subsequently returns to its original shape just like a helical spring. Some helical polymers as well as tubulins arranged as microtubules, some of which are already possible at this stage of evolution, have such properties. When a large number of such molecules are attached to the surface quite randomly, let us say that it produces a net rotation of the entire system, at least, by a tiny amount. Let me depict this as the motion reaction M in Figure 63.1. The source of energy for

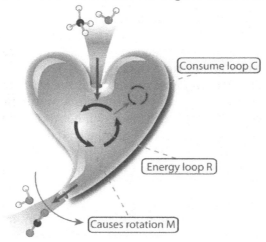

Figure 63.1: Simple SPL system capable of actively searching for inputs. This system has energy loop E, consume loop C and a motion reaction M. The consume loop C requires high activation energy to run. When there is less energy generated by energy loop E, the consume loops C do not run. All of the available energy goes towards rotation M. When excess energy is available through E, consume loop C uses this energy. In other words, the behavior exhibited by this system is: consume inputs when excess is available and rotate or search for inputs when less energy is available.

producing this twist comes from the internal energy loop R.

Let us assume that this system has another chemical loop C (or a set of chemical reactions) that essentially consume all available energy from R. I will assume that the loop C has high activation energy and yet, has highly reactive set of chemical reactions. It does not seem too difficult, at least theoretically, to find some choices for reactions C (experimental justification treated as future work). When there is a lot of energy produced through R, most of it is consumed by C. The reaction C will occur because the amount of available energy is assumed to be sufficiently high to overcome the high activation energy. When C occurs, almost no energy is available for motion reaction M. The system does not rotate under these conditions. This can be seen as if the system is resting when excess food (or energy) is available. However, when there is less energy produced, C reaction slowly stops and is not activated again because of high activation energy. In this case, all of remaining energy will now be consumed by M reaction to produce a motion or twisting instead.

In summary, if there is a lot of food available (i.e., energy R), the system essentially rests and consumes all of it (through C). When the energy level drops to a low amount i.e., the system behaves as if it is 'hungry', it keeps searching (simplified as just rotating) until it finds more food. The effective behavior that nature incorporated with this new design is a *need* to find food by itself. This behavior is achieved from a simple set of looped reactions shown above.

I have used the relative rates of reactions and activation energies of C and M reactions to produce this life-like behavior. The set of reactions for C and M only need to obey the above properties crudely. Note that the C loop can be viewed as a detector of high energy because it runs only when there is high energy available and stops running otherwise. The reason we are expressing reactions like C and R as loops is because we do not want to depend on a constant supply of these chemicals as well. Instead, they will be just re-used. This clear advantage of having looped reactions is maintained with all living beings as well, since this minimizes the total number of external chemicals needed.

This behavior is not necessarily efficient. If this system does not find any input resources through turning, it would simply die down. However, if the food is available on two opposite sides of the system, then it starts by consuming the food on one side. When this source is depleted, it would turn around and consume the other side as well. This is a significant improvement over the previous design (i.e., the original active material-energy looped system). The new design is capable of searching and detecting food on the opposite side as well unlike the previous design. The original material-energy looped system would have consumed food only on one side and would have died subsequently. Therefore, by adding more chemical loops, we have created a new design capable of running its dynamics for a much longer time than the original design.

It is interesting to see that we can get a completely opposite behavior just by reversing the conditions for C and M i.e., when there is more energy, choose chemicals so reaction M would occur instead. The resulting behavior in this case is that when you have a lot of energy, you move around a lot (similar to how kids play a lot when they eat chocolates!). Even though intuitively, this design (with opposite behavior) feels natural, the design described in Figure 63.1 has a better chance of sustaining its dynamics for a long time.

When the evolution reaches a state in which primitive cells are possible and they have an ability to generate modern enzymes, the set of chemicals available for the design of sensors and actuators discussed here are considerably expanded from primitive molecules. We could use a phospholipid membrane based primitive cell with embedded proteins for the intake and exhaust of inputs as the starting point. In the best-case scenario, there are chemicals that are capable of performing mechanical work by harnessing the chemical energy released from the hydrolysis of ATP. For example, we have motor proteins like myosin that are used for muscle contractions, kinesin to move molecules away from the nucleus and dynein to move them towards the nucleus along microtubules, actin polymerization used for propulsion and others. For sensor design, there are chemicals that transduce input signals into a convenient output signal that a cell can use. For example, light sensitive rhodopsin molecules in the eye and chemoreceptors (like olfactory receptor neurons and taste buds) transduce a chemical signal into an action potential in the neuron. Similarly, chemical hormones transported in the blood induce a response in a specific group of cells when it binds to a receptor protein. We are not suggesting that we require all of these chemicals for the designs above. However, some of these chemicals can help with higher efficiencies, though alternate solutions with simpler chemicals can suffice for now.

In the next section, I will discuss how to detect chemical gradient using another simple design. I want to highlight how linking sensors and actuation do help in making the system sustain its dynamics for a much longer time.

63.3 Moving towards a higher concentration food source

Let me now propose another simple feature to the system presented in the previous section by linking its sensing ability with actuation. We want to design a system that detects and moves toward higher concentration gradient of inputs, say, when there are two sources of inputs. In this situation, we need a more precise link between the sensory information (i.e., by detecting the gradient) with actuation. We will see that this feature will also make the system sustain for a much longer time. With the previous design, for two sources of food, if the system turned and discovered a low food source first, it would still consume and finish it before it searches and finds the high food source. The new design of this section will detect this difference in food sources as well and switch to the high food source first before it finds the low food source.

Figure 63.2 shows such a simple system S_H using the same set of reactions described so far, but with two copies of these reactions that occur within two compartments inside this system. You have two sources of input that enter the system at two locations A and B. Let us say that the amount of input at A is higher than at B i.e., the concentration gradient decreases from A to B. From section 63.2, recall that when we have more external inputs (i.e., more energy production), there will be less

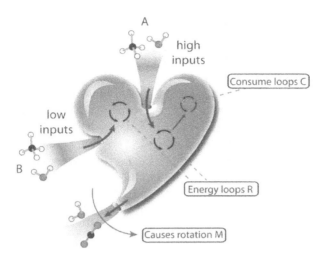

Figure 63.2: An SPL system that rotates towards a source of higher concentration of inputs. Here, we have an SPL system with two openings for receiving inputs. The differential in inputs is detected by the system. The system responds by turning towards the direction of high inputs.

motion and vice versa. As a result, at the high input source A, there is less wiggling motion whereas at the low input source B there is more wiggling motion. The differential in the wiggling between these two regions causes a net rotation of the system from more wiggling location B (and low input) towards the high input source A (see Figure 63.2). Therefore, we have a system that preferentially moves towards the source of higher input concentration.

One point worth noting here is that if we alter the chemicals and the looped reactions in various ways, it is possible to create different behaviors using the same features. For example, nature can create a system S_L that moves from high to low input concentrations, the exact opposite of what the above design does. If we have a collection of S_H and S_L systems, it is most likely that S_L systems will be eliminated as the less fit ones. In other words, the design that self-sustains its dynamics for the longest time is naturally picked as the most-fit design among multiple possibilities. A natural outcome, when several competing designs exist, is to pick a stable parallel looped system that obeys this simple principle (i.e., design that self-sustains for the longest time – see section 55.9).

In summary, the new system S_H is an active one that rests in one place when there are excess inputs. When the food is scarce, it actively engages in fetching new food, picking the highest source of food concentration first. Nature has incorporated a genuine need into the system. If a random external disturbance or a competitive system (say, from similar designs that are adjacent to this system competing for the same set of resources) tries to take its food away, this system would go and seek food by itself. This is an important life-like feature.

Clearly, this design self-sustains for a longer time than the design of the previous section. This is true especially if there is a competition for food from multiple such systems. The current system would pick the highest food source first. The previous systems may have picked lower food source first. Later, when these previous designs search for the next food source, it is likely that the resources are depleted because the newer systems have already consumed them.

Nature has discovered a way to start from simple looped mechanical systems and produced a system with a purpose, namely to search for its own food when it is 'hungry', albeit in a primitive and inefficient way. The active energy source in a material-energy looped system is used in an effective way to achieve this goal without the help of any conscious systems. Nature is creating order from disorder using just subtle variations to the original active material-energy looped systems (unlike the current proposals that require complex and specific-sequenced organic polymers like replicases). We are now beginning to attribute higher life-like features even though we started designing them as purely mechanical systems. In the next section, I will discuss yet another simple and natural design that lets the system wander off randomly while encountering new environmental inputs. This system has an energy storage mechanism as well. This wandering can be viewed as exploration. Its chances of self-sustaining its dynamics will be further improved compared to the current designs because of this exploration.

63.4 Exploration

If the energy loops are efficient or if there are abundant available inputs, there will be excess energy generated internally to trigger both the consume loops C and the movement loops M (see Figure 63.2). In the next design, I will suggest additional chemical loops to store the energy and retrieve it from the internal storage rather than relying entirely on the external inputs. This is advantageous because the system can self-sustain longer even in the absence of external energy. It is possible that the system can be relocated to a different region inadvertently, say, through wind and other disturbances. The new relocated region may not be favorable because of less available external inputs and/or energy. The previous designs would have died in this case, but the new design is capable of consuming its internal energy to relocate to a better region just in time, so it can self-sustain longer.

The energy storage mechanism need not be as efficient as the ATP-ADP mechanism, storing in as fat, in living cells. A simple and crude reversible chemical reaction that can store and release energy to power up other chemical reactions within the system is already an improvement in the existing design. Consider a generalization of Figure 63.2 with additional chemical loops utilizing chemicals like fat for storing excess energy. Even a crude mechanism for storing excess energy already helps the system self-sustain longer than before. This is because when the external inputs are depleted and the searches did not yield new external food sources, the internal energy storage system can power the dynamics.

What is interesting with these systems is that they can use the excess stored energy to wander off for no particular reason. The system can utilize its internal storage mechanism to power the motion whenever the energy level drops. Note that this system is not acting as a gradient sensor anymore but has a new behavior of 'exploring' the environment. Recall that one of the biggest problems before the existence of life is that the transport of molecules from one region to another is purely random. The new design will fundamentally introduce an active and nonrandom way to transport chemicals within wide enough regions. If the systems self-sustain for a while and subsequently 'reproduce' (using the previously discussed primitive replication mechanisms), we will now have these active material-energy looped systems accumulating at the new region. These systems will start creating a similar collection of organic molecules, including polymers, DNA and others at the new region as well. Over a long period, this cumulative effects of this migration through active exploration becomes significant. It will eventually cause all the important organic chemicals and polymers like DNA, RNA and phospholipids to spread to far reaches of the planet in a nonrandom way. Let me now briefly discuss how this system behaves.

When there are enough external inputs available, the system generates energy, consumes some of it and stores some internally. When the energy level begins to drop and there are no external inputs available, it starts to consume its internal storage. When this internal energy is running low, the first state it reaches is one in which the motion stops (because all available energy goes towards the consumption loops). The next state is where all the consume loops stop as well. This causes the motion loops to start again making the system search for food. In other words, the exploration stops first, followed by stopping the consumption of food and subsequently starting the search for food again.

The extreme case is when the energy input is too low. In this case, the chances of self-sustaining the system is significantly diminished. There are, at least, three scenarios with different degrees of self-sustaining ability that are possible. The system can (1) divert the suction into motion instead – this has some chance of self-sustaining if it finds more food immediately or (2) continue the suction instead of the motion – sustains until all resources are depleted or (3) oscillate between suction and motion – this has the highest chance of self-sustaining under diverse external conditions.

The design can now be altered to (a) replace the consumption of food with a mechanism to store the food and/or (b) to oscillate between suction and motion. This change in designs is not hard to achieve by nature itself (to be justified with experiments). You begin to see that the number of possible designs is increasing as we include more and more stable parallel looped reactions. It is no longer obvious which design would self-sustain for the longest time. This is the beginning of diversity of life. Different systems start to develop niche features. The systems diverge in design by incorporating different sets of these features. This gives rise to different amount of self-sustaining time for each of these systems.

In the above discussion, we looked at the change in behavior when the energy transitioned smoothly from a very high value to a very low value or vice versa. There are, however, sudden changes that can occur. They can be either from sudden disturbances in the environment or from explosions when multiple coordinated reactions occur simultaneously (analogous to emotional or realization explosions within our brain – see chapters 49 and 48). These produce discontinuities or abrupt jumps from one state to another. For example, if there is sudden drop in inputs, this causes the loop C to stop suddenly diverting an excess of energy to the motion loop M. This produces a jerky motion. This can be viewed as a sudden life-threatening situation where energy consumption is dropped and instead energy is directed to motion and suction, critical for survival. In more complex systems, this sudden changes (like being attacked by a virus or bacteria and incorporating its genome into the system) can create new behaviors which are memorized to become valid pathways. These newly introduced pathways are quite useful even under normal smooth changes.

As we can see, the stable parallel loop approach has let us design quite abstract and life-like features into the system. A quick summary of what we have done so far is that we have seen how nature can create an increasingly complex network of chemical loops within a system enclosed by a membrane. The different types of disturbances (in the energy or inputs) propagate within this network to yield quite interesting and useful behaviors like an ability to actively search for food, an ability to move towards higher concentration of food sources and an ability to explore and spread the chemical diversity to new regions. Most of these behaviors contribute positively to self-sustain the systems' dynamics for a longer time.

64
Primitive Genetic Code

Having seen how nature can create several life-like features in the previous chapters using variations of active material-energy looped systems, let us now look at how nature can expand the size of the chemical SPL network. For this, an important step is to discover modern enzymes. This is not an easy task. We need to discover a primitive genetic code and also transition from a state with just primitive enzymes (i.e., active material-energy looped systems) to a state with modern enzymes.

I will discuss this problem in a theoretical sense in this and the next chapter. This subject has been studied in considerable detail both theoretically and experimentally (see Vetsigian et. al [158], Woese [171] and others). Several hypothesis have been suggested like a frozen accident (see Crick [27]), adaptor hypothesis (see Crick [26]), a non-Darwinian collective evolution (see Vetsigian et. al [158]) and others. Several of the experimental results would continue to be applicable here, but the analysis would need to be suitably adapted to fit within the SPL framework, which we will do here. The set of assumptions and the proposed mechanisms for creating a primitive genetic code are considerably more involved than the designs discussed in the previous chapters. They do rely on additional randomness. For this reason, experimental justification is quite critical.

Given a DNA sequence, we already know how to create an approximate copy of it using the template-assisted mechanism of chapter 62. Using the same mechanism, we have also seen how to create an mRNA or a tRNA from a DNA sequence. But we do not have a mechanism to create a specific protein using a DNA or mRNA template. We do, however, have a way to create a random protein sequence, with no connection to DNA using monomers-in-polymers-out mechanism. This is clearly not useful. Even if the protein did have beneficial properties, there would be no way to regenerate it reliably after it breaks down.

In this chapter, I will address part of the problem of creating a protein using a DNA or mRNA template. Since copying a DNA and transcribing a DNA into mRNA is already possible, an mRNA template based protein would make it possible to regenerate the same protein as many times as possible. Once this becomes possible, we can say that the construction of a protein is encoded into the DNA, which itself can be successfully copied over to ensure that the construction template is not lost along with the proteins they help create.

One of the first steps towards achieving this goal is to create a genetic code. A genetic code is a well-defined mapping from the set of DNA subsequences to the set of amino acid sequences. It has been shown that the 20 amino acids that are used in all modern enzymes are coded with triplets of genetic sequences (A – adenine, U – uracil, C – cytosine and G – guanine). For example, tryptophan is coded with UGG sequence, glutamic acid with GAA or GAG sequences and leucine with UUA, UUG, CUU,

CUC, CUA or CUG sequences. Before the existence of life, I will refer to such a code as the primitive genetic code. In chapter 65, I will use the primitive genetic code to create modern proteins using primitive enzymes.

The genetic code is enforced in the following way. There are several three-dimensional tRNA molecules with length of about 100 nucleotides, each with a given triplet anticodon arm (like UGG, GAA and GAG). Each such tRNA molecule with the corresponding triplet sequence covalently attaches to a unique amino acid to form aminoacyl-tRNA molecule. If we use the triplet anticodon to identify the tRNA molecule, the genetic code based on this triplet specifies a unique correspondence. A special enzyme called aminoacyl-tRNA synthetase (aaRS), one for each triplet-amino acid pair, facilitates the joining of tRNA to an amino acid. Therefore, to specify a genetic code, we need the following.

1. the set of tRNA molecules with the corresponding anticodon arms,
2. the set of amino acids, and
3. the set of aminoacyl-tRNA synthetases (aaRS's).

Next, to create an enzyme using this genetic code requires, additionally, a mechanism to use the aminoacyl-tRNA's, generated using aaRS, to concatenate the protein chain using an mRNA sequence as a template. This is accomplished by a ribosome. In this chapter, I will only focus on specifying the (primitive) genetic code. In chapter 65, I will address the creation of a modern protein using an mRNA template, specified according to the primitive genetic code.

It is unlikely that the primitive genetic code is the same as the modern genetic code. I will, therefore, discuss a few generic features of the genetic code, irrespective of whether the primitive and the modern ones are the same or not. I will now discuss the problem of existence and uniqueness of primitive genetic code, both of which use similar mechanisms. As an important area of future work, I am very curious to see what the simulation of this iterative process in a computer will yield as the genetic code (see also section 59.6).

64.1 Existence of primitive genetic code

The subject of the evolution of genetic code has been studied in great detail both theoretically and experimentally. Several hypothesis have been suggested. Let me briefly mention a few of them. One of the early suggestions about the universality of the genetic code was that it was a "frozen accident" (see Crick [27]). Since this is not a satisfactory explanation for several, researchers have looked at other possibilities. One of this involved looking at the structure of the code itself to identify special properties. Among these, at least, two were highlighted: (i) synonym order, which specifies how similar the codons are that are assigned to the same amino acid (like, for example, similarity of codons UUA, UUG, CUU, CUC, CUA or CUG assigned to leucine) and (ii) relatedness order, which specifies how related the codons are that are assigned to related amino acids (see Vetsigian et. al [158]). For the latter relatedness order, a notion of similarity among the amino acids is first needed. One measure that seems to express remarkably well is the amino acid polar requirement (see Woese [171], Woese [172], Woese [173] and Woese et. al [170]). The relationship to the codons using the amino acid polar requirement also makes it quite pronounced ("one in a million" compared to randomly generated codes) than without this measure (see Knight [85]). No other known amino acid measure has such a high relatedness order (see Knight [85]).

The above notion of related amino acids and related codons suggest that a given DNA sequence is not translated to a unique protein, rather a family of proteins. However, these proteins should not be viewed as errors at this stage in evolution. In fact, Vetsigian et. al [158] have termed such a family of proteins as 'statistical protein'. This led them to suggest that evolution proceeds in a non-Darwinian manner with no common ancestor. Instead, there is innovation-sharing and a collective evolution of the genetic code with horizontal gene transfer (HGT) as the primary mechanism. They also show the existence of virtuous cycles of cooperation to attain universality and optimality of the genetic code (see Vetsigian et. al [158] for further details).

In this chapter, I will present a similar notion of collective evolution to create a primitive genetic code. However, the main difference is that the primitive genetic code will be formed even in the absence of life using active material-energy looped systems. The underlying mechanism for maintaining and self-sustaining the system is the formation of chemical SPL network within the family of material-energy looped systems.

Let me start by showing the existence of a genetic code. The set of tRNA molecules necessary for the genetic code (see (a) above) can be repeatedly created using the DNA template-assisted mechanism of section 62.2 with some desirable errors. The initial set of DNA sequences are created randomly using the monomer-in-polymers-out mechanism. Each of the randomly created DNA sequences is both copied and transcribed into an RNA sequence with some reliability as well (see section 62.2). Several mutations in the DNA sequences are possible either from entirely new random DNA sequences or through useful errors when copying DNA sequences. Therefore, just from trial-and-error using active material-energy looped systems, nature would eventually discover DNA sequences corresponding to what we can identify as 'primitive' tRNA sequences. Recall that this is not a low probability event because the material-energy looped systems are repeatedly created, behave like primitive enzymes, are active, widespread and abundant in nature (see chapters 59 - 63).

The presence of amino acids (for (b) above) as well as nucleic acids under primordial conditions have been shown experimentally in great detail (see Miller and Cleaves [107] and section 62.1). The chemical reaction for combining tRNA with a given amino acid to create a 'primitive' aminoacyl-tRNA (for (c) above) can occur on the surface membrane of an appropriate active material-energy looped system, as described in section 62.1. This system takes a while to specialize when working with polymers like tRNA (see section 60.8). However, once it is specialized, we can view this active material-energy looped system as the equivalent of a primitive aaRS enzyme.

Given the specialized, active and abundant nature of all of these material-energy looped systems, there is a high enough probability for creating several primitive aminoacyl-tRNA's. Initially, it may happen that a given tRNA will be attached to several amino acids. However, what is natural is that similar tRNA molecules (with different triplets like GAA or GAG) will be attached to a given amino acid (like glutamic acid) on the surface of a given primitive aaRS. This is because of the *similarity* of geometric, physical and chemical properties among these set of tRNA molecules. In the same way, *similar* tRNA's will be attached to similar amino acids on the surface of a given primitive aaRS. We observe these properties with the modern aaRS as well (see Woese et. al [169] and Woese et. al [170]).

We can, therefore, call any system of primitive aaRS's, which are naturally created and which are specialized to one set of tRNA molecules and one amino acid within a given local region as the primitive genetic code. The primitive genetic code does not necessarily have primitive aaRS's for all 20 amino acids yet (as is the case currently). As more tRNA and amino acid molecules are discovered, new material-energy looped systems would need to be created to specialize with pairs of them. Repetitive exposure of the same tRNA and amino acid combination locally helps create a given primitive aaRS. The process of this unique specialization is slow and is prone to errors.

In addition, this primitive genetic code can keep changing over time, in fact, very frequently. A different instance of the material-energy looped system that specializes with a given tRNA, but a different amino acid, say, can be created as often as the original primitive aaRS of the genetic code. This is because the primitive genetic code was created by trial-and-error in the first place. The same randomness can create new primitive aaRS's, altering the code on a regular basis. This makes the primitive genetic code time-varying. Let us see how we can minimize time-varying effects now. In the next section, we will see how to create a universal genetic code as well.

If there are two sets of systems, with one trying to produce the modern enzymes randomly (say, using monomers-in-polymers-out) and the other producing them more reliably using a corresponding set of primitive aaRS's, the latter approach is going to succeed. This is true even if the relative advantage between the two sets of systems is quite small. This advantage combined with the above looped mechanism of creating a genetic code increases the probability of existence of a coding system.

Notice that the above-mentioned primitive genetic code is most likely different from the modern genetic code. It is formed randomly with no obvious relationships between the tRNA sequence and the corresponding amino acid. Yet, it is clear that a *fixed* genetic code is better than a *time-varying* genetic code. A system that develops an advantageous feature (i.e., one that causes its dynamics to continue for a long time – a stable looped feature) using a set of proteins would be able to recreate this feature once again, provided it has a fixed genetic code.

This advantage manifests itself as a purely dynamical phenomenon. The systems with a fixed genetic code (including the very active material-energy looped systems that generate the genetic code itself) will self-sustain much longer than the ones with time-varying code, simply because the repeated creation of favorable proteins and features is much higher with a fixed genetic code. They will continue to evolve several other useful features making them more successful as dynamical systems. For example, a group of material-energy looped systems in a local region that develop an ability to move using, say, flagella will explore farther. They will actively seek food as it starts to deplete in a local region. If there is a fixed genetic code within this group, they will be able to regenerate the flagella if damaged. They will also specialize active material-energy looped systems with higher probability. On the other hand, a group with time-varying code, unable to regenerate such useful features (like flagella), will eventually be eliminated as the less fit systems.

64.2 Uniqueness of primitive genetic code

We have seen that it is natural for a primitive genetic code to exist under the conditions outlined so far (like with the presence of active material-energy looped systems, molecules like nucleic acids, amino acids and others). However, it is not obvious that there is a unique genetic code among different groups of systems across different regions. Since each of these regions are chemically evolving independently, they can all create different primitive genetic codes locally. This should be expected because the primitive aaRS's in one local region need not be the same in a different local region. What this means is that a given DNA sequence will create one protein, say, in one region using its local primitive genetic code while the same DNA sequence will create a completely different protein (or none) in a second region using a different local primitive genetic code. Also, if the primitive genetic code is time-varying in a given local region, it will create a different protein before versus after the changes in the code for the same DNA.

If two sets of subsystems in different regions do not interact with one another, there is no reason for their respective genetic codes to converge to one universal genetic code. However, interesting things happen when these two groups of systems start to interact with one another. In some cases, we can show that both groups of systems will converge to a state where there is only one genetic code. However, there are situations in which systems with more than one genetic code can coexist in a stable manner as well.

As mentioned earlier, this subject of collective evolution and creation of a universal genetic code has already been studied in considerable detail both theoretically and experimentally (see Vetsigian et al [158]). Let us now analyze various possibilities that can occur when two groups of systems at two different regions with two local genetic codes start to interact. This analysis is similar to the one used by Vetsigian et. al [158] with one difference. They were describing it as collective evolution when life existed (using horizontal gene transfer) whereas I am describing it as collective evolution in the context of material-energy looped systems and evolution of the chemical SPL network before life existed.

First, I will assume that both groups of system have 'fixed' genetic codes, though different from each other. There are two reasons for this. Firstly, over a period, the time-varying codes will be eliminated as the less fit feature. The steady-state for the time variations in code is to converge to a fixed genetic code. The systems that use a fixed genetic code would repeatedly and reliably generate the set of all useful chemical features discovered compared to the case with a time-varying genetic code.

A simple non-uniform distribution of systems will naturally create this convergence to a fixed genetic code. Secondly, the mechanisms I will discuss below to show how two groups of systems in two regions will converge to a universal genetic code will be analogously applicable to show how two time-varying codes in a single group in one region will converge to a fixed genetic code.

Next, both groups of systems have stable parallel loops for creating, breaking down, copying, transcribing and translating polymers using appropriate material-energy looped systems. Note that the two groups of systems need not be living beings or have cellular structures. They are simply chemical loops and they contain several variations of material-energy looped systems, modern enzymes created through the primitive genetic code and other chemicals. As discussed in chapter 63, it is possible to create other features like sensing, actuation and exploration using additional looped reactions. In other words, even though these systems do not appear to be living in the traditional sense, they do exhibit a wide variety of interacting and self-sustaining physical/chemical looped reactions.

Now, different enzymes are created within each group of systems. Until now, these enzymes were only used within one group exclusively to create interesting features, like say, flagella for locomotion, energy generation mechanisms and so on. However, when the two groups of systems interact, new stable parallel loops become possible from just the existing enzymes. For example, two systems, one from each group, can together form a loop by exchanging chemicals between them (similar to the feature of plants and animals that exchange CO_2 and O_2). This sort of symbiotic relationship can exist even with different genetic codes. Let me suggest a few possibilities that might occur when the two groups of systems interact.

(i) Dominance: Consider the scenario in which one group of systems becomes more dominant than the other group. The dominance comes from the discovery of a better-evolved set of stable looped features utilizing its fixed genetic code. It is not hard to imagine that some primitive aminoacyl-tRNA's, part of one specific genetic code, are better suited to create complex and more reliable enzymes than the others. This is true even if the primitive ribosome (that helps with the actual translation of an mRNA into a protein sequence) is similar in both systems. For example, some combination of primitive aaRS's and primitive aminoacyl-tRNA's are less prone to errors during the creation of an enzyme. It has been shown that the modern genetic code already exhibits inherent fault-tolerant properties to point mutations through redundancy or degeneracy like, say, fourfold degeneracy in glycine, twofold degeneracy in glutamic acid and threefold degeneracy in isoleucine. Minor errors during replication or transcription of a gene (say, single point mutation) can still yield the same enzyme.

After a long while, the dominant group will take over as the most prevalent systems. The most robust primitive aminoacyl-tRNA's and aaRS's are naturally picked as the successful ones. This is true even if the number of amino acids are different in each coding scheme. The coding scheme that allowed a faster growth of proteins will become prevalent consuming all the resources from other subsystems. The group with a weaker genetic code will be eliminated as the less fit one. The enzymes of the weaker group will also be lost because of the differences in the genetic code. The DNA, on the other hand, of the weaker system is replicated and transcribed using the material-energy looped systems of both groups. We do not yet have living cells. Instead, these are just a series of interacting SPL systems like the material-energy looped systems.

(ii) Peers: Consider another scenario in which both groups have unique niches. Both groups can be considered as peers. They will continue to survive though they tend to get segregated. The relationship can still be either cooperative or competitive. When there is competition, the genetic codes tend to converge to one. The converged genetic code need not be either one. It could also be a new merged genetic code. Typically, this happens when the competition has annihilated both systems considerably that the chemical processes start more or less from the beginning once again.

In the case of cooperative relationship, we can have remnants of both genetic codes present in some form. One example is with mitochondria in eukaryotes, the source of energy generation in cells. Mitochondria has its own genes, different from the cell, and a different genetic code from the standard one. For example, AGA and AGG codes for arginine in the standard code whereas it codes for terminator in the vertebrate mitochondria code. Similarly, AUA codes for isoleucine (standard code) versus methionine (vertebrate mitochondria code) and UGA for terminator (standard code) versus tryptophan (vertebrate mitochondria code). The invertebrates, echinoderms, yeast, protozoans and others have other differences compared to the standard genetic code.

Cooperation does not always mean that you will have two different genetic codes. When working with different groups of *living* beings (as opposed to just nonliving systems like variants of material-energy looped systems), the stable-variant merge loop pattern based on horizontal gene transfer (HGT) discussed in section 55.1 becomes one way to ensure that the beneficial genetic code from each group is itself transferred between the two groups. The aaRS's, and hence the genetic code itself, should be encoded in the DNA for this HGT process to work, unlike the case discussed here with the primitive aaRS. When you transfer these specific genes, you are effectively ensuring that you have a single genetic code.

This is common during the early stages of origin of life. The advantage is that when life begins to diversify, you already have a unique genetic code (with a few exceptions discussed above, say, with mitochondria). This and several other cases, including the above ones, are explored in considerable detail in Vetsigian et. al [158]. They have studied how collective evolution of different groups of living beings can result in a universal genetic code using HGT, or more accurately, stable-variant merge loop based on HGT, as the primary mechanism.

Over a period of time, several different groups of systems interact and evolve into a single universal genetic code (analogous to the mechanisms suggested by Vetsigian et. al [158]). There is a natural iterative mechanism that ensures that as multiple groups of systems interact, they converge to a common genetic code. Let me briefly describe how such an iterative loop works.

1.   Consider an initial condition in which one group *G* has a slightly higher evolved

genetic code (i.e., a code that has generated several useful features).

2. Consider a new group G_1 that has weaker genetic code than the group G. Assume that both G and G_1 interact with one another.

3. At each iterative step m (where $m = 1, 2, ..., N$ iterations):

   a. Since group G is more dominant than G_m (this condition is initially satisfied for G_1 from the assumption in step 2 above), it will have several features that make it adapt and self-sustain for a longer time.

   b. Group G will consume the weaker genetic code system G_m as described using dominance mechanism (i) above. Let us call the resulting combined system as G once again. It will create more new features because of the new genetic material and through the stable-variant merge loop patterns in a directed manner.

   c. This makes the combined system G even more dominant than before. Any new group G_{m+1}, will have higher probability of satisfying assumption (2) above, namely, that G_{m+1} is weaker than G.

   d. The iterative loop now continues. The dominant group G will start consuming other systems one at a time, each time making this combined system G more dominant.

4. After N iterations, the combined system G converges to have a unique genetic code. This can be considered as the universal genetic code.

In this way, the iterative loop based on 'dominance' (as opposed to 'peers' based mechanism (ii) above) has a natural tendency to create and converge to a universal code. Conditions (1) and (2) above are typically easy to satisfy. Given two groups of primitive chemical systems evolving independently in different regions, it is more realistic that one group turns out to be more dominant than the other instead of both groups being equally versatile (i.e., analogous to A ≠ B having higher probability than A = B). As a result, the above dominance iterative loop will be triggered more often in realistic scenarios.

One important assumption with the iterative loop is that the group can undergo several mutations to create a diverse set of features. The diversity of the features is what makes the group dominant. If we have a group that does not undergo such diverse mutations and instead has a symbiotic relationship with the dominant group, there will be less reason for its genetic code to be altered (i.e., peers-based mechanism (ii) above). This is the alternate scenario mentioned above with mitochondria. Mitochondria did not evolve a diverse set of features similar to how cells have evolved. It also has a symbiotic relationship with a cell. As a result, the cell will not consume systems like mitochondria and eliminate them. Therefore, its genetic code continues to deviate from the universal genetic code even now.

In the next chapter, we will discuss how nature can switch over to creating modern enzymes using the primitive material-energy looped systems and the primitive genetic code discussed until now.

65

Transitioning from Primitive to Modern Enzymes

Given that nature can create primitive enzymes and unique primitive genetic code using mechanisms described in the previous chapters, we want to see how nature can create modern enzymes for most organic reactions. The main missing piece is the creation of a primitive ribosome. Once this is achieved, nature can start creating an enormous variety of modern enzymes using primitive material-energy looped systems, which we can call as natures *chemical factory*. With the discovery of a chemical factory for creating modern enzymes, it now becomes possible to take over the functionality of each existing primitive material-energy looped system using an appropriate modern enzyme. This includes taking over the primitive genetic code and replacing it with the modern genetic code as well.

The chemical factory keeps generating an enormous number of modern enzymes using primitive mechanisms and a wide repository of DNA molecules. The errors in DNA replication, transcription and translation using the primitive mechanisms also aid the chemical factory to create a wide range of modern enzymes. This can be seen as a collective evolution as suggested by others like Vetsigian et. al [158]. The above errors at this stage in evolution are not undesirable. They can be regarded as 'statistical proteins', a term suggested by Vetsigian et. al [158].

Without the help of primitive enzymes, the task of creating even a single modern enzyme of a specific functionality using just random collisions has almost zero probability. Even if nature got lucky, it would be unreasonable to claim that nature creates several modern enzymes necessary for, say, replication, transcription and translation. Furthermore, nature should regenerate the same enzymes reliably and repeatedly even when they degenerate. Natural mechanisms described in this book like the active material-energy looped systems and their variations, is a necessary intermediate stage for achieving this goal.

65.1 Creating modern enzymes using primitive versions

In section 62.2, we have shown how a material-energy looped system can start with a DNA and create a complementary mRNA molecule. This is the transcription step. In the same section, we have seen how to make a copy of the DNA itself. This is the replication step. The next step is the translation of mRNA into a modern protein (or enzyme). For this step, we first need 'adaptor' molecules that link a given sequence of nucleic acids to a given amino acid. These are the primitive aminoacyl-tRNA molecules. They are created using primitive aaRS's. This constitutes the primitive genetic code (see chapter 64). Finally, we need a primitive ribosome. The ribosome takes aminoacyl-tRNA's as inputs in a specific order corresponding to a given mRNA sequence. It removes the amino acids in the aminoacyl-tRNA's and concatenates them (forming peptide bonds) in the same order to create an amino acid sequence. This amino acid sequence naturally folds into a three dimensional shape of a protein as the minimum energy configuration. This three dimensional shape is responsible for giving the protein its enzymatic properties.

In the previous chapters, we have already discussed how to create material-energy looped systems for all of the above steps, except for a primitive ribosome. Let us now see how to create a primitive ribosome in a theoretical sense. As mentioned in chapter 62, it is quite critical to obtain experimental verification of this theoretical step. Recall that a primitive ribosome is, by definition, a material-energy looped system that performs the same function as the modern ribosome.

Let us start with a material-energy looped system that we used in chapter 62 for transcription or replication. Let us say that this system latches an mRNA molecule instead of a DNA strand. Since mRNA is also a helical polymer, this system will make the mRNA molecule advance a few turns along the length in every compression-expansion cycle of the material-energy looped system (see section 62.1 for a helical DNA strand).

We now have a primitive aminoacyl-tRNA that matches a triplet of mRNA sequence, namely, those that are created by the primitive aaRS's of the primitive genetic code. These primitive aminoacyl-tRNA molecules have natural affinities to be preferentially located near the corresponding mRNA sequence at one of the intake opening. This is analogous to how the complementary nucleobases have affinities to stay near the opening when working with a DNA strand during replication or transcription (see section 62.2). Therefore, in each compression-expansion cycle of the material-energy looped system, these primitive aminoacyl-tRNA molecules are partially sucked in while the helical mRNA molecule advances as well (similar to the operation of design D of section 59.2). The primitive aminoacyl-tRNA molecules are typically too big to enter the system. Nevertheless, when one of the primitive aminoacyl-tRNA molecules is partially sucked in, another primitive aminoacyl-tRNA molecule corresponding to the exposed new triplet (as the mRNA molecule advanced in a compression-expansion cycle) is naturally attracted because of the triplet base pairing, near the same opening. The above theoretical description appears quite simple in a conceptual sense. However, we need experimental verification to support the argument.

It is now possible for the amino acids in two consecutive primitive aminoacyl-tRNA's to react favorably within the material-energy looped system during an active compression-expansion cycle. In this way, the system continues to grow the amino acid sequence in a specific order during each compression-expansion cycle. The growing protein sequence can be either inside or outside the active material-energy looped system. We can call such an active material-energy looped system as the primitive ribosome. The amino acid sequence formed is clearly correlated to the original mRNA sequence i.e., the primitive genetic code and the mRNA sequence uniquely determines the amino acid sequence. Therefore, the protein can be reliably and repeatedly created in a 'statistical' sense. In other words, nature has an ability to create a modern enzyme entirely from a collection of primitive material-energy looped systems.

I want to emphasize once again the importance of experimental verification of these theoretical processes even though the mechanisms sound feasible. Experimental designs using synthetic materials (like synthetic nanomotors of Fennimore et. al [39] and Laocharoensuk et. al [91]) would be an useful initial step even though we do ultimately need designs that can be created entirely naturally.

The amino acid sequence created using these primitive mechanisms also folds naturally into a unique three-dimensional shape. This corresponds to a minimum energy configuration. Since the protein is constructed serially, it becomes possible to recreate the same three-dimensional shape as long as you reconstruct the same serial sequence. This is the unique advantage with serial construction as opposed to a parallel one (see section 61.3).

Note that it is also possible to view the actual sequence of steps for the creation of a modern enzyme from a given DNA sequence, as disturbance propagation in a network of loops. This was already discussed in chapter 58 for modern processes. However, the same explanation is applicable even with the primitive material-energy looped systems.

65.2 Chemical factory using primitive systems

Assuming that experimental evidence of the above primitive mechanisms will be discovered soon, let me continue the analysis briefly. With each of the above primitive systems in place, nature can start creating a large number of modern enzymes reliably. Random DNA, mRNA and protein sequences can be created using monomers-in-polymers-out mechanism with just nucleic acids or amino acids as the case may be. Replicating the DNA is yet another source of creating new DNA because of the errors in replication (which are desirable at this stage in evolution). Similarly, transcription is another source of creating new mRNA sequences because of the errors (which are desirable for now).

Finally, translation also has errors when creating new proteins. As long as the proteins with errors have similar functionality in a statistical sense, they are useful. A given translation error may not help in catalyzing a given chemical reaction. However, it is possible that this error protein catalyzes a new chemical reaction. As a result, the collection of all proteins (due to these errors) will statistically catalyze a large number of chemical reactions than a few error-free proteins. In other words, *a primitive protein factory which allows nature to create a large variety of proteins through errors is more beneficial than a modern protein factory that creates a smaller set of proteins but with higher accuracy*. Note that both high-accuracy and low-accuracy mechanisms are repeatable, though with varying degrees.

With so many random DNA and mRNA sequences generated (along with reliable and repeatable copies in a statistical sense), nature effectively has an enzyme factory. An enormous variety of enzymes are constantly created (and broken down as well – creation and breakdown of polymers is a looped process in itself that self-sustains for a long time). Even though the enzymes created are quite random, when we take the entire collection, they become extremely useful. Nature would not know ahead of time which set of chemicals a given enzyme would catalyze. The enzyme factory production and the set of chemicals they potentially catalyze are independent processes, at least, initially. In fact, this decoupling is desirable at this stage of evolution primarily because both of them are looped processes – the former is the looped process of creation and breakdown of polymers and the latter is the chemical SPL network in a given region. The two sets of chemicals, namely, the enzyme and the corresponding reactants, begin to match correctly thereby forming a large chemical SPL network analogous to the modern metabolic network (see Nicholson [114]).

Even though the matching problem (of the enzyme with the reactants it catalyzes) has a degree of randomness, it is not improbable at all. This is different from saying that using purely random mechanisms it is possible to create an enzyme like carbonic anhydrase that catalyzes CO_2 and H_2O to create H_2CO_3, primarily because of the existence of a non-random and active mechanisms based enzyme factory (as opposed to passive random chance mechanisms).

Therefore, as there is a continuous production of a variety of enzymes, several of them start to catalyze the input chemicals that are already present in the environment. The discovery of these enzymes and the respective chemical reaction is an inherently slow process. However, there are a large number of such enzyme factories and a large amount of time to help discover several useful proteins. Besides, not every random DNA sequence needs to be tried for each different enzyme. There may be additional inherent structure in the types of DNA sequences and their relationship to the enzymes than we have already discovered.

As the random proteins are generated from the enzyme factory in parallel, several of them start to catalyze chemical reactions immediately. Nature is not waiting to discover a given enzyme like carbonic anhydrase for a given chemical reaction. It simply starts using any of the catalyzed chemical reactions that exist until then. There will be several chemical loops that will be formed with the existing enzymes. Whenever this happens, we know that these chemical loops will self-sustain for a very long time. They will also start to form large and complex SPL network as discussed in the previous chapters. If enzymes like carbonic anhydrase are not yet discovered, other enzymatic chemical loops will be used to achieve the same desired behaviors like locomotion, sensing, actuation and others. Later, when a better and an effective enzyme is indeed discovered for the same chemical reaction, the older enzymes will be naturally replaced by the newer ones.

This is a non-Darwinian collective evolution generating a large chemical SPL network collectively in a given region before the existence of life itself. Several simple organic molecules present in low concentrations until now are converted to create a large variety of organic molecules using the new enzyme factory. There is an explosion of existence of simple organic molecules of different functional groups in a given region (like carbonyl, carboxyl, aldehydes, ethers, esters, alcohols, ketones, amides, amines, sulfhydryl and others) because of the enormous collection of catalysts generated through the enzyme factory.

Additionally, since we have already seen in chapter 58 that any behavioral or functional feature can be expressed as disturbance propagation within an appropriately chosen network of loops, nature begins to create a large variety of functional features as well (like sensors, actuators, exploration, ability to digest and so on – see examples of these features in chapter 63 that depended only on chemical SPL networks of special properties). Expressing a functional feature as disturbance propagation in a suitable chemical SPL network is critical because nature can create a given behavior using just a chemical SPL network without worrying about which specific chemicals and enzymes are part of that network. This result, which is a new outcome of the stable parallel looped architecture highlighted in this book, gives a certain invariance to the problem of origin of life against the discovery of specific chemicals and specific enzymes.

65.3 Transition to modern enzymes and genetic code

Using the above-mentioned chemical factory for creating DNA, RNA and proteins, nature will eventually discover the modern version of DNA polymerases, helicases (for replication), RNA polymerases (for transcription), tRNA's, ribosomes and aminoacyl-tRNA synthetases (for translation) and the corresponding DNA sequences. The process is iterative as shown in Figure 65.1. When this happens, there are two ways to perform the same operation, namely, a primitive way and the modern way to (a) replicate the DNA, (b) transcribe a DNA into an mRNA and (c) translate an mRNA into a protein. Besides, tRNA's and aminoacyl-tRNA's in the primitive and the modern approach could be quite different. Therefore, the primitive and the modern genetic codes could be quite different from each other as well. More importantly, the modern ribosome can only work with the modern aminoacyl-tRNA's, not the primitive ones.

This implies that starting from a given DNA sequence the primitive mechanism might generate the modern RNA polymerase. However, the same DNA sequence would not necessarily generate the same RNA polymerase using the modern mechanism. Therefore, even though nature is able to generate modern RNA polymerase reliably, the primitive mechanisms cannot be eliminated just yet. Nature is only able to create modern enzymes using primitive mechanisms. The transition to modern enzymes and modern genetic code is not yet complete. Nature needs to discover the DNA sequences that generate the modern enzymes using entirely modern RNA polymerases, modern ribosomes, modern aaRS and the modern genetic code.

The modern enzymes discovered are much more efficient than the primitive mechanisms. Strictly speaking, the primitive mechanisms for replication and transcription do become obsolete with the discovery of DNA and RNA polymerases. However, it is important to remember that these very enzymes are still created only using primitive translation mechanism. The DNA sequences for generating these enzymes do not work with the modern ribosome and modern genetic code.

Now, we have two chemical factories for generating modern enzymes – the primitive factory and the modern factory. Both of them generate different enzymes starting from the same DNA sequence. Nature starts to generate an enormous variety of enzymes. The modern factory is faster and more accurate while the primitive factory is slower and has more errors (see Figure 65.1). During the transition period, nature iteratively uses both mechanisms, primitive and modern, to create variety and

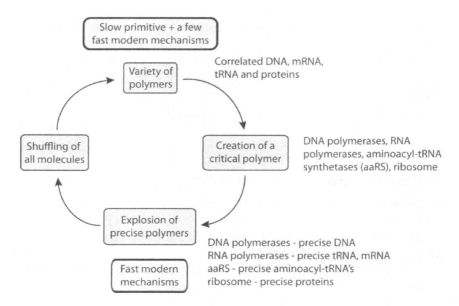

Figure 65.1: Slow-variety followed by fast-specificity mechanism. The iterative process has two phases. In one phase, the primitive mechanisms create a variety of correlated polymers slowly. In this phase, if a critical polymer is discovered, it enters the second faster phase creating several precise molecules until the second law effects slow down the process and brings it back to the first phase.

specific enzymes respectively. Nature resorts to trial-and-error, but with reliable, repeatable and active creation of enzymes using enzyme factories, to discover the correct DNA sequence for a given modern enzyme using the modern genetic code and modern mechanisms.

Once the same set of modern enzymes are generated by trial-and-error using modern chemical factories (i.e., with modern mechanisms), nature has effectively transitioned to the modern system from the primitive system. In other words, DNA replication, transcription of DNA to mRNA, translation of an mRNA into a protein, creation of modern aaRS's and aminoacyl-tRNA's, i.e., the modern genetic code, now uses the modern approach. This discovery is a slow process but it is reliable and repeatable since we have primitive factories. If we were to rely on random self-assembly, the discovery of these chemicals would be almost impossible and non-repeatable. The primitive system can now be eliminated as the less fit mechanism.

Note that in the above discussion, we did not assume the existence of primitive life yet. It is possible that several of these chemical SPL networks may be grouped inside a closed phospholipid membrane to be called as primitive life, but such a requirement is not necessary. Both solutions, with or without primitive life, are valid using the above mechanisms. The evolutionary process at this stage, with or without primitive life, is non-Darwinian and collective in which one of the primary observable changes is the growth of the chemical SPL network over time.

The modern genetic code will also converge to a universal genetic code in an exactly analogous manner as discussed in chapter 64. When multiple modern systems interact, one system will iteratively become more and more dominant as long as it continues to evolve. The genetic code of this system will turn out to be the universal genetic code. There will, however, be other systems with different genetic codes because they do not evolve as much, like with the mitochondria example discussed in section 64.2. It is more precise to say that the modern genetic code will become universal only after the systems evolve into primitive living beings, though not necessary. In that case, the evolution of a large number of features is easier when we have membrane bound stable parallel looped systems (which we can call as primitive life). This allows systems or primitive life to have a notion of evolution to become iteratively more dominant. Given that all the processes discussed here are iterative in nature, it is possible to simulate them in a computer (similar to the discussion in section 59.6). This is an important area of my future work.

65.4 Loss of primitive features

When the primitive material-energy looped systems become obsolete, several primitive features that were useful also disappear. For example, creation of random DNA, RNA and protein sequences using monomers-in-polymers-out mechanism will be lost. The modern enzymes do not have a closed membrane like the primitive system to facilitate such a random polymer generation naturally. Therefore, the modern polymer sequences always start from another existing sequence. The variations or new DNA sequences can only appear from existing DNA sequences through mutations or errors (during replication or transcription). The modern DNA replication requires primases to initiate the synthesis of complementary DNA strand.

The higher fidelity of modern enzymatic mechanisms gives more stability to the features being developed. The primitive system, on the other hand, generated a variety of polymers slowly and with higher inaccuracies. In the modern system, variety is still created through random mutations in the most prolific species like bacteria and viruses. Naturally, the high fidelity is better for directed evolution as we now know.

66
Chemical Diversity

Using all the mechanisms discovered until now, we have seen how nature transitions from primitive material-energy looped systems to modern enzymes, DNA and genetic code to create a wide collection of proteins. In this chapter, I want to discuss how nature creates other important chemicals necessary for life as well. These include different types of polymers beyond DNA, RNA and proteins. The goal is to create an iteratively larger chemical SPL network and hence a diverse set of chemicals. The problem of origin of life starts to become feasible when this happens (see Figure 59.4).

Another common question discussed in relation to the problem of origin of life is homochirality. A molecule is said to be chiral (or an optical isomer) if it is not superposable to its mirror image. This property is described in terms of a molecule' interaction with polarized light. A solution of an optical isomer can turn the plane of a beam of polarized light either clockwise (dextrorotatory) or counterclockwise (levorotatory). This property is represented using different naming conventions, one of which is the D- and L-forms (see McMurry [103] and Solomons and Fryhle [144]). The typical cause of chirality in organic molecules is the presence of an asymmetric carbon atom.

It is observed that a chemical reaction creating chiral molecules always produces racemic mixture (an equal split of L- and D-forms) under no external chiral influences. However, most amino acids observed in nature are L-forms and sugars are D-forms. Proteins made from L- amino acids are referred to as left-handed proteins while those made from R- amino acids are referred to as right-handed proteins. The problem of homochirality is to explain how the asymmetry arose on earth given that the early beginnings of chemicals on earth are actually racemic.

In section 66.5, I will show how a chemical SPL network when combined with second law of thermodynamics has a natural tendency to create an asymmetric distribution (over a period of time) even if the initial collection of molecules is racemic. In fact, we can generalize this result to other situations to show that asymmetry is a natural outcome from a symmetric system as long as the systems are SPL designs subject to the effects of the second law of thermodynamics (like, for example, the presence of more matter than antimatter in our universe).

66.1 Membrane bound system – an early primitive cell

Until now, we have been focusing primarily on polymers like DNA, RNA and proteins that are made of nucleic and amino acid monomers. These polymers carried useful information within their monomer sequences. Therefore, duplicating the polymers' function necessarily required creating a precise sequence of monomers. Without this precision, the polymer would not exhibit its function. The unique sequences and the ability to duplicate them reliably made it ideal to use its static structural sequence to identify the dynamical properties it exhibits. Figure 59.4 summarizes this discussion on how the SPL structure grows even before the existence of life.

Now, other types of chemicals and polymers exhibit useful structural and dynamical properties as well. The creation of these polymers is not as hard as creating nucleic and amino acid based polymers. These polymers have a different type of complexity. The monomers that make up these polymers have useful geometric, physical and chemical properties that enable the corresponding polymer to have several useful features. Any minor variations in the sequence of monomers do not significantly alter the polymers' function. This is unlike the case with variations in amino or nucleic acid sequences. In this section, I will discuss the creation of such polymers and their properties briefly. For more details, I refer the reader to any standard textbook on biochemistry (say, Garrett and Grisham [48], Koolman and Roehm [89] or T. McKee and J. McKee [102]).

Let me arbitrarily classify these molecules into two groups. The first group are polymers based on proteins and RNA sequences. Note that this group is eventually based on nucleic acids and amino acids as well. However, the difference is that these polymers are not directly based on amino acids or nucleic acids as monomers. Instead, this group contains polymers whose monomers are themselves polymers like proteins or RNA sequences. For example, microtubules are helical polymers made of protein polymers themselves (like small globular proteins, α- and β-tubulin). The second group are polymers that are unrelated to nucleic or amino acids. They are, for example, polymers based on lipids or carbohydrates. These two groups of chemicals are important for a number of useful features exhibited by living beings.

The chemistry of these molecules is already studied in detail. They can be found in any standard textbook on biochemistry. What I would like to focus here instead is on how we can create these molecules repeatedly and reliably, just like how we showed this for DNA and proteins in the previous chapters.

66.2 Proteins as monomers

There are several important examples where proteins act as monomers. Actin is an example of globular protein that is a monomer for microfilaments, one of the important components for the cytoskeleton of a eukaryotic cell, and for thin filaments, used in muscle contraction. Intermediate filaments, another component of the cytoskeleton, is also a polymer based on a family of proteins like acidic and basic keratins, desmin, vimentin, lamins and others. Microtubules are yet another component of the cytoskeleton that are polymers based on small globular proteins like α- and β-tubulin. Microtubules also help with the transport of chemicals within the cell. More generally, motor proteins interact with actin (like myosin) or with tubulin in microtubules (like kinesin or dynein) to move along these surfaces.

In order to create these polymers reliably, the first step is to encode the protein monomer as a suitable DNA sequence. The next step is to assemble these monomer units to form the desired polymer. As mentioned earlier, this assembly does not require extreme precision like with DNA or protein sequence. The polymerization process, i.e., the nucleation and the growth of each of these polymers are already studied in detail (see, for example, Garrett and Grisham [48]). Some form of preliminary complex is necessary to initiate the polymerization. For example, ARP (actin-related proteins) complex is needed for actin polymerization. Similarly, γ-tubulin in microtubule organizing center like centrioles forms a ring complex acting as a scaffold to begin polymerization of a microtubule. Once initiated, the growth of the polymer is a spontaneous process until some form of instability sets in to stop the process. Since a high precision is not necessary for its functional role, we can say that the polymer created is sufficiently similar to the desired one.

The sequence of steps for polymerization and depolymerization of these molecules can also be viewed as disturbance propagation in a suitable network of loops. Therefore, the memory (see chapters 61 and 62) of recreating a similar polymer of this type has two parts: (a) the DNA sequences of the monomer proteins and (b) a network of chemical and physical loops to represent the sequence of steps for polymerization.

66.3 Non-protein molecules as monomers

Let me consider polymers that use molecules unrelated to proteins as monomers. For example, these are based on lipids or carbohydrates. The cellular membrane is an important example that is made of phospholipid bilayer. They spontaneously arrange in a spherical shape with the hydrophilic head on the outer side, with one layer facing the exterior and the other layer facing the interior of the cell, and the hydrophobic tail between the two layers. Systems with such a membrane regulate the movement of chemicals in and out of them. If these molecules were present under primitive conditions, we see that they can play a critical role in creating an elastic membrane even for the active material-energy looped systems.

The creation of a phospholipid bilayer polymer is spontaneous and the resulting polymer is not identical every single time. The lipid monomers are neither the same nor are arranged in a similar way each time. In spite of these differences, the resulting enclosed systems are functionally similar. Creation and breakdown of these lipids involve a series of chemical reactions catalyzed by appropriate enzymes (see Alberts et. al [1] for a discussion on the looped reactions for lipid metabolism).

The creation of the monomer sequences can also be seen within the network of loops shown in Nicholson [114]. The polymer creation is partly a spontaneous assembly of these phospholipids. Therefore, we can say that the memory for recreating a similar polymer can be represented as disturbance propagation in the above network of looped reactions (see section 58.2). The situation is similar with other types of polymers like polysaccharides. There are different types of polysaccharides like starch, for storage of energy and cellulose, as a structural component of a plant cell wall. There are specialized proteins for the initiation and the elongation of these polymers. The same applies with the breakdown of these polymers. The metabolic map shown in Nicholson [114] shows the corresponding network of looped reactions for the polysaccharides.

66.4 Directed evolution to create DNA in a primitive cell

As the material-energy looped systems' role as a primitive enzyme starts to become obsolete with the discovery of modern enzymes, the same system begins to simplify itself by taking the role of just a container (i.e., an enclosed membrane) that runs a set of chemical reactions. The set of chemicals in the environment now become compartmentalized within a number of such simplified material-energy looped systems. This has its advantages too. For example, it is now possible to run only a particular set of chemical reactions and in a controlled fashion within these systems. This is the beginning of the formation of a primitive cell. The reactions within this cell are part of the system and the outside is the environment. The surface membrane is made of phospholipids as mentioned above. Lipid bilayers have a *self-sealing* property as well. If there is a tear that opens up the membrane, there is a natural tendency to rearrange the lipids spontaneously to close the tear.

In addition, the membrane-enclosed systems can have an internal structure like a primitive cytoskeleton with polymers like microtubules. The system now starts to attain a structural shape as well. These mechanisms are quite random, but they are reliable and repeatable. As several of these systems start to enclose different types of chemicals, especially DNA and enzymes, there will be some of them that can self-sustain their dynamics as looped reactions. The collection of all such self-sustaining membrane-enclosed systems can have a directed evolution even though they may not have the entire DNA transcription, translation and replication machinery.

The underlying mechanism for this directed evolution is the stable-variant merge loop pattern (see section 53.4). The merge step in this case works quite naturally, as follows. Two simple phospholipid membrane based systems can coalesce and can become one bigger system. When this happens, more looped reactions with the combined set of chemicals become possible. This merging of two systems can be considered as a primitive version of 'sexual' joining or even a primitive version of horizontal gene transfer (HGT). Therefore, in some cases, the combined system is more stable than each system individually. This clearly sets a direction in evolution and can be viewed as a variation of the stable-variant merge loop pattern. The main difference is that the merging of two desirable systems happens by chance, even though a modern enzyme factory exists to assist the process.

As this merging process continues, though randomly, better stable systems begin to form, at least, locally in some regions. If, within this pool of membrane-enclosed systems, you have one system that contains all of the DNA replication, transcription and translation machinery, this system would even develop an ability to reproduce. We can now call this self-replicating system as the first primitive cell. They begin to multiply naturally. The remaining systems will be eliminated as the less fit ones. Several different self-replicating systems will be created each with its own DNA sequences. The net result is that the behavior of this primitive cell begins to deviate considerably from the environment.

The chemical reactions that occur within this primitive cell are no longer random. Just as with the material-energy looped systems, this is an important turning point in the evolution of chemicals and chemical reactions. In the subsequent sections, I will assume the existence of such a membrane enclosed system as the starting point for further designs.

66.5 Homochirality

One common problem discussed in relation to the problem of origin of life is the problem of homochirality. A molecule is said to be chiral (or an optical isomer or an enantiomer) if it is not superposable to its mirror image. This property is described in terms of a molecule' interaction with polarized light. A solution of an optical isomer can turn the plane of a beam of polarized light either clockwise (dextrorotatory) or counterclockwise (levorotatory). This property has several naming conventions, one of which is the D- and L-forms (see McMurry [103] and Solomons and Fryhle [144]). The typical cause of chirality in organic molecules is the presence of an asymmetric carbon atom.

It is observed that a chemical reaction creating chiral molecules always produces racemic mixture (an equal split of L- and D-forms) under no external chiral influences. However, most amino acids observed in nature are L-forms and sugars are D-forms. Proteins made from L amino acids are referred to as left-handed proteins while those made from R amino acids are referred to as right-handed proteins. The problem of homochirality is to explain how the asymmetry arose on earth given that the early beginnings of chemicals on earth are actually racemic.

In this section, I will show how a chemical SPL network when combined with second law of thermodynamics has a natural tendency to create an asymmetric distribution (over a period of time) even if the initial collection of molecules is racemic. This result can be generalized to other situations to show that *asymmetry is a natural outcome from a symmetric system as long as the systems are SPL designs subject to the effects of the second law of thermodynamics*. As an example, the reason why we have more matter than antimatter even if the initial distribution is symmetric can be seen as an application of this result (assuming that the entire universe is an SPL system, which seems reasonable but needs further justification).

The subject of homochirality has been studied in considerable detail. It is suggested that homochirality evolves in three steps: (i) *mirror-symmetry breaking* step in which the enantiomer asymmetry is introduced even from an initial symmetric distribution, (ii) *chiral amplification* step in which the asymmetry introduced starts to get more pronounced over time and (iii) *chiral transmission* step in which the enantiomer asymmetry is propagated to other molecules through an asymmetric distribution of the set of chemical reactions that occur.

Some proposals for the introduction of initial enantiomer asymmetry are: (a) an extraterrestrial origin in the Murchison meteorite (see Meierhenrich [105]). This meteorite was believed to contain a rich set of organic compounds including amino acids (like glycine, alanine and glutamic acid) and nucleic acids (like uracil and xanthine). (b) the Vester-Ulbricht hypothesis that the beta decay from Parity violation (see Weinberg [161]) causes different half-lives of important organic molecules resulting in an asymmetry.

For chiral amplification step, it was shown experimentally that a few autocatalytic reactions (i.e., a type of looped reactions) starting with a small amount of asymmetry can result in a much larger asymmetric set of enantiomers. For example, in Soai reaction (see Shibata et. al [140]), pyrimidine-5-carbaldehyde is alkylated by di-isopropylzinc to produce pyrimidyl alcohol. The presence of 0.2 equivalent of S-enantiomer alcohol produces a 93% excess of enantiomer pyrimidyl alcohol. Other similar experiments have been shown with amino acids and sugars as well (see Kojo et. al [88], Córdova et. al [24] and Mathew et. al [99]). The underlying mechanisms suggested are either autocatalytic reactions or some form of spontaneous symmetry breaking.

In the discussion below, I show that (in a theoretical sense) a chemical SPL network combined with the second law effects introduces asymmetry, which is further amplified through the iterative process of chemical looped reactions (see Figure 66.1). The mechanism suggested for introducing asymmetry from symmetry is generic enough to be applicable to any collection of SPL systems that are subject to second law effects. The iterative process is as follows. The computer simulation of this iterative process is an important area of my future work (see also section 59.6):

1. *Locally non-uniform and non-racemic physical distribution*: The processes described in the previous chapters based on material-energy looped systems are active mechanisms, not passive or random ones. They guarantee that the local physical distribution (not chemical distribution) of molecules in space and time is not racemic or locally uniform even though the total distribution is racemic.

2. *Accumulation of chiral molecules*: Among a given group of chiral molecules (either L- or D-forms), there is a natural tendency to react with each other in a well-defined way. For example, if a D-form molecule A_D reacts with an L-form molecule of B i.e., B_L, then A_D does not react with the D-form molecule of B i.e., B_D. This causes appropriate segregation of molecules. Since the physical distribution is non-uniform and non-racemic (as mentioned in step (1) above), one group will be localized in a small region while another group will be spread out to a larger region (see Figure 66.1). The active mechanisms of material-energy looped systems ensures that the smaller localized region forms a stable chemical looped network.

    This imposes a constraint that the independent amino acids and sugars should take the same chiral forms within each localized group. The chiral molecules that are not part of some loop within the SPL network will disintegrate. In other words, the spread out group of chiral molecules will disintegrate while the localized group of chiral molecules will grow in concentration, similar to what was observed with Soai reaction above and other autocatalytic reactions. The chemical loops within the localized group is the generalization of the autocatalytic reactions.

3. *Self-sustaining nature of chemical loops causing chiral amplification*: Even if both chiral groups evolve evenly, the probability of simultaneous discovery of all molecules necessary for a given chemical loop to occur, within both groups, is close to zero. Therefore, whenever a complete loop is discovered in one group,

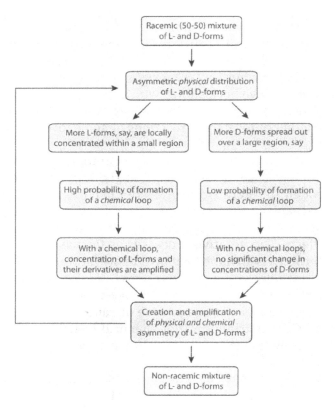

Figure 66.1: Mechanism for homochirality using stable parallel loops. Starting from a racemic (50-50) mixture of L- and D-forms, we can guarantee the creation of a non-racemic mixture iteratively using the asymmetry of physical distribution of molecules and the formation of a specific dynamical structure of chemical loops.

it starts to produce several additional chiral molecules because of the self-sustaining nature of the loop. As the other group does not yet have the same loop, it lacks the corresponding chiral molecules. This creates an initial asymmetry in the collection of chiral molecules.

4. *Creation of chiral asymmetry*: With multiple such active loops, the two types of chiral forms naturally segregate into separate groups, while the second law mixes them together. If the second law effects are slower than segregation, the probability of both chiral groups evolving equally for thousands of years is negligible compared to evolving unequally in which one group becomes dominant (see Figure 66.1). Another state, in which the dominant system becomes weaker and vice versa, resulting in an equal distribution of both chiral forms has near-zero probability. This is expected since the disintegrated molecules of the weaker system (without chiral properties) are reused by the dominant system making the weaker system even weaker and the dominant

system more dominant.

The iterative process now continues as shown in Figure 66.1. The high-probability asymmetric distribution starts to diverge more with time. Eventually, one group remains whenever its active stable parallel looped network becomes sufficiently dense. Therefore, the definite direction in molecular evolution through active stable parallel looped processes causes homochirality.

67

Reproduction

Reproductive ability is a reliable way for a system to make a copy of itself. The copy just needs to appear identical to the original with respect to most of the important observable features. We are not requiring that every single molecule and every single structure is copied over or that these molecules appear in the same numbers. Since the definition of equality of two complex systems introduces this inherent difficulty, let me choose a definition for reproduction that appears reasonable, practical and one that agrees with our everyday usage. I will start with a strict definition by looking at what is critical to recreate a given system and then checking equality of two systems in terms of these features. In this regard, the features and components that make up any system are:

(a) its physical structural components including both its internal and external structure,

(b) its internal dynamics represented through a number of physical and chemical reactions, and

(c) a set of abstract behaviors it exhibits.

System A can be called as a reproduced copy of another system B if and only if the features (a) – (c) of A are similar to B in a satisfactory way. Not every system can be duplicated in the above sense with ease. Strictly speaking, what I mean by reproducing features (a) – (c) is that both systems A and B are 'capable' of exhibiting any given feature, not that each system will actually exhibit it at about the same time. At any given time, both systems may actually be performing different dynamics altogether. However, if and when suitably similar conditions become available at possibly different place and time, both systems should be able to exhibit the same feature.

In this chapter, I want to show that the problem of reproduction, namely, creating the above features (a) – (c) of a stable parallel looped system like a primitive cell can be reduced to:

1. duplicating the DNA,
2. creating the physical structure,
3. ensuring that the SPL network can be regenerated, and
4. having a reasonable initial distribution of chemicals.

I will show that all other features of the reproduced system can be derived from these. One reason is due to the result of chapter 58, namely, that the creation of a given feature via a sequence of steps can be represented as a disturbance propagation in a suitable network of loops. Using the mechanisms discussed in the previous chapters, I will show that nature can perform each of these steps repeatedly and reliably.

Consider the simplest way of creating two systems from one system, namely, by breaking the original system into two parts. If the split creates approximately two halves, then each half will imitate parts of (a) – (c) above compared to the original system, in an obvious way. This split may seem sufficient for systems like simple crystals, simple solids (like sand or rocks) and some liquids. It is not a good mechanism for complex systems like living beings. For example, some of the features that should be present after a cell reproduces are as follows. The physical structure (i.e., (a) above) corresponds to a cell membrane, its internal cytoskeleton and other compartments like nucleus, Golgi apparatus, mitochondria and others. The internal dynamics (i.e., (b) above) has both physical movement and detailed chemical reactions for proteins, lipids, carbohydrates and other chemicals present within a cell. The behaviors (i.e., (c) above) exhibited by a cell are, say, sensing concentration gradients, searching for its own food, resisting death, fighting pathogens and others. Therefore, for a cell, we would like to ensure that all three sets of features (a) – (c) are exhibited by each of the two parts after a cell replicates. Let me discuss this next.

67.1 Structure within the set of reproducible features

At first glance, there appears to be an enormous set of features that should be duplicated when a cell replicates. However, I want to show that the stable parallel looped architecture of cells simplifies the problem of reproduction considerably. A number of looped interrelationships within these systems make it easy to create most of these features. Let me outline some of these simplifications now.

Deriving behaviors from stable parallel loops: All abstract behaviors (i.e., feature (c) above) of cells can be represented within stable parallel looped chemical reactions as disturbance propagation (see chapter 58). This is a nontrivial observation. In each of the designs in chapter 63, we have seen that the abstract behaviors like the simplified versions of resisting death, actively searching its own food, exploration and others emerge from within the interconnected looped reactions. While I have not discussed the same features for actual living beings, I claim that they can still be mapped within a stable parallel looped network.

See chapter 58 for additional examples and the corresponding generalization that all features expressed as a linear sequence of steps can be equivalently represented as disturbance propagation within suitable chemical SPL networks. For bacteria and other living beings without a brain, behaviors like resisting death is just similar to the one discussed in the primitive designs, except that there are a large number of situations where they detect damage (or instability) and generate the equivalent of 'back pressure' for material-energy looped systems to resist or compensate for it. When it comes to beings that have, at least, some degree of awareness, we have already seen that abstract behaviors like consciousness, emotions and others emerge from stable parallel looped physical and chemical dynamics. Therefore, the desire to survive or resist death with these beings is a combination of awareness or consciousness and a set of emotions.

Therefore, any given abstract behavior will be automatically inherited if all physical and chemical stable parallel looped reactions (and the corresponding compounds) critical for this behavior are present in both the parts after a split. In other words, the feature set (c) above is automatically inherited if all of (a) and (b) are inherited in each part after a split. This is because all living beings use stable parallel looped architecture.

DNA replication: Let us see what is needed to replicate the set of chemical reactions mentioned in (b). Most chemical reactions occurring within a living cell are catalyzed either by enzymes (amino acid based catalysts) or by ribozymes (RNA based catalysts). We know that DNA encodes each of these enzymes and ribozymes. Therefore, replicating the DNA along with an initial set of molecules necessary for DNA replication, transcription, translation and the genetic code would provide a capability to create enzymes and ribozymes in the offspring. We have already seen how this can be done in chapter 62 even in a primitive cell.

Additional chemicals and reactions: There are several other chemicals necessary for a cell to function correctly that are not part of the DNA machinery. Some of these chemicals can be obtained from the environment through food as essential amino acids, starch, vitamins, minerals and others. The rest of them can be generated within the cell through a set of looped chemical reactions. For example, Nicholson [114] shows metabolic pathways for generating these chemicals. When a cell splits into two parts, if the initial set of chemicals is unevenly distributed between them, though not in an overly lopsided way, most of the chemicals can still be regenerated via a suitable metabolic pathway as long as the set of enzymes within the pathway can be recreated. For this, we need a supply of external inputs as well as molecules like transcription factors that control the gene expression.

Chained sequence of processes: While the discussion so far addresses how to copy individual or looped set of chemical reactions, we need a way for the system to memorize specific sequences of chemical reactions that corresponds to a given functionality. For example, the 'process' of creation of an enzyme involves a complex sequence of physical and chemical steps. It is not sufficient that each set of these reactions are capable in the reproduced system. The entire sequence should be memorized similar to memorizing an algorithm or rules in addition to the objects themselves. This memory of a given sequence of steps happens naturally through the disturbance propagation in an appropriate network of loops (see chapter 58). Therefore, as long as the same stable parallel looped network of chemical reactions can be reconstructed, the required sequence of physical and chemical reactions will occur in the offspring as well.

Physical structure and dynamics: The cell membrane and the cytoskeletal structures, for example, are critical for the life of a cell. The same applies with a number of other organelles, say, in eukaryotes. While a precise structure at a molecular level is not necessary for its function, the cell does need some level of precision at a macroscopic level. For example, the cytoskeleton and internal structure of cells is made of polymers like actin, intermediate filaments and microtubules. Similarly, the cell membrane is made of a phospholipid bilayer. These polymers can be created approximately to achieve the same function.

If a cell accidentally damages a minor part of its internal structure, membrane, organelles or other structural components, it is necessary to have existing chemical processes to repair these damages. Otherwise, these defects will grow and render that specific structure non-functional. Therefore, without loss of generality, we can say that only those substructures that have processes to repair minor damages would need to be reproducible in the offspring. Furthermore, there should be a sequence of steps (or, equivalently, a network of loops) to create these structures in the first place. Structures that cannot be repaired or even constructed through a network of loops, though advantageous, are only temporary. We can call them as acquired abilities, not heritable qualities. I will discuss these memory mechanisms briefly in chapter 70 when discussing the transition of unicellular to multi-cellular organisms.

The above analysis shows that a reproduction mechanism is one that copies the DNA, rebuilds the physical structure and then ensures that the entire network of stable parallel loops can be regenerated. A cell's SPL network can be considered as highly useful if a mechanism exist that performs these steps. The disturbance propagation within such a useful network of loops would give rise to important processes like creation of the physical structures, creation of enzymes and the very cell reproduction process itself.

67.2 Primitive cell reproduction

In the previous sections, I have already discussed how nature copies the DNA reliably within a primitive cell. If, after this step, nature performs a simple and gradual split of itself into two parts, there will be an approximate outer (not inner) structure created along with an approximately even split of chemicals between the two parts. This may be a reasonable reproductive mechanism for primitive prokaryotes that have no other complex organelles or inner substructures. In fact, the above mechanism is a good approximation of asexual reproduction using binary fission (in bacteria or archaea). This process is surely not enough for eukaryotes and other complex beings.

Therefore, the above discussion implies that we now have a natural reproductive mechanism for primitive prokaryotes or primitive life forms, in general. For the problem of origin of life, we will only focus on mechanisms applicable to primitive organisms like prokaryotes instead of eukaryotes. The problem of diversity of life discussed later in chapter 69 will look into the other specialized organisms and their reproductive abilities. For example, I will discuss the problem of reproduction and even the creation of eukaryotes and multi-cellular organisms for the first time later in chapter 70.

There is obviously one advantage of storing the information needed to create a cell within the DNA sequence and within a network of loops. Under adverse conditions or when there is a lack of energy or resources to run all of the normal reactions of a cell (like during hibernation or ice ages), it is still possible to store the memory of a cell's reconstruction. The minimal energy requirements as well as the static representation of DNA sequence and the network of loops can preserve living beings' capabilities. If there is minimal damage to these small set of chemicals, the remaining chemicals can be reconstructed from looped reactions when favorable conditions reappear. We can say that the dynamical properties of enzymes and other chemical reactions are encoded as geometric or structural properties within a DNA, an initial set of inputs and a network of chemical loops. While the dynamics and behaviors require energy, the geometric structure and the DNA sequences require minimal maintenance.

67.3 On epigenetic mechanisms

The above discussion also highlights several other factors, apart from the DNA replication, that are critical for transmitting features to the next generation. One of them is encompassed into a new feature introduced in this book, namely, the operation of a sequence of steps is memorized equivalently as a disturbance propagation in a suitable network of loops (see chapter 58). More commonly, these non-DNA based memories (like the chemical SPL network itself) are referred to as epigenetic inheritance features (see Jablonka and Lamb [77]), i.e., inheritance of traits other than through genes. For example, if the environment provides only a certain types of inputs to consume, then a corresponding set of enzymes (and hence the genes) will be expressed. It is also likely that some other sets of genes will be actively suppressed through these inputs. If a mother mouse' diet has methyl-rich compounds like folic acid or vitamin B_{12}, its offspring's coat will be brown instead of yellow. The corresponding genes have a high level of methylation (i.e., attachment of CH_3 group to cytosine). Similarly, prions acquired through diet can cause proteins to misfold. The three-dimensional shape of normal proteins switches to a misfolded state causing, say, the mad cow's disease.

The cellular specialization like with skin or liver cells is maintained even though the genome is the same i.e., skin cells reproduce into skin cells and not liver cells even though the genomes are the same. The underlying mechanism is that of the suppression of specific genes through, say, histone modification or methylation (see Jablonka and Lamb [77]). In some cases, it is possible to view these epigenetic mechanisms (i.e., a non-DNA based operation of a sequence of steps) as disturbance propagation in an appropriately chosen network of looped reactions. The disturbance propagation mechanism itself is not part of the genes. Nevertheless, it is a repeatable process whenever transcription factors or other molecules like prions, histones and microRNA's are passed onto the offspring. This repeatability ensures the reproduction of a phenotypic feature like cell specialization even when the corresponding DNA's are identical.

A large number of epigenetic mechanisms that are still being discovered. It remains to be seen if these additional interesting patterns can be formulated within the stable parallel looped network or if new explanations are necessary.

68
Primitive Life

Given the discussion so far, it is now possible to explain the creation of the initial set of primitive life forms. The initial primitive life forms are much simpler than the modern life forms. They can be considered similar to a prokaryote. The more complex eukaryotes or multi-cellular organisms evolve much later. I will discuss them briefly in chapter 70.

Starting from a primitive cell of section 67.2, we have seen that we can incorporate several life-like features, at least in a primitive form. For example, in chapters 59 - 63, we have created simple designs with the following features – actively generating energy, actively seeking inputs, wanting to live, resisting death, searching for food, sensing chemicals and chemical gradients, controlling physical movements, exploring the environment, growth (since it actively seeks inputs), adaptation (since it evolves using stable-variant merge loop based on horizontal gene transfer and others) and reproduction.

It is possible to add more features into the above system naturally using additional loops without altering the existing features. Some examples that are generalizations of the above existing set of features are – efficient ways for motion, more sensors, more energy generation mechanisms, more types of food inputs, storage of food for future use, positional and spatial sensing (like ants that drop pheromones along the path), ways to compete for resources, fighting pathogens that try to destabilize the system and so on. In order to create such new features, nature needs to discover and use several enzymes and other chemicals.

At what point should we say that a system with all of these features is indeed a living being? This system does exhibit all of the common features attributed to life. Our primitive system with the above features already is well-organized, grows, adapts, reproduces, senses and controls the environment, metabolizes and self-sustains its dynamics. If we examine these features closely, we see that while reproduction and self-sustaining the system are well-defined, it is not clear how much growth, adaptation, metabolism, sensing and control of the environment is sufficient in order to call a system as a living being. Nevertheless, this is not a serious issue because we can wait for the system to evolve and accumulate more features, say, using stable-variant merge loop pattern and other directed mechanisms described earlier.

The problem of defining what life is, in precise terms, is quite subjective and debatable. This is especially true when someone attempts to create a synthetic, potentially inorganic based, system that exhibits each of the above features as well. The mechanisms described in this book explain how to create a primitive system with all of these attributes using organic molecules, entirely naturally. However, if we look closely at these mechanisms, we notice that the requirement about organic molecules is not necessary anymore. We can create all of the same features using synthetic and possible inorganic materials, though not naturally, but with help from conscious beings. This creates a fundamental disconnect with our perceived notions of life and organic living beings, which I will not elaborate here, though it is very likely to happen in the near future.

One interesting aspect of a system exhibiting these features is that it will behave as a single entity. The system is not just a collection of either independent or loosely interconnected subsystems. Rather, they are so deeply coordinated that this interrelationship is fundamental. You can say that the system tries to stay stable as a whole. If parts of the system are becoming less stable, there will be feedback mechanisms that will try to restore stability. We see this behavior only when the system has evolved to a threshold state like with a primitive life form. For example, it will feel like parts of the system will be sacrificed at the expense of the survival of the entire system. If the system is unable to recover, the system as a whole will collapse. When this happens, the resulting collapsed system behaves as uncoordinated and independent dynamical subsystems, i.e., the system dies. It is now nontrivial to get back to the previous 'alive' state. In this chapter, I will discuss these, somewhat philosophical, topics in further detail.

68.1 Early life forms

Let me now summarize a possible and natural way of creating primitive life forms. We will see that the feasibility of this approach is justified from the analysis and designs of the previous chapters. I will discuss how these minimal structures will evolve in a directed way to acquire new features. However, keep in mind that life has different evolutionary pathways. The complexity of the network of stable parallel loops of a primitive system naturally gives rise to different ways of actively self-sustaining themselves for a long time. Even after creating the minimal structures as discussed in the previous chapters, we can say that the following discussion is just one of the many possible ways of creating life for the first time. The possible evolutionary pathways to create primitive life for the first time after the creation of the above minimal structures is non-unique.

(a) *Viruses and primitive bacteria*: Let us take the primitive cell of section 67.2 and include the features of chapter 63 into this cell. This system can reproduce as described in chapter 67 using the equivalent process of binary fission of prokaryotes. We can view them as primitive bacteria. Viruses (primitive or otherwise) are much simpler than these systems. They simply have DNA or RNA protected in a protein coat and wrapped inside a fatty enclosure. In order for these viruses to replicate, they should enter, at least, the primitive bacteria and use the replication machinery of this system. The discussion in the previous chapters already shows how this is possible in a natural way.

(b) *Stable-variant merge loop pattern based on horizontal gene transfer*: When we have several such systems with different DNA sequences, they begin to interact either by joining or by competing for the same set of inputs. This naturally leads to a directed evolution of more complex features using, say, stable-variant merge loop pattern relying on horizontal gene transfer. Random mutations in viruses and primitive bacteria act as a factory for creating new enzymes and, hence, other chemicals. The enormous diversity of viruses suggests the possibility of creating as many random DNA sequences as possible, many of which produce fairly stable enzymes that can survive in a host system (like primitive bacteria). Villarreal [159] had already suggested the role of viruses in the evolution of life.

In the beginning, there is no immune system. Therefore, infection by primitive bacteria and viruses are common. It is an effective way to transfer DNA or RNA across systems (cf. stable-variant merge loop pattern using horizontal gene transfer). In fact, it is estimated that 0.03% of human genome is from endogenous retroviruses (see Belshaw et. al [11]). The infection from these DNA or RNA sequences produces a number of proteins in the host cell. These help create new chemical links between existing reactions or a new network of looped reactions. However, when the system evolves to develop additional sensory and control mechanisms, they start to detect and destroy pathogens. While this brings in more stability to the system, the diversity of features slows down significantly as well.

(c) *Regulatory chemical loops*: In order to create more features, it is necessary to have a much wider range of chemicals, enzymes and reactions. The above mechanisms do help with each of these tasks. As these chemicals are discovered, what gives stability to the system is the presence of a number of regulatory chemical loops. We have already seen in chapter 58 that in order to have a coordinated sequence of steps, we do need a network of looped chemical reactions. Since the creation of a variety of enzymes is one of the critical steps in almost all chemical reactions, it is important to be able to regulate their production. Otherwise, either too little or too much of an enzyme will be produced.

Since we have an elaborate sequence of steps to create an enzyme starting from the DNA, it becomes possible to regulate this process at any of these stages. For example, during transcription, there are proteins called transcription factors that bind to specific DNA sequences to control the creation of mRNA. Some proteins act as activators and others as repressors when RNA polymerase tries to bind to the DNA. Similarly, regulation can occur at chromatin when trying to access the gene. They can also happen during post-transcriptional changes, transport of mRNA, translation step, degradation of mRNA and post-translational changes. One of the first studied examples of genetic regulatory mechanism is the *lac* operon. It describes how a cell digests lactose efficiently taking different pathways depending on the presence or absence of glucose. I refer the reader to Alberts et. al [1] for further details on multiple control mechanisms for *lac* operon as well as other mechanisms for gene regulation.

Before these regulatory-looped networks of chemical reactions can be formed, the necessary enzymes and other intermediate chemicals should be discovered using, say, the above stable-variant merge loop patterns. Besides regulation of gene expression, there are a number of other regulatory looped reactions necessary for the catabolism and anabolism of several other chemicals. Allosteric enzymes (those that change their conformation when another molecule binds to it thereby altering its chemical activity) and their regulatory properties are best-known examples of this.

The outcome of the above-directed evolutionary mechanisms is to create a diversity of self-replicating chemical systems with several useful features. These systems can now self-sustain their looped dynamics for a very long time, both individually and together as a 'species'. We can now begin to call these systems as the very first life forms. They were created entirely naturally and starting from inanimate objects without the help of any intelligent being. The transition from nonliving to living beings is now complete using the mechanisms described until now.

In the next section, I will discuss how living beings self-sustain and behave as a single entity instead of just a collection of interconnected subsystems.

68.2 On behaving as a single entity

A system behaving as a single entity should have subsystems that are well connected. If two subsystems are loosely connected, a change in one does not affect the other significantly. In this case, you may disconnect the two subsystems and let them operate independently. Your hand and your other internal organs of your body are an integral part of you because the dynamics in, say, your hand effects and is effected by the rest of your body. When you want to study the dynamics of such a system, you cannot ignore any of its subsystems.

Typically, when a physical system behaves as a single entity, you have some sort of physical connection like rigid rods, wires or strings between the subsystems. These wires or rods help transmit the forces from one subsystem to the next. These tangible links produce a well-defined effect from a cause. However, in chemical systems like a single cell, there are far fewer rigid structures that help carry out a cascading sequence of chemical reactions. Even among the ones that exist, like microtubules and other cytoskeletal structures, they are not well-organized, like axons and dendrites in the brain, to allow a safe passage of signals and chemicals from one location to another. For example, when a hormone like insulin triggers the intake of glucose from the blood in liver or muscle cells, the chained set of reactions happen with minimal and organized physical connections.

Yet, chemical systems achieve a similar level of coordination to behave like a single entity. This is possible when you have a dense network of looped systems, both at a physical and chemical level. The disturbance propagation within such a network of loops involves physical transport of molecules as well as chemical reactions that occur in a specific sequence. In the case of brain, memory of a sequence of steps is stored within physical structures of interconnected network of neurons. This memory gives rise to reliability and coordination when an event repeats multiple times. With chemical systems, the memory is within the network of loops. Unlike the brain, changing the interconnections between the chemical loops is not common because of a lack of organized physical structures.

68.3 On exhibiting primitive brain-like features

Since we also have several sensory-control loops in a chemical system, several of the patterns described for the brain, like the fundamental pattern (see chapter 16) are applicable to these systems as well. I will not discuss these analogies here. Yet, it is possible to translate the sensory-control designs of our brain even at a chemical level. We can use the analogies to explain how simple single-cellular chemical systems also exhibit primitive versions of brain-like features like primitive free will, primitive awareness or primitive sensing of space and time. While we do not say that there are associated feelings for each of these features (because of a lack of self-sustaining membrane of meta-fixed sets that includes space, time and other dynamical features – see chapters 19 - 40), we can observe similar mechanics, albeit at a slower pace or in a lower degree.

For example, the mechanics of free will requires creating an appropriate grouping of dynamical components that self-sustain and, as a whole, have an ability to obey arbitrary laws (within the physical limits) even though each component individually obeys definite physical laws (see chapter 28). A chemical system like a single cellular organism can surely exhibit this property in a mechanical sense. The cell, as a whole, because of its active mechanisms can self-sustain and obey arbitrary laws (within limits) by moving and interacting with other neighboring systems unlike how a nonliving system would behave. This active chemical system is not aware of the feature, but it will oppose the natural laws, within limits, to a degree unlike how nonliving systems of a similar size react. We already see unexpected and unpredictable behaviors in all living beings including single cellular and multi-cellular organisms. Similarly, we can use the mechanics of chapters 43, 35 and 40 to see how a single cellular organism exhibits features like primitive awareness, sensing space and time, though to a very low degree and without any associated feelings.

68.4 Quantum leap in evolution when discovering a complete loop

Once the initial set of life forms (say, bacteria and viruses) are formed, different organisms start to evolve different chemicals. The bacteria and viruses act as chemical factories for creating a large diversity of potential enzymes. These are transferred to other organisms as well using the stable-variant merge loop pattern based on horizontal gene transfer. This causes the network of loops to diverge among living beings. In this section, I want to show that every time all the necessary chemicals for a single chemical loop are discovered, there will be a quantum leap in evolution in terms of the set of features that can be developed from them.

To study this more closely, let me start with a given network of looped reactions. We should note that, unlike a brain's neuronal network, the creation of even a single chemical loop takes several generations. As new enzymes (or ribozymes) are discovered using any of the mechanisms discussed previously, new chemical pathways within this network begin to form. A few possible pathways are:

(a) *New chemical node*: The new enzyme creates a new chemical from existing chemicals.

(b) *New reaction edge*: The new enzyme converts one existing chemical into another when it was not possible before.

(c) *Enzyme regulation*: The new enzyme participates in a chemical reaction corresponding to the above (a) or (b) case. In addition to this, either an existing or a new chemical will affect the new enzyme or a new chemical will affect an existing enzyme. This occurs by trying to aid or prevent its creation or its operation.

(d) *Replace existing enzyme*: The new enzyme is more efficient than an existing enzyme of an existing chemical reaction.

(e) *Physical changes*: A new protein assists with physical mechanisms involving existing or new (from (a) above) chemicals. These include acting as a pump (Na-K pump), as an ion channel (Ca^{2+}, Na^+, K^+), as structural components (actin or tubulin), for ligand transport (oxygen or sugar), as receptor and hormone binding, as antigen binding or as other transport chemicals (motor proteins like myosin and kinesin).

In each generation, one or more mutations help create these pathways incrementally. The memory of these pathways of all the past generations is stored within the DNA and the chemical SPL network. The stable-variant merge loop ensures that one copy maintains the 'stable' history while the second copy brings in new 'variant' changes. In this way, the set of changes become cumulative. The system starts to build a complex chemical network from these pathways accumulating over hundreds of generations of mutational changes.

Initially, the network need not even be connected. The chemicals and the enzymes for the edges and nodes are not necessarily discovered in an orderly fashion that maintains graph connectedness. Therefore, the cascaded chain of a chemical pathway will appear broken in the early stages of evolution. Over hundreds of generations, we begin to see a well-connected network with several intact chemical pathways similar to Nicholson [114]. When we have such a complex connected network, the system starts to exhibit interesting features. One important feature is the formation of chemical loops. Each time the connected network creates a new chemical loop, an explosive evolution occurs.

For example, if the Kreb's cycle (see Figure 2.8) for cellular respiration forms over several generations, the subsequent cells will then have an efficient energy generation mechanism. A single chemical loop results in an 'explosive' jump in the evolution of primitive bacterial cells. If we look closely at how this loop will be created, we notice that in any given generation, typically, only one enzyme or a single chemical, which is part of the Kreb's cycle, will be created. Now, there is no obvious gain for the organism with an incomplete Kreb's cycle. Yet, these partial set of chemical reactions are created and memorized within the DNA and the SPL network.

The order of enzyme discovery over several generations is surely not in the cyclic order of execution of the chemical reactions (of the Kreb's cycle). However, whenever the last missing enzyme anywhere in the cycle, say, for example, aconitase or fumarase, is discovered, suddenly the entire chemical loop can begin to run. This is what I term as an 'explosion', not in the usual sense, but in the sense of generation of a large number of chemical reactions not possible before. For example, large quantities of other chemicals branching from Kreb's cycle within the network can now be generated (see Figure 2.8). This is because the chemicals that were part of Kreb's cycle can now be produced in large quantities due to the continuous looped operation. The lack of a complete Kreb's cycle before had prevented a huge surge in each of these chemicals. Without Kreb's cycle, ATP generation was not consistent and reliable. Therefore, features that depended on a continuous or extended supply of ATP have not occurred previously. However, all of these reactions and, hence, the corresponding features are suddenly possible now.

More generally, when chemicals within a loop or along connected chemical pathways were only intermittently available (as the complete loop is not yet formed), it is not possible to propagate a disturbance in the chemical concentrations effectively. As a result, no concerted action, sensing or any other useful sequence of steps becomes possible. However, when a complete loop exists, it produces these chemicals continuously causing a regulatory interconnected loops with steady-state concentrations. A disturbance in this network of stable loops can now transmit further and reliably to produce an appropriate action like blood clotting or other hormonal effects. If the network is complex enough to have several such interconnected loops, the disturbance propagation can sometimes result in an advantageous feature for the organism. Such disturbance propagation would not have been possible in the absence of the discovery of complete loops.

This is true with several other important chemical loops like Calvin-Benson cycle, glyoxylate cycle, cell cycle and other cycles including those within neurons. A generalization of this can involve multiple input loops simultaneously. Such a multi-input loop pattern does not restrict itself within a single cell or a single organism. It includes a stable loop formed across different species as well. With chemical systems closing the multi-input loop occurs by discovering appropriate enzymes by each of the interacting species. This will result in an explosion of features. Some of these examples include nitrogen fixation cycle, phosphorus cycle, pollination life cycle and other biochemical cycles.

In this way, we have quantum leaps in evolution each time a chemical loop is formed. Sometimes, it may appear that this quantum leap occurs just from a single chemical reaction with the discovery of one new enzyme in the current generation. However, that is because this single reaction is the last missing piece in the formation of one or more chemical loops within the existing SPL network. Complex seemingly hidden links are constantly being created with each mutation and within each generation using (a) – (e) mentioned above. We will not see any observable pheno-typic changes in the organism whenever the organism has only discovered partial set of chemicals or incomplete loops. Yet, the SPL network being built keeps growing within the organism and is being stored in the DNA even though we are unable to notice these changes.

These chemical loops are transferred to other species using stable-variant merge loop pattern based on horizontal gene transfer initially. Therefore, each time all the enzymes necessary for a chemical loop are created over several generations, there is a similar explosive effect for all future organisms of this and other species.

69
Diversity of Life

After the first life forms were created, they will start to spread quite prolifically. The fast enzymatic reactions and the reproduction mechanisms make it easy to replicate these cells. Initially, they will be transported randomly through wind gusts, water waves and other weather patterns. The active mechanisms for transport do exist, but are slower. However, once they reach a given region, the single cellular organisms will occupy and multiply until energy and resource inputs within this region are depleted (or they continue to reproduce by reusing chemicals from the dead cells). There will be replication errors, which initially are advantageous to create diversity of living beings. The stable-variant merge loop pattern will ensure a definite direction in evolution.

In this manner, life diversifies in quite complex ways over billions of years. The topic of diversity of life is too broad to discuss in this book in any detail. In this chapter, I will briefly highlight a few topics. One reason for being brief is that our current knowledge on these topics is already quite extensive. Darwin's theory has provided a framework to allow us to explore these topics in considerable detail. Another reason for being brief is that we need detailed experimental research to understand the mechanisms at a cellular level for each new feature. In this book, we have focused primarily on theoretical ideas. The amount of diversity we can observe across species and across billions of years is too large to be able to discuss just from a theoretical point of view. This, however, was not the case with the problem of origin of life as we have seen in the previous chapters.

69.1 Competition versus cooperation

When we look at most living beings, the diversity we see around us is primarily from the multi-cellular organisms that seem to exist everywhere. A eukaryotic cell has internal cellular structures called organelles and multi-cellular organisms have specialized structures called organs formed from multiple cells. These specialized structures are combined in complicated ways to generate a wide range of body plans. These give rise to enormous diversity among living beings that we observe today. Two of these problems of particular interest are the origin of eukaryotes from prokaryotes and the origin of multi-cellular organisms from single cellular organisms. There are several difficulties when trying to explain how to create these organs and organelles naturally, which I will discuss in the next chapter.

For now, let me briefly discuss one aspect here, which is a central issue for the problem of diversity of life, at least when viewed at a philosophical level. This is about how we can embed the concepts of competition and cooperation naturally within the physical laws. These two concepts are fairly well known to be critical for creating diversity.

I will first present a simple analogy to clarify these concepts. I will use systems and structures created by conscious humans instead of systems created naturally. The world economy is a good example where we see enormous diversity. We have created products and industries specialized in textiles, commodities, housing, farming, banking, finance, technology and so on. Several of these industries are interrelated, in the sense that if one industry fails, then it brings down several others in a noticeable way (like a recession). The world economy a few hundred years ago did not have all the industries that we currently have like for instance, we did not have a similar technology sector hundred years ago. Over the years, the diversity in our economy increased much like the diversity of multi-cellular organisms.

Two of the commonly identified 'forces' shaping our economy are the competitive and the cooperative elements. For example, it is well known that the presence of a healthy competition drives innovation within an industry. A monopoly, on the other hand, has a tendency to suppress innovation (or at least creates less incentive to innovate). We have seen this type of behavior repeatedly in almost all industries. We even use 'survival of the fittest' argument from Darwin's theory as an appropriate analogy in the economy, to capture this notion that competition is critical for creating innovative designs or features. Indeed, the central tenet of Darwin's theory is natural selection and survival of the fittest to create new features within and across species. Similarly, the stable-variant merge loop pattern is a generalization that relies on competition among other things as well.

The second important element for creating diversity in an economy is cooperation between industries. This can be seen as some sort of balance or conservation like the balance between supply and demand. Any given company A depends on a number of other companies for the parts it needs (like, say, a computer company needs parts like transistors, resistors, IC chips, peripherals and others that it does not manufacture itself). Similarly, the product of a given company A can be used by multiple companies in different ways. Yet, each of the companies that the company A depends on does not directly compete with it. It may seem a bit awkward to say that these companies are cooperating especially because they are not developing a common product by working together.

Nevertheless, there are cooperative elements at various stages of product development. They also manifest in different forms like, for example, one company provides a design or a specification while the other company builds a part that meets this specification. One company requires a service with certain quality and another company provides that service. These services could be quite broad like building services, janitorial services, electricity, legal, accounting, email, telephone and so on. The cooperation could be at a quality of service or at a design requirement level. If we look at the entire economy, the cooperative elements of the above type are a lot more prevalent. In fact, the interdependencies between producers and consumers in an economy are so intertwined that it is fair to call our entire economy as a large-scale structured complex system.

With multi-cellular organisms too, cooperation is extremely common. The various organs need to cooperate so the entire organism can survive even at the expense of sacrificing a few components in dire situations. The various organelles cooperate in a eukaryotic cell to perform basic cellular functions. In fact, there is one popular theory to explain the origin of a few organelles in a eukaryotic cell, especially the mitochondria and plastids (like chloroplasts). This is the endosymbiotic theory (see Margulis [133]). Based on detailed microbiological evidence, the theory suggests that some specific primitive prokaryotes were engulfed by other larger prokaryotes to attain a symbiotic relationship. These later became as mitochondria and plastids.

In summary, in large-scale structured dynamical systems created by conscious humans like the economy, we see that both competitive and cooperative forces do drive innovation and diversity. Darwin's theory, on the other hand, seems to focus primarily on competition and less on cooperation.

Let us now see how to extend the above analogy (on economy) to explain the innovation and diversity in living beings i.e., the problem of creation of eukaryotes from prokaryotes and the creation of multi-cellular organisms from single cellular organisms. One thing to note when we discussed competition and cooperation between industries in an economy is that we relied significantly on consciousness. Companies that interact are aware that they are indeed competing and/or cooperating. New relationships between companies are forged by conscious humans who clearly understand the consequences (at least to a degree). There are no physical laws that drive relationships between companies. Instead, they are purely abstract that only conscious humans can understand and build.

If we want to extend the analogy of our economy to explain diversity in living beings (in eukaryotes and multi-cellular organisms), we need to change the language to express competition and cooperation in a form that does not require consciousness. Stated another way, we want to identify the set of physical laws that naturally promote competition and cooperation in a dynamical system.

How do we have competition in physical systems, especially among nonliving and living but non-conscious systems? What is the root cause? Interestingly, this is quite easy. First, we need several similar dynamical systems that have an active source of energy. If all of these systems share a common resource (cf. multiple companies selling a similar product to the same set of customers), then the fact that they are actively consuming inputs suggest that they are competing to get a larger share. This happens quite naturally i.e., without conscious awareness. Some systems consume more inputs than the others do. Since there is no notion of consciousness in these systems, we can say that they are not intentionally trying to compete. Yet, the fittest will survive in such a situation. Even if a monopoly exists (i.e., a single system consumes almost all of the inputs), it is possible to disturb this state to create competition by introducing stresses like unforeseen environmental variations (weather patterns, earthquakes, wind gusts, rainfall, volcanic eruptions and others). Other examples that can disturb a monopoly are systems that are agile and better fit at a smaller scale. The more dynamical the environment is, the higher the chance that the fitness shuffles among these systems. Therefore, competition naturally emerges when multiple systems share a common resource.

How do we have cooperation in physical systems, again especially among nonliving and living but non-conscious systems? If we see streams of water travelling downhill, we never see cooperation between them to create a specific pattern. When multiple clouds coalesce, we never see cooperation between them to form the shape of, say, a human face. Even in cases when there are clear patterns like with tornadoes, hurricanes and others, we rarely see cooperation between components to form this pattern. It seems like a bit of a stretch to say that nonliving beings are cooperating towards a common goal even if we include attractive and repulsive forces within electrical, magnetic, gravitational and other forces. On the other hand, there is a great degree of cooperation among living beings though non-conscious. For example, the mere existence of a multi-cellular organism shows an enormous degree of cooperation between various organs within its body. How can we then explain this difference?

What is the root cause of cooperation within nonliving systems? The answer is the existence of looped dynamical systems. The class of systems studied in this book, namely, SPL systems are precisely the reason why cooperation is possible within nonliving (and living) systems. To see this, let us first pick a simple looped system, like the one in Figure 2.1 with a ball rolling down an inclined plane. Such a looped system can be split into two parts. In one part, the ball rolls down the inclined plane and stops. In the other, a human brings the ball and places it on the inclined plane to complete the loop. When this happens, the first part can occur once again. If this looped dynamics continues for a long time, we can say that both parts are working cooperatively. This is especially true because the natural state is *not* to continue the looped dynamics. Note that we cannot have a perpetual motion machine. In fact, we do need a constant supply of external energy to continue any looped dynamics, which does not happen easily among nonliving systems.

Therefore, given that looped dynamics does not occur naturally often, we would need to have cooperation between two or more systems to create a loop. One system will cause a downhill reaction while another system will cause a corresponding uphill reaction by consuming energy so together they complete the loop (see Figure 53.2). Downhill reactions can occur naturally, but if the corresponding uphill reactions do not occur to form a loop, then the world will settle down to a low energy state with almost no possibility of life. Only if we have looped reactions would we have a possibility of creating life. We have already seen that this was the basis for the origin of life i.e., the physical and chemical SPL systems.

Even within non-conscious living beings, we see cooperation because of the presence of looped processes. Plants consume energy from the sun, CO_2 from animals and H_2O to produce O_2 and starch, which are consumed by animals to produce CO_2 and H_2O needed by plants. Together, they complete a loop so life could continue. This loop, split between two systems (plants and animals), is clearly an example of cooperation. Even within a cell, we have several chemical reactions that consume ATP (an energy source generated in the mitochondria) to produce ADP, while the mitochondria consume ADP to generate ATP, thereby completing the loop. Each such chemical loop in the metabolic network (see Nicholson [114]), though occurs in a single cell, is well regulated. We can identify several chemical reactions managed by different subsystems within our cells for each such chemical loops. We can view all of these subsystems as cooperating to achieve regulation of the chemical concentrations. Even if we disturb the system by introducing additional chemicals, the loops return to the original states by propagating the disturbance within the SPL network, as we have seen before in chapter 58.

Therefore, if a looped chemical reaction, a looped physical process or a combination of both occurs within two or more subsystems causing a loop to occur for a long time, we can say that these subsystems are cooperating to allow a possibility that would otherwise have not happened naturally (namely, the continued occurrence of the looped dynamics). The fact that looped processes *cannot* occur naturally in most cases (since we cannot have a perpetual motion machine) implies that two or more systems must cooperate to make this happen. In cases like earth revolving around the Sun, we accept the looped process as caused by the fundamental force of gravity. In almost all other examples of looped processes, the mere existence of looped dynamics becomes the root cause of cooperation between a few subcomponents.

In this book, we have seen that active looped processes were necessary for explaining both the origin of life and consciousness. They were also sufficient to create synthetic life and synthetic consciousness. What we are seeing just now is that the same looped processes are responsible, at least in a philosophical sense, for creating cooperation between components, which will eventually create diversity of life as well.

69.2 Creating diversity

When studying the problem of diversity of life, we need to answer at least two distinct problems satisfactorily. Firstly, we need to explain the *creation* of specific organs, structures or features like heart, eyes, wings, fins and others. Secondly, we need to understand how these organs, structures or features are used to *perform* appropriate functions as well as how they *cooperate* to perform a new task that helps the organism as a whole even though it may not help each subcomponent individually. The first one is the creation problem and the second one is the usage problem. The laws of nature needed to solve them typically have little in common.

The existing research in developmental biology and other areas primarily focus on the second 'usage' problem in considerable detail. However, all solutions proposed for the first 'creation' problem are based only on random chance. For example, when applying Darwin's theory to explain the existence of diverse features, there are two important stages: (a) the discovery of a feature using random mutations and (b) natural selection with survival of the fittest. In other words, once a feature is discovered that gives an organism a survival advantage over other similar organisms, there is little doubt that in a large population after thousands of years, organisms that remain as offspring will have this feature and the organisms that do not have this feature will be eliminated. This is the natural selection step (b) above. During this step, there is no discovery of a feature. Rather, the feature already discovered is simply used in an advantageous way to survive better and longer.

However, what are the mechanisms for the discovery of a feature itself? The only mechanisms offered are random chance mutations that give an organism just a slight advantage over others. This slight advantage makes them survive better. Hence, they survive while the others are eliminated as less fit. For example, if we want to explain the 'creation' of a heart or an eye, the first step is to argue that the existence of a heart or an eye makes an organism survive better than those that do not have these organs. The next step is to split the creation sequence of these organs into a long sequence of mutations, each of which is small and simple enough so nature can discover them just by random chance. The final step is to argue that over a long period and across large populations, random chance will discover each of the chemicals within the creation sequence. All of these mutations accumulate in a positive manner, linking one-step to the next, to produce the correct cascading chain of chemical reactions necessary to create the organ. For example, if the creation of an eye requires 10 steps, say, $S_1, S_2, ...,$ S_{10}, then it is not sufficient to discover all 10 steps separately through random mutations over a long period across large populations. It is necessary to link them in the correct order (i.e., any permutation of $S_1, S_2, ..., S_{10}$ is not valid).

There are several gaps in the above argument. I will discuss a couple of them here (more details in chapter 70). One issue is on how mutations, if they are purely random, can only produce better and new chemicals instead of deleting existing enzymes (see Figure 54.2). We were able to overcome this issue with Darwin's theory by proposing the stable-variant merge loop (SVML) pattern as an alternative mechanism that guarantees a definite direction in evolution. The second issue is on how to memorize the sequence of steps like, say, the sequence $S_1, S_2, ..., S_{10}$ even after we have discovered the correct chemicals through random mutations. We were able to overcome this issue as well using (a) the SPL network representation of all chemicals and chemical reactions and (b) the disturbance propagation as a way to memorize and run a sequence of chemical reactions in an SPL network.

I will identify a few other gaps later in chapter 70 and attempt at addressing them. In short, I would say that for the most part, the topic of diversity of life needs considerably more future work relative to the other topics discussed in this book. The amount of diversity we observe in the last 2-3 billions of years is too large to rely entirely on random chance at every incremental discovery necessary to create a given feature (i.e., as random mutations spread across generations). The main theme of this book is to replace purely random mechanisms using repeatable or looped processes. It is, therefore, necessary to cast the problem of creation of diversity of life into a suitable SPL framework.

This book introduced several such SPL concepts and mechanisms, laying the necessary foundations, to pursue such an approach. For example, the stable-variant merge loop pattern (SVML) based on sexual reproduction, horizontal gene transfer or inter-species interactions is a generalization of Darwin's theory that additionally guarantees a direction in evolution. The disturbance propagation in an SPL network allows a system to memorize and perform a long sequence of steps, which otherwise requires conscious beings to setup and guide through the steps.

The goal of the new theory is to suggest a shift in the approach we take to study the problem of diversity of life. It is no longer sufficient to understand how a particular organ or a feature works. But it is necessary to ask ourselves how this organ or feature can be built and be integrated with other existing features to provide a specific purpose, without the help of conscious beings and without entirely relying on random chance. Such a constructive approach forces us to identify the simplest and the smallest set of abstractions across multiple features using which most complex features can be built. These were termed as the minimal structures (see section 53.1). We have identified these structures when studying the problem of origin of life and consciousness. We would need to do the same for the problem of diversity of life.

In this book, I will only make an incomplete attempt for just a small set of topics related to the diversity of life. This is both an active area of research in developmental biology (see Nüsslein-Volhard [115]) and is an important area of future work. We would need to identify common abstractions by looking at detailed mechanisms of creation of, say, an eye, a heart and others. We then need to order these steps into simpler ones and ask ourselves how nature could have discovered all of these chemicals and physical structures naturally. Details should be provided from a creation point of view, which is different from providing mechanisms for how an embryo forms into an adult (cf. the mechanisms necessary to create/assemble a computer versus the laws needed to discover a computer for the first time).

The important objective here is to explain how these mechanisms were discovered in the first place, what laws of nature are needed for discovery and why these laws converge to the discovery of an organ instead of just staying in a disordered state. Only when these are understood, both theoretically and experimentally, can a person be convinced of the creation of diversity. I believe the SPL framework is quite suitable here once again, though further work is necessary. I will discuss just a few abstractions later in chapter 70 that are common across creation of multiple organs and features.

69.3 Specialization of cells

When we look at cells as purely dynamical systems, the stable parallel looped reactions self-sustain as long as energy and input resources are available. Since a cell can reproduce, the self-sustaining dynamics will now extend to the offspring. Even though a single organism may self-sustain its dynamics for a short while, if we include all its descendants, the combined dynamics will sustain for hundreds of years reusing chemicals from the past generations, if necessary. The set of looped processes extend beyond a single organism to the entire ecosystem itself. This is a big change in the sustainability of chemical reactions compared to when there was no life. Before the existence of life, the improvement we have seen is that simple looped reactions run repeatable dynamics for long periods compared to non-looped reactions. However, after life exists, cells and species as a whole make the problem of repeatable dynamics much easier.

Over time, these systems are exposed to different environmental conditions. The chemical stresses, the errors in replication and the chemical factories (like bacteria and viruses) give rise to new enzymes. They can be transferred to other cells using stable-variant merge loop based on horizontal gene transfer. The diversity of features created within cells under different environmental conditions start to diverge considerably. The cells, as a result, can be partitioned naturally into groups of similar cells. The amount of specialization across these groups diverges over time. This has already been studied in detail.

The looped dynamics tend to align with the environment and achieve a stable state. This pseudo-equilibrium state between the system and environment can be disturbed either by random environmental changes or by population changes within these interacting systems. The range of possibilities for the diversity cannot be ascertained in precise terms. Yet, this diversity, the direction in evolution and the specialization of cells are well understood using the mechanisms given so far combined with the current scientific knowledge. Therefore, I will not discuss most of them in any detail and instead refer the reader to standard textbooks like Gilbert [54].

69.4 On the origin of two sexes

One specialization that is common among most species is the existence of two sexes – male and female. One possible mechanism that aligns with the theory presented here is the one proposed by Geodakyan [51]. His theory does not refer to future advantages to the species when describing the mechanism unlike the approaches taken by others. He suggests that two sexes evolve eventually from two types of systems based on how they react to different stresses. We can see this as a different type of specialization of cells as their features diverge. While specialization discussed in the previous section typically creates a new species, according to Geodakyan, it creates two sexes within the same species in some situations.

Whenever a set of stable parallel looped systems like living beings are subject to a stress, they naturally fall into different groups based on their degree of fitness to this stress. In approximate terms, we can classify them into two groups. The first group attempts at *conservation* and tries to maintain stable features. This, according to Geodakyan, will eventually lead to the evolution of a female sex. The second group attempts at *change* that involves taking more risks. This, he suggests, will lead to the evolution of a male sex. Furthermore, useful features are passed on to the females from males after several generations.

For more details on his theory, I refer the reader to Geodakyan [51]. The stable-variant merge loop pattern is quite compatible with Geodakyan's theory of the evolution of two sexes. Specifically, the stable systems would correspond to the female species and the variant ones to the male species. As mentioned in section 54.6, when we use Geodakyan's theory, the mechanism for the creation of two sexes for the first time and how it is created or used subsequently (through stable-variant merge loop pattern) become quite similar.

Note, however, that this discussion is not sufficient to explain the origin of two sexes because we need to explain the origin of several cellular mechanisms necessary for sexual reproduction as well. One in particular is the origin of meiosis for sexual reproduction. Disturbance propagation within the SPL network does provide an alternate view of how to memorize the sequence of steps that occurs during meiosis.

In addition, the cellular components needed for meiosis can be discovered using the stable-variant merge loop pattern. However, given that the number of steps occurring during meiosis are too large, the above two theoretical mechanisms will feel convincing only if supplemented by considerable experimental evidence (at a cellular/molecular level). I will leave this as an important area of future work.

70
Intra-and-Inter Cellular Structures

In this chapter, I will consider a few features that gave rise to an enormous diversity of life on earth. One is the creation of special intra-cellular structures like organelles inside a eukaryotic cell. The other is the creation of multi-cellular organisms. At a theoretical level, both problems are similar in that there is (a) specialization of functions either within a cell (like mitochondria, cytoskeleton, Golgi apparatus, nucleus and so on) or across cells in a multi-cellular organism (like heart, digestive system, eyes, ears, lungs and so on) and (b) development of a certain degree of cooperation between these structures even for the basic survival of an organism. The creation of special structures at the intra-cellular level can be seen as the problem of transition from prokaryotes to eukaryotes whereas the creation of structures at the inter-cellular level can be seen as the problem of transition from unicellular to multi-cellular organisms.

We want to understand how nature was able to create the first set of eukaryotes when the world only had prokaryotes or how nature created the first set of multi-cellular organisms when the world only had single cellular organisms. If we look at the present day organisms, the construction of a eukaryotic cell or a multi-cellular organism happens autonomously. Even though there are miscarriages and other defects, nature did become quite skillful at this process now. However, what is unclear is how the same processes were developed for the first time naturally without any assistance from conscious and/or intelligent designers.

It is interesting to note that multi-cellularity occurred independently at least 25 times (see Grosberg and Strathmann [60]). For example, plants, animals, fungi and several others developed multi-cellularity independent from one another. Therefore, we would expect that there are common principles and mechanisms spanning multiple organisms and species. In this chapter, our goal is to identify as many such theoretical mechanisms as possible, most of which will be based on the SPL framework introduced in this book. The discussion in this book cannot be considered as complete (as mentioned in section 69.2). Rather, we try to identify as many gaps in the current theoretical analysis and see if we can provide answers for them. Future work both in the form of theoretical and experimental work is necessary to identify common abstractions.

Let me start highlighting some of the most common sources of difficulties with the problem of origin of multi-cellular organisms. The issues are quite similar for the problem of origin of eukaryotes. I will group them broadly into three categories. The first one deals with creation of a eukaryotic cell or the multi-cellular organism itself. Some issues unique to the creation are: the set of steps to be followed, the sequence in which they occur to create the structures, organs and the body plan (referred to as *creation sequence*), the timing, the locations, how the long sequence of steps are coordinated, how they were discovered and incorporated back into a single cell as memory and others.

The second one is with the issues encountered as the organism uses its components and interacts with the environment. Here the issues are: how the organism knows about the existence of different organs or organelles, how it starts using them effectively for its survival while sacrificing, if necessary, some subcomponents, how different organs and organelles introduced at different times of evolution start to coordinate with one another and others.

The third one is on how new traits are discovered as part of evolution over several generations and how they were introduced back into a single cell i.e., into its genotype. Some organs evolve to become better while other entirely new features are developed over several generations in a directed manner.

In this chapter, I will discuss these three issues in detail focusing primarily on the origin of multi-cellular organisms, though the issues are the same even for the origin of eukaryotes. I will attempt at providing an iterative process that addresses some of these issues. As with the problem of origin of life, an iterative or a looped process is necessary if nature were to discover the solution without help from conscious beings. As mentioned in the previous chapter, theoretical analysis for these topics needs more work, as we need to look at the detailed chemical processes of several of these intra-and-inter cellular structures and then propose suitable abstractions common to most of them. The theoretical analysis of this chapter would also need to be supplemented with experimental research in order to be more complete. This will be an important area of future work.

70.1 Physical and chemical structures

Consider creating cellular structures like organs in multi-cellular organisms and organelles in a eukaryotic cell. When would we accept that we have indeed created the cellular structure ourselves? Let me discuss the steps in simplistic terms by using creation of man-made systems as an analogy.

1. *Creating all parts of the system*: We need to ensure that the shapes of structures and their subcomponents are approximately similar to the one we are trying to create. For example, if we are creating a heart, an eye or an ear, we do need to create most of the 'important' parts and shapes first. What we consider as important parts and shapes are just those minimal set that are necessary for the structure to perform its function (cf. the parts needed for a heart, an eye or an ear to perform their function). The set of materials we use for each of these parts are not that critical and need not be identical to the cellular structure that already exists.

2. *Assembling all parts of the system*: We need to arrange these parts in an approximately similar manner. The hope is that when arranged in a similar manner, the resulting system will perform functions similar to the original cellular structure as well. There is a prerequisite here, which is that we understand the structure and its function in sufficient detail. Here, we are trying to copy an existing structure, which we know is a feasible design. We are

not trying to be creative yet. We are not designing new structures whose feasibility in both construction and operation are still unknown.

3. *Operating the system actively*: We need to let the structure operate in a stable manner for a sufficiently long time. This step may need a one-time jumpstarting either by supplying energy (typically, larger than normal amounts like with activation energy) or by applying forces at a few locations. These are applied just for a short period. The idea is not to be actively involved as the system operates. With man-made systems, we do have static objects that do not seem to operate by consuming energy (like a table, a chair, a house and most furniture in your house). However, with organs and organelles, the interesting sets of structures are not the static ones, but dynamic ones that require active control to function correctly (like heart, eyes and ears unlike skeleton, cytoskeleton and others – though they are interesting to a lesser extent).

4. *Integrating the system within the organism*: We need the structure to interact with the neighboring components in a positive manner so the stability of the overall system is considerably enhanced. This is useful because a system is less interesting if it rarely interacts with the environment. In the case of organisms, the interactions between different organs can be quite involved – a heart, the eyes and the ears are well integrated within our body. It is not enough to build an organ and make it work outside the organism. We do want it to function well within the organism while interacting with other components.

One of our objectives is to explain how nature is capable of doing all of these four steps without the help of any conscious or intelligent beings. These steps are quite critical for the problem of diversity of life (unlike the problem of origin of life) because organisms like eukaryotes and multi-cellular organisms have considerable internal structure.

We have already discussed mechanisms for some of the above steps. As an example, the active control of an organ or an organelle (in step 3) can be addressed using the fundamental pattern (see chapter 16). In section 16.1, we have seen how to auto-focus our eye lens provided the eye has already been created and has simple unstructured mechanisms to make the eye lens spherical or flatter. The other set of assumptions for the fundamental pattern are quite simple and natural for the problem of auto-focusing the eye lens. Even if the organism is born without an auto-focus mechanism, the fundamental pattern allows the system converge naturally towards a mechanism that auto-focuses the eye lens. No help from conscious or intelligent beings were necessary to discover the auto-control mechanism for this task.

Similarly, we have seen in chapter 17 how to turn our eye towards a source of light, how to converge both eyes onto the same object, how to track a moving object with our eyes, how to track a curved boundary with our eyes, how to turn our head towards a source of sound, how to move towards a source of odor and how to learn to put our hand in our mouth. For each of these cases, even if we were born without any of these abilities, we will learn them naturally without the help of any other conscious beings. This is indeed the case with other animals as well. The mechanism that allowed us to learn to auto-control for such diverse set of tasks under a mild set of assumptions (see chapter 16) is the fundamental pattern.

Therefore, the topics that need thorough analysis are the creation, assembling and the integration steps above. One important point to note with these steps is that they are composed of a mixture of physical and chemical mechanisms. However, the boundary between physical and chemical processes is quite vague. Even a simple task may transition between physical and chemical reactions several times. For example, if you recall the operation of a neuron in chapter 5, we see that we require chemical processes to open Na-K ion channels, physical processes like diffusion and active transport of Na-K ions to move them in and out of the neuron, a combination of physical and chemical processes to fire action potentials and to release neurotransmitters to the adjacent neurons, chemical process for the synthesis of neurotransmitters and physical processes for the transport of neurotransmitters along the axon and so on. The presence of both physical and chemical processes even for a single task like the operation of a neuron is quite apparent here.

Our goal is to identify and abstract common mechanisms such as the ones discussed above (for the case of the operation of a neuron) that can be used to create a wide range of functional organs and organelles like heart, eyes, ears, lungs, hands, legs, fins, wings and others. In this book, I will only discuss them cursorily. I will introduce some basic ideas and leave the detailed analysis for future work. The overall approach to create an organ or an organelle (that addresses the above four steps) in a theoretical sense can be stated in simple terms as follows.

a) Propose a rudimentary design that uses a combination of physical and chemical structures. Here, the designs may not resemble the current shapes or structures. They will be the primitive versions of a heart, an eye, an ear and so on.

b) Propose mechanisms that allow the system to improve its designs over time in a directed manner. For this step, one generic solution we have already discussed is the stable-variant merge loop pattern.

Most researchers and textbooks discussing the diversity of life do suggest this approach. However, to make the above approach feasible, we need to provide considerable details, say, by studying most organs and organelles and extracting a common set of abstractions to create them naturally. It is best to provide rudimentary designs similar to how we have done for the problem of origin of life in chapters 59 and 63. This will be an important area of future work.

Let us first look at the physical structures that need to be created. What are some of the basic functions performed by these rudimentary designs? Here are some examples:

- *Structural rigidity*: Most physical structures need to maintain a specific shape to perform their functions. As a result, we need to ensure a certain level of structural rigidity of these components. Examples of subcomponents that provide structural stability are cytoskeleton within cells or skeletal structures using bones in multi-cellular organisms. They help maintain the shape and the structure of the system.

- *Producing motion in the form of transport of chemicals and/or the system as a whole:* Since physical structures do not alter their chemical composition, one of the primary functions of these systems instead is to break, assemble and move components around. Fluidic pressure within blood circulatory system allows us to transport chemicals within our body. Diffusion, osmosis and capillary action are other physical mechanisms for causing motion. Other examples are microtubules to transport neurotransmitters within a neuron, karyopherins called importins and exportins associated to mRNA and proteins to facilitate their transport into and out of the nucleus (nuclear transport – see Alberts et. al [2]), motor proteins (like muscle protein myosin, kinesins, cytoplasmic dyneins and others) to perform a number of such active transport-related tasks. In addition to moving components within a system, we need to move the system as a whole as well. Hands, legs, wings and others can be used to move the entire system about.
- *Sensors:* Most living systems have a large number of sensors capable of detecting different external signals like light, sound, temperature, pressure, vibrations and others. Although most of these sensors require new chemical structures, there still are a few of them that are based on physical structures as well, which we discuss here.
    - o *Sensing light:* Photopsins and rhodopsins allow us to detect light as these new molecules deform differently when light falls on them. As long as repeatability of the signals can be guaranteed, these molecules arranged in an arbitrary but fixed structure can be thought of as a rudimentary design for sensing light in a consistent manner.
    - o *Detecting vibrations:* Eardrum is an example that can detect vibrations and transmit them within a tube of fluid. These vibrations can be sensed by hair cells to transmit auditory information to the brain.
    - o *Sensing pressure and temperature:* Skin sensors using touch is an example to detect pressure as well as temperatures.
- *Actuators:* Living beings use energy in various ways. One of the primary ways physical structures use energy is to generate force and pressure. When a force is applied, it causes the system to move. These structures are the actuators that either push or pull the system and its components.
    - o *Applying forces or pressure either by pushing or by pulling:* Several examples exist that generate force and/or pressure. Muscles as well as the skeletal structures help with applying forces on other external objects. Elastic properties resulting from expansion and contraction of muscles allow us to apply forces. Hands, legs and our body allow us to apply force on external objects to produce motion. Our jaws and teeth allow us to apply force large enough to crush food physically. Heart is another example where fluidic pressure is applied at regular intervals. The active control of these actuation mechanisms use the fundamental pattern in most cases (see chapter 16). Contraction and expansion of the lungs is another

example to apply forces and pressure in a regular pattern. It also aids in the transport of air and other molecules and is actively controlled. Screw-like mechanisms with microtubules and other motor proteins (like muscle protein myosin, kinesins and dyneins) can be used to apply force and cause a motion.

- o *Energy sources:* Some physical mechanisms also help store energy. The energy stored within the system can be released to perform useful work as well as heat. These energy sources (for both storing and releasing) can be actively controlled as well.

- *Active control of actuation and sensing:* When a system has sensors and actuators, such a system has a better survival advantage only when there is an active control of these components. Such active control mechanisms make the system autonomous. Here, one of the primary mechanisms is the fundamental pattern, which we have already seen used for several auto-control scenarios like how to turn our eye towards a source of light, how to converge both eyes onto the same object, how to track a moving object with our eyes and others (see chapter 17). There are other simpler mechanisms discussed in the previous chapters (like Figure 2.12) to allow controlling of sensors and actuators in our body in a stable manner.

Each of these physical mechanisms mentioned here are quite generic and are applicable in a wide range of scenarios. It is not too difficult to create rudimentary designs to perform each of these functions (similar to how we proposed primitive designs for actively searching for food, exploration and others in chapter 63). We do not need to discover a wide variety of chemicals (like proteins and others) in order to create the above primitive functional designs. As long as there is a mechanism like the stable-variant merge loop pattern to improve the designs in a directed manner, nature would be in a position to create a significant diversity of physical features.

On the other hand, the chemical structures exhibiting diversity are much more elaborate than the primitive physical structures discussed above if we look at the present day life forms. The number and variety of proteins, DNA, RNA and other organic chemicals along with the variety of organs and organelles that exist makes us wonder how nature created them without the help of conscious and intelligent beings. Besides, most functional designs have a well-integrated set of physical and chemical structures. If we look closely and separate out the physical from the chemical designs, we will begin to see common patterns across organs and organelles.

Just as we have identified common physical structures above, our task is to list and identify common abstractions across chemical structures as well. We need to show how to create rudimentary chemical structures and show how to improve them over time. In this regard, a few important chemical mechanisms already discussed previously come to our rescue. They are the chemical SPL network, the stable-variant merge loop pattern and the disturbance propagation in an SPL network. Together, they address (i) how to discover new chemicals in a directed manner for the creation of a given organ or organelle, (ii) how to integrate them within the existing SPL network to create new pathways and (iii) how to memorize the operation of long sequence of steps (both creation and operation sequences) using the propagation of disturbance within the stable parallel looped network.

While I will not go over the details here in this book, to make the argument more clear and convincing, it is best to take specific examples like a heart, an eye or an ear and enumerate the precise sequence of steps followed to create and operate the organ. We should then identify how to discover each of these chemicals and the sequence of steps needed to create or operate them. We would not want these steps to require the help of a conscious being or rely entirely on random mutations. In this book, I will not discuss these specifics and instead leave it as future work. For now, let me identify a set of issues we need to answer for creating organs or organelles and try to address one critical aspect, namely, the connection between the creation and the operation of organs and organelles.

70.2 Creational issues

The process of creating a multi-cellular organism typically starts from a single cell like, say, a fertilized egg. It then proceeds along an almost perfectly orchestrated and a long sequence of steps, creating each organ and finally the entire organism (see Gilbert [54] and Nüsslein-Volhard [115]). The entire process is energy and input resource intensive. It also takes several days or months to complete in some cases. Even simple missteps or errors in the chemical sequence of steps leads to severe abnormalities including the possibility of a missing organ itself. One of the biggest challenge then is to explain how nature was able to encode all of this 'meta-information' (i.e., about its own construction) back into a single cell, when there is no notion of awareness, goal or understanding of what, how or why the single cell is building a multi-cellular organism.

Of course, we are looking for designs without a designer similar to how we did in the previous chapters for the problem of origin of life. Nature does not store the meta-information back into a single cell after a feature appears useful. This reverse engineering requires awareness and consciousness. Therefore, we need an alternate explanation that achieves the same purpose, i.e., an alternate way to encode the phenotype information back into the genotype.

The creation of a multi-cellular organism can be seen as one of the most complex processes achieved by life. The first life forms appeared on earth about 700 million years after the formation of earth. However, the first multi-cellular organisms took another 2.5 billion years, i.e., more than 3 times longer, after the first life forms were created. Yet, strangely, multi-cellularity occurred independently at least 25 times (see Grosberg and Strathmann [60]). For example, plants, animals, fungi and several others developed multi-cellularity quite independently from one another.

Our current research showed that the embryogenic processes for organisms as diverse as a fly, mouse or human do share similar set of processes as well as the developmental genes themselves. Surprisingly, the complexity in organs and body shapes between, say, a fly and a human do not come from the differences in the total number of genes. Rather, they come from different ways in which nature combines these genes to influence cellular organization.

The origin of multi-cellular organisms is also uniquely different from the problem of creating life for the first time. Several issues are not only different but are, in fact, absent in single cellular organisms. Let me list a few of them here.

(a) *Discovering new chemicals*: In order to create multi-cellular organisms, nature needs to discover a large set of new chemicals like proteins, polysaccharides, several new genes, cadherins for cell adhesion and others. This situation, however, is not too different compared to the origin of life or the creation of any other feature. The chemical factories and other mechanisms of the previous chapters are the primary mechanisms here as well.

(b) *Sequence of steps for creating a given organ*: Multi-cellular organisms have a number of specialized organs like liver, eyes, lungs, heart and others. Creating each organ requires very distinct and elaborate sequence of steps. This sequence of steps is encoded in a single cell (i.e., the germ cell in an animal that later becomes an egg or a sperm) without actually reverse engineering, as mentioned above. I will call such a sequence of steps for creating any organ or organism as the **creation sequence**. It should be noted that the exact same creation sequence of reactions could occur not just during the birth of an organism, but also during other situations like repair, growth and others. Research has shown that no feedback mechanisms exist to correct any errors in the sequence of steps, at least, in the early stages of the embryonic developmental process (see Gilbert [54] and Nüsslein-Volhard [115]). Therefore, any significant alterations in the chemical gradients will directly result in an appropriate abnormality instead of the developmental process attempting to rectify this error.

(c) *Making the organs work after creating it*: Some organs like eyes do not need to work right after being created. However, other organs like ears, heart, liver and the digestive system should start working correctly. This is partly because several internal processes depend on a functional heart or liver, say. I will call the sequence of steps followed to operate an organ as the **operating or usage sequence**. It should be noted that the creation sequence for an organ is considerably different from the operating sequence for the same organ. For example, the laws, steps or genes needed to create a heart or an eye are considerably different from the laws followed when an eye or a heart operates (cf. the laws to create a car versus the laws to operate a car). In the case of a car, the creation sequence is hidden within a human brain as creativity whereas the operating sequence consists of Newton's laws and other known physical laws. It is necessary to unify or relate these two distinct sequences, at least in the case of multi-cellular organisms if nature were to create organisms without help from conscious beings.

One important aspect during the creation of a multi-cellular organism is that even before it is completely created, we can call it as alive. It has self-sustaining dynamics during the very process of creation itself. For example, a heart starts pumping blood and the digestive system starts working well before the birth of a child. This is in stark contrast with the machines we create like a car, robot or a computer. A car or a computer does not 'operate' as it is being created. Nature never creates the complete organism as a static system first and then let it run its dynamics (unlike a computer that is assembled first and then turned on subsequently).

(d) *Encoding relative location of organs*: The chemical information like the enzymes necessary to run a series of reactions is represented within the DNA. What about the physical and geometric information? How is this information stored in the single cell (like in a fertilized egg)? Your hand should be attached at the shoulder and not at the stomach. This important relative positional information is necessary for both creation and survival of the multi-cellular organism. It is not sufficient to explain how these physical structures are created, say, as spatial and temporal variations of chemical gradients (see Nüsslein-Volhard [115]). It is equally important to explain how this relative positional information is encoded within a single cell in the first place. This is because they should be recreated the same way in every generation.

(e) *Organ creation sequence to facilitate good coordination*: Different organs should coordinate correctly if the system were to be a highly stable one. For example, eyes, ears, nose and mouth are arranged efficiently on the face instead of placing them at different locations on the body. It is not possible to combine organs randomly anywhere on the body and expect to create a viable system. For example, you cannot assume that the organs and features are developed independently as specialized systems and then combined later to create a viable system. The random trial-and-error process is bound to fail easily. In fact, we have seen several abnormalities and miscarriages resulting from, say, untimely and imperfect coordination. We want to ensure that the combined system is better stable and self-sustaining than a group of randomly arranged organs.

Natural selection for placing organs at different random locations of your body and then evaluating if the organism is fittest is quite unsatisfactory. Survival of the fittest as the approach to discover the correct location of an organ may be reasonable for single cellular organisms because of the ease of reproduction and the existence of an enormously large collection of them (it is estimated that there are 10^{30} bacteria on earth – see Whitman et. al [163]) but not for multi-cellular organisms. It is unreasonable to suggest that nature has tried placing hands and legs at every possible location on your body within the population and that the ones that survived (as the fittest) are the precisely the present-day organisms.

Also, note that before nature can evaluate if a multi-cellular organism is the fittest, it should first survive. The basic survival itself is extremely hard to achieve through random means for a multi-cellular organism. Natural selection does help here to ensure that the set of organisms that survive are the ones that continue to reproduce and remain after several generations. However, the type of information that defines how physical coordination and arrangements work for multi-cellular organism in order to survive is so far disjoint from the type of information stored in a single cell, which is later used to create these physical coordination and arrangements.

(f) *Specialization of cells:* The germ cells (which become sperm or egg cells later through gametogenesis) contain all of the information necessary to create an entire multi-cellular organism. After an embryo is formed through fertilization, the cells divide and undergo a sequence of steps to create different organs. The cells within each organ are not identical even though they contain the same DNA. For example, the liver cells cannot be used in the heart or vice versa. We now say that the liver cells are specialized unlike the original germ cells. When a liver cell reproduces, it only creates other liver cells, not heart cells. Stem cells in animals are those cells that can turn into any other type of specialized cells when they reproduce. The specialized cells have uniquely different epigenetic marks. Therefore, the set of genes expressed are quite different between two specialized cells. The processes of when and how this specialization happens during development are interesting problems.

Studying each of these problems is quite an active area of research. Considerable progress has been made in explaining how the sequence of steps works in some of the current multi-cellular organisms (see Gilbert [54] and Nüsslein-Volhard [115]). However, there is very little progress towards explaining how these processes were created for the first time from unicellular organisms.

70.3 Operational issues

A multi-cellular organism should operate or survive without collapsing after it is created. This is not as easy compared to a single cellular organism. The multi-cellular organism is bigger in size and has a lot more moving parts. Achieving simple physical stability is itself not obvious. Yet, we require chemical stability as well. Let me list a few of these operational issues here.

(a) *Importance of different organs for overall stability*: How does nature decide which set of organs are the necessary minimal set to guarantee an overall stability? A conscious being is capable of evaluating this. There may be a few cases when multiple organs can give rise to instability (like several wings, multiple hands and others). One argument would be that nature created some configuration and since the organism survived, that configuration (of one heart, two hands and others) is kept across generations. There is no reason to search for the best configuration unless there is a stress that demanded a change. Yet, the study of relative importance of organs needs a closer look.

(b) *Coordinating various organs*: The problem of coordinating various components is analogous to a similar problem in single cellular organisms. It should be obvious that even when all parts of a car work well separately, it does not mean that, when put together, the entire car will work as expected. The same applies with various organs in a multi-cellular organism. The heart, lungs, hangs and legs can work well, but if the nutrients and oxygen are not properly circulated and in the right amounts, if the hormone levels are not regulated or if your hands and legs do not fight back when attacked, the entire organism dies. In fact, we are not even consciously aware of the complex coordination that happens between our internal organs.

(c) *Cooperation between cells and behaving as a single entity*: Billions of cells that make up an organism behave as a single entity. Each individual cell does not work selfishly as if it were a separate organism. Unlike single cellular organisms, the cells in a multi-cellular organism are considerably constrained. In fact, most of these cells can no longer survive on their own. For example, they rely on nutrients distributed within the blood circulatory system in a fundamental way. There are other essential types of dependencies as well. Yet, the organism, as a whole, does gain higher survival abilities at the expense of independence.

(d) *Sensory-control mechanisms and environmental interactions*: The different components of a multi-cellular organism take on roles like sensors and actuators to control the environment actively for better survival. Single cellular organisms also have sensors and actuators. However, the difference is that a single sensor or an actuator is now composed of millions of cells that should work together to achieve the desired function. When we say that eye is a sensor or that you can control the motion of your hand, we mean it in an abstract sense. We seem to simply assume or forget the millions of interactions to achieve the functionality of the sense organ and instead start treating it as a single abstract entity. We do not explain what causes these millions of cells to coordinate in the first place to give this single entity-like behavior. Working at single entity-level abstraction is much easier, but hides the enormous complexity that exists. The machines we build map these abstractions (like sensors and actuators) faithfully using rigid components. Man-made systems, therefore, avoid millions of fluidic components for each such abstraction. We need careful analysis to understand how to guarantee stability of the feedback looped system.

(e) *On knowing and using its organs for better survival*: The multi-cellular organism behaves as a single entity similar to a single cellular organism. The organism exhibits what appears as a sense of familiarity of its own internal organs even the moment it is born. It starts to use each of these features for its own survival almost instinctively. How does the system know about these features and how to use them? How is this knowledge encoded for the first multi-cellular organism? For example, a snake seems to be familiar about the poison as well as the complex mechanism of how to release it in response to a danger. A spider knows how to weave its web to capture its prey. Similarly, our immune system is actively triggered to fight infections. Organisms naturally use mouth for eating, hands and legs for attacking or running and so on. It is possible to say that the organism is simply executing these tasks mechanically. However, the difficulty is when we try to explain how these unique features offering evolutionary leaps were developed and linked together for better survival for the first time using entirely random means.

(f) *Normal growth of organs*: With multi-cellular organisms, different organs and cells should be regulated carefully to avoid any cancerous growth. This is a unique and serious problem with multi-cellular organisms. As the organism ages, different parts of the body grows, but in a proportionate manner. Each organ like the liver or the heart should create the corresponding specialized cells, but not excessively. This is analogous to a looped mechanism, but at a cellular level. What part of the body coordinates this regulated growth and how? Similarly, when there is damage like a bruise, cut or a fracture, the repair mechanisms should not only be present, but should be regulated as well. It is not reasonable to assume that all types of damages were already foreseen and were encoded ahead of time in its DNA.

Each of these problems have similarities with single cellular organisms or, more specifically, with a eukaryotic cell. Compared to prokaryotes, a eukaryotic cell has specialized internal organelles. How these organelles were created and how they were coordinated to perform specific tasks offer insights when addressing similar problems with multi-cellular organisms.

70.4 Evolutionary issues

A multi-cellular organism does not acquire all of the useful features at once. Furthermore, it is unreasonable to assume no new features will develop in the future. Therefore, the process of developing new features should be quite similar to the process of changing existing features. There are, however, nontrivial issues that should be addressed to clarify this. I will list a few of them here.

(a) *Making existing organs effective*: A multi-cellular organism is not initially perfect, even with all of the existing organs. The various subcomponents can relocate over a period of time to better align with the environmental conditions. This minor reshuffling by adjusting the embryonic developmental process is an important process to achieve better overall stability. It is important to explain how these inefficiencies are indeed eliminated, how chemicals and components are recycled and how the same part takes on different roles as necessary.

(b) *Encoding adaptations back into the single cell*: Several types of changes occur due to environmental stresses. Some of them are chemical changes and others appear to be physical changes (created chemically, however). It seems reasonable to suggest that the chemical changes are encoded within the DNA by way of stable-variant merge loop pattern. However, we need clarification on how and where the physical information is encoded. An example is the shape of the beaks of finches (observed by Darwin). Some have long beaks to punch holes in the cactus fruit while others have short beaks to tear the cactus base. These changes in the physical shape should be represented back within the single cell. This representation, typically, is unrelated to the shape of the beak itself. Rather, current research showed that spatial and temporal variations of specific chemical gradients during embryonic development gives rise to these changes (see Gilbert [54] and Nüsslein-Volhard [115]). However, it is not sufficient to explain how these modified structures are now being created. Rather, it is important to explain how the adaptation problem and the creation problem were synchronized or were made compatible with each other for the first time. This synchronization problem appears nontrivial when we look at the diverse set of body plans like animals with wings, fins and legs.

(c) *Creating entirely new organs and integrating them with the rest of the body*: In addition to reshuffling existing organs, it is necessary to develop new organs as well. This is quite different from modifying an existing feature to better fit in the environment. For example, animals have discovered fins to survive under water, legs and other parts to survive on land, wings to fly in air. Several of these features need considerable knowledge of the laws of physics just to understand how they work (like the aerodynamics of flight). If we need awareness and consciousness to integrate multiple concepts to build a feature like wings, how did nature discover wings and how is a bird using them to actively control its flight? Most of the times, this process appear gradual. However, we have identified a period when there is an immense explosion of diversity of body plans, namely, the Cambrian explosion. New environmental stresses can induce mutations and initiate the process of development of a new feature. How this evolves directionally to result in a functional organ and how it integrates with the rest of the system needs more clarification.

(d) *Vestigial organs*: These organs have essentially lost their original function in one species when we compare it with a similar organ in another species. The usage of an organ in a given lifetime seems to have a connection with its ability to become vestigial or not. Just as with the problem of adaptation, does the organism encode the information that the organ or a structure is not being utilized beneficially, back into the germ cell or does it simply build it according to the original plan, but not use it anymore? The latter seems reasonable.

The gradual evolution of a multi-cellular organism involves the following basic steps: creating new features, using them effectively, modifying them to changing environmental conditions and making some unnecessary organs obsolete. As this process repeats over a long period, the diversity we observe around us begins to appear convincing. Our objective is to make this argument concrete by proposing well-defined mechanisms.

70.5 Challenges with existing theories

Consider the time when there were only single cellular organisms on the planet. How did the first multi-cellular organisms begin to form? If multiple cells aggregate and join to form a group of cells, they are still a random collection of cells that self-adhere to each other. We cannot yet treat this group as a multi-cellular organism. We need to show how the information needed to regenerate the entire group repeatedly is encoded back into a single cell. In addition, we need processes that address the issues mentioned in sections 70.2 - 70.4. Therefore, we are looking for a combined solution that gives multi-cellular organisms their unique features and, hence, the enormous diversity, beyond those achieved by a simple aggregation of cells.

One popular theory suggests that individual cells begin to form colonies of cells either by joining after cell division or through a failure to separate when the cell divides (see Wolpert and Szathmáry [174]). While this theory suggests how to create a colony of multiple cells, the resulting system is not a true multi-cellular organism. For example, there is no guaranteed way to generate the same organ or feature just from random colonization. The colonial theory does not encode the construction of a complex colony. It only states that the colony forms randomly. There is no explanation for the problem of coordination between various cells to perform a specific function. The possibility of cancerous growth exists because each cell is still behaving independently within the colony. Furthermore, there does not exist a mechanism to remember the structural and physical shapes created by the colony, especially if it results in a survival advantage. The colonial random reproduction mechanism cannot recreate these advantageous shapes later. We also cannot guarantee directed evolution of new features starting from a given colony.

The colonial theory relies on arbitrary randomness at its core. Random self-assembly of cells to form colonies needs more details to address issues discussed in the previous sections. For now, it relies only on the discovery of the correct set of chemicals necessary for multi-cellularity like cadherins for cell adhesion, receptor tyrosine kinases for cell signaling and epigenetic marks for cellular differentiation. It is not sufficient to explain the enormous structure using just these chance-based mechanisms just as we could not explain the creation of life from inanimate objects using arbitrary randomness. Other theories like syncytial and symbiotic theories also face similar difficulties with most of the issues mentioned above.

70.6 Sample mechanisms from developmental biology

In this section, I will list a few generic mechanisms that turn out to be useful in understanding developmental processes of organisms (see Gilbert [54] and Nüsslein-Volhard [115] for more details). The list discussed here will be quite brief, however.

Transcription factors: One generic mechanism identified by Jacob and Monad [78] is gene regulation during protein synthesis. They have identified several genes that encode proteins called transcription factors. These transcription factors have an ability to bind to specific DNA sequences. As these sequences are present in front of other protein encoding genes, we now effectively have a way to control the production of these proteins. Some of these transcription factors can promote while others can repress transcription of DNA into mRNA. Therefore, a chemical reaction catalyzed by a given enzyme can be controlled by activating or suppressing the production of the enzyme through appropriate transcription factors.

For the development of a multi-cellular organism, these factors play a much more prominent role. Typically, one transcription factor controls one gene encoding a protein and hence one chemical reaction. However, during the developmental process of a multi-cellular organism, one transcription factor controls the gene of another transcription factor (not just a protein), which in turn controls another transcription factor and so on. This type of hierarchical gene regulation exists even in the developmental process of simple multi-cellular organisms like a *Drosophila* fly. When *Drosophila* develops from an embryo to form specialized body structures, the process requires developing clear segmentation patterns using hierarchical gene regulation as follows. First, a chemical gradient from the anterior to the posterior end of the embryo is generated just from a maternal signal like the bicoid-gene. This gradient triggers a cascading activity of fine-grained zygotic (i.e., both maternal and paternal) genes. This activity, along with a posterior gradient, controls the transcription of gap-genes like hunchback and knirps. This, in turn, triggers activity of pair-rule-genes like fushi-tarazu and even-skip. Subsequently, they trigger segment-polarity-genes like engrailed. The result of these hierarchical layers of genetic control is that the *Drosophila* embryo develops clear segmentation patterns. The next step is a cascade of activity of homeotic selector-genes for creating the necessary specialized structures (see Nüsslein-Volhard [115] for details).

It is also possible that one gene has several transcription factors that activates or represses its activity differently at different times. This helps in reusing several existing chemicals to produce different behaviors at different times.

Chemical gradients: The formation of chemical gradients, both within and between cells, is one way to control spatial and temporal activity. When the correct combinations of signals (like morphogens and hormones) are present in a given region, they trigger specific tissue or organ generation. The timing is controlled through the occurrence of specific events. This involves simultaneous activity of several chemicals that exceed certain threshold concentration. A rather precise spatial location of activity is also maintained by controlling both the set of chemicals and the rates of the appropriate chemical reactions. It has been shown that eyes, antennae or legs of flies can be grown at different regions by introducing these specific set of chemicals at different regions. Other spatiotemporal deviations can make the organism inviable as well (see Nüsslein-Volhard [115] for details).

Looped reactions: Another generic mechanism is the presence of a looped set of chemical reactions, both within a cell and across multiple cells. We have seen several variations of these already throughout the previous chapters. Even with developmental processes, we see a large number of such looped processes. For example, in *Drosophila*, a Delta-Notch signaling pathway creates 'salt-and-pepper' patterns (see Nüsslein-Volhard [115]). These are patterns of arrangement of two different types of cells. Even when all the cells were initially similar, a differentiation into two types of cells, arranging themselves into a 'salt-and-pepper' pattern, starts to occur using a feedback loop in the following way. Activation of Notch by Delta in a given cell lowers Delta protein production resulting in a release of less Delta signals. Similarly, when a cell receives less Delta signals, an internal trigger causes itself to produce more Delta signals. The combination of these two mechanisms results in a feedback loop creating two different types of cells. Eventually, the cells rich in Delta become sensory cells whereas the cells rich in Notch become epidermal cells. Such differentiation into two types of cells using the Delta-Notch feedback loop has been discovered in vertebrates as well. For example, cells with high Delta and low Notch become neurons while cells with low Delta and high Notch become glial cells. Similarly, formation of endocrine cells of the pancreas versus goblet cells of the intestinal epithelium use the same mechanism (see Nüsslein-Volhard [115]).

Another example of looped mechanism discussed in Nüsslein-Volhard [115] is with hedgehog and wingless proteins. Here, the mechanism is a positive feedback loop, which works as follows. A protein called engrailed activates production of hedgehog protein, which binds to neighboring cells, triggering the encoding of a transcription factor Ci causing it to produce wingless protein. This in-turn activates engrailed protein production, closing the loop. It is conceivable that there are several other such looped pathways during the developmental process of most multi-cellular organisms.

The advantages of these generic mechanisms are that an organism can reuse the same chemicals in entirely different scenarios like when creating different organs or subsystems. Of course, in order to make this work, we do need a few specific DNA changes. Fortunately, they do not have to be precise nor do they have to be in the regions encoding a gene. For example, the promoter regions (like the activator or repressor sequences) in one scenario can be copied in front of the genes for an entirely different scenario. Minor errors or mutations in these promoter sequences do not make them ineffective, unlike the case when they do code for a protein.

One example where the same set of chemicals are reused under entirely different scenarios is with Toll as a receptor. In *Drosophila* fly, there is a signal chain involving more than 10 genes, which is partially responsible for producing a maternal dorsal gradient with Toll as a receptor and dorsal protein as the transcription factor. However, in both flies and in vertebrates, a similar pathway involving most of the same 10 genes and a Toll-like receptor triggers an innate immune response instead (see Nüsslein-Volhard [115]).

It is not surprising that the same chemicals are reused in different species with different pathways for different purposes. However, what is interesting is that the entire mechanism involving tens of genes is used with similar pathways, but for different purposes. We achieve this simply by copying the DNA sequences of promoter regions into new locations like, say, in front of other genes giving rise to new functionalities.

70.7 Linking creation to operating sequence of a feature

Among the list of issues identified in the previous sections, the central one is the disjoint nature of the creation sequence and the operating sequence. We can say that creating a subsystem like an organ for a given feature is not useful if we do not know how to recreate it later. The sequence of steps and the set of laws required to create an organ are quite different from the ones required to operate an organ (i.e., after it is created). For example, the set of laws needed to operate a heart are the standard laws of physics (using which we can explain how a pump works). However, the set of laws needed to create a heart are completely different. Here, we need a new set of chemicals, guide them in a specific way for several hours (and sometimes days), which then deposits the required chemicals and/or specialized heart cells in a unique manner to create a fully functional shape. Only a few shapes work as a pump. While the mechanisms that deposit the chemicals and grow the cells obey the standard laws of physics, the laws that coordinate a specific sequence of steps seem to depend on conscious beings with free will. The laws of physics are not directed enough to create specific shaped objects or large-scale structured complex systems. For example, Newton's laws do not autonomously create a table or a chair, but humans do with their intelligence and free will. We, therefore, need an alternate mechanism.

On the other hand, single cellular organisms like prokaryotes do not require detailed mechanisms for the creation of a given feature (as mentioned earlier) because they have limited internal structure. Bacteria have several features like an immune system (from bacteriophages), secretion system and others. The bacteria operate these features by creating the correct set of chemicals and by following a specific sequence of steps. Bacteria discover these chemicals through stable-variant merge loop pattern and represents the sequence of steps as disturbance propagation within an appropriate network of chemical loops. All of the above necessary chemicals are already specified within the DNA. This addresses the usage of the feature. What is needed to create the feature? These bacteria do not need any additional processes, other than the replication of the DNA, stable-variant merge loop pattern and disturbance propagation with its SPL network. Since no special structures are needed for the feature, the network of loops enabled by the DNA already helps with both the creation and the usage of the feature.

The colonial, symbiotic and other theories suggest that multiple cells group together to form a colony that as a whole has an improved chance of survival than if the cells lived independently. This may be a reasonable hypothesis except that it does not address how the formation of colonies and other specializations that occur is encoded back into a single cell so the colony can be recreated reliably in each generation. The answer lies in linking the creation sequence to the operating sequence (and vice versa) of chemical reactions.

The second issue mentioned earlier is that when we use Darwin's theory, the usefulness of an organ is suggested to be important for its creation. However, an organism cannot know an organ's usefulness until after it is designed and created. Except for humans, no other organisms can evaluate the usefulness of an organ because they are not conscious to the same degree as humans. Instead, they simply use the organs they have. To create an organ for the first time without any guarantee of basic survival makes it impossible to relate usefulness of an organ to its creation.

The third issue is the well-known chicken-and-egg problem. An egg (single cell) is needed to create a chicken (multi-cellular organism) while a chicken is needed to create an egg. For each of these problems, the amount of both theoretical and experimental progress is minimal.

In this chapter, I will discuss how to address the above problems. In section 70.6, we have already looked at a few sample mechanisms from developmental biology that is generic enough to be applicable under a wide range of scenarios. The next step is to look at mechanisms that allow us to link creation of a feature (creation sequence) to the usage of the feature (usage or operating sequence) and vice versa. These mechanisms are central to the problem of creating multi-cellular organisms from unicellular ones as well as the creation of eukaryotes from prokaryotes.

If a system adapts and makes changes to the organ to better fit in a new environment, nature should create the adapted organ in the future and not the original less-fit organ (like the different shapes of beaks of the finches suitable for different environments). There is a need to link the creation sequence of steps with the usage sequence of steps for a given feature. The fact that they are all chemical processes does help. However, how can one sequence of steps trigger another seemingly different and disjoint sequence of steps? Linking the two types of sequences of steps imply that there are chemical pathways from one set to another. For this, firstly, both set of reactions should occur in approximately the same region and at about the same time. Secondly, one set of chemicals should affect the other set of chemical reactions i.e., they should be chemically correlated in a reliable and repeatable way.

Let me identify several sources of this link between the two sequences of steps. When an organ or a subsystem is in the process of being created, clearly the usage sequences of steps are not triggered. For example, when the chemical processes during the creation of the heart take place (like when angiogenic cell clusters coalesce), the contraction of the cardiac cells (to pump the blood) does not occur. If it did, the unformed heart may even be damaged. Therefore, in most cases, there is no link in the direction where the usage sequence is triggered during the creation sequence. There are, however, a few cases, especially when an organism is being born, where such a link does seem to exist. Even before the organ attains its optimal functional state, it starts to operate and continue to build itself.

In any case, linking creation sequence to the operating sequence is not critical for the problem of origin of multi-cellularity or eukaryotes. If the multi-cellular system creates the organ first and operates it only after the organ is completely created, it is still an acceptable solution. It is the opposite direction that is both nontrivial and necessary to address when solving the problem of origin of multi-cellularity and eukaryotes.

Now, what is interesting is that the existence of a link in the opposite direction i.e., triggering the creation sequence during the usage sequence is far more common than we would have guessed. Note that a creation sequence refers to the creation of a specific multi-cellular feature or organ – this exact same chemical sequence of reactions can occur not just during the birth of an organism, but also during other situations like growth, repair, adaptation and others. Let me list a few such situations now.

1. *Growth*: During the normal growth of an organism, as the subsystems are being used, parts of the creation sequence that allow them to become bigger in size will be triggered. For example, your hands, your eyes, your ears, your heart and other organs in your body grow bigger. At a young age, the growth is considerably fast. Hence, the link between the creation and usage sequences are quite strong compared to an adult age. The resources from food intake are distributed appropriately between various subsystems. This maintains a steady and a controlled growth. A sort of quasi-equilibrium is ensured to avoid cancerous growth.

These chemical reactions could be transferred to an offspring epigenetically or, in some cases, genetically as well. Even if they are not genetic, it is important to realize that these links present opportunities to transmit the mutated DNA and/or other chemicals as infections to other systems, at least through slow and low probability situations. This is compatible with the fact that major changes in the organs' structure and function happens over thousands of years. The important point to note is that the existence of these links presents opportunities for several chemicals to cross-transfer even across species.

2. *Replacement*: There is a need to recreate some of the existing cells of organs (which are specialized like liver cells, heart cells and others) due to regular wear and tear as the subsystems are used. Old cells are constantly replaced by new cells like with skin cells, red blood cells, hair cells and several others. These situations present opportunities to trigger and link recreation sequence of these cells during the usage sequence of the organ. These could be transmitted to an offspring epigenetically or, in some cases, genetically as well (see above discussion).

3. *Adaptation*: Organisms are constantly under stresses like sudden changes in diet, changes in sources of diet, predatory environments, and others. This forces the subsystems to adapt structurally, triggering and altering parts of the creation sequence. Once again, these could be transmitted epigenetically or, in some cases, genetically as discussed previously. An example of an epigenetic mechanism can be seen as follows with the shapes of beaks of finches as they adapt to changes in the source of food. Darwin had observed and documented the different shapes of the beaks of finches in Galapagos Islands. Their shapes seemed to depend on the type and source of food they consume. The finches that punch holes in the softer cactus fruit had long beaks whereas the ones that eat the insect larvae from the hard cactus base had shorter beaks. The suggestion was that the same initial finch species turned into, say, two different species, with a different beak shape and size, by adapting to the source of food they feed on. Later research suggests that (see Abzhanov et. al [5]) that spatial and temporal changes in the expression of calmodium (CaM) and bone morphogenetic protein 4 (BMP4) is responsible for the change in the shape and size of the beak.

One theoretical process to account for this is as follows. Let us assume initially that all finches only had long beaks. The argument given below would still work even if we assume the opposite i.e., that all finches only had short beaks initially. They feed on different sources of food, some on cactus fruit and others within the cactus base. We now want to explain how, say, after several generations, the population of finches turn out to have both long and short beaks and that these different shaped beaks are, in fact, inherited features, not just random variations.

The initial stress experienced by these finches is the change in food source. From our original assumption, there is already a creation sequence encoded to generate long beaks. As the finches grow to their adult size, their beaks grow accordingly to maintain its long thin shape. This can be seen as a regulatory loop in the expression of BMP4 and CaM chemicals all through their life. Clearly, altering the steady state values of these chemicals will generate different shaped beaks. However, we want to ensure that these altered steady state values are memorized so that it can be inherited. For otherwise, even an advantageous shorter beaked finch, if created through random variations, would be lost in the subsequent generations.

This is when we first notice that the usage sequence enforces a steady state condition in the creation sequence itself, not just in the usage sequence. For example, the long beaks of the finches trying to tear the harder cactus base are not strong enough. Continued exposure to this food source will cause the cells in and around the beak to demand relatively more input energy (analogous to what happens after a workout). This new demand alters the steady state condition of creation and deposition rates of more cells (i.e., the creation sequence) needed for growth and replacement of cells within the beaks. They result in making the beak shorter and stronger. This is the new growth pattern for these finches.

The next important step is to determine how this new steady state chemical condition in the creation sequence can be transmitted to the germ cells. This is required if the offspring were to start using the new steady state chemical conditions. For this, we first need a way to represent the new steady state condition chemically. As mentioned in chapter 58 in more detail, this is represented as a new threshold concentrations of various chemicals involved (i.e., BMP4 and CaM in this case) in the SPL network. The reason is that the older threshold concentrations within the network of loops produce a specific disturbance propagation resulting in the deposition of cells corresponding to a long beak. The newer threshold concentrations will produce different disturbance propagation within the same network of loops resulting in a different output deposition of chemicals, thereby creating a shorter beak.

The new threshold concentration gradients corresponding to the new steady state can leak to the germ cells randomly (say, through maternal effects with changes in the cytoplasm i.e., using epigenetic mechanisms initially). This is not arbitrary randomness, but repeatable randomness in the sense discussed in section 52.3. This is also not guaranteed to occur in every single generation with complete reliability and repeatability. However, once it is transmitted, the memory of the threshold concentrations can be maintained in all future generations. In this way, you start to have short beaked finches as well as long beaked finches after a while.

We can now see that the same iterative process discussed here with beaks of finches can be used for a number of other scenarios of adaptations in organisms.

4. *Repair*: During an organisms' lifetime, it is constantly subject to events resulting in minor damages, including physical damages through accidents and chemical damages through infections. The organism has an extensive set of

mechanisms to repair its subsystems in these situations. These mechanisms trigger parts of the creation sequence in response. As a result, they present opportunities to link creation sequence with usage sequence of chemical reactions, some of which can be transmitted genetically, though they are mostly transmitted through epigenetic mechanisms as discussed previously.

5. *Regeneration*: Some organisms like newts, salamander and planaria (non-parasitic flatworms) have an ability to regenerate several body parts when there are major damages including amputation. In this case, the entire creation sequence for that subsystem is triggered in response. This presents opportunity to link the creation sequence with the rest of the usage SPL network (not just the usage network of the amputated organ).

6. *During embryogenesis*: As mentioned earlier, when the embryo grows and starts to create various organs, several of them start to operate as well (like heart, hands, legs and kidneys) even if the organism is not fully formed. In this case, the creation and usage sequences are triggered in response to each other (especially because there is less interaction with the environment during this period). Furthermore, there are several epigenetic mechanisms (like maternal effects, methylation, histone code and others) that transmit both creation and usage chemical sequences from the parent to the offspring.

In each of these cases, at least a subset of the creation sequence is triggered while the organ or subsystem is being used. Of course, the exact chemical details would need to be identified for each of the organ or subsystem individually. Nevertheless, the existence of a reliable and well-defined correlation is definitely well established.

Conceptually, the mechanism of how the link between creation and usage sequences can be formed is as follows. You have a network A of chemical loops that correspond to usage sequence and their related chemicals. You have another network B of chemical loops that correspond to creation sequence and their related chemicals. In the beginning, there are very few pathways from network A to network B because nature has not yet discovered the necessary chemicals to link the two networks. What nature needs to do now is to discover more and more pathways that link network A to network B using the above processes like growth, replacement, repair and regeneration. This involves discovering new chemicals like transcription factors and other proteins. They can be discovered using stable-variant merge loop pattern to ensure there is a direction in evolution (unlike random mutations followed by natural selection of Darwin's theory). The stress that triggers stable-variant merge loop pattern comes from a lack of mechanisms like growth and repair that try to destabilize the system (or make it less fit). As this process iterates, nature eventually creates a dense set of chemical pathways between network A and network B.

70.8 Transition from unicellular to multi-cellular organisms

Given the existence of several mechanisms to link creation sequence to the operating sequence, let me now present one iterative approach that allows nature to transition from unicellular to multi-cellular organisms. The discussion here will be at a conceptual level to show the viability of a natural process. There will be little discussion, if at all, on the precise set of chemicals and chemical reactions needed to explain the transition. Although there is significant experimental work in the area of developmental biology that outlines several detailed mechanisms (see Nüsslein-Volhard [115]), additional experimental work is necessary to fill in some of the gaps within the iterative process described here. One of the challenges with the problem of origin of multi-cellular organisms is a lack of even a theoretical process that overcomes the conceptual difficulties discussed in the previous sections. The iterative or looped procedure presented here attempts to address at least this issue, although further experimental work is necessary.

The looped procedure outlined here is a significant departure from the existing theories. This is because it guarantees iterative designs with the property that the future designs are more reliable than the past designs. There surely are random mutations even in the new proposal. However, they are combined with other processes and mechanisms iteratively to guarantee a direction in evolution even if they are spread across thousands of generations and millions of years.

The iterative process can be viewed from two angles. The first one concerns with the problem of creating a multi-cellular organism for the first time from unicellular organisms. The second one deals with how nature creates a new multi-cellular feature like an organ after the existence of multi-cellular organisms. You will notice that there are several similarities between the two problems and their corresponding solutions. You will see that the solution proposed here addresses both problems though the primary focus is on the former problem.

To begin the iterative process for the former problem, let us consider the situation when the only organisms on the planet are unicellular. These organisms are constantly subject to different types of stresses like, for example, changes in the environmental conditions (like weather patterns), competition within and across species for input resources, competition manifested as infections like with bacteriophages, normal wear and tear, unexpected damages, growth and others. Let us look at how evolution proceeds under these circumstances.

1. *Stress*: The main cause for developing a new feature or for creating the first multi-cellular organisms is a stress that tries to destabilize the existing system. An environmental stress naturally groups the systems into less fit and already fit systems. The less fit systems will be prone to more errors (say, mutations in the DNA during replication) causing them to introduce variations.

2. *Directed response – a rudimentary unicellular solution*: The adaptation proceeds in a directed manner using one of the stable-variant merge loop patterns described earlier. The unicellular organisms maintain their original set of features and improve themselves in a directed way as they try to discover new solutions. There are multiple ways to alleviate the stress, some physical and some chemical rudimentary solutions. Here are a few of these initial designs, which we refer to as unicellular solutions.

    (a) *Discovering new chemicals*: When an organism is subject to an environmental stress, one of the first responses involves discovering chemicals through mutations to alleviate the stress. The underlying mechanism is a directed process spread across several generations using the stable-variant merge loop (SVML) pattern. The unicellular chemical solution does not guarantee that every subsequent cell will survive the stress. Instead, it offers a higher probability of survival than without the unicellular solution. During this process, there will be no increase in either the number of cells or the cellular complexity of the system. Instead, the solution typically involves random mutations within the DNA giving it a partial advantage.

Given the simplicity of the rudimentary design, namely, the discovery of a few chemicals, the unicellular solution has both the creation and usage sequences. For example, for the development of a feature like an eye, the initial unicellular solution could start with the discovery of chemicals like rhodopsin and photopsins. The goal is to look at how nature can build a multi-cellular rudimentary eye including the muscular control mechanisms. Horizontal gene transfer is possible, at least partially, here and at all stages of development discussed below. In this sense, different species and different populations within the same species compete to create similar unicellular solutions. This results in a divergence of solutions for the same feature.

(b) *Avoiding stress*: A second response is a physical solution that tries to avoid the environmental stress actively. The cells hide from the stress behind other cells (like your skin acting as a layer of protection for cells inside your body) or they move away to a new location where the stress is minimal. A rudimentary solution can be based on the design discussed in section 63.2 on how a primitive cell can actively search for inputs (see Figure 63.1). The probability of survival is now higher than when the cells did not actively hide from the stress.

(c) *Hiding from the stress*: A third response is an improvement to the second one when cells discover chemicals like cadherins that allow cells to adhere. This allows the creation of a random group of cells. The outer cells (like your skin cells) shield the inner cells from the environmental stress. This is just another physical way to hide from the stress. For now, the process of adhesion after cell division is repeatably random and unreliable, with no specific structures created. Each cell within this random grouping is still sufficiently independent that we cannot call the group as a multi-cellular organism yet. We need to turn this process into a reliable one with an ability to create unique specific structures. The probability of survival here is typically higher than the second response mentioned above.

The group of cells formed here could be either homogeneous or heterogeneous. Initially, they are homogeneous colony of cells. However, over time, as the outer cells are exposed to the stress while the inner cells are not, it is likely that the cells become heterogeneous with different cells specialized in different ways.

(d) *Homogeneous colony of cells that have discovered new chemicals to avoid the stress*: A fourth response is a combination of the previous two approaches, namely, a random grouping of cells in which each cell has the unicellular solution to the environmental stress. This solution has higher probability of survival than the previous ones even though the process of creating this system is repeatably random and unreliable (same as before).

This is the formation of a homogeneous colony of cells, all of which have the unicellular chemical solution. Compared to the response in (c), which had a heterogeneous colony, the fittest between the two is a homogenous colony. In order to form a colony of cells, we need to discover a few chemicals. For example, we need chemicals like cadherins for cell adhesion, receptor tyrosine kinases for cell signaling and chemicals like Delta, Notch and bone morphogenetic proteins (BMP) for cellular differentiation. The cell adhesion should happen right after cell division, but not before they separate completely and start to move away trying to live independently from each other.

Now, not every cell within this group uses the unicellular solution to alleviate the stress. The outer cells or the ones closest to the stress react to produce a change that benefits all the other cells within the group as well. The outer cells have effectively shielded the inner cells from the stress. As the colony is exposed to different stresses, there will be a transition from homogeneous colony to a heterogeneous colony and vice versa.

3. *Rudimentary multi-cellular solution*: First, note that not every unicellular solution above will be converted to a multi-cellular solution. Yet, as several homogeneous colonies begin to form, new sets of features will develop over time. For example, some of these colonies of cells can form different types of tissues like connective, muscle, nervous and epithelial tissues. Using these tissues, we can create simple physical structures. Some of these physical structures can apply force and cause motion (like with muscle tissues). Such structures can help create a rudimentary multi-cellular solution for the corresponding unicellular solution of hiding from a stress. Other unicellular solutions discussed in chapter 63 like, for example, mechanisms to search for food, moving towards high concentration of food and exploration can now have multi-cellular generalizations. These should be possible because we can create several other physical structures that apply pressure, generate and detect vibrations, push, pull, bend and so on. These are the rudimentary sensors, actuators, components that can maintain a shape and so on discussed in section 63.1. The active control of these structures can be learnt naturally, in most cases, using the fundamental pattern (see section 16.3). In this way, several primitive multi-cellular designs can be created. However, these designs are useful only if they can be recreated reliably. This is the next step.

4. *Linking usage to creation sequence*: The new physical and chemical multi-cellular structures created above are better suited to fight the environmental stress. The physical and chemical pathways to operate the structure (like, say, for generating force or causing motion) are called as the **usage sequence** of steps. These are, in general, different from the chemical pathways (or the SPL network) needed to create or repair some of the cells within the colony from damages, which I call as the **creation sequence**. When we want to turn the unicellular solution into a multi-cellular solution, our goal is to create links between these two SPL networks and form one unified SPL network.

   The usage of a given structure causes the size of the colony of cells to change based on whether the structure is over-utilized or under-utilized. This forces a need to regulate the creation of the collection of cells within the colony as the structure is used. The stresses that drive the creation of these links between usage and creation sequence are regular wear and tear, damages from environmental effects, unforeseen variations in the environment, competition between multiple colonies and others. The simplest fix is for nature to regenerate new cells to replace the damaged cells. The detailed process of cell division already exists to make this happen. However, the two missing links are (a) a way to trigger the cell division, typically, only when there is a damage or destruction, for the sake of stability and (b) a way to stop or regulate the cell division cycle so we do not have a cancerous growth.

   In section 70.7, we have already seen several natural causes and corresponding solutions discovered for the problem of linking usage sequence with creation sequence of steps. These are, for example, growth, repair, embryogenesis, replacement, regeneration and adaptation. For each of these cases, new chemicals and chemical pathways needed to be discovered.

   How does nature discover solutions to the above two problems (a) and (b)? To understand how to address (a) at a conceptual level, note first that the previous solutions like shielding the inner cells, as a solution to these new stresses are no longer sufficient especially if these stresses persist for a long time. Therefore, we would need to use the stable-variant merge loop pattern using horizontal gene transfer (see section 55.1) to discover chemical links in the SPL network. Even though the outer cells can fight off the original stress, the new stress, namely, the attrition from damages, will force the stable-variant merge loop pattern to discover new chemicals to help with repair and growth. Multiple different types of chemicals solve this problem. Enzymes that catalyze a chemical reaction from the usage network to trigger the creation network are one such example. Transcription factors and the promoter sequences for the genes corresponding to G_1 phase of cell cycle (see textbooks like, say, T. McKee and J. R. McKee [102]) is another example. Hormones, morphogens, growth factors and some other simple chemical pathways that link usage and creation networks are other such examples.

   How does nature learn to maintain a steady state for the growth of cells within the colony in order to avoid a cancerous growth (item (b) above)? This is a regulation mechanism memorized as a steady state condition. This memory is represented chemically as new threshold concentrations of the various chemicals involved (see section 58.4). The older threshold values correspond to older steady state conditions i.e., they produce a specific disturbance propagation within the network of loops. The newer threshold values will produce new disturbance propagation within the same network of loops resulting in a different pattern of outputs. As an example, recall that your behavior is considerably different when you consume alcoholic drinks versus when you do not. This is because of the significant difference in how the disturbance propagates within the neurons of your brain in both cases. When you consume alcoholic drinks, the threshold concentrations of neurotransmitters like glycine, gamma-aminobutyric acid (GABA), aspartate and glutamate have deviated from the normal case. This translates to generating different firing patterns in your brain as the disturbance propagation is altered (see chapter 58).

In a similar way, when a cell memorizes new threshold concentrations in response to the growth of cells, a new steady-state condition is maintained within the chemical network of loops. As an example, the number of cells in your heart, lungs and other organs are lower when you are a baby compared to when you are an adult (i.e., they are proportional to the size of your organs). As a result, the threshold concentrations of various chemicals in the usage sequence of these organs when you are a baby have become different compared to when you are an adult. Therefore, when it is time to regulate the number of cells in each organ (to account for wear and tear, damages and others) when you are a baby versus when you are an adult, the underlying mechanism used is the disturbance propagation within the creation sequence of your SPL network for that organ. This disturbance propagation mechanism needs to deposit new chemicals to create new cells (for repair or growth). The rate of deposition of chemicals will be considerably different for a baby compared to an adult because the corresponding steady-state threshold concentrations have changed. They will be readjusted to correctly account for the size of the organ i.e., it will not deposit chemicals at the rate of a baby's organ size for an adult and vice versa.

Another example where new steady-state threshold concentrations of chemicals are memorized is for the creation sequence of beaks of finches (see section 70.7). Some finches have long beaks and others have short beaks. The rate of deposition of chemicals during the disturbance propagation within the creation sequence of beaks is, therefore, different to create the correct size of beaks. This difference is threshold concentrations is memorized by linking the usage sequence to the creation sequence as discussed in section 70.7. This is a new and useful mechanism for memorizing a sequence of steps in general, i.e., through the steady-state threshold concentrations of chemicals within an SPL network (see section 58.4). In this way, the system succeeds in encoding or memorizing the new creation sequence into a single cell itself that avoids cancerous growth.

After several generations, a cell discovers chemical pathways within its SPL network to (i) trigger creation sequence for cell division in response to damage (through random mutations – cf. Darwin's theory) and (ii) regulate the cell division process using its looped network and using disturbance propagation within its SPL network.

Until now, we have seen that the links between the usage and creation networks were triggered by external factors, which were primarily environmental stresses. However, with genetic mechanisms using hormones, transcription factors and others, it is now possible to trigger them purely internally as well i.e., even without any adverse external conditions. We view these as the normal growth mechanisms for the organism. They are self-regulated to maintain the correct chemical concentrations. Representing these pathways as network of loops ensures that the disturbance propagation within the network would guarantee stability.

Let me briefly summarize the steps discussed until now. We have seen that (a) a single cell discovers a unicellular chemical solution to alleviate a stress to a degree, (b) a colony of cells as a whole alleviate a stress through a combination of physical and chemical means in which the inner cells are effectively shielded, (c) the unicellular rudimentary physical and chemical designs are generalized to multi-cellular rudimentary designs as needed and (d) the initial random way of creating multi-cellular colonies are now made reliable using the links discovered between usage and creation sequence of steps. Several of these mechanisms are present even now. For example, Mladenov [109] has studied how several environmental factors influence asexual reproduction in echinoderms. Triggering asexual reproduction is analogous to triggering the creation sequence in response to a stress mentioned above.

If a single cell separates itself from a group of cells that are stable under a given stress, the single cell first tries to overcome the stress using the unicellular solution. As the unicellular solution is triggered, which is its usage sequence, the existence of links from the usage sequence to the creation sequence causes the cell division process to be triggered. This is analogous to the repair, growth or the regeneration process of an organ (which still exists in some species like newts and salamander). The interesting case is when the repair, growth or regeneration process is triggered even in the absence of a stress. If this happens, we can view this step as asexual reproduction process creating multiple cells in a reliable and repeatable manner. We can now call this solution as the rudimentary multi-cellular solution.

Competition is now effectively being converted into cooperation within each multi-cellular colony. The presence of cooperation is a sign that the colony is beginning to behave like a multi-cellular organism. However, this system does not address all of the issues raised in sections 70.2 - 70.4. A closer analogy of the multi-cellular colony created here is an organ rather than an organism. At this stage, each multi-cellular colony is homogeneous i.e., all cells within a given colony are almost identical (say, with the same DNA). However, it is not necessarily true that cells in different colonies have the same DNA. Together, all of these colonies are heterogeneous. Our next step is to show how several colonies begin to cooperate and become homogeneous (i.e., cells across colonies are also similar).

5. *Cooperation between multi-cellular colonies*: Each cell within a colony has the same DNA i.e., a homogeneous colony, analogous to an organ, but cells across colonies have, in general, different DNA i.e., heterogeneous colonies. Horizontal gene transfer is one mechanism that can help transform heterogeneous colonies into homogeneous colonies as the genetic information is transferred between cells of different colonies. This is an iterative process in itself.

    For example, we have a homogeneous colony A and another homogenous colony B. Cells in colony A and B have slightly different DNA because they were specialized in different ways to different stresses. Using SVML based on horizontal gene transfer, the two colonies A and B start to share a common DNA while still creating each colony reliably (using the links between usage and creation sequences of the combined SPL network). The reliability comes from the fact that SVML pattern is a directed mechanism. These are analogous to having different epigenetic marks causing a stem cell to specialize into heart cells, skin cells and others though with the same base DNA. In this manner, colonies A and B have transformed from heterogeneous state to a homogeneous state (i.e., all cells are similar). If we now pick another colony C with different DNA than the cells in colonies A and B, the SVML mechanism based on horizontal gene transfer can once again be used to form homogeneous colonies. The process is iterative across colonies.

    Where does this iterative process converge? One solution is for multiple heterogeneous colonies to converge to homogeneous colonies. Sexual reproduction is also a possibility to create homogeneous DNA and eventually colonies. The creation sequence, the usage sequence and the links between creation and usage for colonies A, B, C and so on are stored in each cell. Cooperation between colonies starts to form. New links between creation and usage sequence are created based on the cooperation between colonies (different from the links caused by repair, growth and others when the colonies were working independently). The cooperation between colonies alters the links between creation and usage including the threshold concentrations maintained during disturbance propagation within the SPL network.

    The processes discussed here are cumulative and directed. The degree of cooperation within cells in a colony and across colonies starts to increase dramatically. We can now begin to see this as the creation of the first set of multi-cellular organisms from single cellular organisms. The homogeneous collection of colonies with strong cooperative elements seems to suggest that we have a true multi-cellular organism, albeit rudimentary. The links between usage and creation sequence also provide a rudimentary reproductive process of the entire collection of colonies (i.e., of the entire multi-cellular organism).

The iterative process described here is generic enough that the origin of multi-cellular organisms does not have to occur exactly once. Plants, animals and others could have developed multi-cellularity independent of one another using the above iterative process. Some of the key common mechanisms that we relied on were: horizontal gene transfer, SVML, disturbance propagation within an SPL network, linking creation to usage sequence and vice versa, convergence of heterogeneous colonies to homogeneous ones and cooperation between colonies. These are generic enough that they can develop in plants, animals and others independently (as we have observed – see Grosberg and Strathmann [60]).

The creation of a diverse set of body plans is a different problem. There has been considerable research in developmental biology that offers detailed evidence on how different animals share a similar body plan at an embryonic stage (see Nüsslein-Volhard [115]). I have briefly mentioned some of these mechanisms in section 70.6 and referred the reader to Nüsslein-Volhard [115] for more details. The key idea is that once we have a creation sequence using the above mechanism, it is likely that reproduction of the entire homogeneous collection of colonies is not precise each time. The subtle differences at the start of the creation sequence produces large differences in the final body plan as described in Nüsslein-Volhard [115]. At this stage of evolution, errors in reproduction of the rudimentary multi-cellular organism are not undesirable. In fact, they help create diversity of multi-cellular life. The SVML pattern (i.e., a generalization of the survival of the fittest argument) is a directed mechanism that helps narrow down the collection of organisms that are best suited in a given environment.

6. *Iterative process*: In the above discussion, we have several steps that have room to improve using iterative processes. Some of these are (a) improving the unicellular solution, (b) discovering effective ways for multiple cells to form colonies of cells, (c) updating the SPL network to better link the usage sequence to the creation sequence and vice versa and (d) building homogeneous colonies from heterogeneous ones thereby improving cooperation between colonies. The unicellular solution will eventually be discarded, as it will be less used each time. The multi-cellular solution will win because of the presence of repair, growth, regenerative and other processes to update the collection of cells continuously. Furthermore, the environmental and other factors can trigger the equivalent of asexual reproduction (see Mladenov [109]) causing new multi-cellular organisms to be created as offspring. In other words, we also have the first primitive reproductive mechanism to regenerate the same colonies of cells. Eventually,

we will find a large collection of stable multi-cellular colonies, which can be viewed as the first multi-cellular organisms.

The description given here is somewhat vague and theoretical for now. We need considerable experimental evidence to justify each of these steps in the iterative process. The existing body of knowledge in developmental biology does offer considerable evidence for most of the steps discussed here, although I have not presented a large number of examples here. I refer the reader to textbooks like Nüsslein-Volhard [115] and Gilbert [54] for specific examples of chemical processes for the steps outlined here.

While some of the ideas suggested here are similar to existing theories, there are a few new mechanisms like the ability to link creation and usage sequence of steps, rudimentary physical and chemical designs, ability to transform from a unicellular to multi-cellular designs for each rudimentary design, iterative way to create cooperation between colonies of cells and others. Without these mechanisms (and others like the SVML pattern and the disturbance propagation in an SPL network), there would be no iterative process for creating the first multi-cellular organism. Without an iterative process, it is not only impossible to create multi-cellular organisms by random chance processes, but also impossible to explain the origin of multi-cellularity more than once (cf. plants, animals, fungi and several others developing multi-cellularity quite independently from one another – see Grosberg and Strathmann [60]). Random chance is not enough to explain such repeatability. Only a looped or iterative process, such as the one described here, is necessary.

It is important to note that the above iterative procedure can also be applied to design a new multi-cellular feature (like specific organs in our body). In other words, the steps outlined here can also be used to create a new multi-cellular organ iteratively, starting from a multi-cellular organism instead of a unicellular organism. The new organ created would need to be integrated within the creation sequence of the existing organism using iterative mechanism of step (5) above.

70.9 Transition from prokaryotes to eukaryotes

One distinguishing feature between eukaryotic cell and a prokaryotic cell is that eukaryotic cell has several internal organelles and quite a detailed structure unlike prokaryotes. For example, eukaryotic cell has organelles like nucleus, mitochondria, Golgi apparatus, cytoskeleton, smooth and rough endoplasmic reticulum. There is a close analogy between this problem and the problem of transition from unicellular to multi-cellular organisms. For example, the complex structure and organs in a multi-cellular organism is analogous to the complex structure and organelles in a eukaryotic cell while the simple structure in a single cellular organism is analogous to that of the simplicity in a prokaryote. Therefore, we would expect the same iterative process to be used even to explain the transition of prokaryotic cells to eukaryotic cells.

The creational, operational and evolutionary issues discussed in sections 70.2-70.4 are, for the most part, applicable to the current problem as well if we replace the analogies appropriately. For example, we would need to explain how such detailed physical structure of a eukaryotic cell could be encoded back into the DNA so that a cell division recreates all of these substructures. If these precise substructures (like nucleus, Golgi apparatus and others) are not created, the eukaryotic cell cannot survive (similar to failure to recreate substructures like heart, lungs and others in a multi-cellular organism). Similarly, the sequence of steps needed to create each organelle is quite elaborate too. Discovering them for the first time from purely random means is less convincing.

There is one popular theory to explain the origin of a few organelles in a eukaryotic cell, especially the mitochondria and plastids (like chloroplasts). This is the endosymbiotic theory (see Margulis [133]). Based on detailed microbiological evidence, the theory suggests that some specific primitive prokaryotes were engulfed by other larger prokaryotes to eventually attain a symbiotic relationship. These later became as mitochondria and plastids. This idea of engulfing to create detailed substructures within a eukaryotic cell is not, however, a generic process and cannot be extended to multi-cellular cases (like engulfing complex organs like an eye). While the microbiological evidence does point towards this hypothesis for mitochondria and plastids, the endosymbiotic theory does not seem to generalize to all other organelles in a eukaryotic cell.

In this section, I want to suggest that the set of creational, operational and evolutionary issues (similar to the ones in sections 70.2 - 70.4) for a eukaryotic cell can be addressed using a similar iterative process of section 70.8. We can say that the iterative process complements the endosymbiotic theory to yield additional mechanisms that covers organelles beyond mitochondria and plastids to explain the origin of eukaryotes from prokaryotes.

I will not discuss too many details partly because we need experimental evidence, precise set of chemicals and chemical reactions beyond just a theoretical process (which is an important area for future work). Besides, the theoretical discussion here is almost identical to the discussion in section 70.8 for the problem of transitioning unicellular to multi-cellular organisms. One such approach is as follows for creating a given organelle in a eukaryotic cell. Nature starts by discovering a solution to alleviate a stress (like aerobic or anaerobic conditions) using the stable-variant merge loop pattern based on horizontal gene transfer. This initial solution is one without any organelle-like substructure (analogous to a rudimentary unicellular solution). This solution is represented as disturbance propagation in a network of looped reactions. The region within the eukaryotic cell where the solution is localized and most active is itself subject to unexpected stresses from everyday usage. In response to the damages caused by these stresses, the cell starts to discover new membrane based structures within this localized regions using the eukaryotes' existing ability to create phospholipid membrane, say. This uses stable-variant merge loop pattern once again to maintain a direction in evolution.

This is the primitive organelle created as a repair mechanism to these stresses. This is the initial rudimentary physical structure suggested in section 70.7. The creation network for this organelle starts to link to the usage network through the discovery of new chemicals, typically transcription factors and other proteins. As more pathways are established between these two networks, the creation of the organelle starts to become inherited as well. Since there is no germ-cell equivalent, the combined network is already encoded within the DNA. The iterative process continues until a well-defined organelle is created and integrated with the creation as well as operating sequence of steps. The disturbance propagation within this network of loops during cell division triggers the creation of organelle as well. Similarly, the disturbance propagation during usage can repair any minor damages or even regenerate the organelle. Next, cooperation between various organelles begins to form iteratively, similar to step (5) of section 70.8.

In this way, we can see that the transition from prokaryotes to eukaryotes can happen both through endosymbiosis (when the organelle is a viable prokaryote itself) and through the above iterative process. Over a period, multiple organelles will be added in response to new stresses.

71
Deviations from Darwin's Theory and Intelligent Design

The problems of origin and diversity of life on earth have been studied for quite a while now. If we try to classify all of the solutions offered, they fall into two distinct and broad groups. The first one is called the Intelligent Design (ID) with the Creator as the one responsible for creating life. The second one is the Darwin's theory in which nature creates life through some form of random chance or spontaneous processes. There has never been a third approach until now. Yet, for some specific set of problems, there have been a few alternate explanations proposed like with endosymbiotic theory for the creation of several key organelles in eukaryotes. These alternatives, though better suited for the problem in question, have never turned out to be broadly applicable to most problems of living beings, at least, in the same spirit as the above two approaches.

The theory presented in this book is the third approach that provides answers to almost all aspects of living, nonliving and, more generally, all structured complex systems, as we have seen here. In fact, the applicability of the new theory is considerably beyond the above two approaches. For example, the problems of the brain like consciousness, emotions, free will and others seem unrelated to the problems of origin and diversity of life. The former problems are abstract and latter ones are dynamical. Yet, the new theory shows that both sets of problems can be unified within a single framework. In this chapter, I want to outline some of the unique set of differences and show how widely the new theory deviates from the above two popular approaches.

In this chapter, I will begin by describing the key concepts of Darwin's theory and Intelligent Design. I will then raise a number of issues and also list a set of problems for which these approaches yield unsatisfactory answers. The fact that we only have two approaches until now added an extra layer of perceptual uneasiness. If an explanation is unsatisfactory with one approach (say, Darwin's theory), the alternative approach (Intelligent Design) seems to offer no more new insight and vice versa, at least, in the minds of the strong believers of each of these approaches. In these often emotional debates (with respect to which theory is correct), it is mostly forgotten that the issue is not to reject one approach or the other, but to understand and accept the issue at hand and then, subsequently, seek an explanation that is most reasonable and convincing for a given problem.

Given a large list of unsatisfactorily answered or unsolved problems highlighted throughout this book, a third approach deemed necessary. My objective with the new theory presented in this book is to provide detailed and even constructible mechanisms, both natural and synthetic, for each of these problems. In this chapter, I will try to bring out these differences as clearly as possible by comparing them with the two existing approaches and referring to the past sections for further reading, where necessary.

In order to appreciate and evaluate the following discussion objectively, a relevant question that the reader need to answer for himself is the following: what is the structure of the living universe that each of these three approaches (Darwin's theory, Intelligent Design and the new theory described here) reveal?

71.1 Central ideas – Darwin's theory and Intelligent Design

Let me begin by describing the central ideas of both Darwin's theory and Intelligent Design in some detail.

Intelligent Design (ID): The basic premise of Intelligent Design is that the features exhibited by living beings are so complex that they cannot form by themselves naturally or randomly. An intelligent being, who is typically considered as the God or the Creator, should assist in creating any new feature. People working in this area rarely address how this is done in any detail. But the implicit assumption is that all the correct molecules are arranged at the right place, at the right time and in the right order by an intelligent being (like the Creator) so the structured complex system does not have a chance to collapse. Since we, human beings, do not know how to achieve such a level of perfection using the current natural mechanisms, i.e., without factoring the new theory introduced in this book, we can call this proposal as a supernatural mechanism.

While it is suggested that new complex features like the creation of an eye needs assistance from a Creator, small evolutionary changes to existing features are considered acceptable within ID as long as they obey the laws of nature. For example, the chemical variations that result in different shapes of beaks in finches (like thin-long or thick-short beaks) are not, by itself, contradictory according to ID. The unsatisfactory aspect is with the creation of the first beak in any bird itself or the first eye, the first heart, the first ears and others. According to ID, these are too complex to form naturally without assistance from any intelligent being. The term natural mechanism is, unfortunately, used as a synonym to mean random mechanism because the only other known approach, namely, Darwin's theory relies only on randomness. I will discuss this shortly. We will need to fundamentally change this notion in light of the new theory proposed in this book.

To understand the complexity of creating the first life forms, let us order the set of all complex features we encounter into groups. At the top level, we would include all features of living beings. At the second level, we would place all man-made systems like computers, houses and cities. At the lowest level, we would place all other nonliving natural structures like mountains, continents and oceans. What are the mechanisms to create systems at each level? At the lowest level, small gradual and random changes occurring over millions of years obeying the standard laws of nature seems like a valid explanation. If we look at the systems at the second level (man-made systems like computers), even though they are not as complex as living beings, they still cannot be created without the help of a conscious and intelligent being. The argument then (by the proponents of Intelligent Design) is that if the systems at the second level are difficult to occur naturally, how can we be convinced that the most complex systems at the top level, namely, living beings can be created purely by random self-assembly? For example, a computer can never create itself even if all the necessary parts sit on a planet devoid of conscious beings for millions of years. How can a primordial soup of dead components self-assemble and become alive?

While the objections deserve a better explanation, the answer that God created them instead, does not offer any deeper insight either (into the structure of the living universe). I will argue that even Darwin's theory does not meet the necessary standard that we all are honestly seeking for this problem. On the other hand, the new theory presented in this book outlines an enormously more detailed structure of the living, nonliving and, more generally, large-scale structured complex systems. This is the new stable parallel looped structure of our universe that both Darwin's theory and Intelligent Design failed to capture.

Darwin's theory: Let me now state the central idea of Darwin's theory. Darwin's theory has three main principles: (a) random variations, (b) inheritance of variations and (c) natural selection. Consider different species grouped as a population of organisms at a given time instant T_1. Not all of them have the same set of features and even if some do, their ability to use them effectively are not similar. As a result, some organisms are more fit than the others. The objective is to understand the evolution of a given population (consisting of different species) over time, i.e., which organisms or species live and which ones die, why a specific subset dies and so on. To answer this, Darwin's theory keeps track of the 'fitness' within the population as a simpler alternative instead. The assumption is that fitness is a measure of survivability both as an individual and as a species. A more fit organism survives better than a less fit one. In the following discussion, I will use fitness distribution and population distribution patterns interchangeably following Darwin's assumption. Imagine (as a thought experiment) this distribution of fitness within the population at the time instant T_1 as a pattern P_1. Let us now see how each of the three principles affect the evolution of the population pattern.

(a) *Random variations*: There are three main sources of random variations considered within Darwin's theory. Firstly, there are random mutations that occur from errors during DNA replication. Secondly, there are random mutations in the DNA that occur from cross-over during meiosis (during sexual reproduction). Finally, there are random and, possibly, unexpected environmental effects, which may or may not produce changes in the DNA. All three of these variations are unstructured and assumed to be purely random.

The first two variations do not affect the current population fitness pattern P_1 directly. Instead, their effect is felt when these organisms start to have offsprings. When new offsprings are born, those that have a random mutation will have a fitness different from their parents (either more-fit or less-fit). The offsprings that did not have random mutations will continue to have similar level of fitness as their parents, since their DNA's were similar. The resulting population fitness pattern is (i) random in the regions where there were mutations and (ii) approximately similar (to the parents' fitness) in the region where there were no mutations. The contribution from the environmental variations is quite random across the entire region. The final fitness pattern at a future time T_2 (after one generation) will change accordingly. Now, to get the resulting population pattern, we can use fitness pattern directly, as mentioned earlier (i.e., the fittest organisms survive).

(b) *Inheritance of variations*: The first two random variations mentioned above are reliably transmitted to the future generations because the changes are within the DNA. The random environmental effects are, typically, not inherited unless they are serious enough to alter the fundamental chemical structure within the cells (like ultra-violet radiation that produces variations in the DNA itself which can be transmitted to offsprings). The resulting fitness pattern at a future time T_2 is somewhat similar to the one in (a) excluding these random environmental effects. The population pattern can be predicted quite easily once again using Darwin's assumption about the relationship between fitness and survivability.

(c) *Natural selection:* Natural selection is just a restatement of the assumption mentioned earlier about the connection between fitness and survivability. This step essentially takes a fitness pattern and generates an appropriate population pattern. In general, more-fit organisms compete and survive better, while the less-fit organisms die faster. If the features are inherited as-is to their offspring (i.e., less-fit to less-fit and more-fit to more-fit), the population pattern over several generations will naturally filter out the less-fit ones and keep only the more-fit organisms. This process is referred to as the natural selection.

There are two important points to remember here. Firstly, natural selection and changes in the population pattern occur even if no new variations are introduced within these organisms or their offsprings. For example, if all organisms and species replicated asexually with almost exact replicas, we would still observe the change in population pattern just from natural selection.

Secondly, natural selection itself does not directly produce any change in the fitness pattern. The differences in the fitness should have been present already, before natural selection starts to act. For example, if all of the organisms have a similar level of fitness, then natural selection does not start altering the fitness within this population. There will be no significant change in the population pattern in this case. In this sense, natural selection is independent from the other two principles.

The net effect of these three principles is to produce a change in population pattern over a period of time. We can break this down into two parts. In the first part, random mutations in the DNA (i.e., heritable changes) will produce a change in the fitness pattern. In the second part, a given fitness pattern will result in a specific population pattern using natural selection.

71.2 Gaps in Darwin's theory and Intelligent Design

Let us now see if the answers provided by both of these approaches to the questions on how living beings create features, are satisfactory or not. Some sample set of questions are: how did finches develop different shapes of beaks, how were eyes, ears and other organs created, how did humans develop opposable thumbs or start walking upright, how did birds develop feathers to fly, how did fishes develop fins to breathe under water, how did plants develop flowers that bees help by pollinating them and so on. In fact, with almost any feature (from the very simplest to the most complex) about any living being, we can always ask the following simple question: how did this living being develop this feature?

The answer from Intelligent Design is almost always that an intelligent designer, typically the God, created this feature, especially if it is a complex feature. Simple features for which either specialized mechanisms exist or if minor random mutations (like point mutations, simple insertions or deletions) are discovered, both approaches does support this view. The one important requirement they do impose is that the process should be easily verifiable. In other words, a mechanism is considered unreasonable if all it says is that we should wait for millions of years and it will somehow randomly self-assemble and give rise to a specific feature.

Let us now look at the explanation offered by Darwin's theory for the same set of questions. Let me outline this carefully because the argument is a bit subtle and appears quite convincing. The standard approach taken by researchers to explain how a certain feature is developed using Darwin's theory is as follows.

(a) The original feature is first broken down into a large number of tiny subfeatures, which when taken cumulatively results in the entire feature.

(b) At each stage in evolution, Darwin's theory suggests that several random possibilities are tried with no intention of creating a specific tiny subfeature. The mechanism that creates several possibilities is random mutations in the DNA and, additionally, sexual variations.

(c) Natural selection now prunes this population based on their fitness abilities. The organisms that survive are considered the fittest within this group. These fit organisms, as a matter of fact, happen to exhibit one specific small subfeature mentioned in (a). The resulting genetic changes (i.e., the previous random mutations) within these organisms happen to correspond to this small subfeature. Researchers do identify these specific genes and use them as a 'proof' of Darwin's theory by showing that a given genotypic change leads to a given phenotypic change. The claim is that nature did not try to design this small subfeature intentionally. It was just picked naturally from within a set of several random possibilities, some of which were good and others, not so good.

(d) Even if we started with a random distribution of population, the resulting population pattern after one small subfeature is encoded through random mutations, has a definite structure and is not random anymore. For example, it has pockets of more-fit organisms at several regions than the less-fit ones.

(e) Every small subfeature necessarily requires discovering one or more new chemicals, typically, enzymes or transcription factors. No subfeature that is transmitted over millions of years can ever be created without these chemical discoveries. The only mechanism to discover them proposed by Darwin's theory is random mutations in the DNA. Natural selection does not let an organism discover new chemicals.

(f) The memory necessary to transmit a small subfeature to future generations is primarily represented within the DNA.

(g) As the small subfeatures (and, hence, the random mutations) accumulate over millions of years, the future organisms eventually develops the entire feature.

The above process appears quite convincing at first glance. However, there are a large number of serious gaps in the reasoning that simply cannot be addressed easily. Let me list some of them here.

1. *Randomness as the only mechanism to create any feature*: The most serious gap with Darwin's theory is the fact that the only mechanism to create any feature, from the tiniest to the most complex, is just random mutations in the DNA. The reason is the following. In order to create any new feature, even the tiniest one (with one exception discussed below), requires (i) some new chemicals to be discovered and (ii) that they should be reliably transmitted to future generations. It is quite rare that we can use the same chemicals but in different ways to get a new feature. The first problem is that we need the new pathways to be inherited by future generations. Furthermore, we need, at least, a few new enzymes to create these new chemical pathways or to supress existing pathways. For otherwise, the reactions will just proceed in the old ways. Even if the chemicals were discovered elsewhere and the organism is just using them now, the relevant question is: what are the mechanisms to discover such new chemicals using Darwin's theory? The answer is just unstructured random mutations within the DNA. Natural selection only eliminates the less-fit ones when the organisms either compete for the same set of resources or cooperate to survive better. In either case, there is no mechanism within natural selection that allows the possibility of 'discovering' new chemicals.

    There is one exception to creating new features without discovering new chemicals. This occurs during the development of an organism from an embryo. In chapter 70, we have seen that this process is far too complex to rely on randomness. It involves discovering several chemicals as well as re-using existing chemicals in novel ways. No part of Darwin's theory, or any other existing theory for that matter, details any mechanisms for the creation and memorization of these steps. Until now, we only have a few details of how some of these steps work. None of these details include how they were created in the first place, other than just by random mutations (see Nüsslein-Volhard [115]). The new theory proposed in this book is one of the first to outline these mechanisms in much some detail (see chapter 70).

2. *Hidden assumptions with random chance followed by natural selection*: There are several hidden assumptions in the argument given in (b) and (c) above. This argument was that several possibilities are tried in response to a stress and that natural selection subsequently picks the organisms that are the fittest, propagating this information to future generations.

    (i) This approach seems reasonable for small features where you just need a point mutation or a simple insertion or deletion in the DNA sequence. For features that require a specific sequence of steps, like with the developmental process of a multicellular organism from a single embryo or the steps involved in the creation of an enzyme, these random attempts across several generations is not satisfactory. There should be other important mechanisms during these millions of years that are necessary as well.

(ii) For simple and complex features, it is necessary to maintain the original good features while attempting to try random possibilities. For otherwise, there is no guarantee that the process will work at all. Random mutations followed by natural selection as an iterative process is inherently unstable with no guarantees of convergence. It is also a greedy process with no guarantees of success after several generations. There should be other mechanisms to stabilize this process, like, for example, the stable-variant merge loop pattern (see chapter 54).

(iii) The term survival and its inherent abilities seem to capture a strong reason to discover new chemicals. Yet, there is no mechanism given by Darwin's theory (or by any other existing theory) to explain what survival means and why living beings try so hard to survive both individually and as a species. Only living beings when subject to a stress are compelled to survive instead of just dying. Why do organisms even resist death? Why can they not be like nonliving systems? If a nonliving system is pushed downhill, it just falls down and collapses. A living being, on the other hand, will try its best to latch only to anything as it is slipping downhill to prevent its death. Why bother? If not for this innate desire to live, living beings have no reason to even attempt several random possibilities and let natural selection pick the fittest. What is the mechanism that gave rise to this strong innate ability?

(iv) Maintaining stability of an organism seems to be taken for granted before and after random mutations. I am referring to the basic level of coordination necessary to ensure thousands of chemical reactions can occur in a stable way. Basic everyday survival itself is a miracle that neither Darwin's nor any other existing theory answers satisfactorily. For example, just standing, walking, flying or swimming are triggered by a series of chemical reactions in a specific way. This is a nontrivial stability problem (cf. man-made robots attempting similar tasks).

3. *Triggering of natural selection*: In response to a stress, different organisms develop different mutations. Some of them yield immediate advantages after just one mutation. Does this mean the others will be eliminated as less fit? If the less-fit organisms stayed long enough that an accumulation of 4-5 individually less-fit mutations yielded a much better advantage than a good-fit organism with a single mutation. For example, Kreb's cycle requires 8 enzymes. If the less-fit organisms did not stay long enough to accumulate all 8

mutations and instead proceeded with a 'greedy' approach, Kreb's cycle would never have been discovered. What is the threshold number of mutations to wait before eliminating the less-fit? Is it time-dependent, stress-dependent or random elimination of the less-fit? All of these issues inherently introduce instability in the above iterative process of random mutations followed by natural selection.

4. *No relation between physical and chemical survival approaches*: When the survival of an organism is in jeopardy, the organism attempts both physical alternatives (like migrating to a different place) and chemical variations. However, the physical and chemical approaches within Darwin's theory are quite disjoint from one another. For example, the physical changes are not random. There are sensors that detect, say, new sources of food or danger. Then, there are actuators that make the organism move or fight back. This information is used to generate directed nonrandom responses (even without any awareness). On the other hand, the chemical changes suggested by Darwin's theory are purely random.

5. *Detailed structure of living universe different for each feature, especially when creating them for the first time*: The approach proposed with Darwin's theory is the same irrespective of the feature it develops (or what you are attempting to explain). However, the underlying structure of the living universe and the associated mechanisms for most features are much more interesting than pure randomness followed by natural selection. For example, the creation of two sexes for the first time, the creation of first multicellular organisms, the creation of various organs for the first time, the first flowering plants, the first eukaryotes, the first birds, the first brain, emotions and consciousness in humans, the first enzymes and the very first life forms itself, each highlight very uniquely elegant structures (as shown in this book) which are not captured within Darwin's theory.

6. *Direction in evolution*: There is a well-defined direction in the evolution of living beings in spite of the contrary with nonliving beings. As dynamical systems, there is no mechanism specified by Darwin's theory for why living beings should accumulate and integrate more and more features in an enormously complex and stable ways. This problem was discussed in greater detail in chapter 55 with stable-variant merge loop as a mechanism to guarantee a direction in evolution.

7. *Creating life for the first time when several chemicals reactions are thermodynamically unfavorable*: If we take favorable chemicals from, say, crushed dead bodies, creating a single cell using random mechanisms requires setting up conditions (like correct place at the correct time and in the correct order) for thousands of thermodynamically unfavorable chemical reactions. If we do not have even these favorable enzymes to work with, the problem is even more difficult. Darwin's theory does not address this issue clearly, especially because there is no DNA yet to keep a memory of the designs. Requiring all of these processes (like enzyme creation machinery, genetic code,

ribosomes and DNA replication) to occur simultaneously using purely random self-assembly of some sort is unreasonable. The simultaneous discovery is important because you need to recreate the same process for each generation. Random chance discovery of a few of these chemicals just once is not useful.

8. *Sequential versus parallel discovery of new chemicals*: The breakdown of a complex feature into smaller tiny subfeatures is done for our convenience. It is unreasonable to expect that the necessary chemicals were discovered in that specific order, even approximately. In general, a random collection of tiny subfeatures may not offer any obvious advantages to a given organism. A lack of a designer who decides on this order over millions of years within Darwin's theory makes it harder to accept the random mechanism as a way to create any feature. It is not guaranteed that the direction in evolution proceeds favorably with features accumulating with constant improvements. This is because while natural selection proceeds with an improvement in population pattern, random mutations, the only mechanism offered for creation of new features, is not a directed process of evolution.

Additionally, there are several other examples like Kreb's cycle where all the enzymes in the cycle should simultaneously be present for the process to be useful. Breaking down into tinier subfeatures, though possible, does not give any fitness advantages in this case, during partial discoveries. The same applies with the multistep process of creating an enzyme or with the development of a multicellular organism from a single embryo.

9. *Memory within the DNA and the mechanisms to execute specific sequence of steps*: Darwin's theory assumes that the primary way chemical features are transmitted to future generations is through DNA replication. Almost no details on, say, how a sequence of steps involving the creation of a number of enzymes in a particular order, are presented. These set of mechanisms and DNA together highlight the dynamical nature of the cells instead of the usual static view taken by Darwin's theory with just the DNA. For example, in sections 70.7 and 70.8, we have seen how this dynamical view is critical for explaining the origin of multicellular organisms. In chapter 58, we have seen how a long sequence of steps can be equivalently represented as disturbance propagation in a suitable network of loops. Even with the same DNA, by altering the sequence of steps, you can get different features. These are nontrivial mechanisms not addressed by Darwin's theory or any other existing theory.

10. *Competitive versus cooperative interactions with other species*: Darwin's theory does not address how evolution proceeds when multiple species are involved simultaneously to find a solution. For example, with flowering plants, the evolution of colors and pollination by bees go together. Similarly, symbiotic or parasitic relationships are harder to explain using generation of random possible organisms followed by natural selection, when you have cooperation across different species. Collective evolution of the genetic code is another example where the genetic code can become universal through sharing of features across species.

11. *Analysis versus design*: Darwin's theory is not a design-based theory. Rather, it is an analysis-based theory. Its attempt to explain how living beings evolved various features cannot be used constructively, in say a laboratory, to create new organisms or adapt existing organisms to make them more fit. It is probably because the timeframe for random mutations and natural selection to yield any complex features are in the thousands or millions of years. Therefore, Darwin's theory cannot be used to design and create a complex system (not just a living being) that have millions of dynamical subsystems in such a way that it remains stable, even though it is a theory that is attempting to explain how this happened within living beings.

There were experiments performed that make a claim that Darwin's theory is constructive by setting up conditions to let organisms randomly mutate and discover a solution after a few thousands of generations (see Blount et. al [14]). However, these experiments cannot be called as constructive because the solution is not predictive on what features these organisms with discover after mutating for thousands of generations.

When discussing about constructive mechanisms, it is important to address questions like what properties, chemicals and processes are important to create new complex living beings. Can we use completely different materials, even inorganic chemicals, to create living beings? Questions of this sort about designing new life forms starting from the most basic raw materials cannot be answered within Darwin's theory simply because the theory does not specify any detailed underlying structure of large-scale complex systems. This is another important gap within Darwin's theory unlike the new theory proposed in this book and, hence, makes it necessarily incomplete.

12. *Origin and processes of the brain*: Darwin's theory cannot explain the origin of the brain or any of the brains' features like emotions, consciousness, free will, self-awareness, sensing space and passage of time. To be fair, Darwin's theory does not even attempt at explaining the above processes of the brain in any case. Nevertheless, the role of brain is extremely critical for almost all animals both for survival and for adaptive evolution. It cannot be decoupled from the problem of evolution of life even though there are seemingly enormous differences between the two. Unifying both the problems, origin of life and processes of our brain, is necessary. The new theory proposed in this book shows how this can be done using the framework of SPL systems.

In summary, the main force that drives creation of new features in Intelligent Design is a supernatural intelligent being, typically, the God. In Darwin's theory, the main force that drives creation of new features and for the adaption of existing features is pure randomness, i.e., random mutations within the DNA. Waiting for millions of years and working with large populations is constantly invoked as a critical step for the random mutations followed by natural selection mechanism of Darwin's theory to work correctly. It is very confusing when people refer to snowflakes, crystals, cave structures and others as comparative examples of systems as complex as living beings obeying the same principles of random self-assembly. The structured complexity of a living being can never be compared to these nonliving examples even though strangely people seem to relate the two.

Since any living being has several organized structures (like eyes, heart or lungs), the question about how their construction plans are created, represented and modified is constantly under debate. Darwin's theory says that pure random mutations tried across large populations and over millions of years followed by natural selection is sufficient to explain this, while Intelligent Design approach says that these structures are too complex that an external intelligent designer like God is necessary to guide this process, at least, for the first time.

Each of the gaps in Darwin's theory can add up to produce incorrect conclusions sometimes. For example, Darwin's theory makes us believe that there is a single common ancestor to living beings with a tree-like ancestry. However, this is increasingly being recognized as false and that the species forms a graph-like structure instead. This clearly follows from the new theory of stable parallel loops proposed in this book. Another example is the approach Darwin's theory would suggest us take in response to a severe stress like prolonged radiation for, say, thousands of years in human beings. Darwin's theory would suggest that the more-fit individuals be secluded from the less-fit cancer-prone humans if humanity were to survive this stress. Sexual reproduction between these two groups (more-fit and less-fit ones) would be avoided thinking that survival of the fittest would eliminate the less-fit humans anyway. However, as explained in chapter 54, new fit features will most likely be discovered when we allow sexual reproduction between the more-fit and less-fit groups.

71.3 Repeatable randomness – SPL designs in simple terms

For designing new complex features in living beings, if Intelligent Design relied primarily on a designer like God and Darwin's theory relied on pure randomness, what does the new theory proposed here rely on? It seems hard to believe that there could be a third approach for building complex features. Either it was built randomly by itself or someone assisted in building it. The new theory relies on stable parallel looped dynamical systems. But what are the underlying principles used by this class of systems that makes the creation of living beings for the first time by itself easier? Does it ultimately rely, in a philosophical sense, on an intelligent designer or on randomness (if no one is assisting)?

In section 55.8, I have already shown that Darwin's theory was a special case of stable-variant merge loop pattern. In fact, we see that looped systems that *self-sustain their dynamics for the longest time* is already a generalization of the notion of 'survival of the fittest'. This new notion extends beyond living beings to include all dynamical systems. Therefore, we should expect the new theory to rely on randomness, at least in some specialized form. I call the underlying philosophical solution of new theory as *repeatable randomness* (see section 52.3). How is this new idea different from the two existing approaches and how does it work better than the others? In this section, I will express some of the mechanisms of the previous chapters in somewhat vague and simple terms to emphasize the importance of repeatable randomness.

The first point to note is that this new idea, like Darwin's theory, does not require an intelligent designer to build complex systems. Secondly, it does not rely on arbitrary randomness similar to how Darwin's theory does, but instead only on events that are repeatable and, yet, random. Let me discuss a few simple examples first. There are several events around us that produce random outcomes. When there is a wind gust, leaves seem to scatter randomly. When you kick a baseball, it seems to go in random directions. When you toss a coin, you get heads or tails randomly. Among such random events, I want to call some of them as repeatable random outcomes in the following sense.

Definition 71.1: An event is said to be repeatably random if and only if the event is repeatable and is random with a probability sufficiently greater than zero.

I have not specified what a probability close to zero means, but in most cases it will be quite apparent. For example, expecting a heads in a coin toss is a repeatable random outcome because whenever you do this experiment, you do get heads a lot of times (with probability close to half). The repeatability of this event comes from our free will. Expecting a leaf to fall at a given location due to wind gusts is not a repeatable random outcome. On the other hand, expecting the leaf to fall in the northern direction of the tree is a repeatable random event. Similarly, expecting the molecules to move to one corner of a room is not a repeatable random outcome. If we take a biological example, expecting to create a specific DNA sequence (encoding a given protein) using random self-assembly is not a repeatable random outcome.

Therefore, in most examples, we can quite quickly estimate if a given outcome is repeatably random or not. Quite simply, we can check if the following two conditions are satisfied or not:

a) Imagine performing an experiment a large number of times and evaluate if the outcome of interest occurs several times, then the outcome is a good candidate for a repeatable random outcome.

b) Evaluate if the experiment can occur naturally a large number of times. If it cannot occur naturally several times, the event is not a repeatable random event. Otherwise, it is a repeatable random outcome.

The condition (a) is just another way to be sure of the *existence* of a possibility. If you have never (or rarely) seen a particular outcome in your experience, you can be confident that it is not a repeatable random outcome. It is, however, possible that under a different set of external conditions, the same event can become repeatably random. But, in this case, these conditions should be specified clearly enough that the experiment can be performed naturally, in real-time and that the outcome can indeed be observed to repeat several times (condition (b)).

Iterative processes using repeatably random minimal structures: How can we use the above idea to build a complex and highly improbable living being naturally? At first glance, it is quite clear that the creation of the first life-form is 'not' a repeatable random outcome. We want to know how we can convert it into a repeatable random outcome.

For this, we start with outcomes that have high enough probabilities. We then let multiple such events combine in such a way that the resulting probability of the combined outcome is not decreased. What is even better is if the combined probability actually increases. The best case scenario of the existence of such a process is with living beings where the probability has increased from 0% (before life existed on earth) to 100% (after the first life forms existed).

In order to achieve such an increase in probabilities when we combine the outcomes, the most important idea is to pick *looped dynamical systems*. The natural repeatability in a looped system already makes its outcomes repeatably random. The obvious difficulty to overcome here is that among the naturally occurring systems, there are very few looped systems you can pick and *control*. This is especially true during the origin of life or under conditions similar to those on the Moon. For example, you have day-night cycle (from the rotation of the planet), water cycle (if there is water), wind or weather patterns, seasonal cycles (from rotation of the planet around the sun) and slow geological patterns including earthquakes, volcanoes or meteor impacts. To overcome this difficulty (of existence of a few looped systems that can be controlled), formation of some simple repeatably random physical structures, also termed as *minimal structures* in chapter 53, are important.

The simplest example is a looped set of chemical reactions like a simple reversible reaction. Generalizing this, consider the case where we have five chemical reactions together forming a loop. If we treat each reaction individually, the probability of generating a chemical is very low. This is because the chance of the same chemical reaction occurring several times is negligible. However, when all these reactions combine to form a loop, the probability of creation of each chemical has actually improved significantly. In other words, combining multiple chemical reactions together as a loop has actually increased the probabilities of creation of the chemicals, compared to when the reactions were occurring separately. Of course, this particular combination that helped is a rather special structure, namely, a chemical loop. A simple combination as a nonlooped cascade of chemical reactions is not enough to improve the probabilities before versus after.

Therefore, chemical looped reactions clearly satisfy condition (a) above. The question we need to answer next is to identify which set of chemical looped reactions can satisfy condition (b) as well i.e., which set of chemical looped reactions can occur naturally. This is the reason for proposing active material-energy looped systems in chapter 59.

Consider a physical structure that helps in improving the probabilities before versus after as the active material-energy looped system of chapter 59. This is a repeatably random structure that can occur naturally as discussed in chapter 59. Before the existence of this structure (material-energy looped systems), a given chemical reaction occurs purely through random collisions under open conditions. Such chemical reactions, therefore, are not repeatably random processes. However, after the creation of material-energy looped systems, the same chemical reaction has become a much better repeatably random process (see chapter 59). In this way, the probability of the combined system has actually increased i.e., it has become more reliable, compared to when both systems, i.e., repeatable random material-energy system and nonrepeatable random chemical reaction, were operating separately.

Let me mention yet another example that behaves the same way, namely, the stable-variant merge loop (SVML) pattern. Random mutations by themselves are not repeatably random events. They do not necessarily produce higher probabilities and, in most cases, they introduce disorder. How can we make this reliable and repeatably random instead? The clever idea is by using sexual reproduction with one stable parent and another variant parent. The variant parent has random mutations and is not repeatably random, but the stable parent is repeatably random because there are fewer or no mutations. Therefore, an offspring from a sexual combination of one stable and one variant parent has a higher probability of discovering a useful and an entirely new solution to a stressful situation (see chapters 54 - 55).

In chapters 57 - 68, I have outlined more such processes when trying to explain how life can be created for the first time. These details included explaining how nature can add, in a repeatably random way, more and more looped dynamical systems incrementally and iteratively in order to maintain the following property: the combined designs should have higher probabilities (i.e., are more reliable and repeatable) than when they were operating separately. Even small jumps in the improvement of reliability will be clearly noticeable when the probabilities for the system as a whole, are low, until eventually the probabilities cross a threshold value, which causes evolution to proceed in such a way that the reliabilities reach close to 100% as well.

In this way, we see that using stable parallel looped dynamical systems as the core set, it is possible to create systems that are as complex as living beings, entirely naturally. There is no need for an intelligent designer and, yet, not rely on purely arbitrary randomness like Darwin's theory. This important class of looped dynamical systems (which creates simple minimal structures naturally) is the main reason for taking initially unreliable processes 4 billion years ago and incrementally or iteratively converting them into better reliable and repeatable random processes, until finally the randomness disappears when a critical level of complexity is reached. What is quite surprising is that complex features of our brain like consciousness, free will and emotions are also expressed within a similar set of stable parallel looped systems.

72
Conclusions and Future Work

Most of the earlier research focused only on smaller timelines within a total time span of about 4 billion years ago to the present day. These included studies at a single cellular level, aspects of multi-cellular organisms, at the structure of DNA, at a specific area of the brain, emotions, consciousness and others. Let me group them into three important timelines: (*I*) the period when there were only nonliving systems until the first life forms were formed, (*II*) the period when there is great diversity after the first life forms were created, and (*III*) the period when human consciousness was developed. Darwin's theory of evolution was regarded as the most successful theory in timeline (*II*), but it does not extend to the other two timelines. In fact, as discussed in detail in chapter 71, Darwin's theory is not only incomplete but also has several limitations when trying to explain the complexity of chemical dynamics in time period (*II*) itself. For the other two timelines (*I*) and (*III*), we do not even have a satisfactory theory until now. There have been doubts whether it is possible to discover a single theory that applies across all three timelines.

In this book, I have presented a new theory that does cover all three timelines quite well. I have demonstrated how to use the three core principles, namely, loops, parallelism and stability, to provide a more complete understanding of most of the important features across all three timelines. One analogy that has guided, at least, me with several of these problems is the one discussed in section 1.4.2, where we try to build an intricate public transportation looped network (i.e., a physical and chemical SPL network) and using such networks to create features as complex as life. The alternative is to rely on random chance mechanisms to create life-like patterns. Almost all existing theories on life and consciousness rely on latter mechanisms, which are not satisfactory. I did not expand on this analogy intentionally because I realized that when I started providing more details, the analogy started to get weaker (as is the case with most analogies even though they do guide us in unique ways). I have, therefore, decided not to include them in this book. However, I do encourage the reader to explore this analogy in a bit more detail, as it does reveal unique insights into the problem of creating life without the help of other conscious beings.

In this book, as we tried to solve several problems across all three timelines *I-III*, we had to overcome a number of challenges. I did discuss several technical ones in the book. Let me now switch to topics that are either somewhat philosophical or those that highlight the generic approach we should be taking when studying large-scale structured dynamical systems. Since these topics are quite subtle and we tend to overlook them, I will discuss them briefly here.

72.1 Avoiding statements requiring consciousness

When studying cells, animals or life, we, as conscious beings, inadvertently use higher forms of consciousness, realizations and abstract concepts that are logical or that seem reasonable to us, into our reasoning. For example, we use statements like 'an organism wants to survive' or 'an animal decides to fight or run away' in a logical sense without offering further explanation. We build scientific arguments using statements such as these. Unfortunately, they are not only incomplete but are unacceptable when proposing a theory that attempts to answer the very origins of life and consciousness, primarily because we cannot construct systems using such statements.

The theory is also supposed to answer some of these very questions. For example, we should explain why living beings wants to survive longer, why living beings struggle to survive instead of just dying, why they want to sense the environment, why their evolution should proceed in a specific direction that benefits the species instead of just becoming extinct, why and how they are trying to improve instead of just doing the same things repeatedly, and several others questions like these.

These questions deserve a better explanation. For example, why do living beings seem to have a purpose when all nonliving systems simply obey the laws of nature? A rock sliding downhill does just that, whereas a living being falling down will try within its means to survive. What is the origin of such a purpose, namely, to live when all the known laws of nature do not seem to have any? How can we create a system that exhibits desires and purposes (like wanting to live, struggling to survive, resisting death, discovering new features to improve future generations so the species as a whole survives better and others) using the existing laws of nature? If we cannot identify the root cause of these features, it is conceivable that earth could have behaved just like any other life-less planets.

If we pose these questions in terms of *creating* a living organism or a conscious human being instead of just *analyzing* how life and consciousness work, we begin to realize that the above statements and arguments like 'an organism wants to survive' and others are unsatisfactory. A synthetic living system that we would be trying to create would not have these properties. For example, a synthetic 'animal' has no reason to fight, run away or even to survive better unless we know how to impart these life-like features into these systems. In general, we can ask the following question whenever we use the above statements or arguments to justify a theory. Do we know how to build a synthetic system that obeys your statements and arguments? If we do not, then more work is necessary to fill in the gaps.

This is where one of the concepts, namely, loops introduced in this book becomes critical – any dynamical system that forms a loop simply continues its motion for a long time. The energy lost in a loop is minimal, the energy needed to run the loop is minimal, the amount of time a loop runs is sufficiently long that two such looped systems have at least a chance to interact with one another instead of completely missing each other and several other properties that were discussed in this book. If we have a complex network of interacting loops and if we develop the theory carefully as shown in this book (cf. the analogy of building an intricate public transportation looped network), we can address most of these difficult questions clearly and consistently. We can do this without having to state the theory in terms of the above arguments that require consciousness.

72.2 Choosing the correct structure of complex systems

Let us try to distinguish the internal structure of our brain with how we use our brain when performing everyday activities in our lives. In abstract mathematics, there is a subtle difference between the steps you take to solve mathematical problems and the steps you take to understand a theory or solutions to problems that were solved by someone else. When you solve a mathematical problem, you try different approaches with several of them leading to dead-ends as well. Not all of this effort is wasted. In fact, it helps build our intuition about the problem. In several situations, our intuition drives us towards the correct solution. After we get a feel for the solution, we translate our intuition into the rigorous language of mathematics. On the other hand, when you read abstract mathematics like functional analysis to learn it or when you read someone else' solutions, you read the details and try several examples to build an intuition. In other words, even though mathematics is a very rigorous subject with linear logic, our approach to both understand and discover new theorems is mostly by using our imprecise intuition. This is true with several other fields like fashion design, music, art and engineering since they all require a certain level of creativity.

It appears as though there is a fundamental disconnect between how our brain works versus how we work. We work by training ourselves in a very systematic linear (or sequential) and serial way. This is quite apparent in our formal education, our engineering designs (like computers, electromechanical systems) and so on. However, our brain works through intuition, analogies, interrelating different topics and jumping from one seemingly disjoint topic to another. Most of us gain our creative abilities through informal training and from normal day-to-day experiences. The intuition and analogies is what is natural for our brain to remember and work with, while we need the formal training to create the necessary complexity in our brain just to gain this very same intuition.

This suggests that to understand correctly how our brain works, we cannot use linear and serial approaches that we are so trained to do so (which is ironically the intuition I have built after working on this problem for several years!). We need to represent it in a language that is most natural given the brain's internal structure. Our brain's architecture clearly is a massively parallel-interconnected network of neurons. There are two problems with this structure: (1) it is impossible to avoid loops in this network of neurons, and (2) parallel dynamics make it even more difficult to coordinate and synchronize data when you split the problem across many subsystems – the parallel dynamics would need to be combined at some point. In network computing and other fields, we generally consider loops as undesirable because it is not easy to understand when to stop the loop and avoid infinite recursion. If this were the case, how can we possibly hope to organize the two seemingly undesirable features – loops and parallelness – to attain a deeper understanding of how our brain works?

The key breakthrough comes when we realize that loops (and fixed sets – see chapter 14) help in containing the very complexity of parallelism itself. For this reason, we should model memory within the structure and dynamics of loops itself, not in the actual content or strengths of the interconnections between neurons. This latter approach is what was followed with the theory of neural networks. With the new approach, we can exploit sufficient structure from a complex looped dynamical system like our brain (unlike non-looped systems like turbulence). I have analyzed this carefully in the book. Along with a few additional ideas discussed here, I feel we were able to explain the problem of consciousness, realizations, emotions and others as well.

72.3 Physical versus chemical reactions

My goal has been to propose a theory that works universally across the three timelines mentioned earlier: (*I*) transition period from nonlife to life, (*II*) diversity after the first life forms are created, and (*III*) development of human consciousness. There are no existing theories that cover all three timelines. Darwin's theory is the most detailed theory that covers only timeline *II*. The presence of DNA and its role turned out to be very helpful in identifying key structure in the diversity of life in the timeline *II*.

When you try to study timeline *I*, the most important details are the biochemistry of organic molecules. Nature needs to create life entirely by capturing the most innovative features of organic chemicals and chemical reactions, in general. The complexity of physical reactions (like building complex physical structures), on the other hand, is minimal. Amino acids, enzymes, nucleic acids, RNA and DNA, lipids, sugars and other organic chemicals along with a diverse set of chemical reactions give rise to complex life-like behaviors. Some examples of these cellular behaviors are an ability to live, reproduce, detect and adapt with external environment and grow into complex life forms. Nature created all of these behaviors by itself without any assistance from higher forms of intelligence (since there is no life to begin with).

On the other hand, the questions related to how humans gained consciousness in timeline *III* and what it meant, relied on understanding the complex structure of our brain. The structure of our brain is mostly a physical structure even though they were formed and maintained chemically. The chemical reactions needed for the formation of these complex structures, the transmission of chemicals between neurons in response to various stimuli are well-understood processes. However, the true complexity comes from the trillions of interconnections and billions of neurons that perform the same set of chemical reactions repeatedly. These well-coordinated and repeated processes produce complex behaviors like memory, intelligence, emotions, consciousness and others. The physical structure is undoubtedly the most critical aspect of our brain that needed careful analysis instead of the chemical reactions.

In other words, timeline *I* and early life is mostly about chemical reactions, while timeline *III* and the brain is mostly about physical reactions. Timeline *II* and the diversity of life is a mixture of physical and chemical reactions. The existing theories that tried to explain timeline *I* used processes that are quite unique to chemical reactions. Similarly, existing theories that tried to explain timeline *III* used processes unique to physical dynamics. As a result, both of these theories were typically incompatible with each other and are incompatible with theories in timeline *II* like with Darwin's theory of evolution. The need to unify the processes of both physical and chemical reactions was a crucial step in proposing a complete theory for all timelines. This book presented several such principles, namely, stable parallel loops, and covered them in significant detail. The nature of such a universally applicable theory (presented in this book) and the philosophical conclusions are significantly different from any other existing theory that does not realize this similarity and the common set of principles.

72.4 Providing a constructive solution

The research problem that aims at addressing all three timelines covers a vast breadth of topics, as noted earlier. The number of people working on these topics and the amount of scientific knowledge amassed over the years is also enormous. It is quite difficult for a single human being to keep up with this broad area of research. If this were the case, how can anyone, including myself, claim that we have a complete and satisfactory solution to this problem? It is quite possible that an existing research area that the person proposing the new theory is unaware of, which turns out to contradict his theory. Or a future research topic can turn out to contradict the theory as well.

One way to address this concern is by providing a constructive solution to this problem. If the theory outlines, at least, a conceptually well-defined construction of a synthetic system, both for creating the first primitive cell from inanimate objects by nature itself and for designing a suitable brain structure that can lead to consciousness, then it would provide confidence in the completeness of such a theory. Even though this may not address all corner cases yet, the constructive approach can trigger research that will eventually lead to a detailed engineering design. This book outlined such a constructive solution with additional theoretical guarantees.

72.5 Addressing complexity occurring right at this instant

It is quite surprising that the existing research does not emphasize the complexity of life as it is occurring right now compared to emphasis on the three timelines mentioned earlier. I am not referring to questions about how our eyes, ear, heart, lungs and other organs in our body work even though they have sufficient complexity. These organs can be considered as designs that were discovered thousands of years ago, which we are beginning to understand just now. My specific reference is to the complexity, the coordination and the seemingly perfect balance that is present in our current world around us.

There are several well-coordinated sequences of unique events that are occurring in, say, my body right now that are significantly different from your body, plants or an animal's body. Imagine that you had a minor accident with a cut and a broken bone. Your body immediately responds to several parallel events, like say, clotting of blood, attacking foreign particles entering your bloodstream, other healing processes for minor broken ribs and joints and so on. It is likely that several of the accidents that living beings are currently encountering today have never occurred in the past to be memorized in the DNA. How is it then that living beings *discover* a long sequence of steps to heal themselves?

Even a simple event requires cascading effect of parallel responses for which the timing, the order of occurrence are very critical. If any one of this sequence is thrown into disarray, it can have serious consequences. Furthermore, this very accident has implications, not just with our body, but within other external systems as well. How can we explain the entire chain of events that are unfolding right now both within our body and within the external world? The theory presented in this book based on stable parallel loops is the only known one so far that can clearly explain the structure of such large-scale structured complex systems.

It is not just with the problem of origin of life we seem to have difficulty in understanding, but the complexity exists at every instant in all organisms. In fact, it is quite surprising that life works without ever collapsing into a chaotic state. To understand the nature of this complexity, imagine trying to create a complex engineering system with hundreds of dynamical subsystems. We will immediately realize the difficulty in coordinating them, even for a short period. Systems designed as stable parallel loops, on the other hand, offer flexibility to coordinate them for extended periods.

We can appreciate the importance of this perfect coordination if we think about what happens when we are knocked unconscious. Just a moment ago, our entire body has a certain level of stiffness where we can, for example, keep our hand horizontal, walk upright on one foot and others. However, just a moment later, every part of our body has lost this stiffness and we behave externally analogous to a dead body or an inanimate object. If someone raises our hand and lets it go, it drops due to gravity like a dead-weight. We cannot balance even on both legs let alone on one leg.

Similarly, there are thousands of thoughts that occur in our brain every day and thousands of chemical reactions that occur in our cells. Why do they not occur simultaneously? Why do they occur in a coordinated way? Which order should they be coordinated, like say A followed by B followed by C and D (or) B and D followed by A and C? How do they occur in such a way that a given purpose is solved? Who knows this purpose? How do we focus and evaluate even a single thought (or a chemical reaction) long enough? How can we balance ourselves and walk let alone run, play and drive? How can even a single minute of our lives proceed in a coordinated way instead of proceeding in a randomly disorganized hodgepodge of physical events? More interestingly, how are we even aware of these things?

The real requirement of any theory should be to explain these and other such phenomena occurring right in front of our eyes every second with as few assumptions as possible. We cannot simply be content with answering the origin of life or consciousness in human beings. The theory proposed in this book that uses stable parallel loops tried to address most of these questions in sufficiently precise terms.

72.6 Future work

In this book, we have covered a vast breadth of topics. Some of the topics were only discussed cursorily in order to highlight the pervasiveness of the core ideas. For example, the diversity of life from single cellular organisms to complex multi-cellular organisms and the formation of complex organs are quite big topics. We have already done a significant amount of research on these topics and more is needed in the future. The main focus of this book was to show that the central ideas are common across widely disjoint topics and are powerful enough to generalize well beyond what were considered here. Yet, it is clear that more theoretical, experimental work and computer simulations (since most processes described in the book are iterative in nature) is needed in the future.

There have been scattered references to some of these problems that need future work throughout the book. Let me try to list a few additional ones here. We can broadly classify these problems into two groups. In the first group, we need to present additional details and perform experiments (or computer simulations, as appropriate) on known topics, while also discussing several important missed topics. In the second group, we should address new set of problems that were exposed only after we had discovered a viable theory of complex systems.

Some examples of the former type are: ability to design complex and elaborate mechanisms for several organs like eye, ear, heart and others, formation of a diverse set of multi-cellular organisms, further details on more elaborate processes like the creation of a variety of enzymes, transcription factors, cell division and others, an evolutionary theory for sex (Geodakyan's [51] theory seems to fit well within our framework though further details are necessary), further details in developmental biology to explain the formation of an adult from an embryo, specific details for the creation of neurons (though the higher-level processes described in this book are reasonable) and detailed mechanisms to control and sense the external environment. My intuition suggests that these problems do fit well within the framework of stable parallel loops. However, more research and experiments are needed to see if the intuition can be converted into rigorous and constructive arguments.

Some examples of the latter type are: a mathematical theory of loops that clarifies the emergent structures and properties of stable parallel looped dynamical systems, discovering new drugs and systematically analyzing their side-effects within the framework of the complex network of chemical loops, new clinical studies and understanding different types of disorders, building a complex system with life-like features as well as synthetic consciousness and emotions, discovering a new complex system-based solution to engineering designs using SPL architecture, generalizing the theory to nonliving systems, relating the theory of complex systems to existing physical theories including quantum mechanics and general theory of relativity, treating most systems (including electrons, atoms and planets) as complex SPL systems and studying their properties, studying problems like global warming and other ecological imbalances from a complex systems' perspective, creating a sustainable ecosystem on a planet like Mars or the Moon and studying other types of complex systems like economies, weather systems and societies using stable parallel looped architecture.

72.7 Final remarks

It was clear that the most useful ideas introduced in this book when studying large-scale structured complex systems are stable parallel loops. Using these ideas, we were able to gradually guarantee reliability and repeatability of features even when starting from purely random events. This helped create life for the first time from inanimate objects using the active material-energy looped systems. For the features of the human brain, there were a number of other important ideas necessary to explain the full range of questions left unanswered until now. Each of these ideas is generic enough to help with both analysis and design of large-scale structured complex systems.

Given that we have identified a new useful structure within complex systems, is it safe to say that the fundamental feature of 'all' systems, including quantum mechanical particles is complexity? The complexity and stability of every such system comes from the stable parallel looped architecture. Should we no longer treat small-sized particles like electrons and protons as simple systems and astronomical-sized objects as complex? I believe the assumption of simplicity is only a convenience that we conscious beings use. It is not inherent in the external world. Instead, my intuition suggests that every stable object, from microscopic to macroscopic, should be treated as complex that uses a stable parallel looped architecture.

Until now, evolution by nature had been a slow process. Whether any living being existed to understand what nature had been creating or not, the evolutionary processes had been steadily at work. Even after nature developed humans, who for the first time were able to truly see and feel the external world using their consciousness, we were still mere observers in the universe. We did not understand the natural processes of how living beings work, though we did gain considerable insight into the other fundamental physical laws. As a result, nature was the only ultimate designer of features of living beings.

For the first time, I feel we have a theory of structured complex systems, as presented in this book, that helps bridge the gap between nature and us. Very soon, we will be designing features and new species alongside nature itself, typically at a much faster pace. Surely, there will be new surprises and new phenomena that do not fit well with this new theory. We would also need to iterate several times to alleviate the new risks that come from a faster pace of evolution. Nevertheless, this is an important turning point in our history. I feel we have just closed a giant loop in our understanding of nature and structured complex living beings. The more we traverse along this new loop in the future generations, the more stable and perfect this loop will become.

Bibliography

1. R. Albert and A.-L. Barabasi (2002) Statistical mechanics of complex networks. *Reviews of Modern Physics* 74: 47-97.
2. B. Alberts, A. Johnson, J. Lewis, M. Raff, K. Roberts and P. Walter (2002) *Molecular Biology of the Cell*, Garland.
3. M. C. Alliegro and M. A. Alliegro (2008) Centrosomal RNA correlates with intron-poor nuclear genes in *Spisula* oocytes. *Proc. Natl. Acad. Sci. USA* 10: 1073.
4. P. W. Atkins (1998) *Physical chemistry*, Oxford University Press.
5. A. Abzhanov, W. P. Kuo, C. Hartmann, B. R. Grant, P. R. Grant and C. J. Tabin (2006) The calmodulin pathway and evolution of elongated beak morphology in Darwin's finches. *Nature* 442: 563-567.
6. B. J. Baars (1988) *A cognitive theory of consciousness*. Cambridge University Press.
7. S. Baron (1996) *Medical Microbiology*, University of Texas Medical Branch.
8. M. J. Behe (2007) *The Edge of Evolution: The Search for the Limits of Darwinism*, Free Press.
9. V. N. Beklemishev (1969) *Principles of comparative anatomy of invertebrates*, Oliver & Boyd.
10. B. P. Belousov (1959) A periodic reaction and its mechanism, *Compilation of Abstracts on Radiation Medicine* 147: 145.
11. R. Belshaw, V. Pereira, A. Katzourakis, G. Talbot, J. Paces, A. Burt and M. Tristem (2004) Long-term reinfection of the human genome by endogenous retroviruses. *Proc. Natl. Acad. Sci. USA.* 101(14): 4894–99.
12. Y. Bengio (2009) Learning deep architectures for AI, *Foundations and Trends in Machine Learning*, 2(1): 1-127.
13. B. C. Berndt (1985) *Ramanujan's Notebooks: Part I*, Springer-Verlag.
14. Z. D. Blount, C. Z. Borland and R. E. Lenski (2008) Historical contingency and the evolution of a key innovation in an experimental population of *Escherichia coli. Proc. Natl. Acad. Sci. USA.* 105: 7899-7906.
15. C. Borgnakke and R. E. Sonntag (2008) *Fundamentals of Thermodynamics*, Wiley.
16. L. Boroditsky, *Size constancy*. Illustration source from the website: http://www-psych.stanford.edu/~lera/psych115s/notes/lecture8/figures3.html
17. G. Buzsáki (2006) *Rhythms of the Brain*, Oxford University Press.
18. C. Caratheodory (1986) *Algebraic Theory of Measure and Integration*, Chelsea Publishing Company.
19. D. Chalmers (1996) *The Conscious Mind: In Search of a Fundamental Theory*, Oxford University Press.
20. M. H. Chase, P. J. Soja, and F. R. Morales (1989) Evidence that glycine mediates the postsynaptic potentials that inhibit lumbar motoneurons during the atonia of active sleep. *Journal of Neuroscience* 9: 743-751.

21. N. Chomsky (1957) *Syntactic Structures*, The Hague: Mouton.
22. A. Choughuley, A. Subbaraman, Z. Kazi and M. Chadha (1977) A possible prebiotic synthesis of thymine: uracil-formaldehyde-formic acid reaction. *BioSystems*, 9(2–3):73–80.
23. K. M. Cole and R. G. Sheath (1990) *Biology of the red algae*, Cambridge University Press.
24. A. Córdova, M. Engqvist, I. Ibrahem, J. Casas and H. Sundén (2005). Plausible origins of homochirality in the amino acid catalyzed neogenesis of carbohydrates, *Chem. Commun.* 15: 2047–2049.
25. R. Courant and H. Robbins (1941) *What is Mathematics?: An Elementary Approach to Ideas and Methods*, Oxford university press.
26. F. Crick (1958) On Protein Synthesis. *The Symposia of the Society for Experimental Biology* 12: 138-16.
27. F. Crick (1968) The origin of the genetic code. *J Mol Biol* 38:367–379.
28. F. Crick and C. Koch (2003) A framework for consciousness. *Nature Neuroscience* 6: 119-26.
29. G. W. Crile (1998) *Origin and Nature of Emotions*, The World Wide School.
30. C. Darwin (1872) *The Origin of Species by Means of Natural Selection, or the Preservation of Favoured Races in the Struggle for Life*, London: John Murray.
31. P. G. Davison (2008) How to define life, *The University of North Alabama*. Retrieved 2008-10-17.
32. D. Dennett (1992) *Consciousness Explained*, Back Bay Books.
33. J. Dugundji and A. Granas (2003) *Fixed Point Theory*, Springer.
34. G. M. Edelman (1987) *Neural Darwinism: The theory of neuronal group selection*. Basic Books.
35. P. Ehrenfreund, S. Rasmussen, J. Cleaves and L. Chen (2006) Experimentally tracing the key steps in the origin of life: The aromatic world. *Astrobiology* 6(3):490-520.
36. M. Eigen and P. Schuster (1979) *The Hypercycle: A principle of natural self-organization*. Springer.
37. R. J. Ellis and S. M. van der Vies (1991) Molecular chaperones, *Annu. Rev. Biochem.* 60: 321–47.
38. R. J. Ellis (2006) Molecular chaperones: assisting assembly in addition to folding, *Trends in Biochemical Sciences* 31(7): 395–401.
39. A. M. Fennimore, T. D. Yuzvinsky, Wei-Qiang Han, M. S. Fuhrer, J. Cumings and A. Zettl (2003) Rotational actuators based on carbon nanotubes. *Nature* 424:408-410.
40. J. Ferris, R. Sanchez and L. Orgel (1968) Studies in prebiotic synthesis. III. Synthesis of pyrimidines from cyanoacetylene and cyanate. *Journal of Molecular Biology*, 33:693–704.
41. R. Feynman (1970) *The Feynman lectures on physics*, Addison Wesley Longman.
42. R. J. Field, E. Koros and R. M. Noyes (1972) Oscillations in Chemical Systems. Part 2: Thorough analysis of temporal oscillations in the bromate-cerium-malonic acid system, *J. of the American Chemical Society* 94(25): 8649-8664.
43. R. J. Field and R. M. Noyes (1974) Oscillations in chemical systems. IV. Limit

cycle behavior in a model of a real chemical reaction, *Journal of Chemical Physics* 60: 1877-1884.

44. S. Fox and K. Harada (1961) Synthesis of uracil under conditions of a thermal model of prebiological chemistry. *Science*, 133:1923–4.

45. F. Fröhlich and D. A. McCormick (2010) Endogenous electric fields may guide neocortical network activity. *Neuron* **67**, 129-143.

46. I. Fry (2000) *The Emergence of Life on Earth: A Historical and Scientific Overview.* Rutgers University Press.

47. S. Fry, D. L. Kacian, D. R. Mills and F. R. Kramer (1972) A replicating RNA molecule suitable for a detailed analysis of extracellular evolution and replication. *Proc. Natl. Acad. Sci. USA.* 69 (10), 3038-3042.

48. R. H. Garrett and C. M. Grisham (2006) *Biochemistry.* Brooks Cole.

49. M. S. Gazzaniga, R. B. Ivry and G. R. Mangun (2008) *Cognitive Neuroscience:* The *Biology of the Mind*, W. W. Norton & Company.

50. V. A. Geodakyan (2005) Evolutionary theories of asymmetrization of organisms, brain and body. *Usp Fiziol Nauk* Jan-Mar; 36(1): 24-53.

51. V. A. Geodakyan (1991) The Evolutionary Theory of Sex. *Priroda (Nature)* 8: 60-69.

52. A. P. Georgopoulos, J. F. Kalaska, R. Caminiti and J. T. Massey (1982) On the relations between the direction of two-dimensional arm movements and cell discharge in primate motor cortex. *Journal of Neuroscience* 2: 1527-1537.

53. D. G. Gibson, J. I. Glass, C. Lartigue, V. N. Noskov, R. Y. Chuang, M. A. Algire, G. A. Benders, M. G. Montague, L. Ma, M. M. Moodie, C. Merryman, S. Vashee, R. Krishnakumar, N. Assad-Garcia, C. Andrews-Pfannkoch, E. A. Denisova, L. Young, Z. Q. Qi, T. H. Segall-Shapiro, C. H. Calvey, P. P. Parmar, C. A. Hutchison 3rd, H. O. Smith and J. C. Venter (2010) Creation of a bacterial cell controlled by a chemically synthesized genome, *Science* 329(5987): 52–6.

54. S. F. Gilbert (2006) *Developmental Biology*, Sinauer Associates Inc.

55. W. Gilbert (1986) Origin of Life: The RNA world. *Nature* 319:618.

56. E. A. Gladyshev, M. Meselson and I. R. Arkhipova (2008) Massive horizontal gene transfer in Bdelloid Rotifers. *Science* 320:1210-1213.

57. P. R. Gordon-Weeks (2005) *Neuronal Growth Cones*, Cambridge University Press

58. R. B. Graham (1990) *Physiological psychology,* Wadsworth Publishing Company.

59. A. Granas and J. Dugundji (2003) *Fixed Point Theory*, Springer.

60. R. K. Grosberg, R. R. Strathmann (2007) The evolution of multicellularity: A minor major transition? *Annu Rev Ecol Evol Syst.* 38: 621–654.

61. B. G. Hall (1990) Spontaneous point mutations that occur more often when advantageous than when neutral. *Genetics* 126: 5-16.

62. S. Hameroff (2006) Consciousness, neurobiology and quantum mechanics, in Tuszynski, J. (Ed.) *The emerging physics of consciousness,* Springer-Verlag.

63. M. M. Hanczyc, S. M. Fujikawa and J. W. Szostak (2003) Experimental models of primitive cellular compartments: Encapsulation, growth, and division. *Science* 302:618–622.

64. M. M. Hanczyc and J. W. Szostak (2004) Replicating vesicles as models of

primitive cell growth and division. *Curr. Opin. Chem. Biol.* 8:660-664.

65. F. Harary (1994) *Graph* Theory. Westview Press.

66. W. Hargreaves, S. Mulvihill and D. Deamer (1977) Synthesis of phospholipids and membranes in prebiotic conditions. *Nature*, 266: 78–80.

67. T. Hastie, R. Tibshirani and J. Friedman (2013) *The elements of statistical learning: data mining, inference and prediction.* Springer.

68. S. Hathaway, A. Eisenberg and H. Murkoff (2003) *What to expect the first year*, Workman publishing company.

69. I. N. Herstein (1975) *Topics in Algebra*, Wiley.

70. J. Hofbauer and K. Sigmund (1998) *Evolutionary Games and Population Dynamics*, Cambridge University Press.

71. D. Hofstadter (1979) *Gödel, Escher, Bach: An eternal golden braid*, Vintage Books

72. D. H. Hubel and T. N. Wiesel (1962) Receptive fields, binocular and functional architecture in the cat's visual cortex. *Journal of Physiology* 160: 106-154.

73. D. H. Hubel and T. N. Wiesel (1963) Shape and arrangement of columns in cat's striate cortex. *Journal of Physiology* 165: 559-568.

74. D. H. Hubel and T. N. Wiesel (1963) Receptive fields of cells in striate cortex of very young, visually inexperienced kittens. *Journal of Neurophysiology* 26: 994-1002.

75. C. Hurovitz, S. Dunn, G. W. Domhoff, and H. Fiss (1999) The dreams of blind men and women: A replication and extension of previous findings. *Dreaming* 9: 183-193.

76. A. Isidori (1995) *Nonlinear control systems*, Springer.

77. E. Jablonka and M. J. Lamb (2006) *Evolution in four dimensions: genetic, epigenetic, behavioral and symbolic variation in the history of life*, The MIT Press.

78. F. Jacob and J. Monod (1961) Genetic regulatory mechanisms in the synthesis of proteins. *Journal of molecular biology* 3: 318-356.

79. J. Jaynes (2000) *The Origin of Consciousness in the Breakdown of the Bicameral Mind*, Mariner Books.

80. H. Jeong, S. P. Mason, A.-L. Barabasi and Z. N. Oltvai (2001) Lethality and centrality in protein networks. *Nature* 411: 41-42.

81. D. L. Kacian, D. R. Mills, F. R. Kramer and S. Spiegelman (1972) A replicating RNA molecule suitable for a detailed analysis of extracellular evolution and replication, *Proc. Nat. Acad. Sci. USA* 69 (10): 3038–3042.

82. E. R. Kandel, J. H. Schwartz and T. M. Jessell (2000) *Principles of Neural Science*, McGraw-Hill Companies.

83. N. Kanwisher, J. McDermott and M. M. Chun (1997) The fusiform face area: a module in human extrastriate cortex specialized for face perception. *Journal of Neuroscience* 17(11): 4302-4311.

84. H. K. Khalil (2001) *Nonlinear Systems,* Prentice Hall.

85. R. Knight (2001) *Ecology and Evolutionary Biology*, Ph.D. thesis, Princeton University.

86. C. Koch (2004) *The Quest for Consciousness: A Neurobiological Approach*, Roberts and Company Publishers.

87. C. Koch and G. Tononi (2008) Can machines be conscious? *IEEE Spectrum* 45:

55-9.

88. S. Kojo, H. Uchino, M. Yoshimura, and K. Tanaka (2004) Racemic D,L-asparagine causes enantiomeric excess of other coexisting racemic D,L-amino acids during recrystallization: a hypothesis accounting for the origin of L-amino acids in the biosphere, *Chem. Commun.* 19: 2146–2147.

89. J. Koolman and K.-H. Roehm (2005) *Color Atlas of Biochemistry*, Thieme.

90. Y. Kuramoto (2003) *Chemical Oscillations, Waves and Turbulence*, Dover Publications.

91. R. Laocharoensuk, J. Burdick and J. Wang (2008) Carbon-Nanotube-Induced Acceleration of Catalytic Nanomotors. *ACS Nano* 2:1069-1075.

92. S. Laurence and E. Margolis (2001) The Poverty of the Stimulus Argument. *British Society for the Philosophy of Science* 52: 217-276.

93. F. W. Lawvere and R. Rosebrugh (2003) *Sets for Mathematics*, Cambridge University Press.

94. M. Levy, S. Miller, K. Brinton and J. Bada (2000) Prebiotic synthesis of adenine and amino acids under Europa-like conditions. *Icarus*, 145: 609–13.

95. M. Levy, S. Miller and J. Oró (1999) Production of guanine from NH_4CN polymerizations. *Journal of Molecular Evolution*, 49:165–8.

96. M. Lynch (2004) Long-term potentiation and memory. *Physiological Reviews* 84 (1): 87-136.

97. S. Mac Lane (1998) *Categories for the Working Mathematician*, Springer.

98. S. Mac Lane and I. Moerdijk (1994), *Sheaves in Geometry and Logic: A First Introduction to Topos Theory*, Springer.

99. S. P. Mathew, H. Iwamura and D. G. Blackmond (2004) Amplification of Enantiomeric Excess in a Proline-Mediated Reaction, *Angewandte Chemie International Edition* 43 (25): 3317–3321.

100. T. McCollom, G. Ritter and B. Simoneit (1999) Lipid synthesis under hydrothermal conditions by Fischer-Tropsch-type reactions. *Origins of Life and Evolution of the Biosphere*, 29:153–166.

101. C. P. McKay (2004) What is life – how do we search for it in other worlds? *Public Library of Science – Biology* 2(9): 302.

102. T. McKee and J. R. McKee (2002) *Biochemistry: The Molecular Basis of Life*, McGraw-Hill Science/Engineering/Math.

103. J. McMurry (1992) *Organic Chemistry*, Brooks/Cole Publishing Company.

104. R. S. Mehrotra and K. R. Aneja (1990) *An Introduction to Mycology*, New Age International.

105. U. Meierhenrich (2008) *Amino acids and the asymmetry of life*, Springer-Verlag.

106. S. L. Miller (1953) A Production of Amino Acids Under Possible Primitive Earth Conditions. *Science* 117:528–529.

107. S. L. Miller and H. J. Cleaves (2007) *Prebiotic chemistry on the primitive Earth*, In: Rigoutsos, I. and Stephanopoulos, G., Eds. Systems Biology: Genomics, Volume I, Oxford University Press, New York, pp. 3-56.

108. S. L. Miller and H. C. Urey (1959) Organic Compound Synthesis on the Primitive Earth. *Science* 130:245-251.

109. P. V. Mladenov (1996) Environmental factors influencing asexual reproductive processes in echinoderms. *Oceanologica Acta*. 19 (3-4): 227-235.

110. T. M. Montgomery, http://www.tedmontgomery.com/the_eye.

111. F. R. Morales, P. A. Boxer, and M. H. Chase (1987) Behavioral state-specific inhibitory postsynaptic potentials impinge on cat lumbar motoneurons during active sleep. *Experimental Neurology* 98: 418-435.

112. H. Morris and J. P. Schaeffer (1953) *The nervous system – The brain or Encephalon. Human anatomy; a complete systematic treatise*. New York, Blakiston.

113. H. J. Muller (1932) Further studies on the nature and causes of gene mutations. *Proceedings of the 6th International Congress of Genetics*, 213–255.

114. D. E. Nicholson (2003) *Metabolic Pathways 22nd Edition*, International Union of Biochemistry and Molecular Biology (http://www.iubmb-nicholson.org).

115. C. Nüsslein-Volhard (2006) *Coming to life: how genes drive development*, Kales Press.

116. F. Oehlenschläger and M. Eigen (1997) 30 Years Later – a New Approach to Sol Spiegelman's and Leslie Orgel's in vitro evolutionary studies; Dedicated to Leslie Orgel on the occasion of his 70th birthday. *Origins of Life and Evolution of Biospheres* 27, 437-457.

117. J. K. O'Regan and A. Noë (2001) A sensorimotor account of vision and visual consciousness. *Behavioral and Brain Sciences* 24: 939-1031.

118. R. Ornstein and R. Thompson (1991) *The Amazing Brain*, Houghton Mifflin.

119. J. Oró (1961) Mechanism of synthesis of adenine from hydrogen cyanide under possible primitive Earth conditions. *Nature* 191: 1193–1194.

120. L. Pauling (1988) *General chemistry*, Dover Publications.

121. A. Pikovsky, M. Rosenblum and J. Kurths (2003) *Synchronization: A Universal Concept in Nonlinear Sciences*, Cambridge University Press.

122. D. Purves (2007) Neuroscience, Sinauer Associates, Inc.

123. W. K. Purves, D. Sadava, G. H. Orians and H. C. Heller (2003) *Life: The Science of Biology*, Sinauer Associates and W. H. Freeman.

124. V. S. Ramachandran, D. C. Rogers-Ramachandran and S. Cobb (1995) Touching the phantom. *Nature* 377: 489-490.

125. S. Rasmussen, M. A. Bedau, L. Chen, D. Deamer, D. C. Krakauer, N. H. Packard and P. F. Stadler (2008) *Protocells: Bridging Nonliving and Living Matter*. The MIT Press.

126. P. H. Raven and G. B. Johnson (2001) *Biology*, McGraw-Hill Science/ Engineering/Math.

127. M. Ravuri (2012) Stable parallel looped systems: a new theoretical framework for the evolution of order, *Evolving systems* 3(2): 111-124.

128. E. G. Reekie and F. A. Bazzaz (2005) *Reproductive allocation in plants*, Academic Press.

129. L. S. Roberts and J. Janovy Jr. (2008) *Foundations of Parasitology*, McGraw-Hill Science/Engineering/Math.

130. M. Robertson and S. Miller (1995) An efficient prebiotic synthesis of cytosine and uracil. *Nature*, 375:772–4.

131. A. Roorda and D. R. Williams (1999) The arrangement of the three cone classes in the living human eye. *Nature* 397: 520-522.

132. W. Rudin (1986) *Real and Complex Analysis*, McGraw-Hill Science/ Engineering/Math.

133. L. Sagan (1967) On the origin of mitosin cells. *Journal of Theoretical Biology* 14(3): 225-274.

134. R. Sanchez, J. Ferris and L. Orgel (1966) Cyanoacetylene in prebiotic synthesis. *Science* 154:784–5.

135. R. Sanchez, J. Ferris and L. Orgel (1966) Conditions for purine synthesis: did prebiotic synthesis occur at low temperatures? *Science*, 153:72–3.

136. K. Schmitt, http://photographyoftheinvisibleworld.blogspot.com/, Weinheim, Germany, uvir.eu.

137. J. Searle (1980) Minds, Brains and Programs. *Behavioral and Brain Sciences* 3: 417-457.

138. J. Sergent, S. Ohta and B. MacDonald (1992) Functional neuroanatomy of face and object processing: a positron emission tomography study. *Brain* 115(1): 15-36.

139. J. M. Siegel (1999) The evolution of REM sleep. *In Handbook of Behavioral State Control, Lydic, R and Baghdoyan (Eds.)* 87-100, CRC Press.

140. T. Shibata, H. Morioka, T. Hayase, K. Choji, and K. Soai (1996) Highly Enantioselective Catalytic Asymmetric Automultiplication of Chiral Pyrimidyl Alcohol, *J. of American Chem. Soc.* 118(2): 471–472.

141. G. H. Shull (1948) What is heterosis?, *Genetics* 33 (5): 439–446.

142. L. Smolin (2002) *Three Roads to Quantum Gravity*, Perseus Books Group.

143. M. Solms (2000) Dreaming and REM sleep are controlled by different brain mechanisms. *Behavioral & Brain Sciences* 23: 1083-1121.

144. T. W. G. Solomons and C. B. Fryhle (2004) *Organic Chemistry*, John Wiley & Sons, Inc.

145. J. M. Sperling, D. Prvulovic, D. E. Linden, W. Singer and A. Stirn (2006) Neuronal correlates of color-graphemic synaesthesia: a fMRI study. *Cortex* 42(2): 295-303.

146. G. M. Stratton (1896) Some preliminary experiments on vision without inversion of the retinal image. *Psychological Review* 3: 611-617.

147. G. M. Stratton (1897) Vision without inversion of the retinal image. *Psychological* Review 4: 341-360 and 4: 463-481.

148. G. M. Stratton (1899) The spatial harmony of touch and sight. *Mind* 8: 492-505.

149. S. H. Strogatz (2003) *Sync: The Emerging Science of Spontaneous Order*, Hyperion.

150. M. Syvanen (1985) Cross-species gene transfer; implications for a new theory of evolution. *J. Theor. Biol.* 112(2), 333-343.

151. J. W. Szostak, D. P. Bartel and P. Luigi Luisi (2001) Synthesizing life. *Nature* 409:387-390.

152. K. Takakusaki, Y. Ohta and S. Mori (1989) Single medullary reticulospinal neurons exert postsynaptic inhibitory effect via inhibitory interneurons upon

alpha-motoneurons innervating cat hindlimb muscles. *Experimental Brain Research* 74: 11-23.

153. R. H. Thompson and L. W. Swanson (2010) Hypothesis driven structural connectivity analysis supports network over hierarchical model of brain architecture. *Proc. Natl. Acad. of Sci. USA.* 107(34), 15235-15239.

154. G. Tononi (2004) An information integration theory of consciousness. *BMC Neuroscience* 5: 42.

155. M. J. Tovée (2008) *An Introduction to the Visual System,* Cambridge University Press.

156. D. Y. Tsao, W. A. Freiwald, R. B. H. Tootell and M. S. Livingstone (2006) A cortical region consisting entirely of face-selective cells. *Science* 311(5761): 670-674.

157. Y. Ura, J. M. Beierle, L. J. Leman, L. E. Orgel and M. R. Ghadiri (2009) Self-assembling sequence-adaptive peptide nucleic acids. *Science* 325: 73-77.

158. K. Vetsigian, C. R. Woese and N. Goldenfeld (2006) Collective evolution and the genetic code. *Proc. Nat. Acad. Sci. USA* 103(28): 10696-10701.

159. L. P. Villarreal (2005) *Viruses and the Evolution of Life*, ASM Press.

160. A. Voet, and A. Schwartz (1983) Prebiotic adenine synthesis from HCN: evidence for a newly discovered major pathway. *Bioorganic Chemistry*, 12:8–17.

161. S. Weinberg (1995) *The quantum theory of fields,* Cambridge University Press.

162. R. White (1984) Hydrolytic stability of biomolecules at high temperatures and its implication for life at 250 °C. *Nature* 310:430–432.

163. W. B. Whitman, D. C. Coleman and W. J. Wiebe (1998) Prokaryotes: The unseen majority. *Proc. Natl. Acad. Sci. USA.* 95 (12), 6578-6583.

164. A. O. Wilkie (1994) The molecular basis of genetic dominance, *J. of Medical Genetics* 31: 89-98.

165. S. J. Williamson and H. Z. Cummins (1983) *Light and Color in Nature and Art*, Wiley.

166. C. Wills and J. Bada (2001) *The Spark of Life: Darwin and the Primeval Soup*. Basic Books.

167. A. T. Winfree (2001) *The Geometry of Biological Time*, Springer.

168. A. N. Witt, U. P. Vijh and K. D. Gordon (2003) Discovery of Blue Fluorescence by Polycyclic Aromatic Hydrocarbon Molecules in the Red Rectangle. *Bulletin of the American Astro. Society* 35, 1381 (2003).

169. C. R. Woese, G. J. Olsen, M. Ibba and D. Söll (2000) Aminoacyl-tRNA synthetases, the genetic code, and the evolutionary process, *Microbiology Mol Bio Rev* 64(1): 202-236.

170. C. R. Woese, D. H. Dugre, W. C. Saxinger, and S. A. Dugre (1966) The molecular basis for the genetic code, *Proc. Nat. Acad. Sci. USA* 55: 966-974.

171. C. R. Woese (1965) On the evolution of the genetic code, *Proc. Nat. Acad. Sci.* 54: 1546-1552.

172. C. R. Woese (1965) Order in the genetic code, *Proc. Nat. Acad. Sci. USA* 54: 71-75.

173. C. R. Woese (1967) *The genetic code: The molecular basis for genetic expression.* p. 186. Harper & Row.

174. L. Wolpert and E. Szathmáry (2002) Multicellularity: Evolution and the egg. *Nature* 420 (6917): 745.

175. S. H. Wright (2004) Generation of resting membrane potential. *Advances in Physiology Education* 28(1-4): 139-142.

176. Y. Yamane, E. T. Carlson, K. C. Bowman, Z. Wang and C. E. Connor (2008) A neural code for three-dimensional object shape in macaque inferotemporal cortex. *Nature Neuroscience* 11: 1352-1360.

177. A. M. Zhabotinsky (1964) Periodic processes of malonic acid oxidation in a liquid phase. *Biofizika* 9: 306-311.

178. Receptive field. Wikipedia article (Source http://en.wikipedia.org/wiki/File:Receptive_field.png)

179. Growth cone. Wikipedia article (Source http://en.wikipedia.org/-wiki/File:GrowthCones.jpg)

180. Absorption spectrum of cone cells in a human eye. (Source http://hyperphysics.phy-astr.gsu.edu/hbase/vision/colcon.html)

181. Cranial nerves in a human. Encyclopedia Britannica, Inc. (Source http://www.britannica.com/EBchecked/media/46720/The-cranial-nerves-and-their-areas-of-innervation).

182. Functional areas of the human brain. Encyclopedia Britannica, Inc. (Source http://media-2.web.britannica.com/eb-media/32/99532-004-2B7BE4E6.jpg).

Index